2015 27th International Conference on Microelectronics (ICM 2015)

Casablanca, Morocco
20-23 December 2015

IEEE Catalog Number: CFP15473-POD
ISBN: 978-1-4673-8760-6

**Copyright © 2015 by the Institute of Electrical and Electronic Engineers, Inc
All Rights Reserved**

Copyright and Reprint Permissions: Abstracting is permitted with credit to the source. Libraries are permitted to photocopy beyond the limit of U.S. copyright law for private use of patrons those articles in this volume that carry a code at the bottom of the first page, provided the per-copy fee indicated in the code is paid through Copyright Clearance Center, 222 Rosewood Drive, Danvers, MA 01923.

For other copying, reprint or republication permission, write to IEEE Copyrights Manager, IEEE Service Center, 445 Hoes Lane, Piscataway, NJ 08854. All rights reserved.

***This publication is a representation of what appears in the IEEE Digital Libraries. Some format issues inherent in the e-media version may also appear in this print version.*

IEEE Catalog Number: CFP15473-POD
ISBN (Print-On-Demand): 978-1-4673-8760-6
ISBN (Online):   978-1-4673-8759-0
ISSN: 2159-1660

Additional Copies of This Publication Are Available From:

Curran Associates, Inc
57 Morehouse Lane
Red Hook, NY 12571 USA
Phone: (845) 758-0400
Fax: (845) 758-2633
E-mail: curran@proceedings.com
Web: www.proceedings.com

Program

2015 27th International Conference on Microelectronics (ICM)

Circuits and Systems

- Roll Angle Estimation of a Spinning Object: Experimental Setup and Preliminary Results

  Clement Thomas (French-German Research Institute of Saint-Louis, France), Markus Stefer (French-German Research Institute of Saint-Louis, France), Loic Bernard (ISL, France) 1

- Image quality assessment using nonlinear learning methods

  Rshdee Alhakim (University of Grenoble & TIMA Lab, France), Ghislain Takam Tchendjou (TIMA Laboratory, Grenoble University, France), Emmanuel Simeu (TIMA Laboratory, Grenoble University, France), Fritz Lebowsky (STMicroelectronics, France) 5

- Efficient Architectures for HEVC Luma Interpolation Filter

  Ahmed Diefy (Egypt Japan University of Science and Technology, Egypt), Ahmed Shalaby (Egypt Japan University for Science and Technology, Egypt), Mohammed Sayed (Egypt-Japan University of Science and Technology, Egypt) 9

- Dynamic Channel Coding Reconfiguration in Software Defined Radio

  Ahmed Sadek (Cairo University, Egypt), Hassan Mostafa (University of Toronto, Canada), Amin Nassar (Cairo University, Egypt) 13

CAD Tools and Design

- An Efficient Hybrid Power Modeling Approach for Accurate Gate-Level Power Estimation

  Alejandro Nocua Cifuentes (University of Montpellier & Laboratory of Informatics, Robotics and Microelectronics of Montpellier, France), Arnaud Virazel (Laboratory of Informatics, Robotics and Microelectronics of Montpellier, France), Alberto Bosio (Laboratory of Informatics, Robotics and Microelectronics of Montpellier, France), Patrick Girard (LIRMM, France), Cyril Chevalier (STMicroelectronics, France) 17

- Toward Efficient Synthesis Method of Multifunctional Logic Circuits

Vaclav Simek (Brno University of Technology, Czech Republic), Richard Ruzicka (Brno University of Technology, Czech Republic), Adam Crha (Faculty of Information Technology & Brno University of Technology, Czech Republic) 21

- Reducing the Number of Embedded Multipliers in Squaring Large Size Complex Numbers

  Fatima-Ezzahra Guessous (Web Help Company, Morocco), Noureddine Chabini (Royal Military College of Canada, Canada) 25

- General fuzzy models for dynamical systems

  Kheireddine Chafaa (University of Batna, Algeria), Mouna Ghanai (Electronics Department, University of Batna, Batna, Algeria), Lamir Saidi (Laboratoire Automatique Avancée et Analyse des Systèmes (LAAAS), Algeria) 27

- Core and Back-End Enhancements for CUPSHOP HW/SW Partitioning Tool

  Rania Hassan (Faculty of Engineering, Cairo University, Egypt), Ahmed Hamed (Faculty Of Engineering, Cairo University & Cairo University, Egypt), Mohamed B Abdelhalim (Arab Academy for Science, Technology & Maritime Transport, Egypt), Serag E. D. Habib (Faculty of Engineering, Cairo University, Egypt) 31

Smart Sensor and Sensor Network

- Security approaches based on elliptic curve cryptography in wireless sensor networks

  Salah Said (RITM-ESTC / CED-ENSEM, University Hassan II, Morocco) 35

- Enhanced algorithm for QRS detection using discrete wavelet transform (DWT)

  Wissam Jenkal (ENSA, Morocco), Rachid Latif (ENSA, University Ibnou Zohr, Morocco), Ahmed Toumanari (ENSA Agadir - Ibn Zohr University, Morocco), Dliou (ENSA agadir, Morocco), Oussama El b'charri (ENSA Agadir - Ibn Zohr University, Morocco) 39

- A Low-cost design of Transceiver based on DWPT for WSN

  Chehaitly Mouhamad (LGIPM, France), Tabaa Mohamed (EMSI, Morocco), Fabrice Monteiro (University of Lorraine, France), Abbas Dandache (University of Lorraine, France), Ali Hamie (Arts Sciences & Technology University in Lebanon (AUL), Lebanon) 43

- Network Coding for Energy optimization and Throughput Enhancement of WideMac for IR-UWB based WSN

  Tarik Chanyour (ENSIAS - Mohamed V University - Rabat, Morocco), Hanane Chergui Hali (Faculty of Science, Ibn Tofaïl University, Morocco), Rachid Saadane (Ecole Hassania des Travaux Publiques de Casablanca, Morocco), Mostafa Belkasmi (ENSIAS - Mohammed V University - Rabat, Morocco) 47

- Distributed consensus algorithm and its application to detect coverage hole in sensor networks

  Anas Hanaf (University of Reims Champagne-Ardenne & CReSTIC, France), Alban Goupil (Université de Reims Champagne-Ardenne, France), Maxime Colas (DéCom/University of REIMS CHAMPAGNE-ARDENNE, France), Guillaume Gelle (University of Reims Champagne-Ardenne & CReSTIC, France) 51

Circuits and Systems

- Harvesting Energy From Data Lines For Avionics Applications: Power Conversion Chain Architecture

  Maryam Mohajertehrani (Polytechnique Montreal, Canada), Umar Shafique (Polytechnique Montreal, Canada), Yvon Savaria (École Polytechnique de Montréal, Canada), Mohamad Sawan (Polytechnique Montréal, Canada) 55

- Digital image stailization for video images

  Yasser Mohanna (Lebanese University, Lebanon), Sari Malaeb (Lebanese University, Lebanon), Hamze Alaeddine (University of Brest, France), Oussama Bazzi (Lebanese University, Lebanon), Ali Alaeddine (Lebanese University, Lebanon) 59

- Real-time Lane Detection in Different Illumination Conditions

  Ismail El hajjouji (ENSA de Tanger, Morocco), Aimad El Mourabit (National School of Applied Sciences of Tangier & LabTic, Morocco), Zakariae Asrih (ENSA de Tanger, Morocco), Benaissa Bernoussi (ENSA de Tanger, Morocco), Salah Mars (ENSA de Tanger, Morocco) 63

- Comparison Between Analog and Digital Locking MPPT Unit for Micro-scale PV Energy Harvesting Systems

  Nourane G. Tawfik (German University in Cairo, Egypt), Hassan Mostafa (University of Toronto, Canada), Yehea Ismail (Northwestern University, USA) 67

Micro/Nanoelectronics

- Effective Device Electrical Parameter Extraction of Nanoscale FinFETs: Challenges and Results

  Alessandra Leonhardt (Unicamp, Brazil), Luiz Fernando Ferreira (Federal University of Rio Grande do Sul, Brazil), Sergio Bampi (Federal University of Rio Grande do Sul & Microelectronics Group at UFRGS, Brazil), Leandro Manera (Unicamp, Brazil) 71

- Accelerated Lifetime Tests and Failure Analysis of an Electro-thermally Actuated MEMS valve

  Haithem Skima (FEMTO-ST, France), Kamal Medjaher (FEMTO-ST, France), Christophe Varnier (FEMTO-ST, France), Noureddine Zerhouni (FEMTO-ST, France), Eugen Dedu (FEMTO-ST, France), Julien Bourgeois (UFC/FEMTO-ST Institute, France) 75

- Insights for Utilizing the Memristor as a Multi-bit Based Memory

  Mostafa El-Khouly (German University in Cairo, Egypt), Ahmed Madian (NCRRT & Egyptian Atomic Energy Authority, Egypt), Hassan Mostafa (University of Toronto, Canada) 79

- Small-Signal Gain Calculations for a CMOS Analog Amplifier using Short Channel Equations Using nm Technology

  Neda Tawalbeh (Princess Sumaya University for Technology, Jordan), Mahmoud Hassan (Universiy, Jordan), Hazem Marar (Princess Sumaya University for Technology, Jordan) 83

Biomedical

- Pulse Interval Modulation for Biomedical Wireless Sensors

  Décio R. M. Faria (UNIFEI, Brazil), Tales C Pimenta (Universidade Federal de Itajuba, Brazil), Robson Moreno (Universidade Federal de Itajuba, Brazil) 87

- Robust Low Power NB PHY Baseband Transceiver for IEEE 802.15.6 WBAN

  Awny Mohammed Mohsen El-Mohandes (Egypt Japan University of Science and Technology, Egypt), Ahmed Shalaby (Egypt Japan University for Science and Technology, Egypt), Mohammed Sayed (Egypt-Japan University of Science and Technology, Egypt) 91

- Inductive Degeneration Low Noise Amplifier for IR-UWB Receiver for Biomedical Implant

  Maissa Daoud (LETI, Tunisia), Rahma Aloulou (LETI, Tunisia), Hassene Mnif (National Engineering School of Sfax, University of Sfax, Tunisia), Mohamed Ghorbel (ATMS, Tunisia) 95

- A New WSN Transceiver based on DWPT for WBAN applications

  Saadaoui Safa (EMSI, Morocco), Tabaa Mohamed (EMSI, Morocco), Fabrice Monteiro (University of Lorraine, France), Abbas Dandache (University of Lorraine, France), Karim Alami (EMSI, Morocco) 99

Circuits and Systems

- On-chip analog PI controller for calibration of Rogowski coils

  Simon Paulus (University of Strasbourg, France), Jean-Baptiste Kammerer (University of Strasbourg, France), Joris Pascal (University of Applied Sciences Northwestern Switzerland, Switzerland), Calogero Bona (ABB Switzerland Ltd Corporate Research, Switzerland), Luc Hebrard (University of Strasbourg, France) 103

- A Comparative Analysis of Optimized CMOS Neural Amplifier

Ahmed El-Attar (Cairo University, Egypt), Saif Ahmed (Cairo University, Egypt), Youssef Abdelkader (Cairo University, Egypt), Mohamed Badran (Cairo University, Egypt), Ali H. Hassan (Faculty of Engineering, Cairo University, Egypt), Hassan Mostafa (University of Toronto, Canada) 107

- A Circular Patch Antenna Using Two-Dimensional Photonic Crystal Substrate

  Boualem Mekimah (University of Constantine1, Algeria), Abderraouf Messai (University of Constantine 1, Algeria), Abdelkrim Belhedri (University of Constantine1, Algeria) 111

- SBLS: Speed Based Lane Changing System in VANETs

  Jetendra Joshi (NIIT University & NIIT University, India), Kritika Jain (NIIT University, India), Yash Agarwal (NIIT University, India), Manash Jyoti Deka (NIIT University, India), Pravit Tuteja (Niit University, India) 114

CAD Tools and Design

- New Fine-Grained Clustering Algorithm on GPU Architecture for Bias Field Correction and MRI Image Segmentation

  Ait Ali Noureddine (Labo SSDIA, ENSET Mohammedia, Morocco), Cherradi Bouchaib (CRMEF, El Jadida, Morocco), Omar Bouattane (EMSI, Lebanon), Youssfi Mohamed (Labo SSDIA, ENSET Mohammedia, Morocco), Raihani Abdelhadi (Labo SSDIA, ENSET Mohammedia, Morocco) 118

- An Enhancement Transient Response of Capless LDO with improved Dynamic Biasing Control for SoC Applications

  Hatim Ameziane (Sidi Mohamed Ben Abdellah University, Fez, Morocco), Kamal Zared (USMBA & USMBA, Morocco), Hassan Qjidaa (USMBA, Fez, Morocco) 122

- FPGA implementation of Data Encryption Standard using time variable permutations

  Soufiane Oukili (Moulay Ismail University, Meknes, Morocco), Seddik Bri (Moulay Ismail University, Meknes, Morocco) 126

- Simulation of Ground Penetrating Radar Imaging Under Subsurface

  Gamil Alsharahi (Abdelmalek Essaadi University, Morocco), Abdellah Driouach (Abdelmalek Essaâdi University, Morocco), Ahmed Faize (Mohammed 1st University, Morocco) 130

Complex Systems

- Formal Modeling, Verification and Implementation of a Train Control System

Mohammad Hossein Askari-Hemmat (Concordia University, Canada), Otmane Ait Mohamed (Concordia University, Canada), Mounir Boukadoum (Université du Québec à Montréal, Canada) 134

- Maintenance optimization of series-parallel systems operating missions with scheduled breaks

  Abdelhakim Khatab (ENIM/Université Lorraine, France), El-Houssaine Aghezzaf (Ghent University, Belgium), Diallo Claver (Dalhousie University, Canada) 138

- Sampling rate optimization for fault diagnosis of distributed networked control systems

  Dominique Sauter (University of Lorraine & CRAN-CNRS UMR7039, France) 142

- An Optimal Integrated Maintenance Strategy with Switching Policy under Subcontracting Constraint

  Hajej Zied (University of Lorraine, France), Dellagi Sofiene (LGIPM, France), Nidhal Rezg (University of Lorraine, Tunisia) 146

- PMU Based Centralized Adaptive Load Shedding Scheme in Smart Grid

  Hamid Bentarzi (IGEE, University of UMBB, Boumerdes, Algeria) 150

Circuits and Systems

- Differentiated Service for NoC-based Multimedia Applications

  Atef Dorai (Jean-monnet University, Tunisia), Virginie Fresse (Université Jean Monnet, Saint Etienne, France), El-Bay Bourennane (University of Burgundy & LE2I Laboratory, France), Abdellatif Mtibaa (ENIM, Tunisia) 154

- A New 65nm-CMOS 1V 8GS/s 9-bit Differential Voltage-Controlled Delay Unit Utilized for a Time-Based Analog-to-Digital Converter Circuit

  Abdullah Elbayoumi (Valeo Interbranch Automotive Software & Cairo University, Egypt), Hassan Mostafa (University of Toronto, Canada), Ahmed M Soliman (Cairo University, Egypt) 158

- Biomimicry to Network on Chip: Router Heart Rate

  Ahmed El-Naggar (Alexandria University, Egypt), Ahmed Shalaby (Egypt Japan University for Science and Technology, Egypt) 162

- A 14-bit Low-Power Interface Circuit for Piezo-Resistive Pressure Sensors

  Amr Walid (Ain Shams University, Egypt), Ayman Hassan Ismail (Ain Shams University, Egypt) 166

- A practical algorithm using virtual point light for the rendering process

  Khemliche Sarra (University of Biskra Algeria, Algeria), Babahenini Mohamed Chaouki (LESIA, Algeria), Bahi Naima (LESIA, Algeria), Zerari Abd El Moumène (LESIA, Algeria) 170

Micro/Nanoelectronics

- Memristor Emulator Based On Single CCII

  Abdullah G. Alharbi (University of Missouri - Kansas City, MO, USA), Zainulabideen J. Khalifa (King Fahad University of Petroleum and Minerals, Saudi Arabia), Mohammed E. Fouda (Faculty of Engineering, Cairo University, Egypt), Masud H. Chowdhury (University of Missouri Kansas City, Kansas City, MO, USA) 174

- A Novel TSV-Based Power Harvesting System for Low-Power Applications

  Khaled Salah Mohamed (Mentor Graphics, Egypt) 178

- Effect of crystallite size and precursor molarities on electrical conductivity in ZnO Thin Films

  Okba Belahssen (University of Biskra & LPCMA Biskra, Algeria), Said Benramache (University of Biskra, Algeria), Boubaker Benhaoua (University of El-oued, Algeria) 182

- Performance Investigation and Linearity Analysis of New Cylindrical MOSFET for Wireless Applications

  Jay Hind Kumar Verma (University of delhi South Campus & Semiconductor Device Research Laboratory, India), Subhasis Haldar (Motilal Nehru College, University of Delhi, India), R s Gupta (Maharaja Agrasen Institute of Technology, India), Mridula Gupta (Semiconductor Device Research Laboratory, India) 186

Biomedical

- Embedded Agent for medical image segmentation

  Bensag Hassna (LSSDIA, Morocco), Youssfi Mohamed (Labo SSDIA, ENSET Mohammedia, Morocco), Omar Bouattane (EMSI, Lebanon) 190

- Interdigitated Electrodes Biosensor for DNA Sequences Detection

  Ayoub Bourjilat (Université de Lorraine, France), Djilali Kourtiche (Université de Lorraine-CNRS, France), Frederic Sarry (Université de Lorraine-CNRS, France), Mustapha Nadi (Université de Lorraine-CNRS & Institut Jean Lamour, France) 194

Poster Session

- Self-Charging of Medical Instruments Based on Bioelectric Potentials

  Khalifa Elmansouri (FSA, Morocco), Rachid Latif (FSA, Morocco) 198

- Neuro-Space Mapping Technique for microwave nonlinear circuits Modeling

  Taj-eddin Elhamadi (fs Tetouan, Morocco), Mohamed Boussouis (fs Tetouan, Morocco) 202

- Implementation of a 17 bits Pulse Width Modulation Circuit using FPGA

  Lucas Salomon (UNIFEI, Brazil), Robson Moreno (Universidade Federal de Itajuba, Brazil), Tales C Pimenta (Universidade Federal de Itajuba, Brazil) 206

- Low-Power CMOS Variable Gain Amplifier Design in 0.18μm Process

  Fahmi Ayadi (University of sfax Tunisia, Tunisia), Sawssen Lahyani (ENIS, Tunisia) 210

- A SWOT Analysis of TSV: Strengths, Weaknesses, Opportunities, and Threats

  Khaled Salah Mohamed (Mentor Graphics, Egypt) 214

- Shifting the half wave dipole antenna resonance using EBG structure

  Abdenacer ES-salhi (Université Mohamed 1er, Morocco), Mohammad El Ghabzouri (Mohammed First University UMP, Morocco), Paulo Mendes (University of Minho, Portugal) 218

- Direct-Elevator: A Modified Routing Algorithm for 3D-NoCs

  Maha Beheiry (Mentor Graphics, Egypt), Ahmed Aly (Mentor Graphics, Egypt), Hassan Mostafa (University of Toronto, Canada), Ahmed M Soliman (Cairo University, Egypt) 222

- On Getting Energy for Medical Equipment from Human Body

  Yehya Ghallab (Helwan University, Egypt) 226

- Enhancing Power Delay Product in DRAMs Using Resonant Tunneling Diode Buffer

  Ahmed Lutfi Elgreatly (Port-Said University, Faculty of Engineering, Egypt), Ahmed Ahmed Shaaban Dessouki (Port Said University, Egypt), Sayed El-Rabie (Dep. of Electronic and Communication Eng., Egypt) 230

- Low Voltage Low Power Highly Linear OTA using Bulk Driven Technique

  Nejib Hassen (University of Monastir, Tunisia), Karima Garradhi (University of Monastir, Tunisia), Thouraya Ettaghzouti (University of Monastir, Tunisia), Kamel Besbes (University of

Monastir & Centre for Research on Microelectronics & Nanotechnology CRMN Sousse, Tunisia) 234

- Evaluating the Feasibility of Centralized Router for Network on Chip

  Mostafa Khamis (Alexandria University, Egypt), Amir Zaytoun (Alexandria University, Egypt), Ahmed Shalaby (Egypt Japan University for Science and Technology, Egypt) 238

- A Reconfigurable 2-D IDCT Architecture for HEVC Encoder/Decoder

  Ahmed Kilany (Information Technology Institute, Egypt), Maher Mohamed (EJUST University & Benha University, Egypt), Ahmed Shalaby (Egypt Japan University for Science and Technology, Egypt), Mohammed Sayed (Egypt-Japan University of Science and Technology, Egypt) 242

- RSSI optimization method for indoor positioning systems

  Youssef Aiboud (EMSI, Morocco), Hafid Griguer (EMSI, Morocco), M'hamed Drissi UEB (INSA of Rennes, France) 246

- First specifications of Urban Traffic-Congestion Forecasting Models

  Abdellah Daissaoui (EMSI, Morocco), Azedine Boulmakoul (FSTM, Morocco), Habbas Zineb (University of Lorraine, France) 249

- Comparison between Active AC-DC Converters For Low Power Energy Harvesting Systems

  Ehab Belal (Cairo University, Egypt), Hassan Mostafa (University of Toronto, Canada), Mohamed Sameh Said (Cairo University, Egypt) 253

- A comparison of multi-resolution and multi-orientation for breast cancer diagnosis in the full-field digital mammogram

  Abdelaziz Addioui (Faculty of Sciences Ben Msik, Morocco), Faouzia Benabbou (Faculty of Sciences Ben Msik, Morocco), Mohamed El Aroussi (LETI-EHTP & Mohammed V University, Morocco), Sanaa El filali (Faculty of Sciences Ben Msik, Morocco) 257

- Comparison of Control Structures for Variable Speed Wind Turbine

  Salma El Aimani (Ibn Zohr University, Morocco) 261

- Precise electric measurements with temperature using 10-bit embedded system: Application on photovoltaic junctions

  Malaoui Abdessamad (Sultan Moulay Slimane University of Beni Mellal, Morocco) 265

- Simulation and Control of Takagi-Sugeno Uncertain Model of Buck Converter by Linear Programming

Rkia Oubah (EMSI, Morocco), Abdellah Benzaouia (University Cadi Ayyad, Morocco), Ahmed El Hajjaji (University of Picardie Jules Verne & MIS Lab, France) 269

Circuits And Systems

- Power optimization of Decode-and-Forward assisted ARQ relaying

  Ali Kamouch (LEC, Mohammadia School of Engineers, Mohammed V-Agdal University, Morocco), Abdelaali Chaoub (LEC, Mohammadia School of Engineers, Mohammed V-Agdal University, Morocco), Zouhair Guennoun (Ecole Mohammedia d'Ingenieurs & Laboratoire d'Electronique et de Communications, Morocco) 273

- Bearing Fault Diagnosis Based on Alpha-Stable distribution feature extraction and wSVM Classifier

  B Chouri (EMSI, Macao), Mohamed El Aroussi (LETI-EHTP & Mohammed V University, Morocco), Fabrice Monteiro (University of Lorraine, France), Tabaa Mohamed (EMSI, Morocco), Abbas Dandache (University of Lorraine, France), Abderhmane Jarrou (EMSI, Morocco) 277

- Design rules for RF Micro Energy Harvesting under near Field probing considerations

  Hafid Griguer (EMSI, Morocco), Hicham Lalj (EMSI, Morocco), Mohamed amine Benfatah (EMSI, Morocco), M'hamed Drissi UEB (INSA of Rennes, France) 281

- Polarization Insensitive Metamaterial Absorber For Energy Harvesting

  Hicham Lalj (EMSI, Morocco), Hafid Griguer (EMSI, Morocco), Mohamed amine Benfatah (EMSI, Morocco), M'hamed Drissi UEB (INSA of Rennes, France) 284

- A new FPGA-based DPLL algorithm to improve SAT solvers

  Khadija Bousmar (EMSI, Morocco), Fabrice Monteiro (University of Lorraine, France), Habbas Zineb (University of Lorraine, France), Dellagi Sofiene (LGIPM, France), Abbas Dandache (University of Lorraine, France) 287

- Real time EEG compression for energy-aware continous mobile monitoring

  Mohamed Adel Serhani (UAE University, UAE), Mohamed Elmenshawy (Concordia University, Canada), Abdelghani Benharref (University of Wollongong, Australia), Alramzana Nujum Navaz (UAE University, UAE) 291

Micro/Nanoelectronics

- Highly Linear Low Voltage Low Power OTA using source-degeneration Technique and Universal Filter Application

Nejib Hassen (University of Monastir, Tunisia), Karima Garradhi (University of Monastir, Tunisia), Thouraya Ettaghzouti (University of Monastir, Tunisia), Kamel Besbes (University of Monastir & Centre for Research on Microelectronics & Nanotechnology CRMN Sousse, Tunisia) 295

- A Novel High Bandwidth Current Mode Instrumentation Amplifier

  Zineb M'harzi (National Institute of Post and Telecommunication (INPT), Rabat, Morocco), Mustapha Alami (National Institute of Post and Telecommunication (INPT), Rabat, Morocco), Farid Temcamani (Superior National School of Electronics and its Applications (ENSEA), France) 299

- Performance Enhancement Of 0.18μm CMOS OnChip Bandpass Filters Using H-Shaped Parasitic Element

  Nessim Mahmoud (Egypt Japan University of Science and Technology & EJUST, Egypt), Anwer Sayed Abd El-Hameed (Assistant Research & Egypt-Japan University for Science and Technology (E-JUST), Egypt), Adel Barakat (Electronics Research Institute (ERI), Egypt), Adel Bedair (Egypt-Japan University of Science and Technology, Egypt), Ahmed Allam (Egypt-Japan University of Science and Technology, Egypt), Ramesh K Pokharel (Kyushu University, Japan) 303

- A60-GHz double-Y balun-fed on-chip Vivaldi antenna with improved gain

  Anwer Sayed Abd El-Hameed (Assistant Research & Egypt-Japan University for Science and Technology (E-JUST), Egypt), Nessim Mahmoud (Egypt Japan University of Science and Technology & EJUST, Egypt), Adel Barakat (Electronics Research Institute (ERI), Egypt), Adel Bedair (Egypt-Japan University of Science and Technology, Egypt), Ahmed Allam (Egypt-Japan University of Science and Technology, Egypt), Ramesh K Pokharel (Kyushu University, Japan) 307

Renewable Energy

- Design and Performance Analysis of Energy Conversion Chain, from Multilevel Inverter until the Grid

  Chirine Benzazah (University of Sultan Moulay Slimane Faculty of Sciences and Technology & Laboratory of Automatic, Energy Conversion and Microelectronics (LACEM), Morocco), Loubna Lazrak (University of Sultan Moulay Slimane, Faculty of Sciences and Technology, Morocco), Mustapha Ait lafkih (University of Sultan Moulay Slimane, Faculty of Sciences and Technology, Morocco) 311

- Parameters Identification of a Thin-Film Photovoltaic Panels

  Abdellatif Obbadi (Laboratory: Electronics, Instrumentation and Energy Faculty of Science University, Chouaib Doukkali, Morocco), Youssef Errami (Faculty of Science University, Chouaib Doukkali, Morocco), Abdelkrim Elfajri (Faculty of Science University, Chouaib

Doukkali, Morocco), Mustapha Agunaou (Faculty of Science University, Chouaib Doukkali, Morocco), Mohammadi Benhmida (Chouaïb Doukkali University, Faculty of Sciences, Morocco), Smail Sahnoun (Faculty of Science University, Chouaib Doukkali, Morocco) 315

- Validation of a multi-exponential alternative model of solar cell and comparison to conventional double exponential model

  Rachid Bendaoud (Chouaib Doukkali University, Faculty of Sciences, Morocco), M'hamed El Aydi (Regional Center for Education and Training, Morocco), Said Yadir (Faculté des Sciences-Université Chouaib Doukkali, Morocco), Charaf Hajjaj (Chouaïb Doukkali University, Faculty of Sciences, Morocco), Youssef Errami (Faculty of Science University, Chouaib Doukkali, Morocco), Smail Sahnoun (Faculty of Science University, Chouaib Doukkali, Morocco), Mohammadi Benhmida (Chouaïb Doukkali University, Faculty of Sciences, Morocco) 319

- Physical parameters extraction by a new method using solar cell models with various ideality factors

  Said Yadir (Faculté des Sciences-Université Chouaib Doukkali, Morocco), Houssam Amiry (Chouaïb Doukkali University, Faculty of Sciences, Morocco), Rachid Bendaoud (Chouaib Doukkali University, Faculty of Sciences, Morocco), M'hamed El Aydi (Regional Center for Education and Training, Morocco), Ahmed Elhassnaoui (Laboratoire de Génie Industriel, Faculté des Sciences et Techniques Béni - Mellal, Morocco), Abdellatif Obbadi (Laboratory: Electronics, Instrumentation and Energy Faculty of Science University, Chouaib Doukkali, Morocco), Mohammadi Benhmida (Chouaïb Doukkali University, Faculty of Sciences, Morocco) 323

2015 27th International Conference on Microelectronics (ICM)

2015 27th International Conference on Microelectronics (ICM) took place 20-23 December 2015 in casablanca, Morocco.

Committees

General Co-chairs

Mohamad Sawan

Mohamad Sawan (Polytechnique Montréal, Canada)

Honorary Co-chairs

Driss Aboutajdine (CNRST, Morocco)
Mohammed Elmasry (University of Waterloo, Canada)

Technical program Co-chairs

Fouad El Haj Hassan (Lebanese University, Lebanon)
Tabaa Mohamed (EMSI, Morocco)
Fabrice Monteiro (University of Lorraine, France)

Plenary Talk co-chairs

Said Belkouch (Ecole Nationale des Sciences Appliquées, Morocco)
Luc Hebrard (University of Strasbourg, France)

Special Sessions co-Chairs

Abbas Dandache (University of Lorraine, France)
Haitham Zaraket (Lebanese University, Lebanon)

Tutorials co-chairs

Omar Bouattane (EMSI, Lebanon)
Patrick Girard (LIRMM, France)
Ali Siadat (Arts et Métiers ParisTech, France)

Panels co-Chairs

M'hamed Drissi UEB (INSA of Rennes, France)
Hafid Griguer (EMSI, Morocco)

Industry liaison co-Chairs

Nourdine Bouayaakoub (EMSI, Morocco)
Hicham Rhioui (EMSI, Morocco)
Dominique Sauter (Nancy University, France)

Publicity co-Chairs

Abdellah Ailane (EMSI, Morocco)
Emmanuel Simeu (TIMA Laboratory, Grenoble University, France)

Publication co-Chairs

Khadija Bousmar (EMSI, Morocco)
Brahim Elbhiri (EMSI Rabat, Morocco)

Finance Chair

Jaouad Khayate (EMSI, Morocco)

Local arrangement co-Chairs

Karim Alami (EMSI, Morocco)
Zouhair Benabbou (EMSI, Morocco)
Fatmi Bergach (EMSI, Morocco)
Abdellah Daissaoui (EMSI, Morocco)
Lamia Fawzi (EMSI, Morocco)

International Coordinators

Abbes Amira (University of the West of Scotland, United Kingdom)
Mohab Anis (University of Waterloo, Canada)
Falah Awwad (UAE University, UAE)
Mountassar Mamoun (EMSI, Lebanon)
Mohamed Masmoudi (National Engineers School of Sfax, Tunisia)
Abdoul Rjoub (JUST, Jordan)
Khaled N Salama (KAUST, Saudi Arabia)

12665

Kamal Daissaoui (EMSI, Lebanon)

Roll Angle Estimation of a Spinning Object: Experimental Setup and Preliminary Results

Clément Thomas, Markus Stefer, Loïc Bernard

French-German Research Institute of Saint-Louis (ISL)
Saint-Louis, F-68300, France
Email:{clement.thomas, markus.stefer, loic.bernard}@isl.eu

Abstract—**In this paper, a low-complexity concept is presented to estimate the roll angle of a spinning object. The circuitry is based on a radio frequency (RF) power detector in combination with a linearly polarized antenna for signal reception. Due to the utilization of the power detector, the requirement on the sampling rate of the mandatory analog-to-digital converter (ADC) is relaxed significantly. First measurement results utilizing the g-hardened system are introduced and show the feasibility of the concept.**

I. INTRODUCTION

The real time roll position knowledge of a spinning object can be of great interest for many applications. Among others, in the case of gyro-stabilized projectiles of new generation, this parameter is required to trigger thruster and to steer the trajectory into a desired direction. Similarly, the roll position can be useful in the initialization of on-board inertial measurement units (IMU). Earlier work utilized a Doppler-RADAR system positioned close to the gun to estimate the angular speed [1] of a spinning object being impractical to integrate into a spinning object. However, previous results have shown that it is feasible to utilize on-board magnetometers to estimate the roll angle [2]. Based on [2], it has been recently demonstrated that this angular position can be estimated in real time with the help of on-board magnetometers [3]. The work presented in this paper is carried out in this context, the goal is to support the on-board Kalman filter in terms of its convergence by providing additional data for estimating the position. Consequently, it is also possible to utilize the concept as an alternative to the angular position estimation system based on magnetometers.

II. PRINCIPLE

The principle investigated in this paper is based on a reference continuous-wave (CW) signal with a given and known linear polarization. The CW frequency was chosen to $f_{CW} = 2.3\,\text{GHz}$ being part of the telemetry band and enabling the buildup of a compact system. The corresponding free-space wavelength is equal to $\lambda_{CW} \approx 130\,\text{mm}$. A horn antenna is utilized to transmit an electromagnetic wave at radio frequency (RF) ensuring a very small cross-polarization component. Wire antennas like dipoles could have been used but horn antennas possess a higher directivity minimizing the potential reflections in the vicinity of the measurement

setup [4]. At the same time, the link budget is increased due to the higher gain of the horn antenna compared to, e.g., a dipole. The spinning system is equipped at its base with a receiving antenna connected to a power detector. The on-board antenna is a printed antenna whose advantage is the small size at a design frequency of 2.3 GHz. Further advantages of this type of antenna are the easy manufacturing process and its low costs. The drawbacks are a gain of relative low value and a cross-polarization level that depends on design and size [5]. The power detector is a device whose output voltage is proportional to the power provided at its input. When the two antennas are aligned regarding their polarization, the maximum power is received by the power detector. On the contrary, when the polarizations of the two antennas are orthogonal, the minimal power is detected. Due to the spin of the object the output voltage of the power detector alternates continuously between maximum and minimum value. If we consider the spin constant over one or two periods, the roll position can be estimated by estimating the spin velocity from minima instants and by integrating this value from a reference position (minima typically referring to the horizontal plane). In one period of rotation, two minima are observed corresponding to the instant where the antenna polarization is orthogonal to the reference antenna resulting in an ambiguity of 180°. Nevertheless, if the spin is considered to be constant, the same position (between 0° and 360°) is observed every two minima. The remainder of the paper presents the investigations carried out to assess the performance that can be achieved by utilizing a miniaturized on-board system, and the errors induced by different rotational frequencies and azimuthal angular offsets, as it is the case with flying projectiles (pitch and yaw).

III. MEASUREMENT SETUP AND ONBOARD ELECTRONICS

The transmitter consists of the signal generator HP83752B providing a CW signal with a frequency of $f_{CW} = 2.3\,\text{GHz}$ to a horn antenna mounted on a tripod of variable height in the range of 1 m to 2 m. The horn antenna has a gain of $G = 10\,\text{dBi}$ and is the model WR430 by ATM. The height of the antenna is equal to $h = 10.5\,\text{cm}$, the width is $w = 21.0\,\text{cm}$, and the depth of the antenna is equal to $d = 36.5\,\text{cm}$. The cross polarization level measured in the anechoic chamber stays below $-30\,\text{dB}$ at a frequency of 2.3 GHz. The transmitter is placed in an anechoic chamber to avoid reflections at

978-1-4673-8760-6/15 $31.00 © 2015 IEEE

the ground or walls and the transmit antenna is oriented to emit a vertically polarized wave. The antenna utilized in the receiver is a microstrip patch on Rogers RO4350 substrate with relative permittivity of $\varepsilon_r = 3.5$ designed at ISL. It exhibits an external diameter of 46 mm and a total thickness of 4 mm. At a frequency of 2.3 GHz, its measured gain is equal to 3.8 dBi and its half-power beamwidth (HPBW) equals $102°$ in the E-plane while the measured cross polarization level is equal to -40.3 dB. The measured radiation patterns in this plane are depicted in Fig. 1.

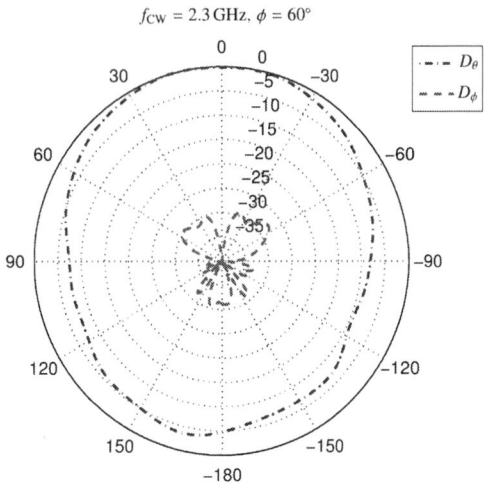

Fig. 1. Measured co-polarized (D_θ) and cross-polarized (D_ϕ) radiation patterns of the circular patch antenna in dB.

Fig. 2 shows the schematic (Fig. 2a) as well as the layout of the circuit (Fig. 2b). The latter fits a PCB of width $w = 60$ mm, height $h = 26$ mm and thickness $d = 1.55$ mm. The circuit in Fig. 2a mainly consists of the miniaturized power detector AD8362 and a balun matching the single-ended patch antenna to the differential input of the power detector. The whole system is integrated into a cylinder with an outer diameter of 60 mm that may represent a projectile. The latter is mounted on a spinning support (horizontal axis), which is itself mounted on a rotating platform (vertical axis) depicted in Fig. 3. The orientation of the spinning object in terms of the pitch angle θ and the roll angle ϕ can be characterized with the help of the measurement setup shown in Fig. 3.

The output signal of the power detector is collected through a slip ring and sampled using an oscilloscope. The slip ring is also utilized to provide the DC power supply to the power detector contained in the spinning object shown in Fig. 4. The whole system that is mounted to the spinning object depicted in Fig. 4 is g-hardened to withstand a high acceleration. Consequently, the potential deterioration of the system performance due to the spinning is mitigated. An optical sensor mounted on the spinning support provides an external reference determining the $0°$-position with respect to the rotating object. The distance between the transmitter and the receiver is equal to 3.9 m $= 29.9 \cdot \lambda_{CW}$, where λ_{CW} is the wavelength of the continuous-wave frequency f_{CW}.

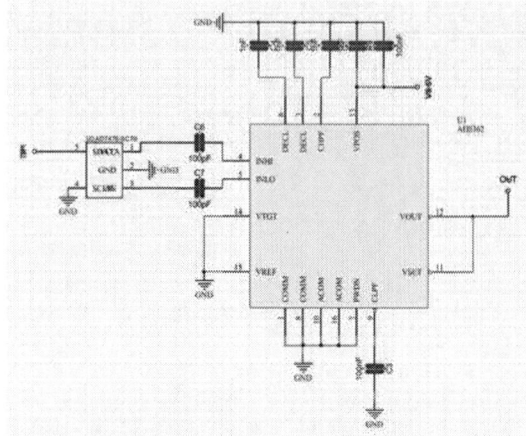

(a) Schematic of the power detector circuit.

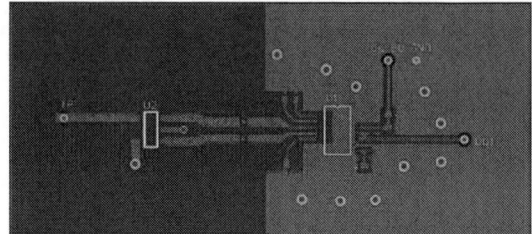

(b) Layout of the power detector circuit.

Fig. 2. Schematic and layout of the power detector circuit.

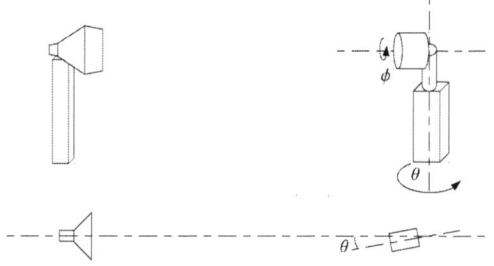

Fig. 3. Side View (top) and top view (bottom) of the measurement setup.

Fig. 4. Photograph of the spinning object with slip ring, optical sensor and g-hardened receive system.

IV. MEASUREMENT RESULTS

First, the relationship between received RF power and the output voltage of the power detector is presented. Secondly, the potential influence of different rotational frequencies of the spinning object is investigated, where the transmitter and the spinning object are aligned for maximum power reception. Finally, the potential influence of different pitch angles between transmitter and spinning object is investigated, where the rotational frequency of the spinning object is kept constant.

A. Power detector characterization

The relationship between input power P_{in} in dB m to the power detector and its output voltage V_{out} has been characterized by applying CW signals with different power levels at a frequency of 2.3 GHz. The evaluation of the measured input-power-output-voltage pairs results in the following linear relationship

$$P_{\text{rx}} = \frac{V_{\text{out}} - 2.8035 \,\text{V}}{0.0483 \,\text{V}} \,\text{dB m} \qquad (1)$$

to compute the received power P_{rx} based on the power detector output voltage V_{out}.

B. Aligned Transmitter and Spinning Object

The pivot-mounted cylinder containing the patch antenna as well as the power detector are rotated at different speeds ranging from 15 Hz to 40 Hz in 5 Hz steps. Fig. 5 exemplarily shows the output signal of the optical sensor (solid curve) and the power detector (dashed curve) depending on the time t in seconds. The two curves have been scaled to make an appropriate comparison feasible regarding the time offsets of the two power detector minima per rotation and the rising edge of the optical sensor output serving as a reference regarding the 0°-position. The optical sensor output signal also enables the estimation of the actual rotational frequency of the prototypical circuit by determining the time period T between two rising edges as indicated in Fig. 5. The latter displays an excerpt of the data acquired with respect to a rotational frequency of 30 Hz. The mean $\mu_{\hat{f}}$ and the standard deviation $\sigma_{\hat{f}}$ of the estimated rotational frequency $\hat{f}$ are listed in Tab. I.

Once the two time offsets Δt_1 and Δt_2 corresponding to the two minima are determined according to Fig. 5, the corresponding angles can be calculated by

$$\alpha_{\text{of},i} = 180° \cdot \left(2 \cdot \mu_{\hat{f}} \cdot \Delta t_i - i\right), \quad i = 1, 2, \qquad (2)$$

where $\alpha_{\text{of},i}$ describes the physical angle between the optical sensor and the H-plane of the antenna. The numerical values of the latter are depicted in Fig. 6 with respect to the different

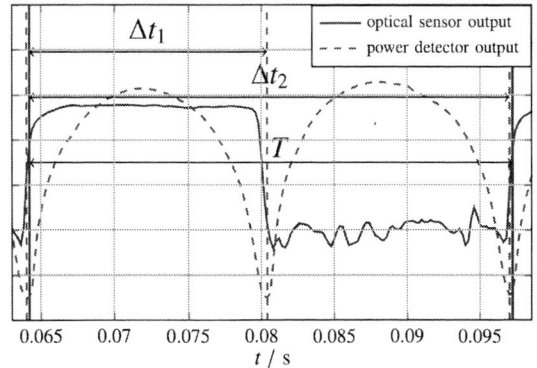

Fig. 5. Determination of the two time offsets Δt_1 and Δt_2 regarding the rising edge of the optical detector output.

TABLE I
ESTIMATED MEAN ROTATIONAL FREQUENCY AND CORRESPONDING STANDARD DEVIATION.

Adjusted frequency f	Est. mean frequency $\mu_{\hat{f}}$	Std. deviation $\sigma_{\hat{f}}$
15 Hz	15.34 Hz	0.03 Hz
20 Hz	20.57 Hz	0.04 Hz
25 Hz	25.56 Hz	0.05 Hz
30 Hz	30.17 Hz	0.05 Hz
35 Hz	35.76 Hz	0.06 Hz
40 Hz	40.34 Hz	0.04 Hz

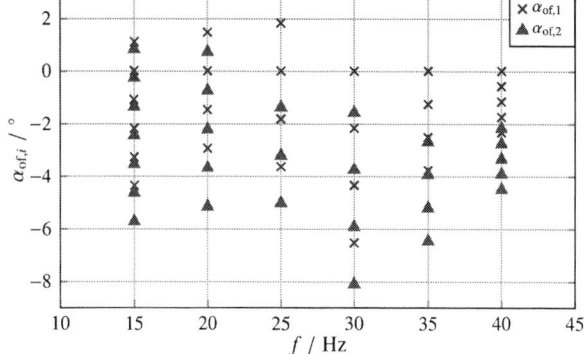

Fig. 6. Offset angle $\alpha_{\text{of},i}$ regarding the optical sensor for the two minima per rotation versus the adjusted rotational frequency f in Hz.

TABLE II
ESTIMATED MEAN OF THE MINIMA-CORRESPONDING ANGLES AND THEIR STANDARD DEVIATION DEPENDING ON THE ROTATIONAL FREQUENCY.

Adjusted frequency f	Mean angle $\mu_{\hat{\phi}}$	Std. deviation $\sigma_{\hat{\phi}}$
15 Hz	−1.55°	1.88°
20 Hz	−2.24°	1.55°
25 Hz	−1.57°	1.61°
30 Hz	−3.59°	1.97°
35 Hz	−3.18°	1.67°
40 Hz	−2.51°	1.92°

rotational frequencies. Regarding the different adjusted rotational frequencies, no azimuthal offset is present, i.e., $\theta = 0°$.

The mean and the standard deviation of the estimated angles $\hat{\phi}_i$ corresponding to the two minima are listed in Tab. II. The mean $\mu_{\hat{\phi}}$ of the estimated angle $\hat{\phi}$ varies significantly and its standard deviation $\sigma_{\hat{\phi}}$ can be considered large, even though the actual rotational frequency can be viewed constant due to its very small standard deviation.

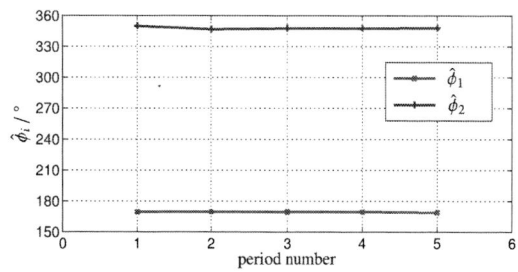

Fig. 7. Computed angles regarding the two minima per rotation for a rotational frequency of 30.17 Hz.

As it can be observed from Fig. 7, the computed angles are constant with respect to the observed periods while their difference equals 180°. An estimator is then defined based on the time period T_{prev} of the previous rotation and the first minimum of the current period $t_{\text{min},1}$ being utilized to extrapolate to the current roll angle $\hat{\phi}$. As the spin can be regarded as constant (see Tab. I), this extrapolation is linear and is given by

$$\hat{\phi} = 360° \cdot \frac{t - t_{\text{min},1}}{T_{\text{prev}}}. \qquad (3)$$

Utilizing the measured data where an excerpt is displayed in Fig. 5, the current angular position is estimated based on (3) and depicted in Fig. 8. Fig. 8 shows that an estimation of the roll angle ϕ based on (3) results in continuous phase values versus time t.

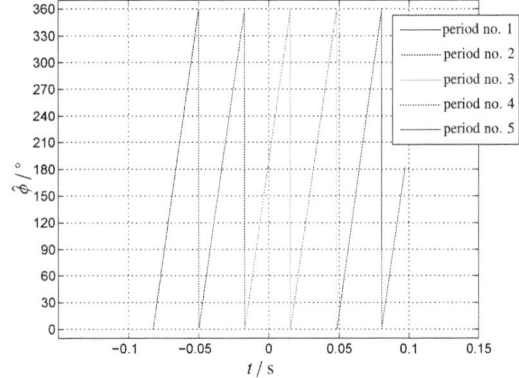

Fig. 8. Roll angle estimation $\hat{\phi}$ versus time t for a rotational frequency of 30.17 Hz.

C. Receiving Spinning Object Tilted Away from Transmitter

Fig. 9 shows the different estimated offset angles versus the pitch angle offset, both in degrees. The rotational frequency was kept constant. The mean $\mu_{\alpha_{\text{of}}}$ and the standard deviation $\sigma_{\alpha_{\text{of}}}$ of the estimated offset angle α_{of} are listed in Tab. III. In conjunction with Fig. 6, the variation of the mean value is

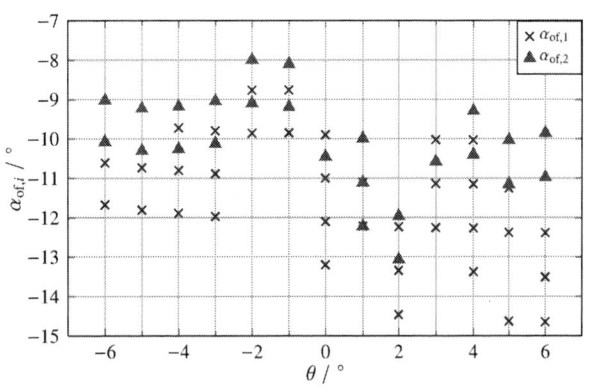

Fig. 9. Offset angle $\alpha_{\text{of},i}$ regarding the optical sensor for the two minima per rotation versus pitch angle offset θ in degree for a rotational frequency of 30.17 Hz.

TABLE III
ESTIMATED MEAN OF THE MINIMA-CORRESPONDING ANGLES AND THEIR STANDARD DEVIATION DEPENDING ON THE ROTATIONAL FREQUENCY.

Azimuthal offset θ	Est. mean angle $\mu_{\alpha_{\text{of}}}$	Std. deviation $\sigma_{\alpha_{\text{of}}}$
−6°	−10.04°	0.90°
−5°	−10.52°	0.56°
−4°	−10.27°	0.80°
−3°	−10.42°	0.84°
−2°	−9.20°	0.67°
−1°	−9.34°	0.52°
0°	−11.01°	0.93°
1°	−11.45°	0.67°
2°	−12.88°	0.90°
3°	−11.23°	0.84°
4°	−10.99°	1.51°
5°	−11.30°	1.32°
6°	−12.09°	1.71°

similar within one experiment, i.e., variation of rotational frequency or altering of the azimuthal offset. However, the system appears to be more robust in terms of the standard deviation regarding the variation of the azimuthal offset compared to the different rotational frequencies.

V. CONCLUSION

In this paper, a low-complexity circuitry is presented and the feasibility of the concept is proven by measurements. It is shown that the estimation of the roll angle yields quasi-continuous values for the roll angle. The static distance does not reflect the dynamically increasing distance during the flight of an object. To cover this scenario, an additional low-noise amplifier stage has to be included to improve the sensitivity of the whole system. To be able to utilize the estimation of the roll angle for guided projectiles, the precision has to be improved. This improvement also influences the reference system regarding the roll angle - a rotary encoder could be used to obtain a high-precision reference. Future work has to consider the potential influence of multi-path propagation as well as the path-loss-related attenuation of the signal that have not been included in the examinations within this paper. Furthermore, it is also relevant to investigate if an additional estimation of the pitch and yaw angle is possible using the presented concept.

REFERENCES

[1] G. Simon, J. P. Mangold, and P. Rateau, "Techniques de mesure de la vitesse de roulis d'un projectile gyroscopé. Étude comparative," French-German Research Institute of Saint-Louis (ISL), Tech. Rep., 1989.

[2] P. Rateau, P. Wernert, and E. Junod, "Détermination de la position en roulis d'un projectile en tir tendu ou courbe," French-German Research Institute of Saint-Louis (ISL), Tech. Rep., 2000.

[3] S. Changey, E. Pecheur, L. Bernard, E. Sommer, P. Wey, and C. Berner, "Real-time Estimation of Projectile Roll Angle Using Magnetometers: In-flight Experimental Validation," in Position, Location and Navigation Symposium (PLANS), Myrtle Beach/SC, USA, Apr. 2012.

[4] C. A. Balanis, Antenna Theory - Analysis and Design, 3rd ed. John Wiley & Sons, 2005.

[5] I. J. Bahl and P. Bhartia, Microstrip Antennas. Artech House, 1980.

Image quality assessment using nonlinear learning methods

Rshdee Alhakim[1], Ghislain Takam Tchendjou[1], Emmanuel Simeu[1], Fritz Lebowsky[2]

[1] TIMA Laboratory, Grenoble University, 38031 Grenoble Cedex, France
[2] STMicroelectronics Grenoble, Grenoble, France
Rshdee.Alhakim@imag.fr, Ghislain.Takam@imag.fr,
Emmanuel.Simeu@imag.fr, fritz.lebowsky@st.com

Abstract—Objective image quality assessment plays an important role in various image processing applications, where the goal of this process is to automatically evaluate the image quality in agreement with human visual perception. In this paper, we propose three different nonlinear learning approaches in order to design image quality assessment models, which serve to predict the perceived image quality. The nonlinear learning approaches used for the aforementioned purpose are nonlinear regression, artificial neural network and regression tree. The largest publicly available image quality database TID2013 is used to benchmark and evaluate the prediction models. The image quality metrics, provided by this TID2013, are not independent and have the redundant information of image quality. This issue might have a negative impact on the training performance and cause overfitting. To avoid this problem and to simplify the model structure, we select the most significant image quality metrics, based on Pearson's correlation measure and principal component analysis. Simulation results confirm that the three nonlinear learning models have high efficiency in predicting image quality. In addition, the regression tree model has low complexity and easy implementation, comparing to the two other prediction models.

Keywords—*Image Quality Assessment; Machine Learning; Neural Network; Non-linear Regression; Regression Tree;*

I. INTRODUCTION

With the continuous evolution of digital visual processing techniques in the acquisition, transmission, storage and display, the digital images/videos have become an important part of everyday life (such as medical imaging, video games, automated traffic control applications and security services) [1-2]. Unfortunately, these visual processes may introduce perceivable distortions and change the quality of visual signals by injecting certain artifacts such as noise, blur, ringing and blocking [3]. Nowadays, measuring the image quality is consider as a hot research topic, since it plays a significant role in enhancing artifact concealment process and in evaluating the performance of many visual processing applications (such as image decoding, lossy compression, watermarking, image denoising, etc.) [4].

Since humans are the final consumers and interpreters of the visual information, the most reliable way for assessing image quality is to ask a group of human observers to watch a test image and rate its quality. The quality scores will then be gathered from all observers and statistically processed in order to obtain Mean Opinion Score (MOS). This method is known as the subjective Image Quality Assessment (IQA) [3]. However, the subjective method requires extensive time and labor to be performed; so it is not applicable to real-time applications [5-6]. Researchers have thus made efforts to develop many objective IQA methods able to automatically assess image quality in agreement with human visual perception. Generally, the objective IQA unit extracts specific metrics (or features) from the image, and then it applies an appropriate score prediction model in order to predict the perceived quality of images. The prediction model is designed based on Machine Learning (ML) techniques, such as artificial neural network, support vector machine, nonlinear regression, regression tree, Bayesian networks, clustering, etc. Thus, ML is employed to define a mapping between image metrics and corresponding desired MOS [3].

Although the subjective IQA methods are time-consuming and rarely feasible, they are usually used for evaluating and benchmarking objective IQA algorithms. In other words, we say that the objective IQA approach has high effectiveness to predict the human perception of images, when the predicted quality scores computed from this objective IQA approach is positively correlated with the MOSs provided by the subjective IQA [3]. To measure the correlation between two quality scores, we propose two coefficients: Pearson's linear correlation coefficient (PLCC) used to measure the degree of linear dependence between MOS and its predicted scores, and Spearman's rank order correlation coefficient (SROCC) used to measure the prediction monotonicity. These two correlation measures give a value between +1 and -1. Large PLCC and SROCC values (close to 1) indicate a strong correlation between objective image quality scores and MOSs [3].

The paper is organized as follows. Section II gives an overview of the image quality database used, and also explains the method applied in order to select the most significant metrics in the database, these selected metrics with the corresponding MOSs will serve to train machine learning models. To predict the perceived image quality three nonlinear ML approaches are studied and analyzed in Section III. Section IV draws a conclusion and highlights future work.

II. IMAGE QUALITY DATABASE AND ITS METRICS

A. Image Quality Database

There are a number of publicly available databases which could be used to evaluate and benchmark the objectives IQA

978-1-4673-8760-6/15 $31.00 © 2015 IEEE

algorithms (such as TID2008, TID2013, LIVE Image, CSIQ, IVC-LAR, Toyoma, etc.) [7-13]. The research work in this paper is evaluated by Tampere Image Database 2013 (TID2013) [8-9]. TID2013 has 25 reference images (see Fig. 1), and 3000 distorted images (obtained from 25 reference images × 24 types of distortions × 5 different levels for each distortion). The MOS was obtained from the results of 985 experiments carried out by observers from five countries (Finland, France, Italy, Ukraine, and USA). TID2013 provides MOS value related to each distorted image, all MOSs range between 0 and 9 [8-9]. The range was rescaled to [0,5] (to be equivalent to the five categories: "Bad", "Poor", "Fair", "Good" and "Excellent") as shown in Fig. 2 [6].

Fig. 1. Reference images of TID2013

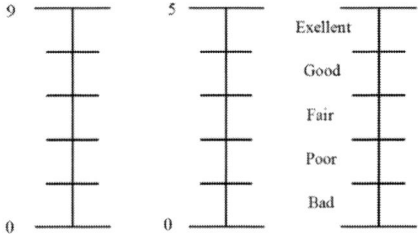

Fig. 2. Range of image quality scale

B. Metric Selection Concept

In addition to MOS, TID2013 also provides 14 image quality metrics related to each distorted image. Thanks to these metrics, we can extract specific information from an image that contributes to predict the perceived image quality. The first column of Table 1 summarizes all TID2013 metrics [8-9].

The objective of our research work, as stated later, is to exploit the image quality metrics with the corresponding MOSs in order to train and validate the proposed machine learning models. To simplify our work, we decided to only select four metrics in order to develop the proposed prediction models. Actually, it might seem that a larger number of metrics always leads to better prediction accuracy. However, this also leads to increase time consumption and model complexity. Moreover, large metric size may increase the risk of having redundant

information of images; this redundancy causes problems at the training stage and reduces the prediction accuracy (overfitting problem)[3]. Therefore, the first task we have to achieve is to filter out unimportant or redundant metrics and to select the four metrics that are the most significant.

We applied the following procedure consisting of two steps for the reduction of the metric space: First step is based on PLCC coefficient to filter out redundant information in the metrics; this step contributes to measure the mutual correlation of all metrics and then eliminate highly correlated metrics without significantly losing information. For example: Table 1 demonstrates that FSIM has high mutual correlation ($R > 0.8$) with FSIMc and MSSIM, so these metrics have redundant information and we could eliminate the two of them without risk of losing information. While the second step is used to filter out unimportant metrics and only select the four most pertinent metrics. This step is based on Principal Component Analysis (PCA) [3, 14].

Suppose the metrics are given by vector X:

$$X = (x_1 \quad x_2 \quad \cdots \quad x_p)^T \tag{1}$$

where X^T denotes the transpose of matrix X, and $p \leq 14$. The assumption is that pertinent metrics correspond to metrics with the highest weighting coefficients in the most significant Principal Components (PCs), represented by the following linear combinations:

TABLE I. PLCC FOR THE SELECTED METRICS FOR TID2013

Metrics	FSIM	PSNR	PSNRHA	SSIM
FSIM	1	0.55	0.74	0.70
FSIMc	0.99	0.55	0.74	0.72
MSSIM	0.95	0.57	0.72	0.79
NQM	0.69	0.71	0.73	0.42
PSNR	0.55	1	0.71	0.58
PSNRHA	0.74	0.71	1	0.64
PSNRHMA	0.74	0.69	0.98	0.58
PSNRHVS	0.63	0.93	0.77	0.48
PSNRHVSM	0.62	0.86	0.74	0.44
PSNRc	0.55	0.78	0.74	0.63
SSIM	0.70	0.58	0.64	1
VIFP	0.68	0.44	0.60	0.84
VSNR	0.02	0.00	0.05	0.02
WSNR	0.59	0.81	0.65	0.36

$$pc_1 = e_{11}x_1 + e_{12}x_{12} + \cdots + e_{1p}x_p$$

$$pc_2 = e_{21}x_1 + e_{22}x_{12} + \cdots + e_{2p}x_p$$

$$\vdots$$

$$pc_p = e_{p1}x_1 + e_{p2}x_{12} + \cdots + e_{pp}x_p \tag{2}$$

where $e_{j1}, e_{j2}, \ldots, e_{jp}$ are weighting coefficients. The variance for the jth principal component is known as the jth eigenvalue λ_j:

$$var(pc_j) = var(e_{j1}x_1 + e_{j2}x_1 + \cdots + e_{jp}x_p) = \lambda_j \tag{3}$$

The first principal component pc_1 has maximum variance (maximum eigenvalue), so it contains more information than other principal components. The second principal component

pc_2 has maximum of the remaining information. Hence $\lambda_1 \geq \lambda_2 \geq \cdots \geq \lambda_p$. As a result, in order to satisfy the requirement to select four pertinent metrics, we suggest selecting two pertinent metrics from each of pc_1 and pc_2; provided that these selected metrics are different.

Having applied the aforementioned procedure, the selected metrics are: PSNR, PSNR-HA, FSIM and SSIM. Table 1 shows PLCC values between the selected metrics and all database metrics. It is worth mentioning that it is possible to directly apply PCA for reducing the metric space without passing through PLCC filter. On the other hand, to ensure that the selection process works well, all image metrics must be scaled and mean-normalized.

III. Machine Learning: Implementation and Simulation

In this section, we seek to deduce an appropriate mapping between the selected image quality metrics and perceived MOSs, to achieve this relationship we proposed three different nonlinear ML approaches: non-linear regression, artificial neural network and regression tree.

Consider that reduced-size TID2013 is composed of N=3000 image datasets on the form $\{(X_1, Y_1), \ldots, (X_N, Y_N)\}$ where X_i is the image quality metric vector: $X_i = (x_{i1}\ x_{i2}\ x_{i3}\ x_{i4})^T$ corresponding to PSNR, PSNR-HA, FSIM and SSIM respectively, and Y_i is the subjective image score, where i represents the index of image set. Thus, all mentioned ML models in this section have four inputs where metric values pass and only one output that corresponds to the predicted MOS. To enhance ML training performance, we scaled and mean-normalized all quality metrics X. The full dataset is then separated randomly into two distinct sets: A training set of 2100 examples (70%) used for the model construction and a test set of 900 examples (30%) for accuracy evaluation of the generated model.

A. Non-Linear Regression Approach

We proposed in [5] the concept and structure of the non-linear regression (NLR) process used to predict MOS. The predicted MOS is represented by the sum of three terms, as follows:

$$\hat{Y}_i = A_i + B_i + C_i \tag{4}$$

where the first term A_i is linear multivariate polynomial (including the constant term), on the form: $a_0 + a_1 x_{i1} + a_2 x_{i2} + \cdots$, and the second term B_i is homogeneous multivariate polynomial of degree 2, on the form: $b_1 x_{i1}^2 + b_2 x_{i2}^2 + b_3 x_{i1} \cdot x_{i2} + \cdots$, while the last term C_i is homogeneous multivariate polynomial of degree 3, on the form: $c_1 x_{i1}^3 + c_2 x_{i1}^2 \cdot x_{i2} + c_3 x_{i1} \cdot x_{i2} \cdot x_{i3} + \cdots$, and the constants $a_0, a_1, \ldots, b_1, b_2, \ldots, c_1, c_2, \ldots$ denote polynomial coefficients. The NLR process, explained in [5], is composed of three main successive iterations: In the first iteration, we apply PLCC algorithm to select the first-degree items in A_i most correlated with MOS score, then the optimal value of the corresponding coefficients is defined by applying least square approach. Similarly, in the second and third iterations, the coefficients of polynomials B_i and C_i are calculated respectively. Fig. 3 presets dependence between the predicted and subjective MOS scores during the test phase with PLCC = 0.9 and SROCC=0.8.

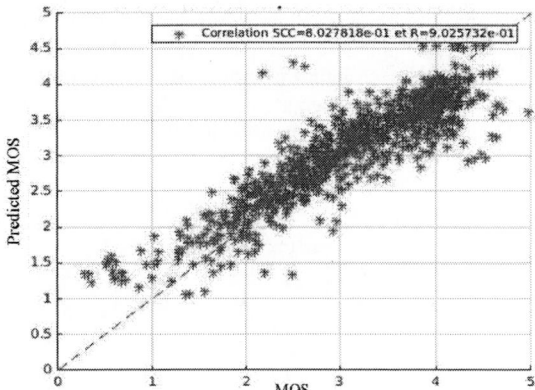

Fig. 3. MOS vs. predicted MOS for NLR method

B. Artificial Neural Network Approach

Artificial neural network (ANN) is one of sophisticated learning algorithms able to model nonlinear relationships between inputs and outputs. We proposed in [6] the concept and structure of ANN process used to predict MOS, where standard Matlab tools were employed for ANN training. Fig. 4 illustrates the proposed structure for ANN with 2 layers: First (hidden) layer contains four neurons and the logistic sigmoid activation function, while second layer contains one neuron and a linear activation function. Fig. 5 presets dependence between the predicted and subjective MOS scores during the test phase with PLCC = 0.9 and SROCC=0.8.

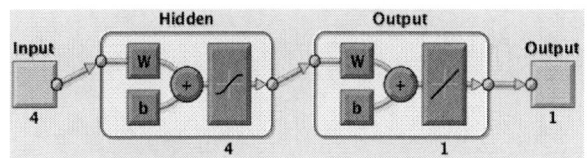

Fig. 4. ANN architecture

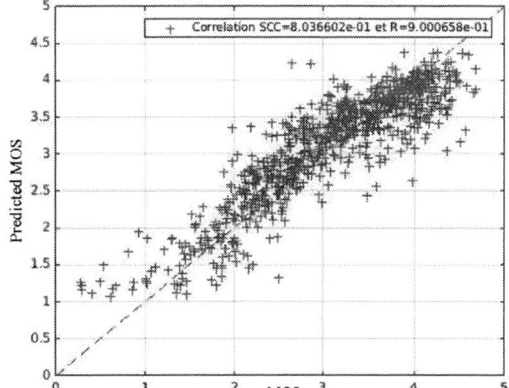

Fig. 5. MOS vs. predicted MOS for ANN method

C. Regression Tree Approach

Regression tree (RT) is nonparametric and nonlinear method, based on recursive binary-partitioning process. RT concept is described in detail by Breiman et al in [15]. We apply RT approach to seek the nonlinear mapping between the quality metrics and MOSs. Standard Matlab tools are used to

978-1-4673-8760-6/15 $31.00 © 2015 IEEE

generate the regression tree, as shown in Fig. 6, where each interior node contains a test on some input metric's value, and the terminal nodes contain the predicted MOS values. Fig. 7 presets dependence between the predicted and subjective MOS scores during the test phase with PLCC = 0.89 and SROCC=0.8.

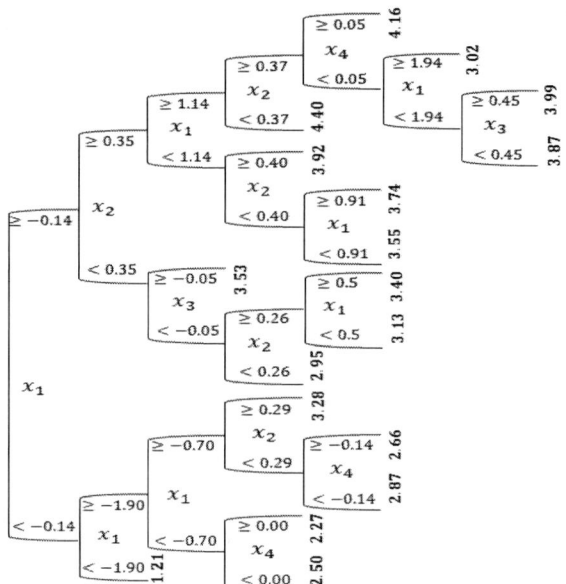

Fig. 6. Regression tree architecture

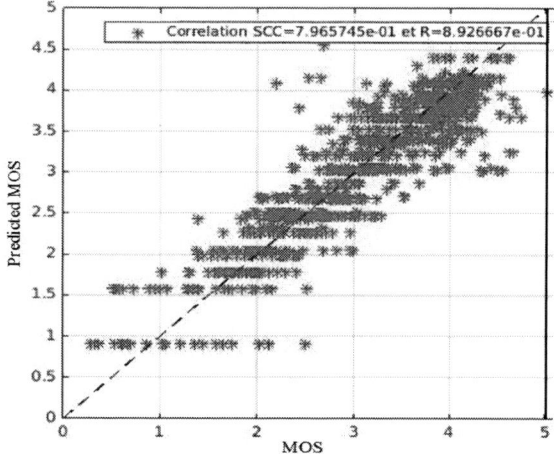

Fig. 7. MOS vs. predicted MOS for RT method

From the previous results, we can notice that the three nonlinear learning methods have almost the same prediction accuracy with high correlation coefficient values (PLCC≈0.9 and SROCC≈0.8). The advantage of RT model is that it has a very low-complexity structure for implementing, compared to NLR and ANN. This is because the RT model consists only of a series of logical if-then conditions (comparators), while NLR and ANN models contain nonlinear functions, such as polynomial or logistic sigmoid respectively.

IV. CONCLUSION

In this paper, we proposed three nonlinear learning methods in order to design an appropriate model, which serves to predict the perceived image quality. The proposed ML methods are nonlinear regression, artificial neural network and regression tree. Moreover, in order to improve model training performance and to avoid redundancy problem, we selected the most significant image quality metrics, based on Pearson's correlation measure and principal component analysis.

Simulation results confirmed that the three nonlinear learning models have high efficiency to evaluate image quality in agreement with human image quality judgments. In addition, the regression tree model has low complexity and easy implementation, compared to the two other prediction models.

In the future research, we plan to exploit these nonlinear learning algorithms on video quality monitoring unit. This unit will extract the quality metrics from the decoder output and then evaluate automatically the video quality. Finally it seeks to adjust in real-time the decoder parameters in order to improve the perceived video quality.

REFERENCES

[1] H.R. Wu, W. Lin, and L. Karam, "An overview of perceptual processing for digital pictures," Proceedings of International Conference on Multimedia and Expo, pp. 113-120, 2012.

[2] A.K. Moorthy, A.C. Bovik, "Visual quality assessment algorithms: what does the future hold?," Multimedia Tools and Applications, Vol. 51, No. 2, pp. 675-696, 2011.

[3] X. Long, W. Lin and C.-C. Jay Kuo, "Visual quality assessment by machine learning," SpringerBriefs in Signal Processing, 2015.

[4] N. Ponomarenko, O. Ieremeiev, V. Lukin, K. Egiazarian, L. Jin, J. Astola, B. Vozel, K. Chehdi, M. Carli, F. Battisti, C.-C. Jay Kuo, A New Color Image Database TID2013: Innovations and Results, Proceedings of ACIVS, Poznan, Poland, Oct. 2013, pp. 402-413.

[5] B.E. Akoa, E. Simeu and F. Lebowsky, "Video decoder monitoring using non-linear regression," IEEE 19th International Conference on On-Line Testing Symposium (IOLTS), pp.175-178, 8-10 July 2013.

[6] B.E. Akoa, E. Simeu and F. Lebowsky, "Using artificial neural network for automatic assessment of video sequences," 27th International Conference on Advanced Information Networking and Applications Workshops (WAINA), pp.285-290, 25-28 March 2013.

[7] TID2008: http://www.ponomarenko.info/ tid2008.htm.

[8] N. Ponomarenko, L. Jin, O. Ieremeiev, V. Lukin, K. Egiazarian, J. Astola, B. Vozel, K. Chehdi, M. Carli, F. Battisti and C.-C. Jay Kuo, "Image database TID2013: peculiarities, results and perspectives," Signal Processing: Image Communication, vol. 30, Jan. 2015, pp. 57-77.

[9] TID2013 Database: http://www.ponomarenko.info/ tid2013.htm.

[10] LIVE Image Quality Assessment Database: http://live.ece.utexas.edu/ research/quality/subjective.htm.

[11] CSIQ Database: http://vision.okstate.edu/ csiq.

[12] IVC-LAR Database: http://www.irccyn.ecnantes.fr/autrusse/Databases/ LAR.

[13] Toyoma Database. http://mict.eng.utoyama.ac.jp/mictdb.html.

[14] S. Larguech, F. Azais, S. Bernard, M. Comte, V. Kerzerho and M. Renovell, "A generic methodology for building efficient prediction models in the context of alternate testing," 20th International Conference on Mixed-Signal Testing Workshop (IMSTW), pp.1-6, 24-26 June 2015.

[15] L. Breiman, J. Friedman, C.J. Stone and R.A. Olshen, "Classification and regression trees: wadsworth statistics/probability," Chapman and Hall/CRC, 1984.

Efficient Architectures for HEVC Luma Interpolation Filter

Ahmed Diefy, Ahmed Shalaby, Mohammed S. Sayed
ECE Department, Egypt-Japan University of Science and Technology (E-JUST), Alexandria, Egypt
{ahmed.diefy, ahmed.shalaby, mohammed.sayed}@ejust.edu.eg

Abstract—Fractional-pel interpolation for motion estimation/ compensation is one of the most computational consuming areas in High Efficiency Video Coding (HEVC). This work presents an efficient design and implementation for luma interpolation filter in terms of hardware complexity and throughput. A new scaling factor for luma interpolation filter is adopted. By applying this modification, remarkable improvement is accomplished on hardware complexity. We propose two different architectures with fewer adders. In addition, optimization is applied on adders' bitwidth, significant improvement in area reduction is achieved; up to 40% compared to the best architecture in the literature.

Keywords—High Efficiency Video Coding (HEVC); Motion Compensation (MC); interpolation filter; hardware implementation.

I. INTRODUCTION

HEVC is the latest video coding standard. It has been developed and standardized by the collaboration of ITU-T and ISO/IEC organizations [1]. HEVC aims to achieve significant compression performance. In comparison with the previous H.264 video coding standard, HEVC provides approximately a 50% bit-rate reduction at the same quality that is essential for ultra-high definition video resolutions [2]. HEVC was designed based on the famous block-based hybrid video coding. A picture is first partitioned into blocks and then each block is predicted by using either intra-picture or inter-picture prediction. In both cases, the resulting prediction error is transmitted using transform coding, scalar quantization and entropy coding of the resulting transform coefficient levels.

Intra-picture prediction exploits the correlation between spatially neighboring blocks. Oppositely, inter-picture prediction exploits the temporal correlation between pictures by Motion Compensation Prediction (MCP). MCP is used to achieve data compression by exploiting symmetry between consecutive frames of video sequence [3]. Each frame is divided into blocks. For each block, a search is applied to find the relative motion between the current block and the best matching block on a reference frame. The relative motion is called a motion vector which is transmitted with the residual signal to the decoder [4]. In order to decrease the residual signal and accordingly increase the coding efficiency; fractional accuracy is utilized alongside integer accuracy for motion vectors. Hence, the reference block has to be interpolated [1].

In HEVC, interpolation filter's accuracy has been improved noticeably over H.264/MPEG-4 AVC. The number of taps is increased and higher precision operations are used in

filter calculations [4]. HEVC interpolation filter is designed to have 25% of pixel accuracy, consequently 15 positions ought to be interpolated. Consequently, motion compensation Interpolation filtering utilizes 20% of encoding time [5]. In addition, increasing filter taps to eight leads to further hardware complexity compared to H.264 filter with six taps. Therefore, to achieve real-time encoding / decoding for high quality videos, several ideas have been proposed to design a proficient architecture for luma interpolation filter. In this paper, we propose a new scheme for the HEVC interpolation filter in order to reduce its complexity. In addition, a new optimized architecture is presented that implements the proposed scheme.

The rest of this paper is organized as follows. In Section II; the luma interpolation filter and related work are reviewed. In Section III; the proposed scheme is introduced. In Section IV; the proposed filter architectures are presented. In Section V; the implementation results are analyzed. In Section VI; the paper is concluded.

II. RELATED WORK

HEVC uses an eight-tap filter for luma half-sample positions and seven-tap filter for quarter-sample positions. Intensity values at fractional-sample positions are calculated by applying these filters in vertical or horizontal directions. Table I shows the coefficients for seven and eight-tap luma interpolation filters according to HEVC draft [6]. Three types of luma filters are tabulated (1/4, 3/4 for quarter-pel and1/2 filter for half-pel).

Finite Impulse Response (FIR) filters are utilized for luma interpolation in HEVC. The coefficients of the FIR filters are designed utilizing a Fourier decomposition of the discrete cosine transform. The resultant interpolation filter is in this way named DCT-based Interpolation Filter (DCTIF). Analytical results of DCTIF by Samsung has been implemented in order to provide detailed complexity and performance analysis and to find out optimal interpolation filter [7]. DCTIF has been tested with three different sizes: six, eight and 12-taps with different bitwidth for coefficient representation. It was clear that longer taps and longer bitwidth filters provide more accurate interpolation and improve motion compensation part. On the other hand, decreasing the number of filter taps and coefficients' bitwidth reduces implementation complexity in terms of multiplication and additions. From these analysis, DCTIF with eight taps and six bitwidth has shown good trade-off between complexity and performance [7]. Many architectures have been proposed

978-1-4673-8760-6/15 $31.00 © 2015 IEEE

for HEVC interpolation filter [8] [9] [10]. In [8], Guo et al. proposed a reconfigurable luma interpolation filter. The filter is composed of 16-adders. The results of the proposed architecture reduced area by 23.8% compared to first 21-adders implementation by Guo, and could support QFHD @60fps. While in [9], the authors proposed a unified filter design for the eight-tap luma filters to optimize area, which uses 13-adders. The design principle adopted an optimized time-multiplexed multiple constant multiplication (TMMCM) for the filter entries. However in [10], Machado et al. proposed parallel hardware architecture for HEVC luma interpolation filter with coarse-grained reconfigurable datapath. It achieves adequate throughput to process ultra-high definition video resolutions QFHD @30fps while reducing area by 50%. The filter is composed of seven adders with pipelining stage to enhance the throughput.

III. Proposed Scaling Factor Modification

The coefficients of the luma interpolation filter are calculated by discrete cosine transform. The filter coefficients are real numbers with magnitude less than one. All filter coefficients are multiplied by a scaling factor 2^{SF} (where SF is six in HEVC) and rounded to the nearest integer. After that, the filter coefficients are corrected so that the gain of the filter is equal to one [4].

This work proposes to reduce the values of scaling factor SF. Thus, hardware implementation complexity is simplified. Table II shows filter coefficients for seven and eight tap luma interpolation filters with different scaling factor, $SF = 4$, 5 and 6 (Nominal one by HEVC). In Table II for each filter, first row shows coefficients without scaling. Then coefficients are presented scaled then rounded to get the final coefficients values. For $SF=4$, first and last coefficients in the final values are zero. This means reducing SF leads to five and six-taps filters instead of seven and eight-taps filters for quarter and half sample positions respectively. This is in contradiction of the proposed modification by the HEVC [4]. As a result, $SF = 4$ was neglected and $SF = 5$ was chosen as a scaling factor to reduce the hardware complexity while maintaining seven and eight-taps filters as required by the HEVC standard.

The effect of the proposed scaling factor modification was studied through several experiments that were executed using HEVC test model version 16.2 (HM16.2) [11] under the common HM test conditions [12]. Evaluation of coding efficiency was completed by measuring the average PSNR and bit-rate difference [13]. The proposed modification for scaling factor was tested with Quantization Parameters (QPs) 22, 27, 32, and 37. With Machine specification Intel Xeon X5690, 3.46 GHz CPU and 96 GB RAM. Table III shows the results of the proposed scaling factor modification (SF = 5) as compared to the original HEVC settings with ($SF = 6$).

From the results, it can be found that coding efficiency loss is trivial. BD_RATE degradation for RA, LB and LP are 0.19%, 0.67% and 1.79% respectively, 0.8833% on average. BD_BSNR degradation for RA, LB and LP are 0.007 dB, 0.024 dB and 0.054 dB respectively, 0.0283 dB on average. So, the proposed modification is practical to be used to reduce

TABLE I.   Luma Filter Coefficients

Filter type	Filter Coefficients							
	A	B	C	D	E	F	G	H
¼	-1	4	-10	58	17	-5	1	
½	-1	4	-11	40	40	-11	4	-1
¾		1	-5	17	58	-10	4	-1

TABLE II.   7-Tap and 8-Tap Luma Interpolation Filters With Different Scaling Factors For Quarter And Half Sample Position

	7- Tap Filter Coefficients						
$SF=0$	-0.013	0.058	-0.157	0.9002	0.2777	-0.084	0.0195
$SF=6$	-0.872	3.7497	-10.10	57.617	17.746	-5.396	1.2533
$SF'=6$	-1	4	-10	58	17	-5	1
$SF=5$	-0.436	1.8748	-5.052	28.808	8.8732	-2.698	0.6266
$SF'=5$	-1	2	-5	29	9	-3	1
$SF=4$	-0.218	0.9374	-2.526	14.404	4.4366	-1.349	0.3133
$SF'=4$	0	1	-2	14	4	-1	0

	8-Tap Filter Coefficients							
$SF=0$	-0.015	0.06	-0.17	0.623	0.623	-0.17	0.06	-0.015
$SF=6$	-0.968	4.24	-11.0	39.87	39.87	-11.0	4.24	-0.968
$SF'=6$	-1	4	-11	40	40	-11	4	-1
$SF=5$	-0.484	2.12	-5.53	19.93	19.93	-5.53	2.12	-0.484
$SF'=5$	-1	2	-5	20	20	-5	2	-1
$SF=4$	-0.242	1.06	-2.7	9.96	9.96	-2.7	1.06	-0.242
$SF'=4$	0	1	-3	10	10	-3	1	0

hardware implementation complexity without notable effect on the coding quality.

IV. Proposed Architecture

The proposed coefficients of seven and eight-tap filters are shown in Table II, at ($SF'=5$). It is very clear the symmetry between the coefficients of ¼ and ¾ types. As a result, these types can be implemented by the same hardware by only inverting the order of the input reference pixels.

A. Proposed Filter Design

The proposed filter is made out of 11-adders. Horizontal and vertical filters have the same structure. The only difference is that the horizontal filter deals with eight-bits input, however, vertical filter deals with 16-bits input. As shown in Fig. 1, the proposed filter is composed of 11-adders and seven multiplexers. All multiplexers are identical with two inputs, one output and controlled with one-bit selector. The selector is used to choose between half pixel or quarter pixel accuracy. For $S_0 = 1$, half pixel accuracy is enabled. For $S_1 = 0$, quarter pixel accuracy is enabled.

978-1-4673-8760-6/15 $31.00 © 2015 IEEE

TABLE III.   RESULTS OF THE PROPOSED SCALING FACTOR
MODIFICATION(SF=5) COMPARED TO HEVC

Anchor (HM 16.2)	BD_RATE [%]			BD_BSNR[dB]		
	RA	LB	LP	RA	LB	LP
Class A (2560x1600)						
Nebuta	0.110	-	-	-0.002	-	-
PeopleOnStreet	0.010	-	-	0.000	-	-
SteamLocomotive	0.510	-	-	-0.009	-	-
Traffic	0.060	-	-	-0.002	-	-
Class B (1920x1080)						
BQTerrace	0.013	1.500	3.561	-0.001	-0.023	-0.059
BasketballDrive	0.030	0.212	1.220	-0.001	-0.005	-0.030
Cactus	0.404	0.708	2.168	-0.008	-0.017	-0.049
Kimono	0.086	0.203	1.792	-0.003	-0.007	-0.058
ParkScene	0.067	1.471	2.415	-0.002	-0.047	-0.076
Class C (832x480)						
BQMall	0.189	0.553	1.686	-0.007	-0.022	-0.065
BasketballDrill	0.343	0.718	0.348	-0.014	-0.030	-0.013
PartyScene	0.183	1.296	1.518	-0.008	-0.055	-0.060
RaceHorses	0.251	0.563	1.281	-0.009	-0.022	-0.049
Class D (416x240)						
BQ Square	0.547	1.697	2.142	0.021	-0.061	-0.078
BasketballPass	0.169	0.179	0.644	-0.008	-0.008	-0.031
BlowingBubbles	0.375	0.438	0.983	-0.015	-0.018	-0.038
RaceHorses	0.227	0.700	1.131	-0.011	-0.034	-0.053
Class E (1280x720)						
FourPeople	-	0.575	3.249	-	-0.021	-0.110
Johnny	-	0.940	6.943	-	-0.018	-0.150
KristenAndSara	-	0.813	3.866	-	-0.022	-0.115
Class F (1280x720)						
BasketballDrillText	0.261	0.545	0.361	-0.011	-0.023	-0.014
ChinaSpeed	0.011	0.000	0.192	-0.001	0.000	-0.009
SlideEditing	0.018	0.354	0.111	-0.002	-0.053	-0.012
SlideShow	0.140	0.096	0.317	-0.011	-0.007	-0.024
Summary						
Class A	0.1700	-	-	-0.003	-	-
Class B	0.1200	0.8190	2.232	-0.003	-0.019	-0.054
Class C	0.2420	0.7820	1.208	-0.009	-0.031	-0.047
Class D	0.3290	0.7530	1.225	-0.140	-0.030	-0.048
Class E	-	0.7760	4.686	-	-0.204	-0.126
Class F	0.1079	0.2490	0.245	-0.006	-0.021	-0.015
Avg.	0.1906	0.6784	1.797	-0.007	-0.024	-0.054

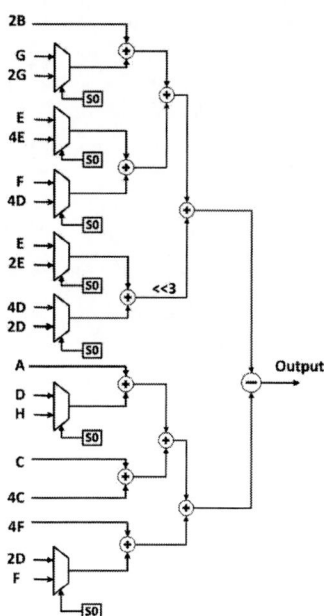

Fig. 1. Proposed Filter Design Using 11 Adders

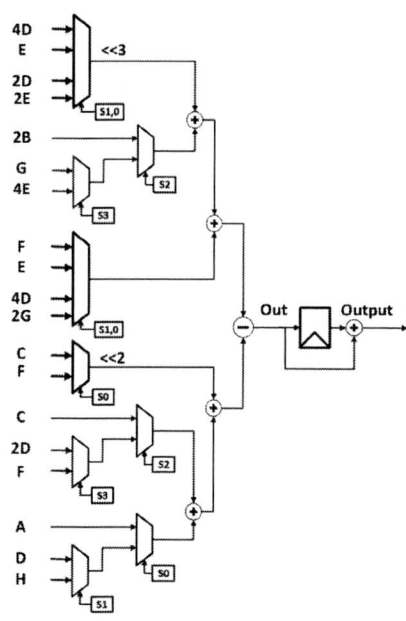

Fig. 2. Proposed Filter Design Using 6 Adders

B. An optimized and reconfigurable Filter Design

Fig. 2 shows another proposed architecture for the luma filter. The difference between this architecture and the previous one is that, the output of this architecture is generated every two clock cycles in contrary to the former architecture that has new output every clock cycle. Table IV shows different configuration for this architecture by changing selector values. For quarter pixel accuracy, intermediate output is calculated in first clock cycle and stored on the register, when selector values ($S_3S_2S_1S_0$= 0100). On the second clock cycle, the reminder portion of the output is calculated and added to the value stored on the register, when selector values ($S_3S_2S_1S_0$=1001). For half pixel accuracy, intermediate output is calculated in first clock cycle and stored on the register, when selector values ($S_3S_2S_1S_0$= 1010). On the second clock cycle, the reminder portion of the output is calculated,

TABLE IV.   DIFFERENT CONFIGURATIONS FOR OPTIMIZED AND
RECONFIGURABLE FILTER DESIGN

Filter Type	Clock Cycle	Selector				Out
		S_3	S_2	S_1	S_0	
1/4	First	0	1	0	0	32D+ G+ F- (4C+ 2D+ A)
	Second	1	0	0	1	8E+ 2B+ E- (4F+ C+ D)
1/2	First	1	0	1	0	16D+ 4D+ 2B- (4C+ C+ A)
	Second	1	1	1	1	16E+ 4E+ 2G- (4F+ F+ H)

978-1-4673-8760-6/15 $31.00 © 2015 IEEE

and added to the value stored on the register, when selector values ($S_3S_2S_1S_0$=1111).

C. Adder-bitwidth size optimization

In order to reduce hardware complexity, we propose to optimize adders' bitwidth. In our architecture, each adder in filter circuits is designed to perform with the minimum number of bits, as shown in (1).

$$Adder\ bitwidth = Input\ bits + log_2\ (Coefficient) \quad (1)$$

This method of optimization has two benefits: first, reducing the delay of each adder; second, decreasing the whole area of the circuit.

V. IMPLEMENTATION RESULTS

The proposed architectures in Fig. 1 and Fig. 2 are implemented in Verilog HDL and synthesized using TSMC 65nm standard-cell library. Table V, VI show the implemented results of the proposed architecture before and after bitwidth optimization. In Table V, the comparison between Huang's [9] work and our proposed architecture is shown. Third row shows the result for the proposed design and, fourth row shows it for the proposed design with adder bitwidth optimization. From synthesis results, reducing the number of adders to eleven achieve area gain of 19% and 13% with respect to no pipelining and fully pipelining designs respectively. In addition, adder bitwidth optimization increases the area gain to 40% and 33% respectively.

Table VI shows comparison between Machado's [10] and our optimized, reconfigurable and pipelined architecture before and after adder bitwidth optimization. Before optimization, our design improves area gain by 9% and frequency gain by 52%. After optimization, area gain is 27% and frequency gain is 120%. Frequency gain shows great gain due to using different technology rather than in [10], 65nm compared to 130nm.

VI. CONCLUSION

We proposed a new scheme and architecture for luma interpolation filter by modifying the scaling factor, in order to reduce hardware implementation complexity. The proposed scaling factor (*SF*=5) is compared with the nominal value (*SF*=6) in HEVC using HEVC test model (HM16.2). Our modification effectiveness has been proved through both average BD_RATE and BD_BSNR degradation, 0.8833%, 0.0283 dB respectively. Additionally, two different implementations are proposed using the proposed scaling factor (*SF*=5), with area gain around 33% and 27% for the new filter and optimized- reconfigurable filter respectively. Furthermore, our results prove significant improvement because of adder bitwidth optimization and using fewer adders compared to previous works.

VII. ACKNOWLEDGMENT

We would like to thank Egypt-Japan University of Science and Technology (E-JUST) for the continuous support and the Egyptian Ministry of Higher Education for funding this work.

TABLE V. COMPARISON BETWEEN THE PROPOSED FILTER DESIGN WITH 11 ADDERS AND THE ARCHITECTURE IN [9]

		No pipelining		Pipelined	
		Timing	*Gate*	*Timing*	*Gate*
	Huang's [9]	4.00 ns	2747	1.80 ns	4928
Proposed	11-Adder	2.41 ns	2225	1.91 ns	4178
	Optimized bitwidth	2.34 ns	1639	1.63 ns	3271
	Gain	41.5%	40.3%	9.4%	33.6%

TABLE VI. COMPARISON BETWEEN THE PROPOSED FILTER DESIGN WITH 6 ADDERS AND THE ARCHITECTURE IN [10]

		Frequency (MHZ)	Gate count
	Machado's [10]	312	1363
Proposed	6-Adder	476	1236
	Optimized bitwidth	689	991
	Gain	120.8%	27.2%

References

[1] G. J. Sullivan, J.-R. Ohm, W.-J. Han, and T. Wiegand, "Overview of the high efficiency video coding (HEVC) standard," IEEE Trans. Circuits Syst. Video Technol., vol. 22, no. 12, pp. 1649-1668, Dec. 2012.

[2] J.-R. Ohm, G. J. Sullivan, H. Schwarz, T. K. Tan, and T. Wiegand, "Comparison of the coding efficiency of video coding standard Including High Efficiency Video Coding (HEVC)," IEEE Trans. Circuits Syst. Video Technol., vol. 22, no. 12, pp. 1668–1683, Dec. 2012.

[3] B. Girod, "Motion-compensating prediction with fractional-pel accuracy," IEEE Trans. Commun., vol. 41, no. 4, pp. 604–612, Apr. 1993.

[4] K. Ugur, et al., "Interpolation Filter Design in HEVC and its Coding Efficiency-Complexity Analysis," Proc. ICASSP, Vancouver, Canada, May 2013.

[5] F. Bossen, B. Bross, K. Sühring, and D. Flynn, "HEVC complexity and implementation analysis," IEEE Trans. Circuits Syst. Video Technol., vol. 22, no. 12, pp. 1684–1695, Dec. 2012.

[6] J. Boyce, et al., "Edition 2 Draft Text of High Efficiency Video Coding (HEVC), Including Format Range (RExt), Scalability (SHVC), and Multi- View (MV-HEVC) Extensions," document JCTVC-R1013, July 2014.

[7] E. Alshina, J. Chen, A. Alshin, N. Shlyakhov, & W. J. Han, "CE3: Experimental results of DCTIF by Samsung", JCTVC-D344, 20-28 Jan. 2011.

[8] Z. Guo, D. Zhou, and S. Goto, "An optimized MC interpolation architecture for HEVC," in Proc. IEEE Int. Conf. Acoust. Speech Signal Process. (ICASSP), Kyoto, Japan, pp. 1117–1120, Mar. 2012.

[9] C. T. Huang, C. Juvekar, M. Tikekar, & A. P. Chandrakasan, "HEVC Interpolation Filter Architecture for Quad Full HD Decoding," in Visual Communications and Image Processing (VCIP), pp. 1-5, Nov. 2013.

[10] C. Diniz, M. Shafique, S. Bampi, and J. Henkel, "High-throughput interpolation hardware architecture with coarse-grained reconfigurable data paths for HEVC," in Proc. IEEE Int. Conf. Image Process. (ICIP), Melbourne, VIC, Australia, pp. 2091–2095, Sep. 2013.

[11] https://hevc.hhi.fraunhofer.de/trac/hevc/browser/tags/HM-16.2.

[12] F. Bossen, "Common HM test conditions and software reference configurations," Document of Joint Collaborative Team on Video Coding JCTVC-L1100, Jan. 2013.

[13] G. Bjøntegaard, Calculation of average PSNR differences between RD curves, document VCEG-M33, 13th VCEG Meeting, Apr. 2.

978-1-4673-8760-6/15 $31.00 © 2015 IEEE

Dynamic Channel Coding Reconfiguration in Software Defined Radio

Ahmed Sadek[1], Hassan Mostafa[1,2], Amin Nassar[1]

[1]Department of Electronics and Communications Engineering, Cairo University, Giza, Egypt,
[2]Center of Nano-electronics and Devices, AUC and Zewail City of Science and Technology, Egypt.
{ ahmd.sadk@gmail.com, hmostafa@uwaterloo.ca, amin.nassar@yahoo.com }

Abstract— **Digital Front End Reconfiguration is considered one of the most promising techniques to implement the Software Defined Radio (SDR) and the Cognitive Radio (CR), allowing the same set of hardware to accommodate Multi-Standard Communication Systems (MSCS). The benefit increases when the reconfiguration is not only dynamic but also takes place in real time without the need to switch off the system. This work shows the advantages of using the Dynamic Partial Reconfiguration (DPR) technique in implementing the SDR convolutional encoders for 2G, 3G, LTE, and WIFI communication standards. Experimental results reveal that the DPR implementations improves the SDR implementation area and power consumption by 67% and 64% compared to the full implementation of these standards. The full implementation design and the DPR-based design are implemented and experimentally tested on Xilinx Virtex 5 design kit XUPV5-LX110T.**

Keywords— **Software Defined Radio; Dynamic Partial Reconfiguration; FPGA; Convolutional encoders**

I. INTRODUCTION

Wireless communication standards are continuously changing and upgrading to support new features and enable new technologies. Therefore, both of the base station and the user terminal need to adopt dynamic communication chains capable of supporting multiple standards which is denoted by Software Defined Radio (SDR) [1]. SDR implementation enables compact system implementation to reduce the Silicon area and the power consumption and lengthen the battery life. Moreover, SDR implementation makes it easy to accommodate new standards with limited hardware modifications which reduces the production time and cost. FPGA is a programmable IC configured to execute a certain application. The utilization of the FPGA in the digital signal processing is increasing due to the fact that it is the best compromise between the configurability and the speed. The flexibility of the FPGA allows it to be used in the hardware implementation of the SDR. The aim of this work is to highlight the benefits of using the DPR capability of the FPGA in the implementation of the SDR compared to the conventional full implementation.

This work compares between two different convolutional encoders system. The first one is denoted by General Encoder Module (GEM) where all encoders exist on the chip and a multiplexer is used to switch among them. The second convolutional encoder system uses the DPR technique where one encoder loaded to the chip at a time and is denoted by Single-Loaded Encoder Module (SLEM). The SLEM module exhibits lower area and power consumption compared to the

GEM module at the expense of more latency delay and memory overheads.

The rest of the paper is organized as follows. Section II gives some background on the dynamic partial reconfiguration feature. Section III shows how the SDR system is used to implement the MSCS system. Section IV explains the GEM and SLEM implementations and simulation results. Some conclusions are drawn in Section V.

II. DYNAMIC PARTIAL RECONFIGURATION

As the complexity of the communications system increases, the upgrading of the fixed and mobile standards is growing. This upgrading requires extra effort and financial resources as well as compatibility with the old standards. One of the future trends is the hardware reusability which means using the same hardware to define multiple standards. This hardware reusability is achieved by using the DPR feature of the FPGA. The benefits of the DPR are its flexibility of redefining system modules, real time functioning while being reconfigured, saving power by replacing unused modules, and reducing the cost of accommodating new designs [2]. Reconfigurable FPGAs are divided into two categories. In the first category, a full image is loaded to the FPGA to run a certain application. This image does not change during run time but new values are set in the registers to change the mode of operation of the application. This approach is called parametrization [3]. The second type is reconfigurablity in which a new image is loaded to the FPGA to execute a new application. FPGA reconfiguration is either full or partial. In the full FPGA reconfiguration, the FPGA has to stop working to download the new bit stream. On the other hand, in the partial FPGA reconfiguration, a part of the bit stream is updated and loaded in the FPGA. The following are the main factors that define the DPR:

A. Configuration mode

Table I shows the different configuration modes for Xilinx FPGA Virtex-5 [4]. There are two methods for loading the configuration data to the FPGA, which is dependent on the way of reaching the configuration plan:

- Internal: Using internal softcore processors such as MicroBlaze or hardcore processor such as PowerPc through Internal Configuration Access Port (ICAP). ICAP primitive is a SelectMap-like protocol that provides access to the internal configuration memory [4].

- **External:** Using external controller DSP or another FPGA as a master. The reconfigured FPGA is used in slave mode and the Serial mode, JTAG or SelectMap are used to access the configuration memory [4].

TABLE I. CONFIGURATION MODE [4]

Configuration Mode	Type	Max Clock	Data Width	Max Bandwidth Bps (bps/8)
ICAP	Internal	100 MHz	32-bit	400 MBps
SelectMap	External	100 MHz	32-bit	400 MBps
Serial Mode	External	100 MHz	1-bit	12.5 MBps
JTAG	External	66 MHz	1-bit	8.25 MBps

B. Reconfigurable Module Style based

Depending on the design size, the needed reconfiguration is either difference based or module based:

- **Difference Based:** used for small designs to edit the connections of few LUTs. The bit stream file contains the difference between the different applications [5].

- **Module Based:** used for large design changes by reconfiguring a complete block with a new one [6].

C. Configuration Memory array type

There are two different methods for reconfiguring the FPGA logic depending on the internal architecture:

- **1D:** A complete array column is reconfigured with the new bit stream such as the DPR feature implemented in Xilinx Virtex II. Fig. 1-a shows how a complete D1 column is reconfigured with a D2 column [7].

- **2D:** It is similar to the memory structure as some specific cells are accessed not a complete column. This type exists in recent FPGAs families such as Xilinx Virtex 4, 5, 6, and 7. Fig. 1-b shows that D1 block is reconfigured with D2 block [2].

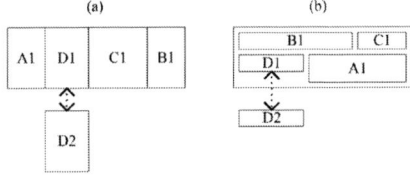

Fig. 1. Configuration memory array types.

D. Reconfiguration type

Reconfiguring part of the FPGA depends on how the data to be altered on the FPGA:

- **Static:** In this type, part of the FPGA is reconfigured while suspending the work of the FPGA. In this reconfiguration type, the communication systems will halt during the reconfiguration process [8].

- **Dynamic:** real-time reconfiguration during the normal FPGA operation, which reduces the reconfiguration time overhead. In this reconfiguration type, the communication systems are functioning during the new image reloading [8].

III. MULTI-STANDARD COMMUNICATION SYSTEM

Recently, increasing the connectivity among people is the main driving force for the wireless technology progress. This progress includes creating distinct standards such as 2G, 3G, LTE, WIFI, and Bluetooth. Unfortunately, the multi-standard communication system results in reducing the battery life due to the increased power consumption and inefficient utilization of the radio frequency spectrum.

SDR is one of the solutions that tackle the multi-standard receiver problem by reusing the same hardware to implement the multi-standard system with software reconfiguration [1]. Another approach that resolves the radio spectrum utilization problem is the Cognitive Radio (CR). CR is a way of using the available spectrum without interfering with the existing users. The implementation of the SDR with the CR concept leads to the MSCS by redefining the hardware blocks to accommodate different communication standards.

Fig. 2 simplifies the general block diagram of the wireless communication system. The smart antenna, RF front-end, Digital to Analog Converter (DAC), Analog to Digital Converter (ADC), and the Digital Signal Processing (DSP) module. The main focus of this work is on how the DPR enhances the area and power consumption compared to the full implementation method in the DSP module.

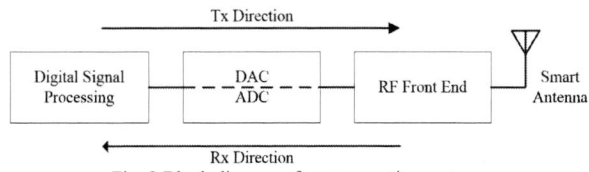

Fig. 2 Block diagram of communication system.

2G, 3G, LTE and WIFI are widely used wireless communication standards. Implementing all these standards on the same Integrated Circuit (IC) results in larger Silicon area and higher power consumption for each channel. However, this area and power consumption can be improved if these standards are implemented by using dynamic and reconfigurable systems.

Fig. 3 demonstrates a simple communication channel adaptation that switches among different channel coding schemes. Applying the same concept to other blocks results in a completely reconfigurable system. A design of modulation chains using the DPR technique has been proposed in [9]. In [10], different implementations of channel coding schemes were compared. This paper investigates experimentally how the DPR technique is efficient in implementing multi-standard systems on the Xilinx Virtex 5 design kit XUPV5-LX110T. Furthermore, a study of the trade-off between the power consumption and the reconfiguration time for different MicroBlaze speeds in the DPR design.

The hardware switching is triggered during the handover between the two different communication systems. However, the handover takes hundreds of milliseconds [11] while the hardware reconfiguration takes much shorter time (i.e., in the microseconds range). On the other hand, many studies have been done on offloading the data traffic between cellular systems, e.g. 3G, and a fixed wireless systems, e.g. WIFI [12].

978-1-4673-8760-6/15 $31.00 © 2015 IEEE

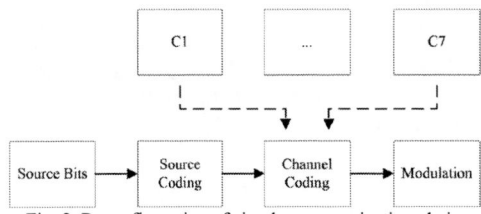

Fig. 3. Reconfiguration of simple communication chain.

IV. IMPLEMENTATION AND EXPERIMENTAL RESULTS

Convolutional encoders are forward error correction coding schemes used to add parity symbols for the data sent over a communication channel. These coding schemes decrease the effect of the noisy channel on the sent data. Three main parameters are used in the convolutional encoders [n, k, l], where n is the number of inputs, k is the number of outputs and l is the constraint length which indicates the number of memory elements used. The convolutional encoder rate is n/k. This rate can be increased by using a puncturing at the output of the encoder. Table II shows the characteristics of the different convolutional encoders used in 2G, 3G, LTE and WIFI technologies. Different convolutional encoders reflect different hardware connections between memory elements.

TABLE II. CONVOLUTIONAL ENCODERS USED IN 2G, 3G, LTE AND WIFI

Conv. Encoders	System	Channel	Rate (n/k)	Constraint length (l)	Generator polynomials (octal)
C1	2G	TCH/FR Speech	1/2	5	G0 = 31 G1 = 33
C2	2G	TCH/HR Speech	1/3	7	G4 = 155 G5 = 123 G6 = 137
C3	2G	Data	1/3	5	G1 = 33 G2 = 25 G3 = 37
C4	3G	BCH, PCH, RACH, DCH, FACH	1/2	9	G0 = 561 G1 = 753
C5	3G	DCH, FACH	1/3	9	G0 = 557 G1 = 663 G2 = 711
C6	LTE	BCH, DCI, UCI	1/3	7	G0 = 133 G1 = 171 G2 = 165
C7	WIFI 802.11a,g	OFDM channel	1/2	7	G0 = 133 G1 = 171

Two designs are presented in this work. First, GEM is shown in Fig. 4, where all the convolutional encoders exist on the FPGA at the same time and the desired one is selected through a multiplexer. Secondly, SLEM, where encoders are stored on external memory and loaded per request, is shown in Fig. 5. A SoC design using an embedded softcore processor MicroBlaze is utilized with simple operating software in both designs. The processing system consists of a MicroBlaze, SysACE controller to load bit stream files form compact flash, ICAP for internal FPGA configuration and UART for interfacing with PC. The test setup used for this work consist of a PC connected to Xilinx Virtex 5 kit XUPV5-LX110T

through a serial cable. A PC terminal emulator Tera-Term is used to communicate with the Kit.

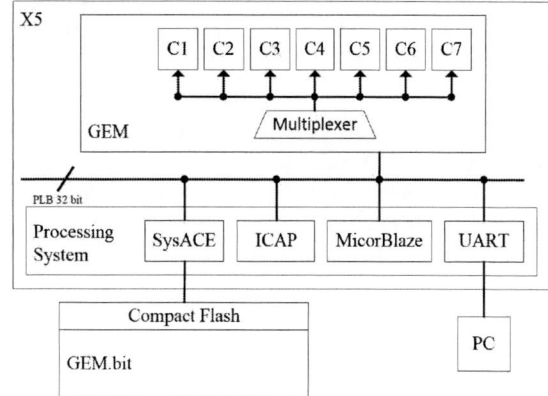

Fig. 4. General Encoder Module (GEM), where multiplexer used to switch among differenet encoders.

In GEM design, a complete image is loaded from the compact flash. The processing system is responsible for the software interface with the PC through UART and controls the multiplexer to switch among the encoders. In SLEM design, part of the FPGA programmable logic is configured as dynamic reconfigurable area using the Plan-Ahead tool. The FPGA is loaded initially by the convolutional encoder C1 in the dynamic part. Encoders (C1 – C7) are stored in compact flash. The needed encoder is loaded to the FPGA dynamic part on demand to replace the existing encoder. The MicroBlaze is responsible for software interface with the PC and manages the switching among the different partial bit stream files (C1 – C7), using ICAP to reconfigure the FPGA dynamic area with the selected encoder. Experimental test results by using the Xilinx Virtex 5 kit XUPV5-LX110T and the above test setup are presented as follows.

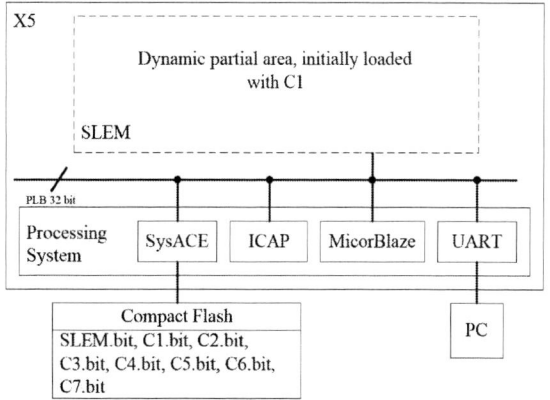

Fig. 5. Single-Loaded Encoder Module (SLEM), one encoder loaded at a time using MicroBlaze.

A. Area

The total area assigned for both designs (i.e., GEM and SLEM) on the Xilinx Virtex 5 kit XUPV5-LX110T is 800 LUTs. The FPGA utilization is calculated with respect to these assigned 800 LUTs. Table III shows the number of LUTs used

978-1-4673-8760-6/15 $31.00 © 2015 IEEE

for both designs and their utilization percentage. It should be noted that the full implementation of the encoders (C1 – C7) should utilize 258 LUTs (i.e., the sum of the utilized LUTs of each encoder). However, to provide a fair comparison, the Xilinx programming tool optimizes the full implementation code which results in a utilization of only 113 LUTs as listed in Table III. It is evident from Table III that the DPR improves the area of the optimized full implementation from 113 LUTs to a maximum of 38 LUTs with area improvement of 66.4%. In addition, the utilization percentage is improved from 14.1% to a maximum of 4.8%.

TABLE III. AREA UTILIZATION FOR BOTH DESIGNS

Conv. Encoders	SLEM							GEM
	C1	C2	C3	C4	C5	C6	C7	
No. of LUTs	36	37	37	37	38	37	36	113
Utilization (%)	4.5	4.6	4.6	4.6	4.8	4.6	4.5	14.1

B. *Memory needed*

The initial file size of the GEM and SLEM designs is 3.8 MB. However, the SLEM requires additional memory for the partial bit stream files of each individual encoder. The size of each partial bit stream file is 60kB and accordingly, the extra needed memory for the SLEM is 420kB (because there are 7 encoders). Thus, the memory overhead of the SLEM is 10.8% compared to GEM.

C. *Power Estimation*

Xilinx Power Analyzer tool (XPA) is used to estimate the power consumption in the convolutional encoder modules. Table IV lists the power consumption of the GEM and SLEM designs for different MicroBlaze frequencies. Once again, it should be highlighted that the calculated power consumption of the GEM full implementation is after performing optimization with the Xilinx configuration tool for fair comparison. It is evident from Table IV that the DPR improves the power consumption up to 64%.

TABLE IV POWER CONSUMPTION

Operating Frequency	SLEM Power Consumption	GEM Power Consumption	Power Improvement
50 MHz	0.08 mw	0.18 mw	56 %
75 MHz	0.12 mw	0.33 mw	64 %
100 MHz	0.17 mw	0.44 mw	61 %

D. *Time overhead*

The initial configuration time for both designs (i.e., GEM and SLEM) is the same because the initial bit stream file has the same size 3.8 MB for both. In the full implementation there is no time overhead added because all encoders exist. In SLEM, partial bitstream files of size 60 kB is generated for each encoder. Correspondingly, this results in a reconfiguration time overhead. Table V shows the configuration and reconfiguration time for both designs. Also, Table V illustrates how the MicroBlaze frequency affects the configuration and reconfiguration time. According to this table, the reconfiguration time overhead in the SLEM design is insignificant and is less than 1.6%.

TABLE V. CONFIGURATION AND RECONFIGURATION TIME

Design	SLEM (C1 – C7)	GEM
Operating Frequency 50 MHz		
Configuration Time (ms)	19	19
Reconfiguration Time (ms)	0.30	0
Operating Frequency 75 MHz		
Configuration Time (ms)	12.7	12.7
Reconfiguration Time (ms)	0.20	0
Operating Frequency 100 MHz		
Configuration Time (ms)	9.5	9.5
Reconfiguration Time (ms)	0.15	0

V. CONCLUSION AND FUTURE WORK

DPR is a flexible and efficient way of realizing SDR in a Cognitive Radio system. Implementing a library of different encoders and switching among them reduces the system complexity and makes it handy and real time upgradable. Future work will include simplifying the communication system chain and generalizing the concept of DPR for other blocks. This would help in offloading between cellular systems and fixed wireless systems such as WIFI, not only on the network level but also on the hardware level.

ACKNOWLEDGEMENT

This work was partially funded by Zewail City of Science and Technology, AUC, Cairo University, NTRA, ITIDA, SRC, ASRT, the STDF, Intel, Mentor Graphics, MCIT, NSERC.

REFERENCES

[1] J. Mitola, "The software radio architecture," IEEE Communications Magazine, vol. 33, no. 5, pp. 26-38, May 1995.

[2] D. Dye, "Partial Reconfiguration of Xilinx FPGAs Using ISE Design Suite," WP374 (v1.2), May 30, 2012.

[3] H. Harada, Y. KAMIO, and M. FUJISE, "Multimode software radio system by parameter controlled and telecommunication component block embedded digital signal processing hardware," IEICE transactions on communications 83.6, pp. 1217-1228, 2000.

[4] Virtex-5 FPGA Configuration User Guide UG191 (v3.11) October 2012.

[5] E. Etto, "Xilinx XAPP290 Difference-Based Partial Reconfiguration (v2.0)," December 2007.

[6] Xilinx Partial Reconfiguration User Guide UG702 (v14.5) April 2012.

[7] P. Sedcole, B. Blodget, T. Becker, J. Anderson, P. Lysaght, "Modular dynamic reconfiguration in Virtex FPGAs," Computers and Digital Techniques, IEE Proceedings - , vol.153, no.3, pp.157,164, 2 May 2006.

[8] Xilinx Partial Reconfiguration Tools & Techniques trining course, "http://www.xilinx.com/training/fpga/fpga31000-ilt.pdf"

[9] K.A. Arun Kumar, "FPGA implementation of PSK modems using partial re-configuration for SDR and CR applications," India Conference (INDICON), 2012 Annual IEEE , vol. 205, no. 209, pp. 7-9, Dec. 2012.

[10] M. Hentati, A. Nafkha, Xun Zhang, P. Leray, J.-F. Nezan and M. Abid, "The study of the impact of architecture design on cognitive radio," Systems, Signals and Devices (SSD), 2011 8th International Multi-Conference, vol. 1, no. 4, pp. 22-25, March 2011.

[11] T. Janevski, "Traffic analysis and design of wireless IP networks," Artech House, 2003.

[12] A. Balasubramanian, R. Mahajan, and A. Venkataramani, "Augmenting mobile 3G using WiFi," Proceedings of the 8th international conference on Mobile systems, applications, and services, ACM, pp 209-222, 2010.

An Efficient Hybrid Power Modeling Approach for Accurate Gate-Level Power Estimation*

A. Nocua, A. Virazel, A. Bosio, P. Girard
LIRMM - CNRS / University of Montpellier
Montpellier-France
Email: <lastname@lirmm.fr>

C. Chevalier
STMicroelectronics
Grenoble-France
Email: <cyril.chevalier@st.com>

Abstract—This paper presents a hybrid power modeling approach based on an efficient library characterization methodology and an effective power estimation flow to accurately assess gate-level power consumption in a faster way. As a case study, we apply the proposed approach on *28nm Fully-Depleted Silicon On Insulator* technology.

Index Terms—FDSOI Technology, Hybrid Power Model, Library Characterization, Power Estimation Technique.

I. INTRODUCTION

High power consumption is a key concern in the design phase of digital circuits. It may cause chip failure, performance issues or an increment in cost or area. To reduce the power consumed by a chip, power optimization techniques are implemented at high abstraction levels. Generally, power estimation at these levels is based on a macro-modeling approach, in which a power model is created using pre-characterized power values as reference.

Usually, chip design uses several high-level components called Intellectual Property (IP), in which general information is saved regarding IP functionality, timing and power information. Mostly, power models are created by computing power consumption at different levels of abstraction [1]. However, in some cases the assessment is not completely reliable as the estimation methodology does not fit the IP characteristics and a detailed information of power consumption per component cannot not be obtained. In addition, time-dependent power consumption is modeled based only on average power results, leaving out the real impact of the instantaneous power dissipation thus reducing the efficacy of the created power model. To ensure the accuracy of these models, power estimation is done at lower levels of the design phase. Then IP power models based on gate-level power estimation are necessary.

Several works have been proposed at gate-level to assess power consumption per component, such as dynamic, short-circuit or leakage power. For instance, the work in [2] presents current/power simulation flow to compute dynamic power. They model the gate current as a triangle shape taking into account different operational conditions (supply voltage, ground voltage) and input switching conditions (rise, fall,

static). Their results shows good accuracy with transistor level simulations. However, their assumptions does not fit quite well in sub-nanometer technologies, due to the impact of non-linear parameters on the current profile. Current-based approaches have been applied to compute short-circuit power by taking into account extra current components due to instantaneous switching currents [3], [4]. Other works consider different conditions independently for power analysis like: under noisy input waveforms, temperature variations [5] or supply variations [6]. Most of these techniques envisage a partial solution to compute accurate power consumption, as they are focused on only one power component and on some simplistic assumptions, i.e. Single-Input Switching (SIS) at gate inputs. Nevertheless, this is not a realistic scenario for almots all the circuits. Regarding the leakage component, the work in [7] developed a probabilistic power model taking into account the transistor-level behavior under static conditions. However, they target obly spatial and temporal independence at the input pins analysis.

In order to handle the above mention issues, we propose a hybrid power model and an effective power estimation flow at gate-level, in which, we are able to exploit the run-time efficiency of logic simulation and the physical accuracy of the transistor-level simulation. In this paper, we present the phases of the characterization methodology and the power estimation flow. We use as case study a 28 nm Full-Depleted Silicon on Insulator (FDSOI) technology from STMicroelectronics [8]. The main contributions of this paper are as follows:

- We present an effective characterization methodology that takes into account multiple conditions at the same time.
- We analyze the use of the characterization results to reconstruct circuit-level current/power trace in an efficient way.
- We study the power consumption of standard cells on a *28nm FDSOI* technology per component, i.e. dynamic, short-circuit, leakage power.

This paper is organized as follows. In section II, the details of our library characterization methodology is described. In Section III, we present the basis of our effective power estimation flow. In Section IV, we present the experimental analysis at gate and circuit-level for the technology under study. Section V, concludes this paper.

---

* This work has been funded by the French Government under the framework of HiCool project: Methods and Tools to Design Low-Power System-On-Chip.

978-1-4673-8760-6/15 $31.00 © 2015 IEEE

II. LIBRARY CHARACTERIZATION METHODOLOGY

In a library characterization, a detailed analysis of basic components of the standard-cell is executed. The characterization results are often saved in a standard maner, like in Liberty (**lib**) format. Generally, it is provided by the foundry and it contains physical, timing and power information for each standard cell. The stored information is then used by the power estimation engine to compute average energy/power per component.

Usually, there is no knowledge about the characterization conditions and if the designer team choose to have its own cell library, then accuracy must be ensured [9]. In addition, lib information is obtained for a small subset of the whole parameter conditions, e.g. only SIS events. Even though, it is a faster analysis at run-time, in some cases it lacks of accuracy.

We propose an effective characterization methodology, based on the physical description of standard cells. With this methodology, we aim to ensure the reliability of data results and to increase the accuracy on the assessment of power consumption. We take into consideration several gate-level parameters that impact the current/power estimation, i.e. Multiple-Input Switching (MIS) events under different environmental and load conditions.

To ensure the quality of the characterization methodology, we need to consider the following aspects: (1) the parameter selection, to include the ones that directly impact the power consumption; (2) the parameter ranges, to highlight non-linear effects on the current trace. (3) How the parameters are set on the electrical simulator, to ensure realistic-case conditions for each gate. In our case, we have analyzed which parameters constraint directly current waveforms, hence the power consumption. Based on a complete analysis of standard cells characteristics, we determine three groups of parameters:

- **Environmental conditions:** Supply Voltage (V_{DD}), Ground Voltage (GND), Body Biasing (BB) and Temperature (T) .
- **Input pin information:** Activity Information of the input pins, i.e. transition type (rise, fall, stable), input transition times (slew rate) and signal arrival times.
- **Output pin information:** Capacitance value of the output pin that takes into account the gates connected to it, i.e. fan-out capacitance.

The characterization methodology is divided into two main phases: a data measurement phase, to create our gate-level current database. And a modeling extraction phase, to develop different models based on current characteristics. We developed an automated characterization flow to set all the parameters, run the simulations and recollect the desired data (Figure 1). As simulation results we obtain current and timing information. The saved currents are: the supply current ($i_{dd}(t)$) and the capacitor current ($i_{cc}(t)$). And the timing information: the propagation delay and the output transition time (slew rate).

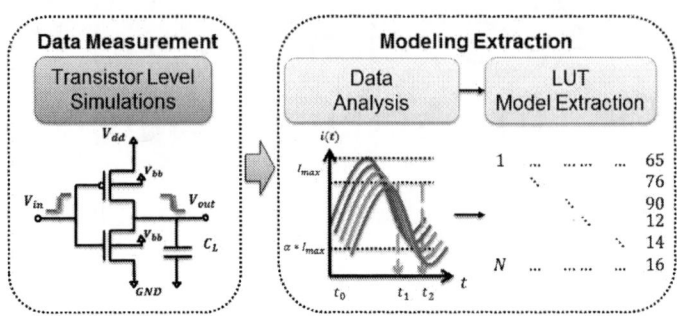

Figure 1: Library Characterization Methodology.

A. Data Measurement

In the first phase of our methodology, we create spice-like decks in which we set all the parameters. We determine how parameters are applied, to ensure that our results are obtained based on conditions close to real circuit scenarios. And we set all simulator parameters, to highlight non-linear behavior of the current profiles, e.g. we save the current values each one picosecond.

To model the input signals, we are able to choose different gate types on the input pins of the gates under analysis. In addition to the current and timing information, a mapping file is also created with the simulated conditions of all the parameters. This file will be used as reference point for the second phase. One of the main advantages of this step, is that we are able to identify and classify all power components at the same time, i.e. we separately are able to compute the dynamic, the short circuit, and the leakage component in a single run.

It is needed to accurately measure the timing characteristics of each gate as this will have a direct impact on the current selection, hence on the power estimation. We use as reference points to compute the slew rate 20-80% and for the propagation delay 40-60%. In our case, we can analyze and save timing information at the same time that our current-characterization, hence there is no increment on the characterization run-time.

B. Modeling Extraction

The modeling extraction phase is subdivided in two steps. In the first one, we analyze all the gate current database and select which information must be conserved. Based on this analysis, we determine characteristics like: current peak value, peak time, current duration time, and then we filter all the extra current information that is not relevant for each condition. In this case, we represent the currents in Look-Up Tables (LUT) for a defined time window, based on the current-trace considering the complete dynamic behavior of the current waveforms. We realize that our LUT are created without any optimization method since our main objective is to show the feasibility of our methodology and their used on the estimation flow. Further effort will be focused on the optimization of the LUT currents and the construction of an equation-based methods.

978-1-4673-8760-6/15 $31.00 © 2015 IEEE

III. POWER CONSUMPTION ASSESSMENT FLOW

To assess the power consumed by a gate or a circuit we use the estimation flow depicted on Figure 2. As input data the following information is necessary: (1) the synthesized gate-level netlist (Verilog format); (2) the activity information of all the internal nodes (VCD File); (3) the input pin capacitance information; and (4) the LUT gate-level current and timing, which are obtained based on the proposed library characterization phase.

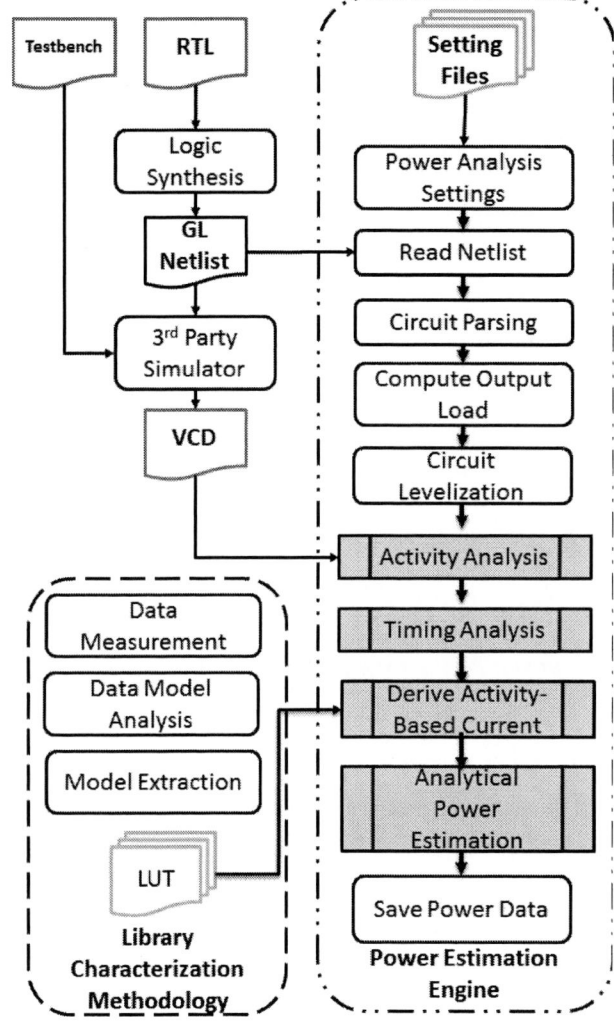

Figure 2: Hybrid Power Estimation Flow.

The estimation engine starts by parsing the circuit to determine the structural information that surrounds every gate, i.e. driving cells and cells driven by each gate. Then, output node capacitance is estimated based on fan-out information and parasitic information if it is available. Afterwards, we run a circuit levelization algorithm to determine the order in which each gate will be analyze and how the timing propagation will be done.

Once we have ordered the gates, we proceed to determine all the switching activity information per net. Then, based on the transition type the slew rate information it is propagated

from primary inputs to primary outputs. Once all the require information is completed, we select the current that is drawn by each gate and add it based on the activation times. Finally, the average power is computed taking into account the simulation period and the derived circuit current.

IV. EXPERIMENTAL RESULTS

In this section, we present experimental results of applying our library characterization methodology and estimation flow on a 28nm FDSOI technology. All the reference simulations are done based on SPECTRE simulator by Cadence [10].

We automate our library characterization using PERL language. We applied it under the same conditions than the ones in the lib file, to directly correlate our results with it. We implemented the power estimation flow presented on last section in C++ and we call it Hybrid Power Estimation Tool (HPET). Using HPET we can easily analyze current/power trace at gate and circuit level. All the analysis are performed on a Linux x86 64bit server with 48 sockets, 1 core per socket, and 158GB of available memory.

A. Validation of the Library Characterization

For the validation phase, we compared the current/power results using our power estimation flow and LUT with the results obtained using the commercial tool PrimeTime-PX [11], in which, dynamic power is computed based on the output load capacitance and the average activity information. Short-circuit power is computed based on liberty data and a single energy value per characterized condition. Leakage power is computed based on a probabilistic approach about the time each net is on a static condition.

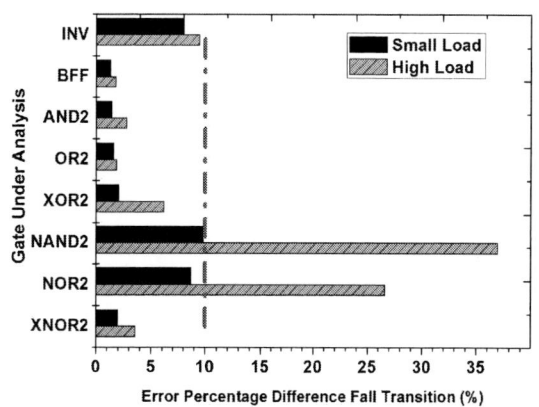

Figure 3: Short-Circuit Power Mean Error.

Dynamic and leakage power are accurately computed based on our characterization results, as mean error difference for both components is lower than 2% for the studied standard cells. On the other hand, we observed a higher difference between our results and the liberty data for short-circuit power component. In general, the mean error percentage is lower than 10% for different gate types and only for two cases we obtained higher differences (Figure 3).

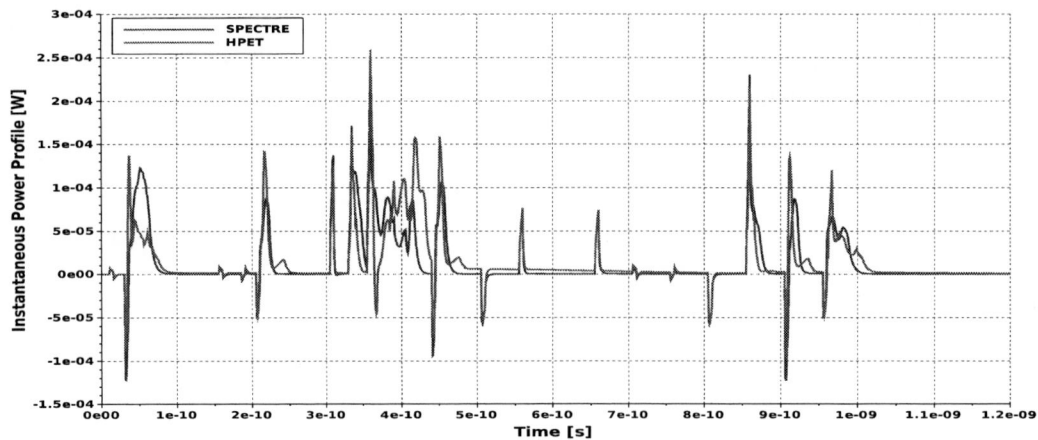

Figure 4: Estimated vs Simulated Current Profile ISCAS C17

We determined that it can be different sources for the obtained percentage error. The main one is that we cannot ensure that simulation conditions were the same than those used to compute the liberty data information. For instance, the use of a different simulator, different driver gate, and different parameter application, will have a direct impact on the current profile when performing the characterization procedure.

Even though we perform the characterization of each gate only once and then use it as many times as necessary, we give some run-time simulation for each gate. In average, for leakage component the LUT creation took less than 5 minutes. For SIS currents, average simulation time was about 1:30h, and for MIS currents it took about 45 minutes.

B. Instantaneous Current Profile Verification

Once our characterization results are validated, we synthesized different circuits using a 28nm technology. As case study, we present the results of the ISCAS85 benchmark circuit C17 to envisage the advantages of the presented work. We derive the current profile using the proposed flow and we compare it with SPECTRE simulation. Estimated versus simulated current profile is presented in Figure 4. Using the Normalized Root Mean Square Deviation (NRMSD) formula, we are able to identify the degree of similarity of the two waveforms, in this case we obtain a 3.4% of difference with a speed-up in computational time of about 5X.

Similar results are obtained on other benchmark circuits with thousands of gates and higher complexity. This demonstrate that we can compute current/power in an efficient way. Future work will be done in ensuring the accuracy of our estimation flow with regard to SPECTRE simulation for instantaneous and average results. And, the complete analysis on real industrial test circuits.

V. CONCLUSION AND FUTURE WORK

In this paper, we proposed a hybrid power model that uses transistor-level current information and gate-level logic simulation, to enhance the power estimation flow. The modeling approach is based on efficient library characterization methodology, in which, we save the current waveforms for multiple parameters conditions at the same time. With this approach we aim to speed-up the estimation run-time while having transistor-like accuracy. Further work will be focused on the LUT improvement and the use of activity information from RTL simulation.

REFERENCES

[1] D. Elléouet, N. Julien, and D. Houzet, "A high level SoC power estimation based on IP modeling," *20th International Parallel and Distributed Processing Symposium, IPDPS 2006*, vol. 2006, pp. 6–9, 2006.

[2] A. Bogliolo, L. Benini, G. De Micheli, and B. Riccò, "Gate-level power and current simulation of CMOS integrated circuits," *IEEE Transactions on Very Large Scale Integration (VLSI) Systems*, vol. 5, no. 4, pp. 473–488, 1997.

[3] H. Fatemi, S. Nazarian, and M. Pedram, "A current-based method for short circuit power calculation under noisy input waveforms," *Proceedings of the Asia and South Pacific Design Automation Conference, ASP-DAC*, pp. 774–779, 2007.

[4] S.-Y. K. Seung-Ho Jung, Jong-Humm Baek, "Short Circuit Power Estimation of Static CMOS Circuits," in *Design Automation Conference, 2001. Proceedings of the ASP-DAC 2001. Asia and South Pacific*, no. 2, 2001, pp. 545–549.

[5] S. Gupta and S. S. Sapatnekar, "Compact Current Source Models for Timing Analysis Under Temperature and Body Bias Variations," *Very Large Scale Integration (VLSI) Systems, IEEE Transactions on*, vol. 20, no. 11, pp. 2104–2117, 2012.

[6] C. Knoth, H. Jedda, and U. Schlichtmann, "Current Source Modeling for Power and Timing Analysis at Different Supply Voltages," in *Design, Automation & Test in Europe Conference & Exhibition (DATE)*, 2012, pp. 923–928.

[7] X. Zhao and K. Wang, "A leakage power estimation method for standard cell based design," *IEEE Conference on Electron Devices and Solid-State Circuits*, pp. 821–824, 2005.

[8] B. S. a. Vitale, P. W. Wyatt, M. Ieee, N. Checka, J. Kedzierski, and C. L. Keast, "FDSOI Process Technology for Ultralow-Power Electronics," *Proceedings of the IEEE*, vol. 98, pp. 333–342, 2010.

[9] J. Jiang, M. Liang, L. Wang, and Y. Zhou, "An effective timing characterization method for an accuracy-proved VLSI standard cell library," *Journal of Semiconductors*, vol. 35, no. 2, 2014.

[10] Cadence, "Cadence® spectre® circuit simulator," 2014.

[11] Synopsys, "Primetime px: Signoff power analysis." 2014.

Toward Efficient Synthesis Method of Multifunctional Logic Circuits

Václav Šimek, Richard Růžička, Adam Crha

Faculty of Information Technology
Brno University of Technology
Brno, Czech Republic
{simekv, ruzicka, icrha}@fit.vutbr.cz

Abstract-**The continuous effort how to achieve space-efficient implementation of a digital circuits may soon arrive at the physical limits behind the scaling of fundamental building components. One possible response to this problem can be recognized in adoption of so called multifunctional logic, which also comes with the intrinsically embedded reconfiguration features. Polymorphic electronics concept with its substantial technological independence opens a way to fulfil this objective through the adoption of emerging semiconductor technologies and advanced synthesis methods. This paper is dealing with a proposal of a novel synthesis method oriented on the exploitation of polymorphic electronics principles. Key part of it is based on Boolean divisor identification and function kernelling technique. Proposed method is evaluated with several test circuits.**

Keywords—Digital circuits, reconfiguration, polymorphic electronics, synthesis methods

I. INTRODUCTION

Nowadays, it is possible to identify a lot of manifold application areas where a digital circuit with the inherent ability to perform a set of different functions at a particular moments in time may prove to be a very efficient means of solution. Obviously the most immediate approach, how to address this specific need, is to design as many different circuits, as the overall number of functions that are actually needed in a given situation. As a next step involved within the execution flow, individual outputs of these circuits are switched in such way that only the presently required function will be taken into account. However, main drawback behind this conception, and its essential limitation as well, will emerge in a direct connection with the overall size of the resulting implementation on a circuit level.

Another possible answer to the outlined purpose lies in the adoption of reconfiguration [1] [2] principles. This course of action clearly enables more flexible way (e.g. circuits that have not been prepared yet during the design phase could be implemented as well, when the conception of evolvable hardware is employed [3]) how to obtain corresponding design with noticeably improved area-aware properties. Nevertheless the chosen procedure may turn out to be less effective in terms

of the necessary processing time – in order to invoke the function change, the structure of the circuit must be adjusted and this takes a bit of time. Moreover, this approach assumes the availability of suitable infrastructure – a reconfigurable circuit with adequate granularity of functional elements and interconnection fabrics.

Recent advancements within the field of digital design techniques and components for digital circuits provide a vital evidence that yet another feasible strategy may be employed – area and time-efficient design of multifunctional circuits based on utilization of individual structural elements with multifunctional features [4]. In this case, the entity of multifunctional circuit is devised as a compact structure involving set of multifunctional components, where their mutual, low-level interconnection scheme remains untouched in all allowable operating modes and only the active function of these components is expected to change intentionally. It's important to note that the gate-level granularity is typically chosen today for design purposes in case of these circuits, while the individual components (gates) are conceived in most cases predominantly at a transistor level. Alternative path with promising outlook for the future is marked by the gradually increasing adoption of unconventional devices that bring significant advantages to this type of circuits – especially in terms of functional characteristics.

A special case related to these multifunctional circuits is based on adoption of polymorphic electronics paradigm [5]. From a technical perspective, circuits with these attributes typically change their function in accordance to the actual state of a target operating environment. The environment in this particular case is represented by a physical quantity that has notable influence on some of the physical parameters of electronic structures – power supply voltage level, voltage amplitude of a signal, temperature etc. It may seem quite impractical on a first sight to consider these properties for any useful circuit behaviour but, in fact, it may help to achieve new (better than existing) solutions for some application classes. Most notable benefit here dwells in the scheme of utterly distributed sensitivity to the well-established factors that bring along the intentional function change. In addition, no

978-1-4673-8760-6/15 $31.00 © 2015 IEEE

configuration network with a global scope or dedicated input pins of these components are required [6].

It's important to point out that change of the active function which is executed by the polymorphic circuit takes place immediately (with no extensive delay associated with this step) and the function change triggering mechanism, that happens due to outlined circuit sensitivity to suitable phenomena from its operating environments, is therefore naturally embedded into the circuit itself. Today's applications are often based on unipolar semiconductor transistors, but the concept of polymorphic electronics has more general nature and allows to conveniently employ new emerging devices like graphene [7] or nanowire structures [8], ambipolar devices utilizing suitable organic polymers with semiconductor-like properties [9], etc., which make it possible to obtain new generation of advanced multi-functional logic elements.

The structure of this paper is organized as follows: the opening section clarifies basic aspects related to multifunctional circuits and their benefits in comparison to conventional approach. Section II is briefly explaining key theoretical aspects behind polymorphic electronics and the implications for digital circuits design. Review of selected circuit synthesis methods and their properties can be found in section III. Then, section IV contains the introduction of novel synthesis method for polymorphic circuits, which is based on adoption of so called Boolean divisors identification and function kernelling technique. The obtained results are demonstrated throughout the section V. Finally, section VI provides the conclusions.

II.   CONCEPT OF POLYMORPHIC ELECTRONICS

The notion of polymorphic electronics [5] determines, in its own essence, a standalone category of reconfigurable circuits, which represents highly appealing perspective how to implement all the required functional properties in a resource-efficient way. In case of these circuits assuming the principles of polymorphic electronics, various modifications in the key physical characteristics of building components (e.g. in a transistor's operation point, usage of ambipolar charge carrier conductivity) are predominantly involved behind the change of their behaviour as a straight response to the influence of external stimuli – temperature, power supply voltage, light intensity, a special signal, etc. However, the structure of the circuit itself remains unchanged on the interconnection level for all the intended functions.

A.   Formal background

From a formal point of view, polymorphic circuit is an electronic digital circuit, which can be described by a graph defined as G = (V, E, φ), where V is a set of vertices (ports of circuit components), E = {(a,b) | a,b ∈ V} is a set of edges (connections in the circuit) and φ is a mapping, which assigns a component from the set K to each that belongs to V, φ: V → K. Then, graph G explicitly determines interconnection of the individual components from set K and, therefore, particular structure of a given circuit, which is able to realize one of the meaningful intended functions from a set $\Phi = \{F_1,..., F_n\}$ a $|\Phi| > 1$.

Furthermore, let X be a physical quantity, assuming values of the real numbers domain **R** and describing an operating environment of the circuit. A mapping $\pi: Y \rightarrow \Phi$, where $Y = \{I_i | I_i \in \mathbf{R}\}$ is a set of intervals of values of the quantity X. If the quantity X has a value $X(t_1) \subset I_k$ at a time t_1, where $I_k \in \mathbf{R}$ is an interval from **R**, then the circuit represented by the graph G performs a function $F_k \subset \Phi$ at the time t_1, briefly $\pi(I_k) = F_k$. If the quantity X has a value $X(t_2) \in I_m$ at a time t_2, where $I_m \in \mathbf{R} \subset I_m \wedge I_k = \cap$, then the circuit represented by the graph G executes a function $F_m \varnothing \Phi$ at the time t_2, briefly $\pi(I_m) = F_m$. Note that even such intervals of X may exist, on which the function of the circuit is not defined.

III.   PROPOSED SYNTHESIS METHOD

In fact, the specification of logic function itself can appear in several, mutually different forms. An elaborate discussion on five of the most common instances using two-level arrangement can be found in [10]. These cases are mostly focused on the various representations of the truth table forms together with disjunctive/conjunctive notation. For the sake of completeness it's important to point out that synthesis and minimization techniques in digital circuit domain are based extensively on multi-level representations as well [11] [12] [13], especially due to reasonable compromise between compact representation and efficient manipulation. Probably one of the most illustrative example here is tied with decision diagrams or, to be precise, BDD (Binary Decision Diagrams) as the widely adopted scheme in various situations.

Those minimization and synthesis techniques could be, as a matter of fact, roughly classified as two-level or multi-level oriented. In case of two-level methods the final circuit composition is delivered as the logic expressions in conjunctive or disjunctive notation. This approach then leads to the situation when input signals will only pass through two logic gates at most. On the other hand, multi-level techniques are generating so called nested expressions with the resulting data path (or interconnection of the individual gates) spanning even far more than two circuit elements within the final circuit arrangement.

A.   Key aspects behind the synthesis method

Searching for the corresponding interconnection graph G (see closer explanation in section 2) may not be an easy task in case of the polymorphic circuitry. Directly related to this observation is the goal to propose novel synthesis method that would be addressing the weak spots of the previous attempts. Main idea behind the novel approach is based on the undeniable identification of common parts across the source circuits, which are virtually shared between them as a so called common divisors, by means of exploiting techniques of function kernelling [12] [14] and Boolean division [12].

Typical execution flow behind the proposed method consists of the following sequence of individual steps, which are further outlined below:

978-1-4673-8760-6/15 $31.00 © 2015 IEEE

1. Minimized expressions in DNF notation (Disjunctive Normal Form) depict the input functions – F_1 and F_2. Both functions are initially provided in two-level PLA format as a truth table.

$$F_1 = ab\bar{d} + b\bar{c}d + \bar{b}cd + a\bar{c} + \bar{a}\bar{b}c\bar{d}$$
$$F_2 = ab\bar{c} + a\bar{c}d + \bar{a}\bar{b} + \bar{b}c\bar{d} + \bar{a}c$$

2. Intersection table at a dimensions given by *m* x *n*, where *m* denotes the number of term groups of F_1 and *n* has the same meaning for F_2. This table is laid out in such way that first column contains terms groups belonging to F_1 and first column holds the number of terms of F_2. Individual boxes within the table are filled up in the following way:

group of terms intersection (remaining terms of F1 | remaining terms of F2)

3. Main task here is to identify those entries that exhibit the mutual intersection of a maximum size. In fact, these so called minterms are basically shared for both input functions. Once the minterm is registered in the final expression, corresponding row and column is eliminated from the table.

4. Second pass through the table constructed in step 2) is commenced. This time, the task is to find the largest intersection. The box fulfilling this requirement is then rewritten into the final expression, whole row and column with this particular box is eliminated from the table.

The final expression obtained at this step is following:

$$F = \bar{b}c\bar{d}(\bar{a}|1)+$$

5. Previous step 4) is continuously repeated until the table contains uncovered boxes with at least some intersection. Once all the intersection are covered, it's possible to proceed with a next step.

6. Now, it's necessary to apply a special functional block called polymorphic multiplexer (labeled as "|" in the expression), which isolates contradictory parts of functions F_1 and F_2.

Decomposition of the obtained expression will be done as a measure towards the best possible mapping onto the set of available circuit components:

$$A_2 = (\bar{a}|a)$$
$$B_2 = (\bar{b}|b)$$
$$C_1 = (c|\bar{c})$$
$$Z = (\bar{d}|1)$$

And now, the resulting expression ready for technology mapping phase will have the following composition:

$$F = \bar{b}c\bar{d}(\bar{a}|1) + b\bar{c}(\bar{d}|a) + \bar{c}d(\bar{b}|a) + A_2B_2Z + A_2C_1$$

7. If both functions F_1 and F_2 have different number of term groups, there will remain for sure certain number of uncovered terms belonging to the function with higher number of term groups. Those uncovered terms are put into the resulting expression by means of using polymorphic operator "|" and neutral element for addition denoted as "0".

II. EXPERIMENTAL RESULTS

The proposed synthesis technique for polymorphic circuits has been tested on several circuits defined by a truth table in two-level PLA format. Detailed specification of these circuits can be found in table 1 and table 2 below. These circuits were either randomly generated or taken from ISCAS benchmark set.

TABLE I.   SPECIFICATION OF TEST CIRCUITS

Test batch#	Example circuits properties		
	Input variables	*Product terms [count]*	*Onset [%]*
1	4	16	25, 50, 75
2	5	16, 32	25, 50, 75
3	6	16, 32, 64	25, 50, 75
4	8	16, 32, 64, 128, 256	25, 50, 75
5	10	16, 32, 64, 128, 256, 512, 1024	25, 50, 75
6	12	16, 32, 64, 128, 256, 512, 1024, 2048, 4096	25, 50, 75
7	16	16, 32, 64, 128, 256	25, 50, 75
8	4, 5, 6, 8, 10, 12, 16	16	25, 50, 75
9	5, 6, 8, 10, 12, 16	32	25, 50, 75
10	6, 8, 10, 12, 16	64	25, 50, 75
11	8, 10, 12, 16	128	25, 50, 75
12	8, 10, 12, 16	256	25, 50, 75
13	10, 12, 16	512	25, 50, 75
14	10, 12	1024	25, 50, 75
15	12	2048	25, 50, 75
16	12	4096	25, 50, 75
17	4, 5, 6, 8, 10, 12, 16	16, 32, 64, 128, 256, 512, 1024, 2048, 4096	25
18	4, 5, 6, 8, 10, 12, 16	16, 32, 64, 128, 256, 512, 1024, 2048, 4096	50
19	4, 5, 6, 8, 10, 12, 16	16, 32, 64, 128, 256, 512, 1024, 2048, 4096	75

TABLE II.   COMPARISON OF THE PROPOSED SYNTHESIS METHOD AND CONVENTIONAL APPROACH

Test batch#	Polymorphic synthesis tool (averaged values)						Espresso – two outputs mode (averaged values)				Improvement [%]
	2-AND	2-OR	INV	P_INV	P_MUX	sum	INV	2-AND	2-OR	sum	
1	9	3	4	3	2	21	5	12	8	25	13,97
2	19	6	5	2	4	35	7	17	20	44	21,21
3	44	10	6	4	7	70	8	28	49	85	17,89
4	164	26	8	9	18	226	10	66	214	290	23,03
5	645	81	10	15	53	803	12	151	844	1008	22,76
6	2443	243	12	29	175	2903	14	409	3200	3623	24,38
7	767	50	16	25	2	859	18	172	1016	1206	27,68
8	60	6	8	7	2	83	10	35	58	103	19,67
9	124	13	10	9	3	159	11	41	155	208	22,77
10	275	27	10	15	6	333	12	71	367	450	24,84
11	599	55	12	20	12	698	14	112	838	963	27,30
12	903	88	11	21	36	1059	13	259	1243	1516	28,27
13	1747	183	11	22	102	2064	13	327	2477	2817	26,26
14	2616	287	11	58	197	3169	13	473	3583	4069	20,54
15	6067	610	12	24	499	7212	14	974	7787	8775	19,25
16	7376	782	13	24	732	8927	14	1119	8902	10035	10,40
17	678	62	10	15	39	804	12	145	1033	1190	28,69
18	1069	106	10	16	73	1276	12	216	1405	1633	25,08
19	1190	130	10	21	82	1434	12	207	1411	1630	15,82

IV.   CONCLUSIONS

The obvious benefits arriving with the proper exploitation of reconfiguration scheme offer significant advantages over the standard or conventional solution. It is even more apparent fact if those procedures are further enriched with the previously outlined principles of polymorphic electronics. However, the currently available approaches within the reconfiguration domain, as well as some of the methods taking use of polymorphic electronics principles, tend to produce relatively inefficient solution in terms of the overall circuit size.

Due the reason, a novel method based on the formal basis has been formulated. The obtained results indicate that it's possible to achieve around 27% improvement especially in comparison with the standard synthesis tool called Espresso. Nevertheless, further improvements are planned from the side of extending the divisors identification procedure also for multi-level circuit specification.

ACKNOWLEDGEMENTS

This work was generously supported by the internal grant Architecture of parallel and embedded computer systems (no. FIT-S-14-2297) and by national COST grant Unconventional Design Techniques for Intrinsic Reconfiguration of Digital Circuits: From Materials to Implementation (no. LD14055).

REFERENCES

[1] Bobda, C.: Introduction to Reconfigurable Computing: Architectures. Springer, 2007, 362 p.

[2] Cardoso, J., Hübner, M. (eds.): Reconfigurable Computing. From FPGAs to Hardware/Software Codesign. Springer-Verlag New York, 2011, 296 p.

[3] Trefzer, M. A., Tyrrell, A. M. (eds.): Evolvable Hardware. From Practice to Application. Springer, 2015, 411p.

[4] Růžička, R., Šimek, V.: More Complex Polymorphic Circuits: A Way to Implementation of Smart Dependable Systems. ElectroScope, Pilsen, 2013, Vol. 7, No. 5, pp. 1-6. ISSN 1802-4564.

[5] Stoica, A., Zebulum, R. S., Keymeulen, D.: Polymorphic electronics. Proc. of Evolvable Systems: From Biology to Hardware Conference, volume 2210 of LNCS, Springer 2001, pp. 91–302.

[6] Sekanina, L., Růžička, R., Vašíček, Z., Prokop, R., Fujcik, L.: REPOMO32 - New Reconfigurable Polymorphic Integrated Circuit for Adaptive Hardware. Proc. of the 2009 IEEE Symposium Series on Computational Intelligence - Workshop on Evolvable and Adaptive Hardware, Nashville, IEEE CIS, 2009, pp. 39 – 46.

[7] Tanachutiwat, S., Lee, J. U., Wang, W., Sung, C. Y.: Reconfigurable multi-function logic based on graphene P-N junctions. Proc. of DAC 2010, pp. 883 – 888.

[8] Weber, W. M., Heinzig, A., Trommer, J., Martin, D., Grube, M., Mikolajick, T.: Reconfigurable nanowire electronics – A review. J. on Solid-State Electronics, Vol. 102, December 2014, pp. 12-24.

[9] Tesař, R., Šimek, V., Růžička, R., Crha, A.: Polymorphic Electronics Based on Ambipolar OFETs. In: EDS 2014 IMAPS CS International Conference Proceedings. Brno: Brno University of Technology, 2014, pp. 106-111.

[10] Wakerly, J. F.: Digital Design: principles and practices, 3d edition. Prentice Hall, NewJersey, USA, 2000.

[11] Darringeret, J. A. et al.: Logic synthesis through local transformations. IBM Journal of

[12] Hachtel, G. D., Somenzi, F.: Logic Synthesis and Verification Algorithms, Kluwer Academic Pub, 1996, 564 p.

[13] Hassoun, S., Sasao, T.: Logic Synthesis and Verification, Boston, MA, Kluwer Academic Publishers, 2002, 454 p.

[14] Brayton, R. K., McMullen, C.: The Decomposition and Factorization of Boolean Expressions, ISCAS Proceedings, April 1982.

Reducing the Number of Embedded Multipliers in Squaring Large Size Complex Numbers

Fatima-Ezzahra Guessous
Web Help Company
Rabat, Morocco

Noureddine Chabini
Department of Electrical and Computer Engineering
Royal Military College of Canada
Kingston, ON, Canada

Abstract—To square large size complex numbers using n×n embedded multipliers as in using FPGAs, one needs to first partition the real and imaginary parts into segments of size less or equal to n and then to multiply the segments using these embedded multipliers. When the size of some segments is small, we show that three multiplications can be carried out using one n×n multiplier thus two n×n embedded multipliers can be saved. The idea is also applicable in carrying out two small squarers using one n×n embedded multiplier.

Keywords—FPGAs; complex numbers; embedded multipliers; large size squarer.

I. INTRODUCTION

Arithmetic on complex numbers is widely used in digital signal processing, in digital image processing, and in scientific computational algorithms [1][2][3]. One important operation is squaring a complex number.

Field Programmable Gate Arrays (FPGAs) have immerged as an important platform to implement various applications due to their low cost and ease of programmability.

Modern FPGAs contain various components like LUTs, adders and multipliers. For instance, Spartan FPGAs from Xilinx Inc. [5] contain a good number of 18×18 embedded multipliers.

In this paper, we are interested in squaring large size complex numbers using n×n embedded multipliers in FPGAs. Once the real and imaginary components of the complex number are partitioned into segments of size less than or equal to n, we show that when some segments are small, one can carry out three multiplications using one n×n embedded multiplier. Thus, two n×n embedded multipliers can be saved. Carrying out more than one multiplication using one embedded multiplier has been proposed in [5]. Our proposed idea is also applicable to squaring small numbers; in this case, two small squarers can be carried out using one n×n embedded multiplier. Papers [2][4] show how squarers are used to carry out multiplications.

II. METHODOLOGY

In the following, we show how to reduce the number of n×n embedded multipliers for the case of n = 16 bits, n = 18 bits and n = 36 bits. The latter two cases correspond to the size of embedded multipliers in some Xilinx FPGAs [5] and some Altera FPGAs [6], respectively.

Let $Z = A + j*B$ be the complex number to be squared, where A and B are unsigned integer numbers.

Assume that $A = AH*2^k + AL$ and $B = BH*2^k + BL$, and the sizes of AH and BH are multiple of n. This corresponds to segmenting A and B from the left to the right for the case of using n×n embedded multipliers.

Case of n = 16 bits:
In this case, we assume that the size of AL and BL is less of equal to 4 bits and k = 4. Let $C = AL*2^9 + BL$. The size of C is less or equal to 13 bits so it can be squared using one 16×16 embedded multiplier. We have:

$$C*C = (AL*2^9 + BL)*(AL*2^9 + BL)$$
$$= (AL*AL)*2^{18}+ (AL*BL)*2^{10} + (BL*BL)$$
$$= T1 + T2 + T3$$

where :

$$T1 = (AL*AL)*2^{18}$$
$$T2 = 2*(AL*BL)*2^9$$
$$T3 = (BL*BL)$$

The terms T1, T2 and T3 do not overlap. So each one of them can be found easily at the output of the n×n embedded multiplier.

Recall that k = 4 bits. We have:

$$(A + j*B)*(A+j*B) = A*A - B*B + j*(2*A*B)$$
$$= (AH*2^k + AL)^2 - (BH*2^k + BL)^2 +$$
$$j*(2(AH*2^k + AL)*(BH*2^k + BL))$$
$$= (AH*AH*2^{2k} +AH*AL*2^{k+1}+\textbf{AL*AL}) -$$
$$(BH*BH*2^{2k} +BH*BL*2^{k+1}+\textbf{BL*BL}) +$$
$$j*(AH*BH*2^{2k+1} + AH*BL*2^{k+1} +$$
$$AL*BH*2^{k+1} + \textbf{2*AL*BL})$$
$$= (AH*AH*2^{2k} + AH*AL*2^{k+1}) -$$
$$(BH*BH*2^{2k} + BH*BL*2^{k+1}) +$$
$$j*(AH*BH*2^{2k+1} + AH*BL*2^{k+1} +$$
$$AL*BH*2^{k+1}) +$$
$$\textbf{(\underline{AL*AL} - \underline{BL*BL} + j*2*\underline{AL*BL})}$$

For the underlined terms in bold in $(A + j*B)* (A + j*B)$, one can save two multipliers by executing the following algorithm:

Algorithm:
1. Construct the term denoted C above,
2. Square C using one embedded multiplier, and
3. Finally take the appropriate terms from the result of C*C since its terms T1, T2 and T3 defined above do not overlap on the output of this embedded multiplier.

Case of n = 18 bits:
In this case, we assume that the size of AL and BL is less of equal to 5 bits and k = 5. Let $C = AL*2^{12} + BL$. The size of C is less or equal to 17 bits so it can be squared using one 18×18 embedded multiplier. Two 18×18 embedded multiplier can be saved as we did for n=16 bits using the algorithm above.

Case of n = 36 bits:
In this case, we assume that the size of AL and BL is less of equal to 11 bits and k = 11. Let $C = AL*2^{24} + BL$. The size of C is less or equal to 35 bits so it can be squared using one 36×36 embedded multiplier. Two 36×36 embedded multiplier can be saved as we did for n=16 bits using the algorithm above.

Besides squaring a complex number, the proposed idea can be used to carry out the two squaring functions AL*AL and BL*BL using one n×n embedded multiplier by constructing the term C above and squaring it.

III. CONCLUSION

In this paper, we focused on squaring large size complex numbers using n×n embedded multipliers. We showed that once the real and imaginary component are segmented and the segments are multiplied using n×n embedded multipliers, some small segments can be multiplied using one n×n embedded multiplier thus saving two n×n embedded multipliers. The idea is also applicable to carrying out two small squarers using one n×n embedded multiplier.

REFERENCES

[1] H. Zaini, and R.G. Deshmukh, "A novel method for arithmetic operations using complex binary number system and the reconversion of the result to the decimal complex number system," *Proceedings of the IEEE SoutheastCon*, 2003, pp. 31-37.

[2] Y.B. Mahdy, S.A Ali, and K.M. Shaaban, "Algorithm and two efficient implementations for complex multiplier," *Proceedings of IEEE Conference on Electronics, Circuits and Systems*, vol.2, 1999, pp. 949-952.

[3] V.G. Oklobdzija, D. Villeger, and T. Soulas, "Considerations for design of a complex multiplier," *Proc. of IEEE Conf. on Signals, Systems and Computers*, vol.1, 1992, pp. 366-370.

[4] A. Skavatzos, "Novel approach for implementing convolutions with small tables," *IEE Proceedings*, Vol.138, No.4, July 1991, pp. 255-259.

[5] Xilinx Inc.: http://www.xilinx.com

[6] Altera Inc.: http://www.altera.com

General fuzzy models for dynamical systems

Kheireddine Chafaa[1], Mouna Ghanai[1] and Lamir saidi[1]

[1] Laboratoire d'Automatique Avancée et Analyse des Systèmes (LAAAS), University Batna 2, Algeria
medjghou.ali@gmail.com; kheireddine.chafaa@univ-batna.dz
kheireddine.chafaa@univ-batna.dz

Abstract—**Fuzzy logic systems can approximate any nonlinear function and they can also be successfully applied to system modeling. This paper describes an efficient fuzzy identification method which can be applied to nonlinear dynamical systems. The proposed method is constituted of three steps: in the first step, an ordinary fuzzy identification technique is applied, steps 2 and 3 are introduced in order to improve the model obtained in step 1, and then a model with higher resolution is obtained. To demonstrate the effectiveness of the proposed method, several examples are presented.**

Index Terms—**Dynamical systems identification, Error modeling, Fuzzy modeling, Fuzzy systems.**

I. INTRODUCTION

Fuzzy logic was successfully used in systems modeling [1, 2, 3]. In this investigation, a simple and efficient fuzzy identification method is proposed. We use the input/output data in order to determine a primary model for the plant. After that, the error between the plant output and the primary model output will be modeled giving us a model for the error. Finally, the primary model and the error model will be interconnected in a parallel configuration which can be considered as a corrected model for the plant. As we will see later, the error modeling will be reduced, and then a high-resolution model is obtained. This modeling strategy has the advantage that it gives a supplementary step to the usual identification methods allowing to obtain more refined models.

II. PROPOSED METHOD

In this section we present the three identification steps in detail. All fuzzy systems considered in this paper have Gaussian membership functions for the premises and singleton membership functions for the consequences.

First stage

In this stage the input/output data (u_1, y_1) are used for the determination of a primary model $\hat{f}_p$ for the plant f to be identified as shown in Fig. 1, where $\hat{y}_{p1}$ the output of the

primary model and $e_1 = \hat{y}_{p1} - y_1$ is the error modeling which excites the adjustment mechanism 1 (*AM1*).

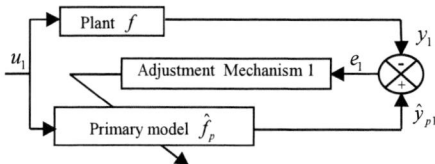

Figure 1. Primary model identification

The design of the primary fuzzy identifier consists of an on-line adaptation of $\hat{f}_p$. *AM1* adjusts the parameters of the fuzzy model $\hat{f}_p$ such as the error e_1 is minimized. The parameters of $\hat{f}_p$ to be trained by *AM1* are the Gaussian centers of premise membership functions (MFs) and the singletons of the consequences denoted $\bar{x}_j^i$ and $\bar{y}^i$ respectively, where i denotes the rule number and j denotes the regressor number. *AM1* is based on the gradient descent technique [4, 5] which we propose to use it with our above parameters as follows:

$$\bar{y}^i(k+1) = \bar{y}^i(k) - \alpha_1 \frac{\hat{y}_{p1} - y_1}{\sum_{i=1}^{M_1} \bar{y}^i \beta_i(x)} \beta_i(x)$$

$$(1)$$

$$\bar{x}_j^i(k+1) = \bar{x}_j^i(k) - \alpha_1 \frac{\hat{y}_{p1} - y_1}{\sum_{i=1}^{M_1} \bar{y}^i \beta_i(x)} (\bar{y}^i - \hat{y}_{p1})*$$

$$\beta_i(x) \frac{2\left(x_j - \bar{x}_j^i(k)\right)}{\sigma_1^2}$$

$$(2)$$

where α_1 is a constant step size, σ_1 is a constant variance for all the Gaussian premise MFs and M_1 the number of fuzzy rules relative to primary model identification.

Second stage

In this stage, and for purposes of generalization, the primary model obtained in stage 1 is validated with a new input data u_2 in order to obtain a general error signal e_2 which will

978-1-4673-8760-6/15 $31.00 © 2015 IEEE

be considered as the output of an error process (see Fig. 2). The error process is defined by the parallel interconnection of the plant and the primary model as shown in Fig. 2, where the input is u_2 and the output is the error signal e_2 given by $e_2 = \hat{y}_{p2} - y_2$. Note that y_2 and $\hat{y}_{p2}$ are the new plant output and the new primary model output, respectively.

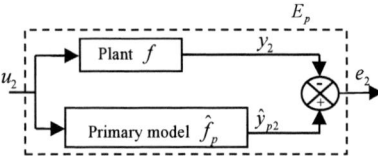

Figure 2. Error process

The problem now is to determine a fuzzy model for the error process by using the new input/output training data (u_2, e_2), i.e., identify the error process. Fig. 3 shows the framework of the proposed algorithm where E_p (the transfer between u_2 and e_2) denotes the unknown function we want to identify (Error process), $\hat{E}_p$ the fuzzy error model of E_p, e_{mo} the output of the error model and e_3 is the error modeling between the error process and the error model which excites the adjustment mechanism 2 (*AM2*).

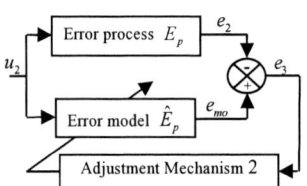

Figure 3. Error process identification

Mathematically, the process to be identified will be given now by the nonlinearity E_p and since $e_2 = \hat{y}_{p2} - y_2 = E_p(u_2)$, then the fuzzy model can be given by $e_{mo} = \hat{E}_p(u_2)$. Now we assume that we have M_2 fuzzy rules describing the unknown nonlinear function E_p of the error process as follows:

$$R_i : \text{If } u_2 \text{ is } U_i(u_2) \text{ Then } e_{omi} = \overline{\varepsilon}^i \quad i = 1, 2, ... M_2 \quad (3)$$

where $U_i(u_2)$ is the ith premise MF for the input u_2 and $\overline{\varepsilon}^i$ the ith consequence singleton MF. The problem now becomes the training of the parameters of $\hat{E}_p$ such that the error e_3 is minimized. The parameters of $\hat{E}_p$ to be trained by the *AM2* are the Gaussian centers of the premise MFs U_i and the singletons of the consequences denoted $\overline{\mu}^i$ and $\overline{\varepsilon}^i$, respectively. To train $\overline{\varepsilon}^i$ and $\overline{\mu}^i$, the *AM2* based on the gradient descent technique is used as follows:

$$\overline{\varepsilon}^i(k+1) = \overline{\varepsilon}^i(k) - \alpha_2 \frac{e_{mo} - e_2}{\sum_{i=1}^{M_2} \overline{\varepsilon}^i \beta_i(u_2(k))} \beta_i(u_2(k)) \quad (4)$$

$$\overline{\mu}^i(k+1) = \overline{\mu}^i(k) - \frac{\alpha_2(e_{mo} - e_2)(\overline{\varepsilon}^i - e_{mo})}{\sum_{i=1}^{M_2} \overline{\varepsilon}^i \beta_i(u_2(k))} \beta_i * \frac{2\left(u_2(k) - \overline{\mu}^i(k)\right)}{\sigma_2^2} \quad (5)$$

where α_2 and σ_2 are relative to error model identification and $\beta_i(u_2(k)) = \mu_{U_i}(u_2(k))$.

Third stage

In this last stage, the primary model $\hat{f}_p$ and the error model $\hat{E}_p$ are interconnected in a parallel configuration (see Fig. 4) in order to obtain the final model $\hat{f}_F$. This configuration allows us to compensate the error modeling obtained in the primary model. Now if we denote $\hat{y}_p$ the output of the primary model, and $\hat{e}_{mo}$ the output of the error model, then the output $\hat{y}$ of the final model can be given by

$$\hat{y} = \hat{y}_p - e_{mo} \quad (6)$$

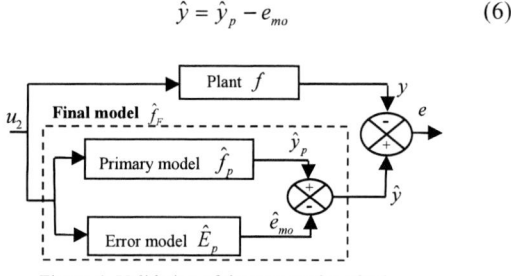

Figure 4. Validation of the proposed method

III.   SIMULATION RESULTS

In this section simulation results for a nonlinear system using our approach are presented. The plant to be identified is governed by the difference equation (7) where f is the unknown function to be identified:

$$y(k+1) = \underbrace{\frac{y(k)y(k-1)\left[y(k) + 2.5\right]}{1 + y^2(k) + y^2(k-1)}}_{f[y(k), y(k-1)]} + u(k) \quad (7)$$

As an identification model, let's use the following model

$$\hat{y}(k+1) = \hat{f}\left[y(k), y(k-1)\right] + u(k) \quad (8)$$

where $\hat{f}$ will be the fuzzy model of f, $\hat{y}(k+1)$ the model output, $y(k)$ and $y(k-1)$ are the model regressors and $u(k)$ is the system input.

The primary model $\hat{f}_p$ is identified according the following rules form

$$R_i : \text{If } y(k) \text{ is } A_{1i}(y(k)) \text{ and } y(k-1) \text{ is } A_{2i}(y(k-1))$$
$$\text{Then } \hat{f}_{pi} = \overline{f}^i \quad i = 1, 2, ... M_1 \quad (9)$$

where $A_{1i}(y(k))$ and $A_{2i}(y(k-1))$ are the premisse membership functions for regressors $y(k)$ and $y(k-1)$, respectively and $\bar{f}^i$ the ith consequence singleton MF.

We simulated the proposed method in all steps with $M_1 = M_2 = 40$ rules, then the primary model identifier will be made up of 40 rules and with $40 \times 3 = 120$ adjustable parameters, for each rule we have 3 free parameters, two parameters for the premise(centers of premise MF) and one parameter for the consequence (singleton position). Parameters α_1 and σ_1 are set to 0.5 and 0.1, respectively.

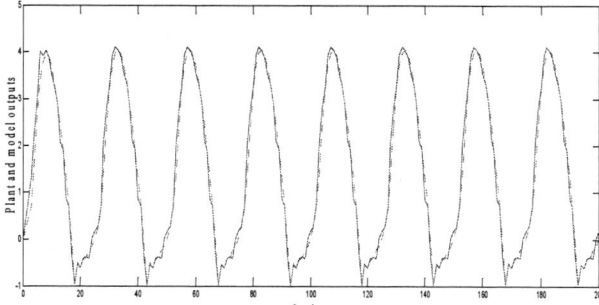

Figure 5. Validation of the primary model

In figures 5 and 6 we give the results of the first stage (see Fig. 1) where Fig. 5 shows the outputs of the plant (dashed line) and the primary identified model $\hat{f}_p$ (solid line), corresponding to the input $u = u_1(k) = \sin(2\pi k / 25)$. Fig. 6 shows a general validation of the obtained model for a new input data $u = u_2(k) = \sin(2\pi k / 25)$ for $1 \le k \le 50$ and $150 \le k \le 200$, $u = u_2(k) = \sin(2\pi k / 10) + \sin(2\pi k / 5)$ for $51 \le k \le 149$ (see Fig. 2). We see from these figures that the output of the identified primary model follows the output of the plant even the input data are changed which confirm the generalization capability of the fuzzy identifiers.

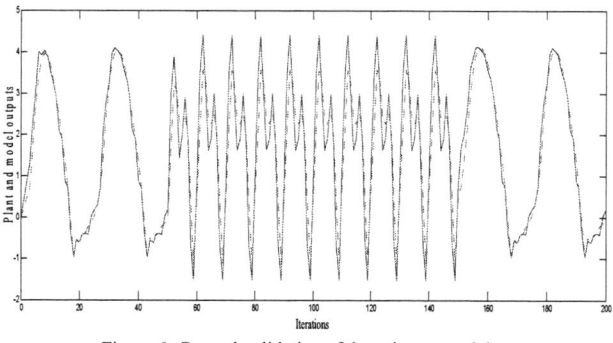

Figure 6. General validation of the primary model

In the second stage, we will build a model $\hat{E}_p$ for the error process E_p. The error process output represented in figure 7 is obtained by the parallel interconnection of the plant (7) with the primary model which gives us the identification data set defined by the set of couples $\{(u_2, e_2)\}$ (see Fig. 2).

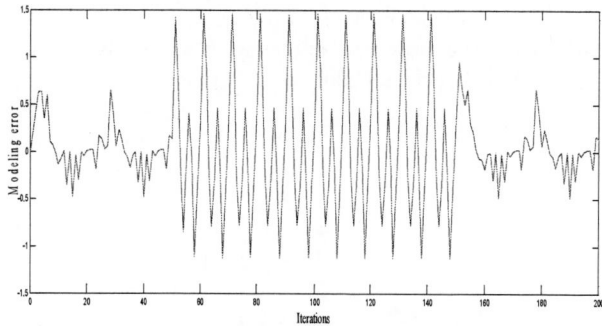

Figure 7. Error process output

The error process E_p is identified according to the rules given in equation 3 and the adjustment mechanism 2 given by equations 4 and 5 with $M_2 = 40$, $\alpha_2 = 0.8$ and $\sigma_2 = 0.1$. In Fig. 8 we present the result of error process identification where we see that we obtain a very good model for this process.

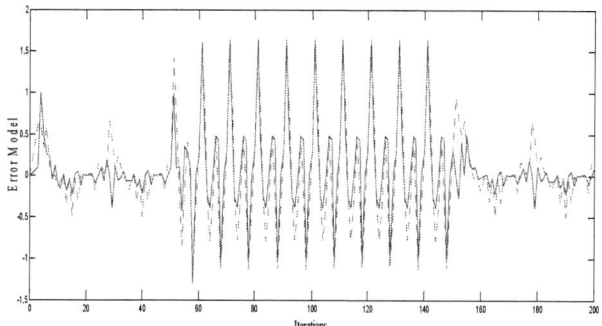

Figure 8. Validation of the error process model

Finally, and to obtain our final model $\hat{f}_F$, the primary model $\hat{f}_p$ and the error model $\hat{E}_p$ are interconnected in a parallel structure as shown in Fig. 4. Simulation results are given in figure 9 where we clearly see that this final model is much better than the primary model (compare figures 6 and 9).

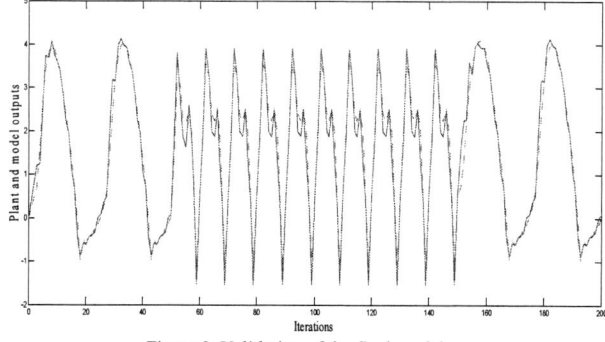

Figure 9. Validation of the final model

We confirm this result by the MSE (mean square error criterion) values and the visual comparison given in figure 10 where we present a superposition of the error curves of the primary model and the final model with their respective MSE. Note that MSE value for the final model (0.0849) is much smaller than the MSE of the primary model (0.3046), which validate our proposed method, and give us a system model with higher resolution.

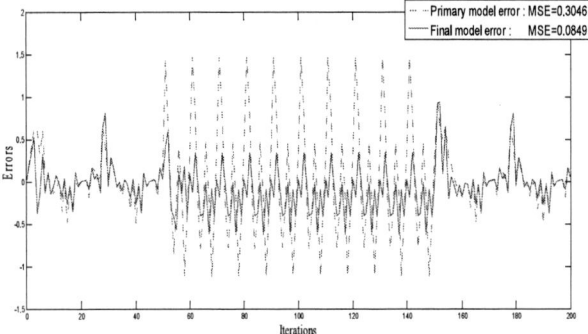

Fig. 10. Comparison between the modelling errors in primary and final models

IV. CONCLUSION

In this paper a new identification algorithm for nonlinear dynamical systems was suggested. This technique allows us to generalize the notion of identification by adding a new identification module which identifies a model for the error modeling. The error model was used to help the identified primary model of the plant to correct its output in order to be more precise. Simulation studies reveal that our suggested method is very effective and gives models with higher resolution.

V. REFERENCES

[1] K. Chafaa, M. GhanaiM, and K. Benmahammed. "Fuzzy modelling using Kalman filter", IET Control Theory and Applications, vol. 1, pp. 58-64, 2007.

[2]. P. Angelov, and D. P. Filev. "An Approach to Online Identification of Takagi-Sugeno Fuzzy Models". IEEE Trans. Syst. Man Cybern., vol. 34, pp. 484-498, 2004.

[3]. Z. Deng, Y. Jiang, F. L. Chung, H. Ishibuchi, and S. Wang. "Knowledge-Leverage based fuzzy system and its modeling". IEEE Trans. Fuzzy systems, vol. 21, pp. 597-609, 2013.

[4]. J. A. Snyman. Practical Mathematical Optimization. Springer, MA, 2005.

[5]. Z. Wanqing, L. Kang, and G. W. Irwin. "A New Gradient Descent Approach for Local Learning of Fuzzy Neural Models". IEEE Trans. Fuzzy systems, vol. 21, pp. 30-44, 2013

Core and Back-End Enhancements for CUPSHOP HW/SW Partitioning Tool

Rania O. Hassan
Electronics and Communications
Engineering Dep.,
Cairo University, Giza, Egypt
rania.osama@aucegypt.edu

Ahmed Hamed Fatehy
Electronics and Communications
Engineering Dep.,
Cairo University, Giza, Egypt
engineerahmedhamed@gmail.com

M. B. Abdelhalim
CCIT-AASTMT,
Cairo, Egypt
mbakr@ieee.org

S. E. D. Habib
Electronics and Communications
Engineering Dep.,
Cairo University, Giza, Egypt
seraged@ieee.org

Abstract — **Hardware/Software co-design has become one of the main methodologies in modern embedded systems. The HW/SW partitioning problem is a central problem in Hardware/Software co-design. The recently-introduced Cairo University PSo-based Hardware/sOftware Partitioning (CUPSHOP) tool advanced the swarm intelligence optimization techniques to solve this problem. In this paper we introduce an augmented version of the CUPSHOP tool. This augmented CUPSHOP version adds an automatic RTL (VHDL) Back-End generation for the HW mapped components. It also complements the different hardware implementations alternatives already supported by the tool with the pipelined hardware alternative. Furthermore, it implements an improved pre-scheduling delay estimation formula.**

Keywords – Embedded Systems, HW/SW Co-designs, HW/SW Partitioning, Intermediate Representation (IR), Control-Data Flow Graphs (CDFG), Front-End, Back-End, FPGAs, High-level synthesis.

I. INTRODUCTION

Hardware/Software co-design proved its efficiency in supporting design cycles starting from behavioral specifications with constraints. The designer of an embedded system is always facing the problem of partitioning the design components into either hardware components; e.g. FPGAs or ASICs; or software components on processor; e.g. DSPs, core processors, or ASIPs. Mapping components to software enhances design flexibility, allows for late design changes, and reduces cost. On the other hand, mapping components to hardware improves performance. This tradeoff between the hardware and software illustrates the optimization aspect of the Hardware/Software (HW/SW) co-design problem [1]. In the last two decades, a large variety of deterministic and heuristic algorithms were advanced in the literature to tackle this optimization problem [2-4].

The Cairo University PSo-based Hardware/sOftware Partitioning (CUPSHOP) is a recently-introduced Hardware/Software partitioning CAD tool. CUPSHOP was developed at Cairo University and its first version was released in 2008 [5]. CUPSHOP was the first to use Particle Swarm Optimization (PSO) technique for solving the HW/SW partitioning problem [2]. CUPSHOP v.1 introduced two bounding hardware alternatives for each design node. Thus, the partitioning has to map each design node between three alternatives: SW, Serial HW, or Parallel HW.

This paper presents an augmented version of CUPSHOP: CUPSHOP v.2; with three added features. Firstly, a fourth alternative, the pipelined HW alternative is implemented. The pipelined alternative is important for high throughput systems with tight delay constraints. Secondly, an improved pre-scheduling delay estimation formula is implemented [7]. Finally, automatic generation of the RTL (VHDL) code for the hardware-mapped components is developed. This automatic RTL (VHDL) code generation supports arbitrary levels of hierarchy.

Fig.1 shows the overall Block diagram of the CUPSHOP v.2. The White blocks refer to the original CUPSHOP v.1 [2]. The core engine of CUPSHOP features a PSO search algorithm and CDFG intermediate representation of the design. The light gray blocks refer to the developed Front-End which accepts ANSI C codes of arbitrary number of hierarchical levels, as is discussed in [7]. The hatched lines blocks present the three added features proposed in this paper.

The outline of the paper is as follows: *Section II* discusses the addition of the hardware pipelined alternative and the improved pre-scheduling delay estimation formula. *Section III* exposes the Back-End of CUPSHOP which is responsible of automatic RTL (VHDL) code generation for the HW-mapped components. Detailed information about VHDL generation process is given in *Section IV*. The case studies and results are shown in Section V. The paper is concluded in Section VI.

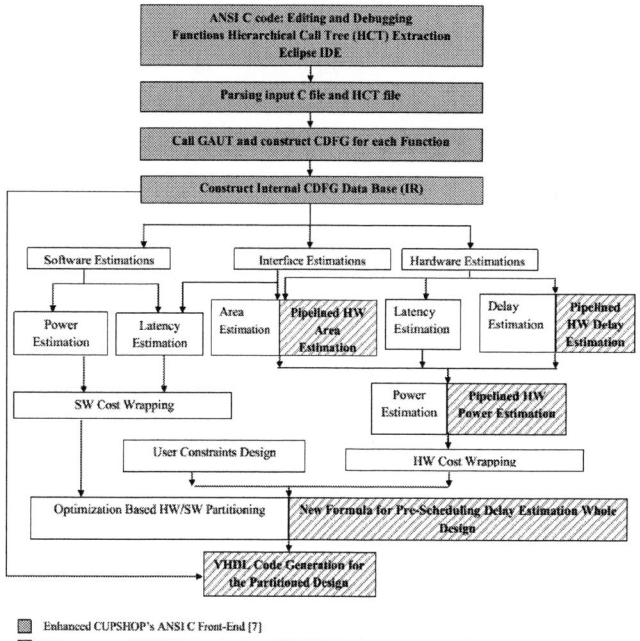

Fig.1 Enhanced CUPSHOP's block diagram

II. CUPSHOP CORE ENGINE ENHANCEMENTS

The HW/SW partitioning problem is compounded by the multitude of possible implementations for the hardware mapped components. Delay, area or power optimized implementations (or combinations thereof) are possible, with different metrics for each implementation. The original CUPSHOP v.1 handled this problem by accounting only for two extreme

978-1-4673-8760-6/15 $31.00 © 2015 IEEE

hardware implementations: the serial implementation with small area and long delay, and the parallel implementation with large area and small delay. The components are thus mapped to one of three alternatives: SW, serial HW or parallel HW alternative.

In the current CUPSHOP version, Pipelining is added as an additional HW alternative. Pipelining targets the maximum throughput. Instead of passing the input and wait for the output to be calculated, new input is passed to the HW resources at each clock cycle, which eventually decreases the average latency to be 1 clock cycle.

Several authors considered the pipelining in the context of HW/SW partitioning [8-12]. Pipelining can be done at four different levels of granularity: at the system level where the Control-Flow Graph (CFG) of behaviors is divided into pipeline stages with equal or non-equal delays; at the behavior level where the hierarchical flow graph of loops within a behavior may be divided into pipe stages; at the loop level where each loop body may be divided into pipe stages; and finally at the operation level where each node in the data flow graph may also be divided into pipe stages. Additionally, pipelining can be implemented in three different ways. First, the pipelining could be done within partitioning. This way achieves the best performance at the cost of design complexity. Second, the pipelining could be done before partitioning. Finally, it can be done after partitioning.

In this paper, we implemented the pipelined HW alternative before partitioning at the component level. The function of the PSO optimization engine is to provide an answer to the question: to which of these four alternatives; namely Software, Serial HW, Parallel HW, and Pipelined HW should we map each component in the system to get the "fittest" partition subject to the specified constraints.

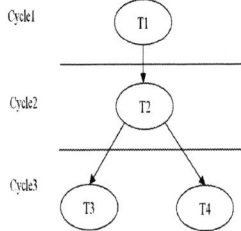

Fig. 2: CDFG for pipelined component

To clarify our procedure, consider the simple CDFG shown in Fig. 2. Assume we have this CDFG as a component, which consists of four basic components T1, T2, T3, and T4. The area, delay and power estimations of these basic components are done offline. The estimation of the area, delay and power metrics of the whole component is done online with the same methodology as in [2,5]. If the whole component is partitioned to the pipelined HW alternative, we calculate these metrics as follows. We calculate the area by adding area of (T1 (Area) + T2 (Area) + T3 (Area) + T4 (Area)). The delay equals the maximum delay of the four tasks (Max (T1, T2, T3, T4)). The power is calculated from power of flip-flops added to power of logic elements. For the parallel implementation we multiply the power of logic elements by the activity rate, which equals the inverse of latency of the full design as each component is scheduled to work for only one cycle for each input processing. In the pipelined case each node works over all cycles so we set the activity rate = 1. Fig. 1 points out the exact block added to the core of CUPSHOP to support the pipeline alternative.

The CUPSHOP v.2 core engine was tested for different practical designs. One of these designs is the JPEG encoder system of Ref. [13]. This design has 47 components to execute the DCT, Quantization, DPCM, Zigzag, Run length coding, and Variable length coding sub-functions of the JPEG encoder. Table I shows how CUPSHOP v.2 engine mapped these 47 components to SW (label 0), Serial HW (label 1), parallel HW (label 2) or pipeline HW (label 3) for different constraints on power,

area, and delay. The constraint on power is entered as a percentage of total power when all the design is implemented in the pipelined alternative. The constraint on area is entered as a percentage of total design area when all the design is implemented in the serial HW alternative. The constraint on delay is entered as a percentage of total delay if all the design is implemented on SW alternative. The results are shown in case 1 after running Re-exited PSO 10 rounds to satisfy the constraints while the results in the other cases satisfied the constraints after only 5 rounds.

TABLE I: DIFFERENT PARTITIONING VECTORS FOR DIFFERENT CONSTRAINTS

% Power Constraint *	% Area Constraint*	% Total Delay Constraint*	Cases	Partitioning vectors for Components **
0.2/0.2	20/12.74	100/82.71	1	0/1/010/010/000/000000 /000010/011/003310/00 0000/101/101/303
5/0.6752	50/46.818	20/18.08	2	3/1/333/010/303/330323 /101011/111/223230/11 1111/111/000/000
2/0.646	70/54.7	10/6.71	3	3/1/333/101/333/033012 /111111/111/233132/31 1111/111/000/000

* Inserted to tool/ Achieved by the Tool

** Components names order as follows:

HEADER/RGB_GEN /RGB_Y, U, V /IN_BUFF_CONT3/IN_BUFF3/CHEN_ADD_SUB/DCT/ MEM_TRANSPOSE/CHEN_ADD_SUB/DCT/QUANT/ZIGZAG/DPCM

Table I depicts the power/area/delay results and the partitioning vectors (last column) for different constraints. The mapping of components is 0 (SW), 1 (Serial HW), 2 (Parallel HW), or 3 (Pipelined HW). For example, the partitioning vector for case 1 indicates that the first component (Header) is mapped to SW alternative (label 0).

The partitioning vector for case 1 has small number of components partitioned to the pipelined alternative (label 3). Relative to case 1, Case 2 has a tight constraint on delay, a loose constraint on area, and a relaxed constraint on power. The components with larger delays are mapped to the pipelined alternative until the delay constraint is satisfied. Case 3 has more tight constraint on delay with relaxation on the area constraint (70%). The resulting partitioning vector achieved better delay and power outputs.

The delay estimation is calculated using our improved pre-scheduling delay estimation formula discussed in detail in Ref. [6].

III. BACK-END AUTOMATIC RTL (VHDL) CODE GENERATION

To generate RTL (VHDL) code for a given component, we need some information about this component. Actually this information is extracted during the conversion of the input C code to CDFG in the recently developed CUPSHOP C-Front-End [7]. This information include the hierarchy inside each component starting from the high level of granularity until we reach the bottom level which is the level of basic operations.

So for each level, we need to know the components working in this level and which components they call and which components are calling them. We need also to specify the inputs and the outputs for each component. The RTL (VHDL) code generation engine needs a library for all possible basic operations for each type (e.g. arithmetic, logic, etc.). First, each level's component should be created from the basic operations and other lower components in an iterative way from bottom up until reaching the main program, which calls the top level component.

Several authors developed automatic HDL code generation from different intermediate representations. We are mainly interested in CDFG to VHDL generation. Nane et. al [14] presented VHDL code generation using DWARV 2.0 which performs standard and custom transformations to the combined CDFG created as an Intermediate Representation (IR) by the CFront engine, then, used this IR to generate the VHDL code. In a sequel publication, the same authors [15] presented a tool for design, integration, and optimization of the DWARV 3.0 HLS compiler. The compiler was designed to accept a large subset of C language as input and to generate synthesizable VHDL code for unrestricted application domains. Menotti et.al [16] developed LALP (Language for Aggressive Loop Pipelining), a novel language to program FPGA-based accelerators, and its compilation framework. The front-end takes as input a program implemented in LALP, and after parsing it, generates the corresponding CDFG. Then, the back end VHDL generator selects the components from the VHDL library and wires them according to the constraints specified in the scheduled CDFG.

IV. VHDL GENERATION PROCESS

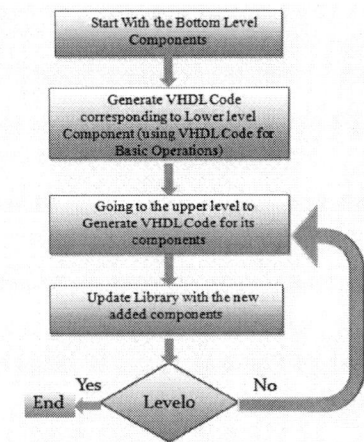

Fig. 3: Flow Chart of VHDL code generation Process

As we stated before, there is a need for constructing a library for basic operations (e.g., arithmetic, logic, compare, resize, multiplexing, and assignment operations). This library should be flexible enough to adapt to different features of the input and output ports of each basic operation (e.g., signed/ unsigned ports, different bit widths, saturated logic operations).

The developed C-based Front-End for the current CUPSHOP tool [7] accepts hierarchal C codes and generates the corresponding hierarchical CDFG. Our methodology for the code generation maintains this hierarchical nature, which is an advantage over other tools that can only handle flat designs. We begin from the bottom level. At this level, each component contains only basic operations, for which an RTL (VHDL) template is stored in a library. Going up to the succeeding upper level, the engine writes its RTL VHDL code using the lower level components, and so on until we write the VHDL for the top design as shown in Fig.3.

In the next sub-Sections, we discuss the details of code generation steps for single-level designs and hierarchal designs.

1. Single Level Designs

Consider the single level C code and its DFG shown in Fig. 4.

```
int main(int a, int b, int* c)
{  *c = (3 * a) + (b / 5); }
```

This code contains three basic operations, namely: add, multiply, and divide. The corresponding VHDL code contains the external inputs and

outputs in its entity, and the components definition of these three operations as shown in Fig.5. Fig. 6 shows the generated RTL VHDL code for the ***main*** function.

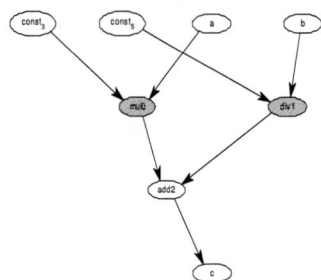

Fig. 4: DFG Generated

```
Entity main is port (
a:in signed(31 Downto 0);      b:in signed(31 Downto 0);    c:out signed(31 Downto 0) );
end main;
Architecture a of main is
Component signed_multiply is
generic  ( DATA_WIDTH1 : natural := 8;  DATA_WIDTH2 : natural := 8;
 DATA_WIDTH3 : natural := 16  );     port (
 a : in signed ((DATA_WIDTH1-1) downto 0); b : in signed ((DATA_WIDTH2-1) downto 0);
 result : out signed ((DATA_WIDTH3-1) downto 0) );  end component;
component signed_adder
generic  ( DATA_WIDTH1 : natural := 8;  DATA_WIDTH2 : natural := 8;
 DATA_WIDTH3 : natural := 8   );   port (
 a : in signed  ((DATA_WIDTH1-1) downto 0);  b : in signed  ((DATA_WIDTH2-1) downto 0);
 result : out signed ((DATA_WIDTH3-1) downto 0) );    end component;
Component signed_divider is
generic  (
DATA_WIDTH1 : natural := 8;   DATA_WIDTH2 : natural := 8;   DATA_WIDTH3 : natural := 8;
port ( a  : in signed ((DATA_WIDTH1-1) downto 0);   b : in signed ((DATA_WIDTH2-1)downto 0);
 result : out signed ((DATA_WIDTH3-1)  downto 0) ); end component;
```

Fig.5 Generated RTL VHDL code of corresponding DFG

```
signal mul0 : signed (31 Downto 0);
signal div1 : signed (31 Downto 0);
signal out_signal: signed (31 Downto 0);
Constant const_3: signed( 31 Downto 0) := CONV_SIGNED(3,3);
Constant const_5: signed( 31 Downto 0) := CONV_SIGNED(5,4);
begin
]L1: signed_multiply Generic map ( DATA_WIDTH1 => 32 ,DATA_WIDTH2 => 3 ,
 DATA_WIDTH3 => 32) port map ( a, const_3, mul0);
]L2: signed_divider Generic map ( DATA_WIDTH1 => 32 ,DATA_WIDTH2 => 4 ,
 DATA_WIDTH3 => 32) port map ( b, const_5, div1);
]L3: signed_adder Generic map ( DATA_WIDTH1 => 32 ,DATA_WIDTH2 => 32 ,
 DATA_WIDTH3 => 32) port map ( mul0, div1,c)      End a;
```

Fig.6 Generated VHDL code of the *main* function

2. Hierarchal Designs

Consider a simple hierarchical set of equations written in C language:

The equation for fun1 is C= 3*a + 5*b
The equation for fun2 is C= 5*a + 3*b
The equation for main is Z= (3*x+5*y) + (5*x+3*y)
fun3 is called to carry out the multiplication by 3 and fun4 called to carry out the multiplication by 5.

The Hierarchical Call Tree (HCT) is shown in Fig. 7. The corresponding DFG graph of the main function is shown in Fig. 8.

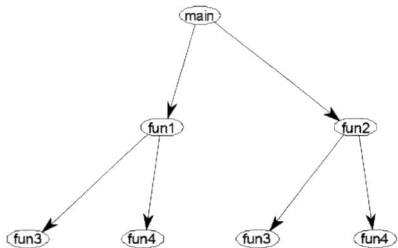

Fig.7 An example Hierarchical Call Graph Tree

The VHDL code generation proceeds as follows:

978-1-4673-8760-6/15 $31.00 © 2015 IEEE   33

(1) Generate the RTL VHDL code for lowest level operations (fun3 and fun4).

(2) Generate fun1 and fun2 codes. Each of them calls fun3 and fun4.

(3) Generate the RTL VHDL code for the main as follows in Fig.9.

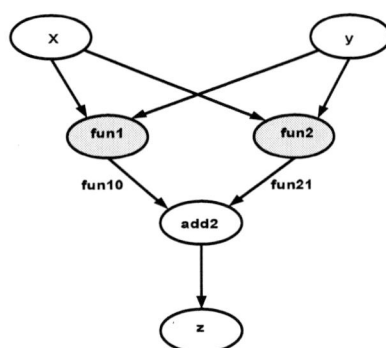

Fig.8 DFG of the main function

```
Entity main is   port (
x:in signed(31 Downto 0);    y:in signed(31 Downto 0);   z:out signed(31 Downto 0) );
end main;
Architecture a of main is
Component fun1 is port (
a:in signed(31 Downto 0);    b:in signed(31 Downto 0);   c:out signed(31 Downto 0) );
end component;
Component fun2 is port (
a:in signed(31 Downto 0);    b:in signed(31 Downto 0);   c:out signed(31 Downto 0) );
end component;
-- component signed_adder definition same as single level example
Signal fun10: signed(31 Downto 0);   Signal fun21: signed(31 Downto 0);
Begin
L1:fun1 port map (x,y,fun10);   L2:fun2 port map(x,y,fun21);
L3:signed_adder  Generic map( DATA_WIDTH1 => 32 ,DATA_WIDTH2 => 32 ,
DATA_WIDTH3 => 32) port map (fun10,fun21,z);   End a;
```

Fig.9 Generated VHDL code of the *main* function

V. CASE STUDIES

We tested the presented Enhanced CUPSHOP version with four practical cases from the GAUT benchmark suite [17].

- YUV2RGB of the MJPEG.
- 8X8Block 16 bits per pixel DCT (3-Level).
- 32 bit CORDIC filter.
- 16 tap Finite Impulse Response filter (FIR16).

The automatically-generated RTL VHDL was verified using Modelsim SE 6.2e simulator and synthesized using Altera Quartus-II 7.2 FPGA synthesizer. Table II summarizes the performance of the CUPSHOP front-end as well as backend.

Table II RTL Back-End Enhanced CUPSHOP test Cases

Parameters		YUV2RGB	DCT	CORDIC	FIR16
Generated VHDL Code Lines		143	925	501	169
L.E. (Logic Element)	Cyclone III FPGA	465	227	3511	1759
Total I/O pins		124	2049	97	65
Embedded 9-bit Multipliers		16	8	0	0

CONCLUSIONS

An Augmented version of CUPSHOP is developed with the following added features: pipelined hardware alternative, improved pre-scheduling delay formula, and automatic RTL (VHDL) Back-End generation that supports unlimited levels of hierarchy. The presented CUPSHOP enhancements are tested via several practical case studies.

REFERENCES

[1] J. Teich, "Hardware/software codesign: The past, the present, and predicting the future," Proc. IEEE, Vol. 100, No. 13, pp. 1411–1430, May 2012.

[2] M.B. Abdelhalim, S.E.-D Habib, "An integrated high-level hardware/software partitioning methodology", Design Automation for Embedded Systems, Springer, Vol. 15. pp. 19–50, 2011.

[3] P. Schaumont, "A practical introduction to hardware/software codesign", Springer, 2012.

[4] S. Gupta, " SPARK: A Parallelizing Approach to the High-Level Synthesis of Digital Circuits", Springer, 2004.

[5] M.B. Abdelhalim, A.E. Salama, S.E.-D. Habib, "Constrained and unconstrained Hardware-Software Partitioning Using Particle Swarm Optimization Technique", Embedded System Design: Topics, Techniques and Trends, The International Federation for Information Processing, Vol. 231, 2007, pp. 207-220.

[6] R. O. Hassan, M. B. Abdelhalim, & S. E.-D. Habib, "Reliable pre-scheduling delay estimation for hardware/software partitioning". In the 56th International Midwest Symposium on Circuits and Systems, MWSCAS, pp. 1246-1250, Columbus, Ohio, USA, 2013.

[7] A. H. Fatehy, "Front-End Enhancements for Cairo University Hardware/Software Partitioning Tool (CUPSHOP)", M.Sc. Thesis, Dept. of Electronics and Communication, Faculty of Engineering, Cairo University, Giza, Egypt, 2015.

[8] M. Yuan, Z. Gu, X. He, X. Liu, and L. Jiang, "Hardware/software partitioning and pipelined scheduling on runtime reconfigurable FPGAs". ACM Transactions on Design Automation of Electronic Systems (TODAES), Vol. 15, No. 2, Article No.13, 2010.

[9] Y. S. Chiu, C. S. Shih, and S. H. Hung, "Pipeline schedule synthesis for real-time streaming tasks with inter/intra-instance precedence constraints". In Design, Automation & Test in Europe Conference & Exhibition, DATE, pp. 1–6, Grenoble, France, 2011.

[10] S. Bakshi and D. D. Gajski, "Partitioning and pipelining for performance constrained hardware/software systems". IEEE Transactions on Very Large Scale Integration (VLSI) Systems, Vol. 7, No. 4, pp. 419–432, 1999.

[11] S. Bakshi and D. D. Gajski, "Performance-constrained hierarchical pipelining for behaviors, loops, and operations". ACM Transactions on Design Automation of Electronic Systems, TODAES, Vol. 6, No. 1, 2001, pp.1–25, 2001.

[12] W. Kim, J. Y. Chang, and H. Cho, "Pipelined scheduling of functional HW/SW modules for platform-based SOC design". ETRI journal, Vol. 27, No. 5, pp.533–538, 2005.

[13] C. Lian Jr, L. G. Chen, H. C. Chang, & Y. C. Chang, "Design and implementation of JPEG encoder IP core". In Proceedings of the Asia and South Pacific Design Automation Conference, ASP-DAC, pp. 29-30, Yokohama, Japan, 2001..

[14] R. Nane, V. Sima, B. Olivier, R. Meeuws, Y. Yankova, and K. Bertels, "Dwarv 2.0: A COSY-based C-to-VHDL hardware compiler," in 22nd International Conference on Field Programmable Logic and Applications, FPL, , pp. 619–622, Oslo, Norway, Aug 2012.

[15] R. Nane, V.M. Sima, C. P. Quoc; F. Goncalves, K. Bertels, "High-Level Synthesis in the Delft Workbench Hardware/Software Co-design Tool-Chain," 12th IEEE International Conference on Embedded and Ubiquitous Computing, EUC, pp.138-145, Milano, Italy, Aug. 2014.

[16] R. Menotti, J. M. P. Cardoso, M. M. Fernandes, and E. Marques, "LALP: A language to program custom FPGA-based acceleration engines", International Journal of Parallel Programming, Springer, Vol. 40, No. 3, pp. 262–289, 2012.

[17] P. Coussy, C. Chavet, P. Bomel, D. Heller, E. Senn, and E. Martin, "GAUT: A High-Level Synthesis Tool for DSP Applications". In High-Level Synthesis, P. Coussy and A. Morawiec (eds.), Springer, pp. 147-169, 2008.

Security approaches based on elliptic curve cryptography in wireless sensor networks

Mohammed Said SALAH

RITM-ESTC / CED-ENSEM, University Hassan II

Km 7, Eljadida Street, B.P. 8012 Oasis, Casablanca, Morocco

salahmedsaid@gmail.com

Abderrahim Maizate*,#, Mohammed OUZZIF#

#RITM-ESTC / CED-ENSEM, University Hassan II

Km 7, Eljadida Street, B.P. 8012 Oasis, Casablanca, Morocco

*STIC Laboratory, Chouaib Doukkali University, El Jadida, Morocco

*maizate@hotmail.com, #ouzzif@gmail.com

Abstract— **Wireless sensor networks are ubiquitous in monitoring applications, medical control, environmental control and military activities... In fact, a wireless sensor network consists of a set of communicating nodes distributed over an area in order to measure a given magnitude, or receive and transmit data independently to a base station which is connected to the user via the Internet or a satellite, for example. Each node in a sensor network is an electronic device which has calculation capacity, storage, communication and power. However, attacks in wireless sensor networks can have negative impacts on critical network applications leading to the minimization of security within these networks. So it is important to secure these networks in order to maintain their effectiveness. In this paper, we have initially attempted to study approaches oriented towards cryptography and based on elliptic curves, then we have compared the performance of each method relative to others.**

Keywords—WSN, Security, ECC, RECC-D, RECC-C, CECKM, AVL...

I. INTRODUCTION

Thanks to wireless sensor networks, we can now monitor and control physical parameters [1]. WSNs have applications in many areas [2], including the observation of nature and the environment, security of buildings and home automation, traffic management, medical monitoring, or military operations.

In many applications of sensor networks, data can be threatened by external events that should not occur during normal network operations [3]. The reliability and security of data carried in a WSN depends on several parameters including energy resources [4], types of protocols used for routing, and transport of data. Ensuring such characteristics is not an easy task to achieve, especially when the nodes are composed of electronic devices with limited hardware capabilities.

Multiple security approaches we will see in detail have the objective of secure communication between nodes in a wireless sensor network. Each approach has advantages, limitations and may be used according to a specific need.

The remaining parts of this paper will focus on the following: In Section II, we introduce wireless sensors and modules of a sensor [5], and the role of each component. Then,

we shall present the characteristics of a wireless sensor networks WSN, the security constraints in WSN [6], and vulnerabilities in a network of sensors. By the end of the section, we will offer an analysis of vulnerable attackers and attacks. In Section III, we shall present the different approaches of cryptography based on elliptic curves. In the last section, we analyze the different approaches and compare the performance of each method by looking at: the memory used by nodes for storing ECC keys [7], the average energy consumed per node and the number of packets exchanged when installing keys.

II. WIRELESS SENSOR NETWORKS

A. Characteristics of Wireless Sensor Network WSN

A wireless sensor network has the following characteristics [8]:

- Lack of infrastructure - sensor networks in particular differ from other networks through the absence of pre-existing infrastructure.

- Important size - a network sensor may contain thousands of nodes.

- Interference - the radio links aren't isolated, two simultaneous transmissions on the same frequency, or on similar frequencies can interfere.

- Dynamic Topology - the sensors can be attached to mobile objects moving freely and arbitrarily [9].

- Limited physical security - Wireless sensor networks are more affected by security settings than by traditional wired networks.

- Constraint of energy [10]- the most critical feature in sensor networks is the modesty of its energy resources because each network sensor has few resources in terms of energy (battery).

B. The vulnerabilities of wireless sensor networks

Vulnerabilities are weaknesses of a system that the attacker exploits to gain privileges. [11] There are two types of vulnerabilities in a sensor network WSN:

978-1-4673-8760-6/15 $31.00 © 2015 IEEE

- Physical vulnerability is a means of attack, which allows the attacker to change in part a sensor, for example by changing its programming code, or by copying protection keys for reuse in a new attack.

- The vulnerability lies in logic programs and protocols. It appears in four forms:
  1. designing defects
  2. implementation defects
  3. configuration errors
  4. resource shortage

C. Description of the attackers and attacks

1) Description of the attackers

The definition of the technical capabilities of the attackers is important in order to know the nature of the threat. For example, an attacker can only receive data transmission, but it can also be introduced as a legal sensor network, and has access to all network services.

Every attacker belongs to a category:

- Passer-by: with spontaneous motivation, resources and limited knowledge,

- Vandal: with resource damage motivation and limited knowledge

- Hacker: access with great motivation, curiosity and interest

- Robber: with great determination and limited resources

- Terrorist: with significant resources and a strong determination

2) Active attacks

The attacker tries to remove, add or change the transmission on a communication channel. An active attacker threatens the integrity and authenticity of data as well as confidentiality. In order to execute the attack, the malicious node is forced to use its energy, emitting a number of packets.

3) Passive attacks

The attacker only monitors the communication channels. Listening occurs when an attacker captures a node and studies traffic without altering the operation.

A passive attacker that threatens the confidentiality of data.

Layer	attack
Physical	Jamming
	sensor forgery
Liaison/MAC	Interrogation
	half-asleep
Network	Modification of control message contents
	Hello flooding
	Homing
Transport	Synchronisation flooding
	dis-synchronisation attacks
Application	Sensor breakdown
	DoS based on a track
	flood attack

TAB. 1. ATTACKS IN WIRELESS SENSORS BY LAYER

- Jamming - Given the sensitivity of wireless media noise, a node can cause a denial of service by transmitting signals at a certain frequency.

- Hello Flooding - The network discovers protocol uses called HELLO type messages to fit into a network and to discover its neighbor nodes. In a so-called HELLO Flooding attack, an attacker will use this mechanism to saturate the network and consume energy.

- DoS - Denial of Service is defined as a malfunction of a sensor-intentioned or malicious action manner. The denial of service may not result from an attack, but a single event preventing the normal functioning of its services.

III. APPROACHES BASED ON ELLIPTIC CURVE CRYPTOGRAPHY

To secure any communication model, it is necessary to encrypt messages exchanged between nodes according to an agreed key management arrangement.

Elliptic curves are mathematical objects [11] used to encrypt with shorter keys than those of public cryptography. This means faster computation and lower power consumption as well as saving of memory and bandwidth.

Because of the small size of the ECC key[12], cryptography based on elliptic curves remains among the best security solutions for wireless sensor networks, for example ECC key [13] 160-bit provides security comparable to RSA keys of 1024 bits.

A. Description Routing Driven Elliptic Curve (RECC)

An elliptic curve cryptographic approach based on the routing protocol GPSR [14] (Greedy Perimeter Stateless Routing) for sensor networks.

The network consists of two types of sensors:

- A small number of powerful sensors (Headers)

- Normal sensors form clusters

All communications goes through the Headers that collect the data delivered by normal nodes and route them to the base station.

Two approaches to establishing key [15], the first is centralized and the other is distributed.

- The centralized approach: Headers are responsible for the establishment and distribution of cryptographic keys.

- The distributed approach: after the creation of clusters each node is pre-loaded with all key neighbors.

B. Cluster Elliptic Curve Cryptography Key Management (CECKM)

This approach has a key management model based on the ECC clustering principle using an appropriate algorithm for deployment associated with a secure data transmission [16] in wireless sensor networks.

CECKM implements asymmetric key systems using ECC on sensor networks and provides dynamic key synchronization mechanism, fast and effective in network nodes without reconfiguration of all nodes when new nodes arrive or leave from the sensor network. This approach is proposed for dynamic wireless sensor networks and on a larger scale.

C. ECC key management based on an AVL tree

Key management should provide a key establishment between all nodes, and must work even if the network topology is not predefined. Unauthorized nodes cannot perform communication with network nodes.

This approach offers management key based on AVL tree system, because in the event of a change of a node (in AVL tree), it can cause a change in a sub-tree, and the key will be also changed at the same time.

- AVL Tree

The AVL tree can perform insert, delete and search in a proportional time to the height of the tree.

Since each membership changes, keys that are along the path of the affected limb at the root must be changed.

Following an addition of a node, the new node goes back to the roots of the tree by calculating the difference in subtree depth of each node encountered. If this difference is equal to two or both less, it balances with the proper rotation.

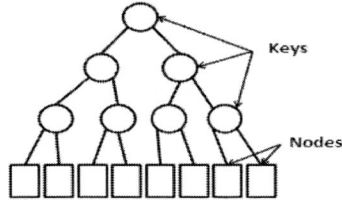

Fig. 1. The representation of keys and nodes in the tree AVL

In Figure 1, the sensors are represented by squares, and the keys are represented by circles.

When messages have to be sent to all the nodes, we use the AVL tree root key because this key is known to everyone[17].

IV. COMPARATIVE STUDY OF DIFFERENT APPROACHES

Each approach has strengths and limitations; in the following we present a comparative study of these three approaches.

To evaluate the performance of each method, we focus on these three metrics:

- The number of ECC keys stored in each sensor.
- Energy consumption per sensor.
- The number of packets exchanged between nodes during key installation.

The following equations show how to calculate the metrics:

- The size of the memory used by a sensor :

$T = (N_{CS} \times 40 \text{ bytes})$

T : Memory size
N_{CS} : number of ECC keys stored
40 bytes : the size of an ECC key

- The energy consumed in the network is:

$(N_{PR} \times E_R) + (N_{PE} \times E_E)$

N_{PR} : Number of received packets
E_R : Energy to receive a packet
N_{PE} : Number of sent packets
E_E : Energy to send a packet

A. *Energy consumption per sensor*

- Consumption by the Headers

CCED-C is the method where in the Header consumes more power because of its role as distributor of ECC keys. However in the AVL-KDC approach, the header does not exchange any message with the nodes of the cluster when installing keys, which accounts for low consumption.

	Energy consumption per Header	consumption depends on
RECC-D	*Average*	*Number of nodes*

RECC-C	*Higher*	*Number of nodes*
CECKM	*Average*	*Cluster size*
AVL-Headers	*Average*	*Number of nodes*
AVL-KDC	*Very low*	*Nothing*

Tab. 2. Average Energy Consumption by Headers

We can see in Table 2 that the AVL-KDC method is the most efficient in energy consumption, because no node participates in the management and distribution of ECC keys. In other methods nodes involved differently and consumption depends on the number of nodes in the cluster.

- Energy consumption per normal sensor

Energy consumption methods CCED-D CECKM and AVL-Headers depends on the number of nodes per cluster and the number of packets exchanged during key installation.

The large number of packet exchanges between normal sensors in a cluster when installing keys, makes the most captivating CECKM consumption method of energy.

	Energy consumption per normal node	Consumption depends on
RECC-D	*Low*	*Nothing*
RECC-C	*Low*	*Nothing*
CECKM	*Large*	*Strongly number of neighbors*
AVL-Headers	*Average*	*Nothing*
AVL-KDC	*Low*	*Nothing*

Tab. 3. Average Energy Consumption by normal nodes

Contrary the method CECKM depends on the number of neighbors, Table 3 shows that normal sensors in the other methods do not participate in the management and distribution of keys, so there is less energy consumption compared to the CECKM method.

B. *Comparison of the number of exchanged packets*

The number of packets exchanged during key installation differs from one method to another.

The simulations show that the approach to key management tree based AVL (AVL-KDC) exchanges fewer packets. However CECKM method remains a method that uses a large number of packages, because of the messages exchanged between all nodes in a cluster.

The AVL-Headers method exchanges more packages between Headers and nodes than AVL-KDC method when ECC key installation.

	Number of packets exchanged	Consumption depends on
RECC-D	*Low*	*Nothing*
RECC-C	*Low*	*Nothing*
CECKM	*Great*	*highly depending on the number of neighbors*
AVL-Headers	*Average*	*Nothing*
AVL-KDC	*Low*	*Nothing*

Tab. 4. number of exchanged packets

We see in table 4 that AVL-KDC method provides smaller communication because it exchanges fewer packets. This guarantees less key installation time and therefore safer, because when communication is prolonged, a hacker is more likely to capture and modify the keys during the exchange.

C. *Comparison of the number of stored keys*

Wireless sensor nodes are characterized by a very limited memory size of about 4KBytes for RAM and flash memory for

512Kbits for Micaz sensor, so memory is the most important constraint and each ECC key a size of 40 Bytes.

Accordingly, we have focused on the number of ECC keys stored in each node type, as operated storage memory size is strongly linked to the number of stored keys.

B. A. Bensaber and H. Boumerzoug [18] compared the memory used for the ECC keys stored in all approaches. The following graphs show the results of comparisons. These graphs show once again that the AVL-KDC method is the least occupying memory either in normal nodes or headers.

- Number of stored keys per Header

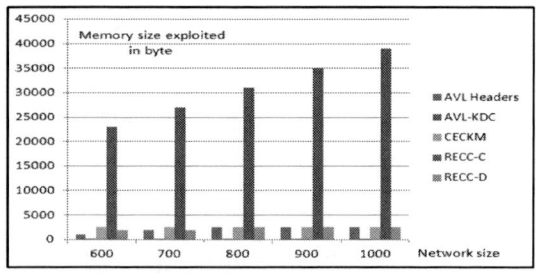

Fig. 2. Number of stored keys per Header

In the centralized approach RECC the Header saves all keys to its cluster, therefore the header uses most of his memory. AVL KDC-method is the best in terms of memory used because ECC keys are stored in the tables of the KDC server.

- Number of stored keys per normal sensor

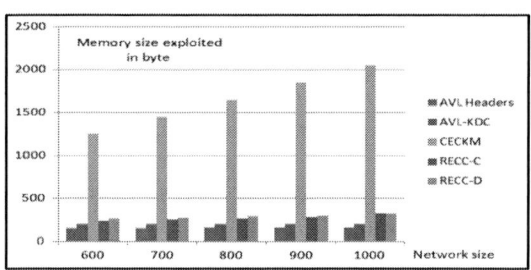

Fig. 3. Number of stored keys per normal sensor

Each node is preloaded with the keys to his neighborhood in the CECKM method which involves large memory size used for storage. Unlike CECKM method, approaches where the header or the KDC server, which ECC key, the nodes do not use their memories therefore long life network.

V. CONCLUSION

In this paper, we compared three security approaches in Wireless Sensor Networks based on Elliptic Curve Cryptography (ECC) that offers good protection, taking into account the limited sensor characteristics that directly influence the overall performance of the network, particularly safety and operating life.

The comparison of the three approaches demonstrated to us the ECC key management method based on an AVL tree offers a significant gain in the storage memory and a huge reduction of packets exchanged during the key installation less calculations while ensuring better security.

These approaches have been designed to reduce consumption of energy, reduce the size of storage, and while minimizing calculation and maximizing performance security.

Finally, the key update method can be further improved with a more study pushed on techniques of encryption and key

update, these techniques must reduce the amount of key transmitted at the start of a node, while ensuring better security.

REFERENCES

[1] A. Bachir, M. Dohler, T. Watteyne, and K. Leung, "MAC Essentials for Wireless Sensor Networks," in Communications Surveys & Tutorials, IEEE, vol. 12, no. 2, pp. 222–248, 2010.

[2] H. Karl and A. Willig, *Protocols and Architectures for Wireless Sensor Networks*, p.p. 200-230, John Wiley & Sons Ltd., 2005.

[3] I. Dietrich and F. Dressler, "On the Lifetime of Wireless Sensor Networks," in ACM Transactions on Sensor Networks (TOSN), vol. 5, no. 1, p. 5, 2009.

[4] P. Boyle et T. Newe, « Security Protocols for Use with Wireless Sensor Networks: A Survey of Security Architectures », in Third International Conference on Wireless and Mobile Communications, 2007. ICWMC'07, 2007, p. 54-54.

[5] D. Roth, J. Montavont, and T. Noel, "MOBINET : gestion de la mobilité à travers différents réseaux de capteurs sans fil," in 12èmes Rencontres Francophones sur les 156 bibliographie Aspects Algorithmiques de Télécommunications (AlgoTel'10)) (M. G. Potop-Butucaru and H. Rivano, eds.), (Belle Dune, France), Juin 2010

[6] E.Munivel, Dr G M Ajit, " Efficient Public Key Infrastructure Implementation in Wireless Sensor Networks ", International conference on Wireless Communication and Sensor Computing, ISBN 978-1-4244-5136-4, February 2010.

[7] Marc Joye, " Introduction élémentaire à la théorie des courbes elliptiques, UCL Crypto Group Technical Report Series ", book in : *http://www.dice.ucl.ac.be/crypto/*.GC-199511, pages (17) (73) (29-52).

[8] H. Karl and A. Willig, Protocols and Architectures for Wireless Sensor Networks. John Wiley & Sons, 2005.

[9] R. Kuntz, J. Montavont, and T. Noël, "Improving the medium access in highly mobile Wireless Sensor Networks," Telecommunication Systems, pp. 1–22, 2011.

[10] O. Landsiedel, E. Ghadimi, S. Duquennoy, and M. Johansson, "Low power, low delay : opportunistic routing meets duty cycling," in Proceedings of the 11th international conference on Information Processing in Sensor Networks (IPSN'12), (New York, NY, USA), pp. 185–196, ACM, Avril 2012.

[11] N. Gura, A. Patel, A. Wander, H. Eberle, and S. C. Shantz, "Comparing elliptic curve cryptography and RSA on 8-bit CPUs," in *Proc. 6th Interna tional on Cryptographic Hardware and Embedded Systems*, Boston, MA, Aug. 2004.

[12] I. Blake, G. Seroussi, and N. Smart, *Elliptic Curves in Cryptography*, London Mathematical Society, Lecture Note Series 265, Cambridge University Press

[13] "IEEE Standard for Information Technology - Telecommunications and Information Exchange Between Systems - Local and Metropolitan Area Networks - Specific Requirement. Part 11 : Wireless LAN Medium Access Control (MAC) and Physical Layer (PHY) Specifications. Amendment 2 : Higher-Speed Physical Layer (PHY) Extension in the 2.4 GHz Band - Corrigendum 1," IEEE Std 802.11b-1999/Cor 1-2001, pp. 0–1, 2001.

[14] I. Dietrich and F. Dressler, "On the Lifetime of Wireless Sensor Networks," in ACM Transactions on Sensor Networks (TOSN), vol. 5, no. 1, p. 5, 2009.

[15] Brad Karp , H. T. Kung, " GPSR: Greedy Perimeter Stateless Routing for Wireless Networks", Proceeding MobiCom '00 Proceedings of the 6th annual international conference on Mobile computing and networking, ISBN:1-581 13-197-6, 2000.

[16] Hua-Yi Lin, " High-Effect Key Management Associated With Secure Data Transmission Approaches in Sensor Networks Using a Hierarchical-based Cluster Elliptic Curve Key Agreement", ncm, pp.308-314, Fifth International Joint Conference on !NC, IMS and IDC, ISBN 978-0-7695-3769-6, Seoul, Korea, 2009.

[17] Yi-Ying Zhang, Wen-Çheng Yang, Kee-Bum Kim, Myong-Soon Park, "An AVL Tree-Based Dynamic Key Management in Hierarchical Wireless Sensor Network ", International Conference on Intelligent Information Hiding and Multimedia Signal Processing, ISBN 978-0-7695-3278-3, August 2008.

[18] B. A. Bensaber, H. Boumerzoug, " A keys management method based on an AVL tree and ECC cryptography for wireless sensor networks", International Conference on Communications, ICC 2011, ISBN 1-4244-0353-7, Glasgow, August 2011

Enhanced algorithm for QRS detection using discrete wavelet transform (DWT)

Wissam Jenkal, Rachid Latif, Ahmed Toumanari, Azdine Dliou and Oussama El B'charri

Laboratory of Systems Engineering
And Information Technology
ENSA, Ibn Zohr University
Agadir, Morocco
Email: wissamjenkal@gmail.com

Abstract—This paper aims at the detection of QRS position from the ECG signal using wavelet coefficients. An enhanced algorithm has been proposed in this work in order to approve the performance of QRS detection. This algorithm is based on three level, namely detection of suspect QRS regions, the discarding of false detected regions and the extraction of the highest peaks of the true QRS regions detected. This method has been tested on some of the MIT-BIH Arrhythmia database signals. The results of this method are promising in comparison with alternative methods recently published.

Index Terms : ECG Signals, QRS complex, wavelets coifficents, enhanced algorithm

I. INTRODUCTION

The electrocardiogram (ECG), as presented in Fig.1, is an important factor for cardiac diagnosis of cardiovascular diseases is, by definition, the representation of the electrical activity of the heart [1]. The ECG signal allows arriving at an accurate diagnosis of cardiac situation. The ECG signal analysing is an important research field either in medical or biomedical domains. Several research subjects have proposed. Those subjects vary among several issues, mention the ECG signal denoising, the QRS detection, and automatic processing methods.

The QRS complex presents the ventricular contractions and it comprises of quick deflections namely Q,R and S waves [2-5]. The Q wave is defined by a low amplitude, which is less than a third of the QRS complex with a short duration less than 0.04 second. The R wave is the first positive wave, whether or not preceded by a wave Q. the S wave is a negative wave that follows an R wave [6-7].

The QRS detection is an essential step for all sorts of ECG signal diagnosis of the ECG signal. The extraction of this complex is a major issue for searchers due to different morphology of ECG signals as well as other influences e.g. noises [8].

Several research works had been proposed to resolve this issue. e.g., digital filters, artificial neural networks [9], derivative based algorithms [10-13], matched filter [9], filter banks [9], wavelet transforms [14-15]. Among these methods, wavelet analysis is an important solution which gives a good estimation

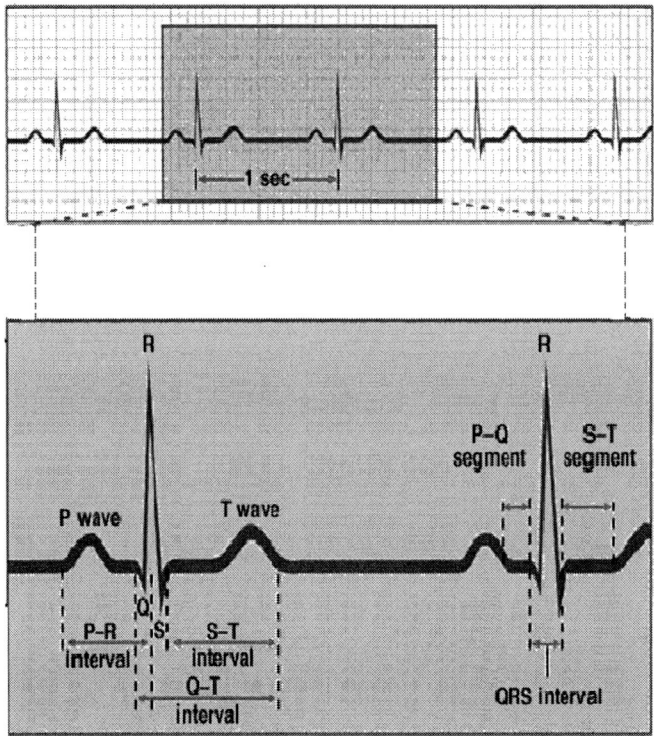

Fig. 1. Normal ECG signal with QRS complex.

of time-frequency localization as well as a flexibility offered by this technique which gives it the possibility to be developed and improved [8].

The purpose of this paper is to introduce an enhanced algorithm for the detection of the QRS complex at ECG signals based on wavelets. The aim of this algorithm is to approve the performance of QRS detection. This algorithm is based on three level, namely detection of suspect QRS regions, the discarding of false regions detected and the extraction of the highest peaks of the true QRS regions detected. This method has been tested on some of the MIT-BIH Arrhythmia database signals.

978-1-4673-8760-6/15 $31.00 © 2015 IEEE

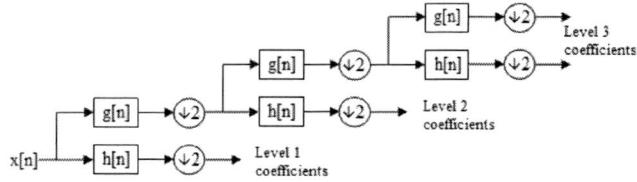

Fig. 2. DWT decomposition

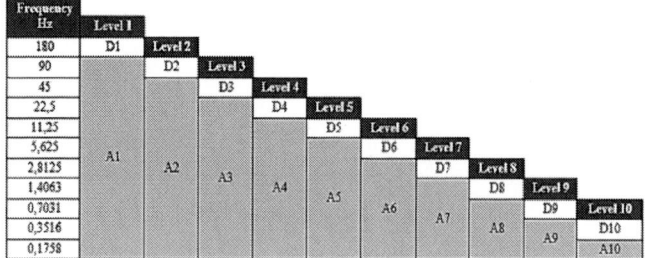

Fig. 3. ECG signal decomposition

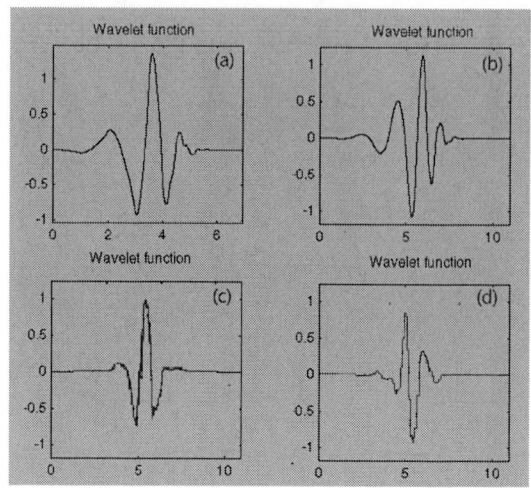

Fig. 4. DWT function. (a) Db4, (b) Db6, (c) Sym6 and (d) Sym7

II. DISCRETE WAVELET TRANSFORM

The discrete wavelet transform (DWT) is a mathematical method widely used throughout the signal processing. The aim of this processing, as illustrated in fig. 2, is to decompose a function of different resolution using high pass filter and low pass filter.

Several high pass and low pass coefficients have been developed to give a larger choice among different scales and translations in order to obtain a different sort of signal analyzing. e.g. Debauchies coefficients (Db), Symlets coefficients (Sym) and Coiflets coefficients.

Figures 2 and 3 shows the DWT decomposition of function where every approximation becomes a novel function to be decomposed in the level that follow.

The Selection of wavelet function is an essential element in the decomposition of the ECG signal [1]. The selected function should be close to the analysed signal. fig. 4 presents some of wavelets function usually used when analysing ECG signals. The symlet 7 (sym7) is the function chosen for this method.

In this study, we shall propose to use The MIT-BIH Arrhythmia signals from the international database Physionet. Those signals are sampled at 360Hz. The maximum range of real frequency components of the signals, as presented in fig. 3, is 180Hz [8].

III. PROPOSED METHOD

As shown in the previous section, the proposed method consist of ECG signal decomposition at several level. this technique allow to extract details and approximation needed for ECG signal denoising as well as extraction of QRS complex as presented in fig.5.

A. Signal denoising

For the ECG signal denoising, we shall propose two levels of filtering, the first for the high frequency noises and the second for the baseline wandering. Detail 1 (D1) and 2 (D2), as illustrated in fig. 6 concentrate the majority of high frequency noises. The elimination of D1 and D2 from the original signal allows reducing high frequency noise from the ECG signal.

For the baseline wandering another method will be proposed in further work. The results of this method are presented in fig. 6.

B. QRS Detection

In this section, two steps of thresholding have been proposed by using details 3, 4 and 5, as shown in fig.7.

For the first step, the threshold value is smaller than normal [8] that allows the possibility of QRS detection even in low amplitude. The purpose of the second step is to remove false QRS localization from the result of the previous step by using a moving window. The window size has been selected based on two parameter, namely the maximum QRS duration , and the maximum beats rate.

After thresholding levels, the QRS localization is assured by finding the maximum beat around suspect QRS localization using a windows size of 160 millisecond (ms).

IV. RESULTS AND DISCUSSIONS

The simulation results have been drawn using MATLAB R2014a. To analyse the efficiency of the proposed method, the use of a standard database is highly recommended. The MIT-BIH arrhythmia database (MITB-DB) is frequently used in the evaluation of different algorithms in several thematic interested by the ECG signal since 1980 [16]. MITB-DB contains 48 records. Each record is of 30 min length with 360 Hz sampling frequency. Some of these records have been proposed to evaluate the proposed method.

Figure 8 shows the application of the proposed method on MITB-DB record n228. This figure illustrates the efficiency of the proposed method over a complex morphology of ECG signals. Table 1 presents results of analysing of this method on the MITB-DB records using statistical parameters, namely true beats (TB), true positive (TP), false negative (FN), False

978-1-4673-8760-6/15 $31.00 © 2015 IEEE   40

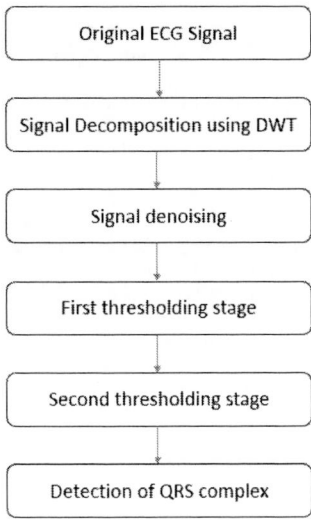

Fig. 5. proposed algorithm

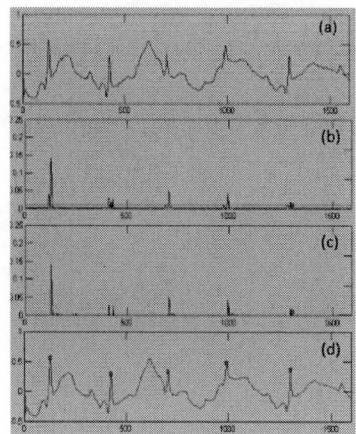

Fig. 8. Application of the proposed method on MITB-DB record n228. (a) original signal, (b) thresholding step, (c) false beats elimination, (d) QRS detection

positive (FP), positive predictivity (Pp)and sensitivity (Se), defined as follows:

$$Se(\%) = TP \div (TP + FN) \times 100 \qquad (1)$$

$$Pp(\%) = TP \div (TP + FP) \times 100 \qquad (2)$$

Where true positive (TP) is the true detected beats index ,False Positive (FP) is the false detected beats index, False negative (FN) is the undetected beats index and total beats (TB) is the total analysed beats.

Table 1 presents results of this method on the 48 MITB-DB records. The maximum errors ration has been shown in the record n203 (2.01%). This is due to variable RR intervals and amplitude of this signal. In spite of that, the total positive predictivity (Pp) of 99.84%, and sensitivity (Se) of 99.67% are very promising. These results, as presented in table. 2, are competitive with a recently published method using wavelets [1,8] or others [17-19]. These results allow concluding that the proposed method presents an efficient technique for QRS detection.

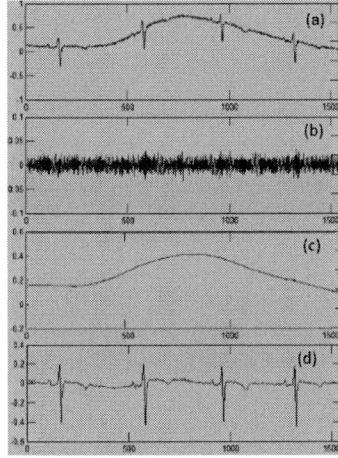

Fig. 6. ECG signal denoising.(a) original signal, (b) D1+D2, (c) Baseling wandering, (d) corrected Signal

V. CONCLUSION

This paper presents an enhanced algorithm of QRS detection using DWT. This method aims at determination of details equations from wavelets decomposition. These details allow filtering the ECG signal as well as detection of QRS regions. Next, two stages of thresholding have been included. The first for selection of suspects QRS regions and the second for elimination of false suspect beats detected from the previous stage. Finally, the last step permits to detect the highest peaks in the QRS regions from the ECG signal. The results of this method, as presented in table.1 and table.2, are promising with the positive predictivity (Pp) of 99.84%, and sensitivity (Se) of 99.71%. Encouragingly, the results of this method are very ambitious compared to recently published methods.

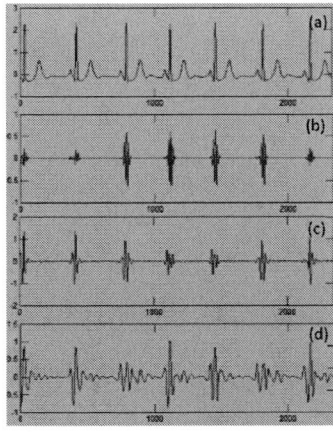

Fig. 7. ECG signal denoising.(a) corrected signal, (b) D3, (c) D4, (d) D5

TABLE I
MITB-DB ANALYZING USING THE PROPOSED METHOD

Rec	N° of beats	TP	FP	FN	Pp (%)	Se (%)
100	2273	2273	0	0	100	100.00
101	1865	1865	1	1	99,95	99.95
102	2187	2188	1	0	99,95	100.00
103	2084	2084	0	0	100	100
104	2229	2230	1	0	99,96	100
105	2572	2561	9	20	99,65	99,23
106	2027	2021	0	6	100	99,7
107	2137	2168	31	0	98,59	100
108	1763	1774	17	6	99,05	99,66
121	1863	1863	0	0	100	100
122	2476	2476	0	0	100	100
123	1518	1518	0	0	100	100
124	1619	1616	0	3	100	99,81
200	2601	2598	0	3	100	99,88
201	1963	1957	10	16	99,49	99,19
202	2136	2131	0	5	100	99,77
203	2980	2920	0	60	100	97,99
205	2656	2637	0	19	100	99,28
207	1860	1858	5	7	99,73	99,62
208	2955	2950	2	7	99,93	99,76
209	3005	3005	0	0	100	100
Total	46769	46693	77	153	99,84	99,67

TABLE II
MITB-DB ANALYZING USING THE PROPOSED METHOD

Methods	Pp (%)	Se (%)
Proposed method	99,84	99,67
Compared method [19] (2014)	99,71	96,28
Compared method [8] (2012)	99.82	99.64
Compared method [1] (2012)	99.5	99.6
Compared method [17] (2010)	99.80	99.66
Compared method [18] (2008)	99.59	98.07

ACKNOWLEDGMENT

We owe debt of gratitude to the National Centre for Scientific and Technical Research of Morocco (CNRST) for their financial support and for their supervision (grant number: 18UIZ2015).

REFERENCES

[1] S. Banerjee, R. Gupta, M. Mitra, Delineation of ECG characteristic features using multiresolution wavelet analysis method, Measurement, Vol. 45, Issue. 3, pp. 474-487, 2012.

[2] X. Liu, Y. Zheng, M. W. Phyu, B. Zhao, M. Je, and X. Yuan, Multiple functional ECG signal is processing for wearable applications of longterm cardiac monitoring , IEEE Trans. Biomed. Eng., vol. 58, no. 2, pp. 380-389, 2011.

[3] A. I. Hernndez, J. Dumont, M. Altuve, A. Beuche, G. Carrault. Evolutionary Optimization of ECG Feature Extraction Methods: Applications to the Monitoring of Adult Myocardial Ischemia and Neonatal Apnea Bradycardia Events, A Comprehensive Framework of Computational Intelligence, Springer, pp. 237-273, 2012.

[4] U. Maji, M. Mitra, S. Pal, Automatic Detection of Atrial Fibrillation Using Empirical Mode Decomposition and Statistical Approach, Procedia Technology, Vol. 10, pp. 45-52, 2013.

[5] R. J. Martis, U. Rajendra Acharya, K.M. Mandana, A.K. Ray, Chandan Chakraborty, Application of principal component analysis to ECG signals for automated diagnosis of cardiac health, Expert Systems with Applications, Vol. 39, Issue. 14, pp. 11792-11800, 2012.

[6] W. Zareba, H. Klein, et al, Effectiveness of cardiac resynchronization therapy by QRS morphology in the Multicenter Automatic Defibrillator Implantation TrialCardiac Resynchronization Therapy (MADIT-CRT), Circulation, vol. 123, no. 10, pp. 1061-1072, 2011.

[7] S. Pal, M. Mitra, Empirical mode decomposition based ECG enhancement and QRS detection, Computers in Biology and Medicine, Vol. 42, Issue. 1, pp. 83-92, 2012.

[8] Z. Zidelmal, A. Amirou, M. Adnane, A. Belouchrani, QRS detection based on wavelet coefficients, Computer Methods and Programs in Biomedicine, Vol. 107, Issue. 3, pp. 490-496, 2012.

[9] B. U. Kohler, C. Hennig , R. Orglmeister, The principles of software QRS detection, IEEE Eng Med Biol, Vol.1 ,pp. 42-57, 2002.

[10] J. Pan, W. J. Tompkins, A real-time QRS detection algorithm, IEEE Trans Biomed Eng, Vol. 32, Issue. 3, pp. 230- 236, 1985.

[11] M. Adnane, Z. W. Jiang , S. Choi, Development of QRS detection algorithm designed for wearable cardiorespiratory system, Comput Methods Programs Biomed, Vol. 93, pp. 20-31, 2009.

[12] M. Paoletti , C. Marchesi, Discovering dangerous patterns in long-term ambulatory ECG recordings using a fast QRS detection algorithm and explorative data analysis, Comput Methods Programs Biomed , Vol. 82, pp. 20-30, 2006.

[13] Y. C. Yeha, W. J. Wanga, QRS complexes detection for ECG signal: the Difference Operation Method, Comput Methods Programs Biomed, Vol. 91, pp. 245-254, 2008.

[14] J. P. V. Madeiro , P. Cortez, F. Oliveira et al, A new approach to QRS segmentation based on wavelet bases and adaptive threshold technique, Med. Eng. Phys. , Vol. 29, pp. 26-37, 2007.

[15] S. W. Chen, H. C. Chen, H. L. Chan, A real-time QRS detection method based on moving-averaging incorporating with wavelet denoising, Comput Methods Programs Biomed ,Vol. 8, pp. 187-195, 2006.

[16] G. B. Moody, R. G. Mark, The impact of the MIT-BIH Arrhythmia Database, Engineering in Medicine and Biology Magazine, IEEE , vol.20, no.3, pp.45-50, 2001.

[17] S. Choi, M. Adnane, G. J. Lee, H. Jang, Z. Jiang, H. K. Park,Development of ECG beat segmentation method by combining lowpass filter and irregular R-R interval checkup strategy, EXPERT Syst. Appl., Vol. 37, Issue. 7, pp. 5208-5218, 2010.

[18] N. M. Arzeno, Z. D. Deng, C. S. Poon, Analysis of first derivative based QRS detection algorithm, IEEE Trans. Biomed. Eng, Vol. 55, Issue. 2, pp. 478-484, 2008.

[19] R. Rodriguez, A. Mexicano, R. Ponce-Medellin, J. Bila, S. Cervantes, Adaptive Threshold and Principal Component Analysis for Features Extraction of Electrocardiogram Signals , International Symposium on In Computer Consumer and Control (IS3C), IEEE, pp. 1253-1258, 2014.

A Low-cost design of Transceiver based on DWPT for WSN

Mouhamad CHEHAITLY [2], Mohamed TABAA[1], Fabrice MONTEIRO [2], Abbas DANDACHE [2], Ali HAMIE[3]

[1]Moroccan School of Engineering Sciences (EMSI), Research and Innovation Department, Casablanca, Morocco
[2] Laboratory of genie industrial and production of Metz (LGIPM), University of Lorraine, France
[3]Art Sciences & Technology University, Beirut, Lebanon

{mouhamad.chehaitly, fabrice.monteiro, abbas.dandache}@univ-lorraine.fr, med.tabaa@gmail.com, ahamie@gmail.com

Abstract — **A novel low-cost design of communication architecture based on discrete packet wavelet transform is proposed. However, a lot of characteristics as demand for wireless communication systems like high bit rate, high capacity for objective to integrate a high speed data in several domains: industrial, medicine, militaries, and others. This paper present a novel design of architecture based on Inverse Discrete Packet Wavelet Transform (IDPWT) in transmitter and Discrete Packet Wavelet Transform (DPWT) in receiver for multichannel mode of communications. We are testing also a transceiver/receiver in different situation of noise from white Gaussian to worse case like random pulse noise.**

Keywords—Discrete Packet Wavelet Transform (DPWT), Transceiver, Wireless Sensor Network (WSN), Multicarrier, Pulse noise.

I. INTRODUCTION

Nowadays, multiband communication is also used in several types of wireless communication applications such as 4G and 5G networks specially the application broadband with high-speed data, WIFI, DVB, UWB and other [1]. The next generations of communication are based at high speed system based generally on frequency division multiplexing scheme for objective multicarrier modulation (MCM) technique. In this later, the bandwidth is divided into sub-bandwidths and each users of architecture can transmitted in different sub-band [3,4,5 and 12].

The Wide-Band or Ultra-Wide Band applications such as the multimedia services based generally on wireless sensors applications. Several techniques are proposed in literature for communication based on two systems: narrow-band and multicarrier, developing different technique in transmitter and receiver like Modulation, Equalization, Filtering, Fading, Multi-path, and Non-Line-Of-Sight (NLOS). OFDM (Orthogonal Frequency Division Multiplexing) has become a preferred proposed in multi-carrier based generally on FFT. This later used addition cyclic prefix on transmission side. By against, wavelet packet based transmission technique has stronger ability and he gives both frequency and time based

information. Recently, wavelet transformation present emerge in digital communication. [9] are first work in this domain presented by Lindsey in 1997 as alternative to communication with multiplex division.

The performances of the DWT algorithms and the multi-resolution analysis authorize a large diversity of applications relying on the wavelets theory. These algorithms are based on several methods among which the filter banks constitute the most often implemented solution. The multi-resolution analysis allows the decomposition of the signal into coefficients of approximation and details, from a scaling function and a wavelet function, respectively [10]. This implementation is based on a set of cascaded low-pass and high-pass digital filters.

Instead, in this paper we propose a novel transceiver for WSN based on Discrete Packet Wavelet Transform (DWPT) cause itself have an orthonormal base characteristic, we use DWPT in several modes: i) multi-channel architecture with two modes of bandwidth allocation: one channel per user in the peer-to-peer mode, several channels per user in the multiusers mode; ii) a single emitter and receiver [3], capable on one hand of detecting the data sent by all the broadcasting stations following both proposed modes, and on the other hand to identify the transmitter.

This paper is organized as follows: in section 2, describe the state of art of different techniques using in research work. The proposed architecture and different form of communication are presented in section 3. Section 4; describe global discussion and results of simulations and finally conclusion and perspectives.

II. DISCRETE PACKET WAVELET TRANSFORM

In the wavelet transform (WT) theory, the wavelet basis functions are obtained from a single prototype function called "wavelet" by translation and dilation or contraction:

$$\Psi_{s,\tau}(t) = \frac{1}{\sqrt{s}} \, \Psi \left(\frac{t-\tau}{s} \right) \quad (1)$$

Where $s \in R^*, \tau \in R$. For large s, the basis function becomes a stretched version of the prototype wavelet, that is a low frequency function, while for small s, the basis function becomes a contracted wavelet, that is a high frequency function.

The Continue Wavelet Transform (CWT) of $x(t)$ is defined as:

$$X(s, \tau) = \frac{1}{\sqrt{s}} \int_{-\infty}^{+\infty} x(t) \Psi^* \left(\frac{t - \tau}{s}\right) dt \quad (2)$$

Where *denotes complex conjugation.

The CWT is too redundancy and to make it practical, we need a finite number of wavelets in the wavelet transform. Therefore, the discrete wavelets transform (DWT) are discretely scalable and translatable. This was achieved by modifying the wavelet representation to create Daubechies (1992) [2,10]:

$$\Psi_{j,k}(t) = \frac{1}{\sqrt{s_0^j}} \Psi \left(\frac{t - k s_0^j \tau_0}{s_0^j}\right) \quad (3)$$

We usually choose $s_0 = 2$ so that the sampling of the frequency axis corresponds to dyadic sampling. In addition, $\tau_0 = 1$ we gave a dyadic sampling in time.

Discretizing the translation and dilation contraction parameters of the wavelet in (1), the dyadic discrete WT of $x(t)$ is:

$$X(j, k) = 2^{-j/2} \int_{-\infty}^{+\infty} x(t) \Psi^* (2^{-j}t - k) \, dt \quad (4)$$

Where j, k $\in$ Z.

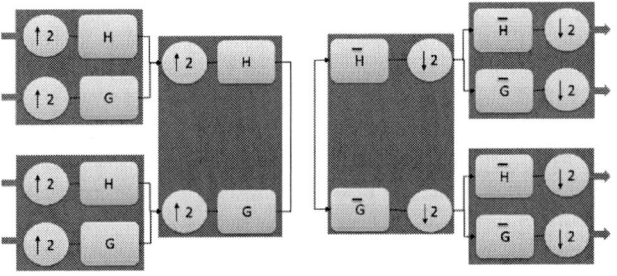

Fig.2. IDWT in transmitter and DWT in receiver

It should be mentioned that WT can be implemented by non-uniform filter banks. Where the input signal separates into low frequency component signal and high frequency component signal. The first is called smooth coefficient, the second part is called wavelet coefficients. The smooth coefficient is separated into two parts repeatedly). In the classical way, we indicate the tow part via low-pass digital filter H and a high pass-filter G. By using the scaling function and there corresponding mother wavelets, we obtain both of the digital filter H and G. We suppose H and G like a FIR filters non-recursive with L length, the transfer functions of H and G can be represented as follows:

$$H(Z) = h_0 + h_1 z^{-1} + h_2 z^{-2} + \cdots + h_{L-1} z^{-(L-1)} \quad (5)$$
$$G(Z) = g_0 + g_1 z^{-1} + g_2 z^{-2} + \cdots + g_{L-1} z^{-(L-1)} \quad (6)$$

Mallat's tree algorithm or pyramid algorithm [2] can be used to find the multiresolution decomposition of DWT, the two scale relations (5) and (6) leads to scaling functions and wavelet functions similar to that in scalar wavelets. But the equations are two scale matrix equations and can be given as:

$$\phi(t) = \sum_n h(n)\phi(2t - n) \quad (7)$$
$$\Psi(t) = \sum_n g(n)\Psi(2t - n) \quad (8)$$

Where $\phi(t) = [\phi_1(t) \, \phi_2(t) \, \dots \, \phi_r(t)]^T$ and $\Psi(t) = [\Psi_1(t) \, \Psi_2(t) \, \dots \, \Psi_r(t)]^T$ forms the set of scaling functions and corresponding wavelets. The suffix r denotes the number of wavelets and is dubbed as multiplicity. The decomposition DWPT at each resolution level can be presented as tree shape as figure 2.

III. SYSTEM DESCRITPION

In this section we present the functioning of our proposed transceiver model. The emitter is based on IDWPT and the receiver is based on DWPT, implemented as a synthesis filter banks Fig.2. The transceiver can be used in several situations and for different application. The wavelet and the depth of the transform can be chosen to adapt the shape of the generated pulse but also, according to the mode of use, to define the maximum throughput and the number of simultaneously exploitable sensors.

The goal of this research work is to propose system multi-user for applications based on star topology. The use of the IDWPT/DWPT is different from what is not novel and it was presented in [9]: the number of users is specified and for every users an IDWPT block is implemented to assure on one hand the activation of only one single input per user, and on the other hand this input corresponds to a given level of details and approximation that will be used as identification label. The unused inputs are all forced to zero. As a consequence of the activation of a single input per user, the throughput per user is reduced, but the overall system's throughput is the same compared with the parallel coding. In [11] DWPT cascade filter is dedicated to the reception and split of the dataflow into separate channels, and a decision block ensures the detection and the correction of data and identification of the emitter. The detection operation consists in studying the amplitude of the received signals and thresholding for decision-making and recognition.

The idea of wavelet packet modulation is to use wavelet packets as the pulse shaping functions, maybe for this raison most of recent research work [4] use an additive Gaussian noise channel and with limit number of channel (2 or 4 channel). So the challenge for us in this work is to use wavelet packets under additive pulse noise with a big number of channel like 8, 16, and 32.

To generate the pulse noise show in fig.3, we use a mixture Gaussian noise generator:

$$X = (1 - a) * f_1 + a * f_2 \quad (9)$$

Where a is a mixing factor ($0<=a<1$), f_1 is the power density function of non-impulsive Gaussian, f_2 is the power density function of impulsive Gaussian, and X is the output impulsive noise signal.

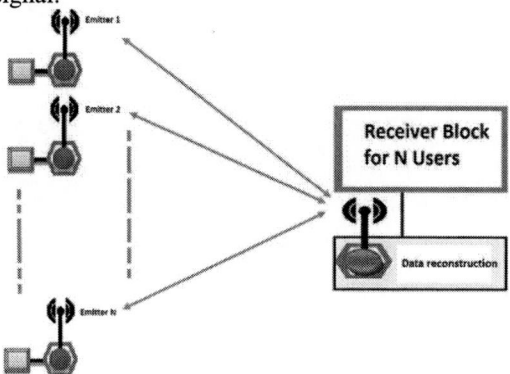

Fig.3. Our proposed

IV. SIMULATIONS AND RESULTS

In this section we show the results of simulations that on one hand allow validating the proposed architecture, and on the other hand, allow defining the characteristics of the target architecture that meets the requirements of the wireless communication. Table 1 gives all characteristics of our architecture proposed.

In table 2, we present the main orthogonal wavelet function (exception biorthogonal spline wavelets is not orthogonal however biorthogonal) and N strictly positive integer represent the order of wavelet function.

Table 1: Characteristics of the architecture

Parameters	Description
Communication	Multi-users
Application	Wide band
TX & RX Sensors	8,16,32
Modulation	Pulse Shape
Transmitter	IDWT
Receiver	DWT
Data	Binary
Frequency band	Multiband

Table 2: Wavelet description

Parameters	Orthogonal	Compact support	Filters length
Daubechies	Yes	Yes	2N
Coiflets	Yes	Yes	6N
Symlets	Yes	Yes	2N
Discrete Meyer	Yes	Yes	102
Biorthogonal	No	Yes	max(2Nr,2Nd)$_{[\bullet]}$+2

The main objective of this research work is proposed a new form of multi-user communication can transformed a data source to a series of pulse series information using several modulation in transmitter and be detected in receiver with complete transmitter. As already presented in section 2, transmitter are based on IDWPT and DWPT on receiver. This structure allowed our proposed multichannel communication model for reconfigurable transceiver input by following the block that this entity IDWPT issue of each user who belongs of architecture, Fig.4.

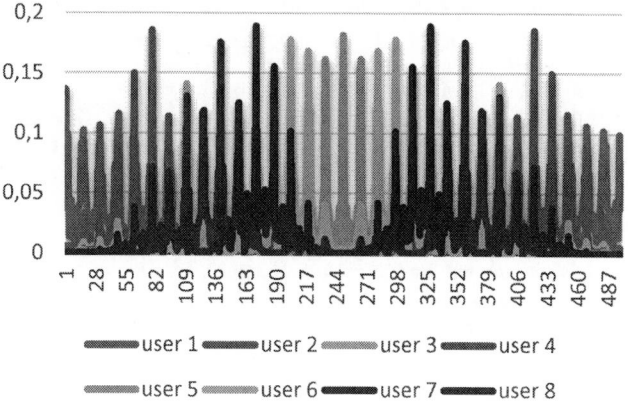

Fig.4. Frequency band for 8 users

We are also proceed on Bit Error Rate study to evaluate the performance of our proposed architecture. To do this, we are using two steps:

In first step, as a performance measure for the proposed technique, we use the BER/SNR with an additive white Gaussian noise. The performances of the proposed system are obtained with different number of users 16 and 32 for multichannel communication. In Fig.6 and Fig.7 we can show a sensible BER with different values of Gaussian noise. The proposed technique for different SNR rate is shown that the BER less than 10% for 5 dB of SNR and less than 20% for 1 dB of the SNR.

In second step, we evaluate the influence of pulse noise on the BER with 6 different wavelets. We adopt the Mallat tree for different depth (for example: depth=4 => 16 users) and to generate the pulse noise we used equation (9). We choose a=0.05, variance1=0, and variant variance, which is a good approximation of industry pulse noise. Using the 20th Daubechie wavelet increases the SNR of 1 dB at BER = $10-2$ and we find a BER reduction capability of the proposed technique depend on the choice of the wavelet. Specifically, the amount of BER reduction increases as the wavelet's order gets larger. We can achieve low BER with the 20th Daubechie wavelet and Discrete Meyer wavelet.

V. CONCLUSION AND PERSPECTIVES

A novel design of transceiver was proposed in this paper, based on IDWT in transmitter and DWT in receiver for multichannel communication with contaminated channel by pulse noise. We are applied a different situation of noise with different wavelet to test a performance of our proposed architecture.

It should be mentioned that the proposed can be implemented in FPGA to profit really of performance of wavelet in some communication systems.

REFERENCES

[1] Fei Hu, Qi Hao (edited by), *Intelligent Sensor Network: The Integration of Sensor Networks Signal Processing and Machine Learning*, Taylor & Francis group, 2013, ISBN: 978-1-4398-9282-4.

[2] Robert X.Gao and Ruqiang Yan, *Wavelets Theory and Applications for Manufacturing*, Springer, 2011, ISBN: 978-1-4419-1544-3.

[3] M. K. Lakshmanan, H. Nikookar, "A Review of Wavelets for Digital WirelessCommunication",*WirelessPersonalCommunications*,Vol. 37, Issue 3-4, pp 387-420, May 2006.

[4] DyonisiusDonyAriananda, Madan Kumar Lakshmanan, HomayounNokookar,"An investigation of Wavelet Packet Transform for spectrum estimation", *Computing Research Repository*, abs/1304.3795, 2013.

[5] M. Ghawami, L. B. Michael, S. Haruyama and R. Kohno,"A Novel UWB Pulse Shape Modulation System", *Wireless Personal Communications*, Vol. 23, pp. 105–120, 2002.

[6] T. Norimatsu et al. A Novel UWB Impulse-radio Transmitter with All-digitally-controlled Pulse Generator. IEEE, 0-7803-9205-1/05, (2005)

[7] I. Dotlic. Design of the Family of Orthogonal and Spectrally Efficient UWB Waveforms. IEEE JOURNAL OF SELECTED TOPICS IN SIGNAL PROCESSING, VOL. 1, NO. 1, JUNE 2007

[8] Y. Kim et C. Shin. Orthonormal Pulses for High Data Rate Communications in Indoor UWB Systems. IEEE COMMUNICATIONS LETTERS, VOL. 9, NO. 5, MAY 2005

[9] AR Lindsey, Wavelet packet modulation for orthogonally multiplexed communication, Signal Processing, IEEE Transactions on, 1997

[10] S. Mallat. A wavelet tour of signal processing. Academic Press, 1999.

[11] Mohamed TABAA, Camille DIOU, A Low-Cost Many-to-One WSN Architecture Based on UWB-IR and DWPT International conference on Control, Decision and Information technologies CoDIT2014 IEEE, Metz France 3-5 Novembre.

[12] Si Chen, Huichang Zhao, Shuning Zhang, Yunxing Yang,"Study of ultra-wideband fuze signal processing method based on wavelet transform", *IET Radar Sonar Navig.*,Vol. 8, Issue 3, pp. 167–172, 2014.

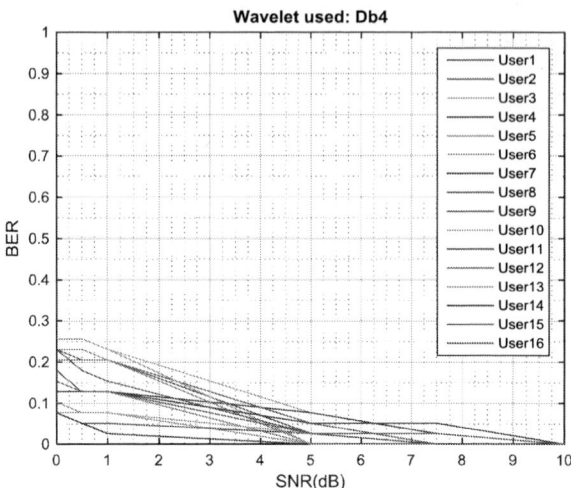

Fig.6. BER as a function of SNR with initial seed=67 for 16 users

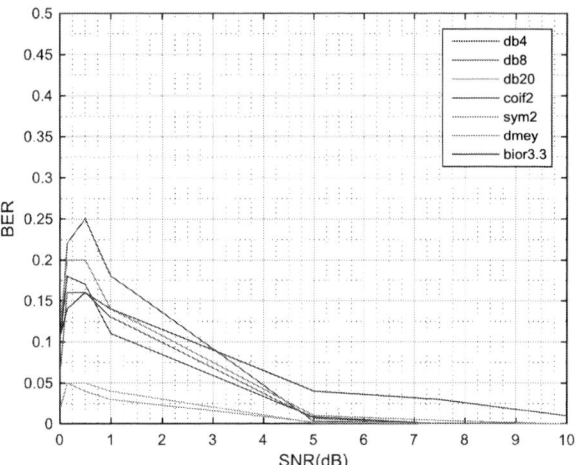

Fig.8. BER as a function of SNR under pulse noise effect of 16 user's communication model with different wavelet family

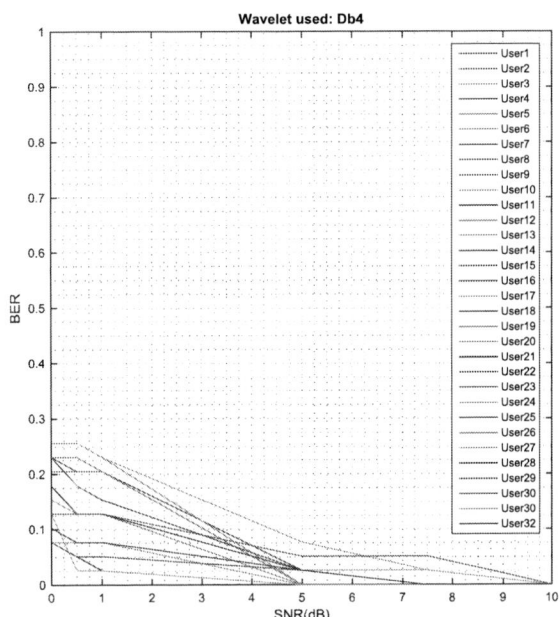

Fig.7. BER as a function of SNR with initial seed=67 for 32 users

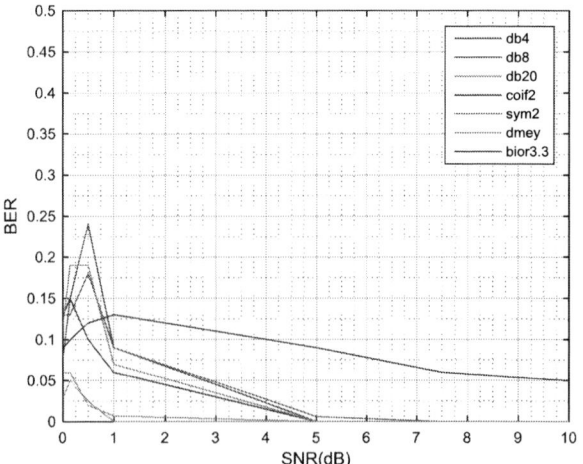

Fig.9. BER as a function of SNR under pulse noise effect of 32 user's communication model with different wavelet family

Network Coding for Energy Optimization and Throughput Enhancement of WideMac for IR-UWB based WSN

Tarik CHANYOUR[*1], Hanane CHERGUI HALI [*2], Rachid SAADANE [3]

[1] Mohammed V-Rabat University
SIME Labo, ENSIAS,
Rabat, Morocco
tarik.chanyour@um5s.net.ma

[2] Ibn Tofaïl University
Faculty of Science
Kenitra, Morocco
hanane.chergui@gmail.com

[3] SIRC2S/LASI EHTP,
Km 7, Oasis, route El Jadida
Casablanca, Morocco
rachid.saadane@gmail.com

Abstract— Energy and throughput optimization occupies an important place in the research area of Wireless Sensor Networks (WSNs). These networks are a class of useful but challenging networks. Combining Network Coding (NC) and efficient scheduling in the MAC layer of such networks gives more efficiency and copes with the problem of energy waste and inefficiency management as well as transmission's lose and collision. Our work's concern is to assess the contribution of the network coding technique's introduction to improve the efficiency and optimize the energy consumption in the MAC layer of IR-UWB based WSNs. Therefore, we tested this new technique using WideMac Protocol; which is a relevant MAC protocols dedicated to WSNs based on IR-UWB. Furthermore, we used XOR coding to fit low computational capabilities requirement in such networks. The obtained results are very encouraging and the proposed MAC scheme's improvements have demonstrated the positive impact of using NC in terms of both used metrics.

Keywords-WSN; IR-UWB; WideMac; ALOHA; network coding; xor coding;

I. INTRODUCTION

According to the results of the work of Ahlswede et al.[1], it is now possible via an appropriate Network Coding (NC) to transmit information in any communication network while achieving the theoretical limits of its capacity. The Network Coding theory has become a discipline in itself which opened a large number of perspectives. Its main principle is to allow intermediate nodes to perform coding operations on packets coming from different sources to produce output flow in order to approach the theoretical limits of the bandwidth to increase throughput.

Another benefit of network coding appears in wireless networks, whether the Ad hoc networks or sensor networks, is the ability to reduce the quantity of consumed energy. Hence, NC becomes a critical technology of the future for wireless networks.

Moreover, linear sensor networks applications are emerging in recent years; and we find various applications: intelligent transportation system and VANETs, monitoring of power-line networks, race circuits' monitoring systems, etc. Besides, the features and advantages of IR-UWB transmission technique make it attractive for a wide range of wireless applications. Furthermore, with these advantages, IR-UWB has been selected by the IEEE 802.15.4a standard as the alternative PHY for the IEEE 802.15.4 standard for mainly its ultra-low power consumption. In addition to that, NC advantages lead the authors of [2] to introduce its use in the field of WSN to improve energy efficiency for unicast sessions. All these points led us to study the contribution of the application of NC in line wireless networks based on IR-UWB.

The remainder of this paper is organized as follows. In Section 2, we present the related works. We briefly introduce WSNs in Section 3. We present ALOHA and WideMac protocols in Section 4. The theoretical fundamentals of the used network coding scheme are presented in Section 5. NC implementation for WideMac is described in section 6. Simulations and results are presented in section 7 and 8 respectively. Finally, the Section 9 concludes the paper.

II. RELATED WORKS

In WSN performance evaluation, modeling and simulating the implemented protocols are the commonly used methods. Several IR-UWB MAC-PHY models have been proposed. Un-Slotted and Slotted ALOHA were the first proposed MAC protocols to go with the new IEEE802.15.4a standard where an analytical modeling is presented in [3], and an additional performance evaluation of slotted ALOHA over IR-UWB using an existing simulation platform is also presented in [4]. New ALOHA-like protocols and their benefits are also presented in [5]. Multichannel distributed protocols: M-ALOHA, MPSMA, BSMA and their performances evaluation are presented in [6].

III. WIRELESS SENSOR NETWORS

The technology of wireless sensor networks (WSNs) has a great improvement in recent years due to the contribution of more research in this area. The ability of a sensor network to use communications without infrastructure has allowed the deployment of sensor nodes with limited resources nearest or

978-1-4673-8760-6/15 $31.00 © 2015 IEEE

far the phenomenon studied. A sensor network is composed of a large number of sensor nodes which consist of sensing, data processing, a power supply unit, and communicating components. With that ability, sensor nodes deployed anywhere will provide intelligence and a better understanding of the environment to the end user.

IV. IR-UWB MAC PROTOCOLS

The IEEE802.15.4a [7] working group targeted the creation and the development of an alternative physical layer of the IEEE802.15.4 standard. The new standard had to offer more location capability and improve performance in terms of data rate, precision ranging and power consumption, and it becomes the physical layer standard amendment for IR-UWB. It consists of two physical layers: an ultra-wideband impulse radio layer and a chirp-spread spectrum layer.

A. ALLOHA

ALOHA like Medium Access protocols for IR-UWB have shown their benefit [8]. A node transmits a frame without sensing whether the channel is busy. If a transmission collides with another one, the frame is retransmitted after a random backoff. After a successful transmission an acknowledgement is received before the expiration of a duration defined by the MAC scheduler.

B. WideMac

WideMac [9] is a MAC protocol designed by Rousselot et al. for ultra-wideband sensor networks which is more compatible with IR-UWB. Its basic principle consists of transmitting a beacon and shortly listens for incoming data right after the end of the beacon transmission. Candidates for transmission wake-up, wait for the beacon of the intended destination, and contend for medium access. Once they receive the destination's beacon for the first time, they know its sampling schedule, thus they are able to save energy by waking-up only at a short time before the beacon is sent. By keeping nodes in their sleep mode most of the time, WideMac saves energy.

V. NETWORK CODING

The inherent sub-optimality of the classic scheme of routing packets, ie store-and-forward, is the source of more than throughput issues. It has repercussions on reliability, delays and power consumption. That leads Ahlswede et al. [1] to view that there is no reason to restrict the function of a node to that of a switch. A node can function as an encoder in the sense that it receives information from all the input links, encodes, and sends information to all the output links.

Network coding is the response of that view; it is used to improve throughput, efficiency and scalability. It enables a node to combine incoming packets then transmit the combination on each of its outgoing links. Considering the network type, the channel properties, and the context of utilization, many variant emerge such as:

A. XOR Network Coding

In XOR network coding, intermediate nodes are allowed to combine the incoming packets by applying the XOR operation.

The basic idea of XOR network coding is illustrated in Figure 1 where the relay node R and the two source nodes S1 and S2 share the common wireless channel. Assume the capacity of network is 1 bit at a time. Due to capacity constraint, in figure 1 node S1 will transmit data packet p1 to R which store in the buffer corresponding to S1. Similarly, node S2 will transmit data packet p2 to R which store in the buffer corresponding to S2. Finally, Node R broadcast the coded packet P1 xor P2. This whole process involves three transmissions instead of four for the normal forwarding process.

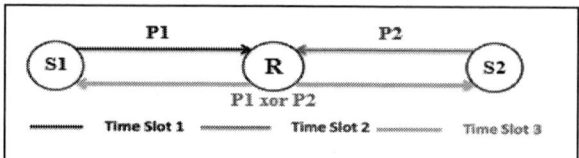

Figure 1. Wireless Network Coding Operation

B. Random linear network coding

Random linear network coding RLNC [10] is a kind of distributed LNC implementation. Its concept is that all nodes other than the receiver nodes perform random linear mapping from inputs onto output over some field. Each network node transmits on each of its outgoing links a linear combination of incoming packets over a finite field, with randomly chosen coding coefficients. The receivers need only know the overall linear combination of source processes in each of their incoming transmissions.

C. Sparse Random Linear Network Coding

Sparse random linear network coding is a special implementation of RLNC [11]. It can greatly reduce the computational complexity at intermediate and destinations nodes. It has been successfully tested as real implementation on various devices to reduce CPU load and extend battery life. Motivated by these promising applications of sparse random linear network coding we adopt it as the main coding process in source and intermediate nodes. In this coding scheme, coefficients are chosen from sample non-zero random values with a threshold probability value, and zero otherwise. The presence of the zero values in the coding vectors leads to the construction of a final coding matrix with large number of zero. Therefore, the decoding process becomes easier with less iteration in the final matrix inversion problem presented in the equation (6).

VI. NC IMPLEMENTATION FOR WIDEMAC

XOR coding is the simple implementation form of NC and has several advantages. It is characterized by the simplicity of its implementation and the low requirements in terms of coding and decoding cost. With these remarks we chose to implement this variant to enhance the performance of WideMac.

The implementation of XOR coding in our WideMac protocol requires, besides the reception buffer, the use of an additional separate buffer for each incoming flow from the neighboring nodes. Furthermore, on the one hand, if the current node is a relay, it uses the following policy: it stores every incoming packet to its corresponding buffer. At its transmission phase, it creates a coded packet by xoring two packets each one from an incoming buffer. And if one of buffers is empty, it transmits uncoded packets only. So, after a successful transmission (acknowledgment reception) we drop the corresponding packet from the corresponding incoming buffer. On the other hand, it only transmits uncoded packets if it is a simple source node.

The synchronization phase in WideMac protocol must result in the assignment of a transmission scheme for the time interval TW for each node. The three possible schemes are: RRS, RSR, and SRR where R denotes 'Receive' and S denotes 'Send'. Figure 2. shows an example of the synchronization process results. As depicted in this figure, RSR scheme was assigned to node A, SRR scheme was assigned to node B, and RRS scheme was assigned to node C, etc.

The node C, for example, wakes up at time t0, sends its beacon, wait for two receptions from it two neighbors for two time intervals Φ. Finally, it sends a packet at time t1 equals to $t_1 = t_0 + 4\Phi$

The parameter Φ is the time interval in which the current node must wait to receive the package from a neighboring node. Additionally, it must take into account the error of synchronization between the moments of awakening of the neighbor nodes.

The wake up interval of a node is thus given by:

$$Ta = t_1 - t_0 + \Phi = 5\Phi \qquad (1)$$

Hence, the activation interval Ta is composed of the following:

- The time required to send the beacon of synchronization

- The time required to receive the packet from the first neighboring node

- The time required to receive the packet of the second neighboring node

- The time required to transmit the packet to the neighboring nodes

- The time required to receive the acknowledgments

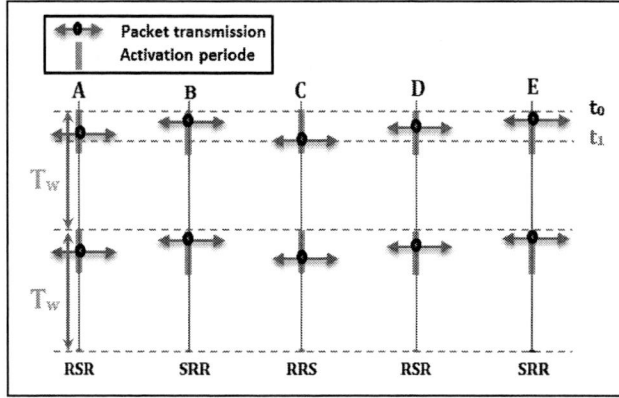

Figure 2.   WideMac optimal transmissions' synchronization with NC

The transmission scheme for WideMac in a line network without NC is characterized by Sending two packets, receiving two packets, and discarding two packets. This scheme is employed as every relay node has to forward exactly one packet to each incoming flow. Therefore, we can notice that the use of NC in this protocol leads the following improvements: collisions avoidance, number of transmissions' minimization, overhearing minimization, and activation period shortening. This last fact is given as the node has to stay in its active state to achieve reception and transmission's operations.

VII.   SIMULATION

The contribution of the introduction of NC in WideMac is evaluated using the MiXiM simulation platform under the OMNet++ simulator [12].

The scenario used to test our proposition is a line network where two source nodes placed at the end of the two directions transmit packets to each other. The total number of nodes varies from 3 to 50 nodes. Also, each source node periodically generates a packet within a period of 1S.

Tow metrics are adopted for the performance evaluation of our proposition: energy consumption and average end-to-end packets delay (latency). Specifically, the packet's latency is the average time it takes from the generating sensor node to the base station (or Sink).

TABLE I.   WIDEMAC AND ALOHA PARAMETERS

Protocol	Parameter	Value
WideMac	Tw	0.037s
	Ta	0,00185s - 0,0074s
	Φ	0.00037s - 0,00148s
	maxTxAttempts	20
	minBE	1
	maxBE	6
ALOHA	maxTxAttempts	30
	minBE	1
	maxBE	10
Both	macAckWaitDuration	0.0003s

VIII.   RESULTS

This section presents simulation results and analysis.

The results concerning the energy consumption are presented in Figure 3. The same figure shows the variation of consumption with the active time Ta from the value Ta=5%Tw to 20%Tw. It illustrates the low consumption of WideMac compared to ALOHA. It shows that the power consumption of both WideMac versions is remarkably less than the ALOHA MAC protocol. This fact of low power consumption of WideMac is the key feature of this protocol. In addition, the introduction of Network coding into WideMac has helped to the decrease in the power consumption of WideMac. This consumption was down 25.1% in case of the case $T_a = 5\% \ T_w$,

978-1-4673-8760-6/15 $31.00 © 2015 IEEE

and 15.2% in case of the case $T_a = 20\% \ T_w$. So it dropped (in mW) from 6.12 to 4.54 and from 20.30 to 17.21 respectively.

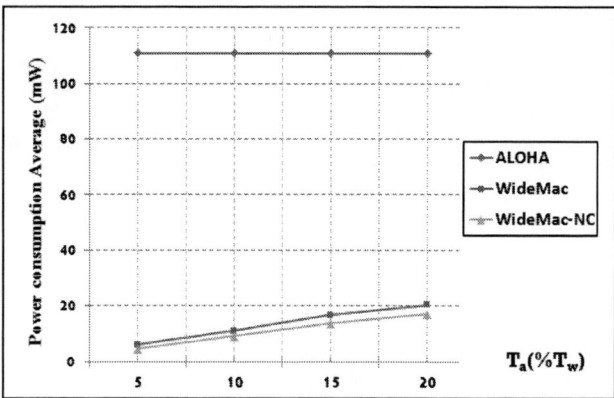

Figure 3. Nodes power consumption average

For many WSNs applications, delay is a critical factor especially for real time applications. It depends surely, in our line network case, on the network size or length. So, obtained results are depicted in figure 4. It shows the variation of the Latency parameter for different nodes' number values: 3, 10, 20, 30, 40, and 50 for ALOHA and the two versions of WideMac. For WideMac the results are obtained with $T_a=15\%T_w$. It shows an arrival delays average for ALOHA case between 44.34ms and 1169.47ms. For WideMac with and without NC, values vary respectively from 74.25ms and 88.68ms to 1888.44ms and 2258.94ms. Therefore, it shows clearly the benefits of the implementation of NC into WideMac. More precisely, NC implementation into WideMac has improved latency between 3.63% for N=50 and 16.40% for N=50. In addition, we can notice that putting nodes in their active periods all the time in ALOHA impacts positively on the average end-to-end packets' delay. But we must not forget the negative impact on energy's consumption.

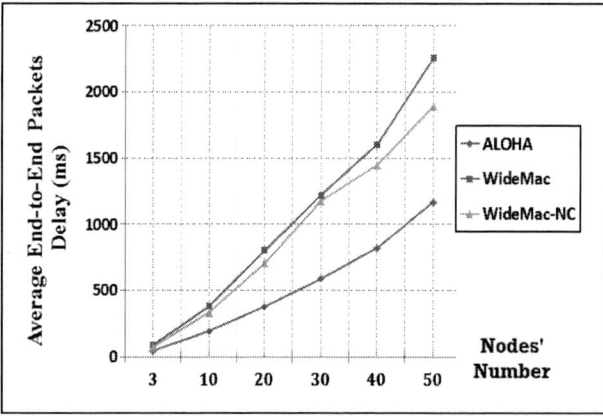

Figure 4. Average End-to-End Packets Delay (Latency)

IX. CONCLUSION

Power consumption was and is an interesting issue that is still a crucial factor in the development of WSN protocols especially in the physical and MAC layers. It is the primary metric to design a sensor node in wireless sensor network. From this perspective, we proposed the use of network coding to improve the performance of IR-UWB based WSNs. Especially; we introduce the implementation of NC at the MAC level using WideMac protocol. Hence, with this new technique, packets' transmission and retransmission are optimized and brought more efficiency in terms of energy consumption as well as latency.

Additionally, the experiments show the superiority for WideMac protocol in terms of energy consumption compared to ALOHA. However, ALOHA achieve good performance in terms of latency compared to the WideMac. We aim, as a future work, to develop a general scheduling scheme at the MAC layer (for both protocols) that fit the NC's implementation requirements to exploit the IR-UWB features.

REFERENCES

[1] Ahlswede, R., Cai, N., Li, S.Y.R., Yeung, R.W, "Network Infomration Flow," IEEE Transactions on Infomration Theory 46(4), 1204–1216 (2000).

[2] D.Lun, M. Medard, and R. Koetter, Network Coding for Efficient wireless Unicast, in Proc. IEEE International Zurich Seminar on Communications, 2006.

[3] M.G. D. Benedetto, L. D. Nardis, M. Junk, and G. Giancola, Performance Evaluation of Uncoordinated Medium Access Control in Low Data Rate UWB networks, Mobile Networks and Application, Vol 2 pp 1129-1135 2005.

[4] H. X. Tan, R. K. Patro, M. C. Chan, P. Y. Kong and C. K. Tham, Performance of Slotted-Aloha over TH-UWB, IEEE ICUWB 2007.

[5] R. Merz, J. Widmer, J.Y. Leboudec, and B. Radunovic, A Joint PHY-MAC Architecture for Low-Radiated Power TH-UWB Wireless Ad-hoc Networks, Wireless Communications and Mobile Computing, John Wiley and Sons Vol. 5 pp 567-580. 2005.

[6] N. J. August, W. Chung, and D. S. Ha, Distributed MAC Protocol for UWB Ad hoc and SensorNetworks, IEEE Radio and Wireless Symposium,pp 511-514, 2006.

[7] IEEE-802.15.4a, "Part 15.4: Wireless Medium Access Control (MAC) and Physical Layer (PHY) Specifications for Low-Rate Wireless Personal Area Networks (WPANs); Amendment 1: Add Alternate PHY," The Institute of Electrical and Electronics Engineers, Inc., Standard, March 2007.

[8] B. Abdoulaye, L. Aubin, D. Daniela, P. Robert," Medium Access Control for Wireless Sensor Networks based on Impulse Radio Ultra Wideband", International Conference – 3rd Edition Electronics, Computers and Artificial Intelligence 3-5 June, 2009, Pitesti, ROMÂNIA.

[9] J. Rousselot, A. El-Hoiydi, J.-D. Decotignie, "WideMac: a low power and routing friendly MAC protocol for Ultra Wideband sensor networks", IEEE International Conference on Ultra-Wideband, Vol 3 pp 105- 108 2008.

[10] S.Y.R. Li, R.W. Yeung, and N. Cai, "Linear Network Coding," IEEE Transactions on Information Theory, Volume 49, Issue 2, pp. 371 – 381, Feb. 2003.

[11] X.L. Li, W.H. Mow, and F.L.Tsang, "Singularity Probability Analysis for Sparse Random Linear Network Coding," Proc. ICC'11, Jun 5–9, 2011,Kyoto, Japan.

[12] A.Köpke, M.Swigulski, K.Wessel, D. Willkomm, P.T.Klein Haneveld, T.E.V. Parker, O.W. Visser, H.S. Lichte, S.Valentin. "Simulating Wireless and Mobile Networks in OMNeT++ The MiXiM Vision" In Proc. Intl. Workshop on OMNeT++, Mar. 2008.

Distributed consensus algorithm and its application to detect coverage hole in sensor networks

Anas Hanaf, Alban Goupil, Maxime Colas and Guillaume Gelle
CReSTIC, University of Reims Champagne-Ardenne, France
{anas.hanaf, alban.goupil, maxime.colas, guillaume.gelle}@univ-reims.fr

Abstract—**In this work, we design a distributed algorithm to detect coverage holes in sensor networks using the consensus algorithm. First, we provide an overview of the work done in consensus algorithm. We assume have no information on the position of the sensors and we work with the methods of algebraic topology to detect coverage holes. Then, we compare the performance of a centralized and a distributed detection and we show with simulations, the performance of our algorithm.**

I. INTRODUCTION

A wireless sensor network is composed of a large number of tiny sensor nodes randomly deployed inside an area or close to it to inspect certain phenomena [1]. There are several applications in wireless sensor networks in such surveillance in hostile environments such as forests, and measures nearby volcanoes. Increasing attention is being devoted in recent years to these problems by different researchers. Indeed, sensor networks are deployed in different ways, ideally, the sensors are made of a manually when it is easy to access environment. However, in reality, the sensors are deployed in hostile environments where human intervention is difficult. There are several factors that can affect the performance of a network, because of some sensors which cease to function. Thus the detection of these areas is important, what will be called coverage hole.

This paper is organized as follows : Section 2 presents the definition of hole and describes a different related work. Section 3 provides details about the distributed consensus algorithm. Section 4 presents the centralized method to detect coverage hole. Section 5 provides details about our distributed algorithms. Section 6 summarizes our algorithm performance and analysis. Section 7 presents our conclusion and futur work.

II. DEFINITION OF HOLE

Detecting holes in sensor networks is one of the major subjects in the field of sensor networks. A hole may be referred to as coverage hole because it is an obstacle to the communication in the network [2]. We suppose a fire in a certain area of the forest, and a set of sensors is destroyed. This area is called hole in the network. We see as the type of environment plays an important role in the formation of the network holes. When the environment is hostile, holes are created easily. If we take the example of the forest, in this environment, it is essential to detect holes for maximum network performance. Fig.1 illustrates an exemple of a sensor network deployed inside a forest. We find different types of holes in [3], but in this work, we study the problem of coverage holes where some areas of the network are not sufficiently covered by the sensors because the random deployment thereof.

Some research projects were carried out in this context. In [2], the author use the positions of the sensors to detect holes coverage. The algorithms in [7], [8] and [9] use the information of connectivity between sensor to detect the boundarty of sensor networks without informations of positions of sensors. The algorithm in [11] use algebraic topology. First, the authors define a notion of the coverage hole in a scalar field, which measures the quality of the estimation by the sensor network without knowing the positions of the sensors.

III. DISTRIBUTED CONSENSUS ALGORITHM

A. Network model

The information flow in a wireless sensor network can be represented using a graph $\mathcal{G}$ where nodes represent the sensors and the edges represent communication links between sensors [10]. The set of nodes $\mathcal{V}$ represents sensors with $i = 1, 2, ..., N$ and the set of edges $\mathcal{E}$ represents the communication links between these sensors. Specifically, if two nodes (i, j) can communicate directly with each other, then we note (i, j) the edge of $\mathcal{E}$ linking. We assume a graph without loops. The set of neighbors of node i is given by $N_i = 1, ..., N$. The adjacency matrix $\mathbf{A}$ of graph $\mathcal{G}$ is given by

$$\mathbf{A} = a_{ij} = \begin{cases} 1, & (i, j) \in \mathcal{E}, \\ 0, & (i, j) \notin \mathcal{E}. \end{cases}$$

The degree matrix $\mathbf{D}$ of $\mathcal{G}$ is a diagonal matrix with the diagonal elements equal to the degree of each node. The Laplacian matrix $N \times N$ is given by

$$\mathbf{L} = \mathbf{D} - \mathbf{A}$$

B. Distributed consensus algorithm

The distributed consensus algorithms aim to achieve a common agreement on a value or a decision between entities constituting a network of information and are divided into two types : Synchronous Algorithms where entities update their status simultaneously and Asynchronous algorithms where entities update their states at different times.

978-1-4673-8760-6/15 $31.00 © 2015 IEEE

Distributed consensus algorithms are iterative algorithms where network entities communicate locally in order to calculate a function (average, maximum, minimum, ...) of a set of initial values.

C. Distributed average consensus algorithm

Distributed average consensus algorithms are algorithms that adress the following problem : Each sensor in the network has a value and wants to learn the average of all the values in the network. However, sensor do not have information about network topology and also about the number of sensors in the network. But, they have information about the identify of their neighbors.

This algorithm is iterative where each entity of a network is programmed to perform a synchronous linear system whose state changes according to the following equation

$$x_i(t+1) = x_i(t) + \sum_{j \in N_i} W_{ij}(x_j(t) - x_i(t)) \qquad (1)$$

Where $x_i(t+1)$ is the state of entity i (nodes or edges), $i = 1, ..., N$ is the number of the entities, $t = 0, 1, ...$ is a number of iterations, $\mathbf{x}(0) = [x_1(0), ..., x_N(0)]^T$ is the initial values of the entities, W_{ij} is a weight associated with the edge and N_i the set of neighbors of i. The algorithm in (1) can be expressesd as

$$\mathbf{x}(t+1) = \mathbf{W}\mathbf{x}(t) \qquad (2)$$

According to [5], the weighting matrix $\mathbf{W}$ is given by

$$\mathbf{W} = \mathbf{I} - \epsilon \mathbf{L} \qquad (3)$$

Where $\mathbf{L}$ is the Laplacian matrix, $\mathbf{I}$ is the indentity matrix, and ϵ is a positif scalar ensuring the convergence. There are different methods to construct the weighting matrix $\mathbf{W}$. According to [5], we find the optimal value method with $\epsilon^* = \frac{2}{\lambda_1(\mathbf{L}_1) + \lambda_{N-1}(\mathbf{L}_1)}$. In this work, we use the maximum degree method with $\epsilon^{MD} = \frac{1}{d_{max}}$.

D. Conditions of convergence

The weighting matrix $\mathbf{W}$ must satisfy the following conditions defined in [6]

$$\mathbf{1}^T \mathbf{W} = \mathbf{1}^T \qquad (4)$$

$$\mathbf{W}\mathbf{1} = \mathbf{1} \qquad (5)$$

$$\rho \left(\mathbf{W} - \frac{\mathbf{1}\mathbf{1}^T}{N} \right) < 1 \qquad (6)$$

Where $\rho(.)$ is the spectral radius (The largest eigenvalue). These conditions are applied to synchronous distributed consensus algorithms.
We give some interpretations of these conditions:

- Condition (4): The average estimate is maintained at each iteration t.

- Condition (5): Stability of consensus.
- So with the conditions (4) et (5), the condition (6) implies with the eigenvalues of the weighting matrix $\mathbf{W}$ the following condition :
$-1 < \lambda_N(\mathbf{W}) \le \lambda_{N-1}(\mathbf{W}) \le ... \le \lambda_2(\mathbf{W}) < \lambda_1(\mathbf{W}) = 1$

IV. CENTRALIZED DETECTION

A. Group of Homology H_k

We will first define a coverage hole in a sensor network using the algebraic topology [4]. Vertices and edges are the elements to construct the topological space defined on the network. In fact, the algebraic topology extends this concept to a simplicial complex, which allows for higher order element having more than three vertices such as triangles. Let $v_i \in V$ a set of vertices. An element with $(k+1)$ vertices is called $k-$simplexe $\sigma_k = [v_1, v_2, ..., v_{k+1}]$. The vertex, the edge, and the triangle are denoted as follows σ_0, σ_1 et σ_2. A set of simplex is a simplicial complex. An example of a simplicial complex is shown in Figure 1. The boundary of an edge σ_1 is two end nodes σ_0 and the boundary of a triangle σ_0 is three edges σ_1 surrounding the triangles. If we note C_k as a chain complex, then the relationship between σ_k et σ_{k-1} is defined using the boundary operator $\partial_k : C_k \to C_{k-1}$. Let $\{ \sigma_k^1, \sigma_k^2, ..., \sigma_{k-1}^q \} \subset C_k$ et $\{ \sigma_{k-1}^1, \sigma_{k-1}^2, ..., \sigma_{k-1}^p \} \subset C_{k-1}$, the linear transormation $\partial_k \in \mathbb{R}^{p \times q}$ from C_k to C_{k-1}.

We will see in detail an examples of boundary matricies in the rest of the article. The Kernel of ∂_k is a set defined by $\ker \partial_k = \{ \sigma_k \in C_k | \partial_k \sigma_k = 0 \}$. The image of ∂_{k+1} is a set defined by $\text{img} \partial_{k+1} = \{ \sigma_k \in C_k | \partial_{k+1} \sigma_{k+1} = \sigma_k, \sigma_{k+1} \in C_{k+1} \}$. Then, The image of ∂_{k+1} is a boundary of σ_{k+1}, which is a cycle $\text{img} \partial_{k+1} \subset \ker \partial_k$. However, a cycle may not be a boundary of σ_{k+1}. Cycles of ∂_k, which are not boundaries of ∂_{k+1} are called group of homology $H_k : H_k = \ker \partial_k \cap \text{img} \partial_{k+1}$. The size of H_k is number of Betti β_k (number of coverage holes).

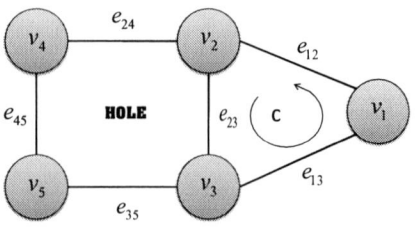

Fig. 1. Simplicial complex

The boundary matrix ∂_1 is calculated as follows:

$$\partial_1 e_{12} = \partial_1[v_1, v_2] = v_2 - v_1,$$
$$\partial_1 e_{13} = \partial_1[v_1, v_3] = v_3 - v_1,$$
$$\partial_1 e_{23} = \partial_1[v_2, v_3] = v_3 - v_2,$$
$$\partial_1 e_{24} = \partial_1[v_1, v_2] = v_4 - v_2,$$
$$\partial_1 e_{35} = \partial_1[v_3, v_5] = v_5 - v_3,$$
$$\partial_1 e_{45} = \partial_1[v_4, v_5] = v_5 - v_4,$$

$$\partial_1 = \begin{matrix} & e_{12} & e_{13} & e_{23} & e_{24} & e_{35} & e_{45} \\ \begin{bmatrix} -1 & -1 & 0 & 0 & 0 & 0 \\ 1 & 0 & -1 & -1 & 0 & 0 \\ 0 & 1 & 1 & 0 & -1 & 0 \\ 0 & 0 & 0 & 1 & 0 & -1 \\ 0 & 0 & 0 & 0 & 1 & 1 \end{bmatrix} & \begin{matrix} v_1 \\ v_2 \\ v_3 \\ v_4 \\ v_5 \end{matrix} \end{matrix}$$

The boundary matrix ∂_2 is calculated as follows:

$$\partial_2 c = \partial_2[v_1, v_2, v_3] = [v_2, v_3] - [v_1, v_3] + [v_1, v_2] = e_{12} - e_{13} + e_{23}$$

$$\partial_2 = \begin{matrix} c \\ \begin{bmatrix} 1 \\ -1 \\ 1 \\ 0 \\ 0 \\ 0 \end{bmatrix} \end{matrix} \begin{matrix} e_{12} \\ e_{13} \\ e_{23} \\ e_{24} \\ e_{35} \\ e_{45} \end{matrix}$$

B. Combinatorial Laplacians

We consider the kth Laplacian : $\mathbf{L}_k = \partial_{k+1}\partial_{k+1}^T + \partial_k^T \partial_k$. The elements in the null space of the kth Laplacian $\ker(\mathbf{L}_k)$ are called k-harmonic.

The method is to define the basic of $\ker(\mathbf{L}_1)$ by calculating the vector in the nul space of $\mathbf{L}_1$: $\mathbf{L}_1 X = 0$ with the goal to find the dimension of $\ker(\mathbf{L}_1)$ and calculating the number of holes in the sensor networks.

The Laplacian matrix L_1 of the example in Figure 2 is

$$L_1 = \begin{pmatrix} 3 & 0 & 0 & -1 & 0 & 0 \\ 0 & 3 & 0 & 0 & -1 & 0 \\ 0 & 0 & 3 & 1 & -1 & 0 \\ -1 & 0 & 1 & 2 & 0 & -1 \\ 0 & -1 & -1 & 0 & 2 & 1 \\ 0 & 0 & 0 & -1 & 1 & 2 \end{pmatrix}$$

The vector X belonging to the null space of L_1 is

$$X = \begin{pmatrix} 0.1741 \\ -0.1741 \\ -0.3482 \\ 0.5222 \\ -0.5222 \\ 0.5222 \end{pmatrix}$$

From the figure 1, we have two possibilities of holes, $e_{12} - e_{13} + e_{24} - e_{35} + e_{45}$ or $-e_{23} + e_{24} - e_{35} + e_{45}$. After calculating $e_{12} - e_{13} + e_{24} - e_{35} + e_{45} = 0.1741 - (-0.1741) + 0.5222 - (-0.5222) + 0.5222 = 1.9118$ and $-e_{23} + e_{24} - e_{35} + e_{45} = -(-0.3482) + 0.5222 - (-0.5222) + 0.5222 = 1.9118$. So from the figure 1, c is a cycle and $-e_{23} + e_{24} - e_{35} + e_{45}$ is a hole.

V. DETECTION USING DISTRIBUTED CONSENSUS ALGORITHM

In section 2, the distributed consensus algorithm was applied to the nodes (vertices), the idea is to apply this algorithm on the simplex σ_1. Thus, instead of assigning random initial values at the nodes (σ_0), we attribute them to the edges (σ_1). So to construct the weighting matrix $\mathbf{W}$, we use the Laplacian matrix $\mathbf{L}_1$ Studied previously.

$$\mathbf{W} = \mathbf{I} - \epsilon \mathbf{L}_1$$

We assign initial values to the edges and steps of the distributed average consensus algorithm is followed taking into account the number of edges.we use the maximum degree method with $\epsilon^{MD} = \frac{1}{d_{max}}$. So, $\mathbf{x}_{consensus} = \lim_{t \to \infty} \mathbf{x}(k)$ is the vector belonging to the null space of $\mathbf{L}_1$

VI. SIMULATIONS

Simulation 1 :

Number of sensors	40
Number of links	146
Connectivity radius	0.3
Graph Type	Random Geometric
Number of iterations	100

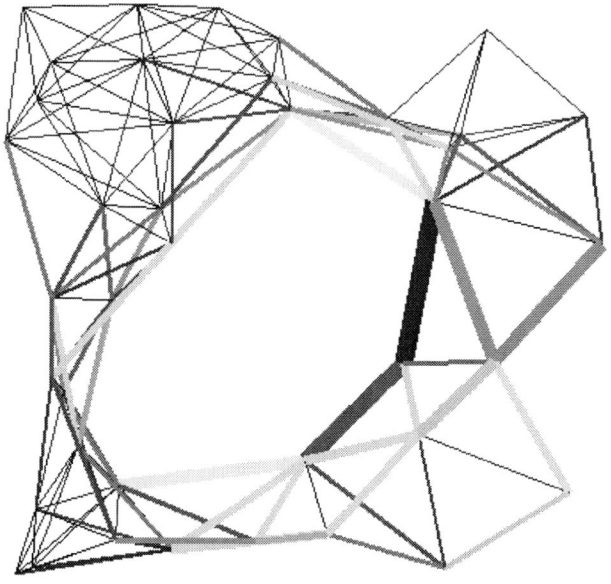

Fig. 2. Distributed consensus method to detect holes

In simulation 1, we use a network with 40 nodes, 146 edges and 208 triangles (two-simplices). We use two methods: centralized and distributed. The originality of this work is the introduction of the concept of distributed consensus algorithms for detecting coverage holes. We note that the performance of distributed method is the same compared to the centralized case, however, this method may be used in the asynchronous case only one sensor is awake, its the futur work.

We start our simulation by assigning random values to the edges, and we apply the algorithm of distributed consensus using the matrix $\mathbf{L}_1$ studied in section 4.

We chose to study in the synchronous case where all sensors woke up and exchange their states simultaneously. After 100 iterations, the values of the edges converge to the vector $\mathbf{x}$ belonging to the null space of $\mathbf{L}_1$ such that $\mathbf{L}_1\mathbf{x} = 0$. The null space of $\mathbf{L}_1$ has a dimension 1.

We note that edges away from the hole have values close to each other. We see it in the graph based on the colors on the edges, the edges away from the hole have the same color as opposed to the closest edges of the hole.

This method allowed us to detect edges surrounding the hole with a high probability, so we can calculate the sum of the values of these edges. After the sum, we will have a zero because according to the example in section 4, when the sum of a cycle is zero, this cycle is a hole.

However, this method only allows a single detection, our future work, we will enable to locate an accurate coverage hole using more advanced routing techniques.

The protocol in [12] uses Clustering by dividing the network into subnetworks, this protocol is useful for the continuation of our futur work to reduced the number of iterations and thus increase the life of the sensors.

Simulation 2 :

Number of sensors	70
Number of links	289
Connectivity radius	0.2
Graph Type	Random Geometric
Number of iterations	100

In simulation 2, we use a network with 70 nodes, 289 edges and 478 triangles (two-simplices). The null-space of the 1-Laplacian has also dimension 1.

VII. Conclusion

A distributed algorithm for detecting coverage holes is presented. The advantage of our algorithm is the distribution side, the information flows between the different sensors bypassing the centralized method which consists in that all sensors send

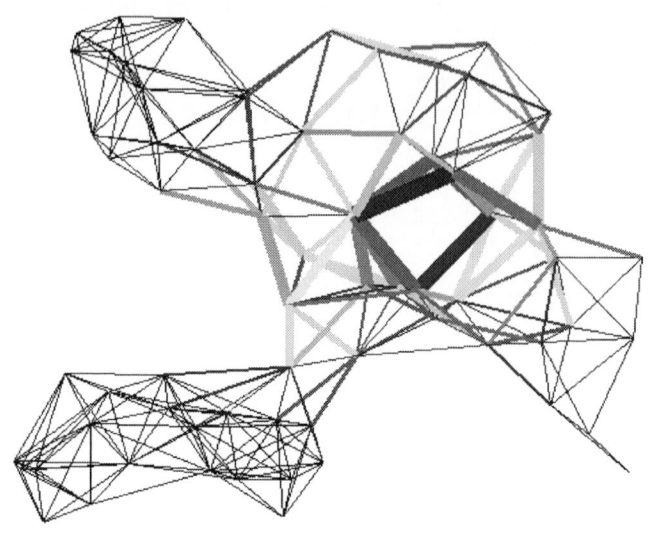

Fig. 3. Distributed consensus method to detect holes

their data to a central unit. Future work includes locating coverage holes using routing methods while exploiting the results of our distributed algorithm generalised in the case of multiple hole in the real time sensor network.

References

[1] I.F. Akyildiz et al., *"A survey on sensor networks"*. Communications Magazine, IEEE, August 2002. 40(8): page(s). 102-114.

[2] Q. Fang, J. Gao, and L. Guibas. "Locating and bypassing routing holes in sensor networks", in proc. Mobile networks and applications, vol.11, page(s).187-200, 2006.

[3] N. Ahmed, S.S. Kanhere, S. Jha. "The holes problem in wireless sensor networks: a survey", in ACM SIGMOBILE Mobile Computing and Communications Review, New York, USA, 2005, page(s).4-18, vol.9, issue.2

[4] Edelsbrunner,H., Harer, J.: *"Computationnal Topology: An Introduction"*. American Mathematical Society Press, 2009.

[5] Reza Olfati-Saber, J. Alex Fax, and Richard M. Murray, *"Consensus and Cooperation in Networked Multi-Agent Systems"*. Vol. 95, No. 1, January 2007.

[6] Lin Xiao and Stephen Boyd, *"Fast linear iteration for distributed averaging"*. Information Systems Laboratory. Stanford University, Stanford, CA, page(s). 94305-9510, USA, 25 February 2004.

[7] R. Ghrist and A. Muhammad, *"Coverage and hole-detection in sensor networks via homology"*, in proc. of 4th International Symposium on Information Processing in Sensor Networks, Los Angeles, 2005, page(s). 250-260.

[8] O. Saukh et al., *"On boundary recognition without location information in wireless sensor networks"*, in proceeding of International Conference on Information Processing in Sensor Networks (IPSN'08), 2008.

[9] K. Bi, K.Tu N. Gu, W.L. Dong and X. Liu, *"Topological Hole Detection in Sensor Networks with Cooperative Neighbors"*, in proc. of International Conference on Systems and Networks Communications (ICSN'06), Tahiti, October 2006, page(s). 31-31.

[10] R. Diestel, *"Graph theory"*, Graduate Texts in Mathematics (173), Springer, 2005.

[11] Mengyi Zhang, Alban Goupil, Maxime Colas and Guillaume Gelle, *"WSN sensing coverage based on correlations"*, In proceedings of International Conference on Wireless Communications and Signal Processing (WCSP), Hangzhou, China, October 2013.

[12] B. Elbhiri, S. El Fkihi, R. Saadane, and D. Aboutajdine, *"Clustering in wireless sensor networks based on near optimal bi-partitions"*, in Next Generation Internet (NGI), 2010 6th EURO-NF Conference on. IEEE, pp. 16, 2010.

Harvesting Energy From Data Lines For Avionics Applications: Power Conversion Chain Architecture

Maryam Mohajertehrani, Umar Shafique, Yvon Savaria, and Mohamad Sawan

Polystim Neurotech Laboratory, Electrical Engineering Dept,
Polytechnique Montreal,
Montreal (QC) Canada, H3T1J4
maryam.mohajertehrani@polymtl.ca

Abstract— **Avionics industry is exploring new techniques to reduce cabling mass in modern aircrafts, leading to lighter and more fuel-efficient aircrafts. New trends are emerging towards the use of self-powered avionics sensors, which are simple to install, requires less maintenance and minimum wiring. A power harvesting circuit interface that uses avionics data lines for power harvesting was previously proposed. This interface can be coupled to an avionics sensor, making it self-powered without disturbing the ARINC 825 field bus. This paper presents a power conversion chain (PCC) module architecture and its circuit implementation required for the energy-harvesting interface. The proposed PCC design features a fast response time (5 msec settling time) and 60% power conversion efficiency. The PCC can be coupled to a compact power storage system consisting of super-capacitors. A 0.18 μm CMOS technology is used for circuit design, and simulations are carried out using Cadence.**

I. INTRODUCTION

Modern aircrafts are routinely equipped with sensors and actuators. Power cables are needed to supply energy and transfer data to and from these sensors. The growing numbers of sensors would require even more cables. These additional cables add weight and occupy space, leading to increased fuel consumption and consequently greater CO_2 emissions [1, 2]. Recently, avionics industry is trying to adopt new schemes to reduce the number of cables required for power and data transmission. These schemes include Wireless communication, power line communication (PLC) and power over Ethernet (PoE) [3]. The PoE technique promises to eliminate the need of separate power lines. In this technique both power and data are transmitted over a single medium. PoE is generally used to power up end systems in conventional Ethernet/AFDX networks [4]. Ethernet is widely used as a standard physical layer solution for networks in many industrial areas [4]. For instance, in aeronautical applications, the Avionics Full-Duplex Switched Ethernet (AFDX) is widely used [3]. The CAN bus standard was also adapted to the requirements of the avionics industry with the ARINC 825 standard [5].

To power the sensors on an avionics sensor networks and to reduce the number of power cables, an approach was proposed by Zhang et al. [1, 2]. They proposed recovering energy from data lines to feed power to the sensors. In this technique, power is harvested from the ARINC 825 data bus during its idle periods.

The architecture of that energy harvesting system comprises isolation blocks, switches, power conversion chain (PCC), voltage distributor, clock generator, control block and a power management block as shown in Fig. 1.

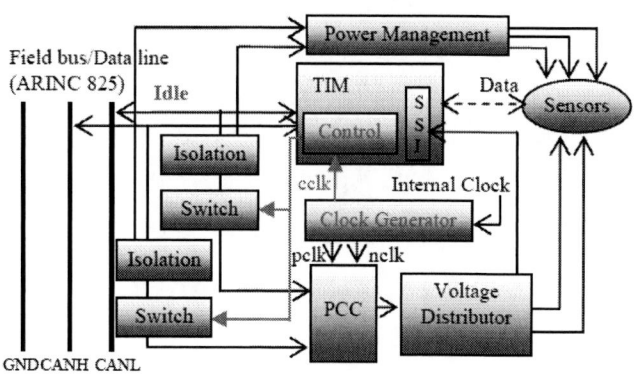

Fig.1. Proposed architecture of energy recovery path by [1]

This paper presents a proposed Power Conversion Chain (PCC) architecture. It consists of a controlled switch that is activated during the harvesting process to receive CAN bus signals in the idle state, a charge reservoir, a voltage regulator, a DC-DC converter to increase the output voltage, a capacitor bank for charge storage, and also the power management block for proper monitoring and distribution of the harvested power. The proposed power harvesting system features simple circuits, short settling time, low-power consumption and consequently, can yield considerable reduction in the cable mass of the cables installed. The remaining of this paper is organized as follows. In Section II, the architecture, description and function of each building block are presented. Simulation results are provided in Section III, which are followed by some concluding remarks in Section IV.

II. THE ENERGY RECOVERY PATH

The architecture of the proposed energy conversion path is presented in Fig. 2. As illustrated, power is harvested from the CAN bus and the power management block manages the power available to the sensors. This harvesting process is developed following the ARINC 825 standard [5]. The idle state of the CAN bus is recommended to be more than 50% in each period T (Major Frame Time).

In order to optimize bus usage for data transmission, a portion of the second half of each period T is used for energy recovery. According to the ARINC 825 standard [5], the non-floating bus provides three rails, CAN-High, CAN-Low, and Ground. The CAN-High is either 2.5 V or 3.5 V, while CAN-low is defined to be 2.5 V or 1.5 V with a differential voltage between CAN-High and CAN-Low of 0 V or 2 V. During the

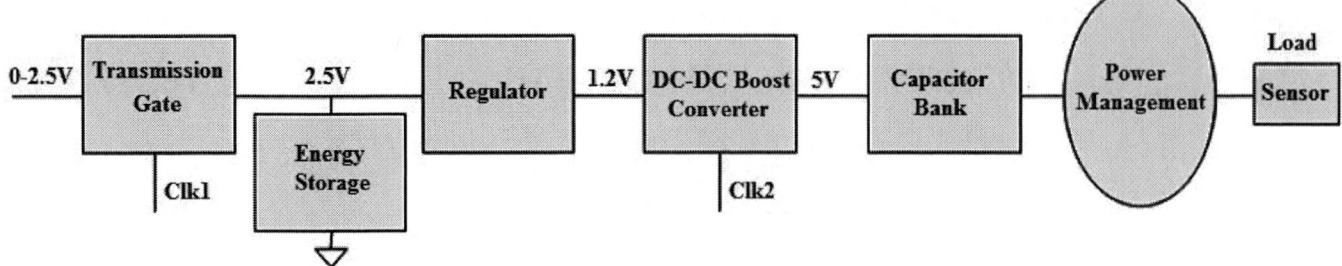

Fig. 2. Architecture of the energy conversion path

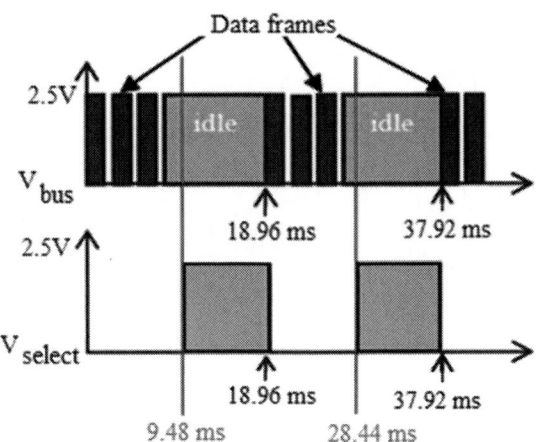

Fig. 3. Idle state in CAN bus

idle state, both CAN-high and CAN-low voltage are 2.5 V (Fig. 3). The maximum current that a CAN transceiver can provide is approximately equal to 10mA.

The energy conversion path comprises, a transmission gate, energy storage, a low drop-out voltage regulator, a DC-DC voltage converter to level up the voltage above the regulated voltage, a capacitor bank and a power management block. The period T is chosen to be 18.96 ms, which corresponds to the maximum data rate of 1Mb/s with ARINC825, at 50% load and 8 byte message payload. The switch is set to open from the middle to the end of each message period T (T corresponding to 52.7 Hz). A clock generator module provides the source clock of 42 kHz to the voltage converter. The main task of the power management module is to properly manage the energy harvested and to distribute the harvested energy. In the following subsections, the various blocks of the energy conversion chain are discussed in detail.

A. Switch

A transmission gate is used to implement the needed switch. A complementary control block will provide control to the switch. The main task of the startup control block is to detect the idle state in the CAN bus and activate the switch to begin the harvesting process.

B. Energy storage

A charge reservoir is used for energy storage during the idle time. The settling time (t_s) grows with the capacity of the

reservoir, while the voltage ripple (V_{ripple}) decreases on the output. A large capacitor is needed to provide enough storage for the harvested energy. It is also useful for supporting operations longer than a frame period.

C. Voltage regulator

There are different types of voltage regulators such as low-dropout (LDO) voltage regulators, linear regulators, switching regulators, and switching capacitor regulators. Each of these regulators has its own characteristics and preferred applications. In this study, a low drop out regulator is used as it can provide fixed and stable output voltage, even in presence of moderate input voltage variations. A basic LDO is composed of an input voltage reference, an error amplifier, a sampling resistor and a series pass element. The LDO's operation is based on the feedback from the amplifier error signal to control the output current flow of the power transistor driving the load. The output voltage is set at a stable level by R1, R2 and the reference voltage. When any change happens on the output voltage, a feedback voltage divided by R1 and R2 and by the reference voltage difference will force the error amplifier to adjust the current flow through the PMOS transistor. Hence, a stable regulated voltage level is provided. The drop-out voltage is the minimum differential voltage between the input and the output when the circuit stops to regulate. The adopted LDO is shown in Fig. 4 [6].

D. DC-DC voltage converter

DC-DC converters are widely used in industrial and commercial applications [7]. These converters are characterized by their voltage conversion ratio, maximum output power, their efficiency, the number of components they comprise, and power density. A DC-DC power conversion process changes the voltage level of a DC source from one level (input) to another voltage level at the output. There are several methods to obtain DC-DC voltage conversion. The choice of the most appropriate technique depends on the requirements [8]. To obtain high efficiency low power consumption while keeping the design simple, an inductive boost converter was selected and used in this study. The boost converter step-up the input voltage to a significantly higher output voltage level. The converter circuit is shown in Fig.6. A clock generator module provides a 42 kHz source clock to the voltage converter that can boost the input voltage (1.2 V) to 5V at its output [9].

E. Capacitor Bank

Energy harvesting systems generally require energy storage. The stored energy can be used when the energy

978-1-4673-8760-6/15 $31.00 © 2015 IEEE   56

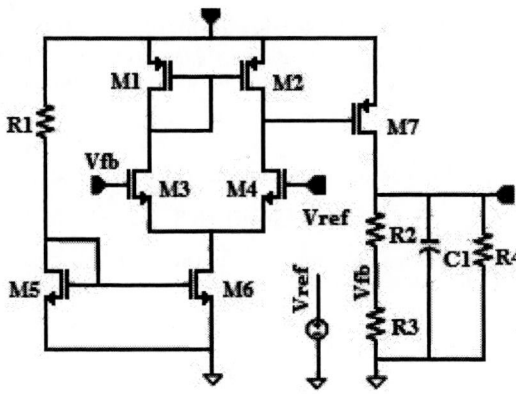

Fig. 4. Low-Dropout (LDO) regulator circuit

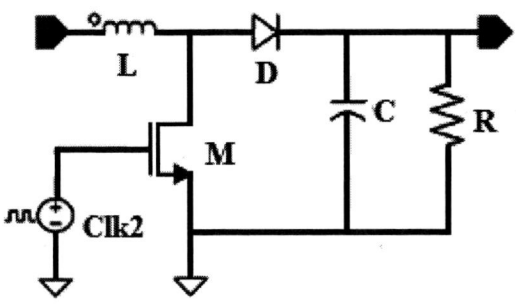

Fig. 5. Converter circuit

source is passive. Numerous types of energy storage systems are available. Rechargeable batteries and super-capacitors are the most portable and popular energy storage components for powering embedded devices. In addition to store the harvested energy, these capacitors also help to reduce the ripple of the current directed to the load [10].

F. Power management

The Power Management block is crucial to the efficiency. It is responsible of the following tasks:

- Monitoring the maximum and minimum voltage against under-voltage (UV) and over-voltage (OV) levels. In case of over-voltage, the harvesting process should be stopped, while for an under voltage level, the harvested power is unable to power the load.
- If the power stored in the capacitor bank is sufficient, it can be used to power the load.
- Monitoring the amount of available charge in the storage (capacitor bank).
- Preventing damage to storage and CAN bus and protection in case of overload [10].

III. SIMULATION RESULTS

Previous report by Zhang et al. showed that a switching frequency of 500 Hz could be used, considering the idle duration during most time frames. This study considers ARIN825 buses with maximum of 30 nodes operating at a data rate of 1Mbps with minimum 50% idle time along with an 8-byte message payload. The switch is set to open from the middle to the end of each message period 18.96 ms,

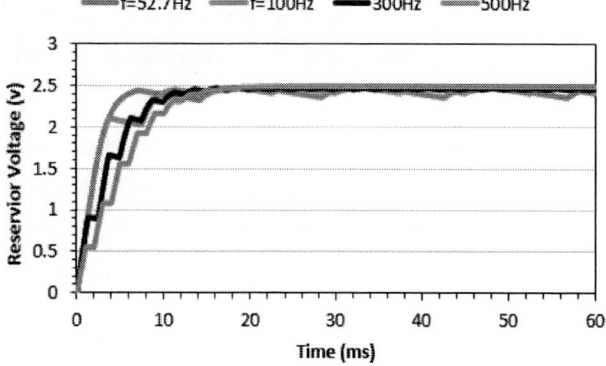

Fig. 6. Effect of switching frequency on the performance of the circuit

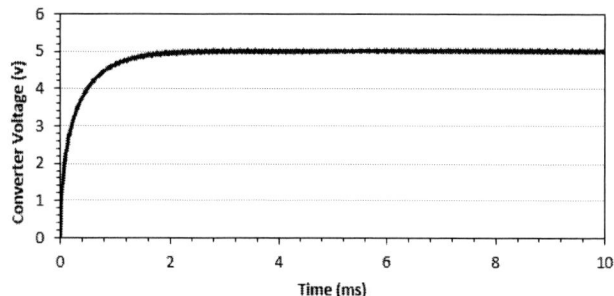

Fig. 7 . Output of the converter corresponding to a 42 kHz clock frequency

corresponding to 52.7 Hz switching frequency). The switch frequency depends on the major and minor time frames, set by the system designer while considering the message frequency of the most active nodes operating over the CAN bus. In general, the switch frequency can be a multiple of the above value when the bus has more than one transmitting active node. Frequency of 52.7 Hz, 100 Hz, 300 Hz and 500 Hz were used for this simulations. Using the above frequencies, the settling time is minimized for the lowest frequency, i.e 52.7 Hz. In that case, a settling time of 5 ms was obtained, while for 500 Hz, a settling time is 16 ms was obtained. However, in spite of settling time differences, the overall power harvested during the steady state is the same as illustrated in Fig. 6.

The converter output voltage is presented in Fig. 7. The Clock frequency on the converter was set at 42 kHz to maximize the conversion efficiency. A 94% maximal efficiency was obtained. The voltage of each stage: input, reservoir, regulator and converter are plotted over a stretch of 50 ms time as shown in Fig. 8. This plot shows that the ripple voltage of the converter is 0.3 V with a 5 ms settling time, t_s. The efficiencies of the switch, regulator and convector are equal to 80%, 80% and 94%, respectively. The overall efficiency of the power recovery system is about 60%. As mentioned above, the voltage level of CAN bus is 2.5 V during the idle state and the bus can maintain a maximum current of 10 mA (CAN transceiver idle state current maximum). Using the second half of each period of the signal T for harvesting energy, the maximum input power is 7.5 mW.

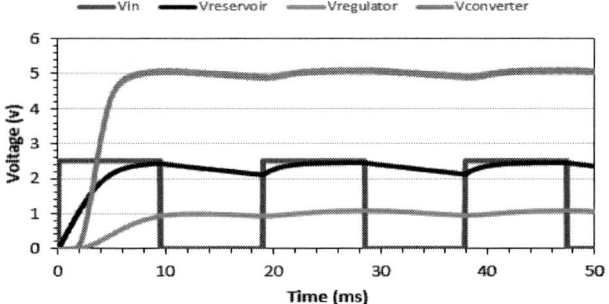

Fig. 8. Transient voltage obtained from the power conversion chain

TABLE 1. Comparison of the simulated results with those obtained by Zhang et al.

	This work	[2]
Input voltage	2.5 V	2.5 V
Switching frequency	52.7 Hz	500 Hz
Reservoir	80 µF	80 µF
Control frequency of converter	42 KHz	10 MHz
Settling time	5 ms	10 ms
Voltage ripple	0.3 V	0.04 V
Total efficiency	60%	30%
Technology	CMOS 0.18µm	Verilog A Modeling

Table 1 summarizes the results of the simulation of the power conversion chain. A comparison is also provided to the results reported by Zhang et al. [2] in the second column of above table. Table 1 shows that the proposed design features half the settling time, making the power readily available while providing a two-fold improvement in the system power efficiency. The PCC block is further integrated to the power management block of the harvesting system to manage power requirements of different loads.

IV. CONCLUSION

A power conversion chain suitable for a harvesting system is designed to recover energy from data lines to feed sensors for self-power applications. The circuits that compose this chain are designed using a 0.18µm CMOS technology. Energy harvesting process takes place during the idle state of a ARINC825 bus system. Enhancements in the circuit design are made to achieve fast recovery time and good system efficiency. Compared to a previous design, the recovery time was reduced by half i.e. from 10msec to 5msec and the output power conversion efficiency was doubled from 30% to 60% as reported. Further work will explore means to improve management of the harvested power for sensitive load operations.

ACKNOWLEDGMENTS

Authors acknowledge financial support from the Canadian Natural Sciences and Engineering Research Council (NSERC, Strategic grant program) of Canada, and thank Thales Canada for providing technical guidance and support. Thanks are also expressed to CMC Microsystems for providing access to technologies as well as design and simulation tools.

REFERENCES

[1] Zhang J, Hashemi, S.; Karimian, M.; Koubaa, Z.; Sawan, M., "A novel power harvesting scheme for sensor networks in advanced Avionic applications," in Electronics, Circuits, and Systems (ICECS), 2013 IEEE 20th International Conference on , vol., no., pp.921-924, 8-11 Dec. 2013 doi: 10.1109/ICECS.2013.6815562

[2] Zhang J. Hashemi, S.; Karimian, M.; Koubaa, Z.; Sawan, M., "Power recovery from data line in avionic applications," in Microelectronics (ICM), 2012 24th International Conference on , vol., no., pp.1-4, 16-20 Dec. 2012 doi: 10.1109/6471394

[3] Heller C. et al., "Powerover- Ethernet for avionic networks," IEEE/AIAA 29th Digital Avionics Systems Conference, pp. 5.A.2-1-5.A.2-11, 2010.

[4] "664P2-2 Aircraft Data Network, Part 2-Etehrnet Physical and Data Link Layer Specification, Aeronautical Radio Incorporated (ARINC) 2009, " Annapolis, MD.

[5] "General Standardization of CAN (Controller Area Network) Bus Protocol for Airborne Use ARINC SPECIFICATION 825-2", Published by Aeronautical Radio Inc. pp. 54-56, 2011.

[6] Torres J., El-Nozahi, M.; Amer, A.; Gopalraju, S.; Abdullah, R.; Entesari, K.; Sanchez-Sinencio, E., "Low Drop-Out Voltage Regulators: Capacitor-less Architecture Comparison", IEEE Circuits and Systems Magazine,, vol.14, no.2, pp.6,26, Secondquarter 2014, doi: 10.1109/MCAS.2014.2314263

[7] Choi B. (2013), "Pulse width modulated DC-to-DC power conversion: Circuits, dynamics, and control designs," Piscataway, New Jersey: IEEE Press.

[8] Zhang.B, Xuemei.W . "Chaos Analysis and Chaotic EMI Suppression of DC-DC Converters," December 2014, Wiley-IEEE Press.

[9] Wens, M.; Cornelissens, K.; Steyaert, M., "A fully-integrated 0.18µm CMOS DC-DC step-up converter, using a bondwire spiral inductor," Solid State Circuits Conference, 2007. ESSCIRC 2007. 33rd European, vol., no., pp.268,271, 11-13 Sept. 2007, doi: 10.1109/ESSCIRC.2007.4430295

[10] Yen-Kheng, T., Yuanjin, Z., & Huey Chian, H. (2012). "Ultralow Power Management Circuit for Optimal Energy Harvesting in Wireless Body Area Network, "In Advanced Circuits for Emerging Technologies (p. 632). John Wiley & Sons

Digital Image Stabilization for Video Images

Y. Mohanna, S. Malaeb, H. Alaeddine, O. Bazzi and A. Alaeddine

Abstract—**The problem of image stabilization in videos is important in many applications, including cameras mounted on vehicles. There are various approaches used for stabilizing the captured videos. This paper explains the subject of digital image stabilization (DIS) and the different steps associated with it. Then three robust DIS techniques are enhanced and developed to stably remove the unwanted shaking phenomena in the image sequences and videos. These methods are the gray-coded bit-plane matching algorithm, feature-based matching algorithm, and target tracking algorithm which is tested as a real-time product.**

Index Terms—**digital image stabilization (DIS), motion estimation, motion compensation, block matching, feature matching**

I. Introduction

Digital videos taken by hand-held cameras such as mobile phone cameras, or cameras mounted in vehicles are usually affected by undesired motion produced by unstable camera holding or platform moving. The unwanted vibrations in a video sequence induced by camera motion will affect the visual quality and obstruct the subsequent processes for various applications such as motion coding, video compression, feature tracking, etc. [1] and [2]. Image stabilization technology was introduced to overcome the mentioned problems and reduce the blur in any shaky video. The challenge of image stabilization systems is how to eliminate the unwanted shaking of the camera without affecting moving objects in the image sequence. Video stabilization can either be achieved by hardware or digital image processing approach. Hardware approach can be classified as mechanical or optical stabilization. Mechanical image stabilization (MIS) uses gyroscopic sensors with two perpendicular spinning wheels to stabilize the entire camera, not just the image. Whereas optical image stabilization (OIS) system employs a moveable lens assembly that moves opposite the shaking of camera to adjust the path length of the light as it travels through the camera's lens system [1], [3], and [4]. Hardware methods are very expensive as they are based on sophisticated sensors that measure camera shakes and then control the jitter acting on the lens or on the sensors, in addition to the large size associated with them [7]. For these reasons, digital approach was applied in our research.

This work was supported by the Lebanese University Doctoral School of Sciences and Technologies.

The authors are with the Physics and Electronics Department, Lebanese University, Hadat, Lebanon (e-mail: yamoha@ul.edu.lb)

II. Digital Image Stabilization

The digital image stabilization (DIS) method is the image post processing method or digital video stabilizer (DVS). The DIS system aims to produce a compensated video sequence so that the image motion due to the camera's undesirable shake can be removed using digital video processing techniques without any mechanical devices such as gyroscopes or fluid prism. The DIS can be performed either as post-processing after the video sequence was acquired, or in real-time during the acquisition process, depending on the application [1], [5], [6] and [8]. The DIS system is generally composed of two processing units: the motion estimation unit and the motion compensation unit. The purpose of motion estimation unit is to estimate the reliable global camera movement between each two consecutive frames in the video. Following the motion estimation, the motion compensation unit generates the compensating motion vector and shifts the current frame according to the compensating motion vector to obtain a smoother image sequence. The DIS system is shown in Fig. 1.

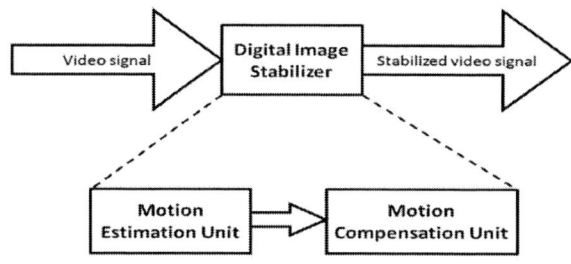

Fig. 1.DIS system, the shaky video should pass through the motion estimation and compensation units to be stabilized.

A. Motion Estimation

Motion estimation (ME) is the process of determining motion vectors that describe the transformation from one 2D image to another; usually from adjacent frames in a video sequence as in our case. Motion vectors can be global (GMVs) or local (LMVs) according to global motion estimation and local motion estimation respectively. The motion vectors may be represented by a translational model, rotational model, both, or many other models that can approximate the motion of a real video camera. The motion estimation step represents the most critical part involving the accuracy and the time complexity of the system [9]. The methods for finding motion vectors can be classified as pixel based methods (direct) [10]-[14] or feature based methods (indirect) [3], [7], [15], and [16]. The feature-based approaches are faster than pixel-based approaches, since they work on a specific chosen feature and not pixel by pixel, but they are more exposed to local effects and there efficiency depends on the feature point's selection. Hence they have limited performance for the unintentional motion. The direct pixel based approach makes optimal use of the information available in motion estimation and image alignment, since they measure the contribution of every pixel. The simplest technique is to pick the search algorithm and try all possible matches, that means do the full search. But this method is very slow and takes a lot of time to complete the motion estimation process [5].

B. Motion Compensation

The motion compensation (MC) unit generates the compensating motion vector and shifts the current frame according to the compensating motion vector to obtain a smoother image sequence. Fig. 2 shows the motion compensation schematics [1], [17], and [18].

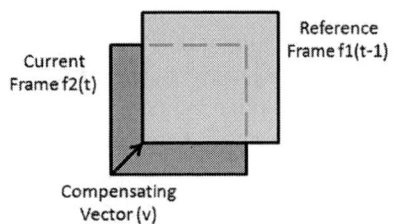

Fig. 2.Motion Compensation Schematics, shift the current frame according to the motion vector to be coincided with the reference frame.

Theoretically, the partial boundary areas will miss when the frames are vibrating. This is because there are no allowable pixels which can correctly compensate such areas. So, four sides of boundary areas are regarded as the missing areas and thus the compensated area is limited in the central part. This requires an interpolation algorithm such as, nearest-neighbor, bilinear, bi-cubic, etc. [16].

III. PRESENTED DIS METHODS

A. Gray-Coded Bit-Plane Block Matching Algorithm

Block-matching algorithms (BMA) are traditional methods used to do motion estimation. The full-search BMA under the mean absolute difference and mean square error criteria can be considered as an optimal solution for motion estimation, but, it requires large amount of computations which causes time delay, and requires complex hardware implementation. To reduce computational complexity, most block-matching algorithms usually divide an image into a small number of blocks and calculate the motion vector of each block. Then, these block motion vectors are utilized together to estimate the global motion vector in order to compensate the movement of the whole image [12] and [13].

Based on the gray-coded bit-plane block matching method proposed in [12] and [13], we developed an algorithm and explained the details of computation used in this method. Also we used the fine-division strategy used in [13]. First, we will explain the bit-plane and the gray-coded bit-plane decomposition of a gray-scale image.

Let us consider the gray level of the pixel at location (x, y)

in an image frame to be $I(x, y)$ which can be represented as

$$I(x,y) = a_{k-1}(x,y)2^{k-1} + a_{k-2}(x,y)2^{k-2} + \dots \\ + a_1(x,y)2 + a_0(x,y) \tag{1}$$

Where $a_{k-1}, a_{k-2}, \dots, a_1,$ and a_0 are the corresponding bit-planes, their values are either 0 or 1, and k is the number of bits needed to represent a pixel, with 2^k gray levels. In our case the frames are 8-bit images (k = 8), each image is composed of eight 1-bit planes ranging from plane 0 to plane 7. The gray-coded bit-planes are defined as

$$\begin{cases} g_i(x,y) = a_i(x,y) \oplus a_{i+1}(x,y) & 0 \le i \le k-2 \\ g_{k-1}(x,y) = a_{k-1}(x,y) \end{cases} \tag{2}$$

Where, $\oplus$ is the exclusive OR operation and a_k is the kth bit plane of the image. This Gray code has the unique property that successive code words differ in only one bit position, i.e., the gray-coded bit-planes tend to have less intensity fluctuation than bit-planes [12] and [13]. We used the 5th gray-coded bit-plane in this algorithm.

Now we start the motion estimation by performing the block-matching process to all blocks. Each block should have its own motion vector. But first we should identify a search region that surrounds each (M x N) block in the gray-coded bit-plane. The search region borders are at distances P and Q from the block borders in the horizontal and vertical directions, respectively. We take P and Q to be 16 pixels. Each motion vector of a block in the current Gray-coded bit-plane image is determined by evaluating GC-BPM over blocks within the search range in the previous Gray-coded bit-plane and selecting the block which yields the closest matching. This approach assumes that all pixels within the block have uniform motion and the range of the motion vector is constrained by the search window [12]. Since the size of each block is (M x N), then the search window size will be (M + 2P) x (N + 2Q). For the proposed GC-BPM, the involved correlation measure is defined as

$$c(m,n) = \frac{1}{MN} \sum_{x=0}^{M-1} \sum_{y=0}^{N-1} g_k^t(x,y) \oplus g_k^{t-1}(x+m, y+n) \tag{3}$$

$$-P \le m \le P, \text{and} -Q \le n \le Q$$

After computation, each block will have a correlation matrix. Each value of the correlation matrix corresponds to the correlation measure between the two blocks in the current and previous frames. To estimate an LMV for each block, we look for the minimum value in the correlation matrix. Minimum value means that the corresponding block is the closest block matched to the block in the current frame. Then the local motion vector can be found from the (m, n) indices of the minimum value in the correlation matrix, this is can be shown by the following equation

$$LMV_j = \arg\min\{c_j(m,n), -P \le m \le P, -Q \le n \le Q\} \tag{4}$$

Where LMV_j is the motion vector of the jth block.

The set of local motion vectors retrieved during block-matching are used to estimate the global motion vector that corresponds to the motion of the whole frame. The GMV

978-1-4673-8760-6/15 $31.00 © 2015 IEEE

estimation is represented by an affine transformation model.

The equations of the affine motion are

$$\begin{cases} \overline{X}_{t+1} = aX_t + bY_t + c \\ \overline{Y}_{t+1} = dX_t + eY_t + f \end{cases} \qquad (5)$$

Where $(\overline{X}, \overline{Y})$ denote the coordinates of the compared frame and (X, Y) denote the coordinates of the reference frame.

To estimate the six coefficients $(a, b, c, d, e, \text{and } f)$ in the affine model, we use the least squares (LS) approach. Let N be the number of motion vectors. We use the standard optimization method to find the "optimal" coefficients that minimize the LS cost function represented by the two following equations [13]

$$J_1 = \sum_{n=1}^{N}(aX_n + bY_n + c - \overline{X}_n)^2$$
$$\qquad (6)$$
$$J_2 = \sum_{n=1}^{N}(dX_n + eY_n + f - \overline{Y}_n)^2$$

After solving (6) we will find the 6 parameters of the motion model. The remaining step is just to compensate the motion in the current frame to obtain a smooth motion between the consecutive frames.

B. Feature-Based Matching

The method adopted in this paper as a feature-based method to do the motion estimation is based on corner detection. Starting with two image frames (reference and shaky frame), the corner detector will collect salient points (corners) from each frame using FAST corner detector. To extract the features of each frame we use the Block method with sum of squared differences (SSD), used as a matching cost to match between the features of each frame and find the corresponding features of these frames. The obtained correspondences are noisy, i.e., contain outliers. Here the RANSAC algorithm is used to get rid of the incorrect correspondences and get the affine model that transforms the motion between the two frames. Last step is compensating the motion to obtain a stabilized sequence of images. Now we will explain more the detailed steps of the algorithm.

1) FAST corner detection:

Features from Accelerated Segment Test (FAST) is an algorithm proposed originally by Rosten and Drummond [19] for identifying interest points (corners) in an image. The algorithm starts testing the pixels of the image to identify the corners. Let us consider a pixel p of intensity I_p in the image which is to be identified. Consider a circle of 16 pixels around the pixel under test (Fig. 3). Pixel p is considered a corner if there exist a set of n contiguous pixels in the circle (16 pixels) which are all brighter than p by a specified threshold t, or all darker than p by the same threshold t. A high-speed test was proposed to exclude a large number of non-corners. This test examines only the four pixels at 1, 9, 5 and 13 (First 1 and 9 are tested, if they are too brighter or darker. If so, then check 5 and 13). If p is a corner, then at least three of these must be brighter than $I_p + t$ or darker than $I_p - t$. If neither of these is the case, then p cannot be a corner.

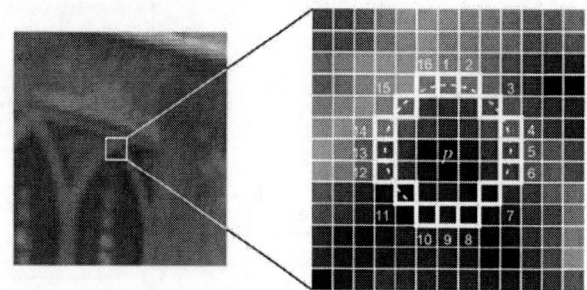

Fig. 3. Circle of 16 pixels and the pixel under test used by FAST to detect the corners in the image [19].

2) Features extracting and matching:

Block method is a simple method used to extract features using the detected corner points. It identifies an (NxN) block corresponds to each corner in each of the two frames. The corner pixel is itself the center of the block. N should be odd positive integer number. Then take the pixels of each block and arrange them in a vector. This vector is called the feature vector. In the matching process, feature vectors from the two frames are compared using SSD as a matching cost, and a certain threshold d. The SSD of each two vectors is computed and compared to d. If SSD is greater than d, matching between the two vectors is rejected, thus their corresponding corners are not matched.

3) Removing outliers and estimating transform:

In this step RANSAC algorithm is used. The motion model we want to estimate is an affine transformation model, this model has 6 parameters have to be estimated, and these parameters were shown in (5). For that we need only 3 points, since each point gives two equations (one for x-direction and the other for y-direction). RANSAC algorithm will robustly fit the best affine transformation to the set of point matches. First, sample randomly a number of sets, each consists of 3 point matches. For each set, compute the unique affine transformation they define. Transform computation is done using Least Squares (LS) approach. Then transform all points from image 1 to image 2 using that computed transformation, and see how many other matches confirm the hypothesis. These steps are repeated N times. Finally, choose the best model that gives the minimum SAD between the two images [20].

C. Target Tracking

Target tracking is a digital stabilization method which requires prior information about the target to track, and position and size of the target-block in the first frame of the video. After taking all necessary information we should indicate a dynamic search region that is used to look for the target in the next frames. Its position is estimated by the last known target location. Searching for the target is done only within this search region, which reduces the number of computations required to find the target. The search region size depends on the strength of vibrations of the camera. After that we take the next frame, apply the block matching inside the search region to specify the location of the target in the current frame and determine how much the target has moved relative to the previous frame. Block-matching is done using

978-1-4673-8760-6/15 $31.00 © 2015 IEEE

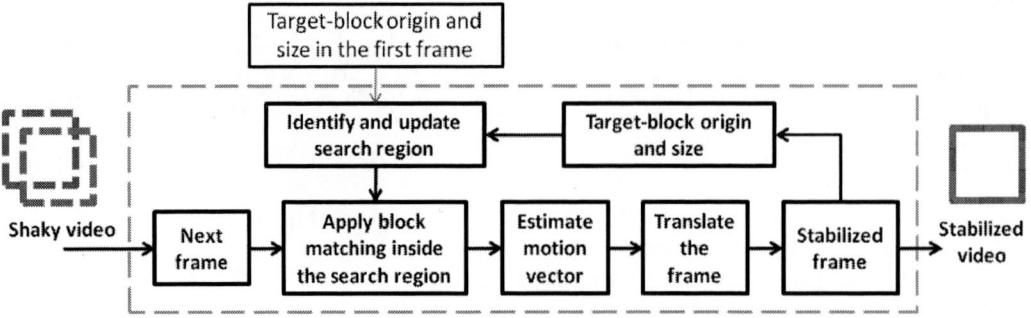

Fig. 4.Target tracking model, first identify the position and size of the target in the first frame, then look for the target in the next frame by applying block matching within the search region, then estimate the motion vector and translate the frame to get a stabilized frame. These steps are repeated until all frames are stabilized, to get finally the stabilized video.

the sum of absolute differences (SAD), by moving the target-block over the search region and computing this cost at each location. The location with the lowest SAD value corresponds to the location of the target in the video frame. After identifying target position the estimated motion vector will be the difference between the position of the origin of the target-block in the first and second frame respectively. Then the motion will be compensated by translating the current frame according to the motion vector. If we identify a target and indicate its size and position while capturing a video, this model can be used as a real-time stabilization algorithm. Fig. 4 demonstrates the model stages.

IV. CONCLUSION

In this paper, we have investigated the subject of image stabilization in its mechanical, optical, and digital aspects. We have explained in details the three main methods in digital image stabilization; Gray-coded bit-plane block matching, feature-based matching, and the target tracking method. Then we enhanced and developed the three methods by matlab.

TABLE I
COMPARISON BETWEEN METHODS

	Gray-coded bit-plane block matching	Feature-based matching	Target tracking
Stabilizing Capability	Robust against translational vibrations. Weak against rotational ones	It can remove large translational and rotational vibrations in addition to scaling effects	Robust against translational vibrations only
Simulation Speed	0.559130 s/frame	1.559727 s/frame	0.097181 s/frame
Prior information	No prior information needed	No prior information needed	Target position and size
Computational complexity	No complex computations since it uses bitwise operations	Some complexity in FAST and RANSAC algorithms	Computational complexity depends on the size of search region

REFERENCES

[1] S.-C. Hsu, S.-F. Liang, and C.-T. Lin, "A robust digital image stabilization technique based on inverse triangle method and background detection," IEEE Trans. Consumer Electronics, vol. 51, pp. 335-345, May, 2005.

[2] J.Narendra Babu, M.Nageswariah, S.Shajahan, and A.Maheswari, "Block Processing Video Stabilization," International Journal of Scientific and Research Publications, Volume 3, Issue 3, March, 2013.

[3] C. Morimoto and R. Chellappa, "Fast electronic digital image stabilization," IEEE Proc. of International Conference on Pattern Recognition, vol. 3, pp. 284-288, 1996.

[4] P. Rawat and J. Singhai, "Efficient video stabilization technique for hand held mobile videos," International Journal of Signal Processing, Image Processing and Pattern Recognition, Vol. 6, No. 3, June, 2013.

[5] P. Rawat and J. Singhai, "Review of motion estimation and video stabilization techniques for hand held mobile video," Signal & Image Processing : An International Journal (SIPIJ) Vol.2, No.2, June 2011.

[6] M. Grundmann, V. Kwatra, and I. Essa, "Auto-directed video stabilization with robust L1 optimal camera paths," in Proc. IEEE Conf. Computer Vision and Pattern Recognition, June, 2011.

[7] S. Battiato, G. Gallo, G. Puglisi, S. Scellato, ''SIFT Features Tracking for Video Stabilization,'' in Proceeding of the IEEE International Conference on Image Analysis and Application, pp. 825–830, September, 2007.

[8] J. Chang, W. Hu, M. Cheng, and B. Chang, "Digital image translational and rotational motion stabilization using optical flow technique," IEEE Transactions on Consumer Electronics, vol. 48, no. 1, February, 2002.

[9] R. Chereau and T. P. Breckon, "Robust Motion Filtering as an Enabler to Video Stabilization for a Tele-operated Mobile Robot," School of Engineering, Cranfield University, UK.

[10] H. Farid and J. B. Woodward, "Video stabilization and Enhancement," TR 2007-605, Dartmouth College, Computer Science, 1997.

[11] H.-C. Chang, S.-H. Lai, and K.-R. Lu, "A robust and efficient video stabilization algorithm," ICME '04:an International Conference on Multimedia and Expo, 29–32, June, 2004.

[12] S. J. Ko, S. H. Lee, S. W. Jeon, and E. S. Kang, "Fast digital image stabilizer based on Gray-coded bit-plane matching," IEEE Trans. Consumer Electron., vol. 45, no. 3, pp. 598-603, August, 1999.

[13] Y. Yeh, H. Chiang, and S. Wang, "Digital camcorder image stabilizer based on gray-coded bit-plane block matching," Optical Engineering, vol. 40, no. 10, pp. 2172–2178, 2001.

[14] S.-H. Lee, K.-H. Lee, and S.-J. Ko, "Digital Image Stabilizing Algorithms Based on Bit-plane Matching," IEEE, 1998.

[15] R. Hu, R. Shi, I.-f. Shen, W. Chen, "Video Stabilization Using Scale-Invariant Features," 11th International Conference Information Visualization (IV'07) IEEE, 2007.

[16] C.-H. Chen, C.-Y. Chen, C.-H. Chen, and J.-R. Chen, "Real-time video stabilization based on vibration compensation by using feature block," International Journal of Innovative Computing, Information and Control, Volume 7, Number 9, September, 2011.

[17] A. Amanatiadis, and I. Andreadis, "An integrated architecture for adaptive image stabilization in zooming operation," IEEE Trans. Consum. Electron. 54(2): 600–608, 2008.

[18] A. Amanatiadis, A. Gasteratos, S. Papadakis, and V. Kaburlasos, "Image Stabilization in Active Robot Vision," In: Ude, A. (ed.) Robot Vision, pp. 261–274. InTech, 2010.

[19] E. Rosten and T. Drummond, "Machine learning for high speed corner detection," in 9th Euproean Conference on Computer Vision, vol. 1, pp. 430–443, 2006.

[20] Y. Shen, P. Guturu, T. Damarla, B. P. Buckles, and K. R. Namuduri, "Video Stabilization Using Principal Component Analysis and Scale Invariant Feature Transform in Particle Filter Framework," IEEE Transactions on Consumer Electronics, Vol. 55, No. 3, August, 2009.

978-1-4673-8760-6/15 $31.00 © 2015 IEEE

Real-time Lane Detection in Different Illumination Conditions

I. El Hajjouji, A. El Mourabit, Z. Asrih, B. Bernoussi, S. Mars

Laboratory of Information and Communication Technologies,
National School of Applied Sciences, UAE University
Tangier, Morocco
eismail89@gmail.com

Abstract—**The paper presents an FPGA implementation of a lane detector able to work at different illumination condition. The proposed approach is based on an adapted Sobel edge detector with a dynamic threshold and Hough transform. Two CORDIC modes vector and rotation are used to implement HT. Moreover, exploiting edge phase information in CORDIC iteration provide an optimal way to increase precision without hardware complexity increase.**

Keywords—*FPGA; Sobel; Hough; Cordic; Dynamic Threshold*

I. INTRODUCTION

Active Driver assistance (ADA) based on an embedded camera is a low cost solution that can be very shortly available on a series-produced car. Recent trend and developments in several technologies (micro-sensor, embedded electronics...) and computer vision plays an important role in the development of these systems to allow big ability to perform embedded intensive calculations with low power, more intelligence and low cost. Several examples in the literature reported applications such lane departure prevention, detection and recognition of road signs, pedestrian detection, parking-aid, automatic cruise control, automatic switching on/off beams, collision risk warning, etc.

There are still several issues that would be solved for the system based -computer vision technology [1] [2]. One main problem is the visibility and illumination environment conditions: sunny, foggy, rainy, cloudy, etc. In such condition camera could give noisy frames which may generate false alarm to the driver or misinterpret a dangerous situation.

At a first approach, ADA system has to satisfy at least to the three following conditions:

1) Hard real-time constraints. This is in fact an important constraint for an embedded functions, since they are going to be integrated into more complex and intelligent vehicles systems which consume a lot of computing resources.

2) Results have to be accurate, especially for segmentation function which is the first processing bloc. Any false detection will affect and erroneous the final result. Segmentation have to be done independently of the illumination road scene [8].

3) The algorithm have to be able to manage with different day scenarios such as: sunny, cloudy, rainy, etc.

In this paper, we propose a processing chain robust to work in spite of the three above conditions. The remainder of this paper is organized as follows. Sections II, present the conventional method of SHT. In section III, we describe the proposed method. Results are reported in conclusion section.

II. STANDARD HOUGH TRANSFORM

Standard Hough transformation (SHT) is widely used in the literature as a way to find different sharps in the image, it is like a better choice for lane detection where parameters number is minimal. The advantage of Hough transform is that the pixel lying on one line doesn't need to be all contiguous. Therefore, it is very useful to detect lines containing short disconnection caused by noise or occlusion. Following steps are used for standard detecting lines with SHT:

1) Edge detection, e.g. using the Canny Edge Detector [3].

2) Mapping of edge points to the Hough space and storage in an accumulator.

3) Interpretation of the accumulator to yield lines of infinite length. The interpretation is done by thresholding and possibly other constraints.

4) Conversion of infinite lines to finite lines.

Canny edge is a gradient-based approach. It is the most used edge detector in machine vision. It is generally superior to the other edge detection methods in terms of detection, localization and single response to a true edge. Canny operator uses two threshold values, to generate binary images from edge images. In the case of detecting the edge in a road scene configuration, the use of global threshold values is hard to be the best solution. In fact, illumination varies during the day. Objects appears differently during sunny/overcast or with the presence of the fog, as a results, non-uniform illumination occurs. Is such a situation, the process of fine tuning one or two appropriate global threshold values becomes a complex task, which requires considerable material resources and computational times. Hence, they are less adequate for real time application with hard constraints.

SHT is not only computation demanding but also memory-demanding. All possible values of θi parameter are used to calculate the corresponding ρi values for a single given feature point. For a given point fully sequential algorithm

978-1-4673-8760-6/15 $31.00 © 2015 IEEE

implementation has complexity θ (N^2xθ_{max}) where θ_{max} is the number of quatizations in θ space and represents the accuracy desired for an NxN image. For example, if the minimal step of θ is 1°, there are 360 times of the multiplication operation and 180 times of the addition operation, just for one edge point.

In the next sections, we present an adaptation of edge and HT operators. The overall goal is to reduce calculus and FPGA complexity.

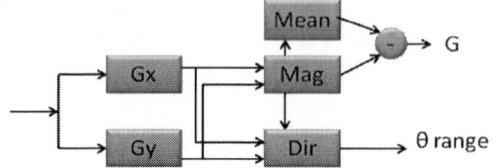

Fig. 2: Gradient and Phase Calculus

III. PROPOSED ALGORITHM

Fig. 1 presents the adopted approach. Edge detection is adapted to meet the required constraint: Robust detection even if changing lighting conditions in the image with optimal FPGA resource. HT is implemented by CORDIC algorithm to meet hard constrains real time.

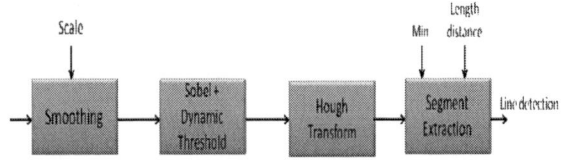

Fig. 1: Proposed Line Detection Chain

A. Proposed Edge Detection

To deal with the constraints reported above, the following algorithm was implemented:

1) Convolution with Sobel vertical and horizontal kernels [3].

2) Gradient calculation at each pixel G.

3) Convolution to generate Mean Gradient Gmean.

4) Thresholding the difference image G - Gmean

$$
\begin{cases}
EM = 1 & if \quad G > G_{mean} \\
EM = 0 & else
\end{cases}
\tag{1}
$$

Applying the some approach of dynamic threshold to Canny detector requires implementation of complex task of fine tuning of two thresholds. The proposed Sobel detector with adaptive thresholding can be an alternative to canny detector especially where the available resources are limited.

To enhance speed processing, horizontal and vertical gradients are calculated in parallel and the total gradient is obtained as the sum of the two gradients absolute values. FPGA implementation of multiplication by 2 is replaced by a right shift operation. The absolute value operator is implemented using comparators to compare the two gradient directions and to decide the range of gradient phase (equation 3). Gradient and phase range are inputs of the next HT stages. Phase range is used in the next stage to reduce rotation number of HT.

B. Hough Transform

The generation sinus and cosines of trigonometric functions is one of the main barriers to implementation of the SHT in hardware [4]. One technique is to employ LUTs [5] [6], but on the FPGA implementation intended herein, LUTs would have a severe impact on on-chip memory usage, given that memory is also required for vote arrays. Another possibility is distributed arithmetic. The Coordinate Rotation Digital Computer offers an elegant way of trigonometric calculus implementation. Thanks to the iterative approximation of an angle by a rotation, the algorithm use only a number of addition, shift and comparison operations, making CORDIC-based algorithm remains attractive for application where real real time hard constrains are critical.

$$
\begin{cases}
x_{i+1} = K_1[x_i - y_i d_i 2^{-i}] \\
y_{i+1} = K_1[y_i - x_i d_i 2^{-i}]
\end{cases}
\tag{2}
$$

atan(2-i) is approximated by 2-i , Ki=cos(atan(2-i)) and di=±1.

CORDIC algorithm can works in two modes, rotation and vector mode. In the vector mode, CORDIC calculates simply the module and the phase of the vector from cartisian Cartesian coordinates x and y. The mode rotation solves, simultaneously from an initial angle, the corresponding sine and the cosine values.

Our implementation of HT is showed in Fig. 3. As for SHT we start from an edge image calculated by the proposed Sobel detector as reported above. θ is obtained through a CORDIC vector mode, sine and cosines are obtained through a rotation CORDIC mode. Accuracy is improved without gain-rotations number trade-off, thanks to restriction of the range of each set of rotations to the corresponding values according the absolute values of Gx and Gy and the resulting sign of their comparison as reported in equation (3). We note here that these values are also computed for gradient stage as reported in parag.2, which simplify material complexity and help timing performance improvement.

$$
\begin{cases}
0 < \theta < \pi/4 & if\ |G_X| > |G_Y|\ and\ G_X > 0 \\
\pi/4 < \theta < \pi/2 & if\ |G_X| > |G_Y|\ and\ G_X > 0 \\
\pi/2 < \theta < 3\pi/4 & if\ |G_X| > |G_Y|\ and\ G_X < 0 \\
3\pi/4 < \theta < \pi & if\ |G_X| > |G_Y|\ and\ G_X < 0
\end{cases}
\tag{3}
$$

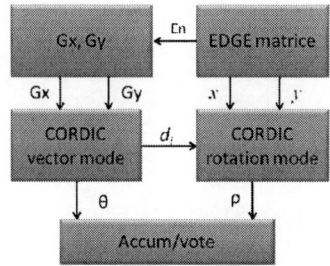

Fig. 3: CORDIC HT principle

Fig. 4 shows the circuit for CORDIC vector mode bloc, used to generate the angle θ from an initial value.

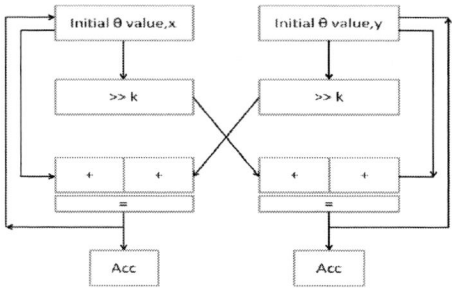

Fig. 4: CORDIC Iteration cell

IV. IMPLEMENTATION AND RESULTS

Fig. 5 presents the set-up setting for testing purpose. Camera Nikon COOLPIX P510 is connected to ALTERA DE2 board through NTSC connector. The last stage of our design is for allowing test purpose through MATLAB facilities. It is constituted by a simple serial liaison RS232 allowing connection between FPGA board and a PC with MATLAB program. Images frames are first recorded to the camera and send to the FPGA to be stored in SSRAM. Edge Detection results SSRAM are sent to the PC for performance analysis and comparison. The module is designed around a dedicated FSM. The design operates using two different clock sources which are 50 MHz and 200 MHz. Operating frequencies of individual design modules are given in Table. 1. The main clock source of the system is the external 50 MHz crystal oscillator. The 50 MHz and 200 MHz clocks are generated internally by a phase locked loop (PLL) circuit from this input clock. The signal transmission between two clock domains are handled by using dual clock FIFO buffers and synchronization stages to prevent timing problems. Regarding FPGA resource utilization, the circuit occupies 45% of the available resources (without taking account VGA and image acquisition functions). In terms of latency, timing analysis obtained during synthesis using the tool Quartus II, the frequency of the FPGA shows that the maximum operating frequency (Fmax) is 23.58 MHz. NTSC (PAL-SECAM) imposes a processing speed of 25 Frame per second which prove that Real time constraint for this standard is granted.

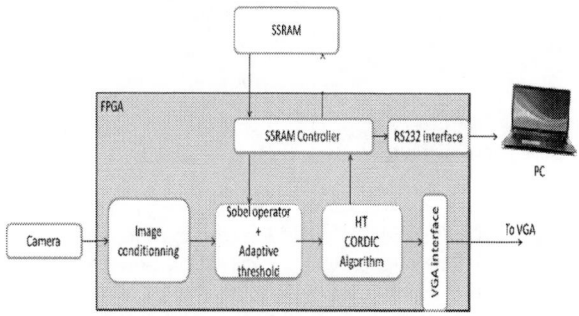

Fig. 5: Test setup setting

TABLE I. FPGA RESOURCES UTILIZATION

Logic Utilization	Used	Available	Utilization
Total logic elements	45,792	114,480	40%
Total combinational functions	45,792	114,480	40%
Total memory bits (Kbit)	1360	3,888	35%
Embedded 18 x 18 multipliers	15	266	6%
Total PLLs	2	4	50%

Table. 2 shows the proposed detector results on multiple images from FRIDA image database [9] degraded by illumination variations. One can see that the proposed edge detector can reliably detect edges even if variation of luminance. We have estimated that the limit of the system in noise-degraded images is down to 20 dB of SNR, where many false edges start appearing.

TABLE II. RESULTS FOR FRIDA IMAGES

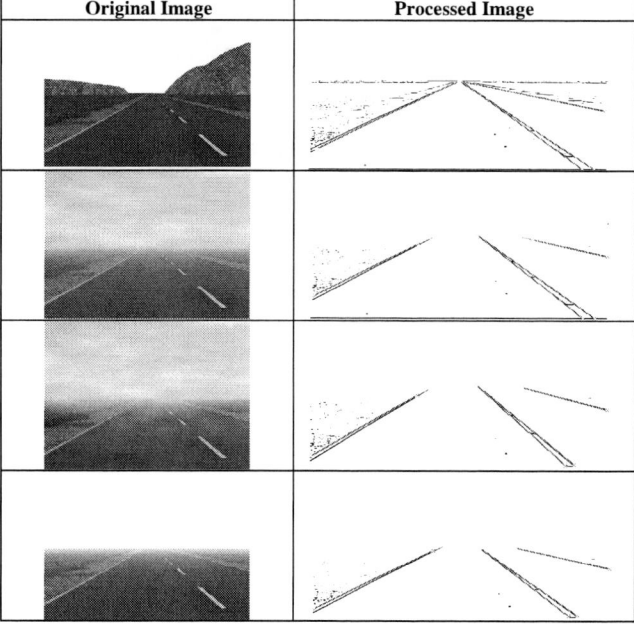

Original Image	Processed Image

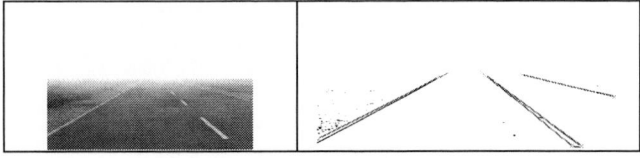

V. CONCLUSION

An implementation of edge detector and Hough transform is presented. The edge detector is based on Sobel kernel with adaptive and dynamic threshold to allow edge adaptation with different illumination conditions. HT is implemented optimally by CORDIC algorithm and by exploiting phase of the gradient calculated early in the edge detection stage. Timing performance can be improved using multiple parallel CORDIC cells.

References

[1] Jack Greenhalgh and Majid Mirmehdi " Real-Time Detection and Recognition of Road Traffic Signs " IEEE TRANSACTIONS ON INTELLIGENT TRANSPORTATION SYSTEMS, VOL. 13, NO. 4, pp. 1498-1506, Dec. 2012.

[2] P-R. Possa, S-A. Mahmoudi, N. Harb, C. Valderrama, "A Multi-Resolution FPGA-Based Architecture for Real-Time Edge and Corner Detection," IEEE TRANSACTIONS ON COMPUTERS, VOL. 63, NO. 10,pp. 2376-2388 Oct. 2014.

[3] H. Maître, Un Panorama de la Transformée de Hough, Traitement du Signal 2 (4) (1985).

[4] S-M. Karabernou, F. Terranti "Real-time FPGA implementation of Hough Transform using gradient and CORDIC algorithm" *Image and Vision Computing*, Elsevier, 2005, 23 (11), pp.1009-1017.

[5] Vijay G., Shashikant L., An improved lane departure method for advanced driver assistance system, International Conference on Computing, Communication and Applications (ICCCA), 2012, 1 – 5.

[6] Stephan S., Sarath K., Alen A., Gamini D., Robust lane detection in urban environments, Proceedings of the IEEE/RSJ International Conference on Intelligent Robots and Systems, 2007, pp.123-128.

[7] Seonyoung L., Haengseon S., Kyungwon M. Implementation of lane detection system using optimized hough transform circuit, *2010 IEEE Asia Pacific Conference on Circuits and Systems (APCCAS)*, 406-409.

[8] Jia H., Hui R., Jinfeng G., Wei H. A lane detection method for lane departure warning system, International ConferenceOptoelectronics and Image Processing (ICOIP), 2010, 28 – 31.

[9] Foggy Road Image DAtabase (FRIDA): http://www.sciweavers.org/datasets/frida-foggy-road-image-database evaluation-database-visibility-restoration-algorithms.

Comparison Between Analog and Digital Locking MPPT Unit for Micro-scale PV Energy Harvesting Systems

Nourane G. Tawfik[1], Hassan Mostafa[2], Yehia Ismail[3]

[1,2,3] Center of Nanoelectronics and Devices, AUC and Zewail City of Science and Technology, New Cairo 11835

[2]Electronics and Electrical Communications Engineering Department, Cairo University, Giza 12613 , Egypt.

nourane.g.tawfik@ieee.org, hmoustafa@aucegypt.edu, hmostafa@uwaterloo.ca, y.ismailg@aucegypt.edu

Abstract—**Energy Harvesting is gaining more attention nowadays due to the urge to have self-powered systems. the biggest challenge when designing the system is having a control unit that does not consume too much power as well as capable of providing a stable and accurate performance. In this work, 2 different implementation approaches for the same control unit logic are investigated and compared to each other to help highlight the advantages of each. All the simulations are done using Cadence Virtuoso Custom IC design tool; the technology used is TSMC 65nm.**

I. INTRODUCTION

Energy harvesting means to collect ambient energy and then transform it into electrical energy through energy transducers; this energy is then stored to be used by an application unit. The fast growing developments in Integrated Circuit (IC) design have caused us to reach ultra-low power Wireless Sensor Nodes (WSN). Even though the majority of these sensor nodes are not power hungry; they require regular change of their power storage element due to the storage units' finite capabilities. Here one can notice the value of having a system capable of providing a standalone performance. When designed carefully, energy harvesting systems can provide the application unit with the energy it needs to maintain a reliable performance. To harvest the ambient energy, we first need an energy transducer; the role of Energy transducers is to transform ambient power to electrical power. The output of the transducer is then connected to a regulator to regulate the energy that is stored in a storage unit. The storage unit is then connected with another regulator to meet the application unit's needs.

The role of these blocks is to deliver a suitable energy level to maintain stable operation mode. A Generic system block diagram is shown in fig.1 In this work, the energy transducer used is a photo-voltaic cell (PV cell). The main reason behind this choice is power density delivered by the PV cell is the highest compared to other transducers as seen in table I.

The most important design consideration that can guarantee a standalone performance is the control unit of the system. The role of the control Unit is to maintain maximum power extraction. This can be done through different famous techniques such as perturb and observe, Hill climbing, Beta method and correlation [2]. A technique in particular will be the scope of this paper; it realizes the relationship in (1) between the optimum charge-pump operating frequency and the optimum

TABLE I: Data for various transducers [1]

Energy Transducer	Condition	Power density
Solar Cell	indoors	$1mW/cm^3$
Solar Cell	outdoors	$15mW/cm^3$
Vibration	continuous vibrations	$110\mu W/cm^3$
Thermoelectric	$10°C$ gradient	$40\mu W/cm^3$
Piezoelectric	inserted in shoes	$300\ \mu W/cm^3$
Acoustic noise	100 dB	$960\ nW/cm^3$

voltage that delivers maximum power. The goal of this paper is to compare between an analog versus a digital implementation of the control unit.Fig.1 shows the generic block diagram of the harvester. The solar cell is a compact verilog-A model [4], the power converter is a 3-stage switched capacitor which is Dickson's linear charge pump [3].

$$f_{clk,opt} = \frac{I_{sat}\left(\exp\left(q\frac{V_{MPP}}{\alpha n_d KT}\right) - \exp\left(q\frac{V_{MPP}}{n_d KT}\right)\right)}{\dfrac{MC\left(MV_{MPP}-V_{EB}\right)}{M-1} + \beta} \quad (1)$$

The analog Control Unit is the one implemented in [5] and shall be re-investigated here can be divided into two parts. The left part interprets the PV voltage change into a current change through the feedback Amplifier. Then, the current is mirrored to the right side using a current mirror. The right part consists of a current controlled ring oscillator. The frequency range is controlled by the delay of the inverter; where the frequency tuning is controlled by the charging current of the capacitor (Cs). The output D-flipflop and the inverter are used to manipulate the duty cycle of the clock.

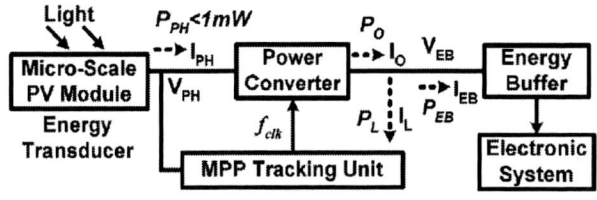

Fig. 1: generic system block diagram [3]

The digital control unit implemented in [6] which follows the same track as the analog control unit mentioned above; monitors the maximum power trajectory of the solar cell and power converter, and then implements this trajectory using a mixed-signal control. The control unit consists of three main blocks as seen in fig.3; An 8- bit Successive approximation register analog-to-digital converter (SAR ADC) used to convert the solar cell voltage sensed into digital bits; exponential decoder that is used to decode the bits into a different digital code that is then fed to a digitally controlled oscillator (DCO); the role of the DCO is to produce the required clock frequency needed by the charge pump to transfer the maximum power of the solar cell to the energy buffer side. System Designers try to increase the percentage of the digital part of any system which is implemented in CMOS. This is because of the CMOS technology scaling in terms of area and gate delay, that's why the digital control unit was implemented and investigated.

Since the charge pump needs two control clock signals. Knowing that the output of either control unit is only one clock terminal, a non-overlap phase generator [7] is added to generate two inverted non-overlapping periods. The need to have non-overlapping clocks is to prevent the switching leakage in the charge pump.

This paper is organized as follows; Section (II) presents an overview over each of the system's blocks. Section (III) presents the design of the analog versus digital control unit. Section (IV) shows the simulation results and section (V) conclusion and future work.

II. SYSTEM BLOCKS OVERVIEW

The system consists of: PV cell which is our energy transducer, 3-stage linear-topology switched-C charge pump, Control Unit and finally our energy storage buffer. A detailed description of system blocks as well as alternatives to blocks used in this work can be found in [2]. However, we will briefly discuss the blocks used in this work.

A. PV cell

The PV cell directly converts the incident light energy into electrical energy and it can be modeled as a voltage limited current source. The output voltage as well as output power of the PV cell depends on several factors such as temperature, aging, but most importantly the input light intensity. I-V curves as well as P-V curves are shown in fig. 2 for different input light intensities.

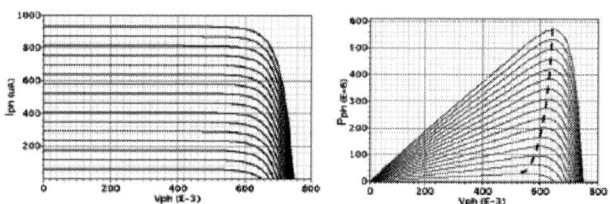

Fig. 2: I-V as well as P-V curves for different light intensities

B. Charge pump

The charge pump that will be used in this work is the Dickson linear-topology charge pump [3]. The ideal voltage step-up ratio for this charge pump is N+1. During each clock cycle, the charge from the previous stage is stored on an internal capacitor of the stage. Then, it's transferred to the following stage. When it reaches the final stage, the harvested charge from the previous stages is dumped into the energy buffer at the output for storage. Linear CP fig. 3

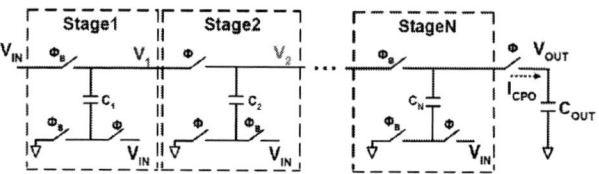

Fig. 3: 3-stage Linear Charge Pump [3]

III. CONTROL UNIT DESIGNS

An efficient control unit design means consuming as little power as possible while delivering accurate tracking results to monitor the MPP. It should verify the transitions from one MPP to another as seen in fig. 4

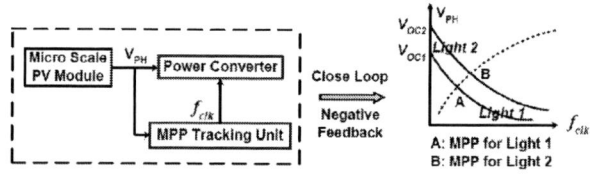

Fig. 4: MPPT block diagram [3]

In [4], a study of [9] was made, a compact relationship between solar cell voltage and charge pump frequency is reached and that is equation (1). The realization of this relationship shows an exponential behavior between frequency and voltage. When implementing the system in [5]; sub-threshold Voltage Controlled Oscillator is designed to match the relation. The design is very efficient in terms of power as well as area, but does not behave well when investigating process corners. This is because sub-threshold operation carries a lot of variations. The control unit design implemented in [6], realizes the same equation but using digital input signals rather than analog input as the analog control unit. As both control units produce only one clock signal, and the charge pump requires 2 non-overlapping cloaks; a non-overlapping 2-phase generator is used after each control unit.

A. Analog Control Unit

The Control Unit designed in [5], [8] approaches the ideal frequency-voltage relationship mentioned above. To understand how that happens; We can see that since the Op-Amp has a high open loop gain the phase voltage approximately equals the feedback voltage node. from the schematic we can also see that transistors $M1$, $M2$ and $M3$ in the design work as potential dividers that deliver $M6$ and $M7$'s gate voltages the role of the first 3 transistors is to guarantee the sub-threshold operation mode and thus consume very little power. The current I_d will be mirrored to the digital side of the control unit. The right side is a ring oscillator with a frequency control $M8$. And the D-flipflop is used to generate a 50 percent clean square wave.

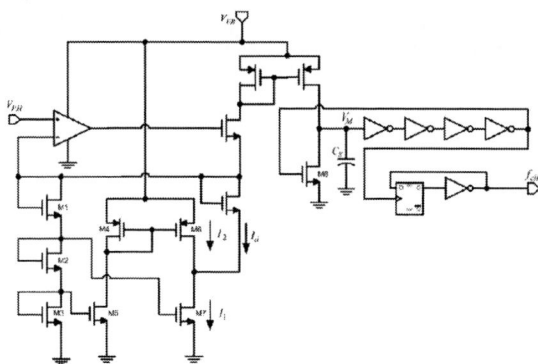

Fig. 5: Analog control unit schematic [5]

B. Digital Control Unit

This control unit design is realized through 3 main blocks: Successive Approximation Register Analog to Digital converter (SAR ADC) followed by an Exponential Decoder and finally a Digitally Controlled Oscillator (DCO). The role of the ADC is to take the PV cell's output voltage and transform it into its digital corresponding 8 bits. The DCO takes as input 20 bits and outputs their corresponding frequency for the charge pump. As for the exponential decoder; it takes the 8 bits generated by the ADC and matches them with their corresponding 20 bits that will realize the optimum voltage frequency relationship.

The SAR ADC is used in this control unit because it can offer high-performance using small amounts of power, which is which is a very important in this design. The SAR ADC basically implements a binary search algorithm to output the correct 8 bits. The ADC implemented in this work is a typical $SARADC$ with a capacitive DAC array. This capacitive array helps us save space because it also operates as the sample and hold unit for the ADC. Dynamic range for this ADC is from 200 mV to 350 mV. a simple block diagram for the ADC is shown in fig. 6

When comparing with a VCO, the DCO is less sensitive to noise and process variations. This DCO is implanted using digitally controlled inverter delay elements connected in ring configuration. The delay is controlled by changing the driving strength of the $MOSFET$ by controlling the current, as well as tuning shunt capacitor loads for different output frequencies.The output frequency is controlled by a binary weighted capacitor array for coarse tuning, as for fine tuning; $PMOS$ as well as $NMOS$ thermometer banks are used. This DCO has 5 stages, each contains a capacitor array as well as thermometer bank with 4 bits as input word for the stage. a DCO unit is shown in fig. 7

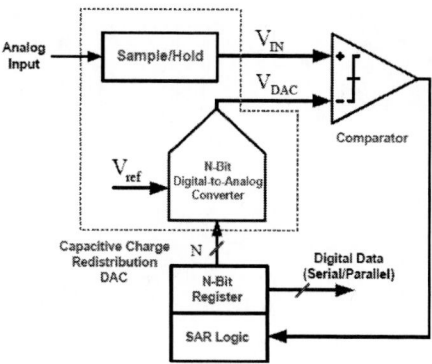

Fig. 6: SAR ADC block diagram [6]

The exponential decoder is a simple Verilog-A code realizing the voltage frequency relationship by matching each 8 bits from the ADC with their corresponding 20 bits for the DCO. A simple block diagram of the control Unit

IV. SIMULATION RESULTS

Over-viewing the system; the PV cell receives solar energy and transforms if into electrical energy. The output of the PV cell is taken as input for input to our charge pump and that output is also sensed by the control unit. The target of the control unit is to determine the optimum clock frequency for the charge pump; this is to maintain maximum power transfer under any condition. Open loop simulations were done to verify the operation of each block and closed loop simulations were done to verify the locking capabilities of each of the control units. The following figures 8, 9 shows different output clock frequency for the charge pump at different input voltage levels.

The following fig. 11 shows the analog control unit's transfer function vs. the ideal relationship. where as fig. 10 shows the analog control Unit transfer function vs. the ideal case. Table II shows the performance metrics for the system

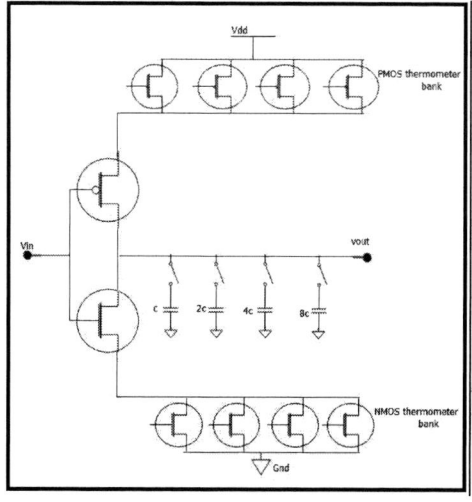

Fig. 7: DCO building Unit

978-1-4673-8760-6/15 $31.00 © 2015 IEEE

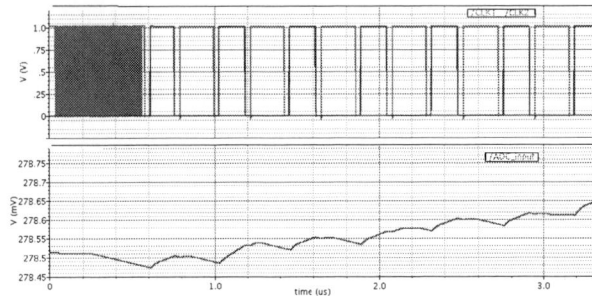

Fig. 8: closed loop simulation for $278mV$ as input voltage

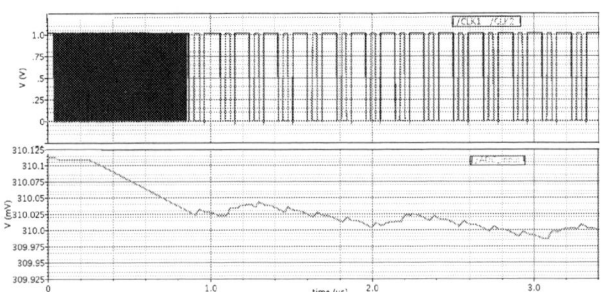

Fig. 9: closed loop simulation for $310mV$ as input voltage

with the digital control unit as well as shows performance metrics for the system with analog control unit installed.

V. CONCLUSION

The capability to track Maximum power is most important aspect when designing a energy harvesting system. Each algorithm can be implemented in different ways as we have seen here. Usually, each implementation method for the same algorithm has its own advantages and disadvantages. This paper presents a comparison between an analog and a digital implementation for the same control unit design.

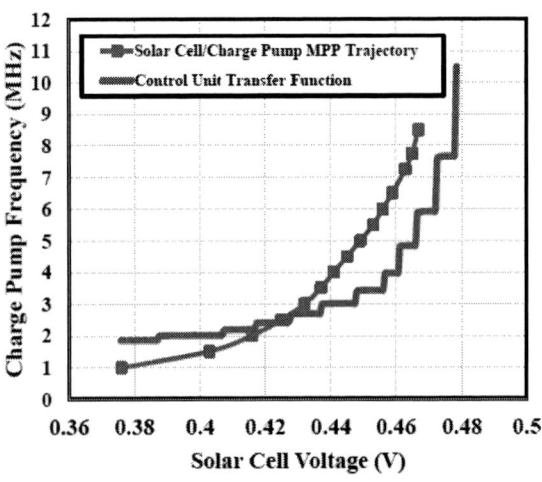

Fig. 10: Digital control unit F-V results versus ideal results [6]

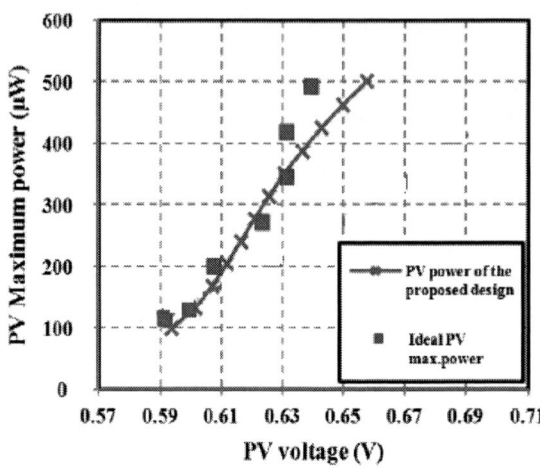

Fig. 11: Analog control unit F-V results vs. ideal results [5]

TABLE II: Comparison between analog and digital control units

Specifications	Analog Control [6]	Digital control
Frequency Range	1.84MHz$\rightarrow$ 5.54MHz	2.3 MHz$\rightarrow$ 7MHz
PV voltage range	638 mV$\rightarrow$ 717	278 mV$\rightarrow$ 320 mV
Max PV power	199 $\mu W \rightarrow$ 1.41 mW	19.7 $pW \rightarrow$ 1.194 nW
Energy buffer power capability	66.8 $\mu W \rightarrow$ 1.08 mW	1.03 $nW \rightarrow$ 1.4 nW
Power efficiency	19.6 $\rightarrow$ 60.6	47 $\rightarrow$ 63
Initial output voltage	2 v	1.8 V
Output capacitor	1 μ	1 μ

ACKNOWLEDGMENT

This research was partially funded by Cairo University, ITIDA, NTRA, NSERC, Zewail City of Science and Technology, AUC, the STDF, Intel, Mentor Graphics, SRC, ASRT and MCIT.

REFERENCES

[1] S. Beeby and N. White, "Energy harvesting for autonomous systems", 2nd ed.,Artech House, 2010.

[2] M. de Brito et al., "Evaluation of the Main MPPT Techniques for Photovoltaic Applications", IEEE Transactions on Industrial Electronics, VOL. 60, NO. 3,, March 2013.

[3] C. Lu, "Efficient Design for Micro-Scale Energy Harvesting Systems", PhD Thesis, 2012.

[4] M. Brinson, S. Jahn and H. Parruitte, "Qucs Report Book", http://qucs.sourceforge.net/docs/reportbook.pdf.

[5] A. Eltaliawy, H. Mostafa, Y. Ismail "Microscale Solar Energy Harvesting for wireless Sensor Networks Based on Exponential Maximum Power Locking Technique", 2013 IEEE 20th International Conference on Electronics, Circuits, and Systems (ICECS), 2013.

[6] A. Eltaliawy, H. Mostafa, Y. Ismail, "A New Digital Locking MPPT control for Ultra Low Power Energy Harvesting Systems", IEEE International New Circuits and Systems Conference (NEWCAS), 2015.

[7] F. Pan and T. Samaddar, "Charge Pump Circuit Design", McGraw-Hill, 2006.

[8] A. Eltaliawy, H. Mostafa, Y. Ismail "Micro-scale Variation-Tolerant Exponential Tracking Energy Harvesting System for Wireless Sensor Networks", Microelectronics Journal, 2015.

[9] AC. Lu, S. P. Park, V. Raghunathan and K. Roy, "Low-Overhead Maximum Power Point Tracking for Micro-Scale Solar Energy Harvesting Systems", 25th International Conference on VLSI Design (VLSID), pp. 215-220, 2012.

Effective Device Electrical Parameter Extraction of Nanoscale FinFETs: Challenges and Results

Alessandra Leonhardt*, Luiz Fernando Ferreira[†], Sergio Bampi[†] and Leandro Tiago Manera*

*UNICAMP, State University of Campinas, Brazil

[†]UFRGS, Federal University of Rio Grande do Sul, Porto Alegre, Brazil

aleonh@dsif.fee.unicamp.br, luiz.ferreira@ufrgs.br, bampi@inf.ufrgs.br, manera@dsif.fee.unicamp.br

Abstract—The accurate evaluation of electrical parameters like effective channel length and series resistance for 10 to 22nm FinFETs is very complex, although extremely important for the correct device modelling. Our paper reviews and applies to measured FinFETs various DC methods for the extraction of effective channel length and source/drain resistance. Experimental devices with fin thickness from 10nm to 20nm were measured and the extraction methodologies explored show consistent values of R_{SD} in the range of 500Ω to 300Ω, respectively, for five parallel fins. All methods herein discussed and applied, however, have limitations in their application to FinFET devices. Our work discusses the challenges and possible options for the L_{eff} extraction on advanced FinFET technologies.

Index Terms—FinFETs, LDD, Parameter Extraction, Series Resistance, Effective Channel Length

I. INTRODUCTION

The MOSFET dimensions have been shrinking constantly over the last five decades. This length reduction faces its limits as short channel and parasitic effects, as well as large statistical device variations, become significant and more influential in the device behaviour. The FinFET architecture manages to further scale the transistor dimensions below 22nm and even 10nm, by providing a better coupling between gate and channel and consequently improving the overall performance. In order to integrate the FinFET technology in modern circuit design, an accurate electrical modelling is essential. This cannot be achieved unless the key electrical parameters are accurately assessed, since they are used as initial parameters when modelling the transistor behaviour. The correct extraction of the FinFET electrical parameters is thus extremely important for the adoption of the technology by designers.

In this paper we explore the extraction of the series resistance R_{SD} and effective channel length L_{eff} in nanoscale FinFETs, discussing the challenges and results obtained. The series resistance is the parasitic element that causes a voltage drop between the source and drain contacts of the transistor and the channel. Only after extracting the parasitic elements, the effective device electrical behaviour can be accurately assessed. The transistor effective channel length L_{eff} is an electrical parameter and its meaning best discussed elsewhere [1]. The effective channel length is given by $L_{eff} = L_{maskk} - \Delta L$, where L_{mask} is the drawn mask gate length and ΔL is the channel length reduction. ΔL is a process dependent parameter, constant for all transistors fabricated in the same

run, and thus used to determine the effective channel length L_{eff} of different transistors with a single extracted parameter.

This work is organized as follows: in section II we describe the fabricated devices, then in the next section the different extraction methodologies are explained. In section IV we discuss the obtained results, with comparisons between the different methods, and section V draws the conclusions and perspectives.

II. EXPERIMENTAL DEVICES

Our FinFET devices were manufactured and measured in IMEC in Belgium. The Ph.D. Thesis [2] has details on device fabrication and DC characterization. The SOI devices have 65nm fin height, 2.2nm effective oxide thickness of SiON dielectric, and gate electrode of 5nm TiN covered with a polysilicon layer. Following the gate formation, the performed steps were source and drain implant, spacer formation, high doping of source and drain, and nickel silicidation on the S/D electrodes. Note that no selective epitaxial growth was performed on the fin extensions towards S/D, thus the measured devices present a higher series resistance. All devices were fabricated with five parallel fins, with a 200nm fin pitch and a 90nm distance between the gate and the silicon pad that connect the fins. Fabricated FinFETs with mask gate length (L_{mask}) ranging from 10μm to 45nm and fin thickness (T_{fin}) of 10nm, 15nm and 20nm were used in this work.

The manufactured FinFETs present lightly doped drain (LDD) structure, used to decrease the hot electron emission [3]. This device architecture creates challenges in the extrinsic parameter extraction, especially regarding the modulation of the source and drain resistance and effective channel length with gate voltage, which will be analysed in this paper.

III. EXTRACTION METHODS

Several extraction methods for the source and drain resistance have been developed [4], [5]. The extraction methodology has to be chosen considering its applicability for the given technology, especially if the assumptions on which it relies can be proven for the devices. Another issue is that the series resistance and the effective channel length have mutual dependence on the device behaviour. Thus, if the extracted value of one parameter is underestimated, the other cannot be properly extracted as well.

In the present work we will explore extraction methodologies that rely exclusively on the DC I-V measurements, since

978-1-4673-8760-6/15 $31.00 © 2015 IEEE

they are more broadly accessible and no C-V characterization of the fabricated devices was available. The limitations of this approach will be exposed and discussed. We explore the Suciu-Johnston [6], Campbell et al. [7] and Torres-Murphy [8] methods for their proposed merits in terms of reproducibility, simplicity or accuracy in extracting the parameters from LDD devices.

A. Suciu-Johnston

The Suciu-Johnston methodology [6] is a traditional method still used to extract both the source/drain resistance and the mobility degradation. The method uses at least two devices of different mask channel length but the results are significantly more reliable if more than two channel lengths are used. The method plots the parameter E (1) as a function of the gate voltage overdrive $V_{GS} - V_{th}$ for the different devices, and extracts $1/\beta_0$ for each. The value of $1/\beta_0$ is given by the linear extrapolation of E, where $V_{GS} - V_{th} = 0$

$$\frac{(V_{GS} - V_{th})}{I_{DS}/V_{DS}} = \frac{1 + A(V_{GS} - V_{th})}{\beta_0} \equiv E \qquad (1)$$

The second step of the method is to plot the slope of E against $1/\beta_0$, since

$$\frac{dE}{dV_{GS}} = \frac{A}{\beta_0} = R_{DS} + \frac{\theta}{\beta_0} \qquad (2)$$

which in turn provides the source and drain resistance R_{DS} as the intercept of the resulting straight line and the mobility degradation factor with transversal electric field (parameter θ) as the slope.

B. Campbell et al.

The methodology proposed in [7] uses the same device in two different – but very close – drain biases in order to obtain the series resistance as a function of the gate voltage overdrive. The extraction relies on the ratio of two $I_D - V_G$ curves

$$\frac{I_{D1}}{I_{D2}} =$$
$$\frac{\mu_{eff} C_{OX} \frac{W_{eff}}{L_{eff}} \left(V_{GS} - V_{th1} - \frac{I_{D1}R_{SD}}{2}\right)(V_{D1} - I_{D1}R_{SD})}{\mu_{eff} C_{OX} \frac{W_{eff}}{L_{eff}} \left(V_{GS} - V_{th2} - \frac{I_{D2}R_{SD}}{2}\right)(V_{D2} - I_{D2}R_{SD})} \qquad (3)$$

where the constants can be cancelled, since the measurements are made in the same device, yielding

$$R_{SD}^2 \left(\frac{I_{D2} - I_{D1}}{2}\right) + R_{SD} \left(V_{th2} - V_{th1} + \frac{V_{D1} - VD2}{2}\right)$$
$$- \frac{(V_{GS} - V_{th1})I_{D2}V_{D1} - (V_{GS} - V_{th2})I_{D1}V_{D2}}{I_{D1}I_{D2}} = 0 \quad (4)$$

The source and drain resistance values are obtained after solving the polynomial equation. The method does not give a single value for the source and drain resistance, but its relationship with the gate voltage. A single value should be extracted from these curves at a very high gate voltage, as discussed in [9].

C. Torres-Murphy

The Torres-Murphy methodology [8] was specifically designed for LDD devices, taking in account the dependence of the series resistance and effective channel length on the gate bias. Its key design is the iterative update of both the channel length reduction ΔL and R_{SD}, calculated from the plot of the total resistance R_T as a function of a parameter K, where

$$K = \frac{1 + \theta \left[V_{GS} - V_T - 0.5 \left(\frac{R_{SD}}{mL_{mask}+b}\right)V_{SD}\right]}{G(V_{GS} - V_T - 0.5V_{SD})} \qquad (5)$$

$$R_T = ML_{mask} + b \qquad (6)$$

where L_{mask} is the mask channel, and M is calculated as

$$M = K + (L - \Delta L)\frac{0.5m\theta R_{DS}V_{DS}}{G(mL_{mask} + b)^2(V_{GS} - V_T - 0.5V_{SD})} \qquad (7)$$

The parameters ΔL and R_{SD}, initially set as zero, are updated every iteration as the slope of the function R_T versus K and the intercept, respectively. This is done for five values of V_{GS} with very small increments, to obtain a linear function for R_T. Since the extracted ΔL and R_{SD} are strongly dependent of the chosen V_{GS} bias, the measurements are repeated for the whole range of V_{GS} biases available. The final value of ΔL and R_{SD} is then assessed by extrapolating at very high V_{GS}.

The Torres-Murphy method depends on the value of the low field mobility μ_0 and mobility degradation parameter θ, however, for the gain factor $G = \mu_0 C_{OX} W$. These constants must be both obtained with different extraction methodologies.

IV. RESULTS AND DISCUSSIONS

The methodologies previously discussed were applied separately on the FinFET measured data, in order to extract the R_{SD} and L_{eff} parameters.

The Suciu-Johnston method, proposed for bulk MOSFETs, may not be applicable to current nanoscale FinFET devices, and more so to the lightly doped drain FETs. The results are highly sensitive to the portion of the curve used for the linear fit of E. Fig. 1 shows the non linearity of the $1\mu m$ mask length transistor, even for higher gate voltage overdrive. The choice of the x-axis interval for overdrive in Fig. 1, for which the linear fitting of the Suciu-Johnston method will be made, impacts directly in the final extraction results. Choosing the part of the curve with higher gate voltage overdrive for the fitting, the extracted R_{SD} is obtained in the graph shown in Fig. 2, for the case of the devices with 15nm of fin thickness. The same was repeated for both 10nm and 20nm devices.

Fig. 3 shows the results obtained with the Campbell methodology, using $V_{DS} = 50mV$ and $V_{DS} = 100mV$. While the series resistance varies for the different mask lengths at low gate bias, it converges to the same value for the larger gate voltage overdrive. It is important to note the dependence of the source and drain resistance with the gate voltage in the measured devices. This is due to the LDD structure, since the electric field generated from the gate changes the carrier density in the lightly doped regions, which in turn changes the

978-1-4673-8760-6/15 $31.00 © 2015 IEEE

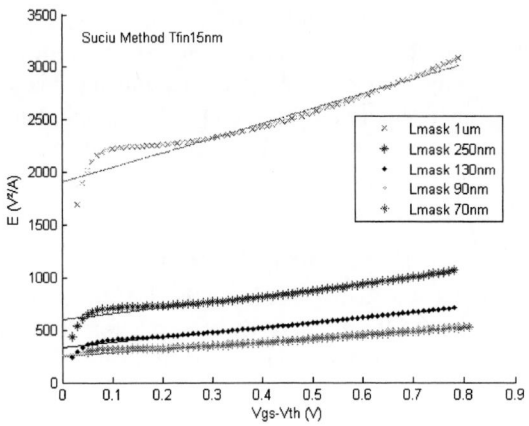

Fig. 1. First step of Suciu-Johnston method, plotting E against the gate overdrive

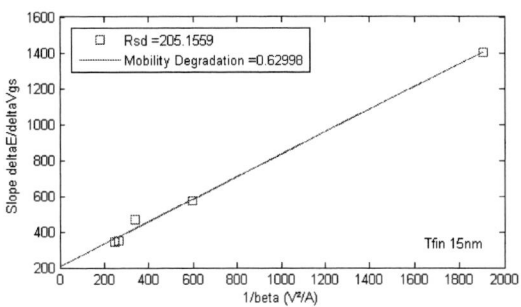

Fig. 2. Suciu-Johnston method final results. R_{SD} is given by the fitting intersection with the axis, while the mobility degradation is obtained from the slope.

resistance [3], [10]. For most modelling purposes, however, a single value of R_{SD} is possible to extract for large gate voltage overdrive, while the exponential curve fitting is applied at low voltage overdrive, as shown in Fig 3.

Since the Campbell method only assumes that μ_{eff}, C_{OX} and geometry parameters are constant for the same device in two very close drain bias, it yields consistent results. The main weakness of the methodology lies in the sensitivity to device-to-device variations and in the lack of a procedure to obtain the effective channel length L_{eff}.

The Torres-Murphy method results are exposed in Fig. 4. The value for the mobility degradation parameter θ used in the Torres-Murphy method was extracted from the Suciu-

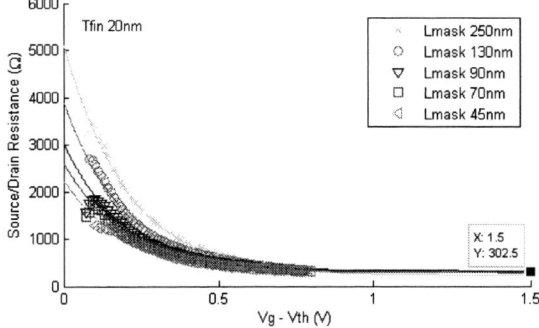

Fig. 3. Campbell et al. method with extrapolation of the curves to yield a single value for the series resistance R_{SD}.

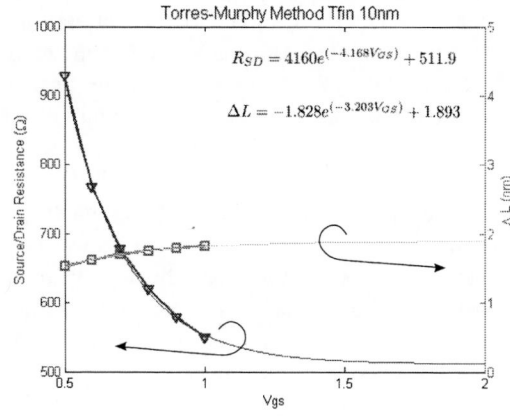

Fig. 4. Torres-Murphy method results, presenting the extracted R_{SD} and ΔL, along with their dependence on V_{GS}.

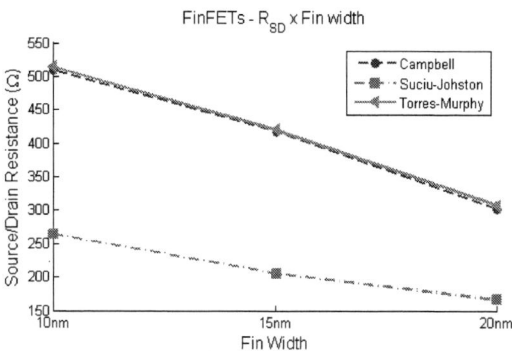

Fig. 5. Source and drain series resistance relationship with fin thickness for the explored methods

Johnston method, 0.53 for T_{fin} 20nm, 0.63 for T_{fin} 15nm and 0.65 for T_{fin} 10nm. The low field mobility parameter μ_0 was extracted from the BSIM-CMG model, $640 cm^2/Vs$, $630 cm^2/Vs$ and $620 cm^2/Vs$ for 20nm, 15nm and 10nm of fin thickness, respectively. Since this method extracts R_{SD} and ΔL as a function of the gate bias, the curves are extrapolated in order to determine the asymptotic values for the parameters.

This method is valuable for the modelling of LDD devices, since it yields R_{SD} and L_{eff} simultaneously. It has an important weakness, however, which is the dependence on auxiliary extractions to determine the gain factor G and the mobility degradation. In nanoscale devices this reliance on external extraction is an issue, since a small error in the evaluation of the parameters creates further inaccuracies. Moreover, the extracted value of ΔL, ranging from 1nm to 2nm for the different groups of FinFETs, is exceptionally small and considered inaccurate by the authors, based on qualitative analysis of the I-V measurements.

The extracted series resistance across different fin thickness for the different methods is shown in Fig. 5. The trend in the graph is expected, since the series resistance is inversely proportional to the fin width, as reported in [9], [11], [12]. The Campbell and Torres-Murphy methods presented very consistent values. The Suciu-Johnston methodology highly underestimates the series resistance, which in turn can lead to an incorrect extraction of other parameters. A 300Ω resistance in the device – as shown in the 20nm fin thickness case – indi-

978-1-4673-8760-6/15 $31.00 © 2015 IEEE   73

cates that each of the five fins has a R_{series} of approximately $1.5K\Omega$. This very high resistance is a result of the narrowness and lack of selective epitaxial growth over the fins. A value of series resistance per fin is important as a figure of merit for the FinFET technology, since the devices may have different number of parallel fins, according to design decisions.

The effective channel length could not be reliably extracted using the purely DC I-V methods studied, however. This is mainly due to the assumptions that the low field mobility μ_0 is constant for all devices of a given technology. For nanoscale FinFETs, this assumption does not hold true, as presented in [13], which in turn gives an underestimation on the extracted value of ΔL. Few options are then left for the extraction of the effective channel length in state-of-the-art devices: to employ DC I-V extraction methods which rely on the transconductance and current relations between transistors of different dimensions and make no assumption regarding the carrier mobility, such as the method presented in [14]; or to perform C-V measurements and perform the effective mobility extraction μ_{eff}, which is then used to assess the channel length reduction ΔL, as discussed in [15]. Another option that can also be explored is to assess the L_{eff} through several rounds of the weighted BSIM-CMG global parameter extraction, as show in [16], [17]. Starting from an initial guess, for every execution of the process, the value of ΔL is updated in order to lower the modelling errors. The final value for the channel length reduction is taken when the modelling errors cannot be further reduced by updating this parameter. This assessment, while providing consistent results – since L_{eff} is refined each iteration – is lengthy and computationally demanding.

The extracted values of the parasitic series resistance R_{SD} – using the Campbell et al. method – and channel length reduction ΔL – assessed by the BSIM-CMG extraction – are given in Table I for each of the three groups of fabricated FinFETs, with fin thickness varying from 10nm to 20nm.

V. CONCLUSIONS

All the methods herein discussed for the extraction of effective series resistance and channel length present weakness when applied to measured FinFET devices. Tests and comparisons have been made using a set of experimental transistors. The reliability of the extraction was assessed comparatively, and the Campbell method was considered the most robust, due to its lower dependence on other parameters and assumptions. The Torres-Murphy method, although giving similar series resistance values, presents a value for the channel length reduction ΔL which is much below the expected. Purely DC I-V methods for the evaluation of the effective channel length

are unsuitable in the current FinFET technology if they assume a constant carrier mobility. In short, the extraction of L_{eff} for nanoscale FinFETs is particularly challenging and the possible solutions either require lengthy iterations of a complete model extraction, I-V methods with fewer assumptions, but prone to fitting issues, or careful C-V characterisations to account for the mobility reduction.

ACKNOWLEDGMENTS

The authors would like to thank the IMEC, Belgium, for the fabrication of the FinFET devices and for providing the processing details and the measurement facilities. Financial support from CNPq is greatly acknowledged.

REFERENCES

[1] Y. Taur, "MOSFET channel length: extraction and interpretation," *Electron Devices, IEEE Transactions on*, vol. 47, pp. 160–170, Jan 2000.

[2] L. F. Ferreira, "Double-gate nanotransistors in silicon-on-insulator: simulation of sub-20 nm FinFETs," Doutorado, Universidade Federal do Rio Grande do Sul. Instituto de Informática. Programa de Pós-Graduação em Microeletrônica, http://hdl.handle.net/10183/65631, 2012.

[3] S. Ogura, P. Tsang, W. Walker, D. Critchlow, and J. Shepard, "Design and characteristics of the lightly doped drain-source (LDD) insulated gate field-effect transistor," *Electron Devices, IEEE Transactions on*, vol. 27, no. 8, pp. 1359–1367, Aug 1980.

[4] D. K. Schroder, *Series Resistance, Channel Length and Width, and Threshold Voltage*. John Wiley & Sons, Inc., 2005, pp. 185–250.

[5] F. G. Snchez, A. Ortiz-Conde, and J. Liou, "On the extraction of the source and drain series resistances of MOSFETs," *Microelectronics Reliability*, vol. 39, no. 8, pp. 1173 – 1184, 1999.

[6] P. Suciu and R. Johnston, "Experimental derivation of the source and drain resistance of MOS transistors," *Electron Devices, IEEE Transactions on*, vol. 27, no. 9, pp. 1846–1848, Sep 1980.

[7] J. Campbell, K. Cheung, J. Suehle, and A. Oates, "A simple series resistance extraction methodology for advanced CMOS devices," *Electron Device Letters, IEEE*, vol. 32, no. 8, pp. 1047–1049, Aug 2011.

[8] R. Torres-Torres and R. Murphy-Arteaga, "An alternative method to determine effective channel length and parasitic series resistance of LDD MOSFET's," in *Devices, Circuits and Systems, 2002. Proceedings of the Fourth IEEE International Caracas Conference on*, 2002.

[9] A. Dixit *et al.*, "Analysis of the Parasitic S/D Resistance in Multiple-Gate FETs," *IEEE Transactions on Electron Devices*, vol. 52, pp. 1132–1140, Jun. 2005.

[10] G. Hu, C. Chang, and Y.-T. Chia, "Gate-voltage-dependent effective channel length and series resistance of LDD MOSFET's," *Electron Devices, IEEE Transactions on*, vol. 34, no. 12, pp. 2469–2475, 1987.

[11] P. Magnone *et al.*, "Gate voltage and geometry dependence of the series resistance and of the carrier mobility in FinFET devices," *Microelectronic Engineering*, vol. 85, no. 8, pp. 1728 – 1731, 2008.

[12] V. Subramanian *et al.*, "Impact of fin width on digital and analog performances of n-FinFETs," *Solid-State Electronics*, vol. 51, no. 4, pp. 551 – 559, 2007.

[13] K. Bennamane, T. Boutchacha, G. Ghibaudo, M. Mouis, and N. Collaert, "DC and low frequency noise characterization of FinFET devices," *Solid-State Electronics*, vol. 53, no. 12, pp. 1263 – 1267, 2009, papers Selected from the Ultimate Integration on Silicon Conference 2009.

[14] A. Cunha, M. Schneider, C. Galup-Montoro, C. Caetano, and M. Machado, "Extraction of mosfet effective channel length and width based on the transconductance-to-current ratio," in *Technical Proceedings of the 2005 Workshop on Compact Modeling*, 2005, pp. 135 – 138.

[15] J. Kim *et al.*, "Accurate extraction of effective channel length and source/drain series resistance in ultrashort-channel MOSFETs by iteration method," *Electron Devices, IEEE Transactions on*, vol. 55, no. 10, pp. 2779–2784, Oct 2008.

[16] A. Leonhardt, "Compact modelling and parameter extraction of nanoscale FinFETs," Graduação, Universidade Federal do Rio Grande do Sul, Instituto de Informática, http://hdl.handle.net/10183/110755, 2014.

[17] A. Leonhardt, L. F. Ferreira, and S. Bampi, "Nanoscale FinFET global parameter extraction for the BSIM-CMG model," in *Circuits and Systems (LASCAS), 2015 IEEE 6th Latin American Symposium on*, 2015.

TABLE I
EXTRACTED VALUES OF THE ASSESSED PARAMETERS.

	Series Resistance R_{SD}	Channel Length Reduction ΔL
T_{fin} 10nm	514 Ω	14 nm
T_{fin} 15nm	440 Ω	15 nm
T_{fin} 20nm	300 Ω	16 nm

Accelerated Lifetime Tests and Failure Analysis of an Electro-thermally Actuated MEMS valve

H. Skima, K. Medjaher, C. Varnier, N. Zerhouni, E. Dedu and J. Bourgeois
University of Franche-Comté (UFC)
FEMTO-ST Institute, UMR CNRS 6174 – UFC / ENSMM
15B av. des Montboucons, Besançon, France
Email: firstname.lastname@femto-st.fr

Abstract—**This paper presents accelerated lifetime tests for an electro-thermally actuated MEMS valve in order to identify and analyze its failures. To perform the different tests, two experimental setup are specially designed. Tests consist in cycling several MEMS valves by changing at each test the operating condition: with unfiltered air, without air and with filtered air. Results show that different failure mechanisms can be detected depending on the operating conditions of the MEMS valve.**

I. INTRODUCTION

A Micro-Electro-Mechanical System (MEMS) is defined as a micro-system that integrates mechanical components using electricity as source of energy in order to perform measurement functions and/or operating in structure having micrometric dimensions. Over the last two decades, MEMS technology has grown from laboratory research projects to global commercialization. Thanks to their miniaturization, low power consumption and tight integration with control and sense electronics, MEMS devices gained wide-spread acceptance in several industrial segments including aerospace, automotive, medical and even military applications to achieve different functions in sensing, actuating and controlling. The most known MEMS are accelerometers for automotive (airbag) applications, gyroscopes for mobiles phones, pressure sensors for engine management and micro-mirror arrays for display applications.

Unfortunately, the reliability of MEMS is considered as a major obstacle for their development [1]. Most of these micro-systems are designed with some basic parts such as cantilever beams, membranes, springs and hinges [2]. These parts are subject to degradation and failure mechanisms which occur during their operation and impact their performances and the availability of systems in which they are utilized. These failures are due to several influence factors [3] such as temperature, humidity, vibration, acoustic noise, etc. Common failure mechanisms identified and known until now concern stiction, wear, fracture, creep, delamination, contamination, adhesion, fatigue, degradation of dielectrics and electrostatic discharge [2], [4]–[8].

One of the main challenges in this field is to identify, analyze and understand the failure mechanism to be addressed so that reliability can be developed. To do so, accelerated lifetime test method for evaluating the reliability of MEMS devices can be considered [9].

Fig. 1. Electro-thermally acuated MEMS valve designed by DunAn Microstaq company.

In this paper, we present accelerated lifetime tests for an electro-thermally actuated MEMS valve in order to identify and analyze its failure mechanisms. Tests are performed in an experimental platform, and inside the Scanning Electron Microscopy (SEM). They consist in cycling several MEMS valves continuously in different conditions: with unfiltered air, without air and with filtered air.

The paper is structured in five sections. After the introduction, Section II presents the targeted MEMS. Section III introduces the experimental setup. Results are presented and discussed in Section IV. Finally, Section V concludes the paper.

II. SYSTEM DESCRIPTION

The targeted system consists in an electro-thermally actuated MEMS valve of DunAn Microstaq, Inc. (DMQ), company designed to control flow rates or pressure with high precision at ultra-fast time response ($<< 100\ ms$). It is currently being used in a number of applications in air conditioning and refrigeration, hydraulic control and air pressure control. Fig. 1 shows a general scheme of the MEMS valve.

This valve is made by using standard semiconductor processes augmented with standard MEMS processes in etching and wafer bonding. It consists of three silicon layers bonded together by using silicon fusion bonding. The center layer is a movable membrane. The other two layers of silicon act as interface plates to either electrical connections or fluid connection ports: common port, normally closed and normally open. The SEM image (Fig. 2) shows the movable membrane through the normally open fluid port.

978-1-4673-8760-6/15 $31.00 © 2015 IEEE

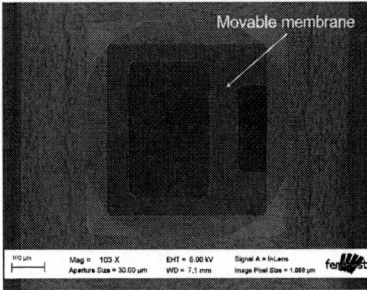

Fig. 2. SEM image of the movable membrane through the normally open fluid port.

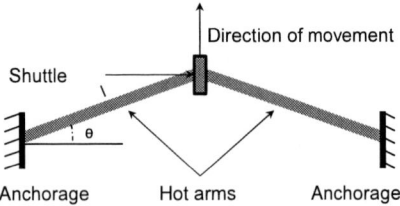

Fig. 3. Schematic view of the Chevron electro-thermal actuator used inside the MEMS valve to move the membrane.

The actuator used in the targeted MEMS to move the membrane, which allows to open or close the fluid ports, is an electro-thermal actuator. This actuator, presented in Fig. 3, is composed of hot arms inclined to the horizontal axis by an angle θ and clamped to the substrate and the freestanding central shuttle. When a voltage difference is applied across the anchor sites, heat is generated along the beams due to ohmic dissipation. The hot arms expand to push ahead symmetrically on the central part of the actuator (the shuttle). This part moves in the direction shown in Fig. 3. The maximum actuation voltage of the MEMS is 12 V and its current can reach 1 A. More details concerning the MEMS can be found in [10].

III. ACCELERATED LIFETIME TESTS

Accelerated lifetime test is an aging of a product that induces normal failures in a short amount of time by applying stress levels much higher than normal ones (stress, strain, temperatures, voltage, vibration rate, pressure, etc.). Reliability results can then be obtained by analyzing the product's response to such tests [9].

The difficulty in MEMS failure analysis arises when structures of interest are not readily exposed for direct observation [11]. In some MEMS, structures that provide the stimulus for motion or actuation are obscured from view.

In our case study, the electro-thermal actuator which allows to move the membrane is obscured from view. However, the movable membrane is accessible through the fluid ports. Thus, failures analysis of the targeted MEMS is based on the health state of the movable membrane (surface state, displacement, response time, etc.).

In order to perform accelerated lifetime tests for the MEMS valve, the two following experimental setup are specially designed.

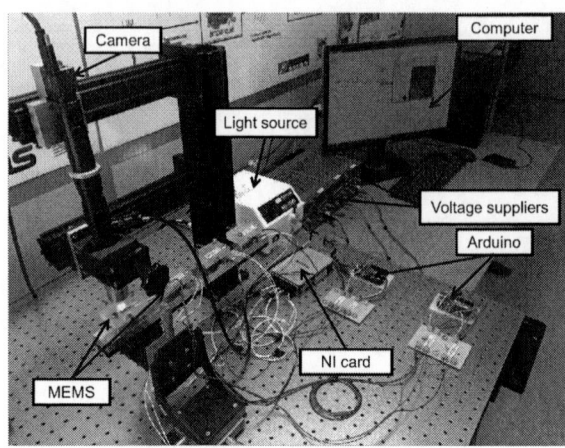

Fig. 4. Overview of the experimental platform.

Fig. 5. Support designed to fix the MEMS vlave.

A. Experimental platform

The platform, presented in Fig. 4, is composed of two Arduino cards, two voltage suppliers, supports for the camera and the MEMS, a light source for the camera allowing to see the movement of the membrane inside the MEMS, an air supply, a pressure regulator, an air filter, a National Instrument card (NI 9216) and PT100 RTD for temperature measurement and a computer.

To fix the MEMS valve on the platform, the support presented in Fig. 5 is specially designed. It is composed of a plastic support made by the 3D printer, a metal plate to allow heat dissipation as the MEMS heats a lot, the input-output of air connected to the fluid connection ports and an electronic card for power supply. The MEMS is bonded on the metal plate under the electronic card by using silicone.

To minimize the mechanical noise, the experimental platform is placed on an anti-vibration table except the air filter.

The acquisition of measurements is the same for all MEMS and for each one of them the following steps are applied: 1) adjust the MEMS below the camera using a 3D positioner until having a very clear image, 2) get the time response by using a Matlab image-processing algorithm, 3) identify the parameters of the system by using the Matlab "system identification toolbox" which leads to the transfer function of

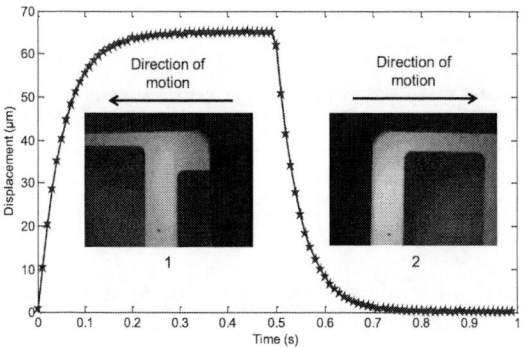

Fig. 6. Time response of the MEMS supplied with a square signal of 8 V magnitude and a frequency equal to 1 Hz. The two images of the membrane, 1 and 2, are taken by the camera through the normally closed port..

Fig. 7. Setup inside the SEM.

the obtained time response, and 4) store the results in different files in a dedicated computer for later use.

The image acquisition is done by using a Guppy Pro F-031 camera with a frame rate equal to 100 frames per second (fps). MEMS are supplied with a square signal of 8 V magnitude and a frequency equal to 1 Hz generated by a voltage supplier and an Arduino card. This voltage value is not too high to not bring up prematurely degradation and not too low to obtain enough displacement. The current consumption of a new MEMS at 8 V is about 0.55 A and the displacement of its membrane is about 65 μm. Fig. 6 shows an example of an obtained time response of one MEMS.

B. Setup in the SEM

This experiment consists in fixing a MEMS valve inside the SEM using the metal plate (Fig. 7). The valve still cycling continuously inside the SEM and it is supplied with the same signal as in the experimental platform. In this experiment, images for the membrane are taken two times per day. These images are used to monitor its health state during cycling and calculate its displacement.

Note that, tests can last long since the manufacturer of the MEMS valve guaranteed 8 million cycles without performance degradation. For that reason, only results of three tests are presented in this paper.

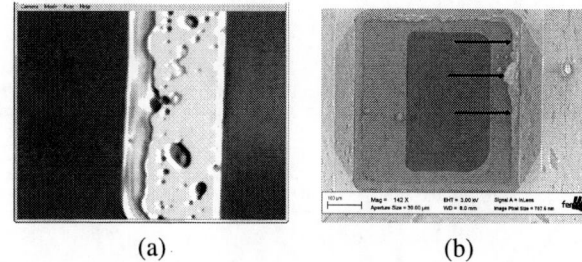

Fig. 8. (a) Image taken by the camera showing the damaged membrane of the MEMS caused by the unfiltered air and (b) SEM image showing contamination at the normally closed fluid port.

IV. RESULTS AND DISCUSSION

Three accelerated lifetime tests are performed in the two previous experimental setup. They consist in cycling MEMS valves and changing at each test the operating condition: A) using unfiltered air, B) without air and C) using filtered air.

Note that, the first and the third tests were performed with the same air pressure in the experimental platform, and the second test in the SEM. Results of the three tests are presented hereafter.

A. Using unfiltered air

This test consists in cycling four MEMS valves with an unfiltered air. Experiments remained running for more than one month. During this period, measurements were collected every day, after 25000 cycles, and at each measurement the displacement of the membrane is calculated.

At the beginning of experiments, the displacement of the membrane was about 65 μm with a good membrane surface state. After 1 million cycles, the displacement decreased to less than 10 μm with degraded membrane surface state (Fig. 8(a)). Therefore, the MEMS valve can be considered as out of service: a loss of performance leading to a very small displacement with damaged surface of the membrane. In fact, this is due to the penetration of unwanted materials coming from the unfiltered air, called contamination, which blocks the moving parts (the actuator and the membrane) and then causes its failure. The SEM image, presented in Fig. 8(b), shows contamination at the normally closed fluid port of a MEMS tested in this experiment.

B. Without air

In parallel to the previous experiment, a second test is performed and which consists in cycling one MEMS valve inside the SEM without air. In this experiment, the displacement of the membrane is calculated by using the SEM images. After 800000 cycles, the displacement decreased from 65 μm at the beginning of cycling to less than 50 μm. The SEM image taken after the end of cycling shows a good state of the membrane (Fig. 9). Therefore, this can be explained by a degradation at the actuator of the MEMS valve. Unfortunately, this degradation can not be analyzed since the actuator is obscured from view.

TABLE I
SUMMARY OF THE THREE PERFORMED ACCELERATED LIFETIME TESTS.

Test	Initial displacement	Performed cycles	Displacement at the end of the test	Membrane surface state
Using unfiltered air	65 μm	1 *million*	10 μm	*degraded*
Without air	65 μm	800000	50 μm	*good*
Using filtered air	65 μm	12 *million*	15 μm	*good*

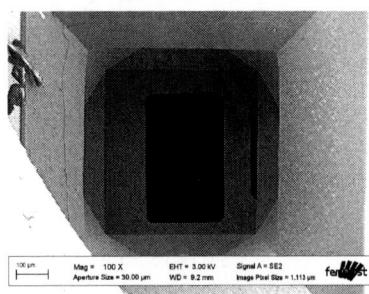

Fig. 9. SEM images of the membrane of the MEMS which cycled inside the SEM without air.

C. Using filtered air

The last experiment consists in cycling four MEMS valves using filtered air. Experiments remained running for more than three months and measurements were collected after approximately 90000 cycles.

All valves tested in this experiment performed 8 million cycles guaranteed by the manufacturer without a significant decrease of the displacement (less than 10% of displacement decrease for some MEMS) or membrane degradation. However, after 12 million cycles, the displacement average for the tested MEMS is about 15 μm (23% of the initial displacement). With a such displacement, the MEMS can be considered as out of service. According to the literature, for the electrothermal actuators, a failure is defined as the point at which the displacement decreases by 20% [12].

Despite the small displacement, membranes of all the tested MEMS still have a good surface state. This can be explained by a degradation at the actuator which is obscured from view.

A summary of the three performed accelerated lifetime tests is given in Table I.

V. CONCLUSION

Accelerated lifetime tests for MEMS valves electro-thermally actuated are presented in this paper. First, a brief description of the targeted MEMS is given. Then, the two experimental setup specially designed to perform accelerated lifetime tests are presented. The obtained results show that operating the MEMS with unfiltered air can cause the contamination at the actuator and the membrane, and then its early failure. Also, we noticed that MEMS operated with filtered air performed 8 million cycles without loss of performance. After that, displacement started decreasing to reach 15 μm at 12 million cycles with a good membrane surface state. This degradation can not be analyzed since the actuator is obscured

from view. In a future work, the evolution in time of the different parameters related to the MEMS (temperature, current, stiffness, response time, etc.) which are collected during tests will be analyzed in order to identify and understand failures at the actuator.

The ongoing experiments consist in performing new tests for new MEMS valves by changing different parameters such as the supply voltage and the operating frequency.

This work is only a step towards the implementation of Prognostics and Health Management (PHM) approaches for MEMS devices. Results of all accelerated lifetime tests will be used to apply a PHM approach for the targeted MEMS in order to predict its health state and estimate its time to failure.

ACKNOWLEDGMENT

This work has been supported by the "Région Franche-Comté" and the Labex ACTION project (contract ANR-11-LABX-0001-01).

REFERENCES

[1] K. Medjaher, H. Skima, and N. Zerhouni, "Condition assessment and fault prognostics of microelectromechanical systems," *Microelectronics Reliability*, vol. 54, pp. 143–151, 2014.
[2] W. Merlijn van Spengen, "MEMS reliability from a failure mechanisms perspective," *Microelectronics Reliability*, vol. 43, pp. 1049–1060, 2003.
[3] R. N. Dean, S. T. Castro, G. T. Flowers, G. Roth, A. Ahmed, A. S. Hodel, B. E. Grantham, D. A. Bittle, and J. P. Brunsch, "A characterization of the performance of a MEMS gyroscope in acoustically harsh environments," *Industrial Electronics, IEEE Transactions on*, vol. 58, no. 7, pp. 2591–2596, 2011.
[4] D. Tanner, "MEMS reliability: Where are we now?" *Microelectronics reliability*, vol. 49, pp. 937–940, 2009.
[5] M. McMahon and J. Jones, "A methodology for accelerated testing by mechanical actuation of MEMS devices," *Microelectronics Reliability*, vol. 52, pp. 1382–1388, 2012.
[6] Y. Huang, A. S. S. Vasan, R. Doraiswami, M. Osterman, and M. Pecht, "MEMS reliability review," *Device and Materials Reliability, IEEE Transactions on*, vol. 12, pp. 482–493, 2012.
[7] Y. Li and Z. Jiang, "An overview of reliability and failure mode analysis of microelectromechanical systems (MEMS)," in *Handbook of performability engineering*, 2008, pp. 953–966.
[8] A. L. Hartzell, M. G. Da Silva, and H. R. Shea, *MEMS reliability*. Springer, 2011, no. EPFL-BOOK-154162.
[9] J. Ruan, N. Nolhier, G. Papaioannou, D. Trémouilles, V. Puyal, C. Villeneuve, T. Idda, F. Coccetti, and R. Plana, "Accelerated lifetime test of RF-MEMS switches under ESD stress," *Microelectronics Reliability*, vol. 49, pp. 1256–1259, 2009.
[10] D. Microstaq, http://dmq-us.com/.
[11] J. A. Walraven, "Failure analysis issues in microelectromechanical systems (MEMS)," *Microelectronics Reliability*, vol. 45, pp. 1750–1757, 2005.
[12] R. A. Conant and R. S. Muller, "Cyclic fatigue testing of surface-micromachined thermal actuators," in *ASME Internation Mechanical Engineering Congress and Exposition*, 1998, pp. 15–20.

978-1-4673-8760-6/15 $31.00 © 2015 IEEE

Insights for Utilizing the Memristor as a Multi-bit Based Memory

Mostafa El-Khouly[1], Ahmed H. Madian[2], *Member IEEE,* and Hassan Mostafa[3], *Senior Member, IEEE*

[1, 2, 3] Electronics Department, German University in Cairo, Egypt,

[3] Electronics and Communications Engineering Department, Cairo University, Giza, Egypt,

[3] Center for Nano-Electronics and Devices, AUC and Zewail City of Science and Technology, Egypt.

Abstract— **The memristor, known as the fourth basic two-terminal circuit element, has attracted many research interests since HP labs were able to develop the device based on what L. Chua predicted theoretically. The memristor is a potential contender for the next-generation memory due to its distinctive characters, such as non-volatility, non-linearity, low-power consumption, good scalability and its ability to store multi-bit values. These electrical characteristics of memristors are mainly determined by the material characteristic and the fabrication process. However, the manufacturing of memristors is facing various challenges due to the difficulty of controlling its process variation, as it is fabricated at nano-scale geometry's size. These process variations lead to deviation in results from the theoretical results which lead to reduction in the reading yield. In this paper, we present the multi-bit memristor and we investigate the effect of the memristor size and the process variations effect on the multi-bit memristor and how we can maximize the reading yield.**

I. INTRODUCTION

Memristor was first theoretically predicted by L.Chua in 1971 [1]. It is considered the fourth fundamental circuit element, to complete the set of passive devices that previously included only resistor, capacitor, and inductor. In 2008, the first physical realization of memristor was demonstrated by HP Lab, in which the memristive effect was achieved by moving the doping front along TiO_2 thin-film device [3].

The memristor developed by HP labs is a TiO_2 thin film of length D, with two metal contacts at each end as shown in (Fig.1) [4]. The memristor can be split into two main parts: I) a high doped region with low resistance (R_{on}), and II) a low doped region with high resistance (R_{off}). Low doped region consists mostly of TiO_2; however the high doped region consists of TiO_{2-x}. The high doped region contains more oxygen vacancies which makes its resistance less than that of the low doped region. (R_{on}) and (R_{off}) notations represent the high doped and low doped regions respectively [6]. Moreover, the memristor has the ability to retain the state for a long time after the current has been switched OFF which means it is a non-volatile device.

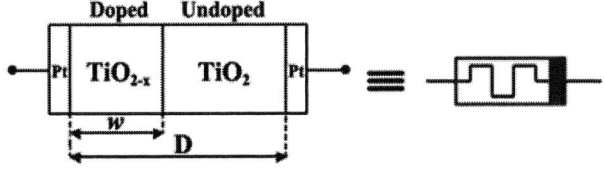

Fig. 1. TiO_2 Memristor of length D and the corresponding Memristor symbol [4].

Due to its ability to remember past charges, an intuitive utilization for it is to be used in memory design [6]. Memristor shows many promising characteristics as the next-generation data storage device [2], such as non-volatility, low power consumption, high performance, high density and excellent scalability [2] [7]. Manufacturing the memristor devices at nano-scale faces many problems which results in critical device parameter fluctuations, incurred by process variations, which is a critical concern affecting the device performance [2] [7]. The process variations sources are line-edge roughness (LER), oxide thickness fluctuation (OTF), and random discrete doping (RDD) [8].

In this paper we attempt to use the memristor as a multi-bit memory and give some insights on the effect of the process variations and dimensions on the reading yield, and how to maximize it through our method of threshold optimization.

II. PRELIMINARIES

A. Yield Concept

The process variations cause the actual results to deviate from its actual results thus we used the yield concept to give us an accurate description of the actual results in storing a value, the yield is the probability of writing a value and reading the same value. We can calculate the yield by calculating the probability of failure, which is the error in reading a different value than the written one, for example in (Fig. 2) is the yield for writing and reading '0' which is denoted by ($Y_0 = 1 - f_{1/0}$) where Y_0 is the probability of writing '0' and reading '0', $f_{1/0}$ is the probability of writing '0' and reading '1'.

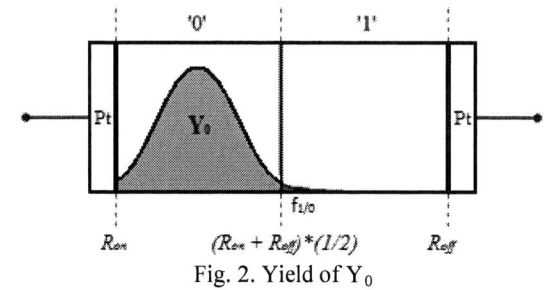

Fig. 2. Yield of Y_0

B. Multi-bit Memristor

The multi-bit memristor memory we can store multi-bit values such as ('00', '01', '10', '11') or ('000', '001', '010', '011', '100', '101', '110', '111') which allows for much better memory storage, however, the storing capacity is limited by the process variations. The multi-bit memristor is divided into four

978-1-4673-8760-6/15 $31.00 © 2015 IEEE

regions, the first regions, for writing and reading '00', is bounded between R_{on} and the first threshold $(R_{on} + R_{off})*(\frac{1}{4})$, the second region, for writing and reading '01', is bounded between the first threshold and the second threshold $(R_{on} + R_{off})*(\frac{1}{2})$, the third regions for writing and reading '10' is bounded between the second threshold and the third threshold $(R_{on} + R_{off})*(\frac{3}{4})$, the fourth regions for writing and reading '11' is bounded between the first threshold to R_{off}.

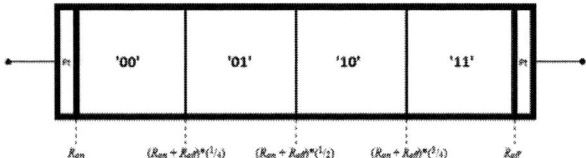

Fig. 3. Multi-bit Memristor storing 2 bit values

III. MEMRISTOR MODEL ANALYSIS

As an external voltage bias is applied across the device, the electric field repels positively charged oxygen vacancies in the doped layer into the pure TiO_2 layer thus changing the length w which is the width of the doped region (TiO_{2-x} layer). Hence, the device's total resistivity changes. If the doped region extends to the full length, that is $w/D=1$, the total resistivity of the device is dominated by the low resistivity region, with a value measured to be R_{on}. Likewise, when the undoped region extends to the full length D, $w/D=0$, the total resistance is denoted as R_{off}. The main parameters of the memristor are R_{on} and R_{off}, the device resistance is bounded between [8]:

$$R_{on} \leq R(w) \leq R_{off} \qquad (1)$$

According to [1][9][10], the memristor has memory effect since the device maintains its resistivity even if the power goes off. The mathematical model for memristive device resistance can be described as [8]:

$$R(w) = (R_{on}.w/D + R_{off}.(1 - w/D)) \qquad (2)$$

The state of the memristor changes according to the amount of flux injection. When the memristor state is controlled by the input flux across the cell, this input flux is given by:

$$\phi = \frac{\phi_D}{R_{off}^2}\left[R_{off}^2 - R_{on}^2\right] \qquad (3)$$

$$\phi_D = \frac{(\beta D)^2}{2\mu_v(\beta - 1)} \qquad (4)$$

Where β denoted the $R_{on} = \beta. R_{off}$ ratio, μ_v is the mobility. In the rest of paper, x and x' are used to represent the design value and the actual value under process variations for any given variable x, respectively [10].

For a given TiO_2 memristor, R'_{off} and R'_{on} are used to denote its actual highest and lowest total memristances, respectively [10].

$$R'_{off} = R_{off}.\eta(\mu R_{off} + \sigma R_{off}.E).(1 + \sigma_y.D) \qquad (5)$$

$$R'_{on} = R_{on}.(\mu R_{on} + \sigma R_{on}).E \qquad (6)$$

Here, two independent random numbers $E \sim N(0,1)$ and $D \sim N(0,1)$ are introduced. E represents the correlation between R'_{off} and R'_{on} due to the same geometry variation sources, D and σ_y represent the impact of RDD. It was found that the actual α', can be modeled as the product of the designed value α, where $\alpha=w/D$ and a coefficient η that represents the influence of process variations as [10]:

$$\alpha' = \eta.\alpha \qquad (7)$$

Here η can be expressed by [10]:

$$\eta = \frac{1}{(1 + \phi.\varepsilon_2 + \phi.\varepsilon_2(\omega_1.E + \omega_2.G)).(1 + \sigma_y.D)} \qquad (8)$$

To avoid overestimating the impact of geometry variations on α', a new random number $G \sim N(0, 1)$ is introduced to offset the impact of LER. ε_1 and ε_2 are two scalars extracted from the actual simulations performed by the device simulator. The coefficients w_1 and w_2 represent the weights of E and G, where $w_1^2+w_2^2=1$. By modeling R'_{on}, R'_{off} and α', the total memristance M' of a TiO2 memristor can be simply calculated by:

$$M'(\alpha) = R'_{on}.\alpha' + R'_{off}.(1 - \alpha') \qquad (9)$$

IV. SIMULATION RESULTS AND DISCUSSIONS

A. Effect of the process variation on the read yield

The process variations effect is different on each region due to change in resistance as the memristance (M) is given by:

$$M(\alpha) = R_{on}\alpha + R_{off}(1 - \alpha) \qquad (10)$$

As an example of how the process variations affect the yield we will assume the values (R_{on} =100 Ω and R_{off} =100 KΩ.), when $\alpha = 0$, $M = R_{off}$ .If α, due to process variation equals 0.001, M= 0.001* R_{on} +0.999* R_{off} ≈99.9 KΩ with an error of 0.1%. However, for $\alpha = 1$, $M = R_{on}$. If α, due to process variation equals 0.999, M= 0.999* R_{on} +0.001* R_{off} ≅ 200 Ω with an error of 100 %. That's why the failure percentage in the case of reading '0' is much lower than that in the case of reading '1'.

A.1. Single bit threshold yield

The region for writing and reading '0' is bounded between Ron to the threshold that is selected to be midway between R_{on} and R_{off}, typically, $(R_{on} + R_{off})$ / 2 [5].

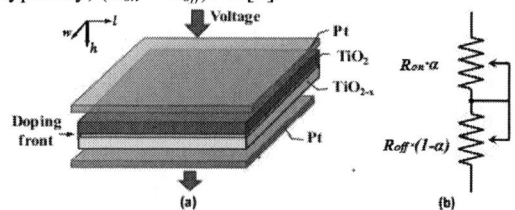

Fig. 4. TiO2 thin-film memristor. (a) Structure, and (b) equivalent circuit. [10].

978-1-4673-8760-6/15 $31.00 © 2015 IEEE

In Table 1 is the reading yield for a memristor with dimensions (h=3 nm, w=l=30 nm and w=l=100 nm), and it's shown how the reading yield Y_0 is lower than Y_1.

Table 1
Single-bit yield for memristor with height 3 nm

Memristor dimensions	Yield	
	Y_0	Y_1
w = l = 30 nm	0.9331	1.0000
w = l = 100 nm	0.9863	1.0000

A.2. Multi bit threshold yield

In the multi–bit memristor the effect of the process variations on the yield are much more apparent due to the fact we are storing more values, which leads to closer regions so the probability of failure is much higher. In Fig.6 is the yield for the different regions for a memristor with dimensions (w=l=100nm and h=10nm), from the figure we can observe how the yield for the multi-bit is much lower than the yield of the single-bit.

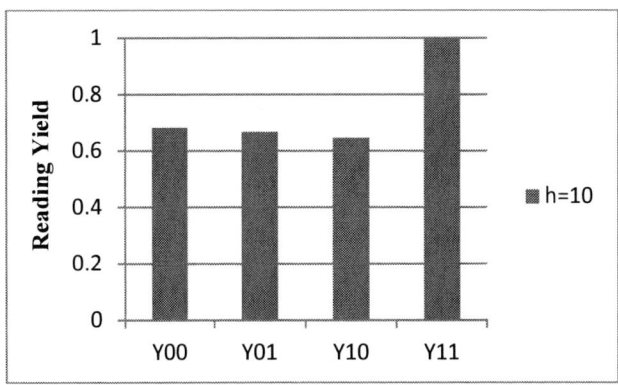

Fig. 5. Multi-bit Yield in case of (w=l=100nm and h=10nm)

B. Effect of the memristor dimensions on the read yield

We attempted to study the effect of varying the memristor dimensions on the reading yield. Thus we compared the yield of different sized memristors and we found that the yield and the size of the memristor are directly proportional, a comparison between memristors with different height (h) but with same width and length (w = l =100nm) for the single-bit memristor is shown in Fig.4 and a comparison between memristors with different height (h) but with same width and length (w=l=100nm) for the multi-bit memristor is shown in Fig.5.

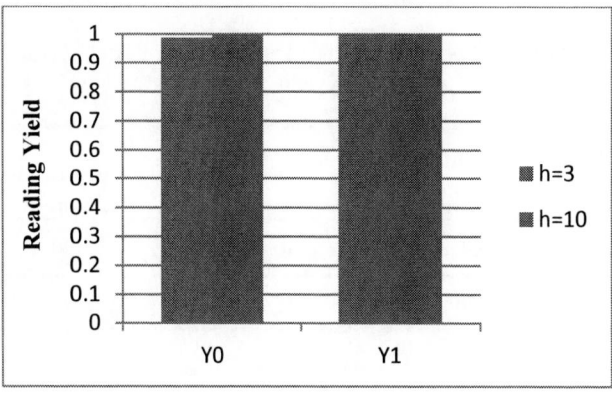

Fig. 6. Single-bit Reading Yield in case of (w=l=100nm)

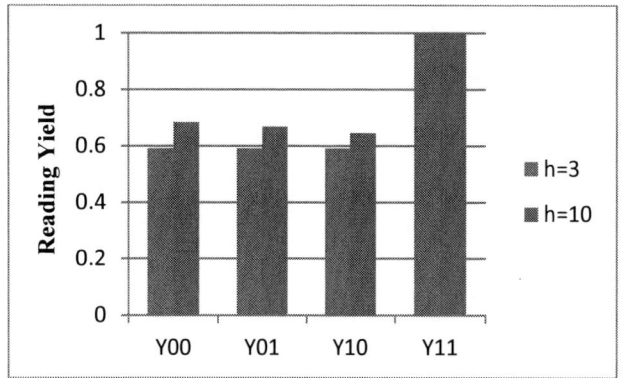

Fig. 7. Multi-bit Reading Yield in case of (w=l=100nm)

C. Yield optimization

Since that process variations effect is different on each region, due to change in resistance, means that choosing to divide the memristor into equal regions, to store the values, wasn't the best decision. In order to maximize the yield we attempted to vary the regions threshold finding the optimal regions for each value that results in maximum reading yield. Taking into consideration the fact that optimizing the region for one value changes the adjacent regions which could lead to lower yield, we calculated the overall yield, denoted by $Y_{overall}$, so that we find the optimal yield for a specific region that leads to a total increase of all the reading yields.

C.1. Single-bit

To optimize the reading yield, the threshold of the single-bit memristor that is $[(R_{on} + R_{off}) / 2]$ has been varied to find the optimal threshold that results in maximum reading yield, the overall yield for the single-bit memristor is ($Y_{overall} = Y_0 * Y_1$). A comparison between the fixed and the optimized threshold overall yield is given in Table 2 for a memristor with height (h=10nm) and with different width and length. As shown in Table 2, optimizing the threshold gives better overall reading yield resulting in 100% yield in some cases, which means there is no probability of failure after optimizing the regions' threshold.

Table 2
Comparison between fixed and optimized threshold overall yield

Memristor dimensions h=10 nm	Overall Yield	
	Fixed threshold	Optimized threshold
w = l = 50 nm	0.9924	1.0000
w = l = 100 nm	0.9980	1.0000

C.2. Multi-bit

In order to verify the method of optimizing the regions threshold, we calculated the overall yield ($Y_{overall} = Y_{00} * Y_{01} * Y_{10} * Y_{11}$) for fixed threshold, then we varied the thresholds of each region to find the optimal region for each value and calculated the overall yield for the optimized threshold. The results from the optimized threshold regions shows clearly the enhancement in the overall yield compared to the results from the fixed threshold region, as shown in Table 3 and Table 4 are a comparison between the overall yield for the fixed and optimized threshold for memristors with h=3 nm and h=10 nm respectively with different dimensions.

Table 3
Comparison between fixed and optimized threshold overall yield for the multi-bit memristor with height 3 nm

Memristor dimensions h=3 nm	Overall Yield	
	Fixed threshold	Optimized threshold
w = l = 50 nm	0.0469	0.8813
w = l = 70 nm	0.0631	0.9167
w = l = 100 nm	0.0842	0.9425

Table 4
Comparison between fixed and optimized threshold overall yield for the multi-bit memristor with height 10 nm

Memristor dimensions h=10 nm	Overall Yield	
	Fixed threshold	Optimized threshold
w = l = 50 nm	0.1474	0.9072
w = l = 70 nm	0.2304	0.9412
w = l = 100 nm	0.2929	0.9543

V. CONCLUSION

In this paper, the design of the multi-bit memristor-based memory is proposed. Also, we observed the effect of changing the memristor size and the process variations on the single-bit and the multi-bit memristor memory reading yield. The "Fast Statistical Model of TiO$_2$ Thin-Film Memristor" had been used for the memristor device simulations, which was done using MATLAB. The paper proposed a method on how to improve the reading yield through optimizing the regions threshold. Simulation results showed that optimization enhances the overall yield from 0.2304 to 0.9412 in some cases. The extension of this work is to try to increase the memory storage of the memristor by using it as a memory for higher multi-bit data.

VI. ACKNOWLEDGMENT

This research was partially funded by Cairo University, ITIDA, NTRA, NSERC, Zewail City of Science and Technology, AUC, the STDF, Intel, Mentor Graphics, SRC, ASRT and MCIT.

REFERENCES

[1] L. Chua. Memristor-the missing circuit element. IEEE Transaction on Circuit Theory, pages507–519, 1971.
[2] D. Niu, Y. Chen, C. Xu, and Y. Xie. Impact of process variations on emerging memristor. In Design Automation Conference (DAC), pages 877–882, 2010.
[3] Budhathoki, R.K., Sah, M.P., Adhikari, S.P., Hyongsuk Kim and Chua, L. Composite Behavior of Multiple Memristor Circuits. IEEE Transactions on Circuits and Systems I: Regular paper, vol 60, no.10, 2013.
[4] Y. Ho, G. M. Huang, and P. Li. Nonvolatile memristor memory: device characteristics and design implications. In International Conference on Computer-Aided Design, pages 485–490, 2009..
[5] Abdallah, M. and H. Mostafa. Yield Maximization of Memristor-Based Memory Arrays. IEEE, International Conference on Microelectronics (ICM'14), Doha, Qatar (In Press)
[6] H. Elgabra, I. Farhat, A. Al Hosani, D. Homouz, B. Mohammad. Mathematical Modeling of a Memristor Device. International Conference on Innovations in Information Technology (IIT), pages 156–161, 2012.
[7] M. Hu, H. Li, Y. Chen, X. Wang, and R. Pino. Geometry variations analysis of TiO$_2$ thin film and spintronic memristors. In Asia and South Pacific Design Automation Conference (ASP-DAC), pages 25–30, 2011.
[8] Y. Ho, G. Huang and P. Li. Dynamical Properties and Design Analysis for Nonvolatile Memristor Memories. 724 IEEE Transactions on Circuits and Systems—I:Regular papers, vol. 58, NO. 4, APRIL 2011.
[9] Y. Ho, G. M. Huang, and P. Li. Nonvolatile memristor memory: device characteristics and design implications. In International Conference on Computer-Aided Design, pages 485–490, 2009.
[10] M. Hu, H. Li, R. Piano. Fast Statistical Model of TiO$_2$ Thin-Film Memristor and Design Implication. The International Conference on Computer-Aided Design, pages 345-352, 2011.

Small-Signal Gain Calculations for a CMOS Analog Amplifier using Short Channel Equations Using nm Technology

Nedaa Al Tawalbeh, Mahmoud Hassan, Hazem W. Marar

Electrical Engineering Department
Princess Sumaya University for Technology

Amman, Jordan

Abstract— **CMOS analog inverter is a basic and simple gain stage for mobile applications. This paper suggests a simple way to calculate the gain of a push- pull inverter which consists only of a one nMOS and one pMOS transistors without additional resistors. This method is based on finding the following two relations for nMOS and Pmos transistors: gm/Ids versus VGS and the channel modulation coefficient λ versus VDS. Then the short channel equations are developed to find gm (transconductance) and gds (output conductance) for both nMOS and pMOS transistors**

Keywords-CMOS analog amplifier; short channel equation; 28nm technology channel length modulation

I. INTRODUCTION

Analog design of integrated circuits is becoming uneasy problem due to good number of the nano technology nonlinearities such as: threshold voltage reductions and leakage current increase when channel length decreases. These changes are due to: drain induced barrier lowering DIBL, carrier mobility degradation and velocity saturation. [1] - [4].

Analog circuits will continue to be essential item in all applications that interface with the real world such as sensors, loud speakers and microphones which are usually combined with mixed signal circuits.

The nature of nano technology has introduced complex issues which need bright ideas to achieve practical analog circuit designs. The short channel equations [2] and other techniques [4] - [16] will be very promising in future analog circuit design for nano technology.

The CMOS analog inverter could be up to three amplifier configurations [17]. These configurations are: active pMOS load inverter, current source load inverter and push-pull

inverter. Each consists only of one nMOS and one pMOStransistors. These simple configurations could give hundreds of small signal voltage gain at MHz range in addition to extremely small size and low power consumption.

Push-pull inverter configuration has been selected for this paper. Figure (1) shows this simple amplifier driven by input signal Vin and 1 V power supply. The output swing of this inverter can be from VDD to ground. Both transistors operate in the saturation region and a large voltage gain can be achieved. Its small signalvoltage gain is given [17] by the following equation.

$$\frac{\text{Vout}}{\text{Vin}} = -\frac{(g_{mn}+g_{mp})}{g_{dsn}+g_{dsp}}(1)$$

whereg$_{mn}$ and g$_{mp}$ are transistors transconductances for the nMOS and the pMOS respectively. Alsog$_{dsn}$ and g$_{dsp}$ are the output conductances for the nMOS are pMOS transistors.

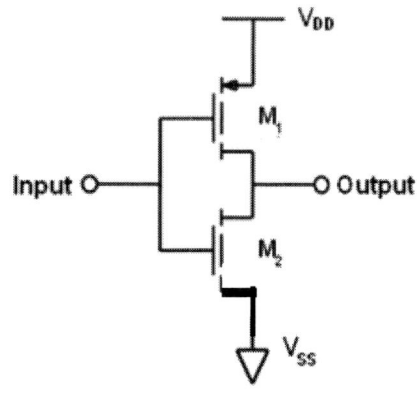

Fig.1: CMOS Amplifier

In the following sections we will use emerging short channel equations [2] to develop equations for gm and gds. The small signal voltage gain will be derived using gm, ID (drain current) and λ (the channel length modulation coefficient). Later the circuit diagrams for measuring above parameters will be introduced, following an example,conclusion will end this paper.

II. G_M AND G_{DS} DERIVATION FROM SHORT CHANNEL EQUATIONS

The emerging current- voltage equations of the short channel nMOS and pMOS transistors operating in the saturation region [2] are:

$$I_{DSn(sat)} = W.v_{satn}.C_{ox} \frac{(V_{GS} - V_T)^2}{(V_{GS} - V_T) + E_C L}.(1 + \lambda V_{DS}) \quad (2)$$

$$I_{DSp(sat)} = W.v_{satp}.C_{ox} \frac{(V_{SG} - V_T)^2}{(V_{SG} - |V_T|) + E_C L}.(1 + \lambda V_{DS}) \quad (3)$$

where
W: channel width. nm
v_{sat}: Saturation drift velocity in the channel.cm/s
C_{ox}: gate capacitance per unit area. F/m
V_{GS}: gate source voltage. V
V_T: thresholds voltage. V
λ:channel length modulation coefficient . V^{-1}
V_{DS}: drain source voltage. V
E_C: channel electric field.V/cm
L: channel length.nm

Define,

$$g_m(sat) = \frac{\partial I_{D(sat)}}{\partial V_{GS}} \quad (4)$$

Then from Equation (2)

$$g_{mn(sat)} = W.v_{satn}.C_{ox} \frac{(V_{GS} - V_T)(V_{GS} - V_T + 2E_C L)(1 + \lambda V_{DS})}{(V_{GS} - V_T + E_C L)^2} \quad (5)$$

Also, define

$$g_{ds}(sat) = \frac{\partial I_{D(sat)}}{\partial V_{DS}} \quad (6)$$

From Equation (2),

$$g_{dsn(sat)} = W.v_{sat}.C_{ox} \frac{(V_{GS} - V_T)^2 \lambda}{(V_{GS} - V_T + E_C L)} \quad (7)$$

Using Equation (2), putg_{dsn} (sat) in term of I_{Dn}(sat)

$$g_{dsn}(sat) = \frac{I_{Dn(sat)}}{1 + \lambda V_{DS}}.\lambda \quad (8)$$

Using Equation (2), (5) and (7), g_{dsn}(sat) can be found in term of g_{mn}(sat) as follows:

$$\frac{g_{dsn(sat)}}{g_{mn(sat)}} = \frac{(V_{GS} - V_T)(V_{GS} - V_T + E_C L)}{(V_{GS} - V_T + 2E_C L)(1 + \lambda V_{DS})}.\lambda \quad (9)$$

Since $E_c L > V_{GS} - V_T$,then

$$g_{dsn(sat)} \cong \frac{g_{mn(sat)}(V_{GS} - V_T)}{2(1 + \lambda V_{DS})}.\lambda \quad (10)$$

Similar equations can be found for g_{mp}(sat) and g_{dsp}(sat)
The small signal voltage gain of Equation (1) can be calculated knowing g_m, ($V_{GS} - V_T$), V_{DS} and λ.

Also using Equation (2) and Equation (5), the following relationcan be found between g_m and I_D.

$$g_m = \frac{I_D(V_{GS} - V_T + 2E_C L)}{(V_{GS} - V_T + E_C L)(V_{GS} - V_T)} \quad (11)$$

Since $E_c L > V_{GS} - V_T$,then

$$g_m \approx \frac{2I_D}{(V_{GS} - V_T)} \quad (12)$$

Equation (1) can be rewritten as follows

$$\frac{Vout}{Vin} = -\frac{g_{mn}/I_D + g_{mp}/I_D}{g_{dsn}/I_D + g_{dsp}/I_D} \quad (13)$$

III. CIRCUITS TO MEASURE ABOVE PARAMETERS

Synopsys simulation software together with the following circuits have been used to measure the following needed parameters to calculate the small signal voltage gain of the circuit of Figure (1).

A. Measuring V_{TO}

Circuit of Figure (2)a is used to measure V_{TO} as explained in Figure (2)b [2].

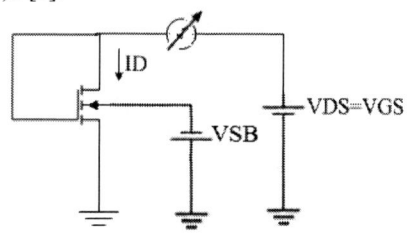

Fig. 2a: Test Circuit Arrangement

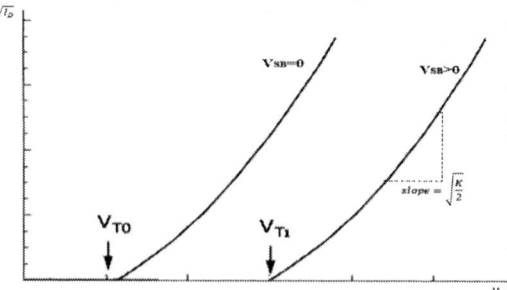

Fig.2.b: Measured data for experimental determination of the parameter V_{To}

Fig. 4: gm/ID Simulation

B. Measuring λ

Circuit of Figure (3)a is used to measure λ as shown in Figure (3) b and using the following equation [2].

$$\frac{I_{D1}}{I_{D2}} = \frac{1 + \lambda V_{DS2}}{1 + \lambda V_{DS1}} \qquad (14)$$

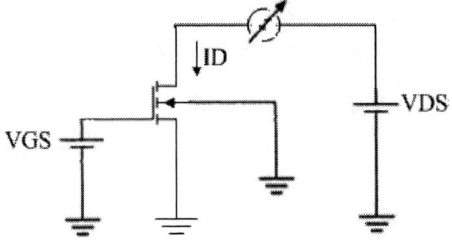

Fig.3a: Test Circuit Arrangement

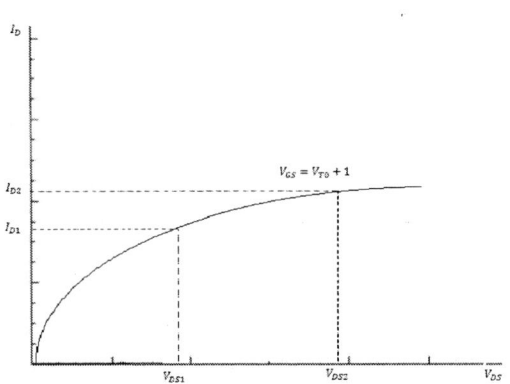

Fig. 3b: Measured data for experimental determination of the channel length modulation coefficient

C. Measuring g_m/I_D

Circuit of Figure (4) [5] is used to measure g_m/I_D. First, Synopsys software has been used to simulate the circuit of Figure (4) in order to have all currents and voltages plots in the waveform window. From the I_D curve we can take the derivation of I_D versus V_{GS} to obtain g_m curve, then simply g_m curve is divided by I_D curve to get g_m/I_D curve.

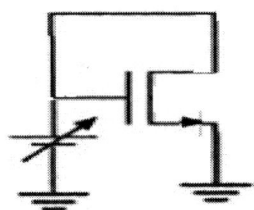

From shown measurements andFigures 5- 10and also similar pMOS Figures(not shown)we can calculate the small signal voltage gain.
3

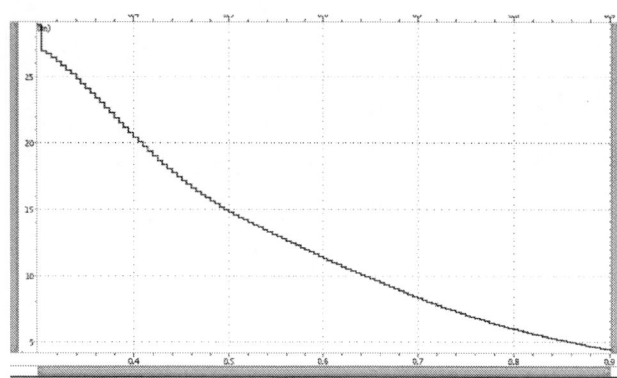

Fig. 5: gm/ID with V_{GS} curve for the nMOS Transistor

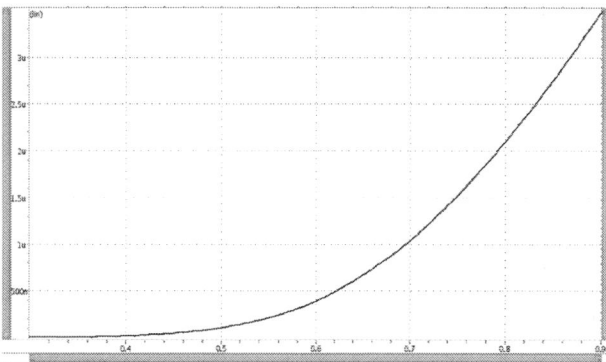

Fig. 6 : The normalized current (ID/ (W/L)) vs Vgs for nMOS Transistor

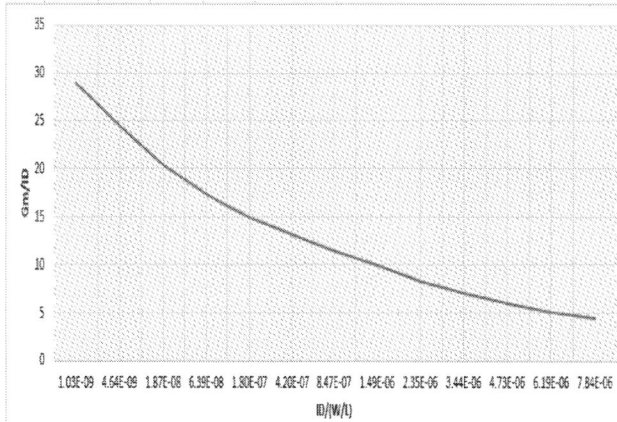

Fig.7:The gm/id vs The Normalized current ID/(W/L)for nMOSTransistor

978-1-4673-8760-6/15 $31.00 © 2015 IEEE   85

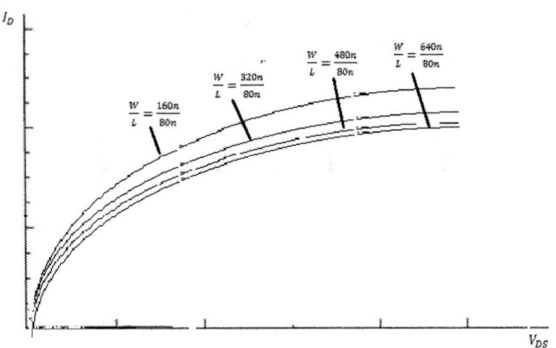

Fig. 8:ID with VDS curve for different W/L ratios

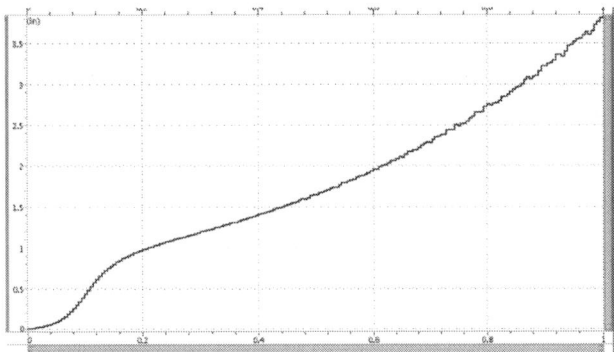

Fig. 9: VA with drain biasV_{DS} when vgs=0.45 (nMOS)

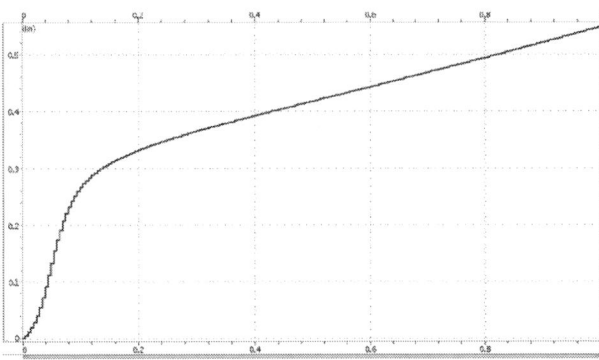

Fig.10: VA with drain biasV_{DS} when vgs=0.45 (pMOS)

IV. CONCLUSION

Small signal voltage gain for a simple CMOS inverter can be calculated using above derived equations based on the emerging short channel MOSFET current- voltage equations and above shown measurements and graphical relations.

The result of a design example confirms that calculated value is very close to simulated value.

Above work will be developed for analog circuits design to more complicated CMOS circuits.

REFERENCES

[1] L.L Lewyn, T.Ytterdal, C.Wulff and K. Martin, "Analog Circuit Design analog in Nanoscale CMOS Technologies" Proceedings of the IEEE, vol.97, no ma10, pp. 1687-1714, October 2009.

[2] S. Mo Kang, Y.Leblebici and C.Kim, CMOS Digital Integrated Circuits Analysis& Design 4th ed., McGraw-Hill 2015.

[3] M.BoLin, Introduction to VLSI Systems A Logic, Circuit, and System Prespective", CRC Press 2012.

[4] S. Pandit, C. Mandel and A. Patra, Nano- Scale CMOS Analog Circuits Models and CAD Technologies for High-Level Design, CRC Press 2014.

[5] P.Jespers, The gm/ID Methodology, a sizing tool for low-voltage analog CMOS Circuits, Springer 2010.

[6] S. Paul, A. Dana and S.Pandit, "An Improved gm/ID Methodology for Ultra-Low Power Nano- Scale CMOS OTA Design" M.S. Gaur et al (Eds): VDAT CCIS 382, PP.128-137, 2013.

[7] R.L. Oliveira Pinto, and F. Maloberti, "Novel Design Methodology for Short-Channel MOSFET Analog Circuits" Proceeding of the 3rd International workshop on System-on- chip for Real-Time Applications, 2003.

[8] A.Tiwari, Design of MOS Amplifies Using gm/ID Methodology, 2014.

[9] M.Goswami and S.Kundu, "Design and Analysis of Semi-Empirical Model Parameter for Short-Channel CMOS Devices, International Journal of Soft Company and Engineering (IJSCE)" vol.4, issue.3, 2014.

[10] K.Gupta, D.Anverkar, and V.Venkateswarlu, "Device Characterization of Short Channel Devices and its Impact on CMOS circuit Design" International Journal of VLSI design of Communication Systems (VLSICS), Vol.3, no.5,2012.

[11] J.Mukhopadhyay, and S.Pandit, "Modeling and Design of a Nano Scale CMOS Inverter for Symmetic Switching Characteristics" VLSI Design, Vol.2012, 13pages, 2012.

[12] A. Fahim, "Challenges in Low-Power Analog Circuit Design for sub-28nm CMOS Technologies" ISLPED,14 CA,USA,2014.

[13] K.Jung, D. Kim, M.Song, W.Jung, and H.IM,"Study of Small Signal Characteristics of Short Channel MOSFETs by Using ∝ -Power Model", Journal of Korean Physical Society, vol.45,pp S924-S927,2004.

[14] R.Baruah, "Design of Low Power Low Voltage CMOS Opamp", International Journal of VLSI design& Communication Systems (VLSICS), Vol.1, no.1, 2010.

[15] H.T,P. Hoang and T.Quan, "A Novel design of Low-Power, high speed OTA in 50nm-CMOS technology", icdrec.edu.vn/45/doc.2012/APP/APP13.pdf, 2012.

[16] O. Ushie, and M.Abbod, "Intelligent Optimization Methods for Analogue Electronic Circuits: GA and PSO Case Study," International Conference on Machine Learning, Electrical and Mechanical Engineering (CMLEME 2014), 2014.

[17] P. Allen, and D. Holberg, CMOS Analog Circuit Design, Oxford University Press, 2009.

Pulse Interval Modulation for Biomedical Wireless Sensors

Decio Renno de Mendonca Faria, Robson Luiz Moreno, Tales Cleber Pimenta
Universidade Federal de Itajuba
Av. BPS, 1303 Itajuba – MG 37500-903

Abstract—**This paper presents a circuit capable of reducing the power used to transmit signals from implanted biomedical wireless sensors. Nowadays, Quasi-Digital sensors are an interesting option that satisfies key performances such as low voltage, low power, low cost, good accuracy and small die area [1]. Those sensors send out analog information using digital signals, either by changing their frequency, period or duty-cycle. In all of them, energy is consumed during the transmitter operation. We propose a power savings by reducing the period that the transmitter is turned on. The proposed circuit changes the square signal of a voltage to frequency converter (VFC) into shorter pulses, but keeping the frequency of the original signal. It generates a Pulse Interval Modulation – PIM signal to be transmitted. We implemented the circuit layout in IBM 130nm CMOS technology. Simulation results demonstrated the circuit work as expected, and the savings may reach over 80 percent.**

Keywords—Wireless Sensor Networks, Voltage to frequency converter, PIM modulation.

I. INTRODUCTION

The technology advancements in the last years, especially in microelectronic, are turning possible the implant of sensors that would be able to measure parameters such as temperature, pressure, heartbeat, glucose and others. The data from the sensors are converted from analog to digital (A/D) and send to a microcontroller outside the body. As an example, Fig. 1 shows the block diagram of a conventional System-on-Chip for Body Sensor Networks [2].

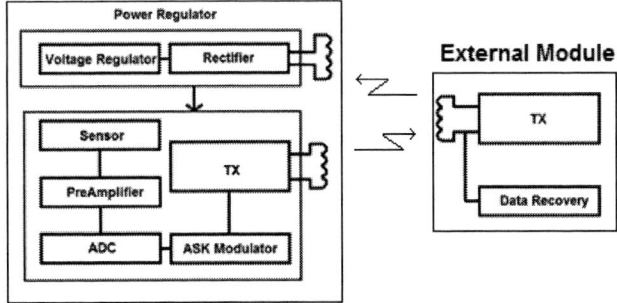

Fig. 1. – Schematic diagram of a system-on-chip architecture.

In this example, the external module consists of a transmitter unit, a coil, and a receiver unit, named Data Recovery. The external module transfers energy to the implanted device via inductive coupling. The implanted circuit

has a transducer sensor and a preamplifier that produce a voltage proportional to the measured parameter (glucose, pressure, Ph and temperature). In many systems, the analogic to digital converter (ADC) employs a microcontroller unit – MCU to convert the data to be transmitted. A simple amplitude shift keying (ASK) modulator and a low power transmitter is used to send the data out via inductive link.

An alternative to ADC technology is the use of voltage to frequency conversion (VFC). The output signal of a VFC can be directly converted to its digital value at the microcontroller in the external module. It is performed by counting the pulses, using its internal timer as reference [3]. As an example of VFC, Fig. 2 shows the basic block diagram of a Quasi-Digital Sensor [4].

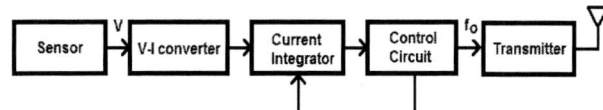

Fig. 2. – Block diagram of a basic VFC circuit.

It can be observed the nonexistence of both ADC and MCU. This is a great advantage since it reflects reduction of size, complexity, and power consumption. In this circuit, the voltage from the sensor is linearly converted into current. The current is then used to charge/discharge a capacitor. The control circuit is a voltage window comparator (VWC). In order to understand its operation, suppose initially the current integrator in charge mode. When the voltage at the capacitor reaches the upper limit, the VWC changes the current integrator circuit into the discharge mode. When the voltage in the capacitor reaches the lower limit, the VWC changes the current integrator again to the charge mode. As a result, the control circuit provides a square wave of a frequency proportional to the voltage of the sensor [5], as indicated in Fig. 3.

II. THE PROPOSED CIRCUIT

Both ADC and VFC systems have digital outputs, meaning zeros and ones. In order to assure low power consumption, miniaturized biomedical devices make use of simple binary modulations, such as amplitude shift keying (ASK), binary phase shift keying (BPSK) and frequency shift keying (FSK). In order to maximize the transmitter overall efficiency, the on-off keying modulation (OOK), the simplest form of ASK can be used. It represents digital data as presence/absence of a carrier wave.

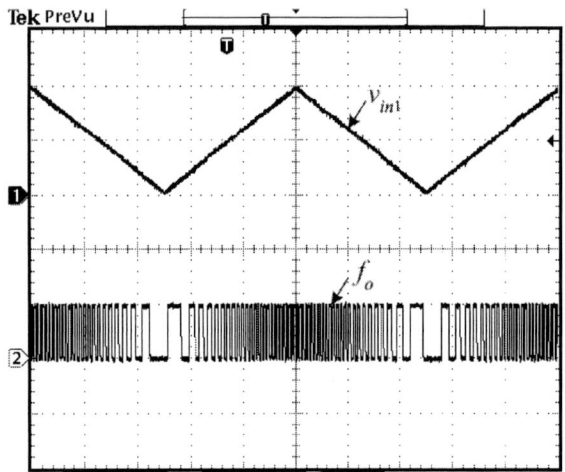

Fig. 3. – A typical output signal of a VFC circuit.

In an OOK system, the transmitter requires power only to transmit the binary 1. The transmitter is off for the binary 0, therefore it requires less power than FSK and BPSK modulations [6].

Nevertheless, when an OOK transmitter is used to transmit VFC information, there is no need to keep the transmitter ON during the whole time of binary 1. This is possible since the information is extracted from the period of the signal [7] (it can be acquired by tracking the rising or falling edge of the signal). Therefore, this paper proposes a circuit that limits the time of the state ON during a transmission. It consists of an equivalent RC circuit in CMOS technology that acts as an edge detector. Fig. 4 shows the pulse interval modulation (PIM) circuit used to validate the proposed approach.

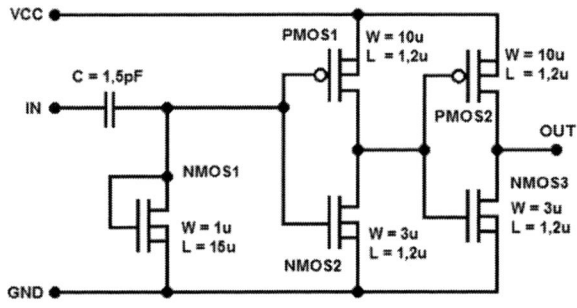

Fig. 4. – A schematic view of the PIM converter.

The capacitor C and the transistor NMOS1 act as an RC circuit that determines the maximum ON time of the transmitter. This time can be adjusted changing the parameters L and W of NMOS1. Initially, PMOS1 is turned on and NMOS2 is turned off, therefore the output is in OFF state due the inverter circuit formed by PMOS2 and NMOS3. When the input change from OFF state to ON, the capacitor is charged and a positive voltage reaches the gate of PMOS1 and NMOS2, thus changing the output state to high. At this time NMOS1, acting as a resistor, discharges the capacitor and after a given time the output returns to the OFF state. When the input state changes from the state ON to OFF, nothing happens.

III. SIMULATIONS

In order to test the circuit, a 10pF capacitor was used as output load, and a symmetrical square signal was applied at the input. Fig. 5 shows the input (dotted signal) and output (full line) signals from the proposed circuit.

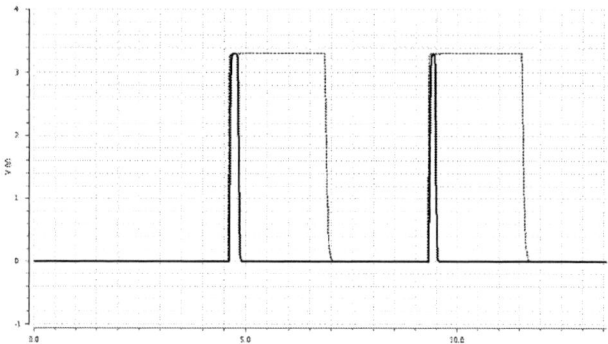

Fig. 5. – Input and output wave signal of the PIM converter.

As it can be observed, the square wave with a short ON state corresponds to the circuit output. This short ON state time is constant and is adjustable by NMOS1 and C parameters, as described.

In order to avoid excessive pulse width shortening, and thus assuring threshold detection at the receiver, the pulse width should be equal to the minimum period of the square wave at the output of the VFC. This procedure will guarantee that only the software must be modified at the receiver. Shorter values can be used for newer systems. A VFC operating in the range of 200 KHz to 700 KHz was used as a test, and the correlation between the frequency and the output of the PIM modulator is presented in Table 1.

TABLE 1 – VFC FREQUENCY VERSUS INTERVAL BETWEEN PULSES.

VFC Frequency	PIM Interval between pulses
200KHz	5.00us
400KHz	2.50us
600KHz	1.66us
800KHz	1.25us

The interval between pulses was directly proportional to the VFC frequency as expected, since it is associated to the rise edge of the signal at the VFC output.

IV. THE POWER CONSUMPTION ANALYSIS

In order to evaluate the energy savings provided by the PIM modulator, the power consumption of the ASK transmitter, when transmitting data provided by a VFC, was analyzed. The PIM pulse length was adjusted to 0.2us. The energy saving is not a fixed value. At 200 KHz, the transmitter is turned on for 2.5us per cycle, as indicated in Fig 6. Since the PIM pulse has a width of 0.2us, the transmitter will be active for 0.2/2.5 = 0.08 or 8% of the time. It means that it requires less than 10% of the energy.

978-1-4673-8760-6/15 $31.00 © 2015 IEEE

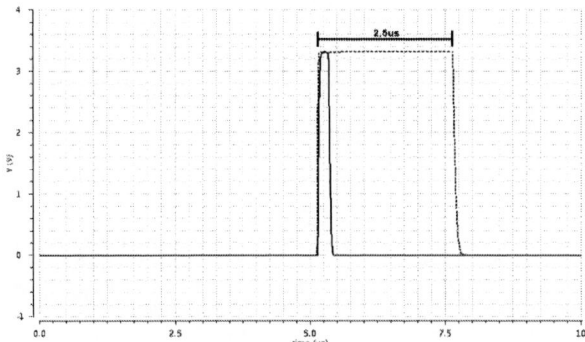

Fig. 6. – ASK transmission at 200 KHz.

At 800 KHz, the PIM pulse will have the same width of 0,2us and the transmitter will be active for 0.2/0.62 = 0.323 or 32.3% of the total time used by the VFC, as indicated in Fig. 7. That is the worst case in our tests.

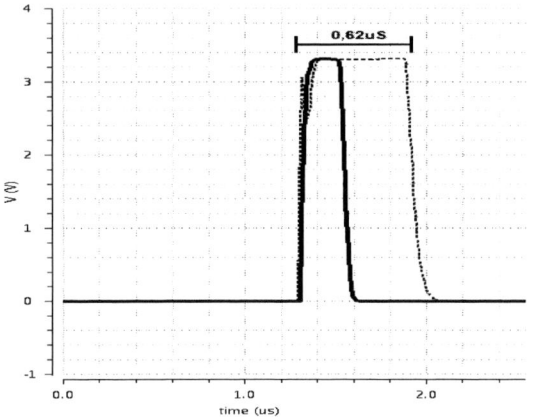

Fig. 7. – ASK transmission at 800 KHz.

Therefore, even the worst case represents an economy of more than 67%. We can derive an expression to obtain the energy saving versus the VFC frequency. The period, TVFC is the inverse of the frequency, FVFC. *A.*

$$f_{VFC} = \frac{1}{T_{VFC}} \qquad (1)$$

The time in which the transmitter remains ON, THalf wave, is half the period of the VFC wave, TVFC.

$$T_{Half\,wave} = \frac{T_{VFC}}{2} \qquad (2)$$

The ratio between the total power used in the VFC transmission and the PIM transmission is given by (3).

$$Ratio = \frac{T_{pulse}}{T_{Half\,wave}} \qquad (3)$$

where Tpulse is the fixed period of the pulse on the PIM modulation. Therefore, the ratio is given as (4).

$$Ratio = 2 \cdot T_{pulse} \cdot f_{VFC} \qquad (4)$$

Expression (4) is a linear equation that express the relation between the VFC output frequency and the ratio of time, which is proportional to the total energy used in the PIM modulation, as shown in Fig. 8.

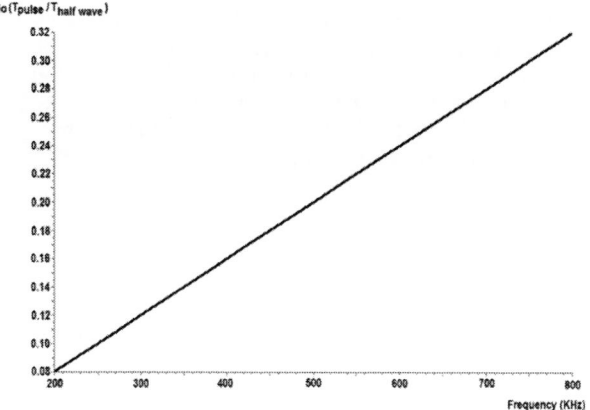

Fig. 8. - Ratio between power used in PIM and VFC.

The PIM modulator was implemented in 130nm CMOS technology. Fig. 9 shows the circuit that takes an area of just $37\mu m$ x $63\mu m$ (0.00233mm2).

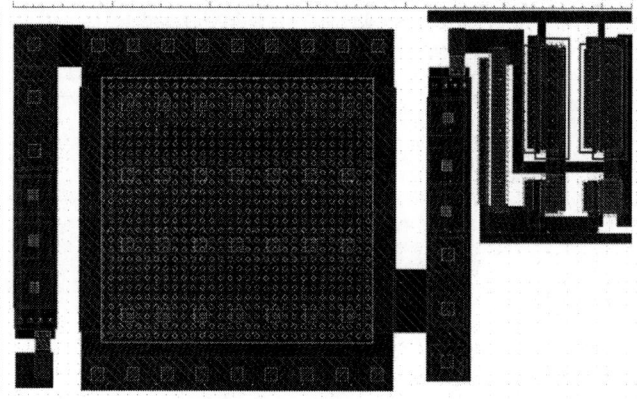

Fig. 9. - PIM modulator layout.

V. RESULTS ANALYSIS

In order to test the PIM modulator circuit, a complete VFC converter was implemented in IBM 130nm CMOS technology using Cadence Virtuoso 6.1.5. The implemented RC circuit using C and NMOS1, as shown in Fig. 4, was initially adjusted to operate with a 5pF capacitor and it worked perfectly. Nevertheless, the implementation using MIM capacitor takes large silicon area. Therefore, it is desirable to use small capacitances and it is the case of this final layout. Another important point to be observed is the transistor sizing of the two inverters PMOS1-NMOS2 and PMOS2-NMOS3, on Fig. 4. They were carefully adjusted to minimize distortions and they take small silicon area. Inadequate W/L ratio or very small transistors may suffer from insufficient current capacity or signal asymmetry. Therefore, although the apparent circuit simplicity, care must be taken into account into its design.

VI. CONCLUSIONS

Since power consumption is one of the most important features of implanted circuits, VFC is a good choice. The absence of a continuous carrier wave reduces power consumption on transmitters using ASK modulation since the transmitter does not work continuously. This paper demonstrated that we can further reduce the time that the transmitter is turned on by using PIM modulation. Therefore, by combining VFC circuit, PIM modulation and ASK transmission we can reach the purpose of sending analog information using low energy budget.

The receiver is the same that is used in regular VFC systems. At the receiver, simple counters track the positive or negative edge of the signal and they can be implemented using discrete components, microcontrollers, field programmable gate array (FPGA), complex programmable logic device (CLPD), and even dedicated IC [8][9]. By keeping the pulse width equal to the minimum period of the square wave at the output of the VFC as suggested, the noise immunity is exactly the same as the worst case of regular VFC.

Additionally, VFC is simple and requires a small number of components, thus the final circuit requires a smaller silicon area.

ACKNOWLEDGMENTS

The authors acknowledge CAPES, CNPq and FAPEMIG for their financial support.

REFERENCES

[1] C. Azcona, B. Calvo, N. Medrano and S. Celma, "CMOS quasi-digital temperature sensor for battery operated systems", Electronics Letters, October 2013, vol. 49.

[2] Kui Zou, Xiuhan Li, Jinbin Hu, Haixia Zhang, "Electronic Design for an Implantable Wireless Power and Data Transmission System", 9th IEEE International Conference on Solid-State and Integrated Circuit Technology – ICSICT, 2008.

[3] Deepthi P, Manju Mohan, S Raghavan, "PSoC Embedded Design for Ultra Low Digital Frequency Meter", International Conference on Circuit, Power and Computing Technologies, ICCPCT, March 2015.

[4] B. Calvo, N. Medrano, and S. Celma, "A Full-Scale CMOS Voltage-to-Frequency Converter for WSN Signal Conditioning", The IEEE International Symposium on Circuits and System – ISCAS, May 30 to June 2, 2010.

[5] Teerawat Thepmanee and Amphawan Julsereewong, "Simple Low-Cost Voltage-Controlled Oscillator", International Conference on Control, Automation and System, October 2010.

[6] Pawel Turcza, Janusz Mlynarczyk, "Design of Wide Band OOK Transmitter for Biomedical Applications", 20TH International MIXDES Conference, June 2013.

[7] Benzhou Jin, Sheng Zhang, Xuan Zhang, Xiaokang Lin, "Pulse Interval Modulation for UWB Communications with Energy Detection", IEEE 3th International Conference on Communication Software and Networks – ICCSN, May 2011.

[8] Setyawan P. Sakti, Dual Channel High Precision 26 Bit Frequency Counter Using CPLD XC95108XL for QCM Sensor System, International Journal of Information and Electronics Engineering, Vol. 4, No. 3, May 2014.

[9] Analog Devices, V/F Converter, Ask The Application Engineer – 3, http : // www.analog.com/library/analogDialogue/Anniversary/3.html in 27/10/2015.

Robust Low Power NB PHY Baseband Transceiver for IEEE 802.15.6 WBAN

Awny M. El-Mohandes, Ahmed Shalaby, Mohammed S. Sayed

Electronics and Communications Engineering Department
Egypt-Japan University of Science and Technology
P.O.Box 179, New Borg El-Arab City, Alexandria 21934, Egypt
{awny.elmohandes, ahmed.shalaby, mohammed.sayed}@ejust.edu.eg

Abstract—The lack of healthcare services in outback areas raises the need for efficient remote health monitoring systems. IEEE 802.15.6 standard defines a new wireless communication standard for healthcare systems to mitigate this problem. Power consumption is the most crucial dilemma in such system. This paper proposes low power architecture of Narrowband Physical Layer transceiver for Wireless Body Area Network based on the IEEE 802.15.6 standard. A reconfigurable architecture was designed to select between DBPSK and DQPSK. Thus variable data rate up to 971.4 kbps is supported with variable packet sizes. The proposed transceiver was implemented and verified by MATLAB. The design was synthesized and implemented on both FPGA and ASIC. The implemented architecture consumes 162 μW, operates at a clock frequency of 6 MHz, and meets all the standard requirements.

Keywords—WBAN; NB-PHY; Transceiver; low power Implementation.

I. INTRODUCTION

The WHO Constitution enshrines the health right as a fundamental right for every human being. The health right means that people have the right to enjoy physical and mental health. While in many counties, people in outback areas have no healthcare services due to the lack of healthcare centers. Today, by dint of Wireless Body Area Network (WBAN), people in these areas can have part of their healthcare services provided remotely without the need of local hospitals. WBAN supports full mobility e-healthcare; so any person can do his daily activities while his health is monitored remotely. WBAN is a wireless sensor network that is placed in or around the human body to collect real-time body physiological signs. WBAN based on IEEE 802.15.6-2012 standard has a lot of unique requirements [1] which make it more specialized in medical applications than other types of IEEE 802.15 standards like Bluetooth and ZigBee. These requirements put conclusive restrictions on reliability, Quality of Service (QoS), low power, data rate, and non-interference [1]-[2].

IEEE 802.15.6-2012 standard defines one common Medium Access Control (MAC) layer that serves three different types of physical (PHY) layers: (1) NarrowBand (NB) layer, (2) Ultra-WideBand (UWB) layer, and (3) Human Body Communications (HBC) layer [2]. For instance, the frequency band for NB-PHY layer is 2.4 GHz, most popular band worldwide, since it is unlicensed industrial scientific medical (ISM) band and has a large usable bandwidth.

Due to the low power requirement [1]-[3] of WBAN and the importance of its application, the need of total hardware system implementation arises. To the best of our knowledge, [4] is the only published work that proposed a hardware implementation of baseband NB-PHY transceiver in 2.4 GHz band. The proposed system in [4] was implemented and tested using Xilinx Virtex 6 (ML605) FPGA Kit. The main work in [4] deals with the digital part in NB-PHY layer, but it was a primitive implementation for idle transceiver missing- symbol mapper, pulse shaping filter, and coarse and fine timing synchronization. Although these blocks are considered from the main parts of the system [1], as they enhance the transmitted spectrum and timing synchronization efficiency, and the overall performance.

This paper proposes low power ASIC transceiver based on IEEE 802.15.6-2012 NB-PHY 2.4 GHz standard with the following specifications: (1) reconfigurable architecture for symbol mapper and demapper, (2) Square Root Raised Cosine (SRRC) pulse shaping, and (3) coarse and fine timing synchronization. Our design targets to enhance the spectral and timing synchronization efficiency and overall transceiver performance, with low power consumption.

The rest of this paper is organized into four sections. Section II presents the IEEE NB-PHY layer specifications, while section III describes our proposed NB-PHY modules and design. In addition, Section IV presents and discusses our results, and Section V concludes the paper.

II. WBAN NB-PHY LAYER

The NB-PHY layer is the link between MAC layer and air channel. PHY layer works to activate and deactivate the radio transceiver, clears the channel assessment within the current channel, and transmits and receives binary data to/from the channel [1]. A 2.4 GHz NB-PHY layer is proposed to fit medical applications, since it has a large bandwidth that supports a quite large data rate transmission, in addition, signals are less degraded by the human body at this frequency band [2]. The NB-PHY layer transmits binary data using certain frame format, which is called PHY layer Protocol Data Unit (PPDU). The PPDU contains three main parts, which are ordered during transmission as: (1) The PHY Layer Convergence Protocol (PLCP) preamble, (2) The PLCP

978-1-4673-8760-6/15 $31.00 © 2015 IEEE

header, and (3) The PHY layer Service Data Unit (PSDU). The receiver uses the PLCP preamble in timing synchronization and carrier offset recovery. The PLCP header contains five main fields. The rate field, with 3-bit length, conveys the information about the type of modulation, the data rate, the pulse shaping, and the spreading factor used to transmit the PSDU. The Scrambler Seed (SS) field contains one bit that indicates the initial value of the scrambler, the Burst Mode (BM) field contains one bit that indicates whether or not the packet is being transmitted in the streaming mode [1]. Four bits Header Check Sequence (HCS) and parity bits generated from BCH (31,19) encoder which are used for error detection and correction at receiver side.

The PSDU contains three parts: (1) MAC header with 7-byte length, (2) MAC frame body with variable length (0 to 255 bytes), and (3) Frame Check Sequence (FCS) with 2-byte length. IEEE 802.15.6-2012 defines a constant symbol rate of 600 ksps for 2.4 GHz band, and two different modulation schemes π/2 Differential-Binary Phase Shift Keying (DBPSK) and π/4 Differential-Quadrature Phase Shift Keying (DQPSK) according to the value of the rate field. The PLCP preamble and header are always modulated using π/2 DBPSK only. The SRRC pulse shaping is used to shape the spectrum of the transmitted signal and eliminates the radiations of the out-of-band side-lobes. The Spreading Factor (SF) takes one value from three possible values, 1 or 2 or 4, according to the value of the rate field, while the PLCP header spread factor has only one value, SF equals four [1]. The transmitted Power Spectral Density (PSD) should keep its out-of-band radiations to -20 dBr at most [1], the latency of the system should be 125 ms at most [3], Packet Error Rate (PER) must not exceed 10% for signal to noise ratio (E_b/N_0) greater than or equal 11 dB [1], and overall power consumption for transceiver is recommended not exceed 30 mW [3].

III. PROPOSED NB-PHY DESIGN

The proposed NB-PHY transceiver, shown in Fig. 1, includes all blocks and modules which were designed. All blocks will be introduced with more details for the major important blocks.

A. Insert and Remove Shortened Bits

The insert shortened bits algorithm depends on BCH (63, 51) input length. It reads data from the MAC PSDU buffer and appends it with zeros to decompose all PSDU data into an integer numbers of codewords with 51-bit length each. The same block is shared with the 19-bit header to insert 32-bit zeros to make its length compatible with the implemented BCH encoder. At the receiver side, this block decomposes the received data to the same number of codewords, as in transmitter, but this time each codeword has 63-bit length which is the length of BCH decoder input codeword. The remove shortened bits performs the reverse operation of the insert shortened bits algorithm. It removes the shortened bits from the output of the BCH encoder/decoder in the transmitter/receiver, respectively.

B. Spreader and Interleaver

The spreader repeats each incoming bit number of times according to the SF which depends mainly on the rate field

content. The output of spreader is applied to the interleaver that reorders the data according to a certain mapping algorithm. In our design we work to optimize implementation so these two blocks were combined in one block by building one buffer with 63-bit length. This buffer stores the output of the removed shortened bits block, and reads data according to interleaver mapping algorithm with a repeated manner, so that each bit is read SF times. At receiver side, the operation is reversed but this time the de-interleaver/de-spreader block writes two bits in a buffer every 2×SF clock cycle.

C. Scrambler

The scrambler randomizes the PLCP header and PSDU to negate the dependency of the energy on data and prevents the appearance of long strings of 1's and 0's in data to improve the performance of coarse and fine time synchronization in the receiver. The scrambler was designed by using 14-bit LFSR with modulo-2 adder according to its polynomial: $g(x) = 1 + x^2 + x^{12} + x^{13} + x^{14}$, that produces serial random bits stream, by using a simple modulo-2 adder circuits data are XORed with the scrambler output to be randomized/de-randomized at transmitter/receiver. MAC layer decides the initial seed of LFSR that takes one of two values related to the scrambler seed for PSDU or channel number for PLCP header.

D. BCH Encoder and Decoder

A binary BCH code is a cyclic error detection and correction code whose code polynomial belongs to Galois Field, $GF(2^m)$. The BCH encoder depends on a generated polynomial with LFSR implementation. BCH (63, 51) is the error detection and correction code that can correct up to two errors per codeword [1]. The encoder was designed using 12-bit LFSR with modulo-2 adders to realize its generating polynomial function: $g(x) = 1 + x^3 + x^4 + x^5 + x^8 + x^{10} + x^{12}$. The register is initialized with zeros at the start of each generation process, and input bits that are applied to the register serially. The contents of the register, after applying the last input to the register, are the parity bits which appended the 51-bit data word to form the 63-bit codeword [5].

The BCH decoder is more complex than the encoder one, the decoding process is composed of three main steps [5]: (i) Syndrome calculation, (ii) Key equation solving, and (iii) Error location finding. Inversion Berlekamp-Massy algorithm is used to design the syndrome calculation and key equation solving modules, while Chein search algorithm is used to find error locations. Due to the iterative nature of these two algorithms, syndrome calculation and key equation solving takes 63 clock cycles to complete its operation and so Chien search, so each codeword takes 126 cycles.

We proposed a new modification, shown in Fig. 2, by adding second buffer to find error locations for the codeword whose syndromes were calculated and key equation were solved, while the incoming codeword is stored in the first buffer to calculate its syndromes and solve its key equation. So; only the first codeword will take 126 clock cycles to appear at the decoder output, and then there is an output at each clock cycle which increases the throughput of the decoder. Also, we design one encoder and decoder which are shared between header and PSDU to reduce the occupied area.

978-1-4673-8760-6/15 $31.00 © 2015 IEEE

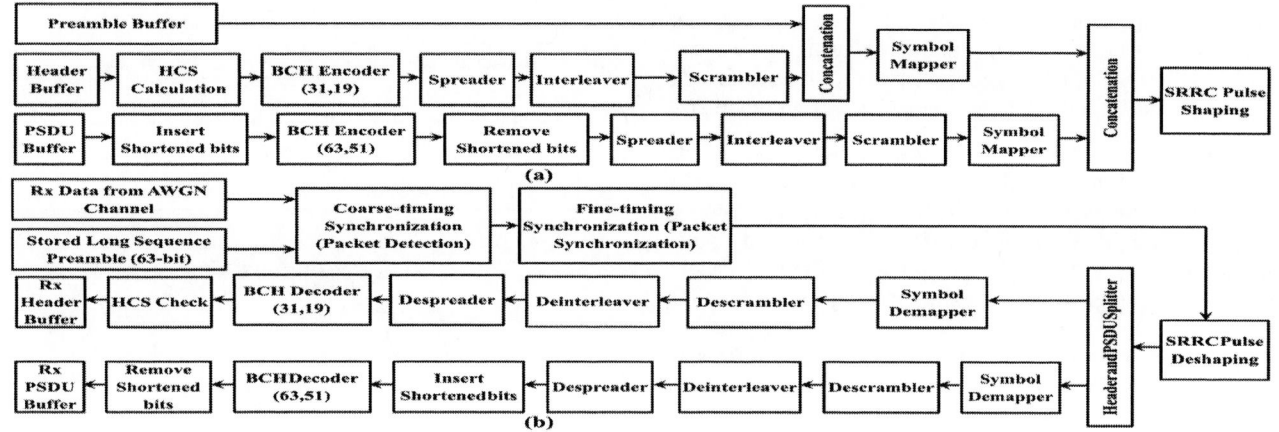

Fig. 1. Proposed Narrowband PHY Transceiver: (a) Transmitter Blocks and (b) Receiver Blocks.

E. Symbol Mapper and De-mapper

The symbol mapper converts each input bit or two bits, according to rate field, to In-phase (I) and Quadrature-phase (Q) values. We designed a reconfigurable architecture symbol mapper, to change its function between π/2 DBPSK and π/4 DQPSK according to the rate field. Hence our proposed transceiver is compatible with all transmission data rates of 2.4 GHz NB-PHY band. The de-mapper is basically the reversed operation of the mapper.

F. Timing Synchronization

The PLCP preamble is used for the synchronization process and is composed of long and short training sequences. The long sequence is a 63-bit m-sequence and the standard defines two unique long sequences in order to mitigate false alarms due to networks operating on adjacent channels [1], while the short sequence, 27-bit long, is common for all channels. The synchronization process is split into a coarse timing estimate based on the long sequence and a fine estimate obtained from the short sequence [1]. There are two common ways for timing synchronization: auto-correlation based algorithms and cross-correlation based algorithms [6]. We proposed to use a cross-correlation based algorithm for coarse time synchronization process, due to good orthogonality properties of the two existing long sequences. We used the long sequence as local reference signal at the receiver and cross-correlates it with the incoming noisy signal according to (1):

$$M_{cl}(i) = \frac{\sum_{k=0}^{N_{cl}-1} r(i+k)L^*(k)}{\sum_{k=0}^{N_{cl}-1} |L(k)|^2}, \quad (1)$$

where M_{cl} is the timing metric, N_{cl} is the length of correlation window, r is the received signal, L is the stored long sequence, and $(.)^*$ stands for the conjugate operation.

However, we proposed to use an auto-correlation algorithm for fine time synchronization, the correlation window has the length of the short sequence and the incoming signal will be correlated with its delayed version by short sequence length according to (2):

$$M_{cs}(i) = \frac{\sum_{k=0}^{N_{cs}-1} r(i+k+N_{cs})r^*(i+k)}{\sum_{k=0}^{N_{cs}-1} |r(i+k+N_{cs})|^2}, \quad (2)$$

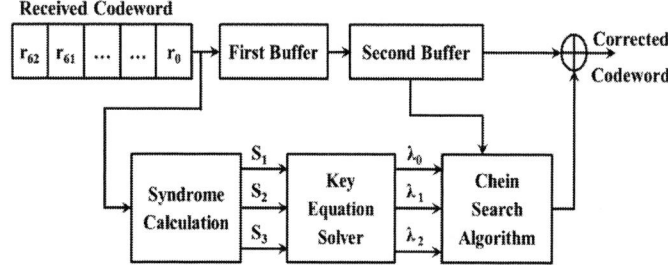

Fig. 2. Proposed Parallel Processing BCH Decoder.

where M_{cs} is the timing metric, and N_{cs} is the length of correlation window that equals the length of short sequence.

G. Pulse Shaping and De-shaping

The SRRC pulse shape with roll-off factor β and symbol period T_s is used to filter the symbols and shape the spectrum. The pulse shaping is very important to keep the out-of-band radiations as minimum as possible. But we proposed to implement this filter in a digital manner to be used for enhancing the overall system performance against channel noise. By sampling the SRRC pulse with 6 MHz sampling frequency and β = 0.5, we can represent each I and Q values, which come from mapper, using 21 samples each with 8-bit. At receiver, we multiplied each incoming 21 noisy samples from the channel with the samples of stored SRRC filter; sum the results in an iterative manner, and then decide the value of I and Q values according to a certain threshold value. So, our proposed SRRC filter acts as a matched filter to enhance the signal to noise ratio of our system.

IV. RESULTS AND DISCUSSION

Our design methodology depends on four main steps: (1) The proposed transceiver was implemented, simulated and verified in MATLAB in two phases: former, floating point transceiver design, and latter, fixed point transceiver design, (2) Then, the IPs were written in Register Transfer Level (RTL) Verilog and simulated by Questa 10.0b, (3) The synthesis, place and route were done on both ASIC and FPGA. For FPGA, we use Xilinx Virtex6 (ML605) Synthesis Technology (XST) for comparison with the proposed work in [4], and (4) the design was implemented and mapped to 65 nm ASIC technology using Cadence Encounter RTL Compiler

978-1-4673-8760-6/15 $31.00 © 2015 IEEE

(RC 11.10). The functionality of our IPs was verified compared to MATLAB implementation by using test vectors from MATLAB as PSDU and header inputs to transmitter and noisy signal inputs, after applying channel effect, to receiver.

To minimize the area, and to reduce hardware complexity and power consumption of the proposed transceiver, reusing and sharing strategy was used for BCH encoder/decoder, scrambler/descrambler, modulator/demodulator, and pulse shaping/de-shaping modules for preamble, header and PSDU. Reconfigurable architecture modulation and demodulation, interleaver, deinterleaver, and spreader were designed to produce a reconfigurable data rate transceiver that supports all transmission rates for 2.4 GHz band. The simulated power spectral density, as shown in Fig. 3, matches the standard requirements as the out-of-band radiations are less than -20 dBr. Also, Fig. 4 shows that our proposed timing synchronization design works properly and gives an accurate coarse and fine time synchronization performance. The performance of the transceiver was tested by using Packet Error Rate (PER) metric vs. signal to noise ratio (E_b/N_0,). As shown in Fig. 5, the PER equals 10% at 6.6 dB and 9.9 dB E_b/N_0 for floating point and fixed point designs, respectively. For a fair comparison with FPGA synthesis results that was done in [4], our synthesis results comparison would done by excluding coarse and fine time synchronization blocks from our design. Synthesis results, in TABLE I, shows an improvement for our proposed design than [4].

ASIC synthesize results are listed in TABLE II. We can notice that the proposed design achieves an efficient implementation regarding area cost and power consumption.

V. CONCLUSION AND FUTURE WORK

Efficient and accurate timing synchronization architecture for WBAN NB-PHY transceiver has been presented. The proposed design operates at different data rates up to 971.4 kbps in 2.4 GHz band and supports variable packet sizes transmission. The overall building blocks of the NB-PHY transceiver were designed and implemented. Our proposed system was synthesized and implemented on both FPGA, for comparison with [4], and ASIC. For ASIC, we used 65 nm technology. Our implemented system consumes 162 µW at 6 MHz clock frequency. The BCH encoder algorithm modules consume about 50% of the total transmitter power and the timing synchronization modules consume about 79% from the receiver power. In future, we will try to minimize the power consumption of these two modules using power optimization strategies.

ACKNOWLEDGMENT

We would like to thank Egypt-Japan University of Science and Technology (E-JUST) for the continuous support and the Egyptian Ministry of Higher Education for funding this work.

TABLE I.   FPGA Synthesize Results Comparison

FPGA Resources	[4]	This Work	Improvement (%)
Number of Slice Registers	2668	846	68%
Number of Slice LUTs	3161	1293	59%

TABLE II.   ASIC SYNTHESIZE RESULTS

Reported Parameters	This Work
Power Consumption	162 µW
Gate Count	18 KGate
Clock Frequency	6 MHz

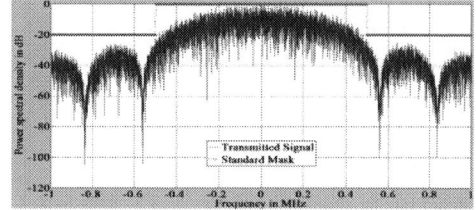

Fig. 3.   Transmitted Signal Power Spectrum Density.

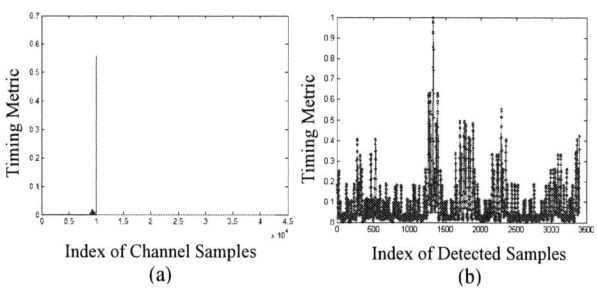

Fig. 4.   Time Synchronization Results: (a) Coarse Timing, and (b) Fine Timing.

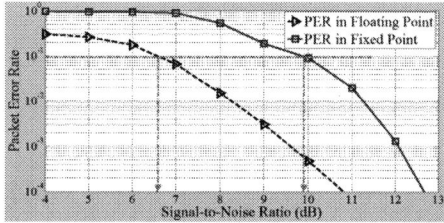

Fig. 5.   Packet Error Rate of the proposed system.

REFERENCES

[1] IEEE Computer Society, "IEEE Standard for Local and metropolitan area networks: Part 15.6 Wireless Body Area Networks", IEEE Standards Association, 29, Feb., 2012.

[2] Ullah, Sana, Manar Mohaisen, and Mohammed A. Alnuem. "A review of IEEE 802.15. 6 MAC, PHY, and security specifications." International Journal of Distributed Sensor Networks 2013 (2013).

[3] Movassaghi, Samaneh, Mehran Abolhasan, Justin Lipman, David Smith, and Abbas Jamalipour. "Wireless body area networks: A survey." Communications Surveys & Tutorials, IEEE 16, no. 3 (2014): 1658-1686.

[4] Mathew, Priya, Lismi Augustine, Deepak Kushwaha, and Vivian Desalphine. "Implementation of NB PHY transceiver of IEEE 802.15. 6 WBAN on FPGA." In VLSI Systems, Architecture, Technology and Applications (VLSI-SATA), 2015 International Conference on, pp. 1-6. IEEE, 2015.

[5] Sutaria, Hardik, and Deepti Khurge. "FPGA based BCH Decoder." International Journal for Scientific Research & Development 1 (2013): 637-640.

[6] Xiao, Pei, Colin Cowan, Tharm Ratnarajah, and Anthony Fagan. "Time synchronization algorithms for IEEE 802.11 OFDM systems." (2009): 287-290.

Inductive Degeneration Low Noise Amplifier for IR-UWB Receiver for Biomedical Implant

Maissa Daoud[1], Rahma Aloulou[1], Hassene Mnif[1], Mohamed Ghorbel[2]

[1]Laboratory of Electronics and Information Technology (LETI), ENIS, University of Sfax, Tunisia
[2]Advanced Technologies for Medicine and Signals (ATMS), ENIS, University of Sfax, Tunisia

Email: daoud_maissa@hotmail.com.

Abstract—**This paper presents the design and simulation of an inductive degeneration low noise amplifier (LNA) for impulsionel radio ultra wide band receiver for biomedical implant. Several techniques was used in this study to improve the LNA features for the [1,5]GHz frequency band. The most important are the use of the diode connected load, the degeneration source and the cascode design. A fully integrated inductive degeneration LNA was designed using 0.18 μm CMOS process. It achieves a minimal noise figure of 1.7 dB, a maximal voltage gain of 15 dB and CP1 of -10 dBm.**

Keywords—low noise amplifier; ultra wide band ; degeneration source; cascade design

I. INTRODUCTION

The receiver unit presents always the most important part of the biomedical implant as it is responsible for the reception of signals carrying critical data related to patients. Several receiver types exist but the two most important are the coherent and non-coherent reception. This last is the most used because it is characterized by its low consumption and small area that respond to the implant design constraints.

Usually the non-coherent receiver includes a low noise amplifier for amplifying the received signal because it is sent with a spectral density close to the noise floor and it is equal to -41.3 dBm for ultra wide band (specified by the IEEE standard 802.15.6), a squarer, an integrator and a comparator as shown in Fig. 1. In this work, we have studied and developed the low noise amplifier, as it is the most critical block of the reception chain.

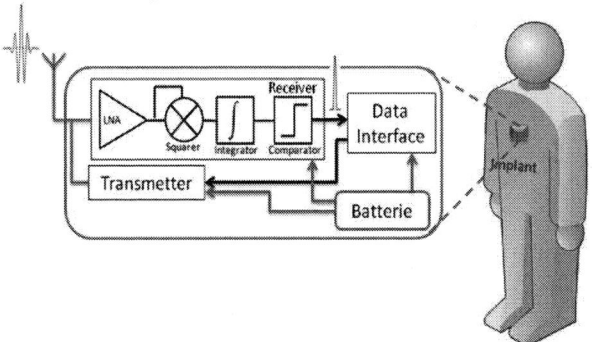

Fig. 1. Biomedical implant schematic

The design of a low noise amplifier, stable, linear, with high gain and not consuming too much energy always remains the most important step for the design of an efficient and high quality receiver. In addition, a best input matching is also an important feature of LNA. A variety of techniques and methods are used in several researches to achieve this goal. The most structure used in the LNA design to obtain a good input matching is the source inductive degeneration architecture (Fig. 2) [1, 2, 3].

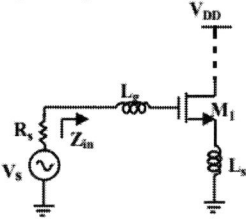

Fig. 2. Inductive degeneration LNA [4]

Its input impedance is expressed by (1). In order to obtain the impedance matching of 50Ω, at the resonance frequency, the imaginary part of (1) is zero, the real part of (1) is 50Ω [4].

$$Z_{in} = jw(L_g + L_s) + \frac{1}{jwC_{gs}} + \frac{g_m}{C_{gs}}.L_s \qquad (1)$$

Because its high gain and small noise figure, this LNA type is used for both single band (narrow and ultra wide band) and multiband.

In [5] a dual band CMOS LNA design with current reuse topology is introduced for the two standards GSM and UMTS at 947.5 MHz and 2.14 GHz frequencies. Its architecture uses an inductive degeneration with dual band low noise amplifier. At 947.5MHz, the LNA exhibits a noise figure of 2.3dB, a voltage gain of 28dB and CP1 of -12dBm. However, the LNA at 2.14 GHz features a noise figure of 2.71 dB and a CP1 of -4.5dBm.

However, in [6] a CMOS low noise amplifier with reconfigurable input matching network is proposed. This LNA has a tuning range of 1.9-2.4 GHz (single band). It uses the inductive degeneration technique as a basic architecture. The LNA has a measured voltage gain of 10-14 dB and a noise figure of 3.2-3.7dB within the band. the LNA consumes 14mA from a 1.2V supply and it is implemented using 0.13μm CMOS technology.

978-1-4673-8760-6/15 $31.00 © 2015 IEEE

On the other hand, this paper proposes an ultra wide band low noise amplifier operated at the [1;5] GHz frequency band, which combines the best input matching from an inductive degeneration topology, the use of the diode connected load, the cascoding architecture and the symmetrical supply [7].

In this paper an inductive degeneration low noise amplifier for IR-UWB Receiver for Biomedical Implant is presented. The remainder of this paper was organized as follows: In section 2, the theory of the proposed LNA was described. The simulation results were reported in section 3. Finally, section 4 was devoted to draw some conclusions.

II. PROPOSED LNA DESIGN

The methodology used for the amplifier design is as follows. First of all we choose the degeneration source architecture because it have an important consequence in the increase in the output resistance of the stage. On the other hand it allows us the freedom to specify the desired band. Secondly we use the diode connected load because in many CMOS technologies, it is difficult to fabricate resistors (with high values) with tightly-controlled values or a reasonable physical size. Consequently, it is desirable to replace resistor with a MOS transistor. This latter can operate as a small-signal resistor if its gate and drain are shorted called diode-connected device in analogy and its impedance is simply equal to $(1/g_m)$. Finally, we use the cascode architecture because it proves beneficial in increasing the voltage gain of amplifiers and the output impedance of current sources while providing shielding as well. The invention of the cascode was motivated by the need for high frequency amplifiers for with relatively high input impedance. The cascode circuit offers the speed of the amplifier by suppressing the miller effect and the input impedance [8].

The proposed LNA is presented in Fig. 3. This amplifier is composed of four transistors two NMOS (NM1 and NM2) for the signal recovery and two PMOS (PM1 and PM2) their loads. The inductance L3 is added in series with the transistor PM2 to provide additional gain and improve the performance of the noise. The use of resistance R3 and capacitor C2 allow to satisfy the constraint on the reflection coefficient at the output S(2,2).

The remainder of circuit includes the matching circuit at the input. Indeed, the use of an impedance matching circuit at the input and the output allows broadening more the frequency band. It is composed of:

- A resistor R2 provides reflection coefficient at the input S (1,1) below -10 dB.

- Inductances L1 and L2 for overcoming the degradation of gain and improving the bandwidth.

- Resistor R1 and capacitor C1 are used to adjust the gain curve on the desired band.

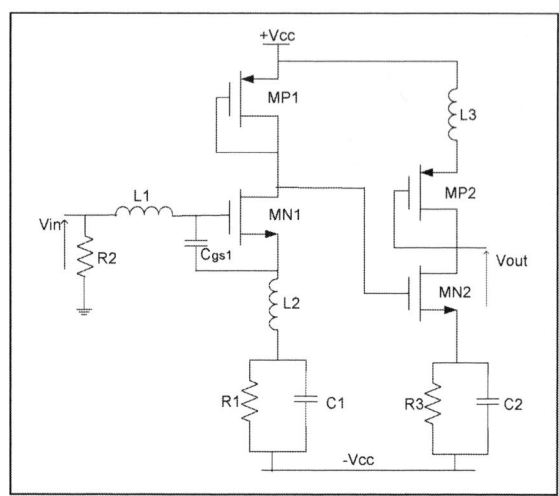

Fig. 3. Proposed inductive degeneration LNA design

The gain expression is achieved using LNA small signal schema (Fig. 4).

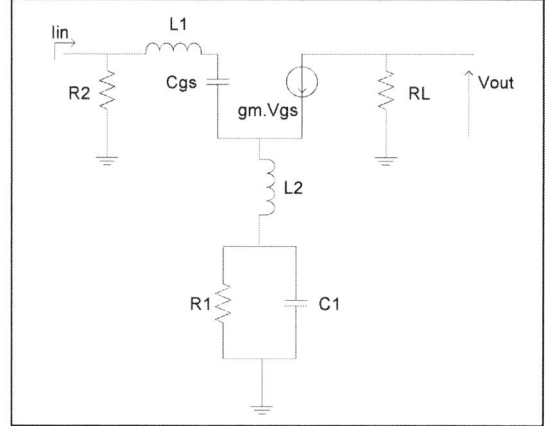

Fig. 4. Small signal Schema of LNA first stage

The gain expression of LNA first stage is then presented by (2).

$$G_1 = \frac{-(sR_1C_1+1)(R_{L1}+r_0)g_{m1}}{(sR_1C_1+1)+(R_1+sL_2(sR_1C_1+1))} \cdot \frac{R_2+sL_1}{R_2} \quad (2)$$

The gain expression of LNA second stage is shown by (3).

$$G_2 = \frac{-(sR_3C_2+1)(R_{L2}+sL_3+r_0)g_{m2}}{(sR_3C_2+1)+g_{m2}\cdot R_3} \quad (3)$$

The total gain of inductive degeneration LNA is the sum of the two mentioned gains. It is expressed by (4) .

$$G_{tot} = G_1 + G_2 \quad (4)$$

We notice from (2) and (3) that the amplifier total gain is directly proportional to the transconductances gm_1 and gm_2. Also, it is based on the inductor "L_1" and "L_3" and the charges of the two transistors NM1 and NM2. Thus increasing the gain is done by varying these parameters. From the same equations (2) and (3), it can be seen that the inductor L2 is in the denominator of the gain equation therefore its increase causes the reduction of the gain.

Table I resumes the component values of the proposed inductive degeneration LNA.

Table. I Component values of the inductive degeneration LNA

PM1 (µm)	L=0.18	W=4
PM2 (µm)	L=0.18	W=45
NM1 (µm)	L=0.18	W=45
NM2 (µm)	L=0.18	W=100
R1 (KΩ)	1	
R2 (Ω)	95	
R3 (Ω)	100	
Cgs (fF)	10	
C1 (pF)	3	
C2 (pF)	3	
L1 (nH)	14	
L2 (nH)	1	
L3 (nH)	2.5	

III. SIMULATION RESULTS

The proposed circuit was simulated with 0.18 µm standard CMOS process from TSMC. In this part we validated the proposed techniques and the LNA specifications through frequency and transient simulation. The first feature to be discussed is the reflection coefficient at the input S(1,1) (Fig. 5). The values of S(1,1) are lower than (-10dB) which reflects a good adaptation to 50Ω.

The second characteristic is the gain, its curve is presented in Fig. 6. The LNA results in a maximum allowable gain of about 15 dB. It is well spread over the 1-5Ghz band, which corresponds to the specification of the UWB system.

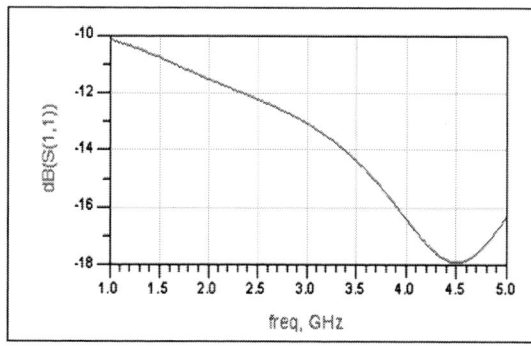

Fig. 5. Curve of the input reflection coefficient

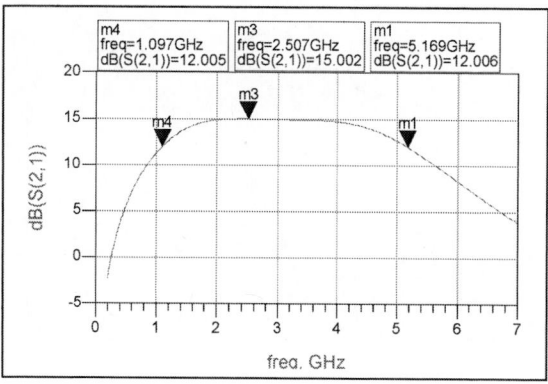

Fig. 6. Gain Curve

To qualify the LNA output signal, it must be measured the noise figurer NF of this block. The best case is to have a small value. According to the curves of Fig. 7, the noise figure of the inductive degeneration LNA at 4.5GHz equal to 1.7dB.

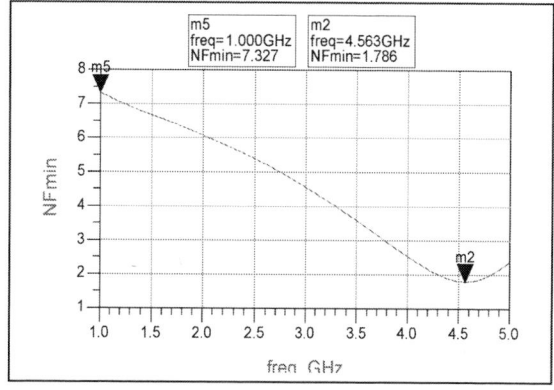

Fig. 7. Noise figure curve

The Fig. 8 shows that the Reverse transmission gain of the inductive degeneration amplifier less than -45 dB which exhibits a very good isolation between LNA input and output. In order to test the linearity of the LNA, we determined the compression point at -1dB (Fig. 9) that it is equal to -10.71 dBm .

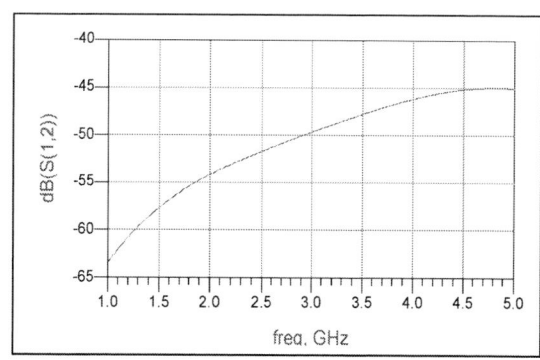

Fig. 8. Reverse transmission gain curve

978-1-4673-8760-6/15 $31.00 © 2015 IEEE   97

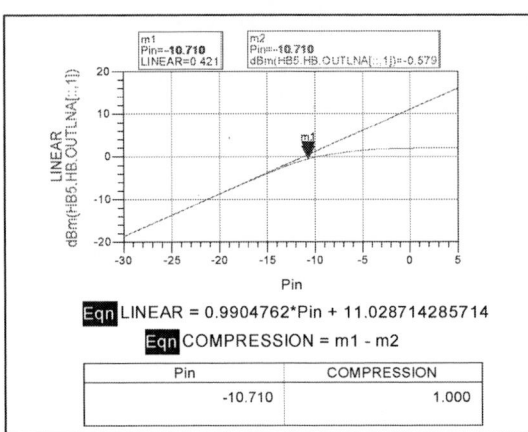

Fig. 9. Compression point (CP1) curve

To test the amplifier output signal an IR-UWB transmitter was simulated where, the received pulses are close to reality. The signal received by the amplifier is illustrated in Fig. 10 (a) and the LNA output signal is presented by Fig. 10 (b). According to the two figures, the signal has undergone a good amplification without degrading its shape.

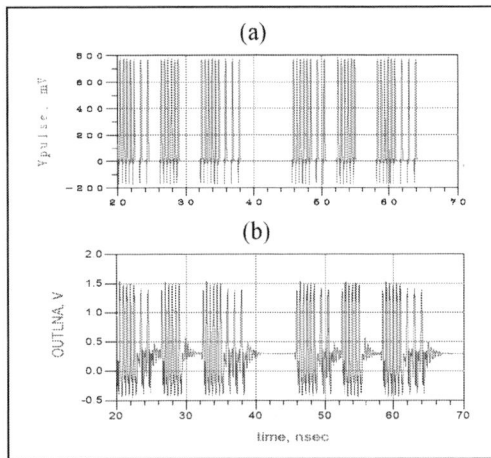

Fig. 10. (a) Received signal, (b) LNA output signal

To show the reliability of the proposed LNA, the novel design is compared to other structures in the literature. These comparisons are summarized in Table II.

TABLE II. PROPOSED UWB LNA PERFORMANCE SUMMERY IN COMPARISON TO RECENTLY PUBLISHED UWB LNAS

	BW (GHz)	Gain (dB)	NF (dB)	S11 (dB)	Process	Supply
[9]	3.1-5	13.8-14.6	5.4-6	<-10	65nm	1V
[10]	3-8	15.2	3.1-6.8	<-8	180nm	1.8V
[11]	3-10	7.9	6	<-10.7	180nm	1.1V
[12]	1.5-8.5	12–13.2	4.2-5	<-7.5	180nm	1.5V
This work	1-5	12-15	1.7-7.3	<-10	180nm	-2/2V

IV. CONCLUSION

This paper presents a scientific research in the biomedical field. In this work we have concentrated on the study of the low noise amplifier because it presents the most critical block in the receive chain. It operates in an UWB as it offers simplified RF architectures due to the emission of baseband pulse enabling equipment with low cost and complexity. the most important in the UWB low noise amplifier is to maintain good performance throughout the band which is presented by our amplifier, and this is achieved by the use of the inductive degeneration topology, the diode connected load technique, the cascoding architecture and the symmetrical supply.

REFERENCE

[1] C. Feng, X. P. Yu, Z. H. Lu, W. M. Lim and W. Q. Sui, "3–10 GHz self-biased resistive-feedback LNA with inductive source degeneration," Electronics Letters, vol. 49, no. 6, pp. 387-388, March 2013.

[2] H.-I. Wu, R. Hu, and C. F. Jou, "Complementary UWB LNA Design Using Asymmetrical Inductive Source Degeneration," IEEE Microwave and Wireless Components Letters, vol. 20, no. 7, pp. 402-404, July 2010.

[3] H. Fouad, K. Sharafw, E. E.-Diwany and H. E.-Hennawy,"An RF CMOS Modified-Cascode LNA with Inductive Source Degeneration," Radio Science Conference (NRSC), 2002, pp.450-457.

[4] S. Mahersi, H. Mnif, and M. Loulou," Novel Control of Input-Match and Output-Match for an Inductively Degenerated Low Noise Amplifier," International Journal of Electronics and Electrical Engineering, vol. 1, no. 3, pp. 199-205, September 2013.

[5] M. Benamor, A. Fakhfakh, H. Mnif and M. Loulou, "Dual band CMOS LNA design with current reuse topology," Design and Test of Integrated Systems in Nanoscale Technology Conference on, 2006, pp. 57-61.

[6] M. El-Nozahi, E.S.-Sinencio and K. Entesari, "A CMOS low noise amplifier with reconfigurable input matching network," Microwave Theory and Techniques, IEEE Transactions on , vol. 57, no. 5, pp. 1054-1062, May 2009.

[7] M. Daoud, M. Ghorbel and H. Mnif, "A Non-Coherent High Speed IR-UWB Receiver for Biomedical Implants," IEEE Advanced Technologies for Signal and Image Processing Conference, 2014, pp. 533-538.

[8] B. Razavi, Design of Analog CMOS Integrated Circuits. Los Angelos, USA: University of California, 2001.

[9] W. Wu, O. Novak, C. T. Charles, Xiaoya Fan, "A low-power 3.1–5GHz ultra-wideband low noise amplifier utilizing Miller Effect," UltraWideband, ICUWB IEEE International Conference on, 2009, pp. 255–259.

[10] A. Meaamar, Boon Chirn Chye; Do Man Anh; Yeo Kiat Seng; , "A 3–8 GHz Low-Noise CMOS Amplifier," Microwave and Wireless Components Letters, IEEE , vol. 19, no.4, pp. 245–247, April 2009.

[11] J. F. Chang, Y. S. Lin, "0.99mW 3-10 GHz common-gate CMOS UWB LNA using T-match input network and self-body-bias technique," Electronics Letters , vol. 47, no. 11, pp. 658–659, May 2011.

[12] A.R. A. Kumar, A. Dutta and S. G. Singh, "Noise-cancelled subthreshold UWB LNA for Wireless Sensor Network application," Ultra-Wideband (ICUWB), 2012 IEEE International Conference on, 2012, pp. 382–386.

A New WSN Transceiver based on DWPT for WBAN applications

Safa SAADAOUI[1], Mohamed TABAA[1], Fabrice MONTEIRO[2], Abbas DANDACHE[2], Karim ALAMI[1]

[1]Moroccan School of Engineering Sciences (EMSI), Research and Innovation Department, Casablanca, Morocco
[2]Laboratory of genie industrial production and maintenance (LGIPM), Lorraine University, Metz, France
med.tabaa@gmail.com, s.saadaoui@gmail.com, fabrice.monteiro@univ-lorraine.fr, abba.dandache@univ-lorraine.fr

Abstract — **Body networks type WBAN (Wireless Body Area Network) are beginning to take a major interest in last years. Generally, modern transmission face three major challenges: energy management for transmission, increase throughput and noise immunity. The use of broadband systems helped answer constraints of modern transmission system, it has questioned the use of complex and cumbersome systems: CDMA and OFDM. This has prompted the search for technical lower layers: physical and MAC. In this paper, we present a new form of communication for WBAN applications based on Discrete Wavelet Packet Transform (DWPT) in transmitter and Inverse Discrete Wavelet Packet Transform (IDWPT) in receiver. We will propose a new form of many-to-one transceiver for medical applications.**

Keywords— WBAN, UWB-IR, DWPT, Multicarrier

I. INTRODUCTION

Nowadays, multiband communication is also used in several types of wireless communication applications such as 4G and 5G networks specially the application broadband with high-speed data, WIFI, DVB, UWB and other [1,5 and 7]. The next generations of communication are based at high speed system based generally on frequency division multiplexing scheme for objective multicarrier modulation (MCM) technique. In this later, the bandwidth is divided into sub-bandwidths and each users of architecture can transmitted in different sub-band.

The use of the ultra-wide band techniques for sensor network is a rapidly growing domain thanks to its characteristics. These systems cover diverse fields of applications such as medical and environment [1,4,5 and 6]. The main characteristic of these systems is doubt less the energy constraint that they must be able to deal with. Indeed, these systems are based on sensors which energy is generally supplied by a limited source such as a battery or a renewable energy source, and whose life expectancy must be able to last for years. In this complex and constrained context, one of the challenges relies on the use of the impulse radio and ultra-wideband communications. Today, the ultra-wideband technology gain in popularity in short-range communications systems [6, 11]: some multimedia transmission systems require a very wide bandwidth, this bandwidth being assured by the impulse techniques.

The performances of the DWT algorithms and the multi-resolution analysis authorize a large diversity of applications relying on the wavelets theory. These algorithms are based on several methods among which the filter banks constitute the most often implemented solution. The multi-resolution analysis allows the decomposition of the signal into coefficients of approximation and details, from a scaling function and a wavelet function, respectively [2,4,6 and 10]. This implementation is based on a set of cascaded low-pass and high-pass digital filters.

In this paper we propose a new architecture for both of many-to-one and one-to-many communications in star topology networks based on DWPT in transmitter and IDWPT in receiver. The rest of this paper is organized as follows: the use of DWPT for UWB applications in part 2, the part 3 is dedicated for details of our proposed. The description of database and simulations in part 4, and finally conclusion and perspectives.

II. DISCRETE PACKET WAVELET TRANSFORM FOR UWB-IR

Spread spectrum techniques for wideband communications often rely on the Discrete Fourier Transform (DFT) that presents some limitations concerning the analysis of the non-stationary signals, because this transform doesn't consider the information in a temporal structure of the signal, what induces a difficulty in localizing the discontinuity of the signal [5,7]. The solution brought at first was the Windowed Fourier Transform (WFT) which compares the signal with a sinusoid of infinite duration within a time-limited window. The Discrete Wavelet Transform (DWT) presents a better solution by bringing a good compromise between time and frequency of analysis. The peculiarity of this transform is that it mixes both notions of time and scale contrary to the Fourier transform which assures the passage from a notion to another [7,10]. The performances of the DWT algorithms and the multi-resolution analysis authorize a large diversity of applications relying on the wavelets theory. These algorithms are based on several methods among which the filter banks constitute the most often implemented solution. The multi-resolution analysis allows the decomposition of the signal into coefficients of approximation and details, from a scaling function and a wavelet function, respectively [2,6,7 and 11]. This implementation is based on a set of cascaded low-pass and high-pass digital filters.

978-1-4673-8760-6/15 $31.00 © 2015 IEEE

In the wavelet transform (WT) theory, the wavelet basis functions are obtained from a single prototype function called "wavelet" by translation and dilation or contraction:

$$\Psi_{s,\tau}(t) = \frac{1}{\sqrt{s}} \, \Psi\left(\frac{t-\tau}{s}\right) \quad (1)$$

Where $s \in R^*, \tau \in R$. For large s, the basis function becomes a stretched version of the prototype wavelet, that is a low frequency function, while for small s, the basis function becomes a contracted wavelet, that is a high frequency function.

The Continue Wavelet Transform (CWT) of $x(t)$ is defined as:

$$X(s,\tau) = \frac{1}{\sqrt{s}} \int_{-\infty}^{+\infty} x(t)\Psi^*\left(\frac{t-\tau}{s}\right) dt \quad (2)$$

Where *denotes complex conjugation.

The CWT is too redundancy and to make it practical, we need a finite number of wavelets in the wavelet transform. Therefore, the discrete wavelets transform (DWT) are discretely scalable and translatable. This was achieved by modifying the wavelet representation to create Daubechies (1992) [2]:

$$\Psi_{j,k}(t) = \frac{1}{\sqrt{s_0^j}} \, \Psi\left(\frac{t-ks_0^j\tau_0}{s_0^j}\right) \quad (3)$$

We usually choose $s_0 = 2$ so that the sampling of the frequency axis corresponds to dyadic sampling. In addition, $\tau_0 = 1$ we gave a dyadic sampling in time.

Discretizing the translation and dilation contraction parameters of the wavelet in (1), the dyadic discrete WT of $x(t)$ is:

$$X(j,k) = 2^{-j/2} \int_{-\infty}^{+\infty} x(t)\Psi^*(2^{-j}t-k) \, dt \quad (4)$$

Where j, k ∈ Z.

It should be mentioned that WT can be implemented by non-uniform filter banks. Where the input signal separates into low frequency component signal and high frequency component signal. The first is called smooth coefficient, the second part is called wavelet coefficients. The smooth coefficient is separated into two parts repeatedly). In the classical way, we indicate the tow part via low-pass digital filter H and a high pass-filter G. By using the scaling function and there corresponding mother wavelets, we obtain both of the digital filter H and G. We suppose H and G like a FIR filters non-recursive with L length, the transfer functions of H and G can be represented as follows:

$$H(Z) = h_0 + h_1 z^{-1} + h_2 z^{-2} + \cdots + h_{L-1} z^{-(L-1)} \quad (5)$$
$$G(Z) = g_0 + g_1 z^{-1} + g_2 z^{-2} + \cdots + g_{L-1} z^{-(L-1)} \quad (6)$$

Mallat's tree algorithm or pyramid algorithm [ ] can be used to find the multiresolution decomposition of DWT, the two scale relations (5) and (6) leads to scaling functions and wavelet functions similar to that in scalar wavelets. But the equations are two scale matrix equations and can be given as:

$$\phi(t) = \sum_n h(n)\phi(2t-n) \quad (7)$$

$$\Psi(t) = \sum_n g(n)\Psi(2t-n) \quad (8)$$

Where

$\phi(t) = [\phi_1(t)\, \phi_2(t) \ldots \phi_r(t)]^T$ and $\Psi(t) = [\Psi_1(t)\, \Psi_2(t) \ldots \Psi_r(t)]^T$ forms the set of scaling functions and corresponding wavelets. The suffix r denotes the number of wavelets and is dubbed as multiplicity. The decomposition DWPT at each resolution level can be presented as tree shape as figure 2.

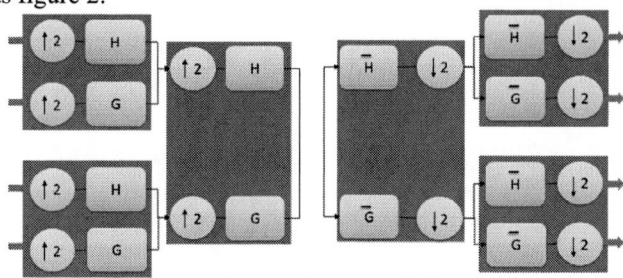

Fig.1. IDWT in transmitter and DWT in receiver

III. OUR PROPOSED ARCHITECTURE

This section presents the proposed architecture. The emitter is based on the IDPWT implemented as a synthesis filter banks, and the receiver is based on the DWPT implemented as an analysis filter banks (Figure 2). This architecture can be configured to suit the targeted application: the wavelet and the depth of the transform can be chosen to adapt the shape of the generated impulse but also, according to the mode of use, to define the maximum throughput and the number of simultaneously exploitable sensors.

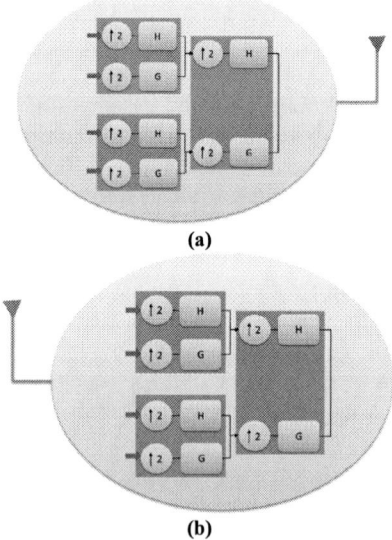

Fig.2. Proposed emitter (a) and received (b)

This architecture is dedicated to many-to-one communications for WBAN applications. From this perspective, the use of the IDWPT / DWPT is different from what was presented in [6,7]: the number of users is specified and for every users an IDWPT block is implemented to assure on one hand the activation of only one single input (sensor) per user, and on the other hand this input corresponds to a given level of details and approximation that will be used as identification label.

The unused inputs are all forced to zero. As a consequence of the activation of a single input per user, the throughput per user is reduced, but the overall system's throughput is the same compared with the parallel coding, as shown in Fig.3.

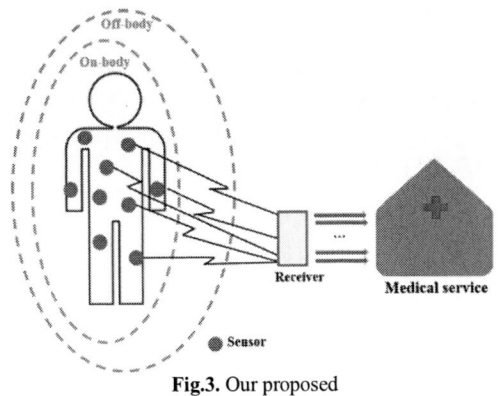

Fig.3. Our proposed

IV. SIMULATIONS AND RESULTS

A. Experimental data

We worked with the data collected by the Antenna & Electromagnetics Group (Body WiSeR) for BAN communications and especially On-body communications. All the parameters for the measurement are presented in Table 1. All the measurements were collected in a room that is 3m high and which geometry is described in Fig.4. More information can be found in [13].

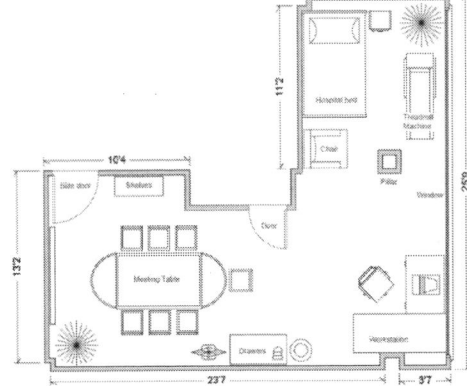

Fig. 4. Dimensions and geometry of the Body Centric Wireless Sensor Lab where the indoor radio propagation measurements are performed

Table 1 : Parameters of Body WiSeR database

Parameters	Values
Frequency range	3-10 GHz
Frequency sampling	4,37 MHz
Maximum time delay	228.8 ns
Maximum observable distance	68.6 m
Frequency span	7 GHz
Maximum temporal resolution	0.14 ns
Maximum spatial resolution	43mm
Time bin size	0.14 ns
Transmit power	0 dBm
IF bandwidth	3 KHz / 101 dBm

B. Global discussions

Over an AWGN channel, the proposed transmission and reception system based on IDWT and DWT algorithms presents very good performances for the whole sensors. Several wavelet functions will be tested to measure the quality of the received signal. The original and received signals with Symlet are presented in Fig.5.

It's clearly noted the influence of the noise on the received signal. Now, it is necessary to see if the various sensors (inputs) will successfully be identified from the received signal after a DWT transform. 3 sensors are considered for simulations; sensor number 1, 7 and 15 noted by S1, S7 and S15 respectively from database.

After DWT, sensors are all identified with various reliability degrees Fig.6. An error signal representing the difference between the received signal and the emitted one will be computed to study the influence of wavelet family on detected signal quality. Table 2 presents the tested wavelet families.

Results of signal error are only presented for sensor S15. For an SNR = 30db, the error signal for tested wavelet families is drawn in Fig.7 and Fig.8.

These results show that for Haar and Symlets wavelets, the error is minimal, so the detected signal corresponds to the original sensor. On the other hand, the rest of tested wavelet families are to be banished due to the important error between original and detected signal. The worst wavelet functions are Meyer and Coiflets.

Table 2 : wavelet families tested in our proposed

Parameters	Orthogonal	Compact support	Filters length
Daubechies	Yes	Yes	2N
Coiflets	Yes	Yes	2N
Symlets	Yes	Yes	2N
Meyer	Yes	Yes	2N
Haar	No	Yes	$\max(2Nr,2Nd)_{[\cdot]}+2$

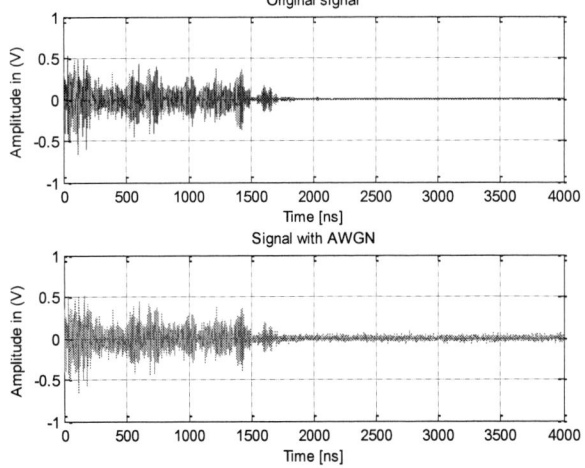

Fig.5. Difference between original and received signal

978-1-4673-8760-6/15 $31.00 © 2015 IEEE

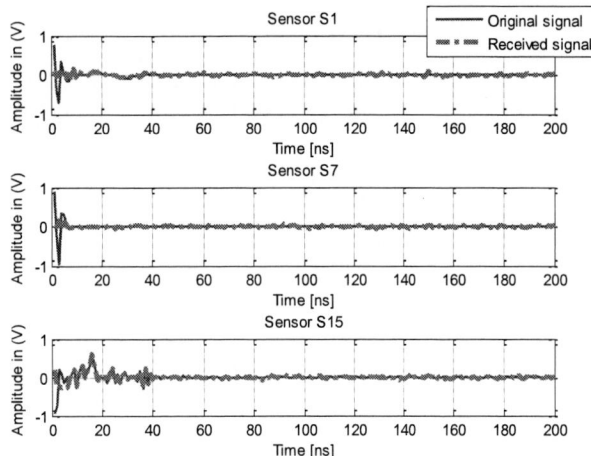

Fig.6. Original and received signal for S1,S7,S15 sensors

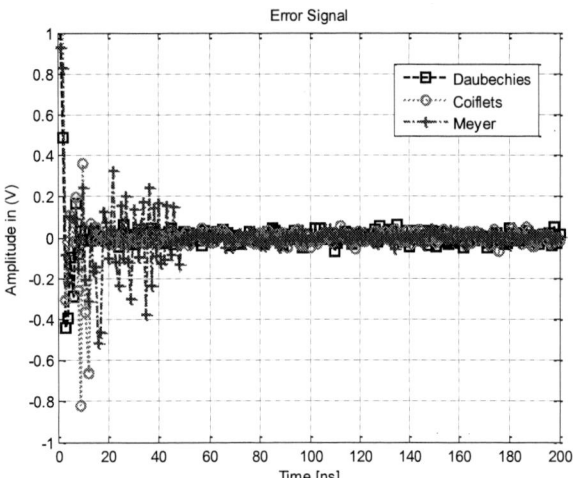

Fig.7. Error signal with Daubechies, coiflet and Meyer

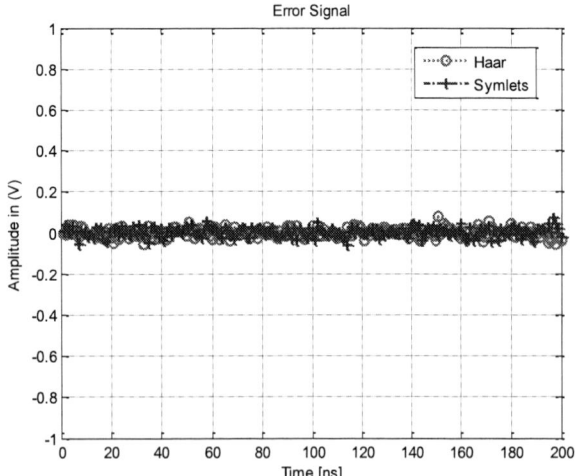

Fig.8. Error signal with Haar and Symlets

V. CONCLUSION AND PERSPECTIVES

A novel form of architecture communication was proposed in this paper, based on IDWT in transmitter and DWT in receiver. The must objective of this research work is to present a new low cost form of communication between sensor *On* or *Off body* for star topology. We are using for this research work the data collected by the Antenna & Electromagnetics Group (Body WiSeR). Global results presented in this paper show that the receiver can detected all data coming the different sensors in architecture in the same time.

ACKNOWLEDGMENT

Special thanks for Dr. Rachid Saadane form EHTP Casablanca, Morocco for his cooperativeness and valuable help.

REFERENCES

[1] Fei Hu, Qi Hao (edited by), *Intelligent Sensor Network: The Integration of Sensor Networks Signal Processing and Machine Learning*, Taylor & Francis group, 2013, ISBN: 978-1-4398-9282-4.

[2] Robert X.Gao and Ruqiang Yan, *Wavelets Theory and Applications for Manufacturing*, Springer, 2011, ISBN: 978-1-4419-1544-3.

[3] R. Dilmaghani, M. Ghavami, "Comparison between wavelet-based and Fourier-based multicarrier UWB systems", *IET Communications*, Vol. 2, Issue 2, Feb. 2008.

[4] Si Chen, Huichang Zhao, Shuning Zhang, Yunxing Yang,"Study of ultra-wideband fuze signal processing method based on wavelet transform", *IET Radar Sonar Navig.*,Vol. 8, Issue 3, pp. 167–172, 2014.

[5] J. Zhang, Z. Sahinoglu, P. Kinney, "UWB Systems for Wireless Sensor Networks", *Proceedings of the IEEE*, invited paper, Vol. 97, No 2, Feb. 2009.

[6] Mohamed TABAA, Camille DIOU, A Low-Cost Many-to-One WSN Architecture Based on UWB-IR and DWPT International conference on Control, Decision and Information technologies CoDIT2014 IEEE, Metz France 3-5 Novembre.

[7] M. K. Lakshmanan, H. Nikookar, "A Review of Wavelets for Digital WirelessCommunication",*WirelessPersonalCommunications*,Vol. 37, Issue 3-4, pp 387-420, May 2006.

[8] Dyonisius Dony Ariananda, Madan Kumar Lakshmanan, HomayounNokookar,"An investigation of Wavelet Packet Transform for spectrum estimation", *Computing Research Repository*, abs/1304.3795, 2013.

[9] J. Zhang, V. Orlik, Z. Sahinoglu, A. F. Molisch and P. Kinney, "UWB Systems for Wireless Sensor Networks", *Proceeding of the IEEE*, Vol. 97, No 2, Feb. 2009.

[10] AbulK. M. Baki and Nemai C. Karmakar, "Improved Method of Node and Threshold Selection in Wavelet Packet Transform for UWB Impulse Radio Signal Denoising", *Progress in Electromagnetics Research C*, Vol. 38, pp. 241-257, 2013.

[11] O. Abedi and M. C. E. Yagoub,"Performance Comparison of UWB Pulse Modulation Schemes under White Gaussian Noise Channels", *International Journal of Microwave Science and Technology*, Vol. 2012, 2012, Article ID 590153.

[12] M. Ghawami, L. B. Michael, S. Haruyama and R. Kohno,"A Novel UWB Pulse Shape Modulation System", *Wireless Personal Communications*, Vol. 23, pp. 105–120, 2002.

[13] Mohammad Monirujjaman Khan, Qammer H. Abbasi, Akram Alomainy, Yang Hao, "Performance of Ultra wideband Wireless Tags for On-Body Radio Channel Characterisation", International Journal of Antennas and Propagation, vol.2012, article ID 232564, 10 pages.

978-1-4673-8760-6/15 $31.00 © 2015 IEEE

On-chip analog PI controller for calibration of Rogowski coils

Simon PAULUS[1], Jean-Baptiste KAMMERER[1], Joris PASCAL[2], Calogero BONA[3], Luc HEBRARD[1]

[1]ICube laboratory – Université de Strasbourg, CNRS – 23 rue du Loess, BP 20, FR-67037 Strasbourg, France
[2]University of Applied Sciences and Arts Northwestern Switzerland FHNW, Muttenz, Switzerland
[3]ABB Switzerland Ltd. Corporate Research, Segelhofstr. 1K, CH-5405 Baden 5 Daettwil – Switzerland

Abstract—**This paper presents an on-chip analog Proportional Integral controller for continuous calibration of contactless Rogowski coil current transducers. Experimental results show that the proposed system ensures a sensor sensitivity deviation lower than ±0.1% whatever the position of the coil around the primary conductor in which the current flows. In addition, it has the ability to ensure an absolute calibration of the sensor. Although this feature is not filled by the system prototype yet, a simple architecture improvement is proposed to solve the issue.**

Keywords: Rogowski coil calibration, CMOS analog Proportional Integral controller, contactless current sensor

I. INTRODUCTION

Rogowski coils are used for contactless current measurement in electrical power applications [1], [2]. They are cheap to fabricate and may be openable and flexible, which allows current measurements in very complex geometrical and environmental conditions. However their accuracy is limited, especially for flexible coils because of the dependence of their sensitivity on the coil geometry and on the relative position of the primary conductor in which the current to measure flows [3] [4]. As a consequence Rogowski coil current transducers (RCCT) are not used for metrology where alternatives are preferred, like shunts or precision current transformers. However, they exhibit many advantages. In addition to their possible flexibility, they are highly linear due to the absence of magnetic core. They are contactless, not intrusive, and light, with a high bandwidth. They are also capable to measure low and high currents with the same coil, and they are cheap. So, strong attention is paid to minimize RCCT drawbacks in order to make them compatible with high performance metrology.

A Rogowski coil is a solenoid wound around a no-magnetic core and bended on itself in such a way that it rounds the primary conductor (Fig. 1). The first way to improve the RCCT metrological performances is to control its fabrication, especially the sensor geometry and the homogeneity of the turns along the coil in order to minimize the off-center error [5] [6]. However, this approach increases the coil price, killing the advantage of the cheap fabrication. In addition, whatever the effort in the manufacturing process, the error is never fully suppressed. The alternative approach we proposed is to implement a continuous calibration of the RCCT. Preliminary results of this approach have been recently presented [7] [8]. Here, more details on the implementation of the control loop in a CMOS chip are given as well as new measurement results obtained with an improved demonstration board. The next section reminds the Rogowski coil sensitivity and presents the calibration principle we proposed. In section III, we describe the chip we designed to implement the calibration control loop. Its limitations and possible improvement are also discussed. Then before conclusion, section IV presents experimental results obtained with our last Rogowski coil calibration demonstrator.

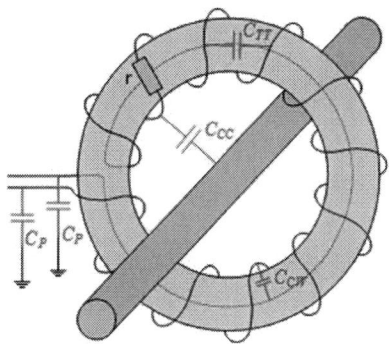

Fig. 1. Sketch of a Rogowski coil and its parasitic capacitances and wire resistance.

II. ROGOWSKI COIL AND CALIBRATION LOOP PRINCIPLE

The RCCT sensitivity, S_{RC}, is given by [7] [8]:

$$V_{rog}(t) = \frac{N}{l_{rc}} \cdot A \cdot \mu_0 \cdot \frac{dI_p(t)}{dt} = M \cdot \frac{dI_p(t)}{dt}$$

$$V_{rog} = M \cdot (2 \cdot \pi) \cdot f_p \cdot I_p = S_{RC} \cdot I_p \tag{1}$$

where $V_{rog}(t)$ is the output voltage of the Rogowski coil (RC), $I_p(t)$ is the primary current to measure, V_{rog} and I_p are respectively the magnitudes of $V_{rog}(t)$ and $I_p(t)$, N is the number of turns in the coil, l_{rc} is the RC length, A is the area of a turn, μ_0 is the magnetic permeability of air, and f_p is the frequency of the current. M is called the mutual inductance between the primary conductor and the RC.

Eq. (1) is valid whatever the position of the primary conductor, only when (i) the area A is small enough to assume the magnetic field induced by $I_p(t)$ uniform over A, (ii) the density of turns is uniform along the RC, and (iii) the coil is perfectly closed, which is never fully the case because room is necessary to output $V_{rog}(t)$. As a consequence in real RC, M and thus the RCCT sensitivity depend on the shape of the RC and on the relative position of the primary conductor.

978-1-4673-8760-6/15 $31.00 © 2015 IEEE

Because of its parasitic capacitances (Fig. 1) the RC sensitivity exhibits a resonance, and is typically derivative up to a few tens, even then a few hundreds, of kilohertz [7]. Since in industrial applications $f_p = 50$ Hz (or 60 Hz) currents have to be measured over roughly the tenth harmonic, the bandwidth of a RC is much higher than necessary. It is why we proposed to benefit from the extra-bandwidth, between 10 kHz and the RC resonance frequency, to add a sinusoidal reference current, I_{ref}, flowing into a second conductor placed coaxially to the primary conductor, and used to continuously calibrate the RCCT thanks to a proportional integral (PI) controller (Fig. 2). Proportional integral derivative (PID) controller was not considered since it is much more sensitive to noise, and because the system is intended to control long term steady-state deviation of the RCCT sensitivity. The coaxiality of both conductors ensures the same M dependence with respect to the conductors position.

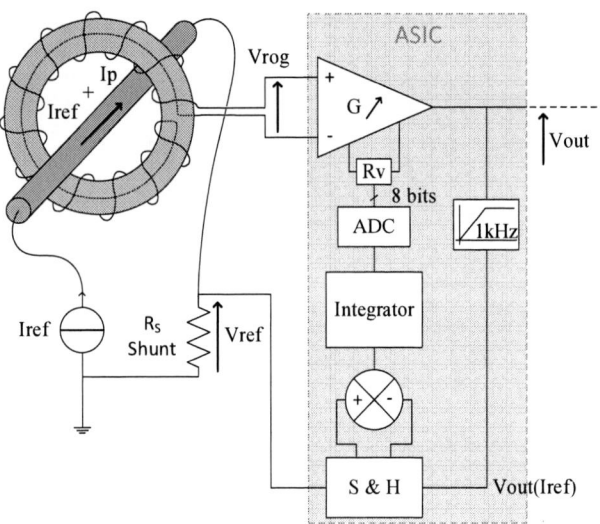

Fig. 2. Calibration loop principle

V_{rog} at the output of the RC is amplified by a variable gain G. Due to the derivative property of the RC, the magnitude of $V_{out}(I_{ref})$ at the output of the G-amplifier is f_{ref}/f_p times higher than $V_{out}(I_p)$. The frequency f_{ref} being typically between 10 kHz and 100 kHz, I_{ref} may be chosen three orders of magnitude smaller than I_p. In our case, it is around 1 A for a targeted nominal I_p of 5 kA. This "small" current flows into a shunt R_S and is compared to $V_{out}(I_{ref})$ in the PI controller (Fig. 2) in order to maintain both signals equal, meaning that:

$$G \cdot M \cdot \left(2\pi \cdot f_{ref}\right) \cdot I_{ref} = R_S \cdot I_{ref} \qquad (2)$$

and thus:

$$V_{out}\left(I_p\right) = \frac{R_S}{f_{ref}} \cdot f_p \cdot I_p = S_{RCCT} \cdot I_p \qquad (3)$$

Assuming that R_S is precise, as well as f_{ref}, which is readily possible with a quartz, the RCCT sensitivity becomes independent of the mutual inductance M, i.e. independent of the relative position of the conductors. In addition, it is constant, i.e. the RCCT is inherently calibrated [9] [10].

III. ON-CHIP PROPORTIONAL INTEGRAL CONTROLLER

The different blocks of the PI controller has been already presented [7]. We focus thus the paper here on details not discussed previously and linked with the main sensitive blocks, i.e. the error sample and hold, and the integrator.

A. Error sample and hold circuit

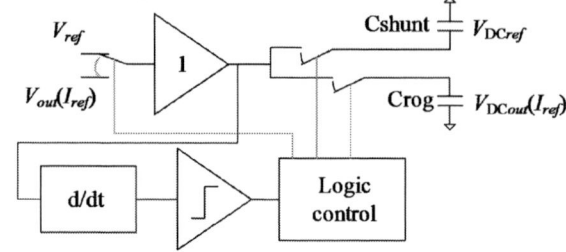

Fig. 3 Error sample and hold circuit (S&H in Fig. 2)

The error sample and hold circuit (Fig. 3) samples the magnitudes of both signals V_{ref}, and $V_{out}(I_{ref})$, and provides them on two capacitances C_{shunt} and C_{rog}. The error signal, which is further integrated, is thus simply the difference of both capacitance voltages. It is mandatory for this error to be free from any offset, otherwise the RCCT absolute sensitivity would be biased. For that purpose, V_{ref} and $V_{out}(I_{ref})$ are sampled alternatively through the same analog path. Note that both signals do not need to be sampled at the same time since we are looking to correct a steady-state error coming from long term sensitivity drift or sensitivity deviation due to off-center conductors. So the first signal is selected by the input multiplexer. The maximum of the signal is detected thanks to a conventional differentiator followed by a comparator with a hysteresis of a few millivolts (Fig. 4). The hysteresis is obtained by cross-coupling two unbalanced single-ended differential stages in such a way that an internal positive feedback occurs at switching [11]. The logic control module samples the signal, switches the input multiplexer, but disables at the same time the sampling for half the reference signal period ($T_{ref} = 1/f_{ref}$) in order to avoid false maximum detection due to unavoidable multiplexer switching noise at the input of the differentiator (Fig. 5). It means that the delay between the sampling of V_{ref} and $V_{out}(I_{ref})$ is $1.25 \cdot T_{ref}$, which is very small comparing to the time scale of possible RCCT sensitivity deviations. In contrast, the analog processing path being the same for V_{ref} and $V_{out}(I_{ref})$, the sampling delay is exactly the same (Fig. 5) leading to an error signal, and thus an RCCT calibrated sensitivity, free from any bias.

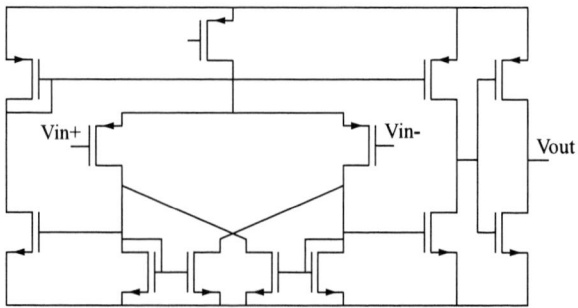

Fig. 4 Architecture of the comparator with hysteresis

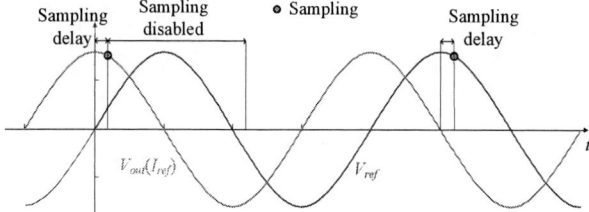

Fig. 5 Timing of the error signal sampling process

B. Chopper stabilized integrator

The integrator (Fig. 6) was implemented from a differential input single-ended output voltage-to-current (V-I) converter whose architecture is given in Fig. 7. It uses a cascode output current mirror in order to have a very high output impedance, and ensure a precise output current $I = (V_{ref} - V_{out}(I_{ref}))/R$, which is integrated over the capacitance C (Fig. 6).

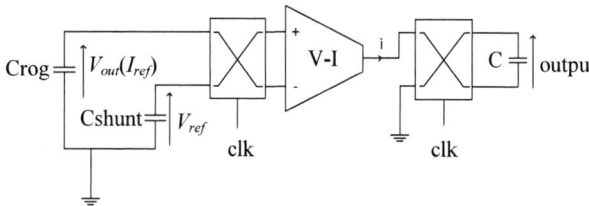

Fig. 6 Chopper stabilized integrator

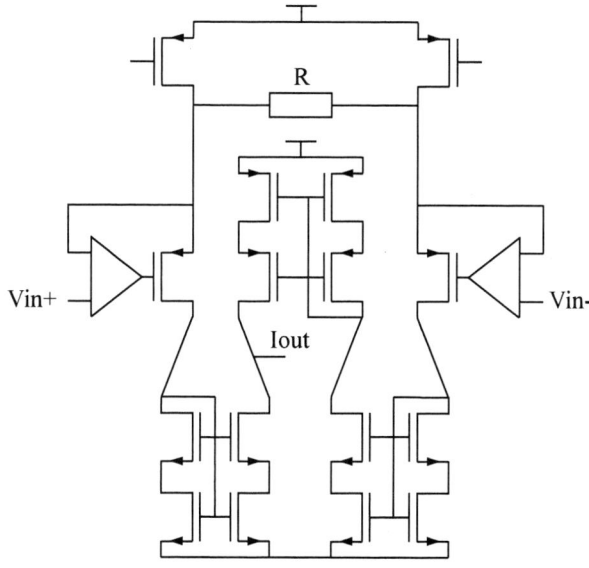

Fig. 7 Architecture of the V-I converter used in the integrator

As for the error sample and hold circuit, the integrator has to be free from offset in order to avoid a random bias of the RCCT absolute sensitivity. It was thus chopper stabilized by inserting switching boxes at its input and output, as shown in Fig. 6. Conveniently, the chopping frequency was chosen to f_{ref}. Unfortunately, when the chopping is activated, the charge injection from the input switching box to the hold capacitances C_{rog} and C_{shunt} leads to a high unavoidable offset. Chopper stabilization was thus disabled during the test.

Adding unity gain buffers to decouple the hold capacitances and the switching box would shift the issue to the buffers. However, a possible improvement would be to suppress the input switching box, and to simply sample successively the error $V_{ref} - V_{out}(I_{ref})$ and its opposite by selecting C_{rog} and C_{shunt}, and then C_{shunt} and C_{rog}, to sample V_{ref} and $V_{out}(I_{ref})$. It can be easily implemented in the logic control block (Fig. 3).

IV. EXPERIMENTAL RESULTS

Fig. 8 Tested flexible Rogowski coil

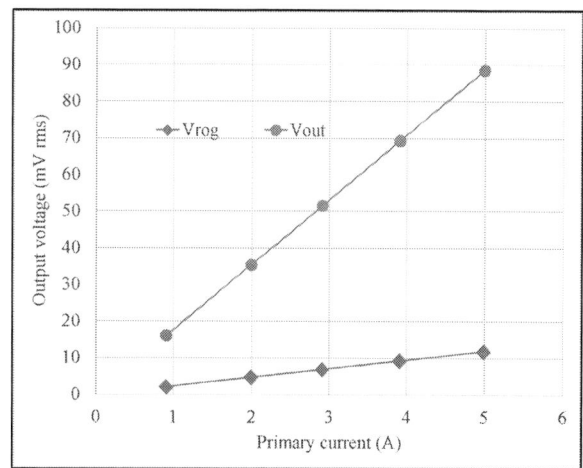

Fig. 9 RC (Vrog) and RCCT (Vout) response

Designed to ensure a ±0.1% accuracy, the PI controller ASIC was integrated in the AMS 0.35µm CMOS technology, and mounted on a Printed Circuit Board (PCB) which also features the reference signal generator whose frequency is 12.68 kHz. As shown in Fig. 8, the primary conductor is made of a 5-conductor gray cable which is inserted 4 times through the tested RC, a commercial flexible RC. All conductors are in series, leading to an effective primary current of 100 A, 20 times higher than the 5 A current provided by the power source (Agilent 6813B). The same way, the reference conductor is made with a basic green cable attached coaxially to the primary conductor and inserted 8 times through the RC. On the PCB, at the output of the RC, an external variable gain amplifier is mounted in order for the calibration system to comply with different RC. The primary current were generated at 60 Hz in order to be free from any ambient 50 Hz noise. Then the signal rms magnitude was measured at 60 Hz and 12.68 kHz, at the output of the RC (Vrog) and at the output of the calibration system (Vout) thanks to a dynamic signal analyzer (Agilent 35670A).

The conductors being centered with respect to the coil (position #3 in Fig. 8), Fig. 9 presents the RCCT response to a primary current varying from 1 A (effective 20 A) to 5 A (effective 100 A). As expected the response is highly linear.

Then the primary conductor was off-centered on positions #1 to #5, and the relative deviation δ of the sensitivity without calibration, i.e. at the output of the RC (Vrog), and with calibration, i.e. at the output of the system (Vout), was determined for each position as:

$$\delta = \frac{S_i - S_3}{S_3} \qquad (4)$$

where $S_i = V_i/I_p$ is the sensitivity, and V_i is the voltage, i.e. Vrog or Vout, at position #i. Note that for each measurement, I_p was monitored since the power source was slightly fluctuating, in the order of ±0.04%.

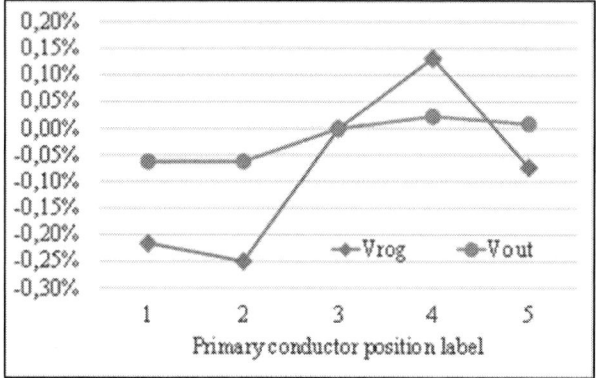

Fig. 10 Sensitivity relative deviation at 60 Hz along the longitudinal axis of the Rogowski coil

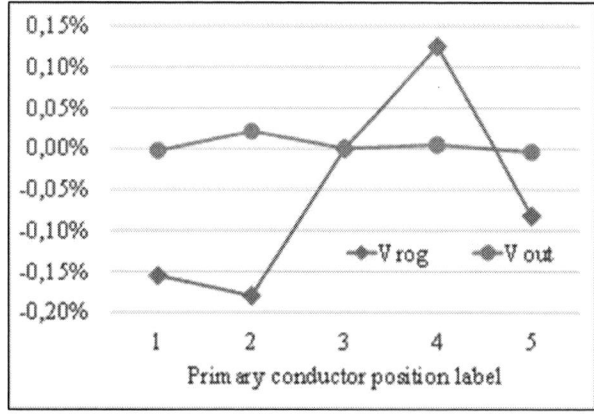

Fig. 11 Sensitivity relative deviation at 12.68 kHz along the longitudinal axis of the Rogowski coil

The commercial RC we used exhibits already high performances since its deviation is between -0.25% and 0.13%. However it doesn't reach a 0.1 class measurement, i.e. ±0.1%. In contrast, with the calibration, the sensitivity deviation is reduced to [-0.06% to 0.02%] and the RCCT becomes class 0.1. Making the same measurements but at the reference frequency (Fig. 11), we see that the deviation at the output of the system is even better, between 0% and 0.02%. The discrepancy between both results comes from the fluctuation of the power source, while the reference generator is much more stable. It means that the system may be close to a 0.01 class, but a better power source would be necessary to prove it.

V. CONCLUSION

An on-chip PI controller designed for continuous calibration of RCCT has been presented. It ensures a 0.1 class measurement whatever the position of the primary conductor, i.e. the sensitivity relative deviation due to primary conductor off-center is below ±0.1%. A way to improve the chopper stabilization of the integrator, a key element of the PI loop, has been proposed. This improvement should allow our system to ensure an absolute calibration of the current sensor, avoiding costly calibration process at the end of the sensor fabrication.

ACKNOWLEDGMENT

The authors would like to acknowledge ABB Switzerland Ltd. – Corporate research, for its financial support under contract ABB-InESS 063108.

REFERENCES

[1] J. D. Ramboz, "Machinable Rogowski Coil, Design, and Calibration", IEEE Trans. on Inst. and Meas., vol. 45, n° 2, pp 511-515, April 1996

[2] W. F. Ray, and C. R. Hewson, "High performance Rogowski current transducers, IEEE Industry Applications Conference, vol. 5, pp. 3083-3090, October 2000

[3] L. Ferkovic, and al., "Mutual Inductance of a Precise Rogowski Coil in Dependence of the Position of Primary Conductor", IEEE Trans. on Inst. ans Meas., vol. 58, no. 1, pp. 122-128, January 2009.

[4] M. Marracci, and al., "Critical Parameters for Mutual Inductance Between Rogowski Coil and Primary Conductor", IEEE Trans. on Inst. and Meas., vol. 60, no. 2, pp.625-632, February 2011.

[5] G. Crotti, D. Giordano and A. Morando, "Analysis of Rogowski coil behavior under non ideal measurement conditions", Fundamental and Applied Metrology, XIX IMEKO World Congress, pp. 876-881, September 2009.

[6] M. Chiampi, G. Crotti and A. Morando, "Evaluation of Flexible Rogowski Coil Performances in Power Frequency Applications", IEEE Trans. on Inst. and Meas., vol. 60, no. 3, pp. 854-862, March 2011.

[7] S. Paulus, J.-B. Kammerer, J. Pascal, and L. Hébrard, "Continuous calibration of Rogowski coil current transducer", IEEE NEWCAS Conference, June 2015

[8] S. Paulus, J.-B. Kammerer, J. Pascal, and L. Hébrard "Integrated front-end for on-line continuous calibration of Rogowski coil current transducer", IEEE ICECS Conference, Dec. 7-10, pp. 391-394, 2014

[9] J. Pascal, Y. Maret, J.-B. Kammerer, R. Disselnkoetter, Current transducer of the Rogowski type and arragement for measuring a current, European Patent, no EP2653875 B1, Published on Sept. 10th, 2014

[10] J. Pascal, J.-B. Kammerer, S. Paulus, Current measurement device and method using a Rogowski type current transducer, European Patent, no WO2015/051983 A1, Published on April 16th, 2015

[11] P. E. Allen, D. R. Holberg, "CMOS analog circuit design", Oxford University Press, 3rd Edition, 2012

A Comparative Analysis of Optimized CMOS Neural Amplifier

Ahmed El-Attar[1], Saif Ahmed[1], Youssef Abdelkader[1], Mohamed Badran[1], Ali H. Hassan[2], and
Hassan Mostafa[3]

[1,2,3]*Electronics and Communications Engineering Department, Cairo University, Giza 12613, Egypt*
[2]*Electronics Department, Faculty of Information Engineering and Technology, German University in Cairo (GUC),
New Cairo 11432, Egypt*
[3]*Center for Nanoelectronics and Devices, AUC and Zewail City of Science and Technology, New Cairo 11835,
Egypt*
Email: {ahmedadelelattar@gmail.com, saifeldeen_murad@yahoo.com, y.abdelkader.ece@gmail.com,
mohammed.mamb@yahoo.com, ali.h.hassan@ieee.org, hmoustafa@aucegypt.edu, and hmostafa@uwaterloo.ca}

Abstract—This paper investigates various implementation techniques of neural amplifiers with emphasis on their design performance metrics and trade-offs such as gain, noise, power consumption, and area. The proposed comparative analysis covers the recently published neural amplifiers in the literature, and proposes a basic optimization method for these amplifiers. These amplifiers are redesigned using UMC 130nm CMOS technology for fair comparison.

Keywords— neural amplifiers, low noise, subthreshold OTAs, self-biased OTAs, and pseudo resistors.

I. INTRODUCTION

The Rapid growth of the electronic industry and neuroscience research led to evolutional milestones in biomedical systems. Based on this growth, scientists spare no effort in developing microsystems that are able to record and stimulate neural brain activities. This helps in treating several neurological disorders such as Epilepsy and Parkinson diseases [1].

Design of a neural recording system faces many challenges, starting from the design of the implantable electrode needed for recording the neural signal, and ended with the feature extraction from the recorded signals. A low noise pre-amplifier (LNA) is needed, because the bio-potential signals' amplitude ranges from a few microvolts to several millivolts, and covers a wide range of frequencies from few millihertz to kilohertz [2]. In order to process these signals, they should be digitized with a high resolution, and low frequency analog-to-digital converter (ADC). The most critical part of this system is the design of the LNA, as it filters the noise without reshaping the signal, adds acceptable gain, rejects the dc offset of the electrode tissue interface, and minimizes the power consumption.

This paper introduces a comparative review and analysis of the different LNA architectures recently published in the literature, and attempts to optimize the performance metrics for lower noise, lower power consumption, better dc cancellation, and lower implementation area. The rest of the paper is organized as follows. In Section II, the design and analysis of the different LNA architectures is presented, followed by the basic optimization technique in Section III, and ended by the simulation results in Section IV. Finally, a conclusion is derived in Section V.

II. LNA ARCHITECTURES

In this Section, different LNA architectures are discussed. The LNA architectures are classified based on (1) differential or single ended, (2) the operational transconductance amplifier (OTA) structure, and (3) the feedback resistors to achieve the required specifications based on the application are presented.

A. Fully Differential Capacitive Feedback LNA

Using a fully differential amplifier applies a great advantage in reducing the noise whether it comes from the circuit itself, or the neighbouring circuits. In Fig. 1, the closed loop LNA circuit is shown, where the mid-band gain is set by the ratio C_{in}/C_f, and the lower cut-off frequency is calculated by $1/2\pi R_f C_f$ [5]. The low cut-off frequency should vary from 0.1 Hz to 10 Hz, and accordingly, we need to implement a very high impedance feedback resistor in range of GΩ [5]. To implement these resistors, PMOS transistors biased in the subthreshold region are used, where we can obtain any frequency in the range stated above (i.e., 0.1Hz to 10Hz) by changing V_{res} [4]. Regarding the higher cut-off frequency, it is controlled by the bias current of the OTA and the load capacitance [5].

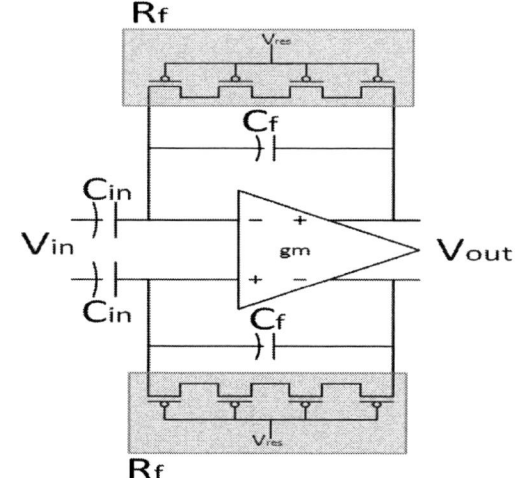

Fig. 1 Closed Loop LNA Schematic [4]

978-1-4673-8760-6/15 $31.00 © 2015 IEEE

A fully differential Telescopic OTA is used to implement this LNA. This OTA provides low power consumption, good noise performance, and a sufficient output swing. The OTA circuit schematic, and its common mode feedback (CMFB) circuit are shown in Fig. 2. Due to the low operating frequency range of neural amplifiers, the impact of the flicker noise and the thermal noise is significant and must be taken into account [4].

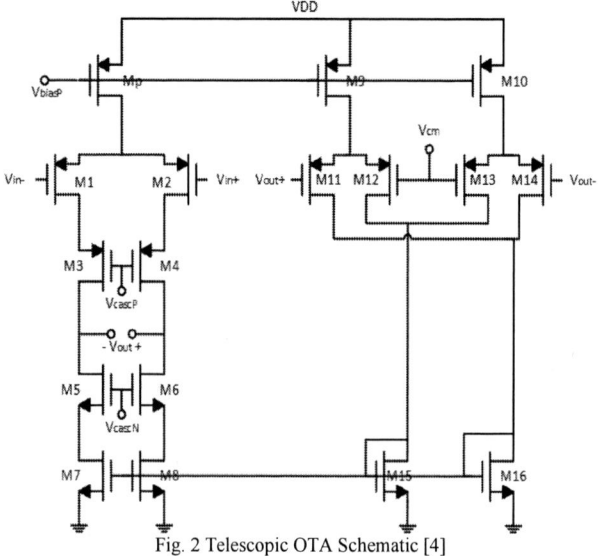

Fig. 2 Telescopic OTA Schematic [4]

B. Single Ended Capacitive Feedback LNA

This LNA uses a single ended OTA with a capacitive feedback topology [3], as shown in Fig. 3. Two coupling capacitors are used to reject the dc offset associated with the sensing electrodes. The high-pass cut off frequency is controlled by the biasing resistance on the positive input terminal of the OTA in parallel with the capacitor. To achieve an extremely low high-pass cut off frequency, a resistance in the range of GΩ is needed. This huge resistance is implemented by using two pseudo resistors connected as shown in Fig. 3. On the other hand, the low-pass cut off frequency is set by the gain of the OTA and the load capacitance connected at the output node. Finally, the gain of the amplifier is controlled by the ratio between the coupling capacitor divided by the feedback capacitor [3].

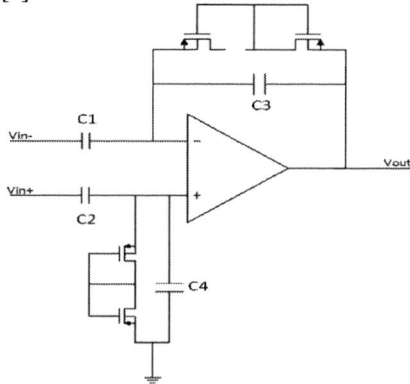

Fig. 3 Single Ended pre-amplifier Schematic [3]

A two stage amplifier with a Miller compensating capacitor C_5 is used for the OTA design as shown in Fig. 4. A cascode stage that is formed by the transistors M3, and M4 is added to the input differential pair transistors (M1, and M2) in order to increase the gain of the first stage, and to enhance the phase margin of the amplifier by cancelling the right half plane zero, which is formed by the miller capacitance, instead of using a compensation resistance [3]. The path formed from the capacitance C_6 with the transistor M10 is used to enhance the settling time of the amplifier. To minimize the flicker noise of the input pair transistors, their gate size is designed large. Thus, increasing their areas means decreasing the input referred noise of the OTA [3].

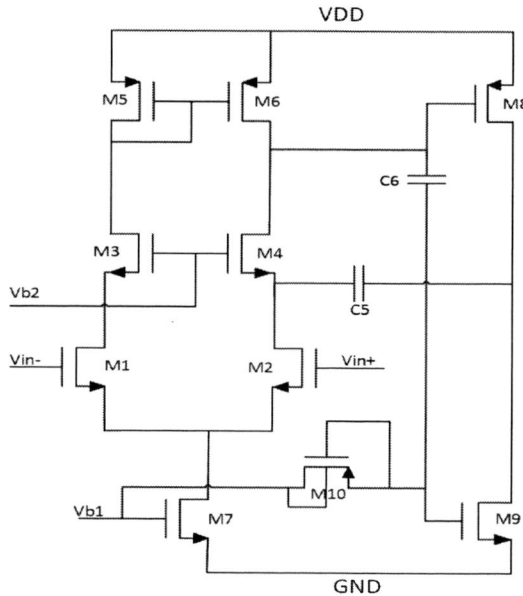

Fig. 4 OTA Schematic [3]

C. LNA with Active Low Frequency Suppression

This LNA with its active low frequency suppression circuit is shown in Fig. 5. This LNA presents an active dc rejection scheme that does not require any AC coupling capacitors. This makes it suitable for the massive integration in implantable neural recording amplifiers, and preserves the LNA high input impedance [6]. This configuration relies on placing an active integrator in the feedback path of the LNA that rejects any DC offset and places the high pass cut off frequency within the transfer function [6].

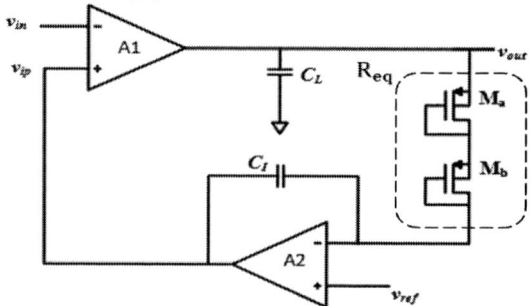

Fig. 5 LNA with the active feedback network Circuit Schematic [7]

978-1-4673-8760-6/15 $31.00 © 2015 IEEE

This LNA circuit consists of a single ended low noise OTA (A1) shown in Fig. 6, where the OTA output is connected to a Miller integrator in the feedback path. The active feedback network is designed from another OTA (A2), a capacitor (C_I) and a high impedance pseudo resistor (R_{eq}) which leads to a high time constant ($\tau = R_{eq}C_I$), and sets the -3db high pass cut off frequency (f_{hp}) as shown in the following equation [7]:

$$f_{hp} = \frac{1}{2\pi}\frac{A_{01}}{R_{eq}C_I}$$

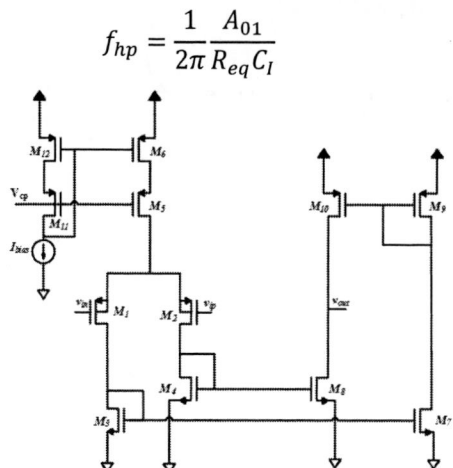

Fig. 6 Main OTA Circuit Schematic [7]

In addition, the midband gain of this LNA is the same as the DC gain of OTA1 (A_{01}), where $A_{01} = g_m/g_0$ ($g_m = g_{m1,2}$ and $g_m = g_{m8,10}$). Finally, the -3db low-pass cut off frequency (f_{1p}) is given by the following equation [7]:

$$f_{1p} = \frac{1}{2\pi}\frac{gm}{A_{01}C_L}$$

For the active feedback network, a two stage OTA topology with a lead compensation was chosen for A_2 as shown in Fig. 7. This OTA is used to provide a high dc gain, and maintain less implementation area. The integrator circuit ensures that node V_{ip} tracks the dc level seen by V_{in}, cancelling any dc offset, or any other low-frequency voltages. Moreover, it introduces a 90° phase shift to the output signal [8].

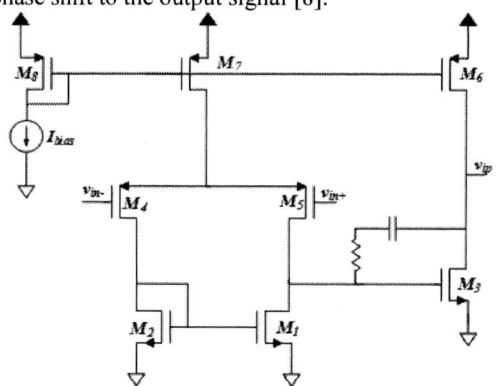

Fig. 7 Feedback OTA Circuit Schematic [7]

III. OPTIMIZATION

In this section, an optimization method for lower power consumption, and better noise performance is discussed. The

pre-mentioned neural amplifiers are optimized by replacing the used OTA by an optimized one.

A. Current-Reuse OTA

The circuit schematic is shown in Fig. 8, where it achieves an extremely low power consumption, and a better noise performance [11]. To minimize the flicker noise, and the thermal noise, the input transistors (P1, P2, N1, and N2) are biased in the subthreshold region. For the CMFB circuit, the transistors (N3, and N4) are used to define the DC output level, and a resistive sensing elements are used [11]. These resistive sensing elements are implemented using the two diode connected PMOS transistors. Thus, they act as a large impedance, and maintain a high gain at the output [11].

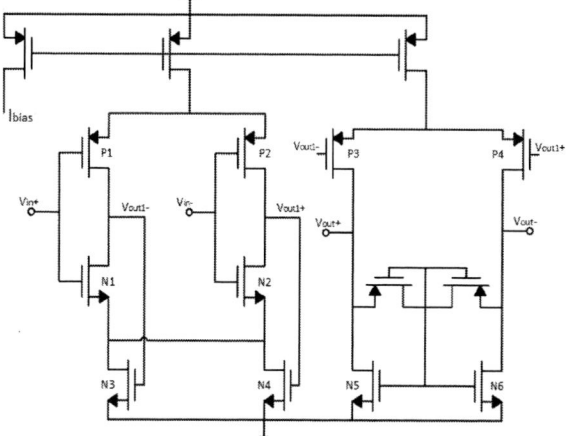

Fig. 8 Fully Differential Current-Reuse OTA Circuit Schematic [11]

B. Bulk-Driven OTA

To reduce the power consumption extensively, bulk-driven OTAs are good candidates for this purpose. The idea is to connect the input voltage to the bulk terminal which offers several advantages such as the removal of the threshold voltage constraints, wider input common mode range spanning from negative voltages to slight positive input voltage, and operation on low supply voltages [9,10]. On the other hand, the bulk driven OTAs have several disadvantages including higher input referred noise, lower transconductance, and lower bandwidth. Moreover, the bulk driven fabrication process usually has higher cost [10].

In Fig. 9, the schematic of the enhanced bulk driven OTA is shown. To increase the transconductance of the transistors, a positive feedback source degeneration is proposed and used to reach a higher gain and relatively a higher bandwidth [10]. Moreover, this enhancement is carried out without any need to increase the power consumption. The positive feedback source degeneration is the only difference from the traditional bulk driven OTAs. This OTA consists of a two stage OTA with a Miller compensation between the two stages, in order to improve the phase margin and stability of the system. The first stage is a folded cascode stage with an input bulk driven differential pair, while the second one is a normal common source OTA [10].

978-1-4673-8760-6/15 $31.00 © 2015 IEEE

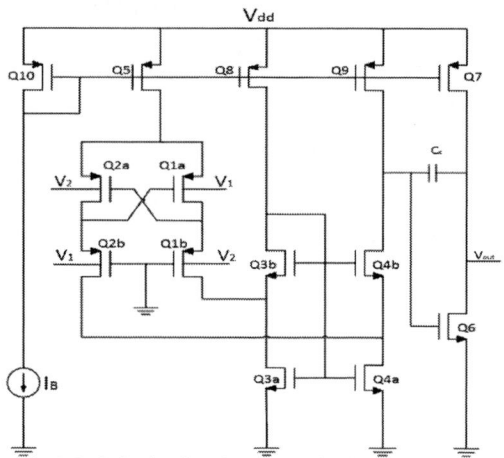

Fig. 9 Optimized Bulk-Driven OTA Circuit Schematic [10]

IV. SIMULATION RESULTS

In this section, all the previous neural LNA circuits are simulated using hardware-calibrated UMC 130 nm CMOS technology under the same environment to guarantee fair comparison as listed in Table I.

Here are the proposed modifications as following:

o The neural amplifiers are re-simulated, and achieved better results due to technology scaling as shown in Table I.

o For the fully differential capacitive feedback LNA, single stage amplifier, the telescopic OTA has been replaced by current re-use OTA for lower power consumption, and better noise performance as shown is shown in Table I.

o For the single ended capacitive feedback LNA, the circuit was resized from the beginning for optimized results, so no need for further investigation, and this obvious in Table I.

o For the LNA with active low frequency suppression, both OTAs replaced by the bulk-driven OTA. This modification has improved the midband gain, the power consumption, but the input referred noise has been degraded. However, this was expected from the bulk driven OTA.

V. CONCLUSION

In this paper, using 1.2 V supply voltage and 130 nm technology, several types of neural amplifiers are designed and simulated using the same technology and the same environmental conditions to guarantee fair comparison. The comparison shows many topologies whether differential or single ended output, wide bands for different applications, and a several dc rejection techniques. The comparison emphasizes the effect of replacing the used OTAs by a more efficient designs reducing the power consumption in such great rate. Moreover, the noise performance, and the bandwidth improved, but not with all designs.

VI. ACKNOWLEDGMENT

This research was partially funded by Cairo University, ITIDA, NTRA, NSERC, Zewail City of Science and Technology, AUC, the STDF, Intel, Mentor Graphics, SRC, ASRT and MCIT.

REFERENCES

[1] R. Shulyzki, K. Abdelhalim, a. Bagheri, C. M. Florez, P. L. Carlen, and R. Genov, "256-Site active neural probe and 64-channel responsive cortical stimulator," Proc. Cust. Integr. Circuits Conf., pp. 5–8, 2011.

[2] R. R. Harrison, "A versatile integrated circuit for the acquisition of biopotentials," Proc. Cust. Integr. Circuits Conf., pp. 115–122, 2008.

[3] X. Zou, X. Xu, J. Tan, L. Yao, and Y. Lian, "A 1-V 1.1-µW Sensor Interface IC for Wearable Biomedical Devices" 2008 IEEE Int. Symp. Circuits Syst., pp. 2725–2728, 2008.

[4] F. Shahrokhi, K. Abdelhalim, D. Serletis, P. L. Carlen, and R. Genov, "The 128-Channel Fully Differential Digital Integrated Neural Recording and Stimulation Interface," IEEE Trans. Biomed. Circuits. Syst., vol. 4, no. 3, pp. 149–161, 2010.

[5] R. Neurostimulation, R. Shulyzki, K. Abdelhalim, A. Bagheri, M. T. Salam, C. M. Florez, J. Luis, P. Velazquez, P. L. Carlen, and R. Genov "320-Channel Active Probe for High-Resolution Neuromonitoring and Responsive Neurostimulation," IEEE Trans. Biomed. Circuits. Syst., vol. 9, no. 1, pp. 34–49, 2014.

[6] B. Gosselin, A. E. Ayoub, and M. Sawan, "A low-power bioamplifier with a new active DC rejection scheme," 2006 IEEE Int. Symp. Circuits Syst., pp. 2185–2188, 2006.

[7] B. Gosselin, M. Sawan, and C. A. Chapman, "A Low-Power Integrated Bioamplifier With Active Low-Frequency Suppression," IEEE Trans. On Biomed. Circuits and Syst., vol. 1, no. 3, pp. 184–192, 2008.

[8] R. J. Baker, CMOS Circuit Design, Layout, and Simulation, Second Edition, New York: Wiley-IEEE, 2005.

[9] L. H. C. Ferreira and S. R. Sonkusale, "A 60-dB Gain OTA Operating at 0.25-V Power Supply in 130-nm Digital CMOS Process," IEEE Trans. Circuits Syst. I Regul. Pap., vol. 61, no. 6, pp. 1609–1617, 2014.

[10] F. Khateb, S. Bay Abo Dabbous, and S. Vlassis, "A survey of non-conventional techniques for low-voltage low-power analog circuit design," Radioengineering, vol. 22, no. 2, pp. 415–427, 2013.

[11] L. Liu, X. Zou, W. L. Goh, R. Ramamoorthy, G. Dawe, and M. Je, "800nW 43nV/$\sqrt{Hz}$ neural recording amplifier with enhanced NEF," Electron. Lett., vol. 48, no. 9, pp. 479–480, 2012.

TABLE I
Comparison between different LNA architectures

Reference	Circuit Topology	Midband Gain (dBs)	High-pass cut off frequency (Hz)	Low-pass cut off frequency (KHz)	Input Referred Noise (µV / $\sqrt{Hz}$)	Power Consumption (µW)	DC Rejection Technique
[4]	Differential Ended Output	33	10	5	3.75	4.76	AC Coupling
[3]	Single Ended Output	45.88	0.05	2.1	1.195	0.1712	AC Coupling
[8]	Single Ended Output	53.78	9	11	4.42	8.1	Active Feedback Sensing
Optimized for [4]	Differential Ended Output	33	5.7	10.5	2.34	0.1953	AC Coupling
Optimized for [8]	Single Ended Output	79.2	6.1	0.143	20	1.52	Active Feedback Sensing

978-1-4673-8760-6/15 $31.00 © 2015 IEEE

A Circular Patch Antenna Using Two-Dimensional Photonic Crystal Substrate

Boualem Mekimah

Electronics Department
University of Constantine1
Constantine, Algeria
b_mekimah@umc.edu.dz

Abderraouf Messai and Abdelkrim Belhedri

Electronics Department
University of Constantine1
Constantine, Algeria
r_messai@yahoo.fr

Abstract—**The Photonic Crystal Substrate (PCS), due to its frequency bandgap, can reduce the surface-wave modes significantly that have direct effects on the gain, radiation efficiency and directivity of the microstrip patch antennas. In this paper, a novel work used a circular patch antenna with 2-D photonic crystal substrate is studied and analyzed using the software package High Frequency Structure Simulator (HFSS). The results show that the circular patch antenna with a PC substrate has a lower return loss that gives us a higher bandwidth and a higher directivity compared to the patch antenna using a conventional substrate.**

Keywords- Two-dimensional Photonic Cristal; bandgap; Circular Patch Antenna; return loss; gain; directivity.

I.   INTRODUCTION

Microstrip Patch Antennas (MPA) have many advantages, in particular low profile and conformity to planar and non-planar surfaces that are highly recommended. However, they have low radiation efficiency and low gain which are highly required. Photonic crystal substrates are one of the solutions to improve these performances by reducing the coupled power into surface-wave modes [1]-[4].

A Photonic Crystal with its frequency bandgap forbid electromagnetic waves to propagate through substrate [5]-[6]. To use PCS with patch antennas, we have to match the bandgap frequency of the PCS with the operational frequency of the patch antenna in order to reduce the excitation of surface wave modes, therefore to improve the antenna radiation efficiency, gain and bandwidth as mentioned previously.

This paper presents a study of a circular patch antenna and a rectangular patch antenna using 2-D photonic crystal substrate. Both designs are analyzed using the finite elements method (FEM) integrated in software package High Frequency Structure Simulator (HFSS). The results show that the circular and rectangular patch antennas with photonic crystal substrate have a lower return loss, which gives us a higher bandwidth and higher directivity compared to patch antennas using conventional substrates.

II.   DESIGN AND RESULTS

2-D Photonic Crystals, with periodic holes in a dielectric substrate, are characterized by lattice constant a: distance between center of holes, r: radius of the holes, ε_r: dielectric constant of substrate, and photonic crystal factor PF which is the ratio of a hole surface area to its unit cell surface area.

Fig. 1 shows the structure of a two-dimensional photonic crystal with periodic holes in a square dielectric substrate. The host material of the dielectric substrate is Arlon AR1000 (relative permittivity ε_r =10).The length and height of the square dielectric substrate are 500 µm and 200 µm, respectively. Holes with radius of 18 µm, lattice constant of 50 µm, then a ratio of 36, and 9x9 holes cylinders are embedded in the dielectric substrate.

The radiating element (patch) and microstrip feed line are both from copper. The length of the square radiating element is 200 µm, and its thickness is 1µm. The width, the length and the thickness of the feed line are 50 µm, 93 µm and 1 µm respectively. To avoid a serious surface waves excitation, we used a ratio: height/wave length, less than 0.03(h/λ<0.03) [7].

We have conserved the same surface of the square patch, and we used a circular patch with a radius (R) calculated from the following equation: πR^2 =200x200 µm^2, in order to keep the same square patch surface, then R=112.84 µm, conserving the same thickness 1 µm, as shown in figure2.

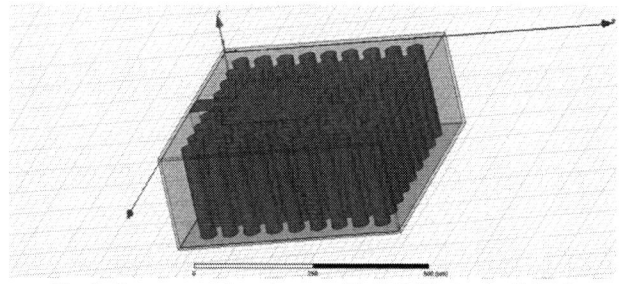

Fig.1: Square patch antenna with 2-D Photonic Cristal substrate.

978-1-4673-8760-6/15 $31.00 © 2015 IEEE

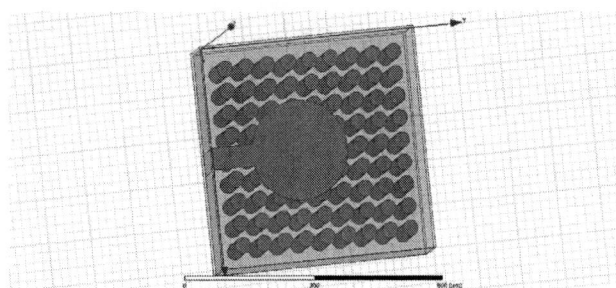

Fig.2: Circular patch antenna with 2-D Photonic Cristal substrate.

In figures 3 and 4, it is clearly indicated that the both circular and rectangular patch antennas with PC substrate, have a lower return loss, and a higher bandwidth. In fig. 3 the return loss (S_{11}) decreased from -10.939 dB to -18.344 dB, with a shift in operational frequency from 32.54 GHz without PC, to 40.34 GHz with PC. Therefore a higher bandwidth is obtained.

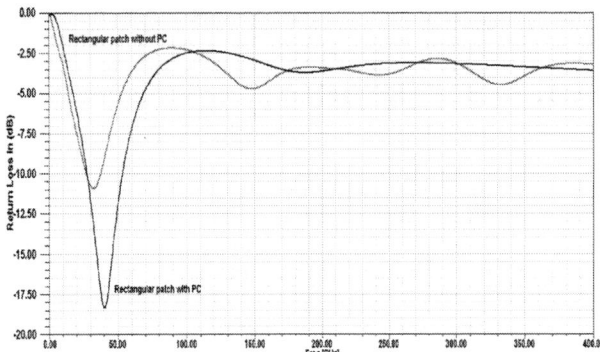

Fig.3: Return loss in (dB) versus Frequency in (GHz); rectangular patch without PC in red, rectangular patch with PC in black.

In the case of circular patch fig.4, the return loss (S_{11}) decreased from -11.136 dB to -18.027dB, with a shift in operational frequency from 32.78 GHz without PC, to 39.42 GHz with PC, which a higher bandwidth is also obtained.

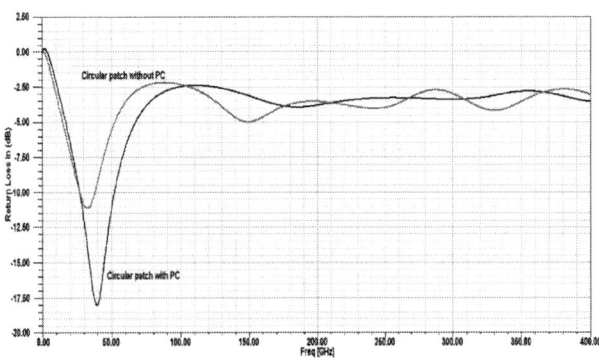

Fig.4: Return loss in (dB) versus Frequency in (GHz); circular patch without PC in red, cirular patch with PC in black.

In figures 5 and 6, Radiation Patterns are presented in the both cases, when phi=0 degree and phi=90 degrees, with PC and without PC. As shown in figures, a higher directivity was obtained with a small variation in gain.

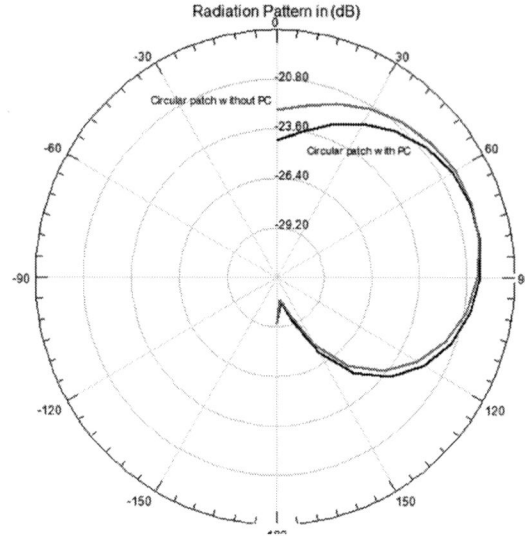

Fig.5: Radiation Pattern in (dB) at 40GHz; cicular patch without PC in red, cicular patch with PC in black; Phi=0 degree.

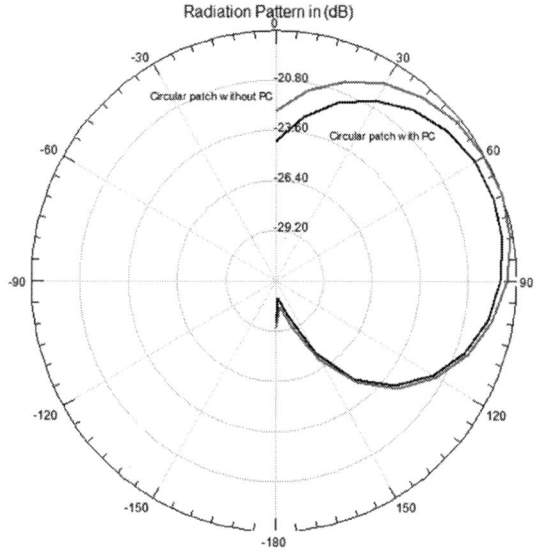

Fig.6: Radiation Pattern in (dB) at 40GHz; cicular patch without PC in red, cicular patch with PC in black; Phi=90 degrees

III. CONCLUSIONS

A comparison between a circular patch antenna with photonic crystal and without photonic crystal substrate has been done. Both designs are analyzed by the finite elements method (FEM) using software package High Frequency Structure

Simulator (HFSS). The results show that the circular patch antenna with a PC substrate gives almost similar performances if we used the same surface of a square patch antenna in the same configuration. Circular patch with PC gives a lower return loss, a higher bandwidth and a higher directivity compared to the circular patch without PC.

REFERENCES

[1] L. Yang, X. Shi, K. Chen, K. Fu, and B. Zhang "Analysis of photonic crystal and multi-frequency terahertz microstrip patch antenna," Physica B 431, pp.11–14 , 2013.

[2] R. Yu-xing, S. Jiang-dong, W. Ge, G. Bo, and T. Xiao-jian, "Design of a rectangular patch antenna with a photonic crystal substrate," The Journal of China Universities of Posts and Telecommunications, 18(Suppl.2), pp.161-163, December 2011.

[3] H. Azarinia, and A. Tavakoli,"Finite difference time domain analysis of a photonic crystal substrate patch antenna," Physica B 370, pp.223–227, 2005.

[4] R. Gonzalo and G. Nagore," simulated and measured performance of a patch antenna on a 2-dimensional photonic crystals substrate," Progress In Electromagnetics Research, PIER 37, pp.257–269, 2002.

[5] R. Gonzalo, B. Martınez, P. DeMaagt and M. Sorolla," improved patch antenna performance by using photonic bandgap substrates," Microwave and Optical Technology Letters , vol. 24, no. 4, February 2000.

[6] R. Gonzalo, P. Demaagt, and S. Mario, "Enhanced patch antenna performance by suppressing surface wave using photonic crystal substrates," IEEE Trains Microwave Theory and Techniques, 47(11), pp.2131–2138, 1999.

[7] J.R. James and A. Henderson," High-frequency behavior of microstrip open-circuit terminations," IEE J Microwaves, Opt, Acoust 3. pp.205-218 ,1979.

978-1-4673-8760-6/15 $31.00 © 2015 IEEE

SBLS: Speed Based Lane Changing System in VANETs

Jetendra Joshi, Kritika Jain, Yash Agarwal, Manash Deka and Pravit Tuteja

Dept. Of Electronics and Communication Engineering

NIIT University

Neemrana, Rajasthan, India

Jetendra.Joshi@niituniversity.in, {Kritika.12.Jain, Yash.Agarwal, Manashj.Deka, Pravit.Tuteja}@st.niituniversity.in

Abstract-- In vehicular ad hoc networks (VANETs) the network services and applications (e.g., safety messages) will require an exchange of vehicle and event location information. Effective lane changing and routing in Vehicular Ad hoc Networks is a challenging task. This paper aims to propose a solution to ensure the safety of drivers while changing lanes on the highways. Efficient and faster routing protocols could play a crucial role in the applications of VANET, safeguarding both the drivers and the passengers and thus maintaining a safe on-road environment. In this paper we propose SBLS: Speed Based Lane Changing System in VANETs, for effective lane changing in the dynamic mobility model. In our approach we present the lane changing based system on speed and minimum gap between the vehicles in VANET. The test bed is created on the techniques used in the proposed system where the analysis takes place in the On Board Embedded System designed for Vehicle Navigation. The designed system was tested on a 4-lane road at Neemrana in India. Successful simulations have been conducted along with real time network parameters to maximize the QoS (quality of service) and performance using SUMO and NS-2.

Keywords-- GPS, RSU, smart cities, V2V, V2I, VANET.

I. INTRODUCTION

Taking the upcoming model of driverless cars and smart cities into consideration, it is very important to build up mobility models and algorithms for safe and efficient environment. With the introduction of vehicular ad hoc network (VANET) in the field of transportation, a new area for research has evolved out of it. VANET is basically a subset of mobile ad hoc network (MANET) where all the moving nodes behave as vehicles. The Vehicle-to-Infrastructure (V2I) and Vehicle-to-Vehicle communications (V2V) [1] play a very important role in this aspect. In the Vehicle-to-Infrastructure model, the information of the traffic is collected at the Road Side Unit (RSU) and is broadcast to the receiver vehicles and then sent to the central server for monitoring the vehicles. Figure 1 shows the communication model for central monitoring of vehicles. In the Vehicle-to-Vehicle model; the exchange of information takes place between the vehicles itself. This exchange of information among the vehicles and the RSU's may influence the movement of these vehicles. The communication between the vehicles, RSU and the central server follow the Dedicated Short Range Communication Protocol (DSRC). In the figure, the vehicles on the road communicate with each other and then the information is sent to the RSU. The RSU further exchanges information with the central traffic-monitoring server with the help of the Internet.

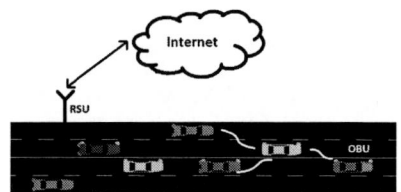

Fig. 1 Communication model (Vehicle-RSU-Central server)

For the implementation of the SBLS algorithm in our hardware and in the scenario as shown in Fig 1, we have assigned the road tracks with particular speed limits for different lanes and the vehicles moving on them will have to follow those particular speed limits. We have taken up a scenario where we have five different lanes on a road with the speed limit increasing from the left most lanes to the right most one. The model is therefore like 120Kmph->80Kmph->40Kmph->0Kmph(stop) lanes. The vehicles have to move according to the speed limits set above. While moving on these lanes, the vehicles can increase or decrease their speed and change the lane accordingly. Changing the lane will be possible only when the minimum gap between the vehicles is as set in the algorithm (SBLS). But if any particular vehicle is not doing so, the hardware implemented OBU (On Board Unit) in the vehicle will display a warning message to the driver to either change the speed or change the lane. If the defaulter vehicle does not follow the speed rules then the emergency warning message will be broadcast to the vehicles and the RSU's in the communication range of this vehicle so as to maintain a safe environment. The other vehicles will also receive this information that will influence their movement in a particular fashion. As soon as the number of defaulter's messages increase, the RSUs transmit the information to the nearest traffic monitoring system. Now the traffic monitoring authority will penalize or bound the faulty vehicle by analyzing the data. The latitude and longitude data will be parsed and then we will check the current position of the vehicle with respect to the lane and then check the correct speed of the vehicle with the lane speed limit. If the driver does not drive at the correct speed then the driver will be warned first in terms of alarm and displaying message on the LCD. After the elapsed time the warning message will be broadcast to the nearby vehicles and RSU's. The information received about the GPS coordinates and velocity readings allows the software implementation for accident prevention system by giving an audible warning tone whenever a vehicle is over speeding or driven in the wrong lane. 'Prioritized' and 'timely' transmission of warning messages for safety

978-1-4673-8760-6/15 $31.00 © 2015 IEEE   114

applications is crucial in VANETs to prevent fatal accidents and warn the drivers beforehand, hence, a SBLS algorithm is presented which is based on the appropriate routing algorithms such as AODV and GPSR as well as Priority Queue and are compared under the given scenario to provide the better Quality of service (QoS). The rest of this paper is organized as follows: In section II we have the related work done in this field. In section III, the system design and problem formulation has been discussed. In section IV, we have defined our protocol. In section V we have shown the performance evaluation, taken a look at the applications of GPS and the hardware implementation of the protocol, traffic model generated with the help of SUMO-MOVE followed by the simulation of the VANET network in NS-2. Section VI, concludes this paper and gives a glance of future work to be done in this model. Lastly, we have listed the references.

II. RELATED WORK

VANET (Vehicular ad hoc networks) is a very vast area for research and has opened gates for new possibilities and better technology in the field of transportation both in terms of safety and efficiency. C. -F. Chiasserini, E. Fasolo, [5] the paper discusses about the smart broadcasting of the messages in VANET. This will help in improving the efficiency and also security up to some extent. In [7] Rajendra Prasad Nayak the thesis proposes a method to calculate the speed of the vehicle based on the position of the vehicle. The vehicles exchange information with RSU and then the RSU sends it to the central monitoring server. Nehal Kassem, Ahmed E Kosba [8] presented the paper proposes a method of vehicle detection and speed estimation based on RF. The main drawback is that a vehicle can proceed to any speed in case of a miss, which can cause an error in the accurate estimation of speed. In [9] Ram Shringar Raw, Manish Kumar presented the paper throws light on various technical applications, advantages, challenges and issues in VANET and methods to improve the network system. Tanvee Kausar, Priyanka Gupta [12], the authors have proposed an approach for the collision avoidance system in VANETS based on the lane changing using GPS based hardware with trans receivers for data exchange. Yong Zhou, Rong Xu [13], the paper discusses about the lane changing and safety warning system based on virtual lane boundary. The safety system alerts the driver based on the width of the lane and the time required to cross that point. In [14, 15], the authors have discussed about the safe lane change assistance system to drivers with the help of cameras and proximity sensors. The basic terminology is to detect the presence of any vehicle in the blind spot region and change the lane based on the speed and the driving style of the driver. In [16,17], the authors have proposed an approach for lane change tracking and vehicle detection in VANET with the help of cameras mounted on the vehicles. The basic process involved is, the camera takes a recording of the surrounding of the vehicle and then image processing is done. Jamal Saboune, Mehdi Arezoomand [18] presented, the paper discusses about the vehicle detection for safe lane changing. The authors have proposed to use cameras mounted on the vehicles for detecting the vehicles in the blind spot region of the vehicle. In [20, 21,

22], the authors have done a comparative analysis on the various routing protocols (AODV and GPSR) for the data packet transmission in VANET.

III. SYSTEM DESIGN AND PROBLEM FORMULATION

Vehicle driving behavior will depend on many factors like situation, speeding, fast lane changing, density of traffic etc. Therefore, driving recklessly will cause accident and also effect the traffic movement. Now if a driver of a vehicle is at fault and does not follow the proper lane changing protocol and does over speeding then quick actions are required to be taken for stopping that vehicle. So what can be the solution to check the abrupt over speeding and lane changing mechanism of vehicle and how to minimize it? Also how to send faster warning message to the neighbor vehicle about the situation? Can we minimize the latency and increase the quality of service by using appropriate routing algorithm?

IV. PROPOSED SYSTEM (SBLS)

The protocol was developed to implement the efficient lane changing based on the speed of the vehicle for the sake of safety and an efficient environment in VANET. In this section we have discussed about the SBLS algorithm to study the behavior of the vehicles with its implementation and is checked with the help of SUMO-MOVE and the simulation of the algorithm in NS2. Before we start with the algorithm, here are a few definitions:

[11] Definition 1 (Road Segment): A road segment is defined by R where $R=\{s(x,y), e(x,y),l,v_l\}$. Here s and e are the starting and end points of the lane respectively with (x,y) as the location (x=latitude, y=longitude). l is the number of lanes (5 according to our scenario). v_l is the speed of the lane. The length of the road segment and the width of the lanes can be estimated with the help of the location of s and e.

Definition 2 (Minimum Gap): It is the minimum gap to be maintained between two vehicles on any particular lane on the road. In this algorithm it is denoted as min_{gap}.

Parameters:

Assume that there are 3 lanes $L_i=\{L_1, L_2, L_3\}$ with speed limits $(V_i)=\{V_1, V_2, V_3\}$ respectively with a tolerance of 10%.

L_{Bi} is the lane buffer ID which will help us in storing the previous lane ID in which the vehicle was moving.

V_{li} and V_{ui} are the considered as the lower and upper limits of the lane L_i. (Each lane has its own upper and lower limit.)The vehicles at the start point initiate the entire path planning process travelling in various lanes L_1, L_2 and L_3 at different speeds $S_i=\{S_1, S_2, S_3\}$ (speed of the vehicles). GPS will monitor L_{Bi} and helps to locate the ID of the last lane, the current position and speed to store in a Buffer of the monitor system. Now suppose that a vehicle is travelling with speed S_{est} in lane L_3 having a upper and lower speed limit lane at a particular point, the GPS based monitor system will check the speed of this vehicle with lane speed limits and give an alert message to the driver to either reduce/ increase the speed or to change the lane, if driving at wrong speed. The system continuously monitors the driving style of the vehicle for a particular time t. If the vehicle changes the lane or reduces the speed in the time t the monitor system gives an OK status to the driver and to the RSU. If the vehicle does not responds to

978-1-4673-8760-6/15 $31.00 © 2015 IEEE

the warning and change the speed or the lane, then the OBU monitoring system broadcasts the emergency message to neighboring vehicles to alert the other nodes and also transmits emergency data packet to nearby RSU.

A. SBLS Algorithm

Position Verification:

Obtain the values of *Sest*, V_{li}, V_{ui}, w.r.t L_i

Pos-label

if ($Sest < V_{ui}$) **AND** ($Sest > V_{li}$)

then

      Alarm OFF;

      Display **"ok speed message to driver";**

      Display **"*Sest* to driver";**

elsif ($S_{est} > V_{ui}$)

      Display **"decelerate or lane change";**

Alarm on;

Timer on ;

      Check for **deceleration, lane change subroutines**;

      Timer ends ;

Greedy forward of the warning message to **RSU**, neighboring **vehicles;**

goto P*os-label*

else($Sest < V_{li}$)

      Check for engine state

if engine off

then

      Display "Vehicle stopped";

else

      Display "Vehicle moving ";

Alarm on ;

Timer on ;

      Check for **acceleration, lane change subroutines**;

      Timer ends ;

Greedy forward the warning message to **RSU**, neighboring **vehicles;**

goto *pos-label*

Deceleration/Acceleration:

1. **if** ($Sest < V_{li}$)

    Accelerate

    Check for ***Sest***

    Check ***min*_{gap}**

goto *Position Verification*

2. **if** ($Sest > V_{ui}$)

    Decelerate

    Check for ***Sest***

    Check ***min*_{gap}**

goto *Position Verification*

Lane Change:

1. **if** ($Sest < V_{li}$) **OR** ($Sest > V_{ui}$)

    Check ***min*_{gap}**

if (min_{gap}) not appropriate

    Display **"Lane changing not allowed"**

    Else

    Display **"Lane changing allowed"**

    Update ***LB*_i**

goto *Position Verification*

V. PERFORMANCE EVALUATION

In this section we have discussed about the working of the on board navigation unit (OBU), the simulation of the algorithm using NS2. The hardware installed in the vehicle will be able to parse the data from that of the data received by the GPS receiver and compare, save them for vehicles and lane ID identification purposes. For example, the location, speed and time information will be sent to the RSU in order keep proper track of the faulty vehicle and pass on information further to the central traffic monitoring server.

A. OBU- GPS enabled monitoring system

It is the hardware that processes the data received by GPS with respect to the moving vehicle. The GPS receiver receives the data packets from the satellites and feeds it to the microcontroller. The microcontroller further parses the required data stores the lane information in Lane ID buffer and checks it with the thresholds set in it. In case of any mismatch, the microcontroller alerts the driver three times. Now if the driver ignores these warnings and continues to break the rules, the warning message is then broadcast to the other vehicles and RSU's in range. The block diagram and implementation of the OBU described above is as shown in the Fig. 2.

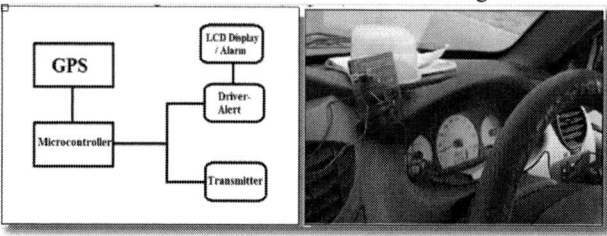

Fig. 2 Block Diagram and implementation of the Hardware Circuit

The hardware of this model comprises of a PIC microcontroller interfaced with a GPS receiver with baud rate of 9600 bps, an LCD display, alarm and a radio module for displaying the warning message, alerting the driver and broadcasting the warning messages respectively.

B. SUMO and MOVE

The mobility model describing our scenario is as shown in the figure below. We have chosen a specific segment of Neemrana, India where we have a 4-lane road and we have implemented the model over there and the simulation results have been recorded. The 4-lane road as shown in the picture has different speed limits assigned to each track. The vehicles moving on these tracks will have to move according to that particular lane speed limit. The lane changing on the tracks is based on the speed of the vehicle and the minimum gap between the vehicles as stated in the SBLS.

Fig. 3 Scenario

978-1-4673-8760-6/15 $31.00 © 2015 IEEE

The lane 1 is the emergency lane where the vehicles like ambulance and police can move in case of emergency or any mishap. The lane 2 has its speed limit set to 120Kmph, lane 3 has the speed limit set to 80Kmph and the lane 4 has its speed limit set to 40 Kmph. If any accident has taken place on the normal lanes, the emergency lane will be open for the vehicles to avoid traffic jam and chaos. The antenna on the top of the road represents the RSU's. The vehicles have their specific range of communication as shown in the figure. They communicate with each other and vehicles in their range and exchange information among themselves. The entire scenario is simulated in NS2 and the SBLS was implemented. The results are described and explained in the next section. SUMO-MOVE gives us the .tcl and sumo.tr file that are the requirements for simulation to be done in NS-2 [3].

C. NS2

The mobility model which we have generated using SUMO-MOVE will be simulated in NS-2 with the help of the .sumo.tr and .tcl files as mentioned in sub section B. The simulation results of the implemented protocol discussed above can be seen in the figure below (Fig. 4.1). In NS-2 the nodes -> 1,2,3,4 are the 4 vehicles on 4 different tracks. In Fig. 4.1 the vehicle no. 2 shifts it lane based on the speed change. Initially it starts from the first lane and then shifts to the third based on the protocol.

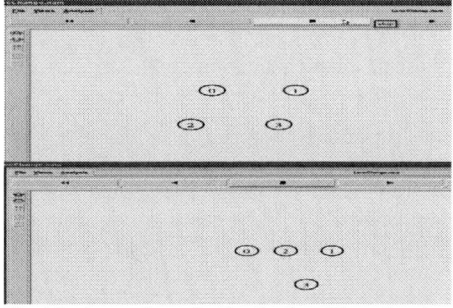

Fig. 4.1 min_{gap} and speed estimation based lane changing

D. Efficiency Analysis

We have checked our algorithm SBLC on the test bed of a flyover in NH-8 Highway of Neemrana, Rajasthan, India and generated the smart flow of vehicles in real time. We have used small range TRX (Transceiver) along with our hardware OBU system in car plying on the four lane highway. Long range TRX has been used as RSU to receive the road segment information and update the congestion status. On for the next test without the SBLS algorithm we deployed four cars with different drivers. During this test three out of four cars were faulty in terms of maintaining their speed on the basis of their lanes.

VI. CONCLUSION AND FUTURE WORK

The speed-based lane changing protocol implemented within the On board system will play a very important role in VANET and will create a semi-automatic environment, in the concept of smart cities and smart transportation. The drivers will get full and authentic information of their surrounding without much effort. The use of GPS based system is more reliable as it is more accurate and the data being provided is real time data. Thus this system is more reliable and efficient in terms of safety and accuracy. In this paper, we have explained the protocol and the first level implementation of the algorithm with the help of SUMO-MOVE, NS-2 and some field trials with the hardware prototype. Further, we will use different packet forwarding strategies for different connection states.

REFERENCES

[1] Kristoffer Lidstrom, "On Strategies for Reliable Traffic Safety Services in Vehicular Networks", Orebro University, 2009.

[2] http://aprs.gids.nl/nmea/#gll

[3] Francisco J. Martinez, Chai Keong Toh, Juan-Carlos Cano, Carlos T. Calafate and Pietro Manzoni, "A Survey and Comparative Study of Simulators for s Vehicular ad hoc Networks (VANET's)", Wireless Communication and Mobile Computing, 2009.

[4] http://nsnam.isi.edu/nsnam/index.php/Main_Page

[5] C. -F. Chiasserini, E. Fasolo, R. Furiato, R. Gaeta, M. Garetto, M. Gribaudo, M. Sereno, A. Zanella, "Smart Broadcast of Warning Messages in Vehicular Ad Hoc Networks".

[6] http://en.wikipedia.org/wiki/Vehicular_ad_hoc_network

[7] Rajendra Prasad Nayak, "High Speed Vehicle Detection in Vehicular Ad-hoc Network", May 2013.

[8] Nehal Kassem, Ahmed E Kosba and Moustafa Youssef, "RF-based Vehicle Detection and Speed Estimation", IEEE, 2012.

[9] Ram Shringar Raw, Manish Kumar, Nanhay Singh, "Security Challenges, Issues, and their Solutions for VANET", IJNSA, Vol.5, No.5, September 2013.

[10] Osama Abumansoor, Azzedine Boukerche, "A Secure Cooperative Approach for Nonline-of-Sight Location Verification in VANET", IEEE Transactions on Vehicular Technology, January 2012.

[11] Zongjian He, Jiannong Cao and Tao Li, "MICE: A Real-time Traffic Estimation Based Vehicular Path Planning Solution using VANETs".

[12] Tanvee Kausar, Priyanka Gupta, Deepesh Arora, Rishabh Kumar, "A VANET Based Cooperative Collision Avoidance System for a 4-lane Highway", Engineering Research and Development, IIT Kanpur Technical Journal, April 2012.

[13] Yong Zhou, Rong Xu, Xiao-Feng Hu, Qing-Tai Ye, "A Lane Departure Warning System Based on Virtual Lane Boundary", Journal of Information Science and Engineering, 2008.

[14] Alrik L.Svenson, Valerie J.Gawron, Timothy Brown, "Safety Evaluation of Lane Change Collision Avoidance Systems using the National Advanced Driving Simulator", unpublished.

[15] Takashi Wakasugi, "A Study on Warning Timing for Lane Change Decision Aid Systems Based on Driver's Lane Change Maneuver".

[16] Gayathiri Somasundaram, Kavitha, K.I. Ramachandran, "Lane Change Detection and Tracking for A Safe-Lane Approach in Real Time Vision Based Navigation System ", CS & IT-CSCP 2011.

[17] Gregory Taubel, Rohit Sharma, Hiann Shiou Yang, "An Experimental Study of a Lane Departure Warning System Based on the Optical Flow and Hough Transform Methods", Wseas Transactions on Systems, Volume 13, 2014.

[18] Jamal Saboune, Mehdi Arezoomand, Lue Martel, Robert Laganiere, "A Visual Blindspot Monitoring System for Safe Lane Changes", unpublished.

[19] Dr. Andreas Festag, Roberto Baldessari, Dr. Wenhui Zhang, Dr. Long Le, Amardeo Sarma, Masathoshi Fukkukawa, "Car-2-X Communication for Safety and Infotalnment in Europe", NEC technical Journal, Volume 3,2008.

[20] Neeraj Shrma, Jawahar Thakur, "Performance Analysis of AODV & GPSR Routing Protocol in VANET", International Journal of Computer Science and Enginering Techology, Vol. 4, No. 2, Febuary 2013.

[21] Abd Al-Razak Tareq Rahem, Mahmood Ismail, Ariff Idris, Aymen Dheyya, "A Comaparative and Analysis of VANET Routing Protocols", Journal of Theoritical and Applied Information Technology, Vol. 66, No. 3, August 2014.

New Fine-Grained Clustering Algorithm on GPU Architecture for Bias Field Correction and MRI Image Segmentation

N. Aitali, B. Cherradi*, O.Bouattane
Labo SSDIA, ENSET
Hassan II University
Mohammedia, Morocco
aitali_noureddine@yahoo.fr

M. Youssfi and A. Raihani
*CRMEF
El Jadida, Morocco
bouchaib.cherradi@gmail.com

Abstract— in this paper, we propose a new fine-grained clustering bias field estimation and segmentation algorithm on Single Instruction Multiple Data (SIMD) architecture (GPU). The goal is to accelerate compute-intensive portions of the sequential version. We have implemented this parallel algorithm using Compute Unified Device Architecture (CUDA) on different NVidia GPU cards. The numerical results in terms of execution time show a gain up to *52x* for GTX 580 versus the sequential implementation.

Keywords— Image segmentation, Bias field correction, GPU, CUDA

I.   INTRODUCTION

In magnetic resonance imaging, intensities inhomogeneities, called bias field, are caused by non-uniformities in the RF field during the acquisition; the result is a shading effect where the pixel or voxel intensities of same tissue class vary slowly over the image domain. This shading can cause severe errors when attempting to segment corrupted images using intensity-based pixel classification methods. It has been shown that this shading is well modelled by the product of the original image and a smooth, very slowly varying multiplier field [1, 2].

Graphic processing units (GPUs) were originally created for rendering graphics. Recently, GPUs has emerged as co-processing units for Central Processing Units (CPU) and has become popular for general-purpose high performance computation (GP-GPU), to accelerate various digital signal processing applications, including medical image processing [3,4]. GPUs are composed of hundreds of processing cores highly decoupled able to achieve immense power of parallel computing. To take advantage of these multi-core architectures, these applications must be parallelized.

Among the few works in the literature on the development of images segmenting algorithms with intensity inhomogeneities correction on GPU architecture, the authors in [5] proposed an extended mask-based version of the level set method, recently presented by Li et al. [6]. They develop CUDA implementations for the original full domain and the extended mask-based versions, and compare the methods in

terms of speed, efficiency, and performance, The GPU implementation of the their version allows a speed up of around 50−100 times for instance, of 512×512×128 slices.

Anderson and al [7] suggested a GPU solution for the Fuzzy C-Means. They have used OpenGL and Cg to achieve approximately two orders of magnitude computational speed-up for some clustering profiles using an NVIDIA 8800 GPU card. Other researchers [8] have focused on accelerating the clustering of the FCM algorithm on GPU using CUDA. They compared their implementation with its C++ sequential implementation, as well as a MATLAB version.

In this paper we propose a conception and CUDA implementation of a new massively parallel fuzzy clustering algorithm for simultaneously Bias field correction and segmentation called PBCFCM based on the work of Ahmed et al [9]. In fact, we propose an efficient implementation of the original version called BCFCM and study his performances in terms of execution time on different NVidia GPU cards with different image sizes.

The rest of this paper is organized as follows. In Section II, we describe the fine-grained parallel used model (GPU of NVidia) and his software development environment (CUDA). In Section III, we review the sequential version of the clustering algorithm proposed by Ahmed et al [9] and called BCFCM. In Section IV, we present our parallel implementation as well as the details of the parallelization. In Section V, we present some findings; compare the results and performance speed-ups. Section VI, concludes the paper and give some perspectives for this work.

II.   PARALLEL ARCHITECTURE MODEL

In this section we will give a summary presentation of the computational model and the software development environment that we choose to implement the algorithm. It is about GPU massively parallel architecture and his SDK CUDA.

Modern GPUs are massively parallel processors that support a large number of elementary processors. They are particularly well suited for exploring the calculations on many data types that have high arithmetic intensity. Currently, GPU architecture is composed of an evolving set of processing units flows (SM, SMX Streaming Multiprocessor or: next-generation Streaming Multiprocessor). Such a multiprocessor contains a number of cores (scalar processor), a multithreaded instruction unit, number of registers, local memory and shared memory.

CUDA is a software development environment based on the C language for GPUs from NVIDIA unveiled in 2007, now in its version 7.5 [10]. CUDA is constituted by a parallel programming model and a set of dedicated instruction. It allows the programmer to define C functions, called kernels (Kernels), which are executed in parallel by multiple CUDA threads (instantiation of a kernel). The programmer organizes these threads into a hierarchy of grids of thread blocks. A block of threads is a set of concurrent threads that can cooperate with each other through barrier synchronization and shared access. During execution, the threads can access data at different levels of hierarchy: registers, shared memory and global memory. The global memory is accessible by all threads, but its access time is about 500 times slower than the access time to the shared memory and registers.

III. BACKGROUND: A MODIFIED FUZZY C-MEANS ALGORITHM FOR BIAS FIELD ESTIMATION AND SEGMENTATION (BCFCM)

As is mentioned in the introduction, the observed MRI signal is modelled as a *product* of the true signal generated by the underlying anatomy, and a spatially varying factor called the gain field.

$$Y_k = X_k G_k \qquad (1)$$

Where X_k and Y_k are the true and observed intensities at the k_{th} pixel, respectively, G_k is the gain field at the k_{th} voxel.

The application of a logarithmic transformation to the intensities allows the artefact to be modelled as an additive bias field:

$$y_k = x_k + \beta_k \qquad (2)$$

Where x_k and y_k are the true and observed log-transformed intensities at the k_{th} pixel, respectively, and β_k is the bias field at the k_{th} pixel.

Ahmed et al. [9] proposed a modification to the standard FCM [11] objective function for partitioning an MRI image containing $\{x_k\}_{k=1}^{N}$ pixels into C clusters by introducing a term that allow the labelling of a pixel (voxel) to be influenced by the labels in its immediate neighbourhood.

The modified objective function is given by:

$$J_m = \sum_{i=1}^{C}\sum_{k=1}^{N} u_{ik}^p \|y_k - \beta_k - v_i\|^2 + \frac{\alpha}{N_r}\sum_{i=1}^{C}\sum_{k=1}^{N} u_{ik}^p \left(\sum_{y_r \in N_k}\|y_r - \beta_r - v_i\|^2\right) (3)$$

$U_{i,j}$: The degree of membership of data x_j in the cluster v_i.

p: Fuzzy weighting exponent.

v_i: The prototypes of the cluster i.

N: The total number of pixels in the MRI image
N_k: Set of neighbor's pixels that exist in a window around x_k.
N_r: Cardinal of N_k.
α : Neighbors effect

The new membership function is given by:

$$u_{ik} = \cfrac{1}{\sum_{j=1}^{C}\left(\cfrac{D_{ik} + \frac{\alpha}{N_r}\gamma_i}{D_{jk} + \frac{\alpha}{N_r}\gamma_j}\right)^{1/p-1}} \qquad (4)$$

Where:

$$\gamma_i = \left(\sum_{y_r \in N_k}\|y_r - \beta_r - v_i\|^2\right) \qquad (5)$$

$$D_{ik} = \|y_k - \beta_k - v_i\|^2 \qquad (6)$$

The cluster prototype updating is done by expression:

$$V_i = \cfrac{\sum_{k=1}^{N} u_{ik}^p\left((y_k - \beta_k) + \frac{\alpha}{N_r}\sum_{y_r \in Nk}(y_r - \beta)_r\right)}{(1+\alpha)\sum_{k=1}^{N} u_{ik}^p} \qquad (7)$$

The estimated bias field is given by the expression:

$$\beta_k = y_k - \cfrac{\sum_{i=1}^{C} u_{ik}^p v_i}{\sum_{i=1}^{C} u_{ik}^p} \qquad (8)$$

IV. PARALLEL FUZZY C-MEANS ALGORITHM FOR BIAS FIELD ESTIMATION AND SEGMENTATION (PBCFCM)

In this section, we describe the implementation of BCFCM in GPU. Fig 1, illustrates the strategy used to distribute the data, at coarse-grained (Blocks) and fine-grained (threads) levels.

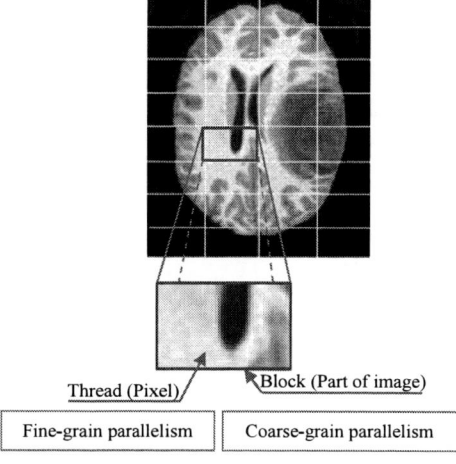

Fig 1: Data distribution strategy for PBCFCM on GPU.

Inputs:
y: Data to be corrected and segmented.
b: Initial bias field matrix (small values e.g: 0.01)
ε *: Termination criterion.*

Parameters:
alpha: Neighbors effect number.
N_r*: Neighbors Cardinal.*
p: Fuzzy weighting exponent.

1. **While** $\left\| v_{new} - v_{old} \right\| < \varepsilon$ **do**

   ### Kernel #1 (in GPU)
   //For each thread ID do

2.   *For all $y_k \in Y_{threadID}$ do*
       // Compute the distance between pixel k and clusters in each thread

3.   *For $j \in [1,c]$ do*
4. $D_j \leftarrow \| y_k - b_k - v_j \|^2$
       // Compute the distance between neighbors r and clusters in each thread
5.   *For all $y_k \in$ neighbors do*
6. $Gamma_j \leftarrow \| y_r - b_r - v_j \|^2 + Gamma_j$
7.   *End for*
8.   *End for*
   //Compute the expression S2 needed to update the clusters
9.   *For (all $y_k \in$ neighbors) do*
10. $s2 \leftarrow y_r - b_r + s2$
11.   *End for*
12. $s \leftarrow y_k - b_k + (alpha/N_r) * s2$
13.   *For $j \in [1,c]$ do*
14. $a \leftarrow D_j + (alpha/N_r) * Gamma_j$
15.   *For $m \in [1,c]$ do*
16. $b \leftarrow D_m + (alpha/N_r) * Gamma_m$
17. $u\_inv \leftarrow (a \, / \, nonzero(b)) \,^\wedge \, (1 \, / \, (p - 1)) + u\_inv$
18.   *End for*
   //Compute the membership function
19. $U_j \leftarrow 1 \, / \, nonzero(u\_inv)$
   // computes expressions needed to update the clusters
20. $Up_j \leftarrow U_j \,^\wedge \, p$
21. $G1_j \leftarrow Up_j * S + G1_j$
22. $G2_j \leftarrow Up_j + G2_j$
23.   *End for*
24. **End for**

End Kernel #1
    //centroids update (in CPU)
25. *For $j \in [1,c]$ do*
26. $v_j = G1_j \, / \, ((1 + alpha) * nonzero \, (G2_j));$
27. *End for*

Kernel #2 (in GPU)
    //Compute the estimated bias Field b_k for each pixel k

28.   *For all $y_k \in Y_{threadID}$ do*
29.   *For $j \in [1,c]$ do*
30. $bfe1 \leftarrow Up_j * classCenters_j + bfe1$
31. $bfe2 \leftarrow Up_j + bfe2$
32. $b_k = y_k - bfe1 \, / \, nonzero(bfe2)$
33.   *End for*
34.   *End for*
35. **End while**

Outputs:
b: Final estimated bias field matrix.
y-b: Corrected image.
$\left\{ v_j \right\}_{j=1}^{c}$ *: Clusters centroids vector.*

U: Membership matrix.

In order to execute the proposed implementation, we reserve necessary threads, allocate and transfer needed data from CPU to GPU, the number of blocks and threads depends on image size *N*, and so each thread becomes responsible of one pixel. In our case we have used 256 threads per block as it is advisable in CUDA occupancy calculator [9], and set the parameters values as below: *alpha =1, p=2, N_r=8.*

First, each thread computes the distance between the data and clusters, then it computes the sum of distance between the neighbors (3 x 3 window) pixels and cluster centers. Each thread computes the expressions that it needs to calculate his membership values to the cluster *Vi*, besides of expressions needed to update the cluster centroids. Finally, each thread computes his new estimated bias field.

The update of clusters and the verification of termination criteria are done on CPU and all the rest of treatment is done on GPU using shared memory and constant memory for data pixels and clusters simultaneously, also we have used registers besides of the efficient library CUBLAS to get more speed up. Notice that for each iteration we have unique data transfer of the new clusters to the constant memory in GPU.

V. RESULTS AND DISCUSSION

We implement both the sequential and parallel algorithms on three NVIDIA devices:
- GT740M with i7 CPU (2.4 GHz, 6 Go),
- GTX 760 and GTX 580, with i7 CPU (3.5 GHz, 16 Go).
Parallel version is compiled with CUDA 6.5 SDK on 32-bit windows 7 platform.

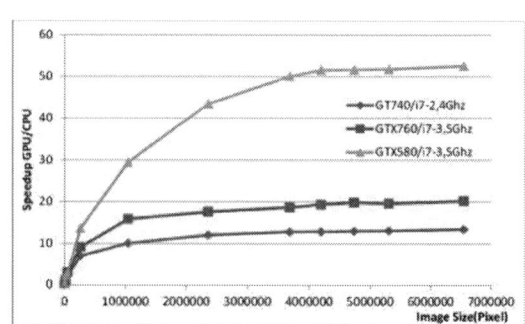

Fig 2: Execution time speedup GPU(s)/CPU(s) versus Data size.

Fig 2 shows graphical representation of execution time speedup GPU(s)/CPU(s) for three cards used in this experiment versus size of the data image. We used lena image grey level with different sizes, the goal is to segment it into his optimal number of cluster 7 [12].

The characteristics of the three curves have the same look and speed-up increases by exponential manner until it reaches a critical value or time behavior to an almost steady state (saturation regime). For GTX 580 the maximum value of the speed-up reached, is around 52x, whereas for GTX 760 and GT740m the speed-up reached is about 20x and 14x respectively. Furthermore, the increased speed-up depends on both the type of the card and that its performance of invested parallel algorithm.

In TABLE I, we present execution times on GTX580 GPU and i7 3.5 Ghz CPU for some data images that we used to validate our algorithm, the results show an important speed up especially when the image size becomes larger. So the use of parallel version is efficient for big data as the case in MRI datasets.

TABLE I EXECUTION TIME ON CPU AND GPU FOR DIFFERENT DATA SIZE

Image Size (Pixel)	i7-3.5Ghz CPU Execution Time (s)	GTX 580 GPU Execution Time (s)	Speedup GPU/CPU(x)
1024	0,09	0,27	0,35
16384	0,27	0,12	2,21
50625	0,80	0,27	2,98
262144	4,13	0,30	13,60
1048576	10,00	0,34	29,41
2359296	30,00	0,69	43,42
3686400	46,61	0,93	50,06
4194304	53,32	1,04	51,52
4734976	45,17	0,89	51,75
5308416	51,73	1,00	51,73
6553600	62,99	1,20	52,49

For image sizes ranging between medium to big data, the efficiency of GPU is greatly investigated, for example the time execution using image size of 16384 pixels in CPU is of 270 ms versus 120 ms using GPU performance. This pertinent feature of GPU becomes more interesting for large image sizes and parallel approach becomes a very valuable advantage. The time execution for image size of 6553600 is 52.49 faster in GPU than using CPU treatment.

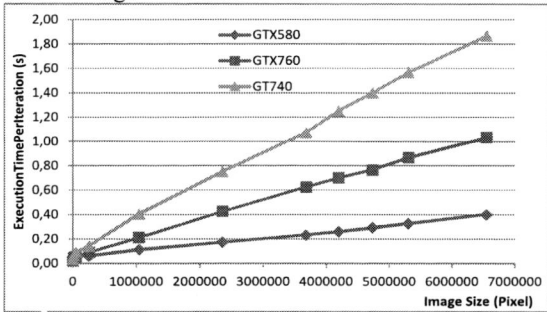

Fig 3: The variation of execution time per iteration versus image size on different GPU devices

The figure above shows the variation of run time per iteration depending on the pixel image size for different GPU cards.

From these graphs it is observed that the behavior characteristics for these three cards vary perfectly linearly.

According to our studies, we deduce that the most efficient is the corresponding card which has the lowest slope. In fact, GTX 580 card is better than GTX 760 which is also better than GT 740m. Our theoretical and practical research is in good agreement with those found in the literature.

VI. CONCLUSION AND PERSPECTIVES

In this paper, we have presented a conception of a PBCFCM algorithm and its implementation on a massively parallel architecture GPU. The use of NVidia GPU and his development environment CUDA is justified by the high performance of this GPU cards and the presence of these devices on most PCs today.

The evaluation of the proposed algorithm execution time shows a significant speedup on GPU, mostly when the data size become more important. The use of parallel version is efficient for big data.

A framework that covers more artefacts correction such as noise and Partial Volume Effect (PVE) will be desirable for correcting and segmenting corrupted MRI images on GPU.

REFERENCES

[1] Rajapakse, J.C., Giedd, J.N., Rapoport, J.L., 1997. Statistical approach to segmentation of single-channel cerebral MR images. IEEE Trans. on Med. Imag. 16, 176-186.

[2] Unser, M., 1995. Multigrid adaptive image processing. In: Proceedings of the IEEE Conference on Image Processing, Vol. I, pp. 49-52.

[3] Eklund, A., Dufort, P., Forsberg, D., & LaConte, S. M. (2013). Medical image processing on the GPU–Past, present and future. *Medical image analysis*, 17(8), 1073-1094.

[4] Pratx, G., & Xing, L. (2011). GPU computing in medical physics: A review. *Medical physics*, 38(5), 2685-2697.

[5] T. Ivanovska, R. Laqua, L. Wang, H. Volzke, and K. Hegenscheid. "Fast Implementations of the Levelset Segmentation Method with Bias Field Correction in MR Images: Full Domain and Mask-Based Versions". In: *Pattern Recognition and Image Analysis*. Springer Berlin Heidelberg, 2013. p. 674-681.

[6] Li, C., Huang, R., Ding, Z., et al.: A level set method for image segmentation in the presence of intensity inhomogeneities with application to MRI. IEEE Trans. on Image Processing 20, 2007–2016 (2011).

[7] Anderson, D., Luke, R., Keller, J. (2007), "Speedup of Fuzzy Clustering Through Stream Processing on Graphics Processing Units", IEEE Trans. on Fuzzy Systems

[8] Rowińska, Z., & Gocławski, J. (2012). Cuda based fuzzy c-means acceleration for the segmentation of images with fungus grown in foam matrices. Image Processing & Communications, 17(4), 191-200.

[9] M.N. Ahmed, N.A. Mohamed, A.A. Farag, T. Moriarty, A modified fuzzy c-means algorithm for bias field estimation and segmentation of MRI data, IEEE Trans. Med. Imaging 21 (2002) 193–199.

[10] https://developer.nvidia.com/cuda-downloads

[11] J. C. Bezdek (1981): "Pattern Recognition with Fuzzy Objective Function Algoritms", Plenum Press, New York

[12] Ouadfel, S., Batouche, M., & Ahmed-Taleb, A. (2012). ACPSO: A Novel Swarm Automatic Clustering Algorithm Based Image Segmentation. In S. Ali, N. Abbadeni, & M. Batouche (Eds.) Multidisciplinary Computational Intelligence Techniques: Applications in Business, Engineering, and Medicine (pp. 226-238). Hershey, PA: Information Science Reference. 5. Chapter 14.

An Enhancement Transient Response of Capless LDO with improved Dynamic Biasing Control for SoC Applications

HATIM AMEZIANE
Laboratory of Electronics, Signals and Information System
Faculty of Sciences Dhar El-Mehraz
Fez, Morocco
hatim.ameziane@usmba.ac.ma

QJIDAA HASSAN
Laboratory of Electronics, Signals and Information System
Faculty of Sciences Dhar El-Mehraz
Fez, Morocco
qjidah@yahoo.fr

ZARED KAMAL
Laboratory of Signals, Systems and Components
Technical and Sciences Faculty
Fez, Morocco
zaredkam@yahoo.fr

ZOUAK MOHCINE
Laboratory of Signals, Systems and Components
Technical and Sciences Faculty
Fez, Morocco

Abstract—*This paper presents an enhancement transient capless low dropout voltage regulator (LDO). To eliminate the external capacitor, the miller effect is implemented through the use of a current amplifier. The proposed regulator LDO provides a load current of 50 mA with a dropout voltage of 200 mV, consuming 14µA quiescent current at light loads, and the regulated output voltage is 1.6 V with an input voltage range from 1.2 to 1.8 V. The proposed system is designed in 0.18 µm CMOS technology. A folded cascode amplifier with high transconductance and high power efficiency is proposed to improve the transient response of the LDO. In addition, multiloop feedback strategy employs a direct dynamic biasing technique to provide a high speed path during the load transient responses. The simulation results presented in this paper will be compared with other results of SoC LDOs demonstrate the advantage of the proposed topology.*

Keywords—Low Dropout Regulator (LDO); active feedback; Integrated circuits (ICs) system-on-a-chip (SoC); MOSCAP.

I. INTRODUCTION

Over the last decade, low dropout voltage regulator (LDO) is widely demanded in portable electronics market. These LDOs require providing strict characteristics such as high PSRR, low noise and fast transient response [1], [2] [3].Furthermore, power-management (PM) design for portable devices imposes new challenges in the areas of I/O interface, energy management and battery lifetime. This has led to increasing demand for higher levels of integration in order to reduce board-space requirements. Utilizing multiple local on-chip voltage regulators is a very promising approach in system-on-chip development [5], [4]. Especially, where the power consumption reduction is required, the latest generation of low drop-out linear regulators (LDO) offers the optimal answer for powering circuitry in many of the portable device applications. In fact, they can provide regulated and accurate supply

voltages for noise-sensitive analog blocks. This advantage makes LDO widely used in portable systems between the switching power converter (SWPC) and RF circuitry to increase battery life and reject the ripple in supply voltage of different RF circuits [9],[10].

In conventional LDO, a large off-chip capacitor (0.47 µF to 4.7 µF) at output is necessary to locate the dominant pole at very low frequency to achieve the frequency compensation and provide a good dynamic performance [7], [9], [13]. The large off-chip capacitor occupies a large chip area, and it is difficult to integrate multiple LDOs on a single chip. In order to design a full on-chip LDO regulator, the number of compensating capacitors must be minimized [8], [11], and [12].

II. PROPOSED LDO

In this work, we present modified miller compensation with insertion sensor Amplifier stage by sensing output voltage in order to inject more transient current in the biasing circuit to feeds the regulator to charge rapidly the gate power PMOS capacitance and improve the fast transient response. Implementing the technique used in [1] in order to reduce the area without influencing stability and realize the full on-chip capacitor LDO.

The proposed LDO, shown in Fig.2, is composed of folded cascode amplifier, a power PMOS transistor and the feedback resistor network. C1, Cf, Cc are the on-chip active MOS capacitances. R1 and R2 construct the active feedback resistive network. C_o is the interconnection lines parasitic capacitor, and typically up to 100 pF. Which affect the accuracy in the ICs.

A. Error Amplifier

The design of error amplifier (EA) is more complex, when a high performance is required to guarantee the stability and transient response, a specific topology is necessary. To move the dominant pole at the output of E.A to low frequencies, low

978-1-4673-8760-6/15 $31.00 © 2015 IEEE

output impedance is designed. To charge rapidly the capacitance seen at the gate of pass transistor (may be as large as 50 pF), EA must provide a sufficient output current [13], [5]. On the contrary, the EA itself should provide very low power dissipation, and its bias currents must be kept as low as possible. In this paper, the proposed EA is the folded cascode amplifier which offers better performances such as high gain, enough load current to drive the power transistor PMOS and improved PSRR characteristic of LDO.

B. Transient response

To enhance the transient response, a novel technique is implemented in this proposed structure as shown in fig .2 the current injected to the load through C1 is given by:

$$ic_1 = SC_1\left(V_{DS4} - V_{out}\right) \tag{1}$$

Where
$$V_{DS4} = \frac{Bg_{m1}V_{out}}{\lambda k' \dfrac{W}{L}\left(g_{m2}V_{out} + V_{th}\right)} \tag{2}$$

The drop of output voltage increases the current flowing through C1, this current injected to the output when the power MOS transistor is not capable to provide the load current.

The current injected by C1 is depends on the sensor amplifier gain. In fact, the drop of Vout decreases the gate voltage of M1. As a result, the current provided by M2 is increased by B2/B1 factor. This latter is amplified again by B4/B3 factor. Consequently the current at the drain of M4 is injected to the load through C1 and the output voltage of LDO recovers his nominal value. In the other hand, when Vout recovers the nominal value the current through M1 takes its minimum value. As result, the current injected by C1 it's also takes the minimum value and the LDO operate in the normal conditions.

C. Power Supply Rejection

Power supply rejection is the LDO ability to suppress power supply noise from its output. Recently, there is a lot of focus on designing high performance, especially low noise and high PSRR. Moreover, the integration level of power management is proportionally increased in time or the dimension of process is decreased rapidly and the parasitic capacitances are up to ten pF. A full on-chip LDO is realized using novel technique to boost the PSRR and transient response. Furthermore, most literature focus in analytic of PSRR on parameters and devices transmit and control the ripple from the supply to the output on the PMOS parasitic capacitance neglecting the effect of the parasitic capacitances at the output of error amplifier and its high output stage gain.

In this analysis, all sources of perturbation are taken into consideration and a new technique is proposed to boost and control the PSRR. The small signal model of PSRR is shown in fig. 2. A small signal input voltage v_{dd} will induce an output voltage v_{out}. The PSRR can be writed in eq. (3)

$$PSRR = \frac{g_{mp}}{sA + B} \tag{3}$$

Where
$$A = \left(C_0 - C_f\right)\left(1 + G_m\right) + C_c$$
$$B = B_1 B_2 g_{m1} - G_m g_{mp}$$

As can be seen from eq. (3), the dc gain of PSRR is controlled by the PMOS gain stage performance. At low frequency, in low load condition, gmp decreases, and also the gain of EA decreases. From eq. (3), gmp in numerator and denominator hence PSRR is not heavily affected by the transconductance variation of power transistor. At full load condition, an increase in gmp and gain of EA, enhance the PSRR|DC as given in eq. (3). As the load current decreases, also the transconductance of PT decreases, the gain of PSRR and the performance of LDO is affected.

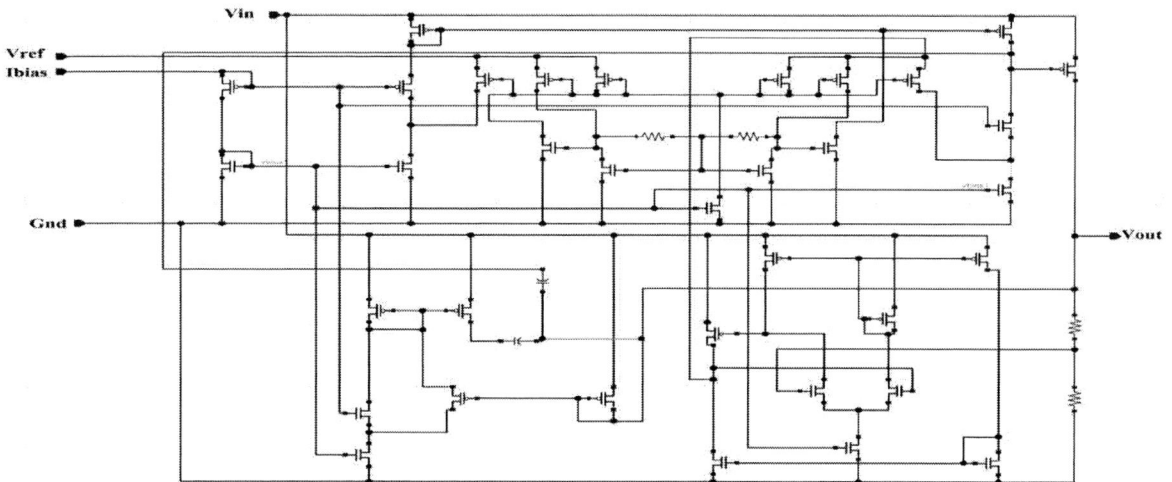

Figure.1. Proposed LDO: schematic of CMOS LDO.

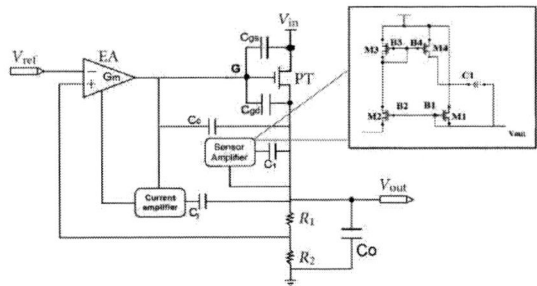

Figure.2. simplified of proposed LDO structure.

D. AC analysis

In load regulation, when the load current suddenly steps from low load to its maximum value, the incapability of the pass transistor to provide the demanded current forces the load capacitor to supply the extra current and the capacitor voltage drops. The active feedback transmits the variation in output voltage of LDO to the EA which in turn downs the gate voltage of the pass element, thus increasing VSGP and providing the output current demanded by the output load. In this structure, the compensation capacitor forms the feed-forward path to inject charges in the load capacitor and decreases the drop of output voltage.

The continuity of the current proves that:

$$\begin{cases} C_L \dfrac{dV_{out}}{dt} = g_{mp} V_{sg} \\ C_g \dfrac{dV_g}{dt} = G_m(V_{out} - V_{ref}) + C'_F \dfrac{dV_{out}}{dt} \end{cases} \quad (4)$$

From eq. (4), the gain of the system can be expressed:

$$A_v = \frac{V_{out}}{V_{ref}} = \frac{w_n^2}{w_n^2 + 2\xi w_n S + S^2}$$

$$= \frac{G_m g_{mp}}{G_m g_{mp} + S\, C_F g_{mp} + S^2 C_g C_L} \quad (5)$$

Where

$$\begin{cases} C_L = +C_1 + C_O + C_c + C_{gd} \\ C_g = C_{gs} + C_{gd} \\ C'_F = C_F + C_c + C_{gd} \end{cases} \quad (6)$$

The damping factor and the poles of the transfer function are given respectively by:

$$\begin{cases} \varsigma = \dfrac{1}{2}\left(C'_F \big/ G_m \right)\sqrt{\dfrac{g_{mp}}{C_L C_g C'_F}} \\ S_1 = w_n(\varsigma - \sqrt{(\varsigma - 1)}) \\ S_2 = w_n(\varsigma + \sqrt{(\varsigma - 1)}) \end{cases} \quad (7)$$

The different capacitors contribute to provide the required current Iload by the output load at ac load transient and the drop of the output voltage of LDO will be small. Moreover, when the load current decreases instantaneously, the over current of the power PMOS transistor charges the all capacitors at output of LDO. As a result, the overshoot of the output voltage will be small.

III. SIMULATION RESULTS

The proposed regulator LDO has been realized in 0.18 μm CMOS technology. The simulation of the proposed LDO was performed with Spectre. The loop-gain simulation has been performed with a total on-chip compensation capacitor Ctotal=10 pF, and the output capacitor Co (up to 100 pF). The proposed LDO is stable with a good phase margin of approximately 66° at full load as shown in Fig.3.

The simulation of PSRR at 50 mA load current as shown in Fig.4, the proposed structure presents a high PSRR at low frequency. Fig 5(a) shows the AC line regulation for supply voltage change from 1.5 to 1.8 V, Fig 5(b) shows the transient response simulation of the proposed LDO with load current switching between 0.1mA and 50mA. The variation of output is about 160 mV with a settling time of 1.21 μs.

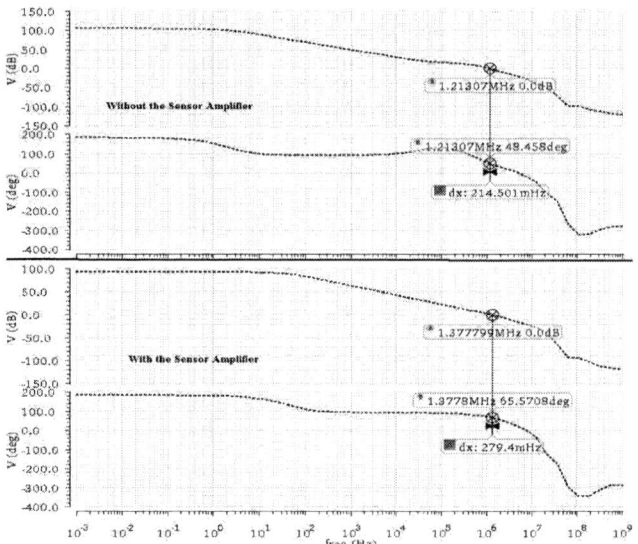

Figure.3. Simulation results of proposed LDO frequency response without and with the sensor amplifier.

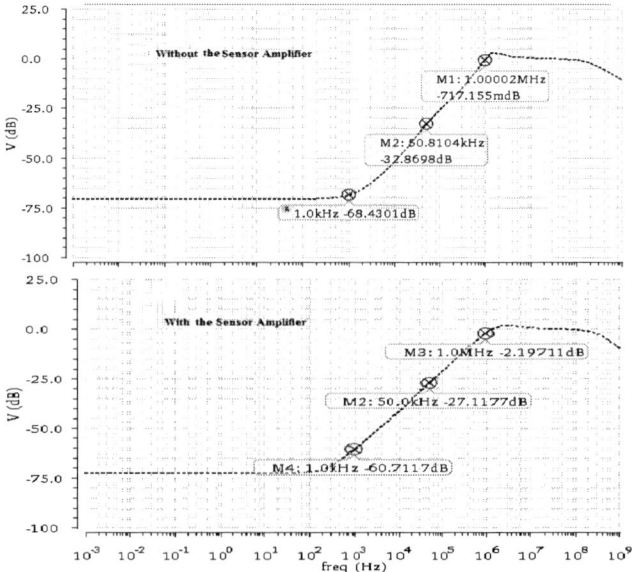

Figure 4. Simulation PSRR performance without and with the sensor amplifier of the proposed LDO.

978-1-4673-8760-6/15 $31.00 © 2015 IEEE

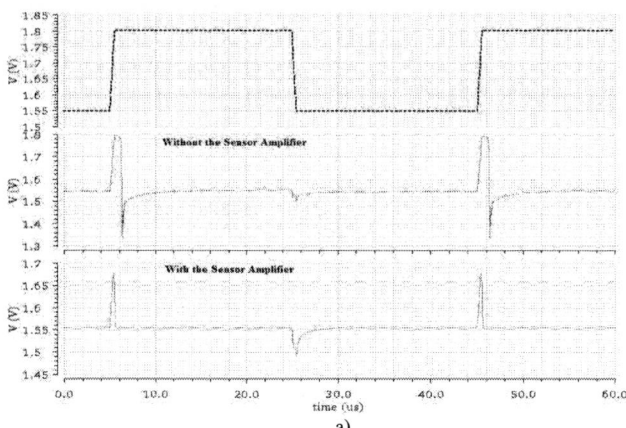

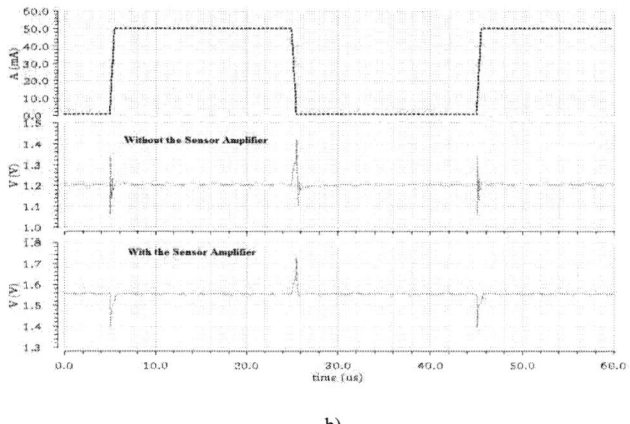

Figure.5. Transient response without and with the sensor amplifier: a) AC line regulation b) AC load regulation.

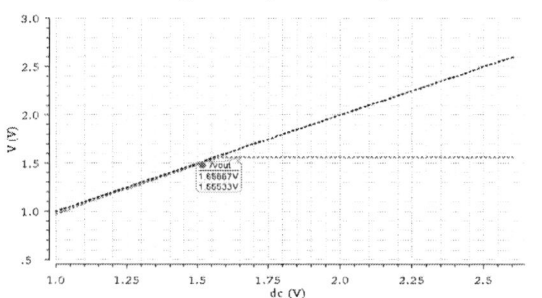

Figure.6. Simulation of DC line regulation.

Parameter	[5]	[6]	This work
CMOS Technology (μm)	0.13	0.18	0.18
VIN (V)	>1.15	1.2-1.5	1.2-1.8
VOUT (V)	1	1	1.6
Drop-out (mV)	>0.15	300	200
Compensation Cap(pF)	N/A	41	10
ILmax (mA)	25	50	50
Line regulation	N/A	0.024%	0.016
Load regulation (mV/mA)	0.048	0.175	0.6
Settling time (μs)	N/A	1.6	1.21
PSSR(dB)	-60dB@100KHz	-70 @ 1 KHz	-60.71 @1KHz

Table 1.Performances and comparison with other works.

IV. CONCLUSION

In this paper, an enhancement transient response of full on chip CMOS LDO using a modified NMC technique has been presented. The regulator circuit design features an active compensation technique, which guarantees the stability through the full load current range of 100uA to 50mA. The high performance is independent of the off-chip capacitor. The technique implemented in this work proves the better transient response without degrading the power consumption of the LDO The detailed analysis of the proposed structure is revealed to justify the performance of the technique utilized. The simulations prove the results theory.

The proposed LDO is capable of providing 50 mA with a drop-out voltage of 200 mV at VDD of 1.8 V. The proposed regulator is mainly used as a regulating power source for wireless applications, RFID and charge pumps.

REFERENCES

[1] K. Zared, H. Qjidaa and M. Zouak, "Full On-Chip CMOS Low Dropout Voltage Regulator With -41 dB PSRR at 1 MHz for Wireless Applications", Journal of Theoretical and Applied Information Technology (JATIT), Dec 2013.

[2] G.A. Rinco-Mora and P.E Allen "Optimized frequency-Shaping Circuit Topologies for LDO's" IEEE trans. Circuits sys.II, Vol. 45, no. 6, Jun 1998, pp. 703-708.

[3] K. Wong and D. Evans, "A 150mA low noise, high PSRR low-dropout linear regulator in 0.13μm technology for RF SoC applications," Proc. of IEEE European Solid-State Circuits Conference, Sep. 2006, pp. 532-535.

[4] S. K. Hoon, S. Chen, F. Maloberti, J. Chen, and B. Aravind, "A low noise, high power supply rejection low dropout regulator for wireless system-onchip applications," Proc. of IEEE Custom Integrated Circuits Conference, Sept. 2005, pp. 759-762.

[5] Gianluca Giustoli, Gaetano Palumbo, and Ester Spitale. "A 50 mA 1-nF Low Voltage Low –Dropout Voltage regulator for SoC Applications". ERTI journal, Vol. 32, Nunber 4, August 2010.

[6] K. N. Leung and P. K. T. Mok, "A capacitor-free CMOS low-dropout regulator with damping-factor-control frequency compensation," IEEE J. Solid-State Circuits, vol. 37, no. 10, pp. 1691–1701, Oct. 2003.

[7] A. Amer and E. Sánchez-Sinencio, "140 mA 90 nm CMSO Low Drop-out Regulator with -56 dB Power Supply Rejection at 10 MHz" ©2010 IEEE.

[8] G. A. Rincon-Mora, "Active capacitor multiplier in miller-compensated circuits" IEEEJ. Solid-State Circuits, Vol. 35, no. 1, pp. 26-32 Jan. 2000.

[9] Sai Kit Lau, Philip K.T.Mok and Ka Nang Leung "A Low-Dropout Regulator for SoC With Q-Reduction" IEEE Journal Of Solid-State Circuits,VOL.42, NO.3, MARCH 2007.

[10] Gabriel A. Rincon-Mora, Phillip E. Allen "A Low-Voltage, Low Quiescent Current, Low Drop-Out Regulator", IEEE Journal Of Solid-State Circuits,VOL.33, NO.1, JANUARY 1998.

[11] W.-J. Huang S.-I. Liu "Capacitor free low dorpout regulators using nested Miller compensation with active resistor and 1-bit programmable capacitor array" IET Circuits Devices Syst., 2008, Vol. 3, pp. 306-3016.

[12] Ma Haifeng, Zhou Feng "Full on-chip and area-efficient CMOS LDO with zero to maximum load stability using adaptative frequency compensation" Journal of Semiconductors, Vol. 31. No. 1 January 2010.

[13] C-C. Wang, C-C. Huang, and U. F. Chio "A linear LDO regulator with modified NMCF frequency compensation independent of off-chip capacitor and ESR" Analog Interg Circ Sig Process (2010) 63: 239-244.

978-1-4673-8760-6/15 $31.00 © 2015 IEEE

FPGA implementation of Data Encryption Standard using time variable permutations

Soufiane Oukili

Materials and Instrumentation (MIN),
High School of Technology, Moulay Ismail University
Meknes, Morocco
Soufiane.oukili@gmail.com

Seddik Bri

Materials and Instrumentation (MIN),
High School of Technology, Moulay Ismail University
Meknes, Morocco
briseddik@gmail.com

Abstract— **The Data Encryption Standard (DES) was the first modern and the most popular symmetric key algorithm used for encryption and decryption of digital data. Even though it is nowadays not considered secure against a determined attacker, it is still used in legacy applications. This paper presents a secure, high-throughput and area-efficient Field Programming Gate Arrays (FPGA) implementation of the Data Encryption Standard algorithm. This is achieved by combining 16 pipelining concept with time variable permutations and compared with previous illustrated encryption algorithms. The permutations change on time by the cryptographer. Therefore, the ciphertext changes too for the same key and plaintext. The proposed algorithm is implemented on Xilinx Spartan-3e (XC3s500e) FPGA. Our DES design achieved a data encryption rate of 9453.47 Gbit/s and 2046 number of occupied CLB slices. These results showed that the proposed implementation is one of the fastest hardware implementations with better area-efficient and much greater security.**

Keywords— Data Encryption Standard (DES) Algorithm; Field Programmable Gate Arrays (FPGA) implementation; Pipelining; time variable permutations, security.

I. INTRODUCTION

The Data Encryption Standard (DES) is an encryption standard for protecting confidential information. It has been developed in the 1970s at IBM and adopted as a Federal Information Processing Standard since 1977 by the National Institute of Standards and Technology [1], [2]. DES has been used pervasively by many applications that require data confidentiality. However, from year 2001, DES has been superseded by the Advanced Encryption Standard AES [3]. But in practice, a lot of hardware or software applications still resort to DES.

The DES is a block cipher that operates on 64-bit blocks of data and uses 56-bit private effective key. Because of its small key size, several attacks against DES algorithm were published [4], [5], [6]. To increase the security of the algorithm, we proposed the Data Encryption Standard based on variable initial and final permutations with time. The proposed scheme uses two permutation boxes, which contain several initial and final permutations to be selected by the sender. Whenever the permutations change, ciphertext

changes also for the same key and plaintext. Therefore, the security is increased because of the time variant behavior.

In this paper, we present an efficient and a secure hardware implementation of 16-stage pipelined DES, based on the variation of the initial and final permutations with time. Data blocks can be loaded at each clock cycle and after an initial delay of 17 clock cycles, the ciphertexts will appear on consecutive clock cycles. The design is implemented on Xilinx Spartan FPGA technology.

The rest of this paper is organized as follows: Section 2 describes the DES algorithm. Pipelining DES and pipelining DES based on time variable permutations are presented in Section 3 and Section 4. Section 5 gives implementation summary. Section 6 compares the achieved results with the previous DES implementations. Conclusion and references are given in Section 7 and 8 respectively.

II. DATA ENCRYPTION STANDARD ALGORITHM

DES algorithm encrypts 64-bit plaintext blocks with 64-bit key and generates 64-bit ciphertext blocks as shown in figure 1. This algorithm uses complicated logical functions such as various types of permutations, XOR and shift functions. One bit in each 8 bits of the key may be utilized for error detection in key generation. Bits 8, 16, 24, 32, 40, 48, 56 and 64 are used in ensuring that each byte of the key is of odd parity and otherwise ignored. Consequently, the effective key length is 56 bits.

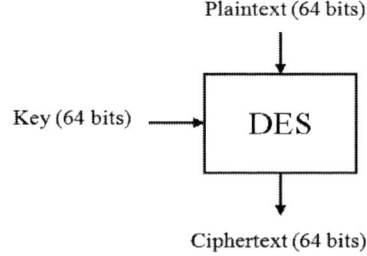

Fig. 1. DES block view

DES is an iterative algorithm as shown in figure 2. For each block of plaintext, encryption is handled in 16 rounds

978-1-4673-8760-6/15 $31.00 © 2015 IEEE

which all perform the identical operation. In every round a different sub-key is used and all sub-keys are derived from the main key. The incoming block of plaintext (64 bits) firstly passes through an initial permutation (IP) and then will be divided into two 32-bit halves, 32 right bits and 32 left bits.

In every round, the right 32 bits are expanded to 48 bits using the expansion permutation (E), by duplicating half of the bits. Then, the result is combined with a sub-key using an XOR operation. The XOR output is divided into eight 6-bit and fed into eight substitution boxes (S). Each of these boxes replaces its six input bits with four output bits, according to a non-linear transformation. The outputs are concatenated and pass through a straight permutation (P). The result is processed through XOR function with the left 32 bits and the output is the right bits of the next round. The left bits of the next round are the right bits of the previous round.

After the 16th iteration, the right and left bits are concatenated and finally pass through a final permutation (IP-1), which is the inverse of the initial permutation (IP). The output is the ciphertext block (64 bits).

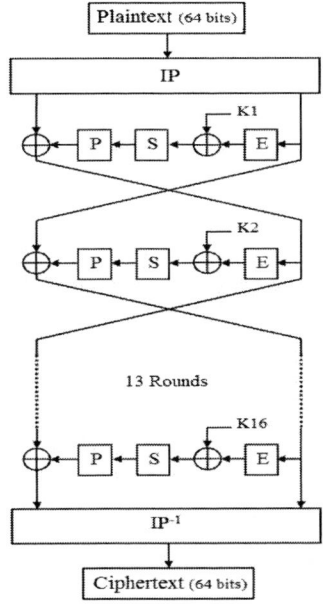

Fig. 2. DES algorithm block diagram

DES is a private key algorithm, in which the same key is used for both encryption and decryption. Figure 3 shows the key schedule generation. It is also an iterative process comprising 16 rounds and generates 16 sub-keys from the main key. The 56-bit effective key gets firstly permutated (PC-1) and splits up into two 28-bit halves; each half is thereafter treated separately. Then, for every round, both halves are shifted left by either one or two bits, depending on the round number. After that, the two outputs go through another permutation (PC-2). The result is the sub-key and it is coded on 48 bits, 24 bits from the left half, and 24 from the right.

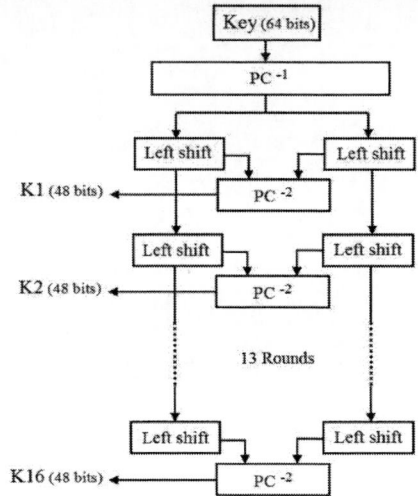

Fig. 3. Key schedule generation

III. PIPELINING DES BASED ON TIME VARIABLE PERMUTATIONS

There have been several approaches to attack DES algorithm. The most popular is the linear cryptanalysis [4], differential cryptanalysis [5] and exhaustive key search [6]. In order to make DES more secure, we developed a new algorithm shown in figure 4. It has an initial and final permutation boxes that contain four permutations in order to be used periodically, in which the final permutations are the inverse of the initial ones. As a result of this, every time the plaintexts are encrypted by different permutations. Therefore, detecting the algorithm will be difficult for the attackers because of the time variant behavior.

We can use several permutations in the boxes. In our design, we have four. Sender specifies how many clock cycles (N1, N2, N3, N4) that he will use each of these four permutations. Constantly, the program checks the Time value. If it is less than sender clock cycle value, the permutations are kept. Otherwise, the next permutations are selected from the permutation boxes and time value is set to zero. Flow chart shown in figure 5 introduces the steps of changing the permutations.

The sender and receiver have the same permutation boxes. They are connected to have the same permutations at any specific time. In order to avoid the disadvantage of the synchronization between them, sender transmit additional data with the ciphertexts to receiver, to indicate the correct choice of permutations from the permutation boxes.

Pipeline is an important technique to increase the performance of a system [7]. The iterative nature of the DES algorithm makes it ideally suited to pipelining and it can be 4, 6, 8 or 16 stages [8]. The pipelining strategy consists in parallelizing the data inputs and outputs with the processing. Basically, it means to process the data that is given as input in a continuous manner without having to wait for the current process to get over.

DES implementation presented in this paper is based on 16 stages pipelining. In order to pipeline the algorithm, registers R and L (32-bit) are placed at the left and right of the outputs of each round of the algorithm to allow the sequentiality of the data.

978-1-4673-8760-6/15 $31.00 © 2015 IEEE

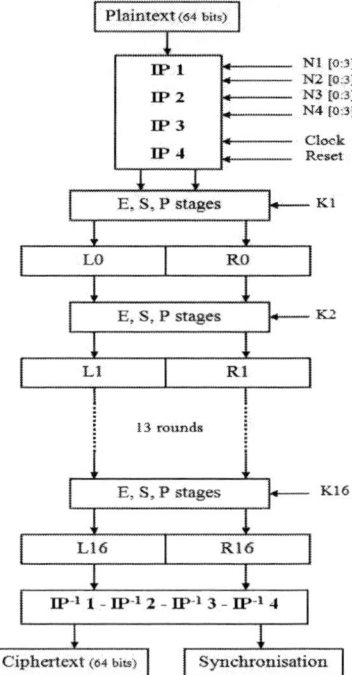

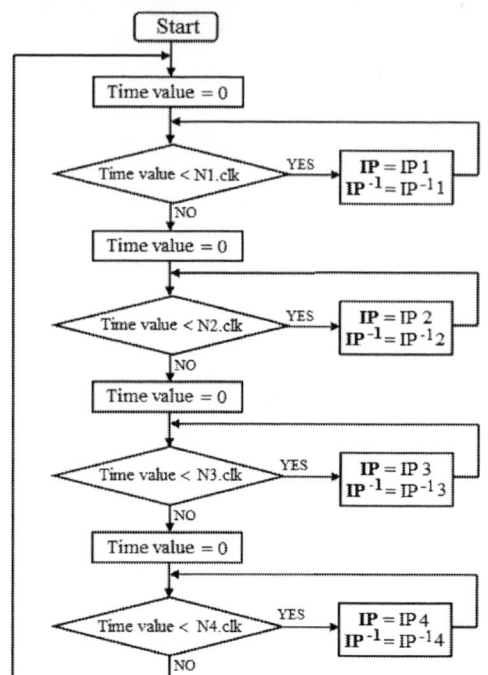

Fig. 4. DES based on time variable permutations

Fig. 5. Flow chart of permutations

IV. IMPLEMENTATION SUMMARY

FPGA implementation of our modified DES algorithm was accomplished on a Spartan-3e device XC3s500e-4fg320 using Xilinx ISE Design Suite 14.7 as synthesis and Modelsim 6.1f as simulation tool. The design was coded using VHDL language. It occupied 2046 (43%) CLB slices, 1143 (12%) slice Flip Flops and 203 (87%) I/Os. The design achieves a frequency of 147.71 MHz. It takes 17 clock cycles latency first time only then encrypts one data block (64 bits) per clock cycle. Therefore, the achieved throughput is (147.71 ×64) = 9.453 Gbits/s. Simulation window is shown in figure 6.

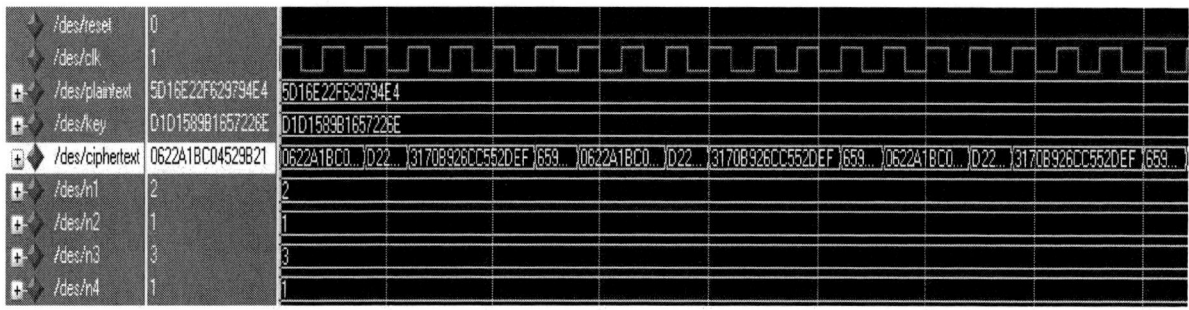

Fig. 6. Simulation Window of our DES design

V. PERFORMANCE COMPARISONS

There are several hardware implementations for the DES algorithm that aim to achieve the most efficient architecture, by improving high-throughput and area-efficient. Table 1 shows the performance figures for some representative hardware implementations of the DES. A Java-base (Jbits) DES implementation [9] achieves the fastest encryption rate of 10752 Mbits/s. The key schedule is computed entirely in software, so all cryptographic key input and sub-key generation logic are removed from the pipelined design. A DES implementation at [10] uses a pipelined design with skew core key-scheduling to load different keys every clock cycle, allowing the possibility of using multiple keys in any one session of data transfer. At [11], the implementation uses a non-standard representation and view the processor same as a SIMD (Single Instruction Multiple Data) computer, as 64 parallel one-bit processors computing the same instruction. A VLSI DES implementation [12] uses 0.6 micron CMOS technology. DES implementation at [13] uses a novel method for implementing the key schedule. This method supports the use of different keys every clock cycle and utilizes permutations to create the sub-keys from the input key. The sub-keys are delayed by the required amount using the necessary array of latches. The Implementation of DES at [14] is based on time variable data permutation. The design uses an initial permutation box that contains several permutations to be selected periodically.

978-1-4673-8760-6/15 $31.00 © 2015 IEEE   128

Table I. PERFORMANCE COMPARISONS

Authors	Device used	CLB slices	System clock (MHz)	Throughput (Mbps)	Throughput per CLB slices (Mbps/slice)	
Biham [11]	Alpha 8400	---	300	127	---	
Kaps, Paar [15]	XC4028EX	741	25.18	402.7	0.543	
McLoone, McCanny [13]	XCV1000	6446	59.5	3808	0.59	
Patterson [9]	XCV150	1584	168	10752	6.78	16-stage pipelined designs
Wilcox, Pierson, Robertson, Witzke, Gass [12]	ASIC	---	---	9280	---	
Patel, Joshi, Saxena [10]	XC3S500E	2814	111.882	7160	2.54	
Abd El-Latif, Hamed, Hasaneen [14]	XC3S500E	2062	124.734	7983	3.87	
Proposed Design	**XC3S500E**	**2046**	**147.71**	**9453.47**	4.62	

From the results in the table, we find that our 16-stage pipelined design gives 1.32, 74.43, 1.01, 2.48 and 1.18 times more throughput than the designs in [10-14], respectively. Looking at the CLB slices area count, our design needs only 0.72 times the CLB slices used in [10], 0.31 times in [13] and 0.99 times in [14]. CLB slices in [11] and [12] are not reported. Also from the table, we note that our proposed design gives 23.47 times more throughput than the pipelined implementation in [15]. But, it needs 2.76 times the CLB slices used, yet our design achieves higher throughput per CLB slice. The proposed design achieves only 0.87 times throughput of the design in [9] and needs 1.29 times the CLB slices used. However, this design is not a single-chip implementation of the full DES algorithm since the key schedule is computed in software. From the comparisons, we notice that our implementation is competitive with the reported implementations. It is more secure because of the time-varying behavior and one of the fastest single-chip FPGA designs with area efficient.

VI. CONCLUSIONS

This paper presents an efficient implementation for the design of a 16-stage pipelined DES algorithm using time variable permutations. The algorithm uses different initial and final permutations periodically with time. Therefore the ciphertext changes by time for the same key and plaintext. As a result of this, the security of the algorithm is increased. The plaintext blocks can be loaded every clock cycle and after an initial delay of 17 clock cycles, the ciphertext blocks will appear on consecutive clock cycles. The implementations of the DES algorithm based on hardware are low cost, flexible and efficient encryption solution. . The implementation of our design is presented by using Spartan-3E (XC3S500E) family FPGAs and is one of the fastest hardware implementations with much greater security. At a clock frequency of 147.71 MHz, it can encrypt or decrypt data blocks at a rate of 9453.47 Mbps.

REFERENCES

[1] National Institute of Standards and Technology (NIST), "Data Encryption Standard," Federal Information Processing Standards Publication 46-3, 1999.

[2] National Institute of Standards and Technology (NIST), "DES modes of operation," Federal Information Processing Standards Publication 81, 1980.

[3] National Institute of Standards and Technology (NIST), "Advanced Encryption Standard," Federal Information Processing Standards Publication 197, 2001.

[4] M. Matsui, "The first experimental cryptanalysis of the data encryption standard", Advances in Cryptology, 14th Annual International Cryptology Conference, California, USA, pp. 1–11, 1994.

[5] E. Biham, A. Shamir, "Differential cryptanalysis of DES-like cryptosystem," Journal of cryptology, vol.4, no.1, pp. 3-72, 1991.

[6] W. Diffie and M. E. Hellman, "Exhaustive Cryptanalysis of the NBS Data Encryption Standard," Computer, vol.10, no.6, pp. 74-84, 1977.

[7] S. Taherkhani, E. Ever, G. Orhan, "Implementation of Non-Pipelined and Pipelined Data Encryption Standard (DES) Using Xilinx Virtex-6 FPGA Technology," 10th IEEE International Conference on Computer and Information Technology, Bradford, UK, pp. 1257-1262, 2010.

[8] V. Patel, R. C. Joshi, A. K. Saxena, "FPGA Implementation of DES Using Pipelining Concept With Skew Core Key-Scheduling," Journal of Theoretical and Applied Information Technology, vol.5, no.3, pp. 295-300, 2009.

[9] Patterson, "High Performance DES Encryption in Virtex FPGAs Using Jbits," In Field-Programmable Custom Computing Machines, IEEE Comput. Soc., Napa Valley, California, USA, pp. 113-121, 2000.

[10] V. Patel, R. C. Joshi, A. K. Saxena, "FPGA Implementation of DES Using Pipelining Concept With Skew Core Key-Scheduling," Journal of Theoretical and Applied Information Technology, vol.5, no.3, pp. 295-300, 2009.

[11] E. Biham, "A Fast New DES Implementation in Software," 4th International Workshop on Fast Software Encryption, Israel, pp. 260-271, 1997.

[12] D. C. Wilcox, L. Pierson, P. Robertson, E. Witzke, K. Gass, "A DES ASIC Suitable for Network Encryption at 10 Gbps and Beyond," First International Workshop on Cryptographic Hardware and Embedded Systems, Massachusetts, USA, pp. 37-48, 1999.

[13] M. McLoone, J. V. McCanny, "High-performance FPGA implementation of DES using a novel method for implementing the key schedule," IEE proc: Circuits, Devices and Systems, vol. 150, no. 5, pp. 373-378, 2003.

[14] K. M. A. Abd El-Latif, H. F. A. Hamed, E. A. M. Hasaneen, "FPGA Implementation of the Pipelined Data Encryption Standard (DES) Based on Variable Time Data Permutation," The Online Journal on Electronics and Electrical Engineering, vol.2, no3, pp. 298-302, 2011.

[15] J. Kaps, C. Paar, "Fast DES Implementations for FPGAs and Its Application to a Universal Key-Search Machine," 5th Annual Workshop on selected areas in cryptography, Ontario, Canada, pp. 234-247, 1998.

Simulation of Ground Penetrating Radar Imaging Under Subsurface

Gamil Alsharahi, Abdellah Driouach
Abdelmalek Essaâdi University
Faculty ofScinces,Tetouan, Morocco
alsharahigamil@gmail.com, adriouach@hotmail.com

Ahmed Faize
Mohammed 1st University
Faculty Polydisiplinarly Nador, Morocco
ahmedfaize6@hotmail.com

Abstract: **This paper is devoted to study the propagation of electromagnetic waves of ground radar (GPR) in geological environments (heterogeneous). GPR is a method based on the analyze of spread the diffraction and reflection electromagnetic waves in high frequency (0.1 MHz to 2.6 GHz). A number of models have been designed to simulate the varieties of geological conditions. FDTD one of these models, which based on the scientific computing software called Reflexw simulation code. During the signal radar spread from the geological environment. We note that the signal decreases due to the absorption phenomenon and reflection. Our results show that the effective permittivities have a great impact on the wave behavior in different geological environments.**

Key words: **Ground Penetrating radar, Propagation, Electromagnetic wave.**

I. INTRODUCTION

The purpose of this part is to make a theoretical study by simulating the propagation of electromagnetic waves from the ground radar in heterogeneous media (geological). The phenomenon of propagation will be considered through the reflected waves: principle on which the GPR work [1, 2]. The software REFLEXW, enabled us the simulation of the ground as a function of its electrical and magnetic parameters. A number of models have been designed to simulate the variety of geological conditions. A rectangular block was used an initial model for simulation. The first model is a simple profile to give an idea on the propagation of electromagnetic waves in different materials and the effects of electromagnetic parameters (σ, ε and μ) on the wave. The second model is used to study the propagation of electromagnetic waves (reflected waves) in geological backgrounds from radargrams.

Ground-penetrating radar (GPR) is a geophysical method that employs an electromagnetic technique. The method transmits and receives radio waves to probe the subsurface. And the method has been extensively used in many applications, such as archaeology, civil engineering, forensics, geology and utility detection (Daniels, 2004).Dielectric constant of the host material plays an important role in GPR technology. Finding out the velocity and the depth of the target dielectric constant is important. In this study it is aimed to identify the behavior of GPR waves under different dielectric constant of the hosting material [12, 9]. The electrical properties of the ground directly beneath a ground penetrating radar (GPR) antenna very close to the earth's surface (ground-coupled) must be known in order to predict the antenna response. In order to investigate changing antenna response with varying ground properties, a series of finite difference time domain (FDTD) simulations were made for a bi-static (fixed horizontal offset between transmitting and receiving antennas) antenna array over a homogeneous ground[13, 6]. The FDTD approach to the numerical solution of Maxwell's equations is to discretize both the space and time continua [14]. Thus,the spatial and temporal discretization steps play a very significant role since thesmaller they are the closer the FDTD model is to a real representation of the problem [14]. However, the values of the discretization steps always have to be finite, since computers have a limited amount of storage and finite processing speed. Hence, the FDTD model represents a discretized version of the real problem and of limited 756 A. Giannopoulos / Construction and Building Materials 19 (2005) 755–762size. The building block of this discretized FDTD grid is the Yee cell named after Kane Yee who pioneered the FDTD method [4, 11]. The interactions of EM waves with physical media can be quite complex and the most exact models known for EM interactions use quantum mechanics [5, 8].

II. MATERIALS AND METHODS

A. Reflexw software

The company MalaGéosciences provides the software Groundvision with radar. This software enables instant viewing of acquisitions, but remains less efficient for their treatments that the Reflexw software. Reflexw is the new Windows version 9X/NT of the Reflx program under DOS. This software is adapted for data processing, originating from various original measures (seismic, GPR, ultrasound). This tool proposes five analysis modules. Software Reflexw offers a diverse range of treatments:

Filtrs 1D and 2D, gains (linear, exponential, local...), interpolation, migration (FK, Kirchoff...).

B. work of GPR (SIR System-3000Manual)

Standard ground penetrating radar systems consist of a transmitting and a receiving antenna. High-frequency (range: 0.1 MHz to 2.6 GHz) electromagnetic pulses are emitted into the ground by the transmitting antenna. The radar wavelet propagates through the soil while the velocity of the wavelet depends on the dielectric properties of the ground. At interfaces, e.g., boundaries of different soil layers or distinct objects, where the dielectric properties of the different media change erratically, the electromagnetic wave is partially reflected. The travel time and amplitude of the wavelet is recorded by the receiving antenna.

978-1-4673-8760-6/15 $31.00 © 2015 IEEE

III. RESULTS AND DISCUSSION

3.1 Simulation of GPR signals

To simulate GPR signals of proposed objects, the simulator (Reflexw) requires a number of parameters, as the frequency of the antenna used, the geometry of the subsoil well as the dielectric permittivity, the magnetic permeability and electrical conductivity the interfering backgrounds in the simulation.

A. Buried iron bar

To make the simulation per Reflexw, the ground, in which the bar is buried, is simulated by concrete whose dielectric characteristics are ($\varepsilon_r = 7$, $\sigma = 0.001$ S/m, and v=0.11m/s). The iron bar, from conductivity ($\sigma = 9.93\ 10^6$ S/m, $\varepsilon_r = 1.45$) is buried at a depth of 0.5 m and is located, on the surface, entre 1.75 m and 2.25 m Ceylon l'axe Ox (Fig.2). The simulation frequency is set at 800 MHz. Each emission and reflection of the simulated signal are recorded on a time window of 30 ns with a space increment of 7 cm.

The results obtained are synthesized and displayed as radargram as in Fig 1b. In this Fig we notice the presence of two diffraction hyperbole that indicates the presence of iron bar around 4 m (i.e exactly the depth at which it was assumed bury this bar).

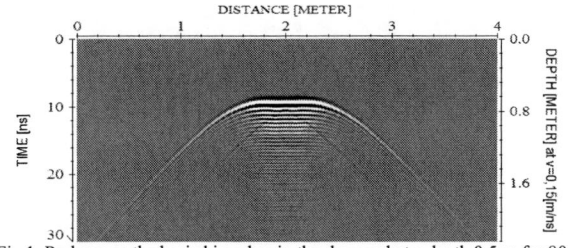

Fig.1. Radargram the buried iron bar in the dry sand at a depth 0.5 m for 800 MHz antenna.

B. Buried plastic tube

The plastic tube is considered very low conductivity $\sigma = 0.00004$ S/m, and dielectric permittivity$\varepsilon_r = 4.5$. One makes the simulation, as in the previous case, in dry sand ($\varepsilon_r = 4$; $\sigma = 0.00001$ s/m $and\ v = 0.15m/s$) at a depth of 0.5 m (Fig .2).

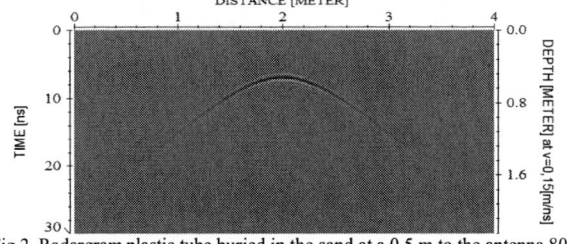

Fig.2. Radargram plastic tube buried in the sand at a 0.5 m to the antenna 800 MHz

C. Simulation signals from a freshwater bottle

Simulation proposed in this section concerns a plastic bottle containing water which is a high electrical permittivity element ($\varepsilon_r = 81$; $\sigma = 0.0005$ s/m). One makes the simulation, as in the previous case, in dry sand at a depth of 1 m. Fig 3 reveals

the presence of several hyperbolas corresponding to several consecutive reflections but at different times for the frequency used (800 MHz). Indeed, the water contained in the bottle, considered a dielectric, has a high permittivity which reduces the propagation velocity of the electromagnetic waves. This accounts for the delay observed when receiving the signal. In this figure we can observe the subhorizontal and sub-parallel interface of the bottle.

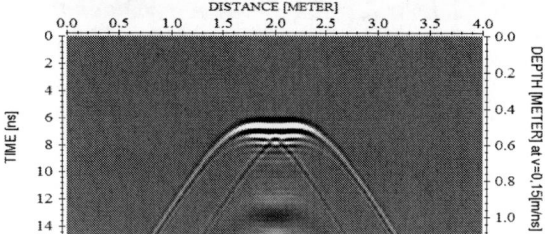

Fig.3. Radargram a plastic bottle filled with water buried in the sand for the 800 MHz antenna.

3.2 Cavity models

This profile is intended to study the propagation of electromagnetic waves through radargram. The simulation of this profile this fact on two models: the first model is the model "layers" consisting of three homogeneous layers (dry sand, saturated sand and limestone) to form a heterogeneous medium, this model can be found in nature in the mountains or massive. The second model is the "cavity", this model consists of a buried cavity in the concrete, depending on the material with which the cavity is full, and this model gives an idea about the form of several geological media.

Table 1: Physical properties of materials

Material	Relative permittivity	Permeability	Conductivity (S/m)
Limestone	4-8	1	0.0005
Saturated sand	20-30	1	0.0005-.002
Dry sand	3-5	1	0.00001
Air	1	1	0
Concrete	7	1	0.001
Clay	4-40	1	0.002-1

A. Cavity model

This model is a geologic medium of 4 m wide and 1m deep. It is constituted by a circular cavity of 0.05 m in diameter within the concrete to a depth of 0.5m from the ground surface. The medium in which the cavity is buried is composed of: The diagram of this structure is given in Fig 4.

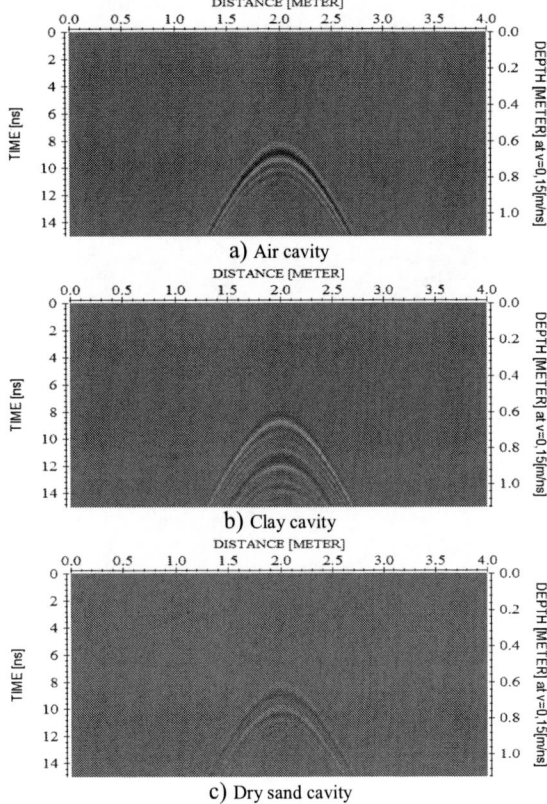

a) Air cavity

b) Clay cavity

c) Dry sand cavity

Fig .4. Radargram the full cavity model.

To study the effect of the geological medium constituting the cavity on the propagation of the electromagnetic wave, we made several simulations using the same materials which were used for profile 1. This allows us to see the response of geological radar in the case of a cavity which is a very close case of reality (pipe, mine, air cavity and other defects in the soil).

IV. COMPARING BETWEEN SIMULATION AND EXPERIMENT

Our calculation have been compared with several experimental results. And shows a good agreement between them.

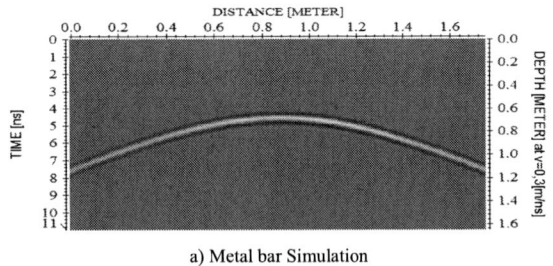

a) Metal bar Simulation

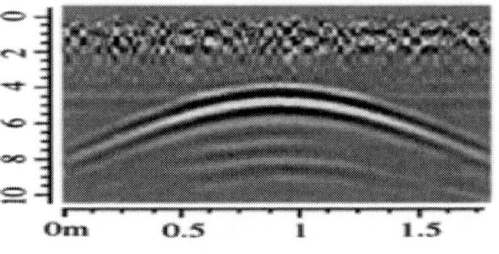

b) Metal bar experiments [15]

Fig.5. Radargram of detect metal bar

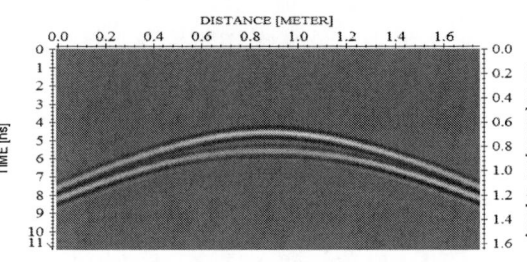

a)Metal bars Simulation

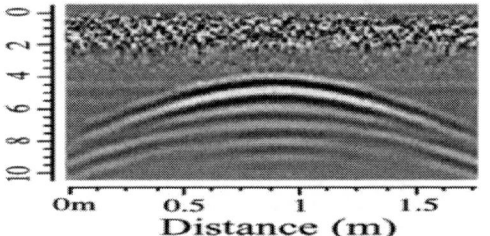

b) Metal bars experiments [15]

Fig.6. Radargram of detect metal bars spacing vertical

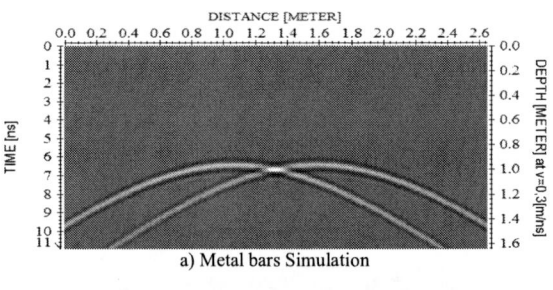

a) Metal bars Simulation

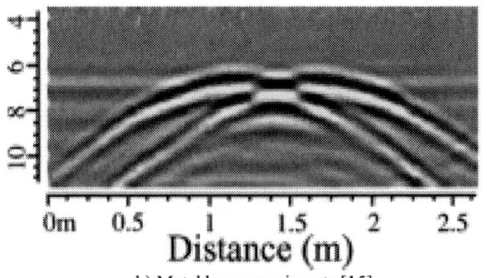

b) Metal bars experiments [15]

Fig.7. Radargram of detect metal bars spacing horizontal

978-1-4673-8760-6/15 $31.00 © 2015 IEEE

The results obtained are summarized and represented by the software, as radargram as in Figure 5(a), for example. In this figure we note the presence of two diffraction hyperbolas that indicate the presence of the iron bar around 0.95 m; this is exactly the depth at which it was supposed to bury this bar. It is possible to compare the simulation results with the experimental results Fig (5b), taking into account the effective parameters of each antenna. The Treaty of the location of the metal (B-Scan) profile is shown in Figure 9b. Hyperbolas that appear between 0.5 and 1, 5 m indicate the presence of the iron rod. For experimental parts: we note the presence of several hyperboles "noise" who could testify to the existence of high conductivity materials.

Figure 6(b) shows the simulation result of this model (Fig 6a). In this figure indicate the presence of several hyperboles that match several consecutive reflections but at different times for the frequency used (800MHz). A higher electromagnetic contrast diminishes the energy of the signal that reaches the second reflector, so a bigger separation is needed in order to detect them as discrete events.

Figure 7 (b) shows Radargram of detect metal bars spacing horizontal.

V. CONCLUSION

Our results, that have been achieved for different objects (iron bar, tube and plastic water bottle plastic) buried, show that the detection of small objects by radar GPR when the process signal lead to production of radargrams. The hyperbole appear indicate that the efficiency of this technique for simulation the small objects. Several simulations with different geological media to understand the phenomenon of propagation electromagnetic waves in the ground by its heterogeneity and texture have been applied. The change of the permittivity causes the reflection of the electromagnetic waves, while the conductivity attenuates these waves. The higher is the permittivity of the dielectric medium, the higher is the amplitude of the reflected waves.

Our calculation have been compared with several experimental results. And shows a good agreement between them.

REFERENCES

[1] A. Benedetto, F. Benedetto, Application Field–Specific Synthesizing of Sensing Technology: Civil Engineering Application of Ground Penetrating Radar Sensing Technology; Volume 13: Sensor Materials, Technologies and Applications, Elsevier 2014

[2] G. Klysz , J.P. Balayssac, X. Ferrières, Evaluation of dielectric Properties of concrete by a numerical FDTD model of a GPR coupled antenna Parametric study, Volume 41, Issue 8, December 2008, Elsevier

[3] Giannopoulos, Antonios (1998) the investigation of transmission-line matrix and finite-difference time-domain methods for the forward GPR and high-frequency EMI data sets 2013 Elsevier

[4] Yee KS. Numerical solution of initial boundary value problems involving Maxwells equations in isotropic media. IEEE T Antenn Propag 1966; 14:49–58.

[5] Nurul Jihan Farhah Bostanudin, University of Portsmouth., Computational
Methods for Processing Ground Penetrating Radar Data, May 2013

[6] A. Shaari, Effects of antenna-target polarization and target-medium dielectric contrast on GPR signal from non-metal pipes using FDTD simulation, 2010 Elsevier, NDT&E International 43 (2010) 403–408

[7] Daniels, David Ground-penetrating radar. - 2nd Ed. © 2004

[8] Sebastian Emanuel Lauro, Estimation of subsurface dielectric target depth for GPR planetary exploration, Journal of Applied Geophysics 93 (2013) 93–100

[9] Michael A. Hatch, The importance of including conductivity and dielectric permittivity information when processing low-frequency GPR and high-frequency EMI data sets 2013 Elsevier

[10] Dr. K.J. Sandmeier. Karlsruhe, REFLEX Version 7.5, 2014

[11] Thomas M. Urban, Jeffrey F. Leon, Sturt W. Manning, Kevin D. Fisher, Catherine M. Kearns & Peregrine A. Gerard-Little Comparing Similar Ground-penetrating Radar Surveys under Different Moisture Conditions at Kalavasos-Ayios Dhimitrios, Cyprus, Archaeological Prospection

[12] A. Benedetto, F. Benedetto, Application Field–Specific Synthesizing of Sensing Technology: Civil Engineering Application of Ground Penetrating Radar Sensing Technology; Volume 13: Sensor Materials, Technologies and Applications, Elsevier 2014

[13] Problem of ground probing radar. PhD thesis, University of York.

[14] Denver Fed. Center, US Geol. Survey, Denver, CO, USA, Ground Penetrating Radar Antenna System Analysis for Prediction of Earth Material Properties, 2005 IEEE (Volume:3B)

[15] Fernando I. Rial, Manuel Pereira, Henrique Lorenzo, Resolution of GPR bowtie antennas: An experimental approach, Journal of Applied Geophysics 67 367–373 2009

AUTHORS BIOGRAPHY

Gamil Alsharahi was born in AmranCity, Yemen in 1979. He received the B.S inmathematics and physic. And Master degrees in Telecommunication and elect- onics from the Abdelmalek Essaadi University, Faculty of sciences,Tetouan.Currently he is workingtoward the Ph.D degree in UniversityAbdelmalek Essaadi,Faculty of sciences "comm-unicatio Systems".

Abdellah Driouach Abdellah Driouach, was born in Al Hoceima city (Morocco) in 1954. He got his license, in electronic physics, at the University of Rabat (Morocco) in 1979, his doctorate (of third cycle) in "spectronomie hertzienne" , at the University of Bordeaux 1 (France) in 1983 and his doctor title (es sciences) at the University of Grenade (Spain). Professor and researcher at the Faculty of Science, Abdelmallek Essaadi,University since 1983. He has participated in several scientific research, including the dispersion of electromagnetic waves on obstacles to arbitrary structures. Currently he is a member of the research team "Communication Systems".

Ahmed Faize was born Morocco in 1985.Currently he is Professor and researcher atmohammed 1stUniversity, Faculty polydisiplinarly Nador

Formal Modeling, Verification and Implementation of a Train Control System

MohammadHossein AskariHemmat, Otmane Ait Mohamed
Dept. of ECE,
Concordia University,
Montreal, Canada
Email: {mo_askar,ait}@encs.concordia.ca

Mounir Boukadoum
COFAMIC, Dept. of Computer Science,
UQAM,
Montreal, Canada
Email: boukadoum.mounir@uqam.ca

Abstract—**For complex system design, it can be very useful to use abstracted models to verify functionality, thereby reducing the number of potential bugs in the implementation phase. In this respect, model checking is a powerful technique for automatically determining whether a design (model) satisfies desired properties. Upon positive results, the design can then be implemented with limited risks of malfunction. In this paper, we show this by modeling a train control system for safe speed and acceleration limits and verifying the correctness of the model properties using the NuSMV model checker. Then, we implement the algorithm of the verified model on an ARM CortexM platform.**

I. INTRODUCTION

The International Council on Systems Engineering (INCOSE)[1] identified Model-Based Systems Engineering (MBSE) [2] as the key driver for effective and efficient system development. For this, the system must be modeled in a way that reflects the correct functionality of the system. Then, model checking [3][4] can be used as a verification technique to automatically determine whether the design (model) satisfies desired properties.

This ability to automatically verify model properties differentiates model checking from other types of formal verification techniques. For instance, in Theorem Proving [5] the system is mathematically modeled with formal mathematical logic. The properties of interest in the system are then derived as theorems that follow from definitions. Various techniques have been developed to automate this process, using computers to handle obvious or tedious steps of the proof. Nonetheless, Theorem Proving is still considered as a non-automated form of formal verification.

Thanks to automation, model checking is the principal form of formal verification practiced in industry [6]. Model checking has another significant benefit: When the system has a bug, that is, its desired properties are not theorems, a decision procedure can provide a counterexample. Such counterexamples can be invaluable in tracing the source of such errors in the system implementation. However, the size of the system state space grows exponentially with the number of processes and variables, ultimately leading to the state space explosion in model checking. The level of expertise of the designer can play a key role to avoid such phenomenon as an expert can potentially reduce the number of states in a model.

In this paper, we present a formal model for a train control system. Then, using the NuSMV [7] model checker, we will verify critical properties of the system and, after successful verification and based on the mathematical model, we will implement the model on an ARM CortexM4 [8] platform to verify the correct functionality of the algorithm.

The rest of the paper is organized as follows: Section II is a brief overview of Model checking tools; Section III describes the train control algorithm and its formal representation; Section IV describes how the model properties are verified using the NuSMV model checker; Section V provides subsequent implementation results on an ARM CortexM plateform and Section VI concludes this paper.

II. PRELIMINARIES

In general, model checking tools are developed to support both CTL (Computational Tree Logic) and PCTL (Probabilistic Computation Tree Logic). For instance, the Symbolic Model Verifier (SMV)[3] system relies on Binary Decision Diagrams (BDDs) to check a finite state system against specifications in temporal CTL. NuSMV [7] is a re-implementation and extension of SMV that aims industrial size designs and can be used as a back-end for other verification tools and as a research tool for formal verification techniques. Another tool, the PRISM model checker [9] was developed to analyze probabilistic models including discrete-time Markov chains, continuous-time Markov chains, Markov decision processes and probabilistic extensions of the timed automata formalism. The properties to be verified against these models are expressed with probabilistic extensions of temporal logic. In this paper, the system that we are modeling does not contain any probabilistic path. Thus, we will use NuSMV to verify the model's formal properties. These are specified in temporal CTL and Linear Time Logic (LTL) for NuSMV. We used AutoFocus3 (AF3) [10] to perform model verification and validation before code generation. AF3 is a model-based tool which provides the means to design a distributed, embedded software systems system from an input model. It is integrated with the NuSMV system to provide model checking, boundary check analysis, reachability analysis and non-determinism checking of the underlying system. It uses NuSMV as the

978-1-4673-8760-6/15 $31.00 © 2015 IEEE

TABLE I: *Dublin* to *Daly City* track properties

Track Name	Segments	Segment Range	Gates	Civil Speed	Grade
Dublin to *Daly City*	DUBL_CAST_E	0km - 16.0km	DUBL (27.6m),WDUB(2447.7m)	36	0.8
	CAST_E_BAYF_S	16.0km - 19.7km	CAST(16129.5m)	80	0.3
	OAKY_BAYF_S	19.7km - 38.1km	BAYF(21396.1m),SANL(25428.4m),COLS(30028.6m), FTVL(33273.5m), LAKE(37714.1m)	70	3.49
	OAKY_SE	38.1km - 38.7km	N/A	50	1.00
	OAKY_DALY	38.7km - 62.6km	WOAK(40577.9m), EMBR(50023.7m), MONT(50577.6m),POWL(51441.5m), CIVC(52306.3m), 16th(53932.1m),24th(55350.1m), GLEN(58038.2m), BALB(59855.1m), DALY(62690.6m)	36	0.8

model checking verification back-end and supports most of the common temporal logic patterns.

III. BART System Modeling

BART (Bay Area Rapid Transit) provides commuter rail service for part of the San Francisco bay area in California. On a typical work day, the BART system serves around 250,000 passengers and during commute hours, over 50 trains can be in service. The system is controlled automatically and the on-board drivers have a limited role to play during normal operation. With a few minor exceptions, BART uses double tracks going in opposite directions and the trains go from a starting point to an ending point (i.e., there are no looping tracks). There are 5 different tracks in the system, and the trains in each one must obey speed and acceleration limits (Other aspects such as communication, error recovery, routing and right-of-way signaling are not covered here). This work is concerned with the design of a control system for these two variables subject to the BART specification constraint.

As case in point, we chose the *Dublin* to *Daly City* track which consists of 5 segments, each one with its speed and acceleration specifications. The track serves 18 gates, each one in an *open* or *close* state. Table I illustrates the *Dublin* to *Daly City* track properties. In it, the distances next to each gate name are with respect to the beginning of the track. The data in the table are provided by the BART Geo-spatial database [11] in Keyhole Markup Language (KML) format. Our objective is to control the speed and acceleration of each train according to the following requirements:

- The train should not get so close to another train that, if the train in front stops suddenly, there will be a collision between the two.
- The train should not enter a closed gate.
- The train should stay below the maximum speed allowed on the track(defined by *CivilSpeed*).

A station computer that is part of the Advance Automatic Train Control (AATC) system controls the trains in its vicinty, using an algorithm that updates the speed and acceleration of each train. We assume that the station computer has direct access to the current train speed, acceleration and position of a train and the control algorithm's output consists of speed (between 0 and 80Kmph) and acceleration (-2 to -0.45 in braking state and 0 to 3Kmphps in propulsion) commands for the train. For model simplicity, we assume that the communication link between

computer and train is ideal and that the interlocking system does not close a gate if it is too late for an approaching train to stop. Moreover, a train is abstracted as a single location on the track. Under these conditions, the operation of each train is modeled as follows:

Listing 1: Modeling the operation of a train

```
1   let delta = 0.5
2   let grade = (−21.9 ∗ currentSegmentGrade)/100
3   then :
4       let n = nosePosition + v × delta + ½ a × delta²
5           + ½ × grade × delta²
6       if (v == 0 and vcm == 0) :
7           nosePosition = nosePosition
8       else :
9           nosePosition = n
10      let speed = v + ½ a × delta + ½ grade × delta
11      if (v == 0 and vcm == 0) or speed <= 0 :
12          v = 0
13      else :
14          v = speed
15      let noseAtNext = nosePosition + v × delta
16          + ½ × a × delta²
17          + ½ × grade × delta²
18      if (v == 0 and vcm == 0) :
19          a = 0
20      elseif ((v > (vcm − 2) and acm > 0) or
21          (v > (vcm − 2) and acm < 0)) :
22          a = (21.9 × grade)/100
23      else :
24          a = acm
```

In this listing, *acm* and *vcm* are the received acceleration and velocity commands, *delta* is the interval between commands in seconds and *grade* is the train acceleration component due to grade. lines 4-9 calculate the next train position based on physical formulas and lines 10-14 calcule its velocity. The train acceleration is calculated in lines 15-24. If the speed reaches the commanded value within $\pm 2Kmph$, the train attempts to maintain it by compensating the acceleration due to grade. It should be noted that a train has no notion of the speed and acceleration limits of the segments. Also, it does not determine whether to stop at an approaching gate. Thus, it is the station computer that guides the trains and prevents potential catastrophes. Listing 2 illustrates a simple control algorithm used by this computer to generate the appropriate velocity and acceleration commands for a train:

Listing 2: Station Computer Algorithm

```
1   let t ∈ trainList
2   let delta = 0.5
3   let grade = (−21.9 ∗ currentSegmentGrade)/100
4   let range = (WCSD(t) × 2 + 230)
5   then :
6      nextStopDistance = calcNextStop(t, trainList, gateList)
7      segment, vcmCivilSpeed = CivilSpeed(t, range)
8      if ((nextStopDistance − t.position) < range) :
9         vcm = 0
10     else :
11        vcm = vcmCivilSpeed
12        d1 = nextStopDistance − t.position
13        if d1 < 0 :
14           acc = train.a() + 0.5
15        else :
16           acc = vcmCivilSpeed² − t.v² / (2×d1) − grade
17        if (acc < 0 and acc > −0.45) :
18           acmCivilSpeed = −0.45
19        else :
20           acmCivilSpeed = acc
21        d2 = nextStopDistance − t.position − WCSD(t)
22        acc = −t.v² / (2×d2) − grade
23        if (nextStopDistance − t.position) > range :
24           acmNextStop = t.a + 0.5
25        else :
26        if (acc < 0 and acc > −0.45) and
27           (d2 > ((t.v × delta) + 0.5 × grade ∗ delta²) :
28           acmNextStop = 0
29        else if (acc < 0 and acc > −0.45) and
30           (d2 <= ((t.v × delta) + 0.5 × grade ∗ delta²) :
31           acmNextStop = −0.45
32        else :
33           acmNextStop = acc
34        if acmCivilSpeed < acmNextStop :
35           acm = acmCivilSpeed
36        else :
37           acm = acmNextStop
```

The Worst Case Scenario Distance (WCSD) used in line 4 serves to set the speed and acceleration safety bounds. It is thoroughly explained in [12]. The *nextStopDistance* is calculated based on the current positions of the train and that in front of it (if any) and on whether a closed gate is within the stopping range. A decision is made according to which of the two is closest. The *civilSpeed* function calculates the next segments within range and returns the lowest *civilSpeed* of all and the corresponding segment object. In lines 8-11, the commanded velocity is computed. The calculation of *acm* (commanded acceleration) is done in lines 12-37. The *acmCivilSpeed* is calculated so that the train reaches a speed 2 Kmph below *vcmCivilSpeed*. If the train is already in that speed segment, *acmCivilSpeed* is set to the current acceleration incremented by 0.5 Kmph. If the resulting acceleration is between 0 and -0.45(which is not allowed), it is rounded off to -0.45. Next, *acmNextStop* is computed. If the next stop is out of range, it is set to the current acceleration incremented by 0.5 Kmph; otherwise it is set to stop the train WCSD feet before the obstacle. Since the WCSD is shrinking while the train gets closer to the obstacle, the train will stop at a reasonable distance. If necessary, the resulting acceleration is rounded off

to be within the allowed acceleration range mentioned before. Finally the commanded acceleration is set to the minimum of *acmCivilSpeed* and *acmNextStop*.

IV. MODEL VERIFICATION

Based on the above algorithms, we modeled the system in AF3 to verify its functionality. Figure1 illustrates the AF3 top model. In this example, we considered two trains moving in the same direction on the *Doublin* track. The Monitor block illustrates the velocities, accelerations and positions of the trains. For simulation and verification, the initial train positions play a key role in avoiding obstacles. In Simulation, the initial positions are calculated in a way that does not violate the safety regulations. For instance, we chose them to be much greater than the WCSD to any heading obstacle. As for verification, the properties are verified only if the initial position is more than the WCSD to the heading obstacle.

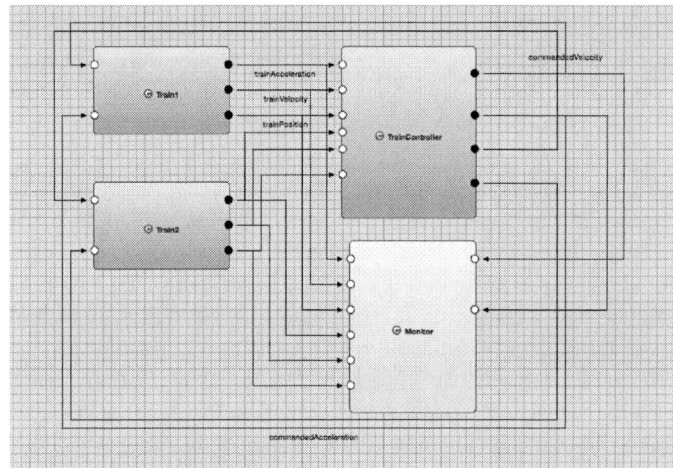

Fig. 1: BART system modeled in AF3

AF3 uses NuSMV for property verification. The following are some of the properties that were verified:

$\forall i \in [1, 2, 3 ..., 18]$
P₁: $AG((T1_{initpos} − T2_{initpos}) > WCSD \& (T1_{initpos} − gate.begin_i) > WCSD)) \rightarrow AF(T1_v < T1.Segment_{CivilSpeed})$
P₂: $AG((T1_{initpos} − T2_{initpos}) > WCSD \& (T1_{initpos} − gate.begin_i) > WCSD)) \rightarrow AF((T1_{pos} − T2_{pos}) < 0)$
P₃: $AG((T1_{initpos} − T2_{initpos}) > WCSD \& (T1_{initpos} − gate.begin_i) > WCSD) \& (T1_a < 0) \& (gate.state == 0)) \rightarrow AF((T1_{pos} − gate.begin_i) < 0)$

Property (**P₁**) shows that if the initial position($T_{initpos}$) is selected as discussed earlier, the train speed will never exceed the segment *CivilSpeed*. The second property(**P₂**) shows that by selecting a proper initial position, the distance between two trains will be greater or equal to WCSD. The third property(**P₃**) shows that if a train is in a braking state ($a \leq 0$) and approaching a closed gate (*gate.state == 0*), by selecting a proper initial position, the distance between the train and the gate will never be negative (thus avoiding collision with the gate).

V. IMPLEMENTATION

This section describes the implementation of the train control algorithm on an ARM Cortex-M4 platform with the proposed tool. The goal is to have the train control system meet all the safety properties that were stated previously. To ensure this, we built a verification platform for the correctness of the output commands. We created a golden system that contains all the required parts in AATC, including the station computer, trains, segments and tracks. Then, we implemented the control algorithm on the ARM Cortex-M4 platform. The control outputs were sent to the PC at the same time that the command were calculated by the golden model. Then, the verification platform compared the two results to decide whether the output of the implemented model was correct or not. We used a FRDM-K64F Freescale freedom platform [13] in our test. As an example, we set the initial position of the train at 50 meters from the beginning of *Doublin* track, and the next stop was the *WDUB* gate located 2447.7 meters from the beginning. Figures 2, 3 and 4 illustrate the position, acceleration and velocity of the train. As shown, the train

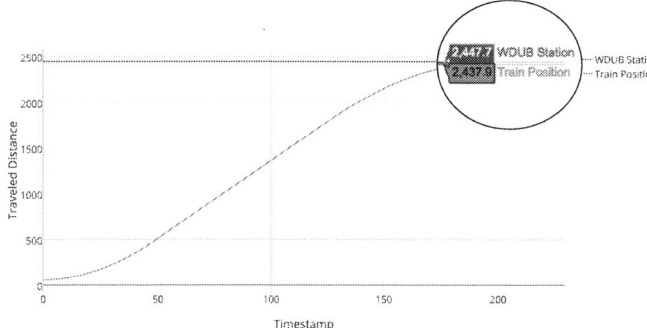

Fig. 2: Train Position

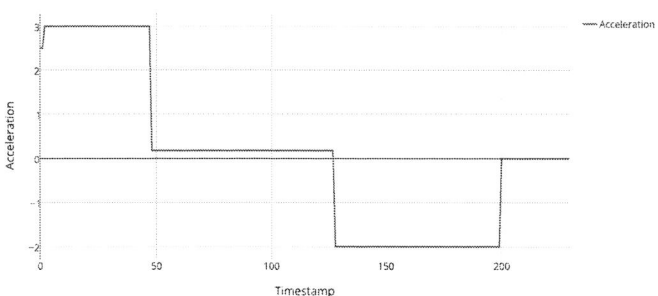

Fig. 3: Train acceleration over time

followed all the safety rules defined earlier. The goal was a full stop within the range of WCSD at the *WDUB* gate located at 2447.7 meters from the beginning of the track. Figure 2 shows that the train successfully stopped before reaching the gate. Figure 3 shows that the train always obeyed the train acceleration limit. Also, Figure 4 shows that the control system successfully controlled the velocity of the train as it never

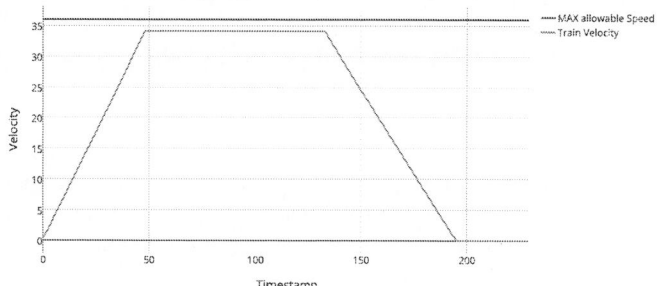

Fig. 4: Train Velocity over time

reached the maximum allowable velocity. The source code for this project is available in a github repository [14].

VI. CONCLUSION

In this paper we provided steps to verify and implement a Train Control System. We started with formally modeling the system. Based on the model and specification, we extracted the essential properties of the system. Using a model checking tool, in our case NuSMV, we verified whether the property held or not. Finally, using the high level model, we implemented the model on an ARM CortexM4 platform. The same approach can be used to verify and implement other systems with high level of complexity.

REFERENCES

[1] The International Council on Systems Engineering (IN-COSE),http://www.incose.org/ , September 2015.
[2] Friedenthal, Sanford, Greigo, Regina, and Mark Sampson, INCOSE MBSE Roadmap, in INCOSE Model Based Systems Engineering (MBSE) Workshop Outbrief (Presentation Slides), presented at INCOSE International Workshop 2008, Albuquerque, NM, pg. 6, Jan. 26, 2008
[3] E. M. Clarke, O. Grumberg, and D. A. Peled. Model Checking. The MIT Press, 2000.
[4] C. Baier, J.P. Katoen, Principles of Model Checking, The MIT Press Cambridge, England, 2008.
[5] M.J.C. Gordon. Mechanizing Programming Logics in Higher-order Logic. In Current Trends in Hardware Verification and Automated Theorem Proving, pages 387439. Springer, 1989.
[6] A. Biere, E. M. Clarke, R. Raimi, and Y. S. Zhu. Verifiying Safety Properties of a Power PC Microprocessor Using Symbolic Model Checking without BDDs. In Proc. of the 11th Inter. Conf. on Computer Aided Verification (CAV99), pages 6071. Springer, 1999.
[7] A. Cimatti, E. Clarke, F.Giunchiglia and M.Roveri,NUSMV: a new Symbolic Model Verifier,International Journal on Software Tools for Technology Transfer, 2000, Volume 2, Number 4, Page 410
[8] ARM Cortex-M processor family, www.arm.com/products/processors/cortexm/, September 2015.
[9] M. Kwiatkowska, G. Norman, and D. Parker. PRISM 4.0: Verification of Probabilistic Real-Time Systems. In CAV, LNCS, pages 585591. Springer, 2011.
[10] AutoFOCUS3 Website. http://af3.fortiss.org/whatn isn AF3.html, September 2015.
[11] Bart Geo-spatial Data, http://www.bart.gov/schedules/developers/geo, September 2015.
[12] Kordon, Fabrice, and Michel Lemoine. Formal Methods For Embedded Distributed Systems. Boston: Kluwer Academic, 2004. Print.
[13] FRDM platform from freescale, http://www.freescale.com/webapp/sps/site/prod summary.jsp?code=FRDMKL25Z, MAY 2014
[14] https://github.com/hossein1387/BART-System-Verification

978-1-4673-8760-6/15 $31.00 © 2015 IEEE

Maintenance optimization of series-parallel systems operating missions with scheduled breaks

A. KHATAB
Laboratory of Industrial Engineering,
Production and Maintenance,
National School of Engineering
Metz, France.
Email: khatab@enim.fr

E.-H. AGHEZZAF
Department of Industrial Systems
Engineering and Product Design,
Faculty of Engineering and Architecture,
Ghent University, Ghent, Belgium.
Email: ElHoussaine.Aghezzaf@UGent.be

D. CLAVER
Dalhousie University
Halifax,Canada.
Email: Claver.Diallo@dal.ca

Abstract—**This paper addresses the selective maintenance optimization problem for a multi-mission series-parallel system. Such a system experiences several missions with breaks between successive missions. To improve the reliability of the system, preventive maintenance actions are performed during breaks. Each preventive maintenance action is characterized by its age reduction coefficient. The selective maintenance problem consists then in finding an optimal maintenance plan which minimizes the total maintenance cost while providing a given required system reliability level for each mission. The fundamental constructs and the relevant parameters of this decision-making problem are developed and discussed.**

I. INTRODUCTION

The growing importance of the maintenance function in variety of industrial environments has lead to an increasing interest in the development and implementation of maintenance optimization models for stochastic degrading Manufacturing production systems. Maintenance operations allow companies to maximize not only their profit, but also rationally exploit their production systems. Indeed, present production systems are often designed in a fully integrated structure with several machines (components). In this kind of design solution, the sudden failure of one component (machine) may lead to the stoppage of the entire production activity. This situation may cause delays with drastic technical and economic consequences. Furthermore, a production system is subject to random deterioration processes resulting from degradation of each of its components. This degradation increases the probability of failures but also impacts production quality. Consequently, to ensure a rational exploitation of such production systems and to meet the need for high product quality as well as market demands and requirements, maintenance activities have been broadly accepted and implemented to assess the degradation of the production system, improve its reliability level, reduce failure, and related economic losses. This have led to an increase in the development and implementation of maintenance optimization models for stochastically degrading production systems. Researchers

have produced many interesting and significant results for a large variety of maintenance optimization models [12, 8].

Selective maintenance optimization problems concern a class of multi-components systems operating an alternate sequence of missions and scheduled breaks. Such systems are those exploited in maritime, nuclear and military industries. In these industrial environments, systems usually benefit from the rapid advance in technology. This lead to highly sophisticated but more complex systems with highly interconnected mechanical, electronic and software parts. To prepare the system to successfully execute its subsequent mission, maintenance activities on the system components are usually planned during a scheduled break. However, due the limited duration of the scheduled break, in addition to the possible budget and maintenance resources constraints, only a limited set of components may be selected to undergo some maintenance actions. Therefore, to meet the predetermined reliability level required to operate the next mission, it is mandatory to select an optimal set of components to maintain as well as the type of maintenance actions to be performed on these components.

Selective maintenance was first introduced by [11] and applied for a series-parallel system with subsystems are assumed to be composed of independent and identically distributed (*i.i.d*) components. The lifetime of each system's component is exponentially distributed and the only one available maintenance option is the replacement of failed components. [1] studied the selective maintenance problem in a series-parallel system, made of components with Weibull distributed lifetimes. Three maintenance actions considered are the minimal repair, the corrective replacement of failed component and the preventive replacement of working component. An enumeration method is used to solve the resulting optimization problem. To deal with large sized systems, [10] proposed four improved enumerative procedures to reduce the computational time. [3] proposed also two heuristic-based methods. Recently, imperfect maintenance actions are considered in the selective maintenance setting [5, 13, 9]. In a more recent work, [2] studied selective maintenance when a mission operated

978-1-4673-8760-6/15 $31.00 © 2015 IEEE

by the system is considered of stochastic duration.

Selective maintenance optimization problems addressed in the above mentioned works are limited to the treatment of only one next mission. However, systems are usually required to perform a sequence of missions, it becomes then important to develop appropriate approaches to manage selective maintenance decisions when the planning horizon considers more than a single mission. In [6], the authors consider a series-parallel system where each subsystem is composed of identical components with exponential distributed lifetimes. The system is assumed to operate a sequence of identical missions and breaks between two successive missions are also of equal durations. The only one available maintenance action is the replacement of failed components. In [4] proposed a modeling approach for selective maintenance of systems operating a series of missions whose durations are possibly non identical. Furthermore, maintenance options allows to deal with actions from minimal repair to replacement, through imperfect maintenance actions.

The present paper revisits the selective maintenance optimization problem initially proposed in [4] for a series-parallel system composed of possibly non-identical components the lifetime of which are generally distributed. The system operates a sequence of missions with possibly different durations and nonidentical breaks are allowed between successive missions. During a given mission, a component that fail undergoes minimal repair while at the end of a mission several preventive maintenance actions are available. The maintenance optimization problem consists then on finding an optimal sequence of preventive maintenance actions the total cost of which minimizes the total maintenance cost while providing the desired system reliability level for each mission. In the present paper, the optimization model is written in a more suitable operational research form resulting in a more comprehensive optimization model.

The remainder of this paper is organized as follows. Section 2 describes the investigated system and defines maintenance time and cost. Section 3 presents a formulation of the selective maintenance optimization model in a multi-mission setting. The optimization model is developed and briefly discussed in Section 4 and conclusions are drawn in Section 5.

II. System's description, maintenance time and cost models

Let us consider a series-parallel system S composed of n subsystems $S_i(i = 1, \ldots, n)$ each of which is composed of N_i independent, and possibly, non-identical components $C_{ij}(j = 1, \ldots, N_i)$. It is assumed that the system and its components experience only two possible states: functioning or failed. We also assume that the system is required to operate a sequence of M missions each with duration $U_m(m = 1, \ldots, M)$. At the end of mission $m(m = 1, \ldots, M - 1)$, the system is turned off during a scheduled break of length D_m for possible maintenance activities. Therefore, the system alternates between up and down states. When in the down state, the system is not operated and becomes available for any preventive maintenance actions.. Such a scenario may arise for systems that operate for some time per a day and then put into the down state for the rest of the day.

To perform maintenance activities, a list of L_{ij} maintenance options (levels) $\{1, \ldots, l_{ij}, \ldots, L_{ij}\}$ is available for each component C_{ij}. Among these maintenance options, there are two particular values $l_{ij} = 1$ and $l_{ij} = L_{ij}$. The former corresponds to the minimal repair maintenance action that when performed brings the component to an *as bad as old* conditions, while the later corresponds to the overhaul after which the component becomes *as as good as new*. Values of l_{ij} where $1 < l_{ij} < L_{ij}$ represent imperfect maintenance actions such that when performed they bring the component's condition between the *as good as new* and *as bad as old* conditions. In the present paper, the age reduction coefficient of Malik [7] is used to model the imperfect maintenance options. According to this model, when an imperfect maintenance action is performed on a component it reduces its age from, say t, to $\theta \times t$ where θ is the age reduction coefficient ($0 \leq \theta \leq 1$). Accordingly, the system becomes *as good as new* (overhaul)if its age is reset to zero ($\theta = 0$) while it becomes *as bad as old* (minimal repair) if the age reduction coefficient $\theta = 1$. In this work, we also assume that minimal repair maintenance option is eligible only for components that fail during a mission. The cost induced by a minimal repair performed on C_{ij} is denoted by MRC_{ij} while its corresponding time is assumed negligible if compared to missions' durations. The cost and time consumed by a preventive maintenance level $l_{ij} > 1$ performed on a component C_{ij} are respectively denoted by $MC(l_{ij})$ and $MT(l_{ij})$.

Let $A_{ij}(m)$ and $B_{ij}(m)$ be the ages of component C_{ij},respectively, at the beginning and at the end of a given mission $m(m = 1, \ldots, M)$. Clearly, one may write $B_{ij}(m) = A_{ij}(m) + U(m)$ and the reliability $R_{ij}^c(m)$ of component C_{ij} to survive mission m, given that its initial age is $A_{ij}(m)$, is such that:

$$\mathcal{R}_{ij}^c(m) = \exp\left(-\int_{A_{ij}(m)}^{B_{ij}(m)} h_{ij}(t)dt\right), \qquad (1)$$

where $h_{ij}(t)$ corresponds to the failure rate of component C_{ij}. From the above equation, it follows that the reliability of subsystem S_i and that of the system S are respectively denoted by $\mathcal{R}c_i(m)$ and $\mathcal{R}(m)^c$ and given as:

$$\mathcal{R}_i^c(m) = 1 - \prod_{j=1}^{N_i}\left(1 - \mathcal{R}_{ij}^c(m)\right), \text{ and} \qquad (2)$$

$$\mathcal{R}^c(m) = \prod_{i=1}^{n} \mathcal{R}_i^c(m). \qquad (3)$$

III. THE SELECTIVE MAINTENANCE OPTIMIZATION PROBLEM

Let us assume that the system has just completed the first mission and has to fulfill the alternate remaining breaks and missions. To meet the required minimum reliability to operate missions, maintenance activities on the system's components are allowed during breaks. However, due to the limitation on both maintenance budget and breaks' durations, not all components may undergo maintenance during a given break. Since, The selective maintenance problem must be solved. For the over all missions, this problem consists then on selecting a subset of components to be maintained, and then determining the maintenance level to be performed on those selected components. The objective of the selective maintenance problem discussed in the present work consists on minimizing the total maintenance cost taking into account the limited duration $\mathcal{D}_m$ of each break $m = 1, \ldots, M-1$, in addition to the required minimal system's reliability $\mathcal{R}_0(m)$ to successfully completing each mission $m = 2, \ldots, M$. The probability of completing a mission is obtained from the conditional reliability $\mathcal{R}^c(m+1)$ computed in the previous section. To evaluate the total cost induced by maintenance actions and the corresponding total time consumed from breaks' durations, we define the following decision variable $z_{ij,m}(l_{ij})$:

$$
z_{ijm}(l_{ij}) = \begin{cases} 1, & \text{if component } C_{ij} \text{ is selected for} \\ & \text{maintenance at the end of mission } m \\ & \text{and maintenance level } l_{ij} \text{ is performed,} \\ 0, & \text{otherwise.} \end{cases}
$$

(4)

Note that, at the end of a mission, only maintenance actions whose levels $l_{ij} \geq 2$ are eligible. In other words, the minimal repair are not allowed at the end of missions and can only be performed during breaks on failed components. Therefore, the total cost of maintenance is computed as the sum of total cost induced by minimal repair actions carried out during missions and preventive maintenance actions performed during breaks. However, since minimal repair time is assumed negligible, the total maintenance time is then computed as the sum of times corresponding to preventive maintenance actions performed during breaks.

Let $PMC(m)$ denote the total cost induced by preventive maintenance actions performed at the end of mission m (m=1,...,M-1) and let $CMC(m)$ (m=1,...,M-1) denote the total cost induced by corrective maintenance actions (i.e. minimal repair) performed during mission m on system's components. The preventive maintenance cost $PMC(m)$ is evaluated as:

$$
PMC(m) = \sum_{i=1}^{n} \sum_{j=1}^{N_i} \sum_{l_{ij}=2}^{L_{ij}} MC(l_{ij}) \cdot z_{ijm}(l_{ij}),
$$

(5)

where only PM actions of level $l_{ij} > 1$ are allowed to be performed on component C_{ij}.

If we assume that system's components fail according to a non-homogeneous Poisson process (NHPP), then the expected number of failures of component C_{ij} during mission m is evaluated to:

$$
\int_{A_{ij}(m)}^{B_{ij}(m)} h_{ij}(t)dt.
$$

(6)

From the above equation, it follows that the corrective maintenance cost $CMC(m)$ is evaluated as:

$$
CMC(m) = \sum_{i=1}^{n} \sum_{j=1}^{N_i} MRC_{ij} \int_{A_{ij}(m)}^{B_{ij}(m)} h_{ij}(t)dt,
$$

(7)

From Equations (5) and (7), the total cost of maintenance incurred during the overall breaks and missions is denoted by TMC and computed as:

$$
TMC = \sum_{m=1}^{M-1} PMC(m) + \sum_{m=1}^{M} CMC(m).
$$

(8)

The total time required to perform maintenance actions during a break is denoted by $PMT(m)$ and evaluated to:

$$
PMT(m) = \sum_{i=1}^{n} \sum_{j=1}^{N_i} \sum_{l_{ij}=2}^{L_{ij}} MT(l_{ij}) \cdot z_{ijm}(l_{ij}).
$$

(9)

In the remainder of this section, we develop the selective maintenance optimization model for a series-parallel system with the objective of minimizing the total maintenance cost to successfully execute a mission with respect to the required minimum reliability level $\mathcal{R}_0(m)$ together with the limited break duration $\mathcal{D}_m$.

IV. THE OPTIMIZATION MODEL

To establish the optimization model, we give a new expression of the system conditional reliability to execute the next mission. This new expression uses the decision variable $z_{ijm}(l_{ij})$ as defined in Equation (4). The conditional reliability $\mathcal{R}_{ij}^c(m+1)$ of component C_{ij} is then given as function of the decision variable $z_{ij,m}$ as:

$$
\mathcal{R}_{ij}^c(m+1) = \frac{\mathcal{R}_{ij}\left(\left[\theta_{l_{ij}} z_{ijm}(l_{ij}) + (1 - z_{ijm}(l_{ij})) \right] B_{ij}(m) + U(m+1) \right)}{\mathcal{R}_{ij}\left(\left[\theta_{l_{ij}} z_{ijm}(l_{ij}) + (1 - z_{ijm}(l_{ij})) \right] B_{ij}(m) \right)}.
$$

(10)

From the above equation, it follows that the reliability $\mathcal{R}^c(m+1)$ of the whole system to operate mission $m+1$ is evaluated from component reliability $\mathcal{R}_{ij}^c(m+1)$ with respect to the system reliability block diagram. In our case, components of the system are arranged in a series-parallel configuration. Therefore, the conditional reliability of subsystem S_i and that of the overall system $\mathcal{S}$ are obtained, respectively, from Equations (2) and (3). The resulting selective maintenance optimization problem (SMP) may now be formulated as follows:

$$\text{Minimize} \quad \begin{aligned} &\sum_{m=1}^{M-1}\sum_{i=1}^{n}\sum_{j=1}^{N_i}\sum_{l_{ij}=2}^{L_{ij}} MC(l_{ij})\cdot z_{ijm}(l_{ij})+\\ &\sum_{m=1}^{M}\sum_{i=1}^{n}\sum_{j=1}^{N_i} MRC_{ij}\int_{A_{ij}(m)}^{B_{ij}(m)} h_{ij}(t), \end{aligned} \quad (11)$$

Subject to:

$$\mathcal{R}^c(m+1) \geq \mathcal{R}_0(m+1), \quad (12)$$

$$\sum_{i=1}^{n}\sum_{j=1}^{N_i}\sum_{l_{ij}=2}^{L_{ij}} MT(l_{ij})\cdot z_{ijm}(l_{ij}) \leq \mathcal{D}_m, \quad (13)$$

$$\sum_{l_{ij}=2}^{L_{ij}} z_{ijm}(l_{ij}) \leq 1, \quad (14)$$

$$z_{ijm}(1) = 0, \quad (15)$$

$$\begin{aligned} A_{ij}(m+1) = &[\theta_{l_{ij}}\cdot z_{ijm}(l_{ij})+\\ &(1-z_{ijm}(l_{ij}))]\cdot B_{ij}(m), \end{aligned} \quad (16)$$

$$z_{ijm}(l_{ij}) \in \{0,1\}; \;\; i=1,\ldots,n;$$
$$j=1,\ldots,N_i; \;\; l_{ij}=1,\ldots,L_{ij}.$$

In the above optimization model, Equations (12) and (13) are, respectively, the required reliability level to operate a mission and the maintenance time constraints. For each component C_{ij}, Equations (14) states that only one maintenance level can be selected if the component is to be maintained at the end of a given mission m. The constraint (15) states that minimal repair is eligible only on a failed component. The constraint (16) allows to update the age of each component at the beginning of the next mission.

The formulation proposed above for the selective maintenance problem is a non-linear combinatorial optimization problem whose exact solution is difficult to obtain with exhaustive search methods. It can be solved using the usual optimization techniques. Among these methods those based on evolutionary algorithms such as degraded ceiling, simulated annealing, genetic algorithm. These metaheusristics have been extensively used to solve large scale nonlinear optimization problem in a reasonable amount of time.

V. CONCLUSION

This paper proposed a selective maintenance optimization model for a series-parallel system. Lifetime of system's components are generally distributed. The system operates several missions and scheduled breaks allotted to perform maintenance actions. The selective maintenance problem was formultaed as finding an optimal preventive maintenance plan to be performed so that to minimize the total maintenance cost while providing, for each mission, the desired system reliability level. The fundamental constructs and the relevant parameters of this resulting decision-making problem are developed and discussed. The future work is of course to solve this mixed integer nonlinear optimization problem.

REFERENCES

[1] C. R. Cassady, W. P. Murdock, and E. A. Pohl. Selective maintenance for support equipement involving multiple maintenance actions. *European Journal of Operational Research*, 129:252–258, 2001.

[2] I. Djelloul, A. Khatab, E.-H. Aghezzaf, and Z. Sari. Optimal selective maintenance policy for series-parallel systems operating missions of random durations. In *Accepted in international confrence on Computers & Industrial Engineering CIE 45*, 2015.

[3] A. Khatab, D. Ait-Kadi, and M. Nourelfath. Heuristic-based methods for solving the selective maintenance problem for series-prallel systems. In *International Conference on Industrial Engineering and Systems Management, Beijing, China*, 2007.

[4] A. Khatab, M. Dahane, and D. Ait-Kadi. Genetic algorithm for selective maintenance optimization of multi-mission oriented systems. In *Annual European Safety and Reliability (ESREL) conference, Amsterdam, Netherlands*, 2013.

[5] Y. Liu and H-Z. Huang. Optimal selective maintenance strategy for multi-state systems under imperfect maintenance. *IEEE Transactions on Reliability*, 59(2):356–367, 2010.

[6] L. M. Maillart, C. R. Cassady, C. Rainwater, and K. Schneider. *Selective Maintenance decision-Making over extended Planing Horizons*. Technical Memorandum Number 807, Department of Operations, Weatherhead School of Management, Case Western Reserve University, 2005.

[7] M.A.K. Malik. Reliable preventive maintenance scheduling. *AIIE transactions*, 11(3):221–228, 1979.

[8] T. Nakagawa. *Advanced reliability models and maintenance policies*. Springer, 2008.

[9] M. Panday, M. J. Zuo, R. Moghaddass, and M. K. Tiwari. Selective maintenance for binary systems under imperfect repair. *Reliability Engineering and System Safety*, 113:42–51, 2013.

[10] R. Rajagopalan and C. R. Cassady. An improved selective maintenance solution approach. *Journal of Quality in Maintenance Engineering*, 12(2):172–185, 2006.

[11] W. F. Rice, C. R. Cassady, and J.A. Nachlas. Optimal maintenance plans under limited maintenance time. In *Proceedings of Industrial Engineering Conference, Banff, BC, Canada*, 1998.

[12] A. Sharma, G.S. Yadava, and S.G. Deshmukh. A literature review and future perspectives on maintenance optimization. *Journal of Quality in Maintenance Engineering*, 17(1):5–25, 2011.

[13] H. Zhu, F. Liu, X. Shao, Q. Liu, and Y. Deng. A cost-based selective maintenance decision-making method for machining line. *Quality and Reliability Engineering International*, 27:191–201, 2011.

978-1-4673-8760-6/15 $31.00 © 2015 IEEE

Sampling rate optimization for fault diagnosis of distributed networked control systems

Dominique Sauter

University of Lorraine, CRAN-CNRS UMR 7039
BP239, 54506 Vandoeuvre Cedex , France

Dominique.sauter@univ-lorraine.fr

Abstract—: **In this paper, the problem of Fault diagnosis is addressed in a particular settings : Autonomous control systems (ACS) sharing a single communication channel (WIFI for example) are considered. The problem is to optimize the sampling rate for the exchange of information between the ACS and the FDI systems in order to have efficient fault diagnosis. In this setting, reducing the number of times that the FDI systems is activated implies a reduction in transmissions and thus a reduction in energy expenditures. The aim is to generate residual signals which, in the fault free case, are supposed to be identical to zero.**

Keywords—Networked Control System; scheduling; Fault diagnosis; sampling rate

I. INTRODUCTION

The functionality of a typical NCS is established by the use of four basic elements in the control loop: sensors to collect information, controllers to provide decision and commands, actuators to perform the control commands and the communication network to enable exchange of information. Compared with conventional point-to-point control systems, the advantages of NCS are less wiring, lower installation cost as well as greater agility in diagnosis and maintenance. Because of these distinctive benefits, typical application of these systems ranges over various fields, such as automotive, mobile robotics, advanced aircraft, and so on.

However, the introduction of communication networks in the control loops makes the analysis and synthesis of NCS complex. There are several network-induced effects that arise when dealing with the NCS, such as time-delays, packet losses and limited communication. Because of the inherent complexity of such systems, the control issues of NCS have attracted most attention of many researchers with taking into account network-induced effects. For instance, the stability and stabilization problems of NCS were investigated in [1, 2, 3, 4, 5] for network-induced delays, [6, 7] for packet losses, and [1, 8, 9, 10, 13]. We refer the readers to the survey [11, 12] for more information about the state-of-the-art of NCS.

The new trends for the realization of fault diagnosis (FDI) and fault tolerant control (FTC) systems that implement supervision functionalities (performance evaluation, fault diagnosis) and reconfiguration mechanisms are to implement them thanks to cooperative functions that are also distributed on a networked architecture [13]. The sensor data and the process information generated from a variety of distributed subsystems are exchanged through the network. Thus, network represents the backbone of integrated communication and control systems which explain why the design of controller and FDI system must take into account the network influence.

However, whereas there are a great number of significant results for the control problems of NECS studies related to FDI of NECS are just at the starting point.

Timing problems in distributed real time control systems is a key issue, in particular when the control designer is faced to limited communication capacity. In this paper, we consider a set of autonomous control systems (ACS) sharing a single communication channel (can be WIFI for example). The problem is to optimize the sampling rate for the exchange of the information between the ACS and the FDI systems in order to have efficient fault diagnosis. It is clear that the reduction of bandwidth necessitated by the communication network in a networked control system is a major concern. In this paper, we explore the effect of reducing the number of data packet exchanges between the sensor and the controller/actuator. The aim of paper as it is illustrated by Fig.1. can be formally stated as follows: Optimize the allocation of the network bandwith among autonomous controlled systems in order to achieve optimal performance with respect to FDI.

II. PROBLEM STATEMENT

We first recall the principle of model based residual generation for fault diagnosis. Let us consider a set of autonomous systems:

$$\Sigma_i \begin{cases} \dot{x}_i = A_i x_i(t) + B_i u_i(t) + E_i(t) \\ y_i(t) = C_i x_i(t) + F_i f_i(t) \end{cases} \tag{1}$$

where $x_i(t) \in \Re^n$ is the state vector, $u_i(t) \in \Re^m$ is the input vector, $y_i(t) \in \Re^p$ the output vector and $f_i(t) \in \Re^q$ represents fault vectors to be identified. All the matrices in Eq. (1) are of appropriate dimensions.

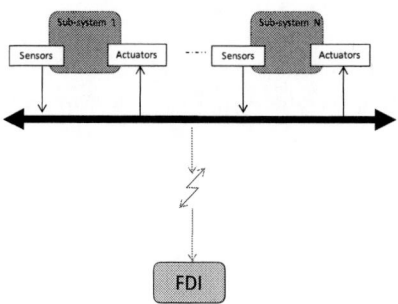

Fig.1.: Communication architecture for FDI of autonomous systems

The aim of failure diagnosis systems [14], [15] is to design a residual which is zero in the fault free case and non zero when a fault occurs. The problem to deal with is to design a linear filter whose inputs are the measurable outputs and control inputs which generates certain outputs called residual signals, one for each possible fault, not affected by other faults and not affected by disturbances. In this paper, we consider a classical observer based residual generator. Therefore, our aim is to find an observer for system of Eq. (2) of the form:

$$O_i \begin{cases} \dot{\hat{x}}_i = (A_i - L_i C_i)\hat{x}_i(t) + B_i u_i(t) + L_i y_i(t) \\ \hat{y}_i(t) = C_i \hat{x}_i(t) \end{cases} \tag{2}$$

and the residual generator :

$$r_i(t) = T_i(y_i(t) - \hat{y}_i(t)) \tag{3}$$

where $L_i \in \Re^{n,p}$, and $T_i \in \Re^{q,p}$ are matrices that are designed in order to fulfill fault detection and isolation requirements. From (1) and (2), the estimation error

$\varepsilon_i(t) = x_i(t) - \hat{x}_i(t)$ and the output of the filter propagate as:

$$\begin{aligned} \dot{\varepsilon}_i &= (A_i - L_i C_i)\hat{\varepsilon}_i(t) + E_i f_i(t) \\ \hat{r}_i(t) &= T C_i \varepsilon_i(t) \end{aligned} \tag{4}$$

Actually, there are various approaches [16] to determine the gain matrices L and T, but we do not discuss this topic in the paper.

III. MULTI RATE SAMPLING

Suppose now, as shown on Fig.2. we have N autonomous sub-systems, monitored at each sampling time by only one FDI systems and sharing a single communication channel.

The aim is to control the sampling rate in order to obtain the best FDI performance for each sub-system. Classical FDI is now revisited, considering multi-rate sampling. We first assume that residual generation algorithms are executed instantaneously at different sampling period h.

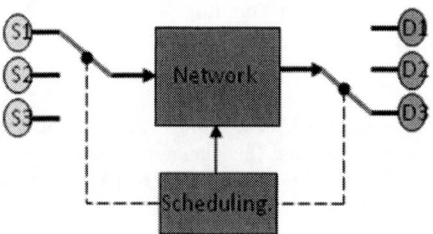

Fig. 2. Allocation of network band-with

The main idea is to feed back the residual generator by updating the estimation using the actual measurement of the plant that is provided by the sensor. The rest of the time the control action is based on a plant model that is incorporated in the controller/actuator and is running open loop for a period of h seconds. The observer based residual generation architecture is shown in Fig.3.

We first assume that residual generation algorithms are executed instantaneously at different sampling period h. We also consider that control input is kept constant (ZOH), during the period $t \in \left[t_{k-1}, t_k^-\right)$, then from the viewpoint of the FDI computer the system behaviour (for simplicity index i is omitted) is described by :

$$\Xi : \begin{cases} x(t) = \Phi_\Sigma(t_k, t) x(t_k) \\ \qquad + \Gamma(t_k, t) u(t_k) + \int_{t_k}^{t} e^{A(t_k - \tau)} f(\tau) d\tau \\ y(t) = C x(t) \end{cases} \tag{5}$$

where $\Phi_\Sigma(t_k, t) = e^{A(t_{k+1} - t)}$ and $\Gamma(t_k, t) = \int_{t_k}^{t} e^{A(t - \tau)} B d\tau$.

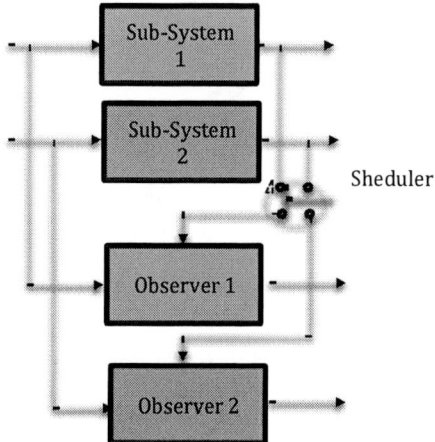

978-1-4673-8760-6/15 $31.00 © 2015 IEEE 143

Fig. 3.: Observer based residual generation architecture.

We suppose that the network used for data transmission is time triggered. Let h be the sampling period of the network and i represents the triggering index of network .

The main idea is to feed back the residual generator by updating the estimation using the actual measurement of the plant that is provided by the sensor. The rest of the time the control action is based on a plant model that is incorporated in the controller/actuator and is running open loop for a period of h seconds. The observer based residual generation architecture is illustrated by Fig.3 where 2 sub-systems are considered..

For the discrete time system described by equation (5) an observer can be designed according to:

$$o: \begin{cases} \dot{\hat{x}}(t) = A\hat{x}(t) + Bu(t) \\ \hat{y}(t) = C\hat{x}(t) \end{cases} \quad t \in \left[t_{k-1}, t_k^- \right) \tag{6}$$

and

$$\dot{\hat{x}}(t_k) = \hat{x}(t_k^-) + \Delta(t_k)L(y(t_k) - C\hat{x}(t_k^-)) \tag{7}$$

with:

$$\hat{x}(t_k^-) = \lim_{\delta t \to 0} x(t_k - \delta t) \tag{8}$$

where L is the state observer gain to be designed according to stability requirements.

From (1), and (6), on the interval $t \in \left[t_{k-1}, t_k^- \right)$ the estimation error $\varepsilon(t) = x(t) - \hat{x}(t)$ and residual generator (3) propagate as:

$$\begin{cases} \dot{\varepsilon}(t) = A\varepsilon(t) + Ef(t) \\ r(t) = TC\varepsilon(t) \end{cases} \tag{9}$$

leading to :

$$\varepsilon(t_k^-) = \Phi_\Sigma(t_{k-1}, t_k)\varepsilon(t_{k-1}) + \int_{t_{k-1}}^{t_k} \Phi_\Sigma(t_{k-1}, \tau)Ef(\tau)d\tau \tag{10}$$

then, with (7), after output updating we get :

$$\varepsilon(t_k) = (I + \Delta(t_k)LC)\varepsilon(t_k^-) \tag{11}$$

Finally, at each sampling time, the state estimation error sequence satisfy to :

$$\varepsilon(t_k) = (I + \Delta(t_k)LC)\Phi_\Sigma(t_{k-1}, t_k)\varepsilon(t_{k-1})$$
$$+ (I + \Delta(t_k)LC)\int_{t_{k-1}}^{t_k} \Phi_\Sigma(t_{k-1}, \tau)Ef(\tau)d\tau \tag{12}$$

The residual generator must be stable when system is operating under no fault conditions. If we assume

$f(t) \triangleq 0, t \in [t_0, t_k]$ have that if at time $t=t0$, $\varepsilon(t_0)$ is the initial condition then :

$$\varepsilon(t_k) = (I + \Delta(t_k)LC)^h e^{Akh}\varepsilon(t_0), \Delta(t_k) \in \{0,1\} \tag{13}$$

If we suppose that on the interval $[t_0, t_k]$ the number of observer updates ($\Delta = 1$) is ℓ, we have :

$$\varepsilon(t_k) = (I + LC)^\ell e^{Akh}\varepsilon(t_0)$$
$$= ((I + LC)e^{Ah})^\ell e^{A(k-\ell)h}\varepsilon(t_0) \tag{14}$$
$$= \Phi_O(t_k, t_0)\varepsilon(t_0)$$

and on the interval $[t_0, t_{k+1}^-]$:

$$\varepsilon(t) = \Phi_\Sigma(t, t_k)\Phi_O(t_k, t_0)\varepsilon(t_0) \tag{15}$$

If, the system described by Equation (5) is stable, that is to say eigenvalues of $\Phi_\Sigma(h)$ are inside the unit circle, then the state observer given by Eq (6 and Eq.(7)) is globally exponentially stable if and only if the eigenvalues of $(I + LC)e^{Ah}$ are inside the unit circle.

We study now the case when system (5) is not stable. A system described by equations : $\varepsilon(t) = \Lambda(t, t_0)\varepsilon(t_0)$ is stable if there exist positive constants c_1, c_2 such that :

$$\|\Lambda(t, t_0)\varepsilon(t_0)\| \le c_1 e^{-c_2(t-t_0)} \text{ for all } t \ge t_0 \tag{16}$$

Taking the norm of the solution described as in Eq. (14):

$$\|\varepsilon(t)\| \le \|\Phi_\Sigma(t, t_k)\| \|((I + LC)e^{Ah})^1\| \|e^{A(k-1)h}\|. \|\varepsilon(t_0)\| \tag{17}$$

we then analyse the successive terms on the right hand side of (17):

If the eigenvalues of $(I + LC)e^{Ah}$ are inside the unit circle, we have:

$$\|((I + LC)e^{Ah})^1\| \le K_2 e^{-\alpha_1 1}, K_2, \alpha_1 > 0 \tag{18}$$

which, with $l = \xi.k$, $0 < \xi < 1$ and $t = kh + \varsigma$, $0 < \varsigma < h$ gives in term of t :

$$K_2 e^{-\alpha_1 \ell} = K_2 e^{-\frac{\alpha_1 \xi(t-\varsigma)}{h}} < K_3 e^{-\alpha t} \tag{19}$$

with $K_3, \alpha > 0$.

The term $\|\Phi_\Sigma(t, t_k)\| \le e^{\bar{\sigma}(A)h} = K_1$ is bounded since here the time difference $t - t_k$ is always smaller than h .

$\bar{\sigma}(A)$ beeing the largest eigenvalue of A , we have also:

978-1-4673-8760-6/15 $31.00 © 2015 IEEE

$$\left\| e^{A(k-1)h} \right\| \leq e^{\bar{\sigma}(A)(k-1)h} = e^{\frac{\bar{\sigma}(A)(1-\xi)(t-\varsigma)}{h}}$$

$$\leq K_4 e^{\frac{\bar{\sigma}(A)(1-\xi)t}{h}} \tag{20}$$

with $K_4 > 0$.

Finally stability conditions are satisfied if :

$$\alpha_1 \xi - \bar{\sigma}(A)(1-\xi) > 0 \tag{21}$$

For a given system dynamics, the stability condition depend on the poles placement of $(I + LC)e^{Ah}$ and the ratio ξ which represent the ratio of observer updates ℓ over the number k of samples.

IV. CONCLUSION

In this paper, the problem of Fault diagnosis in Networked Control Systems (NCS) is addressed in a particular settings. Autonomous control systems (ACS) sharing a single communication channel (WIFI for example) are considered. An optimization approach based on the Pontryagin principle is proposed to optimize the sampling rate for the exchange of information between the ACS and the FDI systems in order to have efficient fault diagnosis.

The stability condition of the residual generator with limited communication has also been studied.

REFERENCES

[1] 1. S.-S. Hu and Q.-X. Zhu. Stochastic optimal control and analysis of stability of networked control systems with long delay. *Automatica*, 39(11):1877–1884, 2003.

[2] 2. Y. Halevi and A. Ray. Integrated communication and control systems: Part I- Analysis. *ASME Journal of Dynamic Systems, Measurement and Control*, 110(4):367–373, December 1988.

[3] 3. J. Nilsson, B. Bernhardsson, and B. Wittermark. Stochastic analysis and control of real-time systems with random time delays. *Automatica*, 34(1):57–64, 1998.

[4] 4. W. Zhang, M. S. Branicky, and S. M. Phillips. Stability of networked control systems. *IEEE control systems Magazine,* 21(1):84–99, February 2001

[5] 5. S. Li, L. Yu, Z. Wang, and Y. Sun. LMI approach to guaranteed cost control for networked control systems. *Developments in Chemical Engineering and Mineral Processing: Special Issue on Advanced Control and Real-Time Systems*, 13(3/4):351–361, 2005.

[6] 6. Q. Ling and M. D. Lemmon. Robust performance of soft real-time networked control systems with data dropouts. *In Proceedings of the 41st IEEE Conference on Decision and Control*, volume 2, pages 1225 – 1230, 2002.

[7] 7. P. Seiler and R. Sengupta. An H∞ approach to networked control. *IEEE Transactions on Automatic* Control, 50(3):356–364, 2005.

[8] 8. P. Zhang, S. X. Ding, P. M. Frank, and M. Sader.Fault detection of networked control systems with missingmeasurements. In *Proceedings of Asian Control Conference*, p 1258-1263, Melbourne, Australia, 2004.

[9] 9. D. Yue, Q.-L. Han, and J. Lam. Network-based robust H∞ control of systems with uncertainty. *Automatica,* 41:999– 1007, 2005.

[10] 10. S. Li, Y. Wang, F. Xia, Y. Sun, and J. Shou. Guaranteed cost control of networked control systems with time-delays and packet losses. *International Journal of wavelets, multi-resolution and information processing*, 2006.

[11] 11. Y. Tipsuwan and M.-Y. Chow. Control Methodologies in Networked Control Systems. *Control Engineering Practice*, 11(10):1099–1111, 2003.

[12] 12. Patton R J, Kambhampati C, Casavola**, Zhang P, Ding S+ & Sauter D "A Generic Strategy for Fault-tolerance in Control Systems Distributed Over a Network" *European Journal of Control, Fundamental issues in Control*, Vol 13, Number 2-3,pp 280-296 , 2007.

[13] Sauter D, Boukhobza T & Hamelin F, (2006), Decentralized and autonomous design for FDI/FTC of networked control systems, *Proc. 6th IFAC- Safeprocess '06, Beijing, Aug 30-Sept 1st*

[14] Patton R.J., P. Frank, R. Clark (Ed.), Fault diagnosis in dynamic systems, *Prentice Hall, International series in systems and control engineering.*, 1989.

[15] Chen J & Patton R J, (1999), *Robust Model Based Fault Diagnosis for Dynamic Systems,* Kluwer Academic Publishers ISBN 0-7923-841-3

[16] Frank, P., & Ding, S. Current development in the theory of FDI. *Proc. IFAC Safeprocess* (16-27), Budapest, Hungary., *2000.*

An Optimal Integrated Maintenance Strategy with Switching Policy under Subcontracting Constraint

Hajej Zied
LGIPM
University of Lorraine
Metz, France
Zied.hajej@univ-lorraine.fr

Dellagi Sofiene, Rezg Nidhal
LGIPM
University of Lorraine
Metz, France
sofiene.dellagi@univ-lorraine.fr
nidhal.rezg@univ-lorraine.fr

Abstract— this paper deals an integrated maintenance strategy for a manufacturing system subjected to a random failure and calling up on two subcontractor machines. The production system composed by a single machine in order to satisfy a random demand during a finite horizon, under service level. That's why it called upon another subcontracting machine. In order to assure an efficiency economical objective, we have a switching choice between two subcontractors having some different data of reliability and cost production. An analytical study showed that the switching choice of the subcontractor machine is trained by the subcontracting availability. The objective of this study is to determine an optimal production plans of principal and subcontracting machines for forecasting demand, which minimizes the total production and inventory cost under a fixed service level and subcontracting constraints. A numerical example confirms the analytical results.

Keywords—Manufacturing system, subcontractor, switching, production plan, random demand, service level.

I. INTRODUCTION

The competitive industrial world is need more cooperation and collaboration between enterprises. In the state, many companies have recourse to the industrial subcontracting which became a very widespread practice to face competition. In this context, it has grown in the industrial world in virtually all domains as noted by Amesse et al.[1]. Andersen, P.H. [2] and Bertrand et al.[4] show that this practice is not always justified by production costs. It is part of cooperation logic and coordination based on technological incentives, to satisfy customers in terms of quantity and delay.

This recourse to subcontracting can be also justified by the will of the company to concentrate on core activities and turn to external sources and collaborate with external partners in order to develop shared technological capabilities Gomes-Casseres, B. [6], Andersen, P.H. [2].
Recently, in the context of integrated maintenance, Dellagi. et al. [5] developed a maintenance strategy integrating subcontracting constraint. They treated a production system represented by a machine producing a single product type to satisfy a constant demand during time. The machine calls

upon the sub-contracting represented by a second machine to complete the entire demand exceeding the maximum machine capacity.

About the production, inventory and maintenance problem optimization, Hajej et al.[7] dealt with combined production and maintenance plans for a manufacturing system satisfying a random demand over a finite horizon. In their model, they assumed that the failure rate depends on the time and the production rate. Ayed et al. [3] dealt with a randomly failing manufacturing system M1 which has to satisfy a random demand during a finite horizon given a required service level. To help meeting this demand, subcontracting is used through another production system M_2. M_1 operates with a variable production rate and its failure rate depends on both time and production rate.

In this paper, we will study a problem of an integrated maintenance policy for a manufacturing system calling upon two subcontractors in the context of the forecasting problem. In fact our approach consists at a manufacturing system composed of one machine which produces a single product in order to satisfy a random demand during a finite horizon. This machine is unable to satisfy the demand, that's why it calls upon another machine, comprising the so-called subcontractor machine, which produces the same type of product.
We have to choice between two subcontractor machines. The subcontractor machines are classified according to their availability rate, and their unit production cost.
The objective of this paper is to determine the economical production planning, during the finite horizon, based on forecasting demand taken into account the availability and unit production cost of subcontracting. In order to assure an economical objective, we have to switching between the offers of the two subcontractor machines.

The remainder of the paper is organized as follows: In Section II we detailed the problem and Section III presents the problem formulation. In section IV we developed the analytic studies. In section V, we presented a numerical example, in order to apply the analytical results. Finally we conclude in section VI.

978-1-4673-8760-6/15 $31.00 © 2015 IEEE

II. PROBLEM DESCRPTION

The present problem consists on a machine M_1 which is unable to satisfy a random demand over a finite horizon H, with a given inventory service level α, and to avoid shortage due to the manufacturing system unavailability, the enterprise has to build a stock, that's why it called upon to another machine, comprising the so-called subcontractor machine. The random demand is characterized by a normal distribution with an average demand $\hat{d}$ and a standard deviation σ.

Points of view reliability, machine M is subject to a random failure. The probability degradation law of machine M is described by the probability density function of time to failure $f(t)$ and for which the failure rate $\lambda(t)$ increases with both time and the production rate U_1. Failures of machine M can be prevented by a preventive maintenance action which is scheduled according to its history. Concerning subcontractor machine we have estimated its availability rate by $\beta_i (i=2,3)$. We have two subcontractor machines which offer their services in order to meet the need of the machine M_1. It's noted that the subcontractor machines differ according their availability rate, and their unit production cost. It is obvious that, more than the subcontractor machine have an increased availability rate more than the unit production cost is increased too. The problem is to make an economical strategy in order to switching between the subcontractor machines.

We noted by β_2, C_{pr2} and β_3, Cpr_3 respectively the availability rate, and the unit production cost of the subcontractor machines M_2 and M_3.

Our objective is to establish an economical production plan satisfying the randomly demand according to the demand forecasting. The aim is to minimize the sum of the production, inventory and maintenance costs.

The problem is illustrated in figure 1.

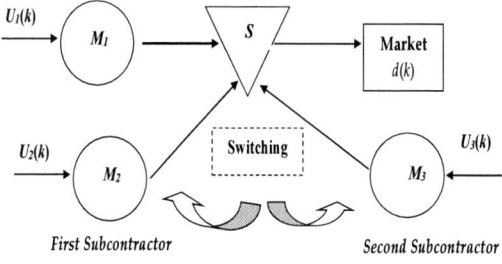

Fig. 1. Problem Description

III. PROBLEM FORMULATION

A. Notations

The following parameters are used in the mathematical formulation of the model:

Δt: length of a production period

H: number of production periods in the planning horizon

$H.\Delta t$: length of the finite planning horizon

$U_{i,k}$: production rate by machine M_i, $i \in \{1,2,3\}$ during period k ($k=0, 1,..., H-1$)

$\hat{d}_k$: average demand during period k ($k=0, 1,..., H$)

$V_{d(k)}$: variance of demand during period k ($k=0, 1,..., H$)

S_k: inventory level of S at the end of period k ($k=0, 1,..., H$)

$\hat{S}_k$: average inventory level of S during period k ($k=0, 1,..., H$)

C_{pri}: unit production cost of machine M_i, $i \in \{1,2,3\}$

Cs: holding cost of product unit during one period

mu: monetary unit

U_1^{max}: maximal production rate of machine M_1

U_1^{min}: minimal production rate of machine M_1

U_2^{max}: maximal production rate of machine M_2

U_2^{min}: minimal production rate of machine M_2

U_3^{max}: maximal production rate of machine M_3

U_3^{min}: minimal production rate of machine M_3

α: probability index related to customer satisfaction and expressing the service level.

β_2: machine M_2 availability rate

β_3: machine M_3 availability rate

y_k: A binary variable that equal to 1 if a subcontractor machine M_2 is performed at period k and 0 if M_3 is performed at period k.

S_0: initial inventory.

B. Production policy

The idea is to minimize the expected production, inventory costs over a finite time horizon $[0, H]$. It's assumed that the horizon is portioned equally into H periods. The demand is satisfied at the end of each period. This kind of problem can be formulated as a linear-stochastic optimal control problem under a stock threshold level constraint, with the production rates corresponding to each period as the decision variables. The stochastic problem as follows:

$$\underset{U(k)}{Min}\left(E\left\{ \sum_{k=0}^{H-1} f_k\left(S_k, U_k\right) + f_H\left(S_H\right) \right\} \right) \qquad (1)$$

$f_k(.)$ denotes functions that represent the production, and inventory costs, and $E\{\}$ denote the mathematical average value operator.

Subject to:

The inventory balance equation for each time period is formulated in this way:

$$S_{k+1} = S_k + U_{1,k} + y_k \cdot \beta_2 \cdot U_{2,k} + (1-y_k) \cdot \beta_3 \cdot U_{3,k} - d_k$$

$$k \in \{0,1,...,H-1\} \qquad (2)$$

The service level requirement constraint for each period is expressed by the following constraint.

$$\text{Prob}\left[S_{k+1} \geq 0\right] \geq \alpha \text{ with } k \in \{0,1,...,H-1\} \qquad (3)$$

The following constraint defines an upper and lower bounds on the production level during each period k.

$$0 \leq U_k \leq U_1^{max} + y_k \cdot U_2^{max} + (1-y_k) \cdot U_3^{max} \quad k \in \{0,1,...,H-1\}$$
$$U_k = U_{1,k} + y_k \cdot \beta_2 \cdot U_{2,k} + (1-y_k) \cdot \beta_3 \cdot U_{3,k} \quad (4)$$

In our problem, we adapted the HMMS model, Holt et al.[9], Silva Filho, and Cezarino [10] to establish an inventory and production policy. The principle characteristic of HMMS model is the use of a quadratic cost function which allows penalizing both excess and shortage in the inventory level.

The expected production and inventory cost for period k is given by:

$$f_k\left(U_{1,k}, U_{2,k}, U_{3,k}, S_k\right) = C_s E\left\{S_k^2\right\} + C_{pr1} U_{1,k}^2 + y_k \cdot C_{pr2} \cdot \beta_2 \cdot U_{2,k}^2$$
$$+ (1-y_k) \cdot C_{pr3} \cdot \beta_3 \cdot U_{3,k}^2$$
$$(5)$$

The quadratic total expected cost of production and inventory over the finite horizon $H.\Delta t$ can then be expressed as follows:

$$F(u) = \sum_{k=0}^{H} f_k\left(U_{1,k}, U_{2,k}, U_{3,k}, S_k\right) = C_s E\left\{S_H^2\right\} +$$
$$\sum_{k=0}^{H-1}\left[\begin{array}{c} C_s E\left\{S_k^2\right\} + C_{pr1} U_{1,k}^2 + y_k \cdot C_{pr2} \cdot \beta_2 \cdot U_{2,k}^2 \\ + (1-y_k) \cdot C_{pr3} \cdot \beta_3 \cdot U_{3,k}^2 \end{array}\right]$$
$$\text{with } k \in \{0,1,...,H-1\}$$
$$(6)$$

C. Maintenance Strategy

It is interesting to develop an optimal maintenance strategy with considering the manufacturing system's degradation according to the production rate. To know the best time intervals when a preventive maintenance action in a manufacturing system must be carried out is very important for minimizing the total cost of maintenance.

The maintenance strategy under consideration is the well known preventive maintenance policy with minimal repair at failure.

The horizon H is partitioned equally into N parts each of length T. Perfect preventive maintenance or replacement is performed periodically at times $i.T$, $i=1,...,N$ and $N.T=H.\Delta t$ following which the unit is as good as new. When a unit fails between preventive maintenance actions, only minimal repair is performed. It is assumed that the repair and replacement times are negligible.

The analytic expression of the total maintenance cost is expressed as follows:

$$\Gamma(N) = M_c \times \phi_{(U,N)} + N \times M_p \quad (7)$$

With :

$$\phi(U_1, N) = \sum_{j=0}^{N-1}\left[\begin{array}{c} \sum_{i=In(j\times\frac{T}{\Delta t})+1}^{In\left((j+1)\times\frac{T}{\Delta t}\right)}\int_0^{\Delta t}\lambda_i(t) + \int_0^{(j+1)\times T - In\left((j+1)\times\frac{T}{\Delta t}\right)\times\Delta t}\lambda_{In\left((j+1)\times\frac{T}{\Delta t}\right)+1}(t)dt \\ \\ + \int_{(j+1)\times T}^{\left(In\left((j+1)\times\frac{T}{\Delta t}\right)+1\right)\times\Delta t}\frac{\left(In\left((j+1)\times\frac{T}{\Delta t}\right)+1\right)}{U_{max}}\times\lambda_n(t)dt \end{array}\right]$$

And
The failure rate in the interval k is expressed as following:

$$\lambda_k(t) = \lambda_{k-1}(\Delta t) + \frac{U_k}{U_{max}}\lambda_n(t) \quad \forall t \in [0, \Delta t]$$

With $\lambda_{k=0} = \lambda_0$ and $\Delta\lambda_k(t) = \frac{U_k}{U_{max}}\lambda_n(t)$ $\quad (9)$

$\lambda_n(t)$ is the nominal failure rate corresponding to the maximal production rate.

The objective is to find the optimal number of preventive maintenance actions N^* $(N=1,2,...)$ which minimize the total cost per monetary unit over a given horizon.

IV. NUMERICAL EXAMPLE

We consider a production system which has to meet a random demand characterized by Gaussian distribution over a finite horizon $H\Delta t$; with a mean $\hat{d}_k$ and a variance V_{d_k} . The number H of periods Δt is equal to 24 months, with $\Delta t=1$. To satisfy the demand under given service level α, the company resorts to a subcontractor machines. We suppose that 2 subcontractor machines, M_i $(i\in\{2,3\})$, propose their offers to the machine M_1 in order to meet the random demand. The maximal production rate of every subcontractor machine is defined by $U_2^{max} = U_3^{max} = 13$ unit/1tu. The unit production cost, the availability rate and the transported delay of every subcontractor machine are defined by:

$C_{pr2}=10mu$, $C_{pr3}=12mu$, $\beta_2=80\%$, $\beta_3=90\%$, $\tau_2=2$ and $\tau_3=1$.

- the other data are presented as following :

$C_{p1}=10$ mu; $C_{s1}=2$ mu/k; $U_{max1}=13$; $S_{1,0}=$ 0; α =90%, $V_{d_k}=1.21$

The average demand is presented in table I below.

TABLE I
MEAN DEMAND

$d_0\rightarrow 15$	$d_1\rightarrow 17$	$d_2\rightarrow 15$	$d_3\rightarrow 15$
$d_8\rightarrow 16$	$d_9\rightarrow 13$	$d_{10}\rightarrow 15$	$d_{11}\rightarrow 14$
$d_{16}\rightarrow 15$	$d_{17}\rightarrow 11$	$d_{18}\rightarrow 16$	$d_{19}\rightarrow 13$
$d_4\rightarrow 15$	$d_5\rightarrow 14$	$d_6\rightarrow 16$	$d_7\rightarrow 14$
$d_{12}\rightarrow 15$	$d_{13}\rightarrow 12$	$d_{14}\rightarrow 15$	$d_{15}\rightarrow 13$
$d_{20}\rightarrow 15$	$d_{21}\rightarrow 12$	$d_{22}\rightarrow 14$	$d_{23}\rightarrow 16$

We used the Numerical Algorithms with MATHEMATICA, in order to realize this optimization. The economically production plan is presented respectively in tables II, III, IV.

In what follows, we interest to find the optimal production plans of principal and subcontracting machines by minimizing the total cost and switching between two subcontractor machines. Using tables II, we can see that the optimal production plan of principal machine. Using tables III and IV, we can note that in the first seven periods, we can call for the subcontractor machine 3 that cost more expensive by cons it has more availability and less transported delay. From period 8, we call the subcontractor machine M_2 which has less cost. This can be explained by the fact that, as demand rates increases, we call the subcontractor machine which has a greater availability and less transported delay despite its high cost. Once the demand rates are reduced, the cheapest subcontractor other is used.

TABLE II
PRINCIPAL MACHINE: $U^*_{1,K}$

$U_0 \rightarrow 9$	$U_1 \rightarrow 7$	$U_2 \rightarrow 10$	$U_3 \rightarrow 6$
$U_4 \rightarrow 3$	$U_5 \rightarrow 5$	$U_6 \rightarrow 7$	$U_7 \rightarrow 3$
$U_8 \rightarrow 10$	$U_9 \rightarrow 6$	$U_{10} \rightarrow 7$	$U_{11} \rightarrow 7$
$U_{12} \rightarrow 8$	$U_{13} \rightarrow 9$	$U_{14} \rightarrow 9$	$U_{15} \rightarrow 7$
$U_{16} \rightarrow 6$	$U_{17} \rightarrow 7$	$U_{18} \rightarrow 8$	$U_{19} \rightarrow 8$
$U_{20} \rightarrow 6$	$U_{21} \rightarrow 10$	$U_{22} \rightarrow 11$	$U_{23} \rightarrow 9$

TABLE III
SUBCONTRACTOR MACHINE: $U^*_{2,K}$

$U_0 \rightarrow 0$	$U_1 \rightarrow 0$	$U_2 \rightarrow 0$	$U_3 \rightarrow 0$
$U_4 \rightarrow 0$	$U_5 \rightarrow 0$	$U_6 \rightarrow 0$	$U_7 \rightarrow 9$
$U_8 \rightarrow 7$	$U_9 \rightarrow 9$	$U_{10} \rightarrow 8$	$U_{11} \rightarrow 4$
$U_{12} \rightarrow 8$	$U_{13} \rightarrow 9$	$U_{14} \rightarrow 0$	$U_{15} \rightarrow 0$
$U_{16} \rightarrow 0$	$U_{17} \rightarrow 0$	$U_{18} \rightarrow 0$	$U_{19} \rightarrow 0$
$U_{20} \rightarrow 7$	$U_{21} \rightarrow 7$	$U_{22} \rightarrow 3$	$U_{23} \rightarrow 0$

TABLE IV
SUBCONTRACTOR MACHINE: $U^*_{3,K}$

$U_0 \rightarrow 4$	$U_1 \rightarrow 10$	$U_2 \rightarrow 7$	$U_3 \rightarrow 6$
$U_4 \rightarrow 8$	$U_5 \rightarrow 7$	$U_6 \rightarrow 6$	$U_7 \rightarrow 0$
$U_8 \rightarrow 0$	$U_9 \rightarrow 0$	$U_{10} \rightarrow 0$	$U_{11} \rightarrow 0$
$U_{12} \rightarrow 0$	$U_{13} \rightarrow 0$	$U_{14} \rightarrow 7$	$U_{15} \rightarrow 8$
$U_{16} \rightarrow 6$	$U_{17} \rightarrow 5$	$U_{18} \rightarrow 4$	$U_{19} \rightarrow 9$
$U_{20} \rightarrow 0$	$U_{21} \rightarrow 0$	$U_{22} \rightarrow 0$	$U_{23} \rightarrow 6$

✓ Maintenance policy:

In this numerical example we suppose that the machine M has a degradation law characterized by a Weibull distribution. The Weibull scale and shape parameters are $\beta = 100$ and $\alpha = 2$. The preventive and corrective maintenance cost are respectively $Mp = 1000$ mu, $Mc = 2500$ mu.

Fig. 2 shows the curve of the average total cost of maintenance according to N (Number of preventive maintenance actions). We conclude that the optimal number of preventive maintenance actions that minimizes the total cost of maintenance is $N^*=3$. Hence, the optimal period for intervention for preventive maintenance $T^*= 8$ months;

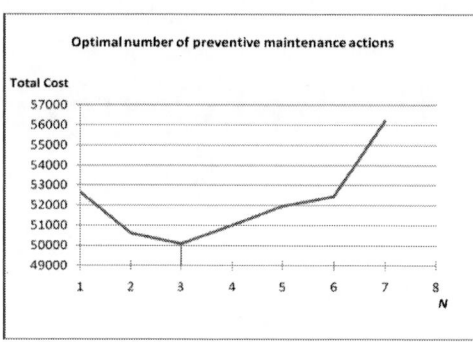

Fig. 2. Total Cost Variation as a Function of N

V. CONCLUSION

In this paper, an optimization integrated maintenance strategy for a manufacturing system subjected to a random failure calling upon a subcontractor machine in order to satisfy a random demand, is presented.

The main contribution of this study is the choice between two subcontractor machines M_2 and M_3. It's noted that the subcontractor machines differ according their availability rate and their unit production cost. It is obvious that, more than the subcontractor machine availability rate increased more than the unit production cost increased too. With an analytical study we proved an economical strategy in order to switch to the subcontractor machine to adopt. A numerical example is treated in order to apply the analytical result proved.

REFERENCES

[1] Amesse, F., Dragoste, L., Nollet, J. and Ponce, S.,."Issues on partnering: evidences from subcontracting in aeronautics", Technovation, vol. 21, p. 559–569, 2001.

[2] Andersen, P. H., "Organizing international technological collaboration in subcontractor relationships: an investigation of the knowledge-stickiness problem". Research Policy, vol. 28, p. 625–642, 1999.

[3] Ayed S., Dellagi S., Rezg N.. " Joint optimisation of maintenance and production policies considering random demand and variable production rate" International Journal of Production Research, vol. 50, Issue 23, p 6870-6885,2012.

[4] Bertrand, J.W.M., and Sridharan, V ., "A study of simple rules for subcontracting in make-to-order manufacturing". European Journal of Operational Research, vol. 128, p. 509-531, 2001.

[5] Dellagi, S., Rezg, N. and Uie, U.,. Preventive maintenance of manufacturing systems under environmental constraints. International Journal of Production Research, 45 (5), 1233, 2007.

[6] Gomes-Casseres, B., Group versus group: how alliance networks compete. Harvard Business Review (July–August), 62–74, 1994.

[7] Hajej Z., Dellagi S., Rezg N.,. "Optimal integrated maintenance/production policy for randomly failing systems with variable failure rate" International Journal of Production Research, vol. 49, Issue 19, p5695-5712, 2011.

[8] Hajej Z., Dellagi S., and Rezg N., 2009. "An optimal production/maintenance planning under stochastic random demand, service level and failure rate". IEEE explore. Issue 22-25 Aug: 292–297. India.

[9] Holt, C.C., Modigliani, F., Muth, J.F., Simon, H.A., Bonini, C.P. and Winters, P.R., 1960. Planning production, inventories, and work force Englewood Cliffs, N.J .: Prentice-Hall.

[10] Silva Filho, O.S. and Cezarino, W., 2004. An optimal production policy applied to a flow-shop manufacturing system. Brazilian Journal of Operations and Production Management, 1 (1), 73-92.

PMU Based Centralized Adaptive Load Shedding Scheme in Power System

H. BENTARZI

Signals and Systems Laboratory, IGEE, UMBB University, Boumerdes, Algeria
sisylab@yahoo.com

Abstract- **One of major contingencies of an electric power system is due to imbalance between loads and generated electric power. This leads to instability in the power system and hence to a blackout. Conventional Under frequency Load Shedding Schemes have been widely used, to restore power system stability. However, the analysis of recent blackouts suggests that voltage collapse and voltage related problems are also important concerns in maintaining system stability The aim of this work is to develop a new approach applied to a load shedding scheme based on both parameters frequency and voltage that can be provided by smart sensor named Phasor Measurement Unit. All generator buses frequencies and voltages are sent to center where a magnitude of disturbance will be calculated and hence the amount of load to be shed as well as its location will be determined.**

Keywords- Smart Grid, Centralized load shedding, Adaptive load shedding scheme, critical frequency, disturbance, Phasor Measurement Unit and voltage sensitivity.

I. INTRODUCTION

An electric power system is a large interconnected system that produces, transmits and distributes an electric energy to different consumers. Stability of the power system is of a great concern, since it is subjected to different disturbances that may cause a local or complete system collapse if no adequate action is taken to prevent it. Therefore, many techniques have been developed to make the power system survives during disturbances and continue to operate. One common disturbance is the imbalance between generation and load due to an overload situation caused by generator outage or loss of transmission lines. Generally, this situation has an effect on frequency behavior of the system and hence the frequency decreases below the rated operating value. The system spinning reserve can compensate small overload, whereas large one requires rapid emergency control actions to be taken by under frequency load shedding schemes that trip temporary certain loads in order to balance the system and consequently recover the nominal operating frequency [1-3].

Under frequency Load Shedding (UFLS) schemes have been widely used, to restore power system stability. However, the analysis of recent blackouts suggests that voltage collapse and voltage related problems are also important concerns in maintaining system stability [2]. For this reason, both frequency and voltage need to be taken into account in load shedding schemes. In the proposed Centralized Adaptive Load Shedding Scheme algorithm, we consider both parameters; voltage and frequency to determine the amount of load to be shed and its appropriate location [4-5].

The most important feature and advantage of this scheme compared to conventional schemes is that bus voltages have an important role in the load shedding scheme. Also the intelligent selection of loads to be shed is another important feature of the proposed scheme. Additionally, the amount of load to be shed is determined adaptively and it is proportional to the magnitude of disturbance. By this way, the problems of under-shedding and over-shedding, existing in the conventional schemes can be avoided.

II. LOAD SHEDDING SCHEME

Under frequency load shedding scheme is the most commonly used control system to balance the generation and load (power demand) and it is the last control step for preventing electric power system from blackouts. It deals with shedding the appropriate amount of load for removing the overload situation. This may be performed in many steps with each step having its own setting frequency and percent of load to be shed. Three main types of load shedding schemes may be distinguished: traditional (conventional), semi-adaptive and adaptive load shedding scheme [3]. In the latter, the amount of load as well as the percent of load to be shed in each step are selected adaptively according to the magnitude of the disturbance. It is determined using the initial rate of frequency decline and is based on the System Frequency Model (SFR) [1, 4]. From the reduced SFR model, the relation between the frequency decline and the size of the disturbance Pd is obtained as follows:

$$\frac{df}{dt} = \frac{P_d}{2 H_{sys}} \quad , \ (\text{Hz/s}) \qquad (1)$$

Where Pd is the disturbance magnitude in per unit,
and H_{sys} is the inertia constant of the system, in seconds.

Load is typically shed by opening the circuit breakers that installed at the terminals of the feeders in the distribution substations. The architecture of load shedding schemes can be: local, distributed or centralized [5]. In this work, centralized load shedding is used where each bus frequency information is sent to a central location where processed by a computer program, which is used to perform the overall load shedding. Decision is then sent from the central station to the distribution substation to trip breakers as selected by the computer control program.

978-1-4673-8760-6/15 $31.00 © 2015 IEEE

II. A PROPOSED ALGORITHM

In this work, a new approach which is applied to the centralized adaptive load shedding scheme is described. The frequencies measured by Phasor Measurement Unit (PMU) [6] are used for calculating the rate of change of frequency as well as the magnitude of the disturbance in the power system.

The advantage of this approach, as compared to the conventional under-frequency load shedding scheme, is to estimate the magnitude of overload occurring from different disturbances and accordingly to determine the necessary amount of load to be shed as well as its location. Therefore, it avoids unnecessary shedding actions.

Once the magnitude of the disturbance is determined, the location and the amount of load to be shed at each bus have to be determined. This determination is based on the voltage sensitivity at each bus measured also by PMU.

2.1 Determination of the disturbance magnitude

After measuring the frequency by PMU at each generator in the power system, the rate of frequency decline of each generator is determined in the center, and then the system mean frequency decline of the system is calculated according to the following formula [7]:

$$\frac{df_c}{dt} = \frac{\sum_{i=1}^{n} H_i \frac{df_i}{dt}}{\sum_{i=1}^{n} H_i} \quad , \quad (Hz/s) \quad (2)$$

Where: df_c/dt is the rate of mean frequency decline,
And df_i/dt is the rate of generator (i) frequency decline,
$\frac{df_c}{dt}$ and H_i is the inertia constant of generator (i).

Then, once the mean rate of frequency decline is known, the size of the disturbance in the system may be determined using the following formula [8]:

$$P_d = 2 \frac{H_{sys}}{f_n} \frac{df_c}{dt}, \quad (pu) \quad (3)$$

Where f_n is the nominal frequency of the system in hertz (50Hz), and H_{sys} is the equivalent inertia constant (in second) of the system given by the following formula [4]:

$$H_{sys} = \frac{\sum_{i=1}^{n} S_i H_i}{\sum_{i=1}^{n} S_i}, \quad (MVA) \quad (4)$$

Where: S_i is the rated apparent power of generator (i), and n is the total number of generators.

2.2 Determination of the amount and location of the load to be shed

Once the magnitude of the disturbance is determined using the above equivalent swing equation (2). The location and the amount of load to be shed at each bus have to be determined. This determination is based on the voltage sensitivity at each bus. Thus the bus with voltage sensitivity very close to the instability limit will have a maximum load shed based on the reciprocal of its sensitivity as a fraction of the sum of the reciprocals of all the load bus sensitivities. Now the QV

analysis is carried out in the following manner. The equation for reactive power is:

$$Q_i = \sum_{j=1}^{n} V_i * V_j * Y_{ij} * \sin(\delta_{ij} - \theta_i) \quad (5)$$

Where: Q_i: is the reactive power of bus (i),
V_i: is the voltage at bus (i),
V_j: is the voltage at bus (j),
Y_{ij}: is line admittance,
δ_{ij}: is the voltage angle difference,
θ_i : is The load angle at bus (i).
Differentiation of Eq.(5) with respect to V_i, gives,

$$\frac{dQ_i}{dV_i} = \sum_{j=1}^{n} V_j * Y_{ij} * \sin(\delta_{ij} - \theta_i) \quad (6)$$

Thus individually, for each bus the relation dQ/dV can be written as:

$$\frac{dQ1}{dV1} = \sum_{j=1}^{n} V_j * Y_{ij} * \sin(\delta_{1j} - \theta_j) \quad (7)$$

$$\frac{dQ2}{dV2} = \sum_{j=1}^{n} V_j * Y_{ij} * \sin(\delta_{2j} - \theta_j) \quad (8)$$

And so on till the nth equation:

$$\frac{dQn}{dVn} = \sum_{j=1}^{n} V_j * Y_{ij} * \sin(\delta_{nj} - \theta_j) \quad (9)$$

The Q-V curve shown in Fig.1 is obtained after plotting the dQ/dV for one such sample bus. The system is said to be unstable beyond the knee point which represents the critical situation. As the knee point is approached, the dQ/dV values become smaller. Thus a system bordering on instability will have a small value of the slope at the knee point.

In Figure (1), the Y axis is the Mvar values of Q and the X axis is the p.u. values of the bus voltage. The plot considers two states of voltage stability represented by two points "A" and "B". It can be noticed that, the point "B" is closer to the knee point as compared to point "A". Thus, the dQ/dV of point "A" will be higher than the dQ/dV of point "B". Thus more load needs to be shed from a bus with the dQ/dV value of point "B". Therefore:

$$\frac{dQ_i}{dV_i} \propto \left(\frac{1}{\text{the load amount be shed from each bus}} \right)$$

Or, $\frac{dV_i}{dQ_i} \propto$ the amount of load to be shed from each bus.

To estimate this load quantity, we consider the reciprocal of the voltage sensitivity as a fraction of the sum of all the reciprocals of voltage sensitivities. The reciprocal is considered because for a higher slope (i.e. a more stable case), the reciprocal will be smaller, hence a less amount of load will be shed from it. Now, the summation of the dV/dQ values of all the load buses is given by:

$$\sum_{i=1}^{n} \frac{dV_i}{dQ_i} = \frac{dV1}{dQ1} + \frac{dV2}{dQ2} + \cdots + \frac{dVn}{dQn} \quad (10)$$

The load that must be shed from each bus is a small portion of the total load required to be shed in order to recover the power balance. This portion of load at each bus is proportional to the fraction of dV/dQ value at each bus with respect to the sum calculated above. This is represented as:

Table I The generation unit parameters of the system

Generator / Parameters	Gen. 1	Gen.2	Gen.3
Volt Ampere (MVA)	247.5	192.0	128.0
Power factor (PF)	0.9	0.85	0.85
Inertia constant (H)	9.55	3.33	2.35
Droop factor (R)	0.06	0.05	0.08
High Pressure Power Fraction (FH)	0.2	0.3	0.4
Reheat Time Constant (TR)	8	7	5
Damping Factor (D)	1	1	1
Mechanical Power Gain factor (Km)	0.95	0.9	0.85

$$\frac{\frac{dVi}{dQi}}{(\frac{dV1}{dQ1}+\frac{dV2}{dQ2}+\cdots+\frac{dVn}{dQn})} \quad \text{for each bus (i)} \quad (11)$$

This is the fraction of the total voltage sensitivities. Thus, the closer bus 'i' is to the knee point of the Q-V curve the higher value of the above fraction will be. Hence, when this fraction is multiplied by the total amount of load to be shed, the load to be shed from each bus is obtained.

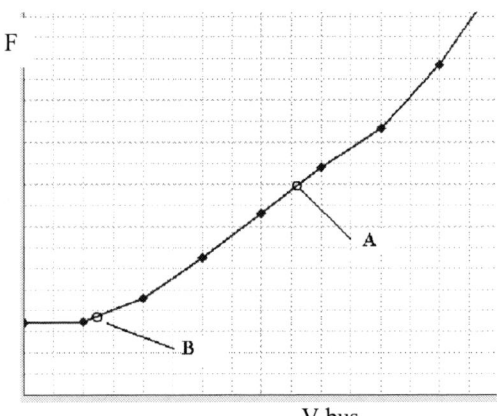

Figure (1): Q-V analysis.

Therefore, the formula to be tested is:

$$Si = \frac{\frac{dVi}{dQi}}{\sum_{i=1}^{n}\frac{dVi}{dQi}} * P_d \quad (12)$$

IV. SIMULATION RESULTS AND DISCUSSION

The model of IEEE-9 bus power system that is shown in Figure (2) is used to test and evaluate the proposed algorithm whose flow chart shown in Figure (3). The power flow from one bus to another is indicated as well as the size of each load and the generating capacity.

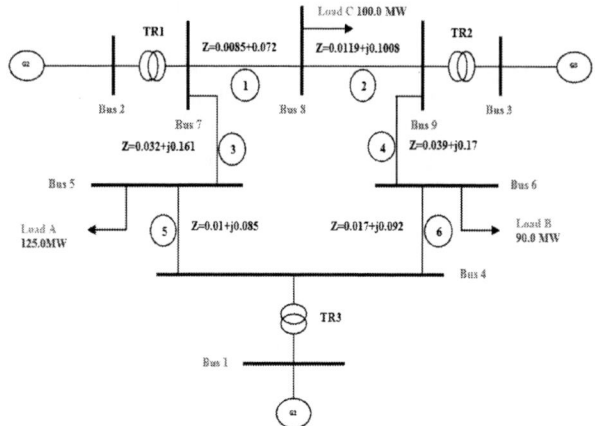

Figure (3): IEEE 9 Bus Power Network Model

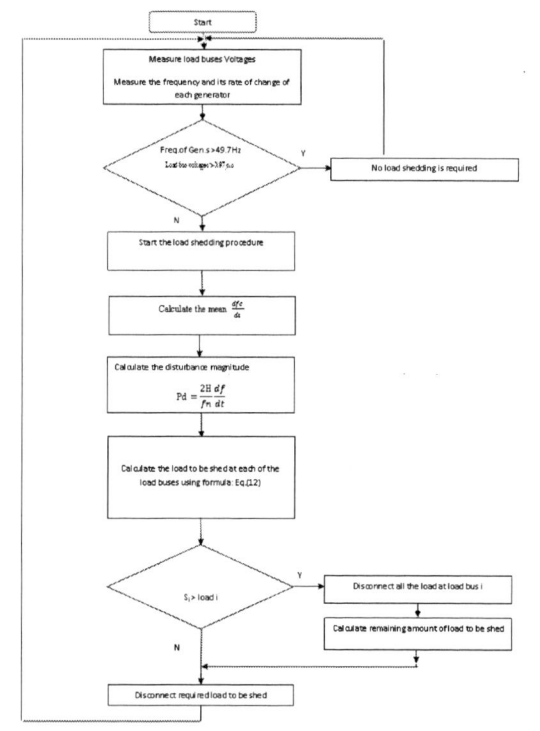

Figure (3): The flowchart of the proposed Algorith.

The table I gives the different parameters of the three generators.

The PSSE software has been used in the simulation for two cases causing disturbances which are: Loss of transmission line (3) and Generator G3 Outage with Increase of loads A B and C by 50% (Over load). Then its results have been used for testing our algorithm programmed by Matlab.

It can be noted that before the use of the load shedding scheme, the load bus voltages decrease below their predetermined standards and become stable at the lower values as given in table II. The voltage sensitivities; dV/dQ values are calculated individually for each load bus and the results are also listed in the table II.

978-1-4673-8760-6/15 $31.00 © 2015 IEEE

The summation of all of the dV/dQ values is 0.001880236.This value is used in the voltage sensitivity formula to calculate the load to be shed at each bus. The obtained results are also given in table II. The total load to be shed is 196.696 MW. After applying the load shedding algorithm, the frequency generators and the load bus voltages plots are shown in Figure (4) and Figure (5) respectively.

The generator bus frequencies have been improved above the specified lower limit. They settle at a value very close to 50 HZ. The load bus voltages also show an improvement due to the appropriate amount of shedding.

Table II: Voltage Sensitivities at each Load Bus (Case 2)

Load Bus number	Voltage (pu)	dQ/dV	dV/dQ	Load to be shed (MW)
5	0.95477	1080	0.000926756	96.946
6	0.97328	9691	0.000103185	10.8
8	0.98175	1176	0.000850300	88.95

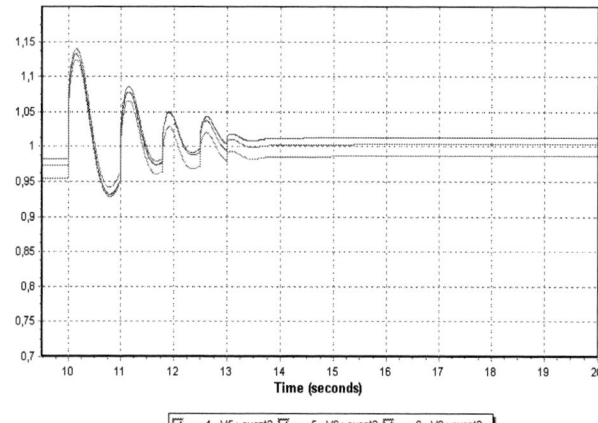

Figure (4): Generator bus Frequency (After applying Load shedding).

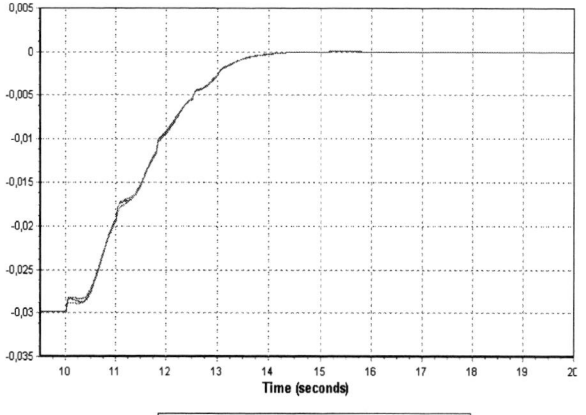

Figure (5): Load Bus Voltages (After Load Shedding).

V. CONCLUSION

We have considered several cases here. In the first case study, the disturbance was very small. The values of frequency and voltage were in the safe margin and the disturbance can be removed by the spinning reserve with no need to perform any further load shedding. However, in the second case, the disturbance was very large; as a result both frequency and voltage settle at low unacceptable values.

The obtained simulation results for any disturbance sizes are really satisfactory. It can be said that the load shedding based on the voltage sensitivities definitely improves the frequency and voltage of the power system. The load shedding algorithm is activated at the threshold values; frequency of 49.7 Hz and voltage of 0.97 p.u for a 50 Hz power system.

The average time required to start shedding load is 0.3 seconds. After load shedding has been applied an improvement in the frequency and voltage profiles is observed. In totality, it takes around 20 seconds for the system to reach an acceptable value.

The advantage of this method is that the amount of load to be shed is not large for all the disturbances unlike the conventional one. Therefore, unnecessary shedding is avoided which allows both a better service to different consumers and the system collapse prevention.

REFERENCES

[1] H. Bentarzi, A. Ouadi, N. Ghout, F. Maamri and N. E. Mastorakis, "A New Approach Applied to Adaptive Centralized Load Shedding Scheme controllers", in Proc. 8th WSEAS Int. Conf. on (CSECS), Tenerife, Spain, December 14-16, 2009.

[2] H.Seyedi and Sanaye-Pasand, "New Centralized Adaptive Load-Shedding Algorithms to Mitigate Power System Blackouts", IEEE Paper, Published in IET Generation, Transmission & Distribution, 2008.

[3] B. Delfino, S. Massucco, A. Morini, P. Scalera, and F. Silvesto, "Implementation and comparison of different under frequency load shedding schemes", Proc. IEEE Power Engineering Society Summer meeting, **Vol.1**, 2001.

[4] P.M. Anderson, *Power System Protection*, IEEE press, New York, 1990.

[5] P.M. Anderson and A.A. Fouad, *Power System Control and Stability*, IEEE Press, Sec. Ed., New York, 2002.

[6] A. Ouadi, H. Bentarzi and J. C. Maun, "A new computer based phasor measurement unit framework", (IEEExplore), SSD'09, Pp.1-6, Djerba, Tunisia, 2009, DOI: 10.1109/SSD.2009.4956773

[7] H. Seyedi, M. Sanaye-Pasand, M. R. Dadashzadeh, "Design and simulation of an adaptive load shedding algorithm using a real network", IEEE Power India Conference, April 10-12, 2006.

[8] V. Terzija, "Adaptive underfrequency load shedding based on the magnitude of the disturbance estimation", IEEE Transactions on Power Systems, **Vol.21**, No. 3, Aug. 2006.

Differentiated Service for NoC-based Multimedia Applications

Atef DORAI* [†‡], Virginie FRESSE*, El-Bay BOURENNANE[†], Abdellatif MTIBAA[‡]

*Hubert-Curien Laboratory, UMR CNRS 5516, University of Lyon, 42000 Saint-Etienne, France
[†]Le2i Laboratory UMR CNRS 6306, University of Bourgogne, 21078 Dijon, France
[‡]EmicroE Laboratory LR99ES30, University of Monastir, 5019 Monastir, Tunisia

Abstract—**As communication on-chip evolves toward the global multi-service network, applications with different service requirements have emerged. A key factor is guaranteeing the quality of support services. Differentiated service with double physical planes is seen as the key technology to achieve this goal. It focuses on the control of traffic and on recognizing the need for aspects for the management plan achieved by the bandwidth broker. In this paper, a novel QoS architecture for multimedia application via NoC is proposed. The gains in latency and in resource are possible due to the simplicity of the NoC architecture.**

Index Terms—**NoC, Quality-of-Service, Multiple physical planes.**

I. INTRODUCTION

Network-on-chip (NoC) dedicated to multimedia applications typically requires more bandwidth that the NoC can provide, leading to congestion and varying performance in terms of latency, throughput and propagation delays. Traditional IPs (Intellectual property) connected to NoC provide a best-effort service that treats any request for access to a network resource equally. Nevertheless, different works have focused on improving the quality of service (QoS).

Two approaches have been proposed to provide QoS in NoC: *Soft* QoS based virtual channels (VC) and *hard* QoS based circuit-switching. Because of the complexity of hard QoS, scalability has become a major issue. Consequently, we propose a differentiated communication service (DiffServ), which is a distinct way to achieve scalability. There is an absolute guaranteed QoS and low resource utilization for *hard* QoS based flows. The timing and area performances of a given system depend on the network topology and on the category of traffic. For smaller systems with more streaming based computation, a flexible combination of routing algorithm and a specially designed structure can ensure high performance.

The NoC topology and routing algorithms provide some degrees of support for QoS. Implementing deterministic routing algorithms is easy but they are not able to balance the load in the links. With adaptive routing algorithms, a packet can take several paths when traveling between a source and a destination. This increases the probability of moving packets outside congested regions. Thus, there is an increasing need for a NoC architecture that treats and recognizes different level services. The key technical contributions of the architecture presented in this paper are the following:

- Keep traffic level separate and protect the high-priority data from the low-priority data.

- It provides controlled access to physical planes between routers using the double-XY routing algorithm.
- It borrows and adapts the streaming multimedia application in NoC based the DiffServ technology.

This paper is organized as follows. Section II reviews related works. Section III provides a description of DiffServ and their deployment in NoC multimedia. Section IV presents the experimental results for timing and resource utilization. Finally, Section V presents our conclusions and future outlook.

II. RELATED WORK

This section presents some examples of NoC that offer different QoS strategies to allow applications to meet their performance requirements. Some works propose streaming multimedia data through NoC. These proposals are based on very complex architecture containing virtual channels, extra controls and routing tables.

The QoS supported by NoCs are usually classified in two groups: guarantee of service (GS) or best-effort (BE). BE is the most commonly found option in NoCs because of its simplicity. However, BE only guarantees traffic accuracy and completion, whereas GS offers additional guarantees such as commitment at several levels, accuracy of result, and so on. A combination of BE and GS is thus desirable, since BE improves resources utilization, while GS is important in real-time systems [1].

The architecture presented in [2] links the hardware level of the NoC to the software level based-MPSoC application levels. These authors propose the development of a communication API that exposes the communication services offered by the NoC to the application developed. To demonstrate the proposed approach, mixtures of real and synthetic applications in different MPSoCs were executed using different NoC communication services. The real applications are a video application using a MJPEG decoder and an audio application composed of an ADPCM decoder. Audio and video flow has the highest priority in the mapping tasks.

A novel structure for congestion-aware and QoS-aware NoC architecture is proposed in [3]. It maintains quality network transmission, the implementation cost is reasonable and the traffic within the network is well balanced. Using differentiated service for different classes of traffic (GS and BE), routers with fully adaptive routing and multiple parallel buffers, bandwidth allocation is managed to meet the service quality requirement.

A novel NoC communications structure is presented in [4]. It supports several QoS based on applying the MPLS technology inspired by computer networks and adapted to the NoC environment. The original MPLS protocol is used for large scale, wide area networks, and its implementation is extremely complicated. More research is thus required to find simplifications and additional improvements to this new proposal.

III. QoS ARCHITECTURE

Considering the scenario of future systems with multiple IPs, different bandwidths will be needed. For this reason, in our architecture, a specific topology is fundamental to improving performance without using expensive interconnects chips. A NoC topology has many advantages for complex designs that can exploit the locality of communication systems. The basic NoC architecture used in this work is based on the comparative study presented in [5], where network performance, power consumption and area occupation was compared in many physical and virtual channels. The final architecture is generated in the following steps:

- An existing multimedia application is extracted and separated into several units.
- Once the differentiated service is specified, the NoC based multiple physical planes are integrated.
- An adapter is incorporated between the local port of router and the separate units.

A. Differentiated Service based priorities

The flow of traffic is received from the multimedia file in the application layer. The classified communication services are supported at the middle abstraction levels. The data streams are extracted to different units in the network. Each QoS unit has its own priority in the MAC layer. Each unit can be accessed based on to its access-category (voice, video, best-effort and background). The priority of each packet in the current traffic is checked in the system.

To meet the demands of computation-intensive application, four IPs components and processor/DSP cores are connected to each local port of the router NoC through the wrapper module. Packets in all access-categories have the same structure, with a header and a data payload. Table I details the sub-fields of the packets. In our work, we consider that the multimedia data are extracted and each access-category is generated by management interfaces.

The allocation field in header is given below.

- Prio: Denotes priority information. Its size is 4-bits
- DEST: Denotes the destination address. Its size is equal to the half the first flit.
- LENG: denotes the size of the packet.
- SRC: Represent the source of packet.
- MSG: Is the number of packets in each message.

In the computer network, some use 8 (from 0 to 7) priorities for 4 access-categories. The number priority 8 is mapped in four priorities depending on the number of access-categories. The Prio field is encoded in 4-bits to indicate the priority

TABLE I
STRUCTURE OF THE HEADER OF A DATA PACKET

HEADER					Payload DATA
Flit0	Flit1		Flit2	Flit3	
Prio	DEST	LENG	SRC	MSG	

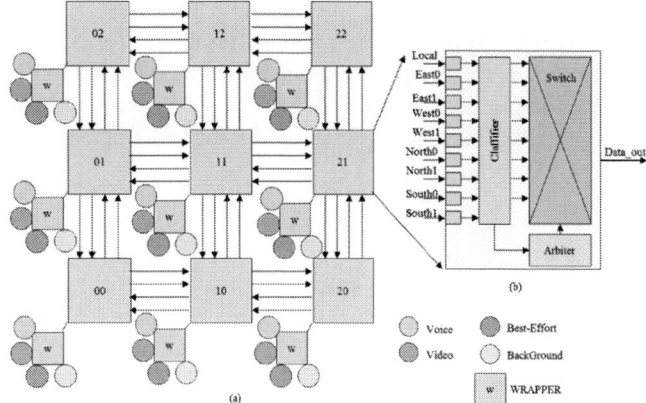

Fig. 1. The NoC architecture: (a) Topologie, (b) structure of the router

of the access-category, where "0001" is the highest priority, followed by "0010", then by "0011" and finally by the lowest priority "0100". Packets belonging to different sources in the same class can be dispatched to different output ports. Thanks to this dispatch, the load injected into the network is balanced.

B. Support for communication services

The system multimedia illustrated in Figure 1.a is composed of a global and local architecture: the basic components of the global part are the units of the access-categories, a wrapper (W), and NoC architecture. The local architecture for NoC includes the routers that are interconnected by multiple physical plane links. These links are used to increase the bandwidth and give the service and the chance to recover time lost due to the delay. The basic router architecture used in this NoC structure is illustrated in Figure 1.b. It contains a DiffServ process, deploys a double-XY routing algorithm, and an arbiter to route the packet in the appropriate direction. The final architecture adopted for NoC multimedia has the following features: a 2D mesh network, a handshake protocol, a double-XY routing algorithm, eight bidirectional ports with one local unidirectional port connected to the wrapper.

1) Classifier: The task of the classifier is to select a packet in a message stream based on the information contained in the packets header, assign it to one of the service classes in the NoC. This is the first step in processing packets and routing them to a free port.

2) Wrapper: The wrapper generates packets from the IPs according to their type. This is done by a priority-based arbiter that schedules different types of access to NoC. The use of the arbiter is justified because it is flexible and does not consume a lot of resources. This allows multiplexing service communications with local ports in each router and enables management of priorities between access-categories.

978-1-4673-8760-6/15 $31.00 © 2015 IEEE

Each wrapper is configured and has the same communication interface whose requests are scheduled according to their priority. The highest or lowest priority is detected automatically.

To remedy the problem of priorities, we propose a solution using dynamic allocation mechanism priorities that is based on the type of traffic to be transmitted. When a new type of traffic arrives, the packets are arranged in order of priority in the wrapper. The order of priority is chosen by referring to the types of flow in the local port. Voice has the highest priority. In the absence of voice, video has the highest transmission priority, and so on.

3) The double-XY routing algorithm: The architecture of NoCs uses a simple wormhole approach and the double-XY routing algorithm. The double-XY routing algorithm can support deadlock-free data and can give high priority packets several paths to recover timer lost in the delay, which allows high priority packets to be accelerated. Using a double-XY routing algorithm is one advantage of this proposal.

The double-XY algorithm uses an extension of the XY algorithm. To avoid cyclic dependencies of channels, each channel in the 2-D mesh topology is doubled and the network is partitioned to provide eight directions: $N_{+E0}, N_{-E0}, N_{+E1}, N_{-E1}, N_{+W0}, N_{-W0}, N_{+W1}, N_{-W1}, N_{+N0}, N_{-N0}, N_{+N1}, N_{-N1}, N_{+S0}, N_{-S0}, N_{+S1}, N_{-S1}$. The choice of the output port is specified based on the destination address and the availability of a port. x0-dir, x1-dir, y0-dir and y1-dir are parameters to determine the shortest path to the destination address based on the current address. According to the double-XY routing algorithm, the first physical plane among the Y directions to be used when the packet is moved toward the east or west is north0 or north1, south0 and south1. Similarly, the first physical plane among the X directions used when the packet is moved toward the north or south is east0, east1 and west0, west1. However, when there is a delay in the routing of a high priority packet, it can use the first available port by overtaking a lower priority to reach its destination. If the second port in same direction is blocked, the double-XY can change the direction of the packet. The physical links can support any type of traffic.

As illustrated in Figure 2, where voice, video and background are competing for the outgoing port of router 1, which can only allow the highest priority access-category to obtain the port resource and transmit to the west region via router 2. In this case, voice is granted while the other port in the same direction serves video with a lower QoS requirement than voice. There will be a high chance of background traffic being blocked in router 1. When voice arrives at router 2, port N_{+W0} is blocked and N_{+W1} is available, so voice uses the second port to follow the path to reach its destination. To get video to the outside of router 2, as both west ports are busy, it will be blocked in the east port N_{-E1} on the same routers. Inter-router arbitration receives information from the classifier to decide which traffic has the higher priority to use the physical output plane between a pair of neighboring routers.

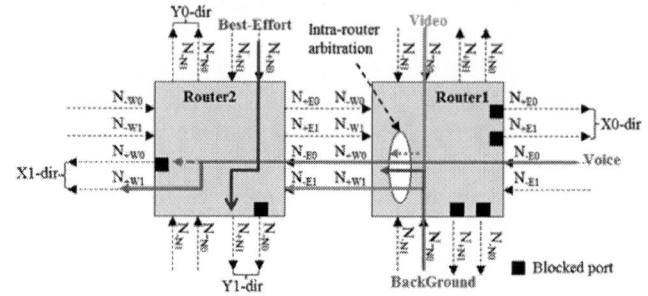

Fig. 2. Intra-router arbitration for different QoS requirement in NoC

IV. EXPERIMENTS

In these experiments, we assume that each packet consists of 32 flits and that the size of each flit is 32 bits. The size of NoC is 4x4. The buffer size of the router is 32 flits. A cycle-accurate NoC simulation environment was implemented in VHDL with different traffic patterns used to evaluate the performance characteristics of the NoC multimedia. Traffic generators (TG) and traffic receptors (TR) are connected to the wrapper and replace the generated data to emulate the NoC architecture. TR and TG are based on models that are widely used in the performance evaluation of communications. These traffic patterns include bit-complement, bit-reverse and transpose. Configuring different access-categories is based on the parameters defined by the IEEE Standard 802.1D-2004 [6]. Table II lists the characteristics of the traffic injected in the network for each access-category.

Synthesis results were obtained on a FPGA platform using ML605. The ML605 board contains a Virtex-6 FPGA. The simulation and the generation of the bitstream use respectively ModelSim 6.5 and Xilinx ISE 14.7 design Tools. The synthesis results include FPGA resources use (registers and LUTs).

TABLE II
PARAMETER OF INJECTION RATE

Service	Priority	Rate
Voice	1	368 kbps
Video	2	1.4 Mbps
Best-Effort	3	1 Mbps
Background	4	1.2 Mbps

A. Timing results

This section analyzes the timing performances obtained with the architecture we propose. The objectives are to analyze the impact of the DiffServ architecture on the different access-categories (high level and low level). In the bit-complement traffic, each router *(i, j)* communicates only with router *(M-i, N-j)*, where the mesh size is $M \times N$. For bit-reverse and transpose, the destination routers are in positions *(N-j+1, N-i+1)* and *(j,i)*, respectively. The set of simulation tests were performed by varying the pattern of traffic generation and the rate of injection. The load injected ranged from 10% to 90%.

Figure 3, Figure 4 and Figure 5 show the average total latency for the multimedia application with the bit-complement

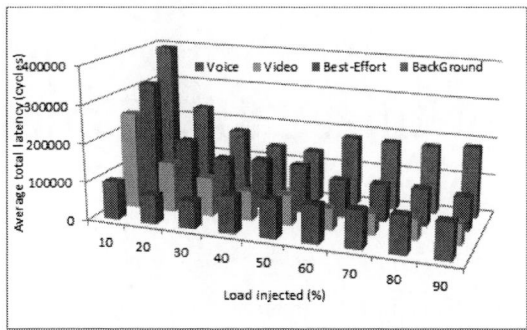

Fig. 3. Bit-complement synthetic traffic

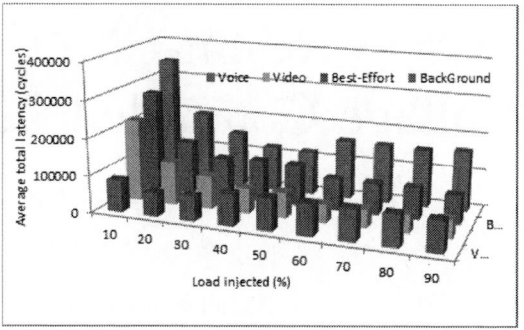

Fig. 5. The Transpose traffic synthetic

traffic pattern. It is clear that the voice and video access-categories have low latency compared to best-Effort and background. All the access-categories except voice have a descending curve. Latency voice remains constant because it has the highest priority, even if there is a low priority access-category advance before the high priority, high priority can overtake and recover priority via the double buffer. From 40%, the video becomes faster than the voice; this is due to the quantity of data transmitted by video. Latency background is high whatever the load injected.

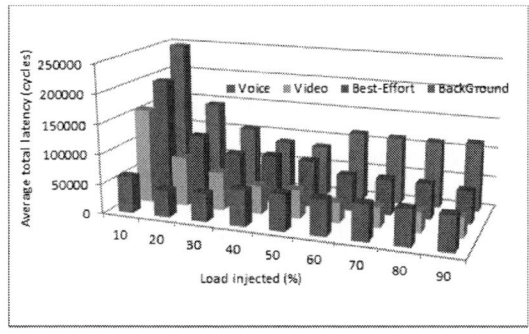

Fig. 4. Bit-reversal synthetic traffic

The average gain in latency during the load injected with all traffic patterns is 7% between latency voice and video, 50% between voice and best-effort, and 122% between voice and background. The gain in latency between video and best-effort is 53% and 141% between video and background.

B. Resource results

Table III lists the percentage of resources used. The number of registers is small compared to the number of resources available on the FPGA. On the other hand, the number of LUTs becomes high with a size of 6x6. The advantage becomes clear with an increase in the size of the network. Hence, the percentage of register utilization increases from approx.0.4% for a 2x2 to 5% for a 6x6. The percentage of LUTs increases from approximately 4% to 76% for NoCs. The number of linear resources depends on the size of the NoC irrespective of the architecture used. The percentage of resources is important in the case of a big NoC.

TABLE III
NUMBER OF RESOURCES ON VIRTEX 6 FPGA

Network size	Synthesis to Xilinx virtex 6 Evaluation ML605	
	registers	LUTs
2x2	1240 (0%)	6152 (4%)
4x4	6914 (2%)	45395 (30%)
5x5	11410 (3%)	76176 (50%)
6x6	17012 (5%)	115512 (76%)

V. CONCLUSION

This paper is a contribution to NoC multimedia structure using service differentiated and doubled physical planes. Thanks to the adaptive strategy, it also has features that can support different traffic patterns. The contributions of our proposed architecture are illustrated in the use of local mechanisms with limited FPGA resources, composed of routers that use the double-XY routing algorithm to the QoS. Experimental results using the three traffic patterns and synthesis resources can significantly improve the performance in terms of latency for voice and video compared to best-effort and background and low consumption resources. The future work will target to model the average latency for each access-category.

REFERENCES

[1] E. Rijpkema, K.G.W. Goossens, A. Radulescu, J. Dielissen, J. van Meerbergen, P. Wielage, and E. Waterlander, "Trade Offs in the Design of a Router with Both Guaranteed and Best-Effort Services for Networks on Chip". Design, Automation and Test in Europe Conference and Exhibition, 2003.

[2] E.A.Carara, N.L.V. Calazans, F.G. Moraes, "Differentiated Communication Services for NoC-Based MPSoCs". IEEE Transactions on Computers,Vol. 63, NO.3,MARCH 2014.

[3] C. Wang, N. Bagherzadeh, "Design and Evaluation of a High Throughput QoS-Aware and Congestion-Aware Router Architecture for Network-on-Chip". 20th Euromicro International Conference on Parallel, Distributed and Network-based Processing, 2012.

[4] M. Kim, D. Kim, G.E.Sobelman, "Network-on-chip quality-of-service through multiprotocol label switching". Symposium on IEEE International.Circuits and Systems. ISCAS 21-24 may 2006. Island of kos.

[5] Y. J. Yoon, N. Concer, M. Petracca, L. Carloni, "Virtual Channels and Multiple Physical Networks: Two Alternatives to Improve NoC Performance". IEEE Transactions on computer-aided design of integrated circuits and systems, vol. 32, NO. 12, December 2013.

[6] IEEE 802.11 WG, "IEEE 802.11e/D6.0, Draft Supplement to Standard for Telecommunications and Information Exchange between Systems LAN / MAN Specific Requirements". December 2003.

978-1-4673-8760-6/15 $31.00 © 2015 IEEE

A New 65nm-CMOS 1V 8GS/s 9-bit Differential Voltage-Controlled Delay Unit Utilized for a Time-Based Analog-to-Digital Converter Circuit

Abdullah El-Bayoumi[1], Hassan Mostafa[2] and Ahmed M. Soliman[3]

[1]Valeo InterBranch Automotive Software, Smart Village, Cairo-Alex Road, Giza, Egypt,

[1,2,3]Electronics and Electrical Communications Engineering Department, Cairo University, Giza, Egypt,

[2]Center for Nanoelectronics and Devices, AUC and Zewail City of Science and Technology, New Cairo, Egypt.

{ abdullah.elbayoumi@pg.cu.edu.eg, hmostafa@uwaterloo.ca, asoliman@ieee.org }

Abstract—A new differential Voltage-Controlled Delay Unit (VCDU) is proposed. The VCDU converts an input voltage into a pulse delay, and delivers it to a Time-to-Digital Converter (TDC) which outputs a digital word. Both circuits form a Time-Based Analog-to-Digital Converter (ADC). In scaled CMOS technology, the Time-Based ADC is a substantial block in designing Software Defined Radio (SDR) receivers, as it exhibits high speed and low power. The new manually-calibrated differential VCDU circuit operates on a high sampling frequency of 8GS/s in 65nm CMOS technology, with a supply voltage of 1V. It achieves a wide dynamic-range of 0.56V at a 3% linearity error and effective-number-of-bits (ENOB) of 8.9 bits. Additionally, it consumes an area of 742µm² and a power consumption of 1.6mW. A metal-insulator-metal capacitor is used to minimize the process-voltage-temperature variations. The simulation results are compared to single-ended VCDU results and to state-of-the-art analog-part ADCs results to show the strength of the proposed design.

Keywords—Nanometer CMOS technology; voltage-controlled delay unit; time based analog-to-digital converter; software defined radio; metal-insulator-metal capacitor; effective-number-of-bits; linearity .

I. INTRODUCTION

The significance of Analog-to-Digital Converters (ADCs) becomes more crucial in emerging applications such as Software Defined Radio (SDR) receivers [1], [2] for future ultra-wide-band wireless communication services. A single Integrated Circuit (IC) of SDR receiver can acquire several chains of operating blocks of different wireless standards. It can configure and control the required chain, otherwise all chains will be switched on. This overcomes the high dissipated power and the wasted area of other receivers. This leads to exploit a new reconfigurable high-speed low-power less-area ADC. In SDR receiver, the received RF signal is applied to an ADC, followed by a digital signal processor (DSP).

The revolution of the deep sub-micron CMOS technology shows the improvement of conventional ADCs is highly challenging compared to digital systems [1]–[4]. In high frequencies, the time resolution of digital signals is so efficient. Consequently, the ADC digital part percentage will be increased in order to: (1) get the full use of the DSP, (2) reduce area

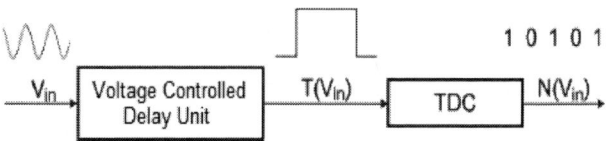

Fig. 1. Time-based ADC architecture.

and power consumption of the analog blocks and (3) speed the ADC up. In Time-Based ADC architecture, shown in Fig. 1, a Voltage-Controlled Delay Unit (VCDU) samples the signal amplitude into stream of pulses in the time-domain, then a Time-to-Digital Converter (TDC) quantizes the pulses into digital words [3].

Several VCDU circuits in CMOS 0.18µm technology have been introduced in the literature [5]. The basic VCDU circuit controls proportionally the delay of the input clock edge with respect to a sampled input voltage. There are superb advantages in the differential design. Firstly, the even-order harmonic components caused by a single-ended VCDU non-linearity will be suppressed. Secondly, the differential input offers a doubling of the signal amplitude resulting in an improvement in the SNR. Finally, the common-mode noise will be rejected. In this paper, a new manually-calibrated differential VCDU circuit is presented and compared to its single-ended design in the CMOS 65nm technology node.

This paper presents a new VCDU circuit, in which promising results are achieved in linearity, dynamic range (DR), effective-number-of-bits (ENOB), sensitivity and Figure-of-Merit (FOM) with a higher operating sampling-rate at the expense of extra area/power overheads. A metal-insulator-metal capacitor (MIMCAP) is used to degrade the process-voltage-temperature (PVT) variations. The rest of the paper is organized as follows. Section II dis-cusses design and analysis of the proposed circuit. Simulation results are demonstrated for the proposed design with its single-ended design and state-of-the-art analog-part ADCs in Section III. Finally, the presented work is summarized in Section IV.

II. PROPOSED DESIGN AND ANALYSIS

The new proposed VCDU architecture is based on the diff-

978-1-4673-8760-6/15 $31.00 © 2015 IEEE

erential approach as shown in Fig. 2. The upward VCDU core converts the positive input voltage into a time-difference variable (ΔT_2) measured with respect to a reference clock event (Φ_{CLK}). The downward VCDU core converts the negative input voltage into a time-difference variable (ΔT_1) measured with respect to Φ_{CLK}. Each time-difference variable is considered the delay equation of a single-ended VCDU circuit. So, the delay equation of the proposed differential VCDU design (ΔT_0) is the time difference between ΔT_2 and ΔT_1. Fig. 3 illustrates the timing diagram of the proposed design. Due to the fact that the VCDU circuit is considered a buffer delay circuit, the time-difference variables are measured between the rising edges of both Φ_{CLK} and their corresponding output (Φ_{O2} and Φ_{O1} respectively).

Fig. 4 portrays a single VCDU core circuit schematic. During the logic '0' of Φ_{CLK}, the capacitor (C) is reset via M6. When Φ_{CLK} raises up to logic '1', C charges by the current (I_{IN}) through the transmission gate switch formed by transistors M4 and M5. I_{IN} is generated by a Wilson current mirror [6] formed by M1-M3. M2 has a high gate voltage to decrease the high sensitivity of the current value to neglect gain-source voltage variations. The current-steering amplifier, constructed by M7-M13, senses the difference between the input voltage (V_{IN}) and the capacitor voltage (V_A) and lets the latching circuit M14-M17 make a logic decision. The output inverter M18-M19 provides a high linear output delay. Once V_A reaches the desired V_{IN}, the VCDU core output switches to logic '1'.

In fact, C is a MIMCAP which is a parasitic capacitor between the higher metal layers (i.e. Metal 7 and Metal 8). It has a higher density and linearity and smaller parasitic capacit-

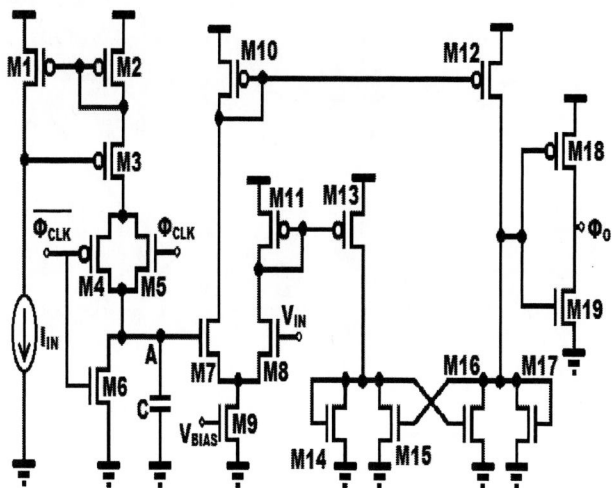
Fig. 4. The single VCDU core circuit schematic.

ances. Generally, it can highly overcome the PVT variations that happen in the drain-source voltage (V_{DS}) of a MOSFET capacitor especially at high frequencies [7]. The main disadvantage of the MIMCAP is that it consumes a larger area as the field-oxide insulator is thicker than the gate-oxide of the MOSFET capacitor. Also, there will be few process-corner variations as an expense of the insulator thickness over several wafer lots [8].

III. SIMULATION RESULTS

Design simulations were conducted on Cadence Virtuoso using industrial Taiwan Semiconductor Manufacturing Corporation (TSMC) 65nm CMOS technology. These results are tested using MATLAB. Optimal bias conditions have been selected using a manual calibration technique by which the largest linear range can be achieved. The applied DC voltage for the proposed differential design and the single-ended design is 0.5V and 0.55V, respectively. The supply voltage is 1V, while the operating clock frequency which has a 50% duty cycle is 8GHz for both designs. The capacitor's value is 167fF.

A. Linearity and Dynamic Range

The method of sweeping the input voltage in order to get a wide linear range during calibration has been discussed in [9]. Fig. 5(a) and Fig. 5(b) show the dynamic range of the proposed differential VCDU design and the single-ended VCDU design, respectively, where the linear input voltage range of the proposed design is clearly enhanced. Fig. 6(a) and Fig. 6(b) show the linearity error check for both designs, respectively.

Table I shows all specifications of both designs at an 8GHz sampling frequency (F_S). It illustrates that the proposed design has a better dynamic range, sensitivity and FOM at the expense of the area, power dissipation and the root-mean-square (RMS) noise than the single-ended design.

B. Voltage Sensitivity

Among the linear range, any two points will be chosen to evaluate the slope between the delay (y-axis) and V_{IN} (x-axis).

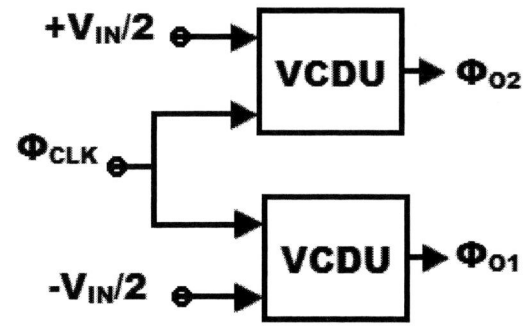
Fig. 2. The proposed differential VCDU architecture.

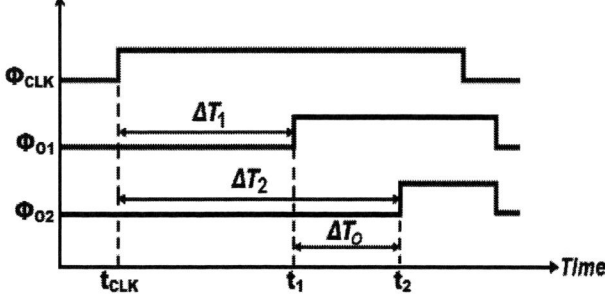
Fig. 3. The proposed design timing diagram.

978-1-4673-8760-6/15 $31.00 © 2015 IEEE

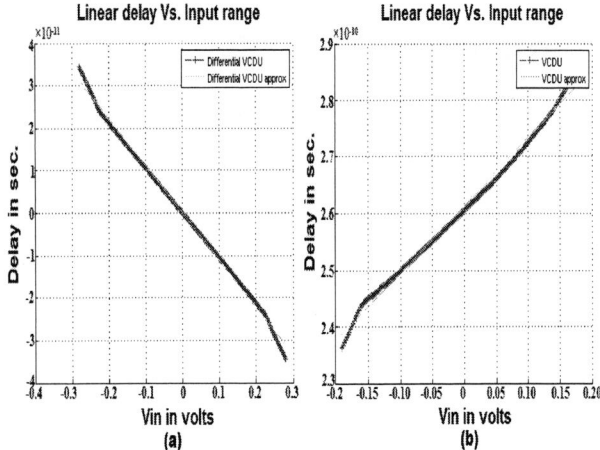

Fig. 5. Curve fitting of the linear range using MATLAB. (a) The proposed differential VCDU circuit. (b) The single-ended VCDU circuit.

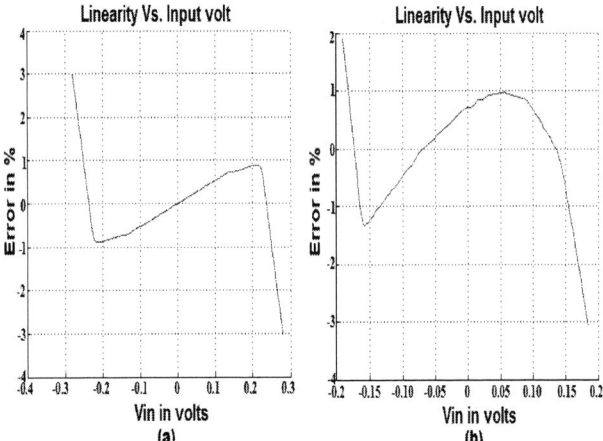

Fig. 6. Linearity error check of the linear range using MATLAB: (a) The proposed differential VCDU circuit. (b) The single-ended VCDU circuit.

The slope represents the sensitivity of a circuit. The sensitivity equals to $0.9ps/m$V for the proposed design, while it equals to $0.3ps/m$V for the single-ended design. The proposed design shows a higher sensitivity as the sensitivity is a function of the delay which is enhanced due to the differential approach.

C. Maximum Sampling Frequency and Power Consumption

Due to the fact that high frequencies distort the signal linearity, low-frequency applications can get higher dynamic range. In fact, high frequency applications have larger power dissipation (P_D) [3]. This power can be minimized by decreasing transistors' size. Hence, the flowing current will get higher resistance from electrons and it will be decreased.

D. Noise Simulation

Noise represents a random fluctuation in any electrical signal [10]. Thermal noise is the main dominant noise parameter of a MOSFET transistor represented in equation (1). Where V_N is the RMS voltage due to thermal noise generated in a resistance over a bandwidth in a room temperature. The

main contributing noise parameters in the proposed design are the flowing current of the steering amplifier, the output inverter and the latching circuit of each VCDU core circuit. They contribute with a percentage of 41.3%, 32.2% and 25.6% from the overall total output referred noise ($V_{N,RMS}$), respectively. $V_{N,RMS}$ is represented in Table I for an 8GHz wide frequency which is larger in case of the proposed design due to its larger number of transistors.

TABLE I: Performance summary of the proposed differential design and the single-ended design at an 8GHz F_S

Parameter	This work	The single-ended design
V_{IN} Range	-0.28V: +0.28V	-0.19V: +0.19V
Dynamic Range (V)	0.56	0.38
Input DC Bias (V)	0.5	0.55
Linearity Error (%)	3	3
Sensitivity (ps/mV)	0.9	0.3
P_D (mW)	1.6	0.8
Area (μm^2)	742	360
FOM ($\times 10^{12}$)	1.6	1.3
Noise ($nV/Hz^{1/2}$)	14.5	10.2
ENOB (bits)	8.9	5.7

The total RMS noise should be less than a single step conversion of the ADC as in equation (2). Where V_{FS} and N are the full scale voltage and the number of bits, respectively. So, this design can be part of a 9-bit ADC due to an $1.3m$V $V_{N,RMS}$ which lets the proposed design have a good ENOB.

$$V_N = (4KTR\Delta F)^{1/2} \qquad (1)$$

$$V_{N,RMS} < V_{FS} / (2^N-1) \qquad (2)$$

E. Area

The area of the proposed design, shown in Fig. 7, is 742 μm^2 which is bigger than by a factor of 2X that of the single-ended design due to the circuit metal wire connections. The area of the MIMCAP is $156\mu m^2$ which can be added on the differential VCDU layout due to its higher metal layers.

F. Effective-Number-of-Bits

ENOB is considered the main metric that tests all different types of errors (including noise and distortion errors) that a circuit practically faces. The ENOB is calculated at an input frequency of 3.5GHz in order to ensure that Nyquist conditions are achieved [4]. In equation (3), SNDR represents the signal to noise and distortion ratio. The proposed design shows a higher ENOB than the single-ended design due to the utilized differential architecture and the MIMCAP that enhanced the linearity.

$$ENOB = (SNDR - 1.76) / 6.02 \qquad (3)$$

G. Figure-of-Merit

The Figure-of-Merit (FOM) approach has been discussed, in [11], as in equation (4). The FOM represents the efficiency of using the power to increase the dynamic range and/or the maximum frequency. The proposed design has a higher FOM due to the higher DR and F_S and the lower dissipated power.

$$FOM = (DR^2 \times F_S) / P \qquad (4)$$

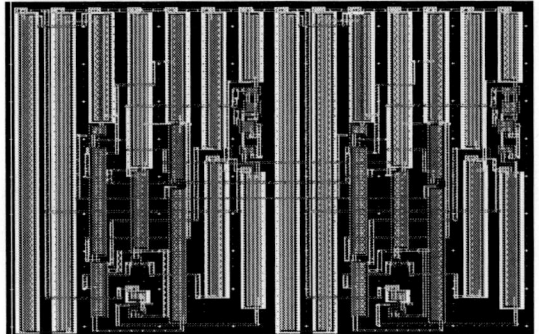

Fig. 7. The proposed design layout using Calibre

TABLE II: Comparison of 65nm CMOS state-of-the-art analog-part ADCs at a 3% acceptable error

Parameter	This	[12]	[13]	[14]	[15]
Supply Voltage (V)	1	2.5	1	1.2	1.2
Dynamic Range (V)	0.56	0.8	0.2	0.13	0.5
Sensitivity (ps/mV)	0.9	-	0.3	0.4	-
ENOB (bits)	8.9	11.2	3.5	3.7	4.9
P_D (mW)	1.6	7	4	21.4	3.7
Area (mm^2)	0.00074	0.075	0.0008	-	-
F_S (GHz)	8	2.4	5	6	2
FOM ($\times 10^{12}$)	1.6	0.2	0.1	0.01	0.1

Table II demonstrates a comparison between the proposed design and state-of-the-art analog-part ADCs [12]–[15]. According to Table II, this work provides a higher FOM, sensitivity and sampling speed that a circuit can operate on and a lower power consumption and area. Also, it provides a reasonable dynamic range and ENOB. Optimization of the proposed design is an active current research work to: combine the proposed design with a new differential TDC circuit to form a complete differential time-based ADC circuit, and add an automatic calibration circuit for the whole system to omit the PVT variations resulted from the TDC block.

IV. CONCLUSION

In this paper, a new manually-calibrated differential VCDU circuit is proposed which achieves a higher operating sampling frequency, linearity, ENOB, dynamic analog input range and sensitivity on 65nm technology. The novelty in this design is emerged from the dependency on three major factors. The first is the power of the CMOS technology which provides a high-speed, low-area and low-power design. The second is the differential architecture, in which the even order harmonics are suppressed as well as the input voltage noise. The third is depending on the MIMCAP which degrades the PVT variations.

The proposed design is compared to the single-ended VCDU circuit and state-of-the-art ADCs to show its simulation results strengths. It provides an 8GS/s sampling-speed, 0.56V dynamic range, 8.9bits ENOB, 0.9ps/mV sensitivity, 1.6mW power, 742μm^2 area, 1.6×10^{12} FOM and 14.5nV/Hz$^{1/2}$ RMS

noise. This work is a part of a differential time-based ADC and it is suitable for high-accuracy high-frequency applications especially the SDR application.

V. ACKNOWLEDGEMENT

This research was partially funded by Cairo University, ITIDA, NTRA, NSERC, Zewail City of Science and Technology, AUC, the STDF, Intel, Mentor Graphics, SRC, ASRT and MCIT.

REFERENCES

[1] M. Palkovic, P. Raghavan, M. Li, A. Dejonghe, L. Van Der Perre, and et al., "Future Software-Defined Radio Platforms and Mapping Flows," IEEE Signal Processing Magazine, vol. 27, no. 2, pp. 22–33, 2010.

[2] R. Saad, D.L. A.-Ramirez, and S. Hoyos, "Sensitivity Analysis of Continuous-Time ΔΣ ADCs to Out-of-Band Blockers in Future SAW-Less Multi-Standard Wireless Receivers," IEEE Transactions on Circuits and Systems I: Regular Papers, vol. 59, no. 9, pp. 1894–1905, 2012.

[3] P.K. Yenduri, A.Z. Rocca, A.S. Rao, S. Naraghi, M.P. Flynn, and et al., "A low-power compressive sampling time-based analog-to-digital converter," IEEE Journal on Emerging and Selected Topics in Circuits and Systems, vol. 2, no. 3, pp. 502-515, 2012.

[4] T. OH, H. Venkatram, and U.-K. Moon, "A Time-Based Pipelined ADC Using Both Voltage and Time Domain Information," IEEE Journal of Solid-State Circuits, vol. 49, no. 4, pp. 961–971, 2014.

[5] C.S. Taillefer, and G.W. Roberts, "Delta–Sigma A/D Conversion Via Time-Mode Signal Processing," IEEE Transactions on Circuits and Systems I: Regular Papers, vol. 56, no. 9, pp. 1908–1920, 2009.

[6] B.A. Minch, "Low-Voltage Wilson Current Mirrors in CMOS," IEEE International Symposium on Circuits and Systems (ISCAS'07), pp. 2220–2223, May 2007.

[7] C. Ng, C.-S. Ho, S.-F. Chu, and S.-C. Sun, "MIM capacitor integration for mixed-signal/RF applications," IEEE Transactions on Electron Devices, vol. 52, no. 7, pp. 1399–1409, 2005.

[8] P. Chiu, and M. Ker, "Metal-layer capacitors in the 65 nm CMOS process and the application for low-leakage power-rail ESD clamp circuit," Elsevier Microelectronics Reliability, vol. 54, no. 1, pp. 64–70, 2014.

[9] A. El-Bayoumi, H. Mostafa, and A.M. Soliman, "A New Highly-Linear Highly-Sensitive Differential Voltage-to-Time Converter Circuit in CMOS 65nm Technology," ISCAS'2015, pp. 1262–1265, 2015.

[10] J. Cheon, and G. Han, "Noise Analysis and Simulation Method for a Single-Slope ADC With CDS in a CMOS Image Sensor," IEEE Transactions on Circuits and Systems I: Regular Papers, vol. 55, no. 10, pp. 2980–2987, 2008.

[11] F. van der Goes, C.M. Ward, S. Astgimath, H. Yan , J. Riley, Z. Zeng, and et al., "A 1.5 mW 68 dB SNDR 80 Ms/s 2 Interleaved Pipelined SAR ADC in 28 nm CMOS," IEEE Journal of Solid-State Circuits, vol. 49, no. 12, pp. 2835–2845, 2014.

[12] G. Taylor, and I. Galton, "A reconfigurable mostly-digital delta-sigma ADC with a worst-case FOM of 160 dB," IEEE Journal of Solid-State Circuits, vol. 48, no. 4, pp. 983–995, 2013.

[13] A.R. Macpherson, J.W. Haslett, and L. Belostotski, "A 5GS/s 4-bit Time-Based Single-Channel CMOS ADC for Radio Astronomy," 2013 IEEE Custom Integrated Circuits Conference (CICC), pp. 1–4, 2013.

[14] A. Hussein, M. Fawzy, M. Ismail, M. Refky, and H. Mostafa, "A 4-Bit 6GS/s Time-Based Analog-To-Digital Converter," 26th IEEE International Conference on Microelectronics (ICM), pp. 92–95, 2014.

[15] J. Yang, Y. Chen, H. Qian, Y. Wang, and R. Yue, "A 3.65 mW 5 bit 2GS/s flash ADC with built-in reference voltage in 65nm CMOS process," 2012 IEEE 11th International Conference on Solid-State and Integrated Circuit Technology (ICSICT), pp. 1–3, 2012.

Biomimicry to Network on Chip: Router Heart Rate

Ahmed El-Naggar
EE Department, Alexandria University
Alexandria, Egypt
ahmed.m.el-naggar@ieee.org

Ahmed Shalaby
ECE Department, Egypt-Japan University of Science and
Technology (E-JUST), Alexandria, Egypt
ahmed.shalaby@ejust.edu.eg

Abstract— **The growing complexity of systems-on-chip creates the need to replace the bus-based architecture. Network-on-chip has been proposed to address the communication bottleneck of system-on-chip. Router is the key component of network-on-chip architecture. Router frequency is one of the critical parameters, which has direct impact on network-on-chip performance. This paper proposes an adaptive scheme for controlling the router frequency based on biomimicry. A complete evaluation for the proposed scheme over various network-on-chip sizes and different evaluation parameters are performed. Results show improvement in throughput and latency. Moreover, it saves up to 75% of buffer storage, up to 60% of dynamic power and achieving load balance for all routers in the network.**

Keywords— System-on-chip (SoC); Network-on-chip (NoC); Biomimicry; router frequency, router architecture.

I. INTRODUCTION

System-on-Chip (SoC) is becoming more complex and multi-functional; consisting of tens or hundreds of integrated IPs supporting one or several applications. Connecting such large numbers of IPs (processor, DSP, memory, etc.) per chip through point-to-point or bus architecture become a complicated problem, owing to the constraints of fan-outs tree and difficulties related to submicron process technology such as non-scalable wire delays, errors in signal integrity and clock skew [1]. Network-on-chip (NoC) is proposed as a promising solution to address the communication challenges of SoC. NoC connects IP blocks through interconnection routers, resembling the computer network, where data is routed through network infrastructure using packet switching rather than global buses [2]. In comparison to bus-based architectures, NoC has many advantages like scalability, modularity, reliability and performance. Over the last decade, many techniques have been proposed in order to improve NoC performance in terms of latency and throughput, trading-off its area utilization and power consumption. However, many research challenges remain to be solved.

Network-on-Chip architecture is defined by: topology, routing algorithm, flow control protocol, and router architecture. Router is a critical component for NoC, steering and coordinating the data flow. The function of the router is to receive packets, determine their destination based on the routing algorithm, and then forwarding packets to the appropriate output. Router architecture is composed of registers, switches, function units, and control logic. It implements the routing algorithm and the flow control protocol to buffer and routes packets to their destinations [3]. Router

frequency is one of the critical factors that has direct influence on NoC performance. For instance, increasing router frequency can improve performance in terms of latency and throughput but at the expense of power consumption. Thus, controlling router frequency can improve NoC performance. Our idea, to control router frequency, is inspired by biomimicry.

Biomimicry is an approach that seeks sustainable solutions for human challenges by emulating the strategies of nature [4]. Proceeding from this approach, we thought of the router frequency as analogous to the heart rate. Heart rate varies according to the physical needs and activities of the body. Physical needs include the need to absorb oxygen and excrete carbon dioxide and activities such as physical exercise, sleep, stress, etc. In this context, we propose adaptive control for router frequency according to network congestion level. We provide a precision performance analysis in terms of latency, throughput, and power consumption. This paper is organized as follows: the next section sheds the light on background and related work. Section III describes the proposed scheme and implementation. Section IV presents the simulation and synthesis results followed by their discussion. Finally, section V concludes the paper.

II. BACKGROUND AND RELATED WORK

A. Introduction to Biomimicry

Biomimicry is a recent field of interdisciplinary research that brings biology together with engineering. The basic principle is to mimic nature's systems to solve engineering problems. There are huge efforts in different fields like energy, transportation, architecture, etc. One of the famous examples is the Mercedes-Benz bionic car [5]. It was modelled after the yellow boxfish, a highly skilled swimmer which, despite its box-like body, it has outstandingly low flow resistance. Designs were managed to achieve greater rigidity with low weight, undiminished levels of safety and the lowest possible fuel consumption and exhaust emissions. Another famous example is the Japanese Shinkansen bullet train. The problem was the changes in air pressure every time the train emerged from a tunnel. It produced large thunder claps causing residents one-quarter a mile away to complain. The solution was to model the front-end of the train after the beak of kingfisher, which dives from the air into water with very little splash to catch fish. The result was not only a quieter train, but 15% less electricity use even when the train travels 10% faster [6].

978-1-4673-8760-6/15 $31.00 © 2015 IEEE

B. Dynamic Voltage Frequency Scaling

Operating frequency is a critical design parameter that directly affects NoC performance. Dynamic Voltage/ Frequency Scaling (DVFS) techniques have been proposed to allow the network to operate at a lower frequency with the intention to reduce power consumption. Former, DVFS has been proposed for microprocessors [7]. DVFS exploits the variance in processor utilization, according to processor load, and adaptive frequency and voltages are tuned. For instance, when the processor is heavily loaded, the frequency is decreased and vice versa. Shang et al. proposed DVFS for links, where the frequency and voltage of links are dynamically adjusted to minimize power consumption based on past utilization [8].

In [9], the authors continued to develop DVFS for links. They proposed a DVFS link, which adopts a clock boosting mechanism and DVFS scheme, which includes the link utilization estimator, DVFS algorithm and link controller. [8] - [9] focused specifically on link latency and power consumption. In [10], DVFS is proposed to reduce power consumption for NoC by lowering the frequency. However, reducing NoC frequency leads to an increase in latency and a reduction in throughput, which in turn degrades overall system performance. This degradation leads to longer duration for cores to achieve the same functions. Consequently, it consumes more leakage power. In this context, we were motivated to find out a solution inspired by biomimetic approach.

III. BIOMIMICRY TO NETWORK ON CHIP

A. Router Heart Rate

Imagine an organism that pumps blood with equal pressures to the vast majority of its organs and limbs regardless their sizes or required energy. That is, it possesses a relatively static blood pressure and hormone levels, paying no heed to dangerous or safe situations. Such an animal would surely be at an evolutionary disadvantage to other animals.

One of the basic principles in animal physiology is "supplies are proportioned to needs". It is performed by an accurate control mechanism in the cardiovascular system; for example, during stressful conditions neural and hormonal responses occur to cope with the increased needs for energy supplies. Sympathetic nervous system stimulation, increase in heart rate, blood pressure, increase in adrenaline, opening of energy stores., etc. Another example is controlling the blood flow to various organs according to activity; for instance, muscular blood flow increases during exercises, gastrointestinal blood flow increases after meals, all of these are through accurate moment to moment regulatory mechanisms that is created in our physiology to adapt to continuously changing situations [11].

Our idea is to mimic the momentarily adaption of such an animal system. It is indeed inefficient for a system to distribute its activity capability (allowed power) equally among all its parts all the time regardless of their individual activity. This work focuses on creating a system that resembles the parts of the biological system of advanced creatures. It is apparent that heart rate and blood flow are, in a very loose way, analogous to the system frequency. Hence, similar to the control of blood flow, we propose the control of the system frequency according to the activity required by each module of the system.

To implement the desired control, we propose tuning the operating frequency of each router in the network according to its congestion level, based on buffers utilization. However, proposed frequency tuning differs from DVFS since it is not scaled by percentage up to 20% like the prior work, but the proposed tuning is alternating the clock between two frequencies: f and $2f$. By this means, all routers on the network can continue operating within a fully synchronous domain. Furthermore, switching between f and its doubled $2f$ can considerably enhance the performance in a short duration. In the evaluation section, more details will be shown regarding results.

B. Router Architecture

Figure.1 demonstrates the clock manager module, which is implemented at every router in the network. The sum of utilization of all FIFO buffers (north, east, south, west and local) in the router is calculated. Then, according to utilization level, is compared to a certain threshold, clock manger is responsible for tuning the router frequency between f and $2f$. Several thresholds for buffers utilization were implemented and tested. The threshold is set as a ratio of the sum of all buffers sizes. The best threshold was 12.5%-25% of buffers utilization, which achieved the best performance in terms of latency, throughput and power consumption.

Clock Manger

*Calculate the **sum** of buffers utilization*

***if** (total utilization) > threshold **then** clk = 2f **else** clk = f*

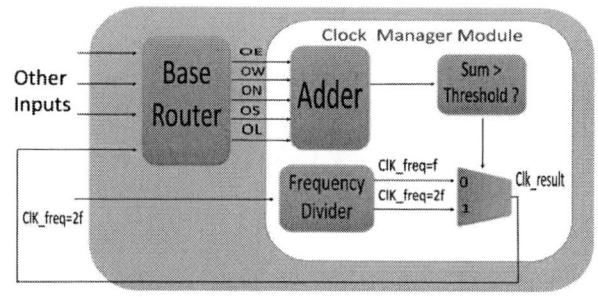

Fig. 1. Modified Router architecture: clock manager.

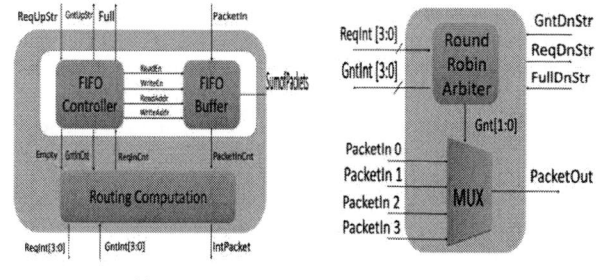

(a)   (b)

Fig. 2. Base router architecture: (a) input port module, (b) output port module.

The base router, without clock manager module, consists of five input/output ports, connected together using the intermediate crossbar. The main two modules at each port are: Input port module Fig.2 (a) and output port module Fig.2 (b).

1) Input Port Module:

It receives the packets from the upstream routers using four-phase handshaking (request/grant) scheme. As shown in Fig.2 (a), the main components are:

- FIFO buffer: Stores the packets received at the input port until it is routed to the suitable output port.
- Input FIFO Controller: Controls read/write processes from/to the FIFO. Moreover, it handles the communications between the upstream router and output port.
- Routing Computation: Applies the routing algorithm according to the destination address and selects the appropriate output port for the packet.

2) Output Port Module:

It sends the packets to the downstream routers using four-phase handshaking (request/grant) scheme. As shown in Fig.2 (b), the main components are:

- Arbiter: Receives incoming requests for the output port and then grants one of them according to arbitration logic function. Round Robin is implemented due to its simplicity and fairness
- Output Controller: Handles the communications between the downstream router and input port.
- MUX: Drives packets to the downstream router according to the arbiter selection.

IV. PERFORMANCE EVALUATION AND DISCUSSION

A. Simulation Framework

We evaluate the performance of the proposed scheme for frequency tuning by a cycle accurate simulator. The router is implemented in RTL using Verilog. Then a 2-D mesh network is implemented with different sizes 2×2, 4×4 and 8×8 and various buffer sizes 4, 8 and 16 for each router port. Dimension-order algorithm is implemented for routing function. The data link width between any two routers is equal to the packet size. Each router is connected to a processor element, which consists of two modules: injector and collector. The injector works to feed packets into the network according to a random-uniform traffic pattern (synthetic workload). Packets travel through the network and terminate at the packet collector. The collector works to log the fields of the packets. Time stamp is added to the packet to calculate the overall performance of the network in terms of latency and throughput.

The packet is composed of different fields: destination, source, packet identification and payload as shown in Fig. 3, where *x_Dst and y_Dst* represent x and y coordinates of the destination node on the network. Similarly, *x_Src* and *y_Scr* represent x and y coordinates of the source node. *Packet_ID* is an ID unique to each individual packet. Finally,

Packet_Payload represents the data in the packet, including the transmission time.

x_Dst	y_Dst	x_Src	y_Scr	Packet_ID	Packet_Payload

Fig. 3. Packet Structure.

B. Simulation Results

In this subsection, the simulation results in terms of throughput, latency and power consumption are presented and discussed. For fair evaluation, base router architecture and the modified one (with clock manger) are implemented. Both are simulated by ModelSim for latency and throughput evaluation. To calculate average latency and throughput, the simulator is run under different injection rates. In addition, both are synthesized on TSMC 65 nm standard library on Synopsys DC for power consumption evaluation.

Although the evaluation is performed with different 2-D mesh network sizes 2×2, 4×4, and 8×8, only results for 8×8 are presented due to space. Nevertheless, as network size increase, the performance improvement by proposed scheme is proportionally increased, owing to the fact that NoC performance degrades with network size due to capacity limitation [12]. In general, the results show a comparison between networks with: 1) routers run at *f* frequency, 2) routers run at *2f* and 3) routers run with proposed scheme (switching between *f* and *2f*).

Fig. 4 demonstrates the throughput performance with different injection rates for different buffer sizes (8 and 16) per port. For the proposed scheme, different thresholds have been examined. It can be noticed that the proposed scheme achieved performance between *f* and *2f* networks. In addition, it can be noticed that as threshold decreases, the performance improves. This is because as the threshold decreases, the average number of packets stored within the router decreases, leading to the release of any congestion in a short time. Similar observation is shown in Fig. 5 for average latency performance.

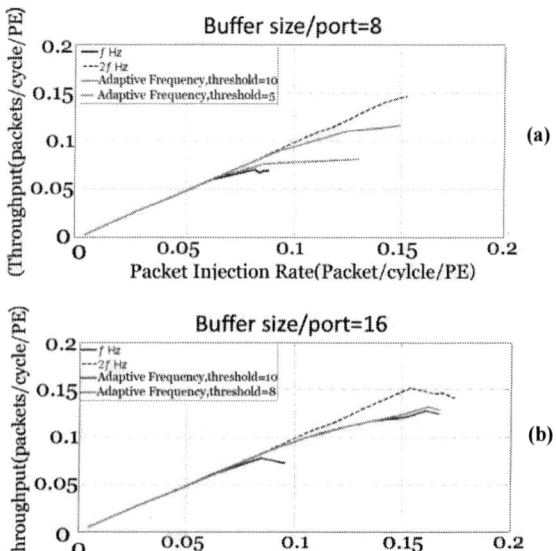

Fig. 4. Throughput versus injection rate for 8×8 NOC with different buffer size: a) 8 and b) 16 for each port.

978-1-4673-8760-6/15 $31.00 © 2015 IEEE

Since the proposed scheme handles the congestion situation adaptively according to predefined threshold that is observed as it decreases, performance improves. Hence, by applying the proposed scheme, buffer sizes can be reduced to less than 25%, saving more than 75% of the total buffers on the network. Moreover, better performance can be achieved. In order to verify that, the average number of packets stored by each router at certain injection rate is measured during the simulation time as shown in Fig. 6. It can be observed that the number of packets is reduced vastly and a load balance between all routers is achieved. Moreover, notable congestion mitigation in the network is achieved, allowing a quantitative boost of packets into the network, consequently improving performance in terms of throughput and latency.

From Fig.4 and Fig.5, it can be noticed that the network with routers run at *2f* achieves better performance than the proposed scheme, but it comes at the cost of power consumption. Dynamic power consumption is measured for networks with *f* and the adaptive scheme (with 50% of total size of all FIFOS). It differs according to router position and activity. For instance, in 2-D mesh the router at the center of the network consumes 700 mW (*f*) and 283 mW (adaptive), while the corner router power consumes 689 mW (*f*) and 428 mW (adaptive). So we can notice that running network at adaptive frequency decreases the dynamic power by 40%-60% according to router position and activity. This is due to the saving in buffers. Since FIFO buffers alone can consume about 50% of the total power consumed by the router [13].

V. CONCLUSION

In this work, we propose an adaptive scheme inspired by biomimicry to adaptively control the router frequency in NoC. It works according to router congestion level based on buffers utilization. The router architecture is introduced. The proposed scheme improves NoC performance in terms of throughput, latency. In addition, it attains save in storage up to 75% of total buffer sizes leading to 40%-60% saving in dynamic power consumption according to router position and activity.

REFERENCES

[1] A. Agarwal, C. Iskander and R. Shankar, "Survey of Network on Chip (NoC) Architectures and Contributions," Journal of Engineering, Computing and Architecture Volume 3, Issue1, 2009.

[2] Dally, W. J., & Towles, B, "Route Packets, Not Wires: On-chip Interconnection Networks," In Design Automation Conference Proceedings, 2001, pp. 684-689.

[3] Keckler, Stephen W., Oyekunle Ayinde Olukotun, and H. Peter Hofstee. Multicore processors and systems. Springer, 2009 .

[4] Benyus, Janine M. Biomimicry. New York: William Morrow, 1997.

[5] "Concept Cars: Bionic Car" [Online]. Available: http://www2.mercedes-benz.co.uk/content/unitedkingdom/mpc/mpc_unitedkingdom_website/e n/home_mpc/passengercars/home/passenger_cars_world/innovation_ne w/concept_cars.0008.html

[6] S. Sheppard. (2012, Apr. 23). "Eiji Nakatsu: Lecture on Biomimicry as applied to a Japanese Train" [Online]. Available:
http://labs.blogs.com/its_alive_in_the_lab/2012/04/biomimicry-japanese-train.html

[7] T. Burd and R. Brodersen, "Design Issues for Dynamic Voltage Scaling," In Proc. Int. Symposium on Low Power Electronics and Design, 2000, pp. 9–14.

[8] L. Shang, L.-S. Peh, and N. K. Jha, "Dynamic Voltage Scaling with Links for Power Optimization of Interconnection Networks," in Proc. Int. Symp. High-Performance Computer Architecture, 2003, pp. 91–102.

[9] S. E. Lee and N. Bagherzadeh, "A Variable Frequency Link for a Poweraware Network-on-Chip (NoC)," Integration, the VLSI Journal, vol. 42, no. 4, pp. 479–485, 2009.

[10] A. K. Mishra, R. Das, S. Eachempati, R. Iyer, N. Vijaykrishnan, and C. R. Das, "A Case for Dynamic Frequency Tuning in On-chip Networks," in Proc. Int. Symposium on Microarchitecture, 2009, pp. 292–303.

[11] Hall, John E. Guyton and Hall Textbook of Medical Physiology. Elsevier Health Sciences, 2010.

[12] Dally, William James, and Brian Patrick Towles. Principles and Practices of Interconnection Networks. Elsevier, 2004.

[13] Y. Hoskote, S. Vangal, A. Singh, N. Borkar, S. Borkar, "A 5-GHz mesh interconnect for a teraflops processor," Proceeding of the IEEE MICRO, vol. 27, pp. 51–61, 2007.

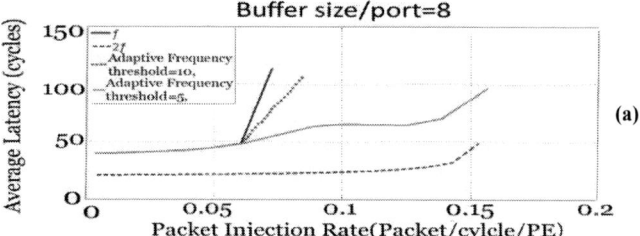

(a)

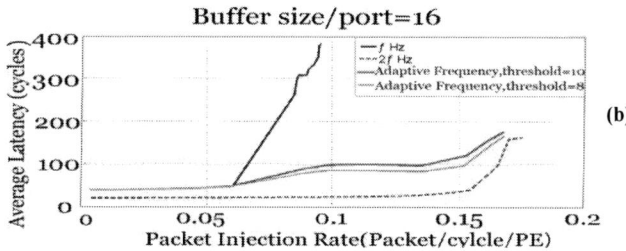

(b)

Fig. 5. Average latency versus injection rate for 8×8 NOC with different buffer size: a) 8 and b) 16 for each port.

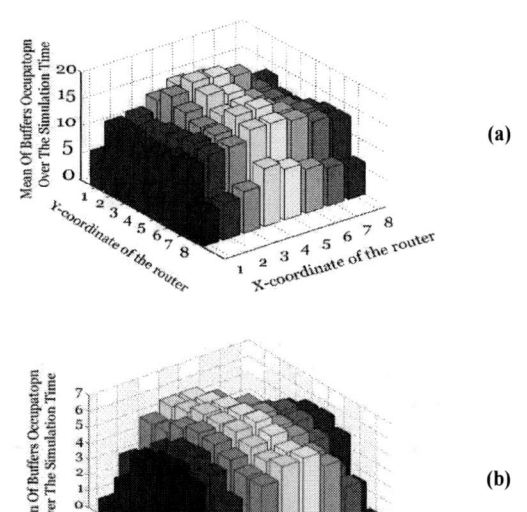

(a)

(b)

Fig. 6. The average number of packets in each router when the all routers work: a) with frequency *f*, b) with the adaptive frequency, threshold= 5

A 14-bit Low-Power Interface Circuit for Piezo-Resistive Pressure Sensors

Amr Walid
Integrated Circuit Lab.
Ain Shams University, Cairo, Egypt
amr.walid.1989@ieee.org

Ayman H. Ismail
Integrated Circuit Lab.
Ain Shams University, Cairo, Egypt
ayman.hassan@eng.asu.edu.eg

Abstract—In this work, a new low-power interface circuit for Wheatstone bridge piezo-resistive pressure sensors is introduced. Different than conventional interface circuits, it is proposed to periodically switch the resistive bridge to save power. To reduce the interface power even further, the turn ON time of different system blocks is optimized. The proposed interface is composed of a sample-and-hold circuit, a front-end-amplifier and a $\Delta\Sigma$ analog-to-digital-converter. The interface circuit achieves a high resolution of 13.82 bits at a power consumption of 221.8μW, including the bridge power, in 0.13μm CMOS technology, with a supply voltage of 1.2V. Hence, a very low figure-of-merit of 5.076 pJ/Conversion is attained.

I. INTRODUCTION

In the recent years, there has been a growing demand on high resolution pressure sensors for new applications such as mobile altimeter, air speed measurements, control systems, medical instruments and accuracy enhancement of the GPS receivers [1], [2]. Many of these applications are hand-held or battery operated, therefore the power dissipation of the pressure sensors remains as a main concern.

In piezo-resistive sensors, the pressure to be measured changes the resistance value of piezo-resistive element. Typically, piezo-resistive sensors are realized in one of two configurations, either as a single resistor with current excitation, or in the form of Wheatstone bridge as shown in Fig. 1. Although a single resistor configuration can potentially achieve high signal-to-noise ratio (SNR) [3], the Wheatstone bridge is the most widely used readout configuration, due to its immunity to common mode noise and interference from supply and environmental disturbance such as temperature.

Wheatstone bridge piezo-resistive sensors can be voltage driven or current driven as shown in Fig. 1a and Fig. 1b, respectively. In the voltage driven configuration, the change in the output voltage of the bridge, corresponding to a change in pressure, in the case of using four varying piezo-resistors is [4] $\Delta V_{ob} = V_s \frac{\Delta R}{R_o}$, where V_s is the supply voltage, R_o is the nominal value of the piezo-resistors and ΔR is the change in the piezo-resistor corresponding to the applied pressure.

In the current driven bridge configuration, the change in bridge output voltage is $\Delta V_{ob} = I_{bridge} \times \Delta R$, where I_{bridge} is the bridge excitation current. The bridge sensitivity is defined

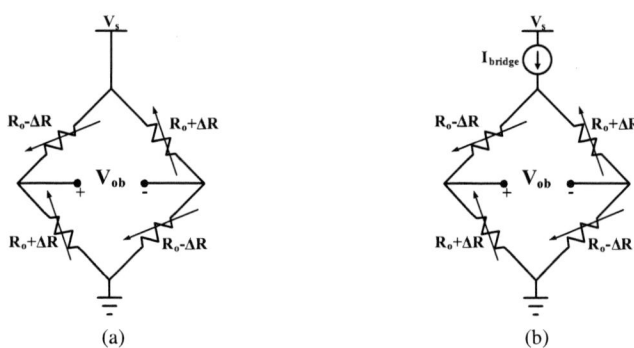

Fig. 1. (a) Voltage driven Wheatstone bridge, (b) Current driven Wheatstone bridge

as [4] $S = \frac{\Delta V_{ob}}{\Delta R}$. Hence, the sensitivity of the voltage driven configuration is given by

$$S_V = \frac{V_s}{R_o}, \qquad (1)$$

and the sensitivity of the current driven configuration is

$$S_I = I_{bridge} \qquad (2)$$

It is clear from (1) and (2) that piezo-resistive sensors exhibit a tight power-sensitivity trade-off and that higher sensitivity is obtained at the expense of higher dissipated current in the bridge. The objective of this work is to introduce a low-power interface circuit for piezo-resistive sensors. Different than the conventional and the state-of-the-art Wheatstone bridge sensors implementation [5] - [6], the Wheatstone bridge current is periodically switched to reduce bridge current consumption without degrading sensitivity. Furthermore, the turn ON times of system blocks are optimized for minimum average power.

This paper is organized as follows. In Section II, the architecture of the proposed system is presented. Section III introduces the interface sample-and-hold circuit (S/H) and the front-end-amplifier. The design of the $\Delta\Sigma$ analog-to-digital converter (ADC) of the interface is detailed in Section IV. In Section V, the simulation results of the interface are presented. The paper is concluded in Section VI.

978-1-4673-8760-6/15 $31.00 © 2015 IEEE

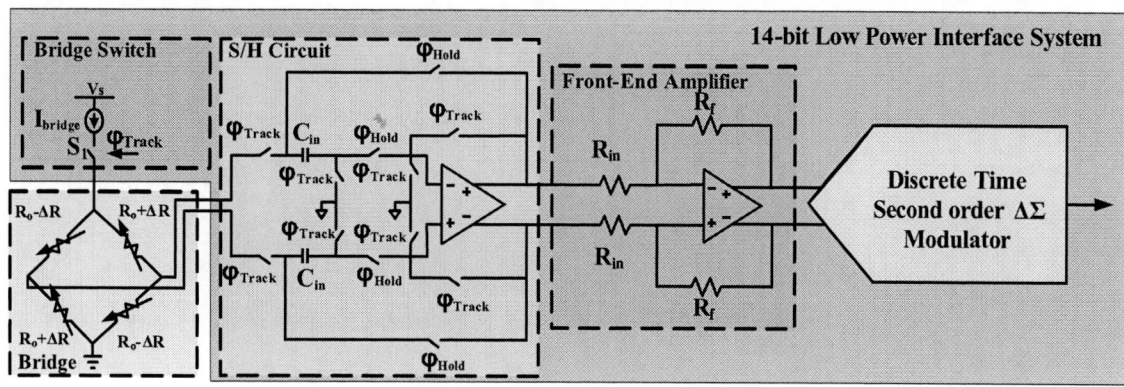

Fig. 2. The proposed interface system

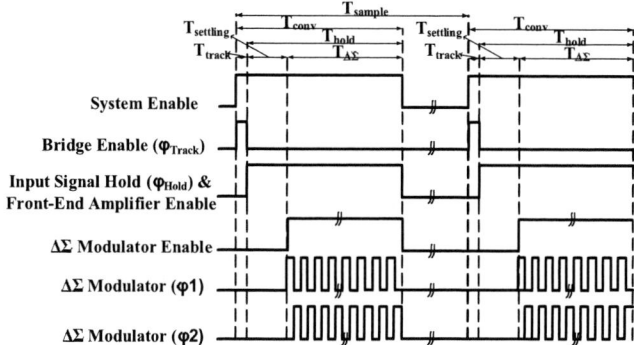

Fig. 3. Timing diagram of the proposed system

II. SYSTEM ARCHITECTURE

The proposed interface system is shown in Fig. 2. It is composed of sample-and-hold circuit (S/H), a front-end-amplifier and a 2nd order $\Delta\Sigma$ modulator. A current driven bridge is adopted because its power dissipation is independent on the sensor R_o and is fully defined by the interface system. Different system blocks are periodically switched to reduce the power dissipation as shown in the timing diagram of Fig. 3. The system samples the sensor output every period of time equals to T_{sample}. The interface converts the sampled signal to the digital form in a time T_{conv}. The bridge is turned ON (S_1 closed) for a time T_{track} and then is switched OFF for the rest of the conversion time T_{conv} to save power.

The time T_{track} is set by the minimum time required for settling to the target resolution and is given by

$$T_{track} = (0.5R_o + 2R_{sw}) \times C_{in} \times (N_{bit} + 1) \times ln(2) \quad (3)$$

where R_{sw} is the sampling switch resistance, C_{in} is the S/H sampling capacitance and N_{bit} is the target resolution. The front-end-amplifier is given a time $T_{settling}$ to settle. The $\Delta\Sigma$ modulator is operated after the settling of the amplifier for a time $T_{\Delta\Sigma}$. The average power of the interface can be expressed as

$$P_{avg} = P_{conv} \times \frac{T_{conv}}{T_{sample}} \quad (4)$$

where P_{conv} is given by

$$P_{conv} = V_s \times (I_{bridge} \times \frac{T_{track}}{T_{conv}}$$
$$+ I_{S/H} + I_{amplifier} \times \frac{T_{hold}}{T_{conv}}$$
$$+ I_{\Delta\Sigma} \times \frac{T_{\Delta\Sigma}}{T_{conv}}) \quad (5)$$

where $I_{S/H}$ is the S/H operating current, $I_{amplifier}$ and $I_{\Delta\Sigma}$ are the operating currents of the amplifier and the $\Delta\Sigma$ modulator, respectively. Furthermore, a figure-of-merit (FOM) for the interface is defined as [5]

$$FOM = \frac{P_{conv}[W] \times T_{conv}[s]}{2^{ENOB}} \quad (6)$$

where $ENOB$ is the effective number of bits of the system. The FOM represents the energy per conversion per resolution.

It is straight-forward to conclude from (4) and (5) that periodic switching of the bridge results in scaling down the bridge power by a factor of T_{track}/T_{sample}. The bridge average power becomes $(V_s.I_{bridge}.T_{track}/T_{sample})$, while T_{sample} is set by input signal sampling requirements (Nyquist rate), the bridge energy per conversion term $(V_s.I_{bridge}.T_{track})$ needs to be minimized for low power operation.

Decreasing the current I_{bridge} by a factor x, reduces the sensitivity of the bridge by the same factor. Hence, the maximum input signal to the interface is reduced by x and signal power is reduced by x^2. For a certain target signal-to-noise ratio (SNR), the noise power is reduced by the same factor and S/H C_{in} is increased by x^2 to reduce thermal noise accordingly. However, this results in an increase in T_{track} by a factor of x^2. Therefore, the overall effect is that the energy per conversion for the bridge $(V_s.I_{bridge}.T_{track})$ increases by the factor x. Hence, for low-power operation and for a switched bridge where the turn ON time is set by the settling requirements of the bridge and the subsequent S/H circuit, I_{bridge} should be increased. In Fig. 4 the bridge energy per conversion vs T_{track}, which corresponds to smaller C_{in} and smaller T_{track}, set by settling requirements of the bridge S/H.

I_{bridge} is generated by a current source stacked to the resistive bridge. Increasing I_{bridge}, increases the voltage drop

978-1-4673-8760-6/15 $31.00 © 2015 IEEE

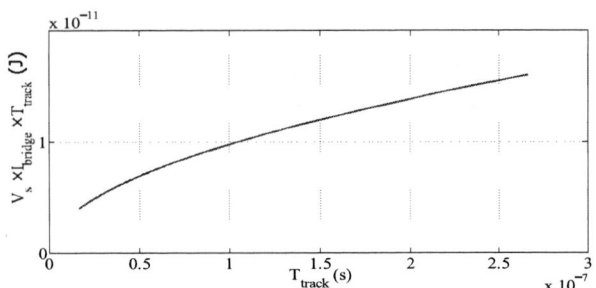

Fig. 4. Bridge energy per conversion vs T_{track}, which is limited by the settling time of the bridge - S/H

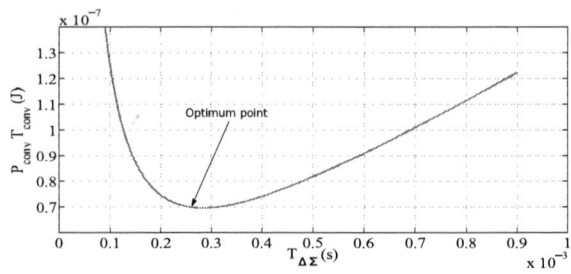

Fig. 5. Energy per conversion of the interface system vs $T_{\Delta\Sigma}$

across the bridge and reduces the voltage headroom available for the current source. Therefore the increase in I_{bridge}, and the corresponding reduction in the required T_{track}, is ultimately limited by the minimum required voltage headroom for the current source.

To further reduce the power dissipation of the interface system, the turn ON time for the $\Delta\Sigma$ modulator ($T_{\Delta\Sigma}$) is optimized for minimum power dissipation. Eq. (4) and (5) indicate that the $\Delta\Sigma$ power dissipation is ($V_s.I_{\Delta\Sigma}.T_{\Delta\Sigma}/T_{sample}$). The energy per conversion term ($V_s.I_{\Delta\Sigma}.T_{\Delta\Sigma}$) needs to be minimized for low power operation. Reducing $T_{\Delta\Sigma}$ requires increasing the speed of the $\Delta\Sigma$ modulator to finish conversion in a shorter time, which in turn imposes higher $I_{\Delta\Sigma}$. Therefore, an optimum value for $T_{\Delta\Sigma}$ that minimizes the $\Delta\Sigma$ energy per conversion exists as discussed in details in section IV. The same argument applies for the amplifier current $I_{amplifier}$ and T_{hold}.

III. THE SAMPLE-AND-HOLD CIRCUIT AND THE FRONT-END AMPLIFIER

The S/H circuit can be either implemented using an open-loop topology or a closed-loop topology. The open-loop topologies can operate at high speed, but the closed loop S/H circuits can achieve higher accuracy [7]. Therefore, a closed-loop topology is adapted for this work. The S/H circuit for the interface system is shown in Fig. 2. The bridge output is sampled when ϕ_{Track} is high and then held during ϕ_{Hold} for the amplification and analog-to-digital conversion.

The front-end amplifier is a resistive feedback amplifier employing a two-satge opamp (Fig. 2).

IV. THE $\Delta\Sigma$ MODULATOR

The target resolution for the proposed interface system is 14-bits. Therefore, $\Delta\Sigma$ analog-to-digital conversion technique is adopted. Since the operation of the interface is based on periodic switching of the system blocks, an incremental type $\Delta\Sigma$ is used, since incremental $\Delta\Sigma$ is suited for intermittent operation. A 2nd incremental $\Delta\Sigma$ requires M number of cycles to resolve N_{bit}, where M is given by [8]

$$M \approx (\sqrt{2} \times 2^{N_{bit}/2})/\sqrt{V_{in-max}/2V_{ref} \times (l-1)} \quad (7)$$

where V_{in-max} is the maximum input voltage to the modulator at which it remains stable, V_{ref} is the reference voltage of the modulator and l is the quantizer number of levels, which is two for the proposed design. Therefore, the system $T_{\Delta\Sigma}$ is defined as

$$T_{\Delta\Sigma} = M/f_{s_{\Delta\Sigma}} \quad (8)$$

where $f_{s_{\Delta\Sigma}}$ is the $\Delta\Sigma$ modulator clock frequency, which defines the modulator integrators unity gain frequency according to [9]

$$f_u \approx (\frac{(N_{bit}+1) \times ln(2)}{\pi}) \times f_{s_{\Delta\Sigma}} \quad (9)$$

From the circuits point-of view f_u can be only increased at the expense of higher current dissipation as f_u is given by

$$f_u = \beta \times \frac{g_m}{2\pi C_{Ltot}} \quad (10)$$

where C_{Ltot} is the equivalent load capacitor of the OTA, β is the feedback factor, g_m is the transconductance of the amplifier, $g_m \propto \sqrt{I}$ and I is the current consumed to achieve a specific g_m.

Eq. (8), (9) and (10) indicate that in the case $T_{\Delta\Sigma}$ is reduced, the current dissipation of the modulator $I_{\Delta\Sigma}$ needs to be increased. For the energy per conversion term ($P_{conv}.T_{conv}$) an optimum value exists that minimizes the term and hence leads to low power operation, as shown in Fig. 5. A similar design approach is used to find the optimum turn ON time for the front-end amplifier to minimize its energy per conversion contribution ($V_s.I_{amplifier}.T_{hold}$).

The $\Delta\Sigma$ modulator is implemented as a cascaded integrators with distributed feedback (CIFB) as shown in Fig. 6.

V. SIMULATION RESULTS

The proposed interface circuit of Fig. 2, is designed in $0.13\mu m$ CMOS technology with a 1.2V supply. T_{sample} equals to $50ms$. To achieve a resolution of $N_{bit} = 14bits$, T_{track} is $0.5\mu s$ and the front-end conversion time, T_{conv}, is $331\mu s$.

The bridge maximum variation $\Delta R/R_o = 10\%$ and $R_o = 5k\Omega$ and $I_{bridge} = 200\mu A$. The maximum output voltage from the bridge (V_{ob-max}) is $100mV$. The $\Delta\Sigma$ modulator consumes $102.5\mu A$ and has $f_{s_{\Delta\Sigma}} = 1MHz$. The average current for the whole system equals to $184.86\mu A$ resulting in power $\approx 221.8\mu W$.

The output signal spectrum of the whole interface system is shown in Fig. 7, considering circuit noise, in addition to

TABLE I

COMPARISON OF THE STATE-OF-THE-ART RESISTIVE PRESSURE SENSOR INTERFACES

Reference	Supply [V]	Circuits Power [μW]	Bridge Power [μW]	Total Power [μW]	Conversion Time [ms]	Linearity Error [ppm]	ENOB	FOM including bridge power [pJ/conv]
[4]	1.5	270	Not Used	270	3.3	2300	8.7	2014
[5]	1	124.5	50	174.5	0.05	2093	8.9	18.26
[6]	5	1350	N/A	1350	100	5	17.6	679
[10]	1.2	366	Not Used	366	1	61	14.13	20.4
[11]	2.5	27.5	Not Used	27.5	5	1000	9.965	137.5
This Work	1.2	221.32	.48	221.8	.331	55.4	13.82	5.076

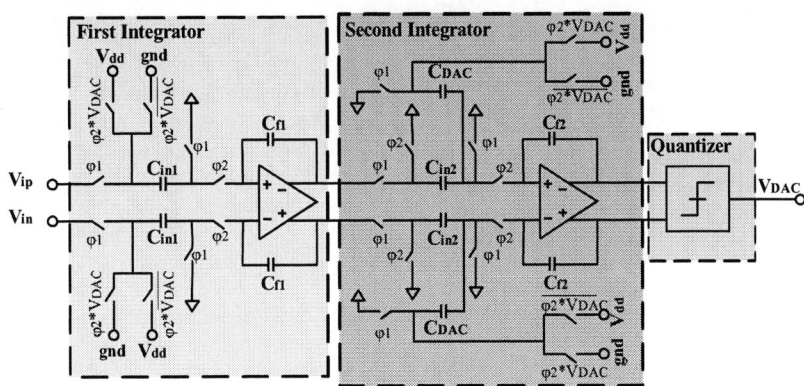

Fig. 6. The 2nd order $\Delta\Sigma$ modulator

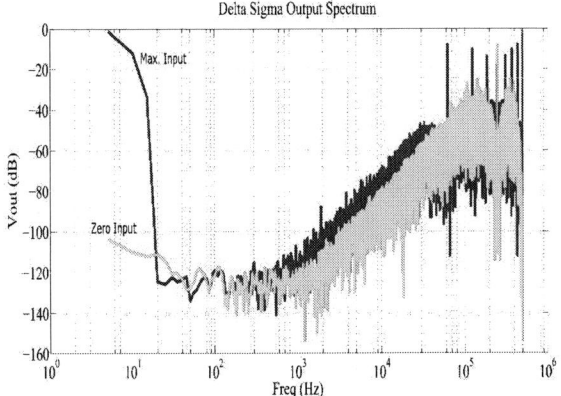

Fig. 7. The Output Spectrum of the interface system

quantization noise. The interface achieves a signal-to-noise ratio of 94 dB (ENOB=13.82 bits).

The front-end total conversion time, T_{conv}, in Fig. 3 is $331\mu s$. The whole interface achieves a very low FOM of 5.076 pJ/Conversion, compared to the state of the art as listed in Table I.

VI. CONCLUSION

A new low-power interface circuit for Wheatstone bridge piezo-resistance pressure sensors is introduced, where it is proposed to periodically switch the sensor bridge, in addition to other system blocks. Furthermore, the turn ON time of the system blocks is optimized for minimum power dissipation. The proposed interface is implemented in $0.13\mu m$ CMOS technology. It consumes a 221.8 μW from a 1.2V supply

and completes a single conversion in 331 μs. The interface achieves an ENOB = 13.82 bits and a FOM of 5.076 pJ/Conv.

REFERENCES

[1] *PC Board Mountable Pressure Sensor Combination Temperature and Pressure Analog Output and IC or SPI Digital Output Gage, Differential, Absolute, Compound, Vacuum 3.3 or 5.0Vdc Supply Voltage*, Measurement Specialties.

[2] *BMP085 Digital pressure sensor*, Bosch Sensortec.

[3] E. Boujamaa, B. Alandry, S. Hacine, L. Latorre, F. Mailly, and P. Nouet, "A low power interface circuit for resistive sensors with digital offset compensation," in *Circuits and Systems (ISCAS), Proceedings of 2010 IEEE International Symposium on*, May 2010, pp. 3092–3095.

[4] A. Thanachayanont and S. Sangtong, "Low-voltage current-sensing cmos interface circuit for piezo-resistive pressure sensor," in *ETRI Journal*, Feb 2007, pp. 1–9.

[5] J. Van Rethy, H. Danneels, V. De Smedt, W. Dehaene, and G. Gielen, "An energy-efficient bbpll-based force-balanced wheatstone bridge sensor-to-digital interface in 130nm cmos," in *Solid State Circuits Conference (A-SSCC), 2012 IEEE Asian*, Nov 2012, pp. 41–44.

[6] R. Wu, J. Huijsing, and K. Makinwa, "A 21b ±40mv range read-out ic for bridge transducers," in *Solid-State Circuits Conference Digest of Technical Papers (ISSCC), 2011 IEEE International*, Feb 2011, pp. 110–112.

[7] M. Waltari, "Integration of high-speed cmos sample-and-hold circuits," Licentiate Thesis, Helsinki University of Technology, 1999.

[8] J. Mrkus, "Higher-order Incremental Delta-Sigma Analog-to-Digital Converters," Ph.D. dissertation, Budapest University of Technology and Economics, 2005.

[9] G. C. T. Richard Schreier, *Understanding Delta-Sigma Data Converters*.

[10] B. Jayaraman and N. Bhat, "High precision 16-bit readout gas sensor interface in 0.13 μm cmos," in *Circuits and Systems, 2007. ISCAS 2007. IEEE International Symposium on*, May 2007, pp. 3071–3074.

[11] J.-M. Park and S.-I. Jun, "A resistance deviation-to-time interval converter for resistive sensors," in *SOC Conference, 2008 IEEE International*, Sept 2008, pp. 101–104.

A practical algorithm using virtual point light for the rendering process

[1] khemliche sarra, [2]Babahenini Mohamed Chaouki, [3]BahiNaima, [4]ZerariAbd El Moumène

LESIA laboratory

Mohamed Kheider University

Biskra, Algeria

[1]s.khemliche@univ-biskra.dz

[2]chaouki.babahenini@gmail.com

[3]naima.bahi@yahoo.fr

[4]a.zerari@live.fr

Abstract— Developing robust visual tracking algorithms for real-world is still a major challenge today in many fields such as design, manufacturing, ecommerce and other fields are used to simulate the appearance of objects and scenes. Realistic looking is often critical, and calculating global illumination (GI) is an important contributor to image reliability; in which we take into account all interreflection effects in the scene.GI approximation methods, such as virtual point light (VPL) algorithms, are efficient, but they can produce image artifacts and distortions of object appearance. In this paper we present a practical algorithm for rendering diverse frequency of interreflection effects. Our method builds upon a standard representation of the scene, based on which a new mathematical formulation of the interreflection equation is made. We systematically study simulation of complex effects on image quality and material appearance of global illumination approximations made by VPL algorithms to provide a fast and smoother complex illumination .Efficiently, the computational effort is heavily reduced.

Keywords—global illumination; interreflection effects ; light source; virtual point light(vpl).

I. INTRODUCTION

The goal of computer graphics is to create images according to the purpose of the user. Photorealistic rendering is used when images indistinguishable from real photographs are desirable. This issue has a big impact on a wide range of various industry applications such as simulation, film production, and video games. One of the essential aspects of creating photorealistic images is that objects look as though made of real materials. Additionally, appearance of objects should match the scene lighting and be in accord with the scene configuration. This match requires accurate, physically-based simulation of light transport in the scene, carried out by, so-called, global illumination (GI) algorithms, which describes the light propagation through a 3D environment and its interaction with all the scene geometry. In contrast to direct or local illumination where only one bounce of light is considered, global illumination considers several bounces.

In our paper we introduce new system which aims at solving some problems mentioned before in less time consuming .our model proposal inspired by the concept of *Instant Radiosity* [1]. Instant Radiosity is a solution to approximate global illumination by tracing rays from the light source to the scene, letting them bounce around to a specified depth and creating secondary light sources at each intersection point which called virtual point light (VPL) this latter is at the core of our rendering system; The final results are then placed into the scene when doing the rendering at the client side. However, we extend it to simulate non- diffuse light paths as well.

Indirect Illumination is a key element to achieve realistic rendering. Unfortunately, since computing this effect is costly, there are few methods that render it with real-time frame rates. In this paper we present a new method based on virtual point

lights and topological information about the scene to render indirect illumination in real-time.

Our main contribution includes:

❖ Approximation of indirect illumination using VPL.

❖ Simulating light transport of different environment.

❖ Giving an automatic setting of scene standard.

The remaining paper is structured as follows: Section 2 gives an overview of related work done in the field of real time global illumination and other recent called many light approaches. In Sections 3 we describe our method in detail. Results and performance measurements are presented in Section 4. We will conclude our paper and also give outlooks into possible future work in Section 5

II. STAT OF THE ART

The problem of global illumination was first formalized mathematically by Kajiya [8]; Rendering Equation (RE) that calculates the equilibrium distribution of light energy in a scene. The rendering equation links light sources, surface reflectance (BRDFs), and visibility (fig.2). It says that at each particular point that has a position and direction on a surface, the outgoing light L_o at a surface location x in direction ω is the sum of emitted radiance L_e and reflected radiance L_r (Eq.1).

$$L_o(x, w) = L_e(x, w) + L_r(x, w) \qquad (1)$$

The reflected radiance is computed as

$$L_r(x, w) = \int_{\Omega^+} Li(x, w_i) f_r(x, w_i \to w) <N(x), w_i>^+ dw_i \quad (2)$$

Where Ω^+ is the upper hemisphere oriented around the surface normal $N(x)$ at x, f_r the bidirectional reflectance function (BRDF) and$< >^+$ a dot product that is clamped to zero. To determine the incident radiance, L_i, the ray casting operator is used to determine from which other surface location this radiance is emitted and reflected. The goal of global illumination algorithms is to compute $L_o(x, \omega)$ for a given scene; materials and lighting L_e [14]. It is a recursive equation which makes it unsolved analytically by exact method for this end several methods have been developed to solve the rendering equation.

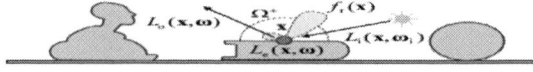

Figure.2 rendering equation

Global illumination is a set of techniques for approximating real-life lighting situations in computer graphics. State-of-the-art reference solutions of algorithms that are used for realistic image synthesis usually rely on; Monte carlo methods (e.g., ray tracing [16], metropolis light transport [15]), are general enough to allow light transport simulation in arbitrary scenes with all possible material types. However, those algorithms are too slow. Finite element methods such as radiosity [2], that can perform the rendering in real-time and also creates

978-1-4673-8760-6/15 $31.00 © 2015 IEEE

realistic images, but unfortunately the fast rendering is gained by an expensive pre-processing step, an enormous amount of storage and a limitation to static scenes, and the realism just includes indirect illumination, another category is density estimation methods (e.g., photon mapping [7]) that can be adapted to achieve interactive results under some conditions.

In contrast to those traditional algorithms, another category to do the rendering is the so-called **Many light rendering (MLR) [3].**They are attractive because they offer a simple solution to many difficult rendering problems. Their main idea is that the general light transport problem can be approximated by the simpler problem of calculating the direct illumination from many virtual light sources. MLR derived from the instant radiosity algorithm of Keller [1] that offers a unified mathematical formulation to simulate the light transport in a scene by computing the direct illumination from many virtual points light sources (VPLs), it models light as a source of energy which distributes its energy outward to several surfaces and replaces the computation of indirect diffuse illumination by direct diffuse illumination from these virtual point lights by the end summing up their contributions and accumulating them to render low-noise images covering a wide range of performance and quality goals .This basic idea was then formulated as the many-lights problem and inspired a series of works .While some VPL generations have low importance because they are in a very little region of interest i.e. .Energy conservation; in general, each VPL does not contribute equally, scalable algorithm tries to exploit important VPLs and unimportant ones to reduce computation cost. In order to do this, some methods have tried to cluster unimportant VPLs together (e.g., Lightcuts [17]) but However, visibility based on ray tracing has to be evaluated for every element in the cut and remains the computational bottleneck. One way to quickly create VPLs is by rendering a so called Reflective Shadow Map (RSM) [4]. The RSM technique builds on the classical shadow mapping technique, but augments the shadow map with material and surface normal information. However, RSMs do not provide a solution for secondary visibility: They allow computing surface elements that send the first bounce, but for every VPL (hundreds) another shadow map is required for secondary visibility.

The idea of Virtual Point Lights (VPLs) is to create a large number of point lights, each with a shadow map(SM), to simulate indirect illumination. So we need to render all these SMs for each vpl and this cause a big problem by choosing poor performance or temporal inconsistencies. Incremental instant radiosity [10] deciding which VPLs need to be removed and inserted to minimize temporal inconsistency. This approach allows for moving lights, but moving geometry leads to indirect shadows that lag behind the actual geometry. Depending on the frame rate and quality desired this can or cannot be acceptable.

The idea of IR was extended further by rendering so called imperfect shadow maps **[13]** for each VPL. Their idea is to create small and partial shadow maps for each VPL each frame, which they demonstrated gave little visual error. To do this, they created a point representation of the scene geometry, and for each imperfect shadow map they choose a subset of these points. This subset was then rendered to a texture, creating a coarse approximation of a shadow map, which was then improved in a push and pull step. By creating all these shadow maps in the same pass on the same texture resource, they were able to do this very quickly. These imperfect shadow maps were then used as shadow maps for the VPLs; in the same manner as the classical Instant Radiosity paper. This method can produce nice results in a few seconds but use hierarchy on light to reduce. Shading point or visibility testing

problem was interpreted as a matrix problem. Matrix row column sampling ([5], [6],[12]) samples sparsely rows and columns of the lighting matrix to reconstruct the image (rows represents shading points and columns represents lights).

VPLs have two inherent limitations. First When connecting the camera to a VPL, the BRDF is much larger than the probability density of generating the path. This limitation leads to spikes ([6], [9]). The closer a VPL is, the more the artefact will be. Even it was treated by using clamping way but it causes energy lose, which means bias. [9] Introduced an unbiased path tracing approach to compensate for the loss of energy. However, large computation penalty was also introduced. The other limitation is the difficult path When glossy materials exist in the scene, S*DS paths are inevitable since the paths are connected from a surface sample to a VPL. The main reason for these difficult paths is that the probability of generating a proper path is too small. By increasing the size of the path, we can increase the size of the path [3] or increase the size of the light **[8].** Of course, bias is also introduced. These approaches considerably reduce the computation cost, but are limited to a small number of VPLs and they are very inefficient at representing glossy light transport. It is only suitable for primarily diffuse materials; this is because a VPL on a glossy surface illuminates a small fraction of the scene only, and a huge number of VPLs might be necessary to render acceptable images.

III. PROPOSED APPROACH
A. System Overview

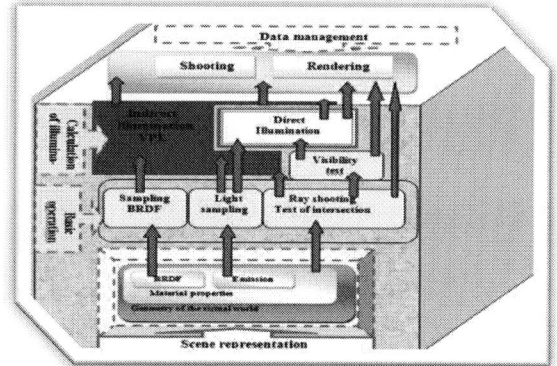

Figure.2.A scheme illustrates our system

Figure 2 illustrates the architectural of our system.

- **Scene representation :** All algorithms operate on a numerically defined virtual world that represent our scene, an internal model consisting of the geometry of the virtual world, optical material properties and the description of the lighting in the scene emission and reflection (BRDF) properties; Scene representation, which is described using Extensible Markup Language (XML) file. Sensors, light sources, objects and also the rendering technique (integrator) are described. In this way we are trying to make a standard script that model 3D scene which is alternate by the user in need.

- **Basic operations**: The basic operations the virtual world representation has to support are light source sampling, directional sampling according to the reflection properties (BRDF sampling), and the computation of the intersection of a light ray with the geometry (ray shooting).

- **Computation of illumination:** The computation of illumination can be separated into two parts direct and indirect. Direct illumination requires light source sampling and then a visibility test for every sample has to be performed. Unlike, indirect illumination is computed by generating multi-bounce light paths using random walks. This requires light source sampling, directional sampling and ray shooting. And computation of direct illumination is also required. As indirect illumination is usually more expensive and the most consuming time to evaluate, this is the part that is typically pre-computed; we transform each indirect light to a direct inspired by the idea of vpl. The defiance is how identifying those parts of the computation that are constant, or can be rapidly adapted to changes happening in the virtual world (scene).

- **Data management:** While finishing preparing all data (direct illumination evaluation, light tracing, ray shooting) that had been pre-computed which are needed to finally be gathering to render images. The main role of final gathering, however, is the validation and adaption of pre-computed information. This typically means checking whether a moving object has interrupted a light path: that is, visibility testing. This latter and other points are the most critical points that contribute to the image synthesis algorithm are ; approximation for visibility testing , ray shooting in general, ray tracing of visible reflective or refractive objects, improving sampling in random walk algorithms and in direct illumination, respectively should be evaluated efficiently to acquire a good, effective system to render smooth images .

B. Detailed description

1. Scene representation

Before light paths can be calculated, a scene and a viewpoint must be defined. Our scene is described using Extensible Markup Language (XML) file. Sensors, light sources, objects and also the rendering technique (integrator) are described. In this way we are trying to make a standard script that model 3D scene which is alternate by the user in need.

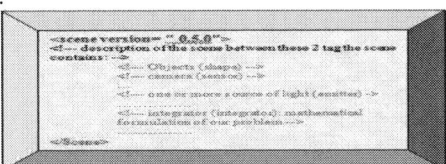

Figure.3. scene representation

After loading the scene which is well presented by an xml file, we trace rays from each light source to all of the scene objects, points, and creating virtual point lights (VPLs) at these points**(generating vpls)**. Randomly choose one of the primary light sources in the scene, sample a random position x and direction w (create a VPL at this location if direct illumination is not handled otherwise).Trace the ray if it intersects a surface then create a VPL at this intersection location. Decide randomly whether or not to terminate the path using Russian roulette. If continued, sample outgoing direction, update path throughput based on BRDF and direction, and continue tracing. Then, during the final shading pass, we evaluate each light (actual and virtual), calculating its weighted contribution to the scene, and shade the fragments appropriately .Every actual source of light will have unit intensity, which we distribute equally to all of the virtual light sources created from it. Thus, we obey the principle of transfer of energy, which is the basis of radiosity. The intensity of each light dictates the share of its contribution to the final shading.VPL generation pre-process, is equivalent to the direct illumination of the first camera hit point by the VPLs.

2. Basic operations

a) Sample generator

When rendering a scene, we have to solve a high-dimensional integration problem that involves all properties that make up the scene; the geometry, materials, lights, and camera. Because of the mathematical complexity of the integral used , it is generally impossible to solve them analytically, instead, they are solved numerically by evaluating the function to be integrated at a large number of different positions referred to as samples. Sample generators are an essential ingredient to this process: they produce points in a (hypothetical) infinite dimensional hypercube $[0, 1]^\infty$ that constitute the canonical representation of these samples. To do its work, a rendering algorithm, or integrator, will send many queries to the sample generator. (That is what arrows means in the diagram of our system that is mentioned before). The gain by using Halton sequence QMC is to decreases the effect of clumping in samples by eliminating randomness completely. Samples are deterministically distributed as uniformly as possible that can lead to a higher order of convergence in renderings. Because of the deterministic character of the samples, errors will manifest as grid armoire patterns rather than random noise, but these diminish as the number of samples is increased[12].we consider a set of points P. Consider each possible, axis-aligned box with one corner at the origin. Given a box of size B_{size}, the ideal distribution of points would have NB_{size} points.

The star discrepancy measure computes how much the point distribution P deviates from this ideal situation,

$$D^* N_{(P)} = \sup_B \left| \frac{N\,umPoints(P \cap B)}{N} - B_{size} \right|$$

where $NumPoints(P \cap B)$ are the number of points from the set P that lie in box B.

The Halton sequence (HS) in particular provides a very high quality point set that unfortunately becomes increasingly correlated in higher dimensions. HS based on the radical inverse function and are computed as follows. Consider a number i which is expressed in base b with the terms a_j :

$$i = \sum_{j=0}^{\infty} a_j(i)b^j$$

The radical inverse function Φ is obtained by reflecting the digits about the decimal point:

$$\phi\, b(i) = \sum_{j=0}^{\infty} a_j(i)b^{-j+1}$$

b) Ray shooting

Ray shooting is one of the expensive operations in the algorithm. The number of rays that needs to be shot in order to compute the distribution of energy in the scene to given accuracy with given confidence. In this case we refer to ray tracing is to trace many light paths using random walks starting from the sensor (camera). A single random walk entails casting a ray associated with a pixel in the output image and searching for the first visible intersection. A new direction is then chosen at the intersection, and the ray-casting step repeats over and over again (until one of several stopping criteria applies).At every intersection, the path tracer tries to create a connection to the light source in an attempt to find a complete path along which light can flow from the emitter to the sensor.

3. Calculation of illumination

Always referring to the rendering equation to equate the contribution for all vpl using for the rendering. For estimating the radiance that represents the light transport from L lights to S surface samples.

$$L(x, w_o) = \sum_{x_i \in \Omega} ghater.VPL(i) \quad (3)$$

$$L(x, w_o) = \sum_{s=1}^{\Omega} C_{LS} \quad (4)$$

Each vpl is represented by its incoming light (L) to hit a surface (S).

L(x, w_o) is the outgoing radiance of sample point and C is the contribution of light L to sample S, which can be decomposed Into $\qquad C_{LS} = M_{LS}G_{LS}V_{LS}I_{LS}$ (5)

Where M_{LS} , G_{LS} ,V_{LS} ,I_{LS} are material, geometry, visibility and intensity of light respectively.

To calculate the contribution for all vpls two operations should be done **sampling** and **shading** which need surface sampling, light and scene geometry. These are the main component that we are showed on our system which are represented by blocks. To fulfill a good process to create image with global illumination in real time, we are going to start with surface samples and light in which we use the best representation of lights for each sample, and then we prepare geometry to evaluate visibilities of sample-light (ray intersection), all these data are pre-computed for the final rendering (data management). To reduce the time in rendering pre-computation is needed, while loading geometry of the virtual world i.e. material properties (light, geometry) L and G, in relative to time we have the notation, TL and TG respectively.

$$\begin{cases} Light = L \rightarrow \sum L \\ Geometry = G \rightarrow \sum G \end{cases} \text{ including the factor of time}$$

$$\begin{cases} L = \sum TL \\ G = \sum TG \end{cases}$$

Light and geometry are dependent to represent material properties (MP)

$$PM = \sum TL + \sum TG \qquad (6)$$

The optimal data management is to minimize the time consuming we deal with MIN_{PM} (Eq. 7)

$$MIN_{PM} = min_{n_L,n_G}(\sum n_L TL + \sum n_G TG) \qquad (7)$$

Where n_L , n_G are the number of times that the algorithm loads light and geometry respectively . In order to achieve this we should take all computation with surface samples as well as all visibility tests before unloading it. Visibility testing may increase the occurrence of geometry because visibility is between geometry and light if G is visible or occluded from the camera point of view, it is preferable to test visibilities with as many sample-light rays as possible which will incur high number of light.

IV. RESULTS AND DISCUSSION

A. System used

our implementation is based on visual studio C++ language using the Visual Studio 2010 IDE and Python. Our system rendering output took place on a computer with i7-3517U processor with 4GB RAM and NVIDIA GT 635M/2GB. A table illustrates results of different integrator which is compared according to various property values

B. Results

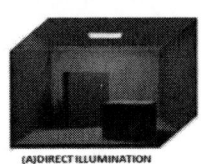  

(A)DIRECT ILLUMINATION   (B) PHOTON MAPPING   (C) INSTANT RADIOSITY

OUR INTEGRATOR

C. Evaluation and discussion

1. Used metrics

a) *Shadow map resolution :*Shadows are created by testing whether a pixel is visible from the light source, by comparing it to *depth* image of the light source's view, stored in the form of a texture

b) *Maximum depth:* Specifies the longest path depth in the generated output image (where -1 corresponds to ∞). A value of 2 will lead to direct-only illumination.

c) *Clamping factor:* used to control the rendering artifact discussed below

d) *Sample:* The number of samples per pixel specified to the sampler is interpreted as the number of VPLs that should be rendered.

e) *Filter :* Image reconstruction filters are responsible for converting a series of radiance samples generated jointly by the sampler and integrator into the final output image.

	Property value					
	Shadow Map resolution	Max depth	Clamping factor	Sample /pixel 64	Filter	Time
Only direct illumination	512*512			Low discrepancy	Gaussian	1.8 m
Photon mapping	512*512	∞		Low discrepancy	Gaussian	24.6 m
Instant radiosity	512*512	∞	0.1	Low discrepancy	Gaussian	1.07 m
Our approach	512*512	∞	0.5	Halton QMC	Gaussian	0.6 m

V. CONCLUSION

We have showed a detailed system aiming at approaching the various phenomena of interaction matter/light by inexpensive processing and less time consuming. As future work, we will investigate for improving other complex phenomena.

References

[1] A,Keller. (1997, August). Instant radiosity. In *Proceedings of the 24th annual conference on Computer graphics and interactive techniques* (pp. 49-56). ACM Press/Addison-Wesley Publishing Co..

[2] COHEN, M. F., WALLACE, J., AND HANRAHAN, P. 1993. Radiosity and realistic image synthesis. AcademicPress Professional, Inc., San Diego, CA, USA

[3] Dachsbacher, C., Křivánek, J., Hašan, M., Arbree, A., Walter, B., & Novák, J. (2014, February). Scalable Realistic Rendering with Many-Light Methods. In *Computer Graphics Forum* (Vol. 33, No. 1, pp. 88-104).

[4] Dachsbacher,C. and Stamminger, M. (2005). Reflective shadow maps. In I3D '05: Proceedings of the 2005 symposium on Interactive 3D graphics and games, pages 203–231, New York, NY, USA. ACM.

[5] Davidovič, T., Křivánek, J., Hašan, M., Slusallek, P., & Bala, K. (2010, December). Combining global and local virtual lights for detailed glossy illumination. In *ACM Transactions on Graphics (TOG)* (Vol. 29, No. 6, p. 143). ACM.

[6] Hašan, M., Pellacini, F., & Bala, K. (2007, August). Matrix row-column sampling for the many-light problem. In *ACM Transactions on Graphics (TOG)* (Vol. 26, No. 3, p. 26). ACM.

[7] Jensen, H. W. (1996). Global illumination using photon maps. In *Rendering Techniques' 96* (pp. 21-30). Springer Vienna.

[8] Kajiya, J. T. (1986, August). The rendering equation. In *ACM Siggraph Computer Graphics* (Vol. 20, No. 4, pp. 143-150). ACM.

[9] Kollig, T., & Keller, A. (2006). Illumination in the presence of weak singularities. In *Monte Carlo and Quasi-Monte Carlo Methods 2004* (pp. 245-257). Springer Berlin Heidelberg.

[10] Laine, S., Saransaari, H., Kontkanen, J., Lehtinen, J., and Aila, T. (2007). Incremental instant radiosity for real-time indirect illumination. In Proceedings of Eurographics Symposium on Rendering 2007, pages xx–yy. Eurographics Association.

[11] Ou, J. & Pellacini, F. (2011). Lightslice: matrix slice sampling for the many-lights problem.In Proceedings of the 2011 SIGGRAPH Asia Conference, SA '11, 179:1–179:8, ACM.37

[12] P ,Dutre. Bekaert, P., Bala, K., & (2006). Advanced global illumination.

[13] Ritschel, T., Eisemann, E., Ha, I., Kim, J. D., & Seidel, H. P. (2011, December). Making Imperfect Shadow Maps View-Adaptive: High-Quality Global Illumination in Large Dynamic Scenes. In *Computer Graphics Forum* (Vol. 30, No. 8, pp. 2258-2269). Blackwell Publishing Ltd.

[14] T,Ritschel. C ,Dachsbacher.T, Grosch, & Kautz, J. (2012, February). The state of the art in interactive global illumination. In *Computer Graphics Forum* (Vol. 31, No. 1, pp. 160-188). Blackwell Publishing Ltd.

[15] Veach, E., & Guibas, L. J. (1997, August). Metropolis light transport. In *Proceedings of the 24th annual conference on Computer graphics and interactive techniques* (pp. 65-76). ACM Press/Addison-Wesley Publishing Co..

[16] WALD I., KOLLIG T., BENTHIN C., KELLER A SLUSALLEK P.: Interactive global illumination using fast ray tracing In *Proc EGRW*(2002) pp15–24

[17] Walter, B., Khungurn, P., & Bala, K. (2012). Bidirectional lightcuts. *ACM Transactions on Graphics (TOG)*, 31(4), 59.

Memristor Emulator Based On Single CCII

Abdullah G. Alharbi[1], Zainulabideen J. Khalifa[2],Mohammed E. Fouda[3] and Masud H. Chowdhury[1]

[1]Computer Science and Electrical Engineering, University of Missouri Kansas City, Kansas City, MO 64110, USA
[2]Electrical Engineering, King Fahad University of Petroleum and Minerals, Saudi Arabia, 31261
[3]Engineering Mathematics and Physics Dept., Faculty of Engineering, Cairo University, Egypt, 12613
Email: a.g.alharbi@ieee.org, zainkhalifa@kfupm.edu.sa,m_elneanaei@ieee.org, masud@ieee.org

Abstract—With the introduction of the memristor design by the HP lab, there has been a surge of interest to perform different theoretical and experimental works to investigate various potential applications of the memristor. However, since the memristor is not available commercially, it is essential to design the emulator circuit for the memristors from the available electronic components. In this paper, we present the validation of a previously published memristor emulator through experimentation and introduce a simpler memristor emulator circuit that is built using only one second generation current conveyer (CCII), two diode-connected transistors, a capacitor and some resistors. The proposed circuit satisfies the fingerprints of the memristor in the I-V plane accurately. The circuit is very simple to implement in the lab for educational purpose. Furthermore, the mathematical model and analysis of the proposed emulator circuit are introduced. SPICE simulation and experimental results are also provided.

I. INTRODUCTION

Memristor is the fourth two terminal passive element after resistor, R, inductor, L, and capacitor, C. Memristor was introduced theoretically by Leon Chua in 1971 [1]. He showed that no combination of the well-known three two terminal passive elements (R, L and C) could be able to duplicate the characteristics of memristor. Even though having this unique property, memristor was restricted only in theory till 2008, when HP team successfully built the physical structure of memristor using $TiO2$ [2]. Memristor acts like a memory since it is charged by applied voltage and doesn?t discharge with removal of voltage and it preserves its pre-charged state. The unique fingerprint of the memristor is its pinched hysteresis loop in the I-V plane comparing with the other three two terminal devices such as resistor, inductor and capacitor (R, L, C). Furthermore, some substantial fingerprints, also known as the signature of a memristor are required for any device to call it memristor [3].

Now-a-days, several papers are introduced from the research community to study the dynamic nature and applications of memristor. Memristor has been used in different applications including implementing high speed memory arrays like RRAM, analog and digital circuits, sinusoidal and relaxation oscillators, neuromorphic circuits and adaptive filters [4], [5].

Because of the high cost and structural complexity of the components, the physical memristors are not available in the market yet. Hence, simple and precise emulator circuits and SPICE macro models are required to imitate the characteristics of real memristor. Even though, numerous SPICE macro models and emulator circuits have been proposed, some of these

models and circuits do not exhibit the three fingerprints of memristor. Several micro SPICE models have been proposed [6]–[8]. SPICE model of memristor with nonlinear dopant drift is reported in [6]. However, these models have many drawbacks those need to be fixed. So, many different SPICE models have been introduced to mimic the behavior of the fabricated memristors [5].

Recently, several emulator circuits have been proposed and constructed using off-the-shelf active and passive devices which are commercially available in the market [9]–[12]. For instance, the memristor emulator presented in [9] uses Microcontroller, ADC and DAC blocks. This emulator is topologically complex which limits their application in connecting with active and passive devices. In addition, a memristor emulator circuit based on Operational Transconductance Amplifier has been proposed in [10]. However, the experimental results of this emulator circuit do not satisfy the condition of memristor. Also, the hysteresis loop area does not decrease as frequency increases. Also, a CMOS based memristor emulator has been introduced in [13]. Whereas the limitation and main drawback of these emulator circuits are they are complex and not suitable for all real world hardware applications. So, here, we are validating a simple memristor emulator and introducing a new simpler one which is built based only on one CCII and two transistors.

This paper is organized as follows. In section II, the emulator circuit, the mathematical modeling and the experimental results are introduced. In section III, mathematical analysis and PSPICE simulation for reduced memristor emulator circuit are presented. Finally, Section V concludes the paper and gives the future work.

II. MEMRISTOR EMULATOR: ANALYSIS AND VALIDATION

The emulator has been designed with an integrator and an exponential amplifier. The integrator is built using CCII+, capacitor and resistor as shown in Fig.1. In addition, the exponential amplifier is used to achieve the required nonlinearity for memristive behavior.

A. Mathematical Analysis of The Proposed Emulator

In the integrator circuit, the input current, i_{in} , is created by subtracting the feedback voltage, V_{fb} from the input voltage,

978-1-4673-8760-6/15 $31.00 © 2015 IEEE

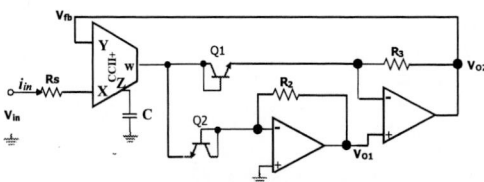

Fig. 1. The proposed emulator circuit.

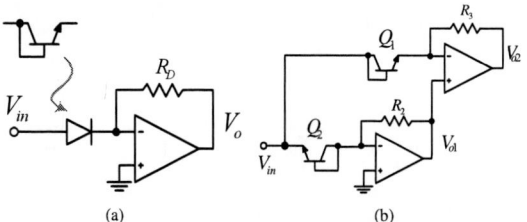

(a)   (b)

Fig. 2. Exponential amplifier (a) single-sided, and (b) double sided.

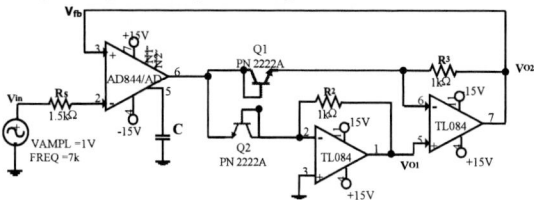

Fig. 3. The schematic diagram of the proposed memristor emulator.

V_{in}. Then, this current is imposed in the capacitor. So, the voltage across the capacitor is given by (1)

$$V_c = \frac{1}{C}\int_0^t i_{in}(\tau)d\tau = \frac{q_{in}}{C} = \frac{1}{CR_s}\int_0^t (V_{in} - V_{fb})d\tau \quad (1)$$

A single-sided exponential amplifier is usually built as shown in Fig. 2(a), which consists of one Op-Amp, resistor and diode. The output voltage is exponentially proportional to the positive input voltage and zero for the negative input voltage. The input-output relation of the exponential amplifier is given by (2)

$$V_o = I_{ES}R_D\left(1 - e^{\frac{V_{in}}{V_T}}\right) \approx -I_{ES}R_D e^{\frac{V_{in}}{V_T}} \quad (2)$$

where I_{ES} is the reverse saturation current, which is in order of $10^{-13}A$ and V_T is the threshold voltage. In result, in order to design symmetric non-linear circuit two single-sided exponential amplifiers with reversed connected diode are used as shown in Fig. 2(b). The output of the second Op-Amp, V_o2, is given by (3) for $R_2 = R_3 = R_D$.

$$V_{o2} \approx -I_{ES}\left(R_3 e^{\frac{V_{in}}{V_T}} - R_2 e^{\frac{-V_{in}}{V_T}}\right) \quad (3)$$

The circuit is connected as shown in Fig. 1 using the integrator and exponential amplifier. So, the voltage across the capacitor is applied to the exponential amplifier, which creates the feedback voltage, V_{fb}. By substituting (1) into (3), the feedback voltage can be given by (4)

$$V_{o2} = -2I_{ES}R_D \sinh\left(\frac{q_{in}}{CV_T}\right) \quad (4)$$

Moreover, the input voltage is a function of the feedback voltage and the input current, which can be expressed as is $V_{in} = V_{fb} + i_{in}R_s$. Hence, from (4) we can derive the expression of the input voltage as in (5).

$$V_{in} = i_{in}R_s - 2I_{ES}R_D \sinh\left(\frac{q_{in}}{CV_T}\right) \quad (5)$$

The memristance, $R_m = V(t)/i(t)$, can be given by (6).

$$R_m = R_s - \frac{2I_{ES}R_D}{i_{in}} \sinh\left(\frac{q_{in}}{CV_T}\right) \quad (6)$$

The $sinh$ function can be first-order approximated as $sinh(x) = x + ...$ So, the memristance can be simplified to

$$R_m = R_s - \frac{2I_{ES}R_D}{CV_T}\left(\frac{q_{in}}{i_{in}}\right) \quad (7)$$

It is obvious that the memristance R_m is a function of the total input current and the charge $q(t)$.

III. EXPERIMENTAL VALIDATION

The proposed emulator circuit shown in Fig.3 has been implemented in the lab with one AD844 (current feedback operational amplifier) as second-generation current conveyors (CCII+). In addition, the exponential amplifier is implemented using two conventional Op-Amps TL084 and two transistors (implemented using PN2222A). Here, the used values of the passive elements are $R_s = 1.5K\Omega$, and $C = 10nF$, and the DC supply voltages = $\pm 12V$. The applied voltage signal has an amplitude of 2V. In order to draw the I-V curves, the

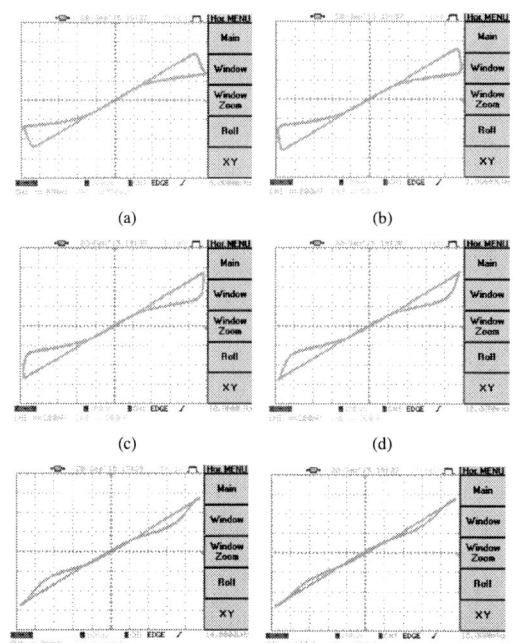

(a)   (b)

(c)   (d)

(e)   (f)

Fig. 4. Experiment results of the pinched hysteresis loop of proposed memristor circuit with different frequencies (a) 6KHz, (b) 8 KHz, (c) 10 KHz, (d) 12 KHz (e) 14 KHz and (f) 15 KHz.

978-1-4673-8760-6/15 $31.00 © 2015 IEEE

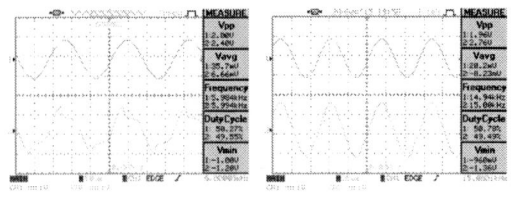

(a)   (b)

Fig. 5. Waveform of input voltage v(t) (yellow) and the input current i(t)(blue) for the proposed memristor at (a) 6KHz and (b)15KHz

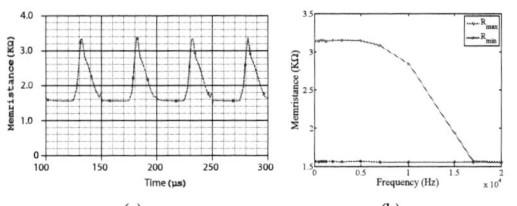

(a)   (b)

Fig. 6. (a) Transient memristance at 10KHz,(b) Maximum and Minimum achievable resistance with the frequency of the proposed emulator.

current was sensed using an instrumentation amplifier sensing the differential voltage across the resistance R_s with gain of (1+50k/300).

Previously, in [11], we showed the PSPICE simulation results of this emulator. Here, we are introducing the experimental results as shown in Fig. 4. by using the off-the-shelf components which are available in the market. It is worthy to mention that the input current has been integrated non-ideally since we have used a resistor R_1 having value of $2K\Omega$ in parallel with the capacitor C. Our circuit is still valid even though the integrator is non-ideal.

It is clearly evident that the proposed emulator circuit can mimic the behavior of memristor as shown in Fig.4. The signature of memristor which is pinched hysteresis loop shows large area at low frequency in the I-V plane. However, when we keep increasing the frequency, the lobe area starts to shrink, as a result, the emulator circuit tend to behave as pure resistor. This phenomena is known as one of the three signatures of memristor [14].

Fig.5 demonstrates the time domain waveforms of the voltage and current of memristor where the current and voltage has the same phase shift. Hence, the obtained memristor is purely resistive element. Also, Fig. 6(a) shows the change in the memristance because of the applied sinusoidal signal, where the memristance varies from $1.7K\Omega$ up to $3.45K\Omega$. From Fig. (6b) it can be noticed that initially, the value of R_{max} was very high (approximately $3.2K\Omega$) at a very low frequency. Then, it started to decrease sharply as the frequency of the applied signal is increased and at around 1.7 KHz, it became very low (almost $1.6K\Omega$) to merge with R_{min}. R_{min} was almost constant at $1.6K\Omega$ for a range of frequency from 0 to 2 KHz.

IV. REDUCED PROPOSED EMULATOR CIRCUIT

The proposed emulator circuit, shown in Fig.7, is the modified and improved version of the original emulator circuit

discussed in the previous. Here, we reduce the number of components in our previous work to make the circuit simple and easy to implement [12].

A. Mathematical Analysis

The current passing through the upper and lower diode-connected transistors are I_1 and I_2 respectively.

$$I_1 = I_{ES}\left(e^{\frac{V_x - V_{fb}}{V_T}} - 1\right), I_2 = I_{ES}\left(e^{-\frac{V_x - V_{fb}}{V_T}} - 1\right) \quad (8)$$

The total current passing through R_1 creating the feedback voltage V_{fb}

$$V_{fb} = (I_1 + I_2)R_1 = 2I_{ES}\left(R_1 \cosh\left(\frac{V_x - V_{fb}}{V_T}\right) - 1\right) \quad (9)$$

It is seen that this relation is implicit relation of V_{fb}, in order to separate the variables to obtain an explicit relation of V_{fb}. The second-order approximation of $cosh$ is used. So,

$$V_{fb} = 2I_{ES}R_1\left(1 + \frac{V_x^2}{V_T^2} - \frac{2V_x V_{fb}}{V_T^2} + \frac{V_{fb}^2}{V_T^2}\right) \quad (10)$$

Using the quadratic formula then

$$V_{fb} = \frac{V_T^2}{4I_{ES}R_1} + V_1 \pm 0.5V_T\sqrt{\frac{V_T^2}{4I_{ES}R_1} + \frac{2V_x}{I_{ES}R_1}} \quad (11)$$

The input current is injected to the first capacitor C to generate V_C. Thus,

$$V_x = \frac{1}{C}\int_0^t i_{in}(\tau)d\tau = \frac{q}{C} \quad (12)$$

By substituting (12) into (11), the input voltage is given as follows

$$V_{in} = i_{in}Rs + \frac{V_T^2}{4I_{ES}R_1} + \frac{q}{C} \pm 0.5V_T\sqrt{\frac{V_T^2}{4I_{ES}^2R_1^2} + \frac{2q}{I_{ES}R_1C}} \quad (13)$$

Then, the memristance is given by (14)

$$R_m = Rs + \frac{1}{i_{in}}\left(\frac{V_T^2}{4I_{ES}R_1} + \frac{q}{C} \pm 0.5V_T\sqrt{\frac{V_T^2}{4I_{ES}^2R_1^2} + \frac{2q}{I_{ES}R_1C}}\right) \quad (14)$$

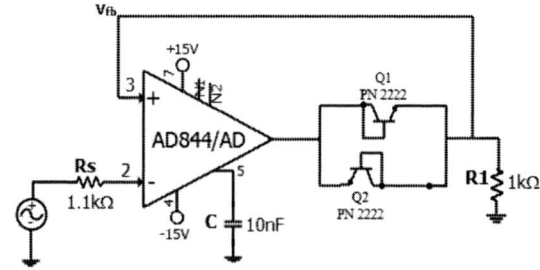

Fig. 7. The schematic diagram of the proposed memristor emulator

978-1-4673-8760-6/15 $31.00 © 2015 IEEE

B. SPICE Validation of The Circuit

The proposed circuit has been implemented with AD844AN (constructed by the commercial AD844 current feedback operational amplifiers) as second-generation current conveyors (CCII+), two diode-connected transistors which is built using PN2222A and some passive elements as shown in Fig. 7. The used values of the passive elements are $R_s = 1.1K\Omega$, $R_1 = 1K\Omega$, and $C = 10nF$. It is important to say that we use a resistor in parallel with the capacitor and the value of this resistor is 900Ω. As we mentioned before the integrator is non-ideal. From the results of Fig. 8, we observe that the proposed emulator can mimic the behavior of memristor perfectly. It is remarkable to mention that the hysteresis loop is restricted in the first and third quadrants with symmetrical behavior. Hence, we can conclude that our proposed emulator acts a passive element, which satisfies Chua's condition in [1], [14].

Moreover, the pinched hysteresis lobe area clearly seems larger at low frequencies whereas the area of lobe disappears at high frequencies. Furthermore, we observe that when the frequency tend to infinity, the pinched hysteresis loop becomes a straight line. This behavior validates the unique property of a memristor, which is known as the frequency dependence of pinched hysteresis loops. Fig.9. illustrates that the memristance varies with time for applied sinusoidal signal with amplitude 1V and frequency 10 KHz. The memristance changes from $0.7K\Omega$ up to $1.12K\Omega$.

V. CONCLUSION AND FUTURE WORK

In this paper, we have introduced the experimental validation of our previously published emulator. In addition, a simpler emulator circuit was introduced. Moreover, these emulator circuits have been comprehensively studied via theoretical analysis, experimental and simulation realization. As noticed that our circuits are very simple and practical comparing with other emulator circuits those have been reported in the literature. Furthermore, these emulator circuits can be used for designing memristor application circuits such as in implementing various analog circuits and educational purposes. In our future work, we will address the reliability and scalability of these circuits in some applications such as relaxation and chaotic oscillators.

REFERENCES

[1] L. O. Chua, "Memristor-the missing circuit element," *Circuit Theory, IEEE Transactions on*, vol. 18, no. 5, pp. 507–519, 1971.

[2] D. B. Strukov, G. S. Snider, D. R. Stewart, and R. S. Williams, "The missing memristor found," *nature*, vol. 453, no. 7191, pp. 80–83, 2008.

[3] S. P. Adhikari, M. P. Sah, H. Kim, and L. O. Chua, "Three fingerprints of memristor," *Circuits and Systems I: Regular Papers, IEEE Transactions on*, vol. 60, no. 11, pp. 3008–3021, 2013.

[4] A. G. Radwan and M. E. Fouda, "On the mathematical modeling of memristor, memcapacitor, and meminductor," 2015.

[5] R. Kozma, R. E. Pino, and G. E. Pazienza, *Advances in neuromorphic memristor science and applications*. Springer Science & Business Media, 2012, vol. 4.

[6] Z. Biolek, D. Biolek, and V. Biolkova, "Spice model of memristor with nonlinear dopant drift," *Radioengineering*, vol. 18, no. 2, pp. 210–214, 2009.

[7] A. G. Alharbi, M. E. Fouda, and M. H. Chowdhury, "Memristor emulator based on practical current controlled model," in *Circuits and Systems (MWSCAS), 2015 IEEE 58th International Midwest Symposium on*. IEEE, 2015, pp. 1–4.

[8] H. Abdalla and M. D. Pickett, "Spice modeling of memristors," in *2011 IEEE International Symposium of Circuits and Systems (ISCAS)*, 2011.

[9] Z. Kolka, V. Biolkova, and D. Biolek, "On hybrid emulation of memsystems," in *Proceedings of the 2014 European Modelling Symposium*. IEEE Computer Society, 2014, pp. 490–494.

[10] M. Kumngern and P. Moungnoul, "A memristor emulator circuit based on operational transconductance amplifiers," in *Electrical Engineering/Electronics, Computer, Telecommunications and Information Technology (ECTI-CON), 2015 12th International Conference on*. IEEE, 2015, pp. 1–5.

[11] A. G. Alharbi, M. E. Fouda, and M. H. Chowdhury, "A novel memristor emulator based on an exponential amplifier and a ccii+," in *2015 IEEE International Conference on Electronics, Circuit and System (ICECS)*. Cario, Egypt, 2015, Accepted.

[12] A. G. Alharbi, Z. J. Khalifa, M. E. Fouda, and M. H. Chowdhury, "A new simple emulator circuit for current controlled memristor," in *2015 IEEE International Conference on Electronics, Circuit and System (ICECS)*. Cario, Egypt, 2015, Accepted.

[13] A. Hussein, M. E. Fouda *et al.*, "A simple mos realization of current controlled memristor emulator," in *Microelectronics (ICM), 2013 25th International Conference on*. IEEE, 2013, pp. 1–4.

[14] L. Chua, "If its pinched its a memristor," in *Memristors and Memristive Systems*. Springer, 2014, pp. 17–90.

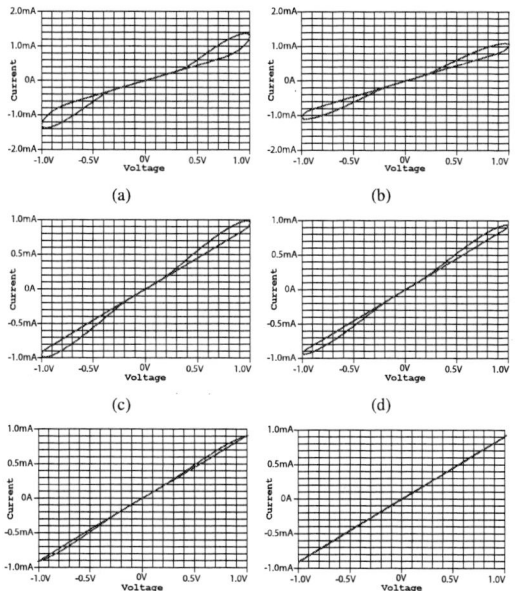

Fig. 8. Simulation of the current-voltage relation of proposed memristor circuit with various frequencies (a) 3 kHz, (b) 7 kHz, (c) 10 KHz, (d) 12 KHz, (e) 15 KHz and (f) 20 KHz.

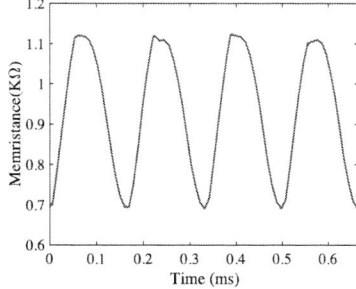

Fig. 9. Transient memristance of the second emulator circuit at 10KHz

A Novel TSV-Based Power Harvesting System for Low-Power Applications

Khaled Salah

Mentor Graphics

Cairo, Egypt

Khaled_mohamed@mentor.com

Abstract—This paper presents a Novel TSV-Based power harvesting system for low-power applications. The proposed system can harvest the energy from the coupling power between the signal through silicon vias (TSVs) in the 3D-SoC. The proposed system consists of three modules: energy source, energy harvesting which is based on TSV-based patch antenna, and power management module. The energy harvester module consists of three sub-modules: TSV-based patch antenna, 8-stage voltage multiplier and regulator, and energy storage. TSV technology is used to build an antenna on high resistivity substrate to be used as a transducer for the energy harvester system. The characteristic of this antenna is a function of TSV diameter, TSV length, and silicon resistivity. Compared to conventional on-chip antennas which suffers from low gain and low radiation efficiency, our novel antenna provides better performance parameters such as higher radiation pattern efficiency and higher gain as the antenna delivers a high gain of 5.8 dBi and the radiation efficiency is 86% over the prescribed range of frequencies. The proposed antenna is centered at 2.4 GHz with 200 MHz bandwidth. The overall area of the proposed antenna is 400µm×100µm. Building the antenna using TSV technology not only improves the performance, but also improves the isolation between the antenna and other active circuits. Simulations show that we are able to obtain dc voltage up to 2.5 V which is suitable for many low-power applications.

Index Terms — Three-Dimensional ICs, Through Silicon Via, TSV, Antenna, Insertion Loss, Energy, Harvesting.

I. INTRODUCTION

For decades, semiconductors manufacturers have shrinking transistor size in ICs to achieve the yearly increases in performance described by Moore's Law, but interconnect delay dominate with scaling. The TSV is an innovative vertical electrical connection passing completely through a silicon substrate. By using TSV technology, three-dimensional (3D) integrated circuits can be built with better performance and small area, where it reduces

parasitics and losses and occupies less die area by making use of the z-direction [1]- [5].

Antenna is very important and critical element in many RF applications, wireless communications, integrated passive devices (IPD), and power harvesting systems. High-gain antenna is required to overcome the limited output power of the silicon-based transmitter and the great free-space propagation loss at very high frequencies as the output power of transmitter antenna and the receiver antenna sensitivity are limited by the breakdown voltage and noise factor of CMOS transistors [6]- [10].

Batteries improvement is relatively slowly. Sensor size is limited by battery size and recharging is an issue. Clever power management and circuit design techniques such as clock scaling and adiabatic computing can reduce power requirements. We want to move from Battery-based solution to "Forever"-based solution. This can done by extracting power from ambient sources which is known as energy harvesting, or energy scavenging.

Mainly, there are many ambient sources for energy harvesting: **Light-based**: such as solar cells, **RF-based**: such as Rectenna which extracts power from cellular telephones, Wireless internet (WiFi), AM and FM radio, **Thermal-based**: body-heat based harvesting, **Motion-based** (including fluids, wind, and mechanical): such as dams, energy-harvesting shoes, **Acoustic-based** energy harvesting. Choice of harvesting approach is application-dependent as there are different power requirements for different applications [11]- [12]. Self-powering devices are a hot topic in VLSI industry to replace the need for batteries.

Our contributions as compared to previous related publications are (i) introducing a new concept of using already-existing coupling energy between TSVs as a source of energy, (ii) introducing a new architecture for an on-chip patch antenna based on TSV technology with smaller area and better performance which can be used in a power harvesting system. This antenna is used, as a core for power harvesting as portable and long-lasting devices have a huge market demand. The source of the energy that is being harvested is the coupling energy between TSVs in any 3D-SoC.

This paper is organized as follows. In Section II, the proposed architecture for the TSV-based power harvesting system is analyzed. In Section III, simulation results are discussed. Conclusions are given in Section IV.

978-1-4673-8760-6/15 $31.00 © 2015 IEEE

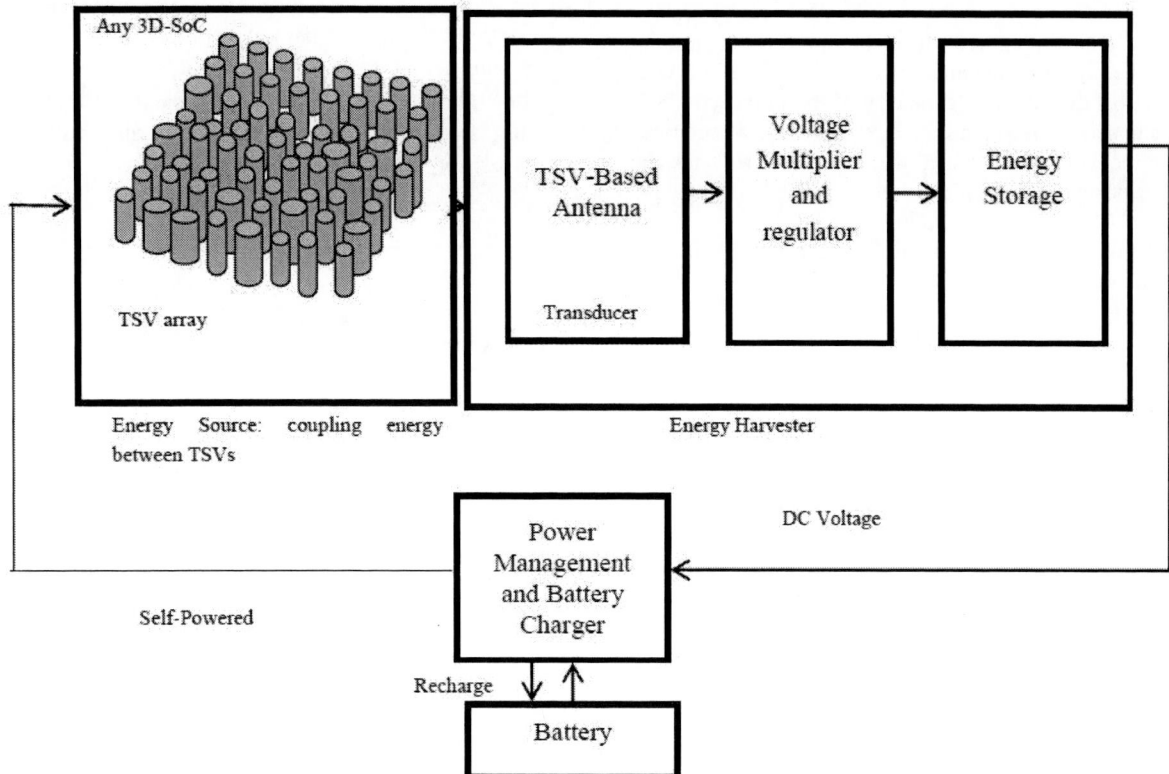

Fig. 1 TSV array is a part of a 3D TSV-based SoC. TSV-Based Antenna is placed close to this array to work as transducer to catch the RF. This power is then entered to voltage multiplier to convert it to DC and then it is stored to be used in the power management block of the Soc. Battery is needed for two reasons: for power on reset and if the energy is not sufficient to support the required operation, the gap is filled by the battery.

II. THE PROPOSED TSV-BASED POWER HARVESTING SYSTEM

In this section, the proposed TSV-based power harvesting system is introduced and analyzed. The submodules of these architectures are also analyzed. Moreover, the simulation results are discussed.

A. The Proposed Architecture

The power harvesting system consists of three modules as depicted in Fig. 1: energy source, which is the coupling energy between TSVs in any 3D-SoC, energy harvesting which is based on TSV-based patch antenna, and power management module. The energy harvester module consists of three sub-modules: TSV-based patch antenna, 8-stage voltage multiplier and regulator, and energy storage.

TSV array is a part of 3D TSV-based SoC. TSV-Based Antenna is placed close to this array to work as transducer to catch the coupling energy between the TSVs.

TSV is used to build the antenna as it provides smaller area and better performance. This power is then entered to voltage multiplier to convert it to DC and then it is stored to

be used in the power management block of the SoC. We need energy storage element to prevent energy from back-feeding into the transducer. Battery is needed for two reasons: for power on reset and if the energy is not sufficient to support the required operation, the gap is filled by the battery. Moreover, the battery can be recharged by the harvested energy. Some power needs for small electronics are shown in TABLE I. The details of the designed components are shown in the subsequent sections.

TABLE I
POWER NEEDS FOR SMALL ELECTRONICS

Application	Needed power
Desktop	100 W
Laptop	10 W
GSM	1 W
MP3	100 mW
Bluetooth	10 mW
Hearing aid	100 μW
RFID Tag	10 μW
Watch/ Calculator	1 μW
Quartz oscillator	100 nW

B. The Proposed TSV-Based Antenna

WiFi is the band which have the most power density around us. So, The proposed antenna is centered at 2.4 GHz with 200 MHz bandwidth. The geometry of the proposed TSV-based antenna is shown in Fig. 2, where a TSV penetrates through the silicon substrate, and connects *M1* and the ground beneath the silicon substrate. The TSV dimensions from the ITRS are summarized in TABLE II [13].
The insertion loss for different TSV diameter, TSV length, and silicon resistivity is studied.

TABLE II
ITRS ROADMAP FOR TSV

Parameters	2009-2012	2012-2015
Minimum TSV diameter	4-8 µm	2-4 µm
Minimum TSV pitch	18-16µm	4-8µm
Minimum TSV depth	20-50µm	20-50µm
Minimum TSV aspect ratio	1:10	1:20
Minimum contact pitch	20 µm	10 µm
Number of tiers	2-3	2-4

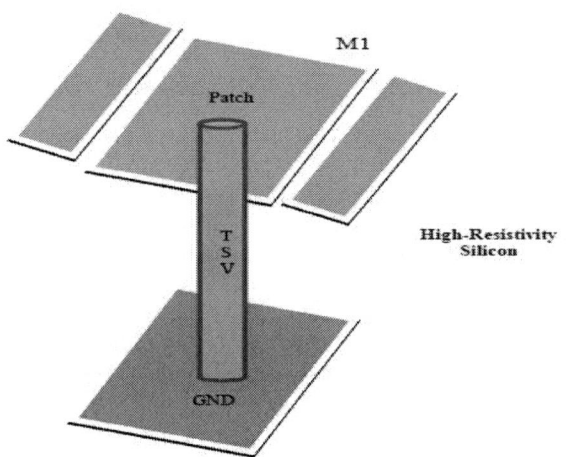

Fig. 2 The proposed TSV-based patch Antenna, with high resistivity substrate and a TSV penetrates through the silicon substrate, and connects M1 and the ground beneath the silicon substrate.

C. The Voltage Multiplier

For voltage multiplier, we need a passive solution to convert RF power to DC power, such as Half-wave voltage multiplier (Villard cascade). Voltage doubler has Schottky diodes due to their low open junction capacitance, low threshold voltage, low drop voltage, and fast switching capabilities, which speeds the high-frequency RF signal rectification process. These diodes direct current flow between successive capacitors. Cascading these sections can yield significant gains. The proposed voltage multiplier is shown in Fig. 3. The DC output voltages obtained through simulation is 2.5 V. When the number of stages increase, the output voltage also increases. The capacitor value used here is 1 nF.

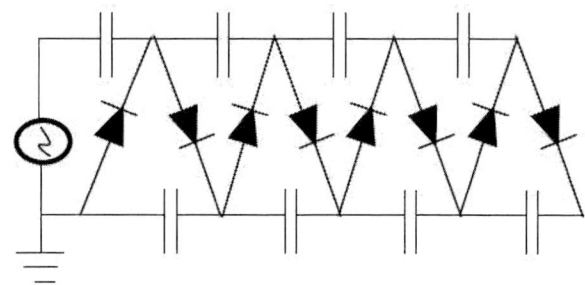

Fig. 3 N-stage Voltage multiplier.

D. Voltage Regulator

Since the output voltage of the voltage multiplier depends of the load current and the RF input at each moment, a regulator is necessary, to assure that the device powered by this system has the correct level of voltage to have a proper functioning. The voltage regulator cannot be linear, since linear voltage regulators use Zener diodes which are dissipative which is not allowed in energy harvesting systems as these losses will be very significant, so the voltage regulator must be a switching voltage regulator [19].

E. Energy Storage Element

Considerations must be made about using energy storage elements, like batteries or supercaps. To use these storage elements as efficient as possible, care must be taken when choosing the properties of them. The supply voltage level of the application in comparison to the voltage level of the storage element itself is very important. Especially when using batteries, high battery voltages together with low supply voltages of the circuits should be avoided.

F. Power Management

The output of the harvester is not directly suited as power supply for circuits because of variations in its power and voltage over time, so a power management circuit is required. This power management module should be able to handle very low feeding power and be able to adapt its input to the energy harvester and its output to the load. It should also be self-starting and this is can be done using a battery [14] - [16].

978-1-4673-8760-6/15 $31.00 © 2015 IEEE

III. RESULTS

Some of the important results from electromagnetic simulations using ANSOFT HFSS, [17], are summarized in the followings:

1) The Antenna is providing acceptable insertion loss (~ 0.5 dB) and return loss (~ -25 dB) if the silicon resistivity is higher than 1000 Ohm.cm (Fig. 4).
2) The radiation efficiency is 86% and the antenna delivers a high gain of 5.8 dBi.
3) The proposed antenna is centered at 2.4 GHz with 200 MHz bandwidth.
4) The overall area of the proposed antenna is 400μm × 100μm.

Moreover, the coupling power between the TSVs of a typical chip is sufficient to merit harvesting and simulations show that we are able to obtain dc voltage up to 2.5 V with a power of 10 mW which is suitable for many low-power applications.

There is no cost for using this system, as we assume that the coupling network between the TSVs is already existing in most of the modern 3D-SoC.

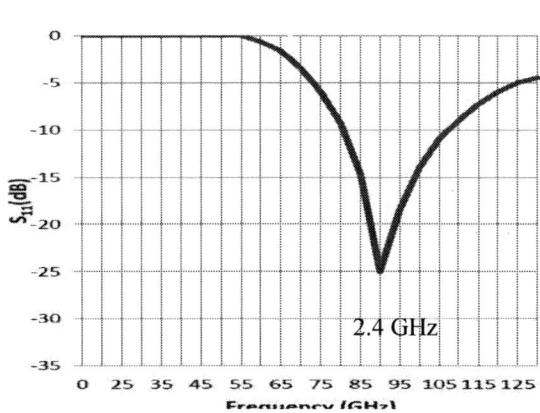

Fig. 4 The proposed antenna return loss S_{11}, where the Antenna is providing acceptable return loss if the silicon resistivity is higher than 1000 Ohm.cm.

IV. CONCLUSIONS

Energy harvesting has become one of the prominent design techniques to keep high battery life. In this paper, a Novel TSV-Based power harvesting system for low-Power applications. A high-efficiency, high performance, and high-gain On-Chip TSV-Based antenna is proposed and developed. The proposed system can harvest the energy from the coupling power between the signal TSVs. The primary advantage of using TSVs for antenna design is that a reduction of the silicon footprint of the antenna is possible by making use of the third dimension extending into the silicon substrate. The EM simulation results show that 200 MHz bandwidth, 5.8dBi gain and 86% efficiency can be achieved with small footprint. These good characteristics of

the proposed antenna indicate that TSV/3D integration is a promising solution for RF applications. Simulations show that the proposed architecture is able to obtain voltage up to 2.5 V which is suitable for many low-power applications.

REFERENCES

[1] W. R. Davis, J. Wilson, S. Mick, J. Xu, H. Hua, C. Mineo, A. M. Sule, M. Steer, and P. D. Franzon, "Demystifying 3D ICs: The Pros and Cons of Going Vertical", IEEE Design and Test of Computers, 22(6):498–510, Nov. 2005.

[2] N. Miura, H. Ishikuro, T. Sakurai, and T. Kuroda, "A 0.14pJ/b Inductive-Coupling Inter-Chip Data Transceiver with Digitally-Controlled Precise Pulse Shaping", In Proceedings of the International Solid-State Circuits Conference (ISSCC'07), pages 358–359, Feb. 2007.

[3] K. Salah, Y. Ismail, and A. El-Rouby, Arbitrary Modeling of TSVs for 3D Integrated Circuits. Springer Publishing Company, Incorporated, 2014.

[4] K. Salah, H. Ragai, Y. Ismail, and A. El Rouby, "Equivalent lumped element models for various n-port Through Silicon Vias networks," 16th Asia and South Pacific Design Automation Conference (ASP-DAC), 2011, pp. 176-183.

[5] K. Salah et al., "Compact lumped element model for TSV in 3D-ICs", in proc. IEEE International Symposium on Circuits and Systems (ISCAS), Rio de Janeiro, 2011, pp. 2321-2324.

[6] C. Jin, V.N. Sekhar, X. Bao, B. Chen, B. Zheng, and R. Li, "Antenna-in-Package Design Based on Wafer-Level Packaging With Through Silicon Via Technology " IEEE Transactions on Components, Packaging And Manufacturing Technology, vol. 3, no. 9, September 2013.

[7] Y. Lamy, L. Dussopt, O. El Bouayadi, C. Ferrandon, A. Siligaris, C. Dehos, P. Vincent "A compact 3D silicon interposer package with integrated antenna for 60GHz wireless applications" IEEE, 2013.

[8] C. Jin, Boyu Zheng, Liang Ding, Rui Li and Kafai Chang "High Performance Integrated Passive Devices (IPDs) on Low Cost Through Silicon Interposer (LC-TSI) " IEEE, 2013.

[9] P. Li, Y. Shang, H. Yu , M. Yu "A Wideband 150GHz Antenna by 3D-TSV based Composite Right/Left Handed Transmission Line for Sub-THz Biomedical Imaging" IEEE MTT-S International Microwave Workshop Series on RF and Wireless Technologies for Biomedical and Healthcare Applications (IMWS-BIO), 2013.

[10] E .J. Marinissen, D. Y. Lee, J. P. Hayes, C. Sellathamby, B. Moore, S. Slupsky, and L. Pujol" Contactless Testing: Possibility or Pipe-Dream?" EDAA, 2009.

[11] Le, T. T., "Efficient power conversion interface circuits for energy harvesting applications," PhD Thesis, Oregon State University, USA, Jun. 2008.

[12] Asefi, M., S. H. Nasab, L. Albasha, and N. Qaddoumi, "Energizing low power circuits by using an RF signal harvester," 16th Telecommunications Forum TELFOR, Nov. 2008.

[13] http://www.itrs.net/

[14] Doms I, Merken P, Van Hoof C. Comparison of DC–DC-converter architectures of power management circuits for thermoelectric generators. In: Proceedings of 2007 European conference on power electronics and applications, p. 1–5.

[15] Dickson JF. "On-chip high-voltage generation in MNOS integrated circuits using an improved voltage multiplier technique". IEEE J Solid-State Circ 1976; 11(3): 374–8.

[16] Shao H, Tsui C-Y, Ki W-H. "An inductor-less micro solar power management system design for energy harvesting applications". In: Proceedings of ISCAS; 2007. p. 1353–6.

[17] http://www.ansoft.com/products/si/hfss.

Effect of crystallite size and precursor molarities on electrical conductivity in ZnO Thin Films

Okba Belahssen*, Said Benramache

Material Sciences Department, Faculty of Science,
University of Biskra, Biskra 07000, Algeria

Boubaker Benhaoua

VTRS Laboratory, Institute of Technology, University of
El-Oued, El-Oued, Algeria

Abstract— **In present paper, ZnO thin films were deposited using the simple, flexible and cost-effective spray ultrasonic technique at different precursor molarity values. The films deposited on glass substrate at 350 °C. The electrical and structural properties were studied as a function of precursor molarity. The as sprayed films exhibit an hexagonal structure wurtzite and (002) oriented with the maximum value of crystallite size G = 31.82 nm is measured of ZnO film sprayed with 0.1 M. The crystallites size in the thin films depend by the defect (less defects), where the minimum defects confirmed the high crystallinity. The maximum value of electrical conductivity of the films is 7.96 $(\Omega.cm)^{-1}$ obtained in ZnO thin film for precursor molarity 0.125 M. The correlation between the electrical and the structural properties with the precursor molarity suggests that the electrical conductivity of the films is predominantly influenced by the crystallite size and the precursor molarity. The measurement in the electrical conductivity of the films with correlation it is equal to the experimental with the error is about 0.3 % in the higher conductivity.**

Keywords— *ZnO; Thin films; Electrical conductivity; Correlation; Ultrasonic spray technique.*

I. INTRODUCTION (*HEADING 1*)

Zinc oxide (ZnO) has a wurtzite (WZ) structure, this is a hexagonal crystal structure (lattice parameter: a=0.325 nm, c=0.521 nm), belonging to the space group P63mc, and is characterized by two interconnecting sublattices of Zn^{2+} and O^{2-}, such that each Zn ions is surrounded by a tetrahedral of O ions, and vice-versa [1,2]. Zinc oxide (ZnO) which is one of the most important binary II–VI semiconductor compounds has a hexagonal wurtzite structure and a natural n-type electrical conductivity with a direct energy wide band gap of 3.37 eV at room temperature, a large exciton binding energy (~60 meV) [3,4]. The resistivity values of ZnO films may be adjusted between 10^{-3} Ω cm and 10^{-4} Ω cm by changing the annealing conditions and doping [5]. This material is receiving considerable attention due to its broad range of applications such as microelectronic devices [6], antireflection coatings transparent electrodes in solar cells, thin film, gas sensors, varistors, spintronic devices, surface acoustic wave devices, light emitting diodes and lasers [7-11].

For the deposition of ZnO thin films various methods like reactive evaporation, molecular beam epitaxy (MBE) [12], magnetron sputtering technique[13], pulsed laser deposition (PLD) [14], the sol-gel technique, chemical vapor deposition, electrochemical deposition [15] and spray pyrolysis [16], have been reported to prepare thin films of ZnO. Among these, we will focus more particularly in this paper on the spray ultrasonic technique that is a low method suitable for large-scale production, it has several advantages in producing nanocrystalline thin films, such as, relatively homogeneous composition, a simple and deposition on glass substrate because of the low substrate temperatures involved, easy control of film thickness and fine and porous microstructure. It is possible to alter the mechanical, electrical, optical and magnetic properties of ZnO nanostructures [16, 17].

The aim of this paper is study the possibility of the correlation between the electrical and structural properties of ZnO thin films with precursor molarity. Ramana et al [18]. They found that the grain size of V2O5 thin films produced by pulsed laser ablation strongly influences their optical characteristics. Tudose et al [19]. They studied the correlation of ZnO thin film surface properties with conductivity. There is a limited amount of literature dedicated to systematic and detailed studies on the surface evolution with growth parameters and their effect on the ZnO surface conductivity under reduction/oxidation. However, same works were investigated the dependence of physical properties of ZnO thin film as a function of parameters conditions such temperature, thickness, oxidizing conditions, nitrogen addition and doping for characterizing the thin films [20–31].

This paper is to present a new approach to the description of correlation between electrical conductivity and crystallites size with precursor molarity of ZnO thin films. ZnO thin films were deposited on glass substrate at various precursor molarity between 0.05 to 0.125 mpl/l-1 at a substrate temperature of 350 °C, the films deposited by spray ultrasonic technique. We have studied the effect of the spraying concentration on electrical conductivity and structural properties.

II. 2. EXPERIMENTAL PROCEDURE

A. Experimental procedure

The spray solution were prepared by dissolving $(Zn(CH_3COO)_2, 2H_2O)$ in the solvent containing equal volumes absolute ethanol solution (99.995%) purity, then

---

* Corresponding author. Tel.: +213774637626.
E-mail address: belahssenokba@gmail.com (*O. Belahssen*).

added a drops of HCl solution as a stabilized, the mixture solution was stirred at 50 °C for 180 min to yield a clear and transparency solution. The samples were deposited keeping the substrate temperature TS equal to 350 °C, and with the precursor solution concentrations M of 0.05, 0.075, 0.1 and 0.125 mol.l−1. The substrate cooling was prevented by spraying a solution jet of 2 min pulses with each new pulse delayed by 10 min from the previous one. The film forms as the solution's atomized aerosol droplets vaporize on the heated substrate. This procedure has led to highly adherent ZnO films. The distance between the substrates and the spray gun nozzle was fixed at 3.5 cm [32,33]. The Crystallographic and phase structures of the thin film were determined by X-ray diffraction (XRD, Bruker AXS-8D) with CuKα radiation (λ = 0.15406 nm) in the scanning range was between 2θ = 25° and 55° and the electrical conductivity of the films was measured in a coplanar structure obtained with evaporation of four golden stripes on film surface. All spectra were measured at room temperature (RT).

B. Methods and model

The correlation between the electrical and structural properties were studied for the electrical conductivity (σ) as a function the crystallite size (G) and precursor molarity M of ZnO films. These parameters correlates were resulting from the following equation:

$$\sigma_{(*)} = \frac{\sigma_{(e)}}{\sigma_{(e)Max}} \tag{1}$$

$$G_{(*)} = \frac{G_{(e)}}{G_{(e)Max}} \tag{2}$$

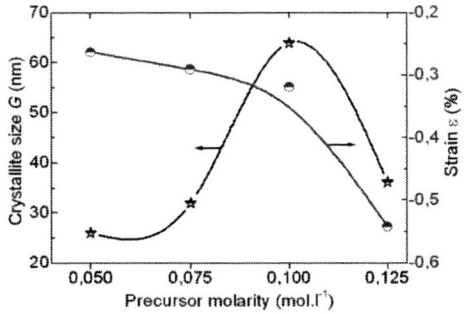

Fig. 1. X-ray diffraction spectra of ZnO thin films as a function of spray concentration.

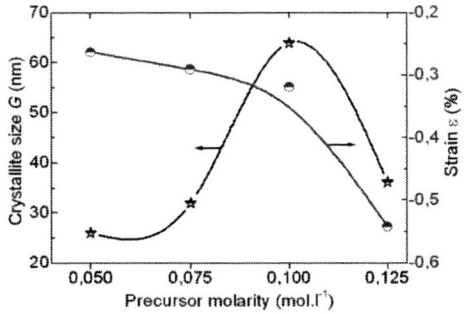

Fig. 2. Variation of crystallite size and strain of ZnO thin films with spray concentration.

$$\text{Fig. 3.} \quad M_{(*)} = \frac{M_{(e)}}{M_{(e)Max}} \tag{3}$$

where $\sigma_{(e)}$, $G_{(e)}$ and $M_{(e)}$ are the experimental values; $\sigma_{(e)Max}$, $G_{(e)Max}$ and $M_{(e)Max}$ are maximal experimental values and $\sigma_{(*)}$, $G_{(*)}$ and $M_{(*)}$ are the first values have been consisted in the correlate relationships.

III. Results and discussion

A. Structural properties

The XRD patterns for ZnO thin films as a function of precursor molarities [32] are shown in Fig. 1. The film exhibit polycrystalline structure that belongs to the hexagonal wurtzite type of ZnO [15,34]. The obtained XRD spectra matched well with the space group P63mc (186) (No. 36-1451). There are four peaks that could be observed in the film at diffraction angle are 31.7, 34.5, 36.3 and 47.5°. The peak at position 34.5° corresponding to the (002) plans is very sharp, the film obtain with 0.1 M has higher and sharper diffraction peak indicating an improvement in (002) peak intensity, revealing that the films are nanocrystalline and a preferred orientation with the c-axis perpendicular to substrate. The crystallite size was calculated using the Scherer's equation [35]:

$$G = \frac{0.9\lambda}{\beta \cos\theta} \tag{4}$$

where G, λ, β and θ denote the crystallite size, the X-ray wavelength, FWHM and the Bragg angle of (002) peak, respectively. The variations are shown in Table 1.

The variation of crystallite size of ZnO thin films as a function of precursor molarity as shown in Figure 2, as can be seen, the maximum value of crystallite size in the ZnO on glass substrate was 63.99 nm at 0.1 M, indicate that the enhancement of the crystallinity and c-axis orientation of ZnO thin films [36].

The average uniform strain for the lattice along the c-axis in the randomly oriented ZnO films deposited on glass substrate has been estimated from the lattice parameters using the equal below [37]:

$$\varepsilon = \frac{c - c_0}{c_0} \times 100 \ \% \tag{5}$$

where ε is the mean stress in ZnO thin films (Table 1), c the lattice constant of ZnO thin films and c_0 the lattice constant of bulk (standard $c_0 = 0,5206$ nm).

According to the hexagonal symmetry, the lattice constant can be calculated by the following formula [38,39]:

$$d_{hkl} = \left(\frac{4}{3} \frac{h^2 + hk + k^2}{a^2} + \frac{l^2}{c^2} \right)^{-\frac{1}{2}} \tag{6}$$

where a, c are the lattice parameters, (h,k,l) are the Miller indices of the planes and d_{hkl} is the interplanar spacing. The

TABLE I. BRAGG ANGLE 2θ, THE FULL WIDTH AT HALF-MAXIMUM FWHM β, THE GRAIN SIZE G, LATTICE PARAMETERS c AND a, THE STRAIN ε FOR (0 0 2) PLANE AND ELECTRICAL CONDUCTIVITY σ FOR ZnO THIN FILMS WERE MEASURED AS A FUNCTION OF CONCENTRATION.

Concentration (Mol/l)	2θ (deg)	β (deg)	G (nm)	c (A°)	a (A°)	ε (%)	σ $(\Omega.cm)^{-1}$
0.05	34.52	0.32	25.99	5.192305	3.245197	- 0.263	0.24
0.075	34.53	0.26	31.99	5.190847	3.244279	- 0.291	2.38
0.1	34.54	0.13	63.99	5.189391	3.243369	- 0.319	7.65
0.125	34.62	0.23	36.18	5.177764	3.236103	- 0.542	7.96

variation of lattice parameters c of ZnO thin films were estimated in Table 1. As can be seen, the minimum value of strain was equal to - 0.542 % measured in ZnO thin film for precursor molarity 0.125 M. The decreases of the strain (less defects) with precursor molarity indicated the increases of the crystallinity and c-axis orientation of ZnO thin films. We found that the crystallite size along height direction. Negative value of the strain in the film means that the sample is in a tensile condition.

B. Electrical conductivity

Figure 3 shows the variation of the electrical conductivity σ of ZnO films as a function of precursor molarity. As can be seen, the electrical conductivity increases from 0.24 to 7.96 (Ω.cm)-1 with increasing of precursor molarity of ZnO thin films, which mean an increase in the carrier concentration. The increase in conductivity of films with increasing precursor molarity has been explained by decreasing of the strain (less defects) (Table 1) in the films hence the potential barriers decreased and decreases defects, which resulted in an increased carrier density [40-42].

C. Correlation between the electrical and optical properties

We have described previously the experimental data; one can be seen from this data, the electrical conductivity of ZnO thin film is varied nonlinear the precursor molarity. The model proposed of ZnO thin film with precursor molarity is discussed.

We have estimated the relationships between the electrical conductivity and the crystallite size and the precursor molarity in all films. We found the following empirical relationships:

$$\sigma_{(c)} = a \times b^{E_{g(*)}} \times M_{(*)}{}^{c}, \qquad (7)$$

where $\sigma_{(c)}$ is the correlate electrical conductivity; a, b and c are empirical constants as $a \approx 0.5276$, $b \approx 3.118$ and $c \approx 2.466$. The results are collected in Table 2.

As shown in figs. 4 and 5. The variation experimental electrical conductivity and correlation of ZnO thin films as a function of precursor molarity and crystallite size, respectively. On can be seen, this correlation indicates that the measurement in the electrical conductivity of the films from the crystallite size and the precursor molarity was equal; it is predominantly influenced by the crystallinity and the chemical composition of undoped ZnO thin films.

TABLE I. THE EXPERIMENTAL CORRELATION OF THE CRYSTALLITE SIZE G AND THE ELECTRICAL CONDUCTIVITY σ OF ZnO THIN FILMS WERE MEASURED AS A FUNCTION OF CONCENTRATION.

Molarities (mol/l)	0.05	0.075	0.01	0.125
σ Experimental $(\Omega.cm)^{-1}$	0.24	2.38	7.55	7.96
σ Crrelation $(\Omega.cm)^{-1}$	0.695	2.110	7.553	7.988

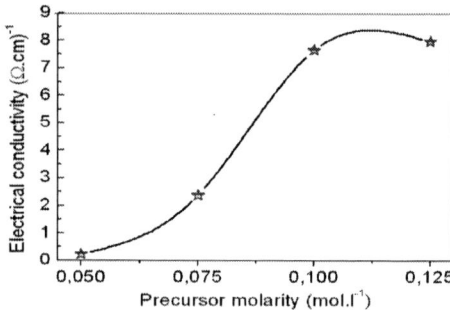

Fig. 4. Variation of electrical conductivity of ZnO thin films with spray concentration.

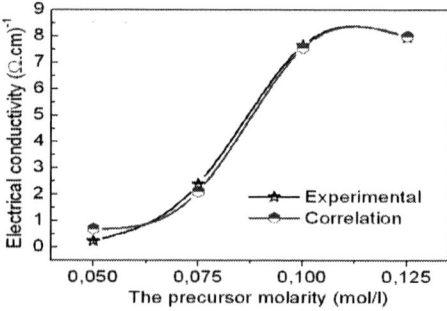

Fig. 5. The experimental electrical conductivity and correlation of ZnO thin films as a function of the precursor molarity.

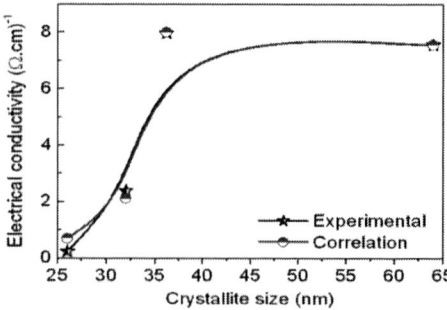

Fig. 6. The experimental electrical conductivity and correlation of ZnO thin films as a function of the crystallite size.

In our experience there was no evidence for significant changes in electrical conductivity with correlation upon varying the crystallite size; the main effects have been observed upon variation of the modifying the precursor molarity. This difference could be caused by the different compositions. The correlation between the electrical conductivity and the crystallite size with the precursor molarity was investigated.

IV. CONCLUSION

In summary, high-quality transparent ZnO thin films were grown on glass substrates at room temperature by ultrasonic spray technoque; the influence of precursor molarity on structural, electrical and structural properties was investigated. The as sprayed films exhibit an hexagonal structure wurtzite and (002) oriented with the maximum value of crystallite size $G = 31.82$ nm is measured of ZnO film sprayed with 0.1 M. The crystallites size in the thin films depend by the defect (less defects), where the minimum defects confirmed the high crystallinity. The maximum value of electrical conductivity of the films is 7.96 $(\Omega.cm)^{-1}$ obtained in ZnO thin film for precursor molarity 0.125 M. The correlation between the electrical and the structural properties with the precursor molarity suggests that the electrical conductivity of the films is predominantly influenced by the crystallite size and the precursor molarity. The measurement in the electrical conductivity of the films with correlation it is equal to the experimental with the error is about 0.3 % in the higher conductivity.

REFERENCES

[1] V.A. Coleman, C. Jagadish, Zinc oxide bulk thin films and nanostructures processing properties and applications gainesville. FL, USA, 2006, p. 1-20.

[2] M. Ohtsu, Progress in nano-electro optics VII. Tokyo, 2002, p.73-108.

[3] L. Ma, X. Ai, X. Huang, S. Ma, Effects of the substrate and oxygen partial pressure on the microstructures and optical properties of Ti-doped ZnO thin films. Superlattices and Microstructures 50, 2011, s.703–712.

[4] S. Benramache et al. Effect of the Dip-coating Speed on the Structural and Optical Properties of ZnO Thin Films IJSEI 1, 2012, s. 28-31.

[5] Y.D. Ko, K.C. Kim, Y.S. Kim, Effects of substrate temperature on the Ga-doped ZnO films as an anode material of organic light emitting diodes. Superlattices and Microstructures 51, 2012, s. 933–941.

[6] Z. Zhang et al. Influence of deposition temperature on the crystallinity of Al-doped ZnO thin films at glass substrates prepared by RF magnetron sputtering method. Superlatt Microstruct. 2011; 49: 644–653.

[7] X. Zhu et al. Optical and electrical properties of N-doped ZnO and fabrication of thin-film transistors. J. Semiconductors 2009, 30(3): 033001

[8] H. Zhang et al. Low-temperature deposition of transparent conducting Mn-W co-doped ZnO thin films. J. Semiconductors 2010, 31(8): 083005

[9] Z. Ben Ayadi, L. El Mir, K. Djessas, S. Alaya, Effect of the annealing temperature on transparency and conductivity of ZnO:Al thin films. Thin Solid Films. 2009; 517: 6305–6309.

[10] P. Nunes et al. Influence of the annealing conditions on the properties of ZnO thin films. I J Inorganic Mate. 2001; 3: 1125–1128.

[11] G. Babita et al. Development and characterization of sol-gel derived Al doped ZnO/p-Si photodiode. J Mater Sci Techn. 2010; 26: 223-227.

[12] S. Singhal, A.K. Chawla, H.O. Gupta, R. Chandra, Influence of cobalt doping on the physical properties of Zn0.9Cd0.1S nanoparticles. Nanoscale Research Letters. 2010; 5: 323–330.

[13] W. Buguo et al. Hydrothermal growth and characterization of indium-doped-conducting ZnO crystals. J. Crystal Growth. 2007; 304: 73–79.

[14] SJ Kang et al. Effect of substrate temperature on structural, optical and electrical properties of ZnO thin films deposited by pulsed laser deposition. J Mater Sci: mater Electro. 2008; 19: 1073–1078.

[15] D. Vernardou, G. Kenanakis et al. The effect of growth time on the morphology of ZnO structures deposited on Si (100) by the aqueous chemical growth technique. Journal of Crystal Growth. 2007; 308: 105-109.

[16] Z.B. Bahsi, A.Y. Oral, Effects of Mn and Cu doping on the microstructures and optical properties of sol–gel derived ZnO thin films. Optical Materials. 2007; 29: 672–678.

[17] X. Yan et al. Nanostructure and optical properties of M doped ZnO (M=Ni, Mn) thin films prepared by sol–gel process. Physica B. 2011; 406: 3956-3959.

[18] C.V. Ramana et al. phys stat sol A 199 (2003): R4–R6.

[19] I.V. Tudose et al. Correlation of ZnO thin film surface properties with conductivity, Applied Physics, 2007, 89: 57-61.

[20] M. Ivill et al. Journal of Applied Physics, 2005, 97() 53904.

[21] W. Vollmann, W. Berger, et al. Relations between the morphology and conductivity of thin films of tetrathiofulvalinium tetracyano quinodimethane. Thin Solid Films, 1984, 111(1) 7–16.

[22] A. Asadov et al. Correlation between structural and electrical properties of ZnO thin films. Thin Solid Films, 2005, 476(1): 201–205.

[23] C. Ton-That et al. Correlation between the structural and optical properties of Mn-doped ZnO nanoparticles. J Alloys and Compounds, 2012, 522(1) 114–117.

[24] O. Kappertz, R. Drese, M. Wuttig, Correlation between structure, stress and deposition parameters in direct current sputtered zinc oxide films. Journal of Vacuum Sciences Technology A, 2002, 20(6): 2084–2095.

[25] I.V. Tudose, P. Horvath et al. Correlation of ZnO thin film surface properties with conductivity. Applied Physics A, 2007, 89(1) 57–61.

[26] B. Joshi et al. Correlation between electrical transport, microstructure and room temperature ferromagnetism in 200 keV Ni2+ ion implanted zinc oxide (ZnO) thin films. Applied Physics A, 2012, 107(2) 393–400.

[27] S.D. Wang et al. Correlation between grain size and device parameters in pentacene thin film transistors. Appl Phy Lett, 2008, 93(4) 043311–1.

[28] T. Minami et al. Correlation between film quality and photolumine-scence in sputtered ZnO thin films. J. Mater Sci 1982, 17(5) 1364–1368.

[29] J.J. Zhu, L. Vines, T. Aaltonen, A.Y. Kuznetsov, Correlation between nitrogen and carbon incorporation into MOVPE ZnO at various oxidizing conditions. Microelectronics Journal, 2009, 40(2): 232–352.

[30] Y. Nakanishi, K. Kato, H. Omoto, M. Yonekura, Correlation between microstructure and salt-water durability of Ag thin films deposited by magnetron sputtering. Thin Solid Films, 2013, ().

[31] R.S. Ajimsha et al. Correlation between electrical and optical properties of Cr:ZnO thin films grown by pulsed laser deposition. Physica B, 2011, 406() 4578–1583.

[32] S. Benramache et al. J Sci Eng 2013;2:97–104.

[33] S. Benramache, B. Benhaoua, F. Chabane, Effect of substrate temperature on the stability of transparent conducting cobalt doped ZnO thin films. Journal of Semiconductors. 2012; 33: 093001-1.

[34] S.K. Pandey et al. Effect of growth temperature on structural, electrical and optical properties of dual ion beam sputtered ZnO thin films, J Mater Sci: Mater Electron (2013) DOI 10.1007/s10854-013-1130-5

[35] S. Benramache et al. A comparative study on the nanocrystalline ZnO thin films prepared by ultrasonic spray and sol–gel method. Optik. (2012). http://dx.doi.org/10.1016/j.ijleo.2012.10.001.

[36] C. Zhang, High-quality oriented ZnO films grown by sol–gel process assisted with ZnO. J. Physics and Chemistry of Solids, 2010, 71(2): 364

[37] S. Benramache, B. Benhaoua, F. Chabane, F.Z. Lemmadi, Influence of growth time on crystalline structure, conductivity and optical properties of ZnO thin films. Journal of Semiconductors. 34 (2013) 0203001-1.

[38] M. Mekhanache et al. Properties of ZnO thin films deposited on (glass, ITO and ZnO:Al) substrates. Superl.Micros, 2011, 49(3): 510

[39] S. Benramache et al. Elaboration and Characterisation of ZnO thin films. Matériaux & Techniques. 100 (2012) 573–580.

[40] A. Khorsand et al. X-ray analysis of ZnO nanoparticles by Williamson–Hall and size–strain plot methods. Solid State Sci 2011; 13: 252-255.

[41] S. Benramache et al. Influence of annealing temperature on structural and optical properties of ZnO: In thin films prepared by ultrasonic spray technique. Superlattices and Microstructures. 2012; 52: 1062–1070.

[42] S. Benramache, B. Benhaoua, Influence of substrate temperature and Cobalt concentration on structural and optical properties of ZnO thin films prepared by Ultrasonic spray technique. Superlattices and Microstructures. 2012; 52: 807–815.

Performance Investigation and Linearity Analysis of New Cylindrical MOSFET for Wireless Applications

Jay Hind K. Verma, Mridula Gupta
Department of Electronic Science
University of Delhi South campus,
New Delhi, India, 110021
jkv.electronics@gmail.com
mridula@south.du.ac.in

Subhasis Haldar
Department of Physics
Moti Lal Nehru College,
University Of Delhi, New Delhi,
India, 110021
subhasis_haldar@rediffmail.com

R. S. Gupta
Department of electronics and
Communication Engineering
Maharaja Agrasen Institute of
Technology, Delhi, India, 110085
rsgupta1943@gmail.com

Abstract—**This paper presents linearity and analog performance of new cylindrical MOSFET for wireless application. It is based on the incorporation of an inner core gate in the Cylindrical Surrounding Gate (CSG) MOSFET. Inner gate incorporation improves the electrical performance of the device in the nanoscale region. Linearity parameter like higher order transconductance, second and third order voltage intercept point and third order intercept input power are calculated using ATLAS device simulator. ITRS roadmap has been taken in consideration as 22 nm gate length for the analysis.**

Keywords-ATLAS; linearity; cylindrical gate (CGT) MOSFET; wireless applications; CMOS.

I. INTRODUCTION

Modern Communication system requires CMOS devices with higher analog performances and linearity [1-2]. It requires maintaining linear operation even when receiving a weak signal with high accuracy. International Technology Roadmap for Semiconductor (ITRS) provides the roadmap of the technology node for future CMOS devices and their limitation [3]. The main intention of the roadmap is to support the semiconductor industry in extending the rapid pace of improvement in semiconductor products into the future [4]. To the end, a lot of different targets, e.g., for device size, device structure, electrical performance, etc., are defined, and possible roadblocks that may impede the progress. For a long time, the Si MOSFET has been considered a slow and noisy device not suited for RF applications. However due to continuous scaling of MOSFETs for digital applications, the RF performance of Si MOSFET has been improved considerably [5]. Thus for future CMOS devices and improved RF performances, multiple gate structure i.e. double gate [6], tri-gate [7] and cylindrical-surrounding gate [8] has been proposed with more enhanced gate controllability over the channel. The presence of the gate electrode on more than one side of the device effectively improves the electrostatics control over the channel and reduces short-channel effects [9]. The controllability over the channel to reduce SCEs due to scaling can further be increased with innovative MOSFET design [10]. During the last decades, the RF performance of Si MOSFETs has improved at virtually breathtaking pace.

Therefore, to-date the Si MOSFET is a well accepted device for low-performance applications in the lower GHz range, such as WLAN transceivers, and further more advanced applications requiring enhanced device performance seem to be realistic in the near future [5].

Tekleab et al. in 2012 [11] patented a silicon nanotube MOSFET. This silicon nanotube MOSFET structure has tubular channel controllable by an inner and outer gate [12]. The structure looks like CSG MOSFET [8], which consists of only an outer gate surrounding the silicon substrate. In 2011 Hussain et al. introduced the concept of silicon nanotube field effect transistor with core-cell gate stacks for enhanced high performance operation and area scaling benefit [13].

In this paper, a new cylindrical dual gate MOSFET [14-15] has been investigated with improved performance in the nanoscale region for wireless application. The proposed cylindrical shaped dual gate MOSFET has shown the potential to fulfill the requirements for future CMOS device as well as RF application. It is the cylindrical version of double gate MOSFET in which two gates, outer and inner gate control the silicon channel at same voltage. The outer gate acts as CSG and the inner gate acts as the second gate for enhanced charge control. Apart from these, there is a need to estimate the RF and linearity distortion analysis to realize the device feasibility and stability for high frequency applications and RFIC design. Linearity is an essential requirement in all RF systems, ensuring that inter-modulation and higher order harmonics are minimal at the output. Linearity is mainly determined by metrics such as second order Voltage Intercept Point (VIP2), third order Voltage Intercept Point (VIP3), Third order intercept Input Power (IIP3) and higher order transconductance coefficients [16-17]. In RF and wireless communication systems, device must maintain linear operation even while receiving a weak signal in the presence of a strong interference.

II. DEVICE ARCHITECTURE AND SIMULATION

Figure 1 (a) shows schematic view of CSDG MOSFET with two gates named outer gate and inner gate (Charge Control Gate). The CSDG MOSFET operates in two distinct modes– separate inversion and volume inversion. In separate inversion two conduction channels forms one is at the interface of outer gate oxide-silicon substrate and other at the interface of inner gate oxide-silicon substrate. In volume inversion inner and outer channel merge in to whole silicon region. Thus in this mode of operation, the carrier number and their mobility are increased and therefore performance improves significantly. Fig. 1 (b) shows a schematic view of conventional CSG MOSFET with same diameter of silicon substrate equal to CSDG MOSFET. Fig. 2 (a) shows a circular view of CSDG MOSFET with only inner gate inside is shown. It also has an outer gate surrounding the silicon substrate. Fig. 2 (b) is the circular view of CSG MOSFET. In the present study CSDG and CSG MOSFETs electrical parameter are compared by device simulator. Simulation incorporates Newton-Gummel method with Lombardi CVT mobility model with Concentration dependent mobility model. Shockley-Read-Hall (SRH) recombination model with fixed carrier life time (1x10^{-7} s) along with Auger model have been included to account for the minority and high carrier density recombination [18]. Various device parameters used in the simulation are tabulated in Table 1 Threshold voltage is obtained as gate to source voltage for which drain current is 5 x 10^{-8} A using constant current method. The threshold voltage is optimized by tuning the metal gate work function. In the analysis the following three cases have been studied.

TABLE I
UNITS FOR PARAMETERS USED IN THE SIMULATION FOR THE THREE CASE OF CSG MOSFET.

Symbol	Parameter	Value		
		CSDG		CSG
		(Case: 1)	(Case: 2)	(Case: 3)
L	Channel length	22 nm	22 nm	22 nm
D	Diameter	24 nm	10 nm	24 nm
Na	Channel doping	10^{17} cm^{-3}	10^{17} cm^{-3}	10^{17} cm^{-3}
t_{ox1}, t_{ox2}	Oxide thickness	2 nm	2 nm(t_{ox})	2 nm (t_{ox})
ϕ_m	Work function	4.65 eV	4.60 eV	4.83 eV
IG	(Inner Metal Gate) ⇔Radius	5 nm	-	-

Case: 1 (CSDG MOSFET) Diameter 24 (nm) (including the diameter of metal = 10 nm, two oxide thickness = 4 nm and two silicon thickness = 10 nm) Fig. 1 (a).

Case: 2 (CSG MOSFET) Diameter 10 (nm) only silicon thickness is equal to CSDG MOSFET Fig. 1 (b).

Case: 3 (CSG MOSFET) Diameter 24 (nm) (Diameter of Silicon = 24 nm) same diameter equal to CSDG MOSFET Fig. 2 (b).

III. RESULTS AND DISCUSSION

The higher order derivatives of trans-conductance are

$$gmn= \frac{d^n I_{ds}}{dV_{gs}^n} \qquad \text{Where n = 1, 2, 3} \qquad (1)$$

The device Figures of Merit are evaluated as

$$VIP2=4X\frac{gm1}{gm2} \qquad (2)$$

$$VIP3 = \sqrt{24X\frac{gm1}{gm3}} \qquad (3)$$

$$IIP3=\frac{2}{3}X\frac{gm1}{gm3XR_s} \qquad (4)$$

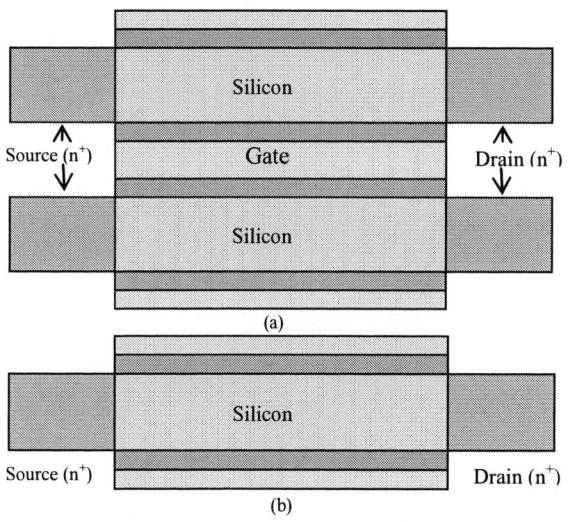

(a)

(b)

Fig. 1 (a) schematic view of CSDG MOSFET and (b) schematic view of CSG MOSFET of gate length 22 nm for case 1 and case 2.

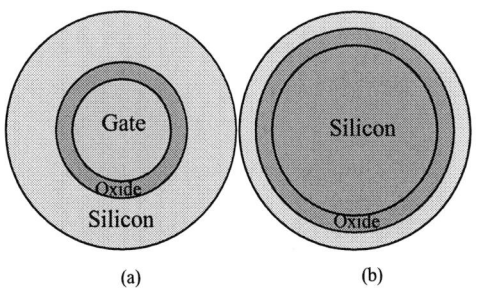

(a)   (b)

Fig. 2 (a) circular view of CSDG MOSFET for case 1 (Outer gate is not shown here) and (b) circular view of CSG MOSFET for case: 3 at diameter of 24 nm.

Figure 3 (a) shows the variation of transconductance with respect to gate to source voltage at fixed drain voltage, V_{ds}=1. It shows that the proposed two cylindrical gates device MOSFET has higher transconductance than the other two CSG MOSFET. Higher value of transconductance is required for the device to increase the sharpness of drain current with voltage and improvement in cut off frequency. Case: 3 have the higher drain current so that its transconductance is between Cases: 1 and case: 2 even though it has higher leakage current also. Second order trans-conductance in Fig. 3 (b) shows that case: 1 has higher than the case: 2 and case: 3 of CSG MOSFET.

978-1-4673-8760-6/15 $31.00 © 2015 IEEE

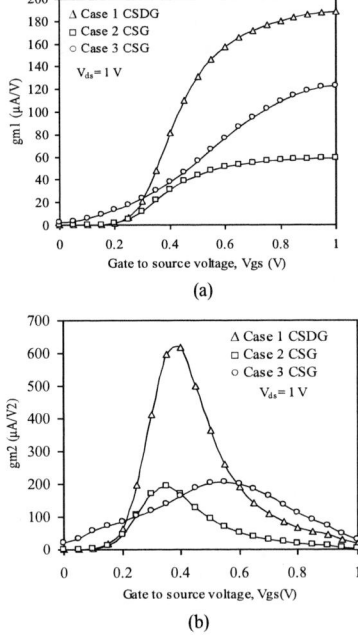

(a)

(b)

Fig. 3 (a) First order of transconductance (gm1) with respect to gate to source voltage for the three case of transistor and (b) Second oredr of transconductance (gm2) with respect to gate to source voltage for the three case of transistor.

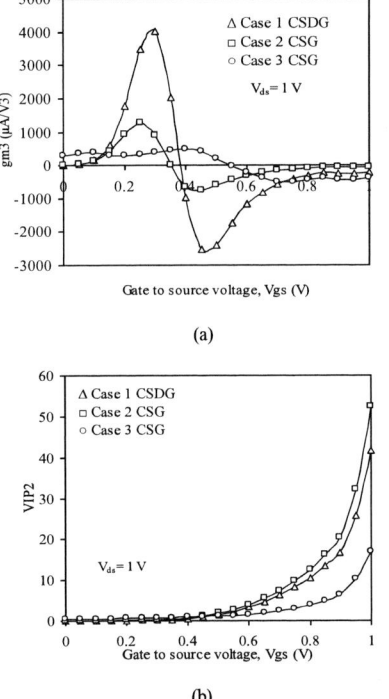

(a)

(b)

Fig. 4 (a) Third order of transconductance (gm3)with respect to gate to source voltage for the three case of transistor and (b) Second order voltage intercept point (VIP2) with respect to gate to source voltage for the case of three transistor.

University Grant Commission (UGC), Government of India with references no. Schs/JRF/F.15-9/4010/2013-14 provides the necessary financial support to carry out this work.

Fig. 4 (a) shows the variation of third order of trans-conductance with variation in gate to source voltage. Case: 1 has much value of gm3 than the other two. For wireless communication coefficient of higher order transconductance should be low. The zero crossover point of gm3 determines the dc bias point for optimum device operation, and hence, the nonlinear behavior of gm3 determines the dc point for optimum device operation. Zero crossover point of gm3 in case: 1 slightly higher than case: 2 but lower than the case: 3.

Fig. 4 (b) shows the variation of second order of voltage intercept point with respect to gate to source voltage at fixed drain bias. VIP$_2$ is a FOM which can determine the distortion characteristics for different dc parameters. For high linearity performance and low distortion operation, a higher value of VIP2 is required. Figure shows that case: 1 has slightly lower than the case: 2 but higher than the case: 3. Case: 2 of CSG MOSFET shows bettter performance for linearity but it has low drive current. Fig. 5 shows the VIP3 and IIP3 with variation in gate to source voltage at fixed drain voltage. These values are approximately equal for case:1 and case: 2 and lower for case: 3. So the CSG MOSFET with incarpotation of inner gate shows higher drain current and transconductance due to increase of current surface area. Its linearity performance are between case: 2 and cas:3.

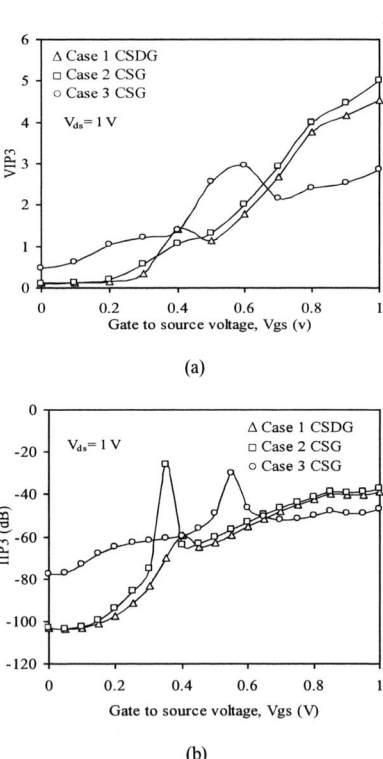

(a)

(b)

Fig. 5 (a) Third order voltage intercept point (VIP3) with respect to gate to source voltage for the three case of transistor and (b) Third order intercept input power (IIP3) with respect to gate to source voltage for the case of three transistor.

IV. CONCLUSION

In this paper, a Cylindrical Surrounding gate (CSG) MOSFET with impact of an inner Gate is analyzed for suppression of short channel effects and harmonic distortion. Linearity performance (VIP2, VIP3 and IIP3) and higher order transconductance (gm2 and gm3) are analyzed with comparison to two case of CSG MOSFET on the basis of their diameter and silicon substrate thickness. It is concluded that performance of inner gate CSG MOSFET reside between case: 2 and case: 3 of CSG MOSFET.

REFERENCES

[1] P. H. Woerlee, M. J. Knitel, R. van Langevelde, D. B. M. Klaassen, L. F. Tiemeijer, A. J. Scholten, and A. T. A. Zegers-van Duijnhoven, "RF CMOS performance trends", *IEEE Trans. Electron Devices*, vol. 48, no. 8, pp. 1776–1782, Aug. 2001.

[2] C. Yu, J. S. Yuan and H. Yang, "MOSFET linearity performance degradation subject to drain and gate voltage stress", IEEE Trans. Device Mater. Rel., vol. 4, no. 4, pp. 681–689, Dec. 2004.

[3] International Technology Roadmap for Semiconductor. online [http://public.itrs.net].

[4] F. Schwierz, "RF transistors: Performance trends versus ITRS targets" Devices, Circuits and Systems, Proceedings of the 6th International Caribbean Conference on. IEEE, 2006.

[5] F. Schwierz, H. Wong and J. J. Liou, Nanometer CMOS. Pan Stanford Publishing, 2010.

[6] F. Balestra, S. Cristoloveanu, M. benachir, J. birini and T. Elewa "Double-gate silicon-on-insulator transistor with volume inversion: A new device with greatly enhanced performance" IEEE Electron Device Lett., no. 9, pp. 410-412, Sept. 1987.

[7] M. C. Lemme, C. Moormann, H. Lerch, M. Moller, B. Vratzov and H. Kurz "Triple-gate metal–oxide–semiconductor field effect transistors fabricated with interference lithography" Nanotech. pp. 208-210, 2004.

[8] B. Cousin, M. Reyboz, O. Rozeau, M. A. Jaud, T. ernst, J. Jomaah "A unified short-channel compact model for cylindrical surrounding-gate MOSFET" Solid-State Electron. no. 56 pp. 40-46, 2011.

[9] I. Ferain, A. C. Cynthia and J. P. Colinge. "Multigate transistors as the future of classical metal-oxide-semiconductor field-effect transistors", Nature, pp. 310-316, 2011.

[10] I. Hiroshi "Future of nano CMOS technology" Solid-State Electronics 2015.

[11] D. Tekleab, H. H. Tran and J. W. Slight, "Silicon nanotube MOSFET", U.S. Patent, Aug. 30, 2012.

[12] D. Tekleab "Device Performance of Silicon Nanotube Field Effect Transistor", *IEEE Electron Device Lett.,*; no. 5, pp. 506-508, 2014.

[13] H. M. Fahad, C. E. Smith, J. P. Rojas and M. M. Hussain "Silicon Nanotube Field Effect Transistor with Core-Shell Gate Stacks for Enhanced High Performance Operation and Area Scaling Benefits" Nano Lett., vol. 11, pp. 4393-4399, Sept., 2011.

[14] Y. Chen and W. Kang "Experimental study and modeling of double–surrounding-gate and cylindrical silicon–on–nothing MOSFETs", *Microelectron. Eng.*, no. 97, pp. 138-143. 2012.

[15] J. H. K. Verma, S. Haldar, R. S. Gupta and M. Gupta "Modeling and simulation of subthreshold behaviour of cylindrical surrounding double gate MOSFET for enhanced electrostatic integrity" Superlattices and Microstructures 2015.

[16] R. Gautam, M. Saxena, R. S. Gupta, and M. Gupta "Effect of localised charge on nanoscale cylindrical surrounding gate MOSFET Analog performance and linearity analysis", *Microelectronics Rel.*, no. 6, pp. 989-994, 2012.

[17] P. Ghosh, S. Haldar, R. S. Gupta and M. Gupta "Investigation of linearity performance and inter-modulation distortion of GME CGT MOSFET for RFIC design" *IEEE Trans. Electron Devices*, no. 12: pp. 3263-3268, 2012.

[18] SILVACO Int. Santa Clara, *ATLAS: 3D Device Simulator*, 2015.

978-1-4673-8760-6/15 $31.00 © 2015 IEEE

Embedded Agent for medical image segmentation

Hassna BENSAG, Mohamed YOUSSFI, Omar BOUATTANE

Laboratory SSDIA, ENSET Mohammedia, University Hassan II Casablanca, Morocco

h.bensag@gmail.com, med@youssfi.net, o.bouattane@gmail.com

Abstract—**In last years, a lot of efforts have been made to deploying the intelligent devices like agent to enjoy their performance. Multi-agent systems (MAS) are simple and strong software suitable for complex and dynamic environment like simulation of economies, societies, biological and others. The aim of this paper is to present a new embedded agent especially for Cardiac Magnetic Resonance Image (MRI) segmentation. With this work, we focused on the first to implement the parallel algorithm of classification using C-means method in embedded systems and on the second to propose a new concept of distributed classification using multi-agent systems based on JADE and Raspberry PI 2 devices.**

Keywords—MAS, JADE, Raspberry PI 2, Segmentation, Cardiac RMI.

I. INTRODUCTION

In last years, grid computing has been widely used to solve complex and critical problems in the areas of science and engineering such as drug discovery, earthquake simulation, climate modeling and other [1,2,7,8 and 9]. Today, the need for use of devices such as embedded devices becoming more and more common, they are not only equipped with powerful processors but also a ubiquitous connectivity and convergence of technology have resulted in hardware/software systems being embedded within everyday products and places. These are prompting scientists to consider these intelligent devices as a fertile platform for use in grid computing. For that, the developing of intelligent middleware attired attention a lot of research laboratory which combines grid and embedded computing, supports mobile and users and resources in a seamless, transparent, secure and efficient way [3,4 and 5].

The multi-agent systems are used in several domains like simulation of economies, renewable energy, computer science domain, and others [3,8,9,10 and 11]. These distributed systems used in medical and healthcare application and also for image segmentation. This last area presents a main problem in majority of medical application [5,6]. The complexity of image segmentation can be reduced by using parallel and distributed computing. In order to integrate available resources and parallel computing capacity it necessary to use the middleware that manage the complexity of distributed systems and simplify the development process. They are several method of classification used for

segmentation, we are particularly interesting on C-means using also in a lot of research work [6].

In this paper, we propose a new method for classification which is a distributed c-means algorithm. It is applied to cardiac MRI image segmentation in order to be implemented on an embedded agents executed on Raspberry PI devices using the JADE middleware.

The remainder of this paper is organized as follows: section 2 describes the related work, all methods and tolls using in our proposed. Some obtained results on the proposed model are presented in section 3, section 4 concludes this paper with a remarks and perspectives.

II. RELATED WORK

This section introduces selected methods, approaches, and tools that increased the acceptance of MASs and distributed systems for our proposed.

A. Muti-agent systems (MAS)

An agent is an encapsulated computer system with an environment in which it is capable of performing a flexible and autonomous action compatible with the design objectives. In particularly, agents are characterized by properties such as autonomy, reactivity, proactivity, intelligence, adaptability, collaboration, mobility, and mobile agents have the characteristics to move from one machine to another [2,4,9,10 and 14]. Multi-agent system is applied in different domains, offer strong models for representing complex and dynamic environments. MAS can also be used to simulate the behavior of complex computer systems, this simulation models can help designers and developers of complex computational systems. So, the simulation based multi-agent provides a good set of tools to manage complex systems for online resource allocation environments.

B. JADE Agent platform

In this section, we are looking to present all information about open source platform JADE [9]. This software is distributed by Telecom Italia the copyright holder, in open source under the terms and conditions of the LGPL (Lesser General Public License Version 2) license.

JADE (*Java Agent DEvlopment*) is a most popular open source framework for development of multi-agent systems, is

978-1-4673-8760-6/15 $31.00 © 2015 IEEE

a software framework fully implemented in the JAVA language. He is a FIPA (Foundation for intelligent Physical Agents) compliant agent platform, composed of multiple containers which host and execute agents. The main goal of platform JADE is especially to simplify development while ensuring standard compliance through a comprehensive set of systems for agents and services [5,9]. The starting of JADE platform consists of at least one container called MainContainer. After, all agents said are incremented and registered with the main container, the figure 1 present the JADE architecture.

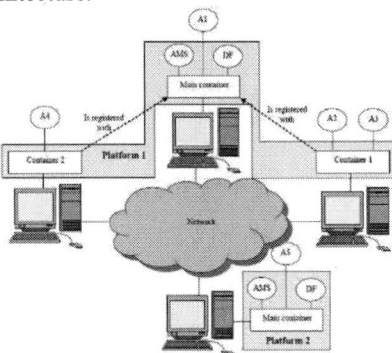

Fig1.JADE Agent platform

C. Raspberry P

This card provide high speed, better accuracy, good flexibility and low cost solution for development of embedded system equipped by ARM. Using this last board as development platform speed up the process of development. Raspberry pi Model B (as shown in Fig.2) is currently most popular ARM board. It has a Broadcom BCM2835 system on a chip SoC, which includes an ARM1176JZF-S 700 MHz processor, VideoCore IV GPU, and is shipped with 512MB of RAM .It does not include a built-in hard disk instead it uses an SD card for booting and long-term storage. It comes with two USB ports, RJ45 Ethernet port, HDMI port and RCA output on board.

Fig2. Raspberry PI 2

D. Distributed C-Means Algorithm

The C-means method of classification defined in [6], is an algorithm for segmentation of image which consists on a partitioned groups of set S of n attribute vectors into c classes (clusters C_i , i= 1,..., c), generally based on different criteria segmentation : gray levels, texture, or shapes. The main goal of the algorithm is to find the class centers for objective to minimize the cost function by using:

$$J = \sum_{i=1}^{c} J_i = \sum_{i=1}^{c} \quad \sum_{k \in c_i} d(x_k, C_i)$$

Where:

C_i is the center of the ith class

$d(x_k, C_i)$ is the distance between i^{th} center C_i and the kth data of the set S

We use the Euclidean distance to define the objective function as follows:

$$J = \sum_{i=1}^{c} J_i = \sum_{i=1}^{c} \quad \sum_{k, x_k \in C_i} ||x_k - C_i||^2$$

The partitioned groups can be defined by a binary membership matrix U(c, n), where each element u_{ij} is formulated by:

$$\begin{cases} 1 \ if \ ||x_k - C_i||^2 \leq ||x_k - C_k||^2, \forall k \neq i \\ 0 \ otherwise \end{cases}$$

(i=1 tp c, j=1 to n; n is the total number of points in S).

Since a data must belong to only one class, the membership matrix U has two properties which are given in the following equations:

$$\sum_{i=1}^{c} u_{ij} = 1, \forall j = 1, \dots, n$$
$$\sum_{i=1}^{c} \sum_{j=1}^{n} u_{ij} = n$$

The value C_i of each class center is computed by the average of all its attribute vectors:

$$C_i = \frac{1}{|C_i|} \sum_{k, x_k \in C_i} x_k$$

$|C_i|$ is the size or the cardinal of C_i.

The C-means classification is achieved using the following algorithm stages:

Table 1.All stages of C-means algorithm

Stage	Description
1	Initialize the class centers C_i (i= 1, … , c). This carried out by selecting randomly c points in the gray level scale [0 … 255]. For each iteration i
2	Determine the membership
3	Compute the objective function J
4	Compare the obtained objective function J to the one computed at iteration i-1 and stop the loop (i.e. go to stage 6) if the absolute value of the difference between the successive objectives functions is lower than an arbitrary defined threshold (S_{th})
5	Calculate the new class centers, and return back to perform stages 2,3 and 4
6	End

III. SIMULATIONS AND RESULTS

This section presents the implementation of the proposed distributed algorithm in our proposed middleware based on embedded agent using the cardiac MRI. This intelligent platform is a distributed and parallel embedded agent executed on raspberry pi devices on which we implemented the image segmentation algorithm. The proposed is based on all concepts: intelligent devices, JADE Middleware, agent and distributed C-means algorithm for objective to perform a high performance computing.

The experimental evaluation was done on the test-bed with three Raspberry PI model B which host agent, see Fig.3. We obtain the following results (see Fig.4), the figure (a) corresponds to a input image, and the figures (b),(c),(d), are the segmented output images where each of them corresponds to class centers (c1=0, c2=127, c3=255).

All features of this distributed method for data image classification where we are usingthe initial class centers (c1=0, c2=127, c3=255) is shown for dynamic convergence analysis, where we see clearly in Table 2 the convergence of the algorithm after 8 iterations to the final class centers (c1,c2,c3)=(13.00,99.00,220.00) and which are illustrated in fig5.

Iteration	c1	c2	c3	$|J_n\text{-}J_{n-1}|$
1	0,00	127,00	254,00	5,29E+05
2	15,00	110,00	234,00	8,50E+04
3	15,00	106,00	226,00	4,21E+03
4	14,00	103,00	223,00	9,36E+03
5	14,00	101,00	221,00	1,22E+03
6	13,00	100,00	220,00	7,79E+03
7	13,00	99,00	220,00	6,81E+02
8	13,00	99,00	220,00	0,00E+00

Table2.Differents states of classification algorithm

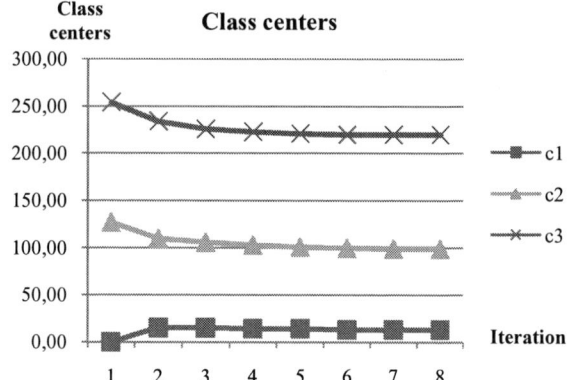

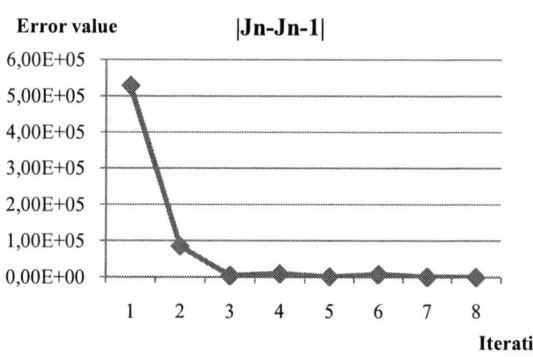

Fig.5.Class centers (c1, c2, c3)=(0,127,254)
And Error of the cost function

The process of our proposed is presented in Fig.6. The processes begin by activation of Agent Task Dispatcher (ATD) for identification of workers agents. The ATD send image and the class centers to Agent Local Worker (ALW) for objective split image based on number of available Embedded Remote Agents (ERA). After preparation of data, ALW send the elementary image to all ERAs, in this architecture we are using three ERA respectively: 1, 2 and 3. All ERA used perform the distributed C-means classification and send results to ALW. This last, create the c output segmented images where c corresponds to the class number.and liberate both of ALW and ERA.

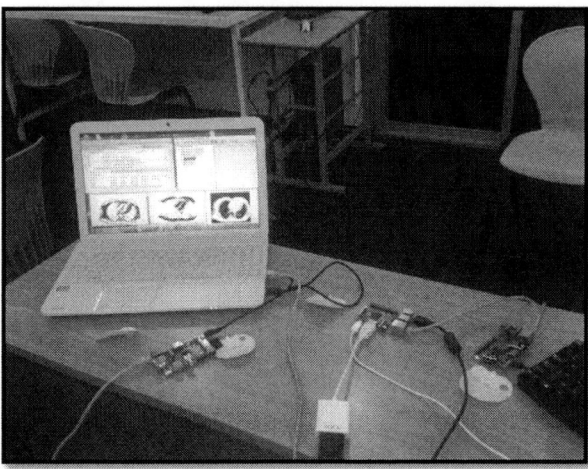

Fig.3. Testbed

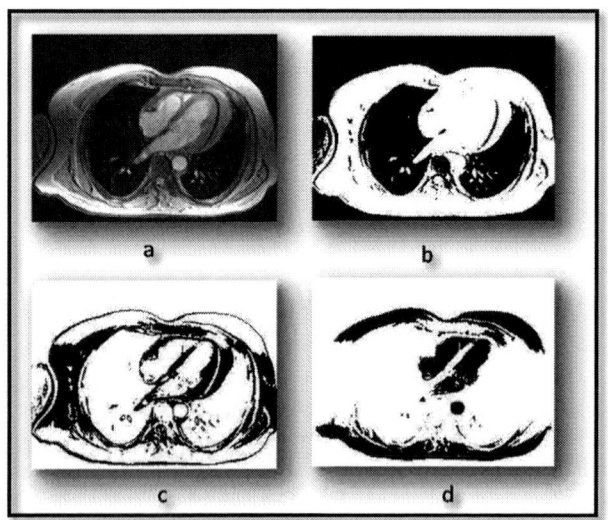

Fig.4. Segmentation results

IV. CONCLUSION AND PERSPECTIVES

This research work presented in this manuscript proposed a method to fill the gap in the design methodology for MASs related to the performance and responsiveness. The approach is for objective segmentation of medical image using JADE platform and embedded agent based on Raspberry PI model B. The technique provides many opportunities for future work. First of all, the standardization of our proposed for using in others domains, load balancing for all agents and integrated of Internet Of Things (IoT).

REFERENCES

[1] S.Sotiriadis, N.Bessis, Y.Huang, P.Kuonnen, N.Antonopoulos, A JADE Middleware for Grid inter-cooperated infrastructures, international conference on advanced information networking and applications, 978-7695-4338-3, 2011.

[2] L, Zhang, Q.Wang, X.Shu, A mobile-Agent-Based Moddleware for wirless Sensor Networks Data fusion, International Instrumentation and measurement technology, 978-1-4244-3353-7, May 5-7 Singapore 2009.

[3] L.Chunlin, L.Layuan, A multi-agent model for service-oriented interaction in a mobile grid computing environment, Pervasive and mobile computing 7, 270-284, ELSEVIER, 2011.

[4] M.Youssfi, O.Bouattane, J.Bakkoury, M.O.Bensalah, A new massively parallel and distributed virtual machine model using mobile agents, international conference on multimedia computing and systems, 978-1-4799-3823-0, April 14-16 Marrakech, Morroco 2014.

[5] M. Youssfi, O. Bouattane, and M.O. Bensalah " On the Object Modelling of the Massively Parallel Architecture Computers", Proceedings of the

IASTED Inter.Conf. Software engineering, Innsbruck, AUSTRIA, pp 71-78, February 16 - 18, 2010.

[6] O.Bouattane, B. Cherradi, M. Youssfi and M.O. Bensalah "Parallel cmeans algorithm for image segmentation on a reconfigurable mesh computer" ELSEVIER. Parallel computing, 37 pp 230-243, 2011.

[7] F.Bellifemine, A.Poggi, G.Rimassa, Developing Multi-agent systems with JADE, Intelligent Agents VII, pp.89-103, speinger 2001.

[8] M.Higashino, T.Hayakawa, K.Takahashi, T.Kawamura, K.Sugahara, Management of streaming multimedia content using mobile agent technology on pure P2P-based distributed e-Learning system, international conference on advanced information networking and applications, 978-0-7695-4953-8, March 25-28 Barcelona 2013.

[9] F. L. Bellifemine, G. Caire, and D. Greenwood, "Developing Multi-Agent Systems with JADE",Wiley, 2007.

[10] I.Satoh, Mobile Agent Middleware for dependable distributed systems, international conference on informatique technology interfaces, june 27-30, Cavtat, Croatia 2011.

[11] R.Abidar, K.Moummadi, H.Medromi, Mobile device and multi agent systems: an implemented platform of real time data communication and synchronization, international conference on multimedia computing and systems, 978-1-61284-730-6, 7-9 April Ouarzazate, Morocco.

[12] F.Bergenti, G.Caire, D.Gotta, Agenst on the move : JADE for android devices, CEUR workshop proceeding voal-11260, sepy. 25-26 catania, Italy 2014.

[13] Petr Kadera1, Petr Novak1, Vaclav Jirkovsky, Pavel Vrba1, Performance models preventing multi-agent systems from overloading computational resources, Automation, Control and Intelligent Systems, 2(6): 105-111, 2014.

[14] AbhilashKantamneni, Laura E. Brown, Gordon Parker, Wayne W. Weaver, Survey of multi-agent systems for microgrid control, Engineering Applications of Artificial Intelligence 45, 192–203, ELSEVIER, 2015.

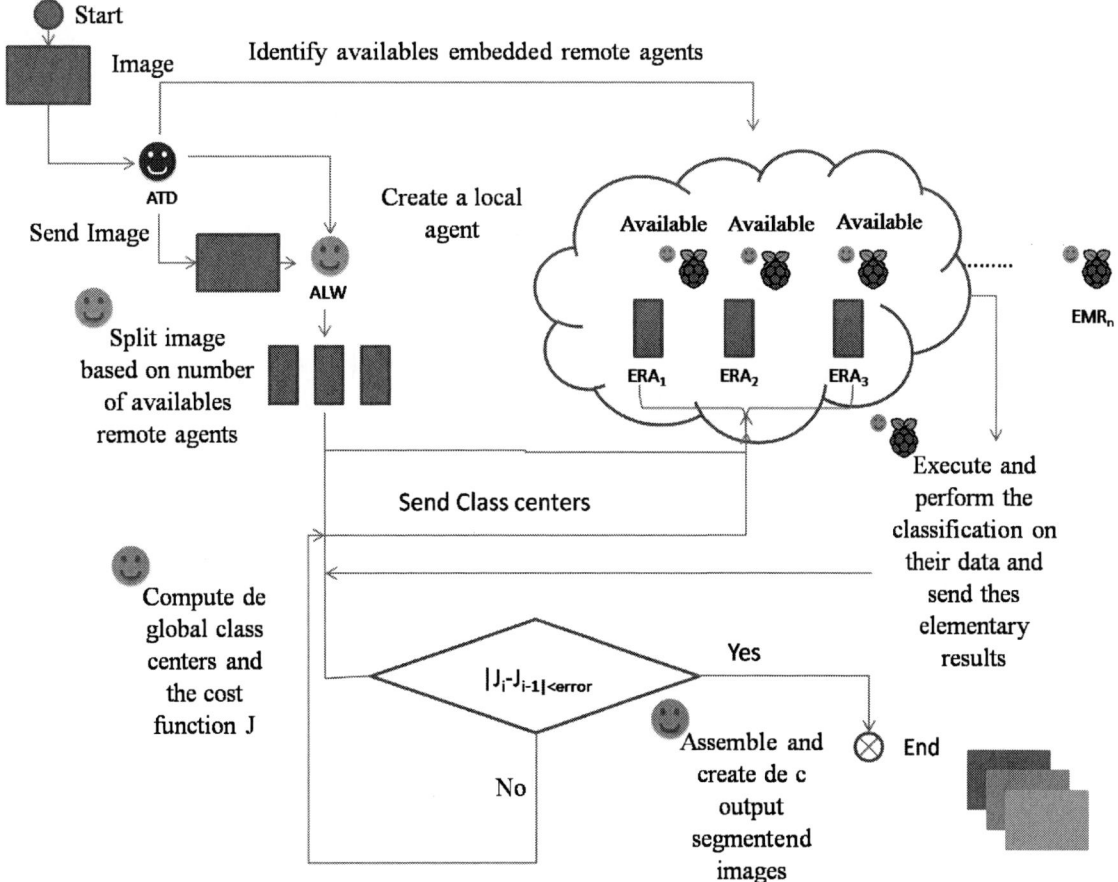

Fig.6. Our proposed process of image segmentation

Interdigitated Electrode Biosensor for DNA Sequences Detection

Ayoub Bourjilat[1], Djilali Kourtiche[1], Frederic Sarry[1], and Mustapha Nadi[1]

[1]Université de Lorraine – CNRS, UMR 7198, Institut Jean Lamour (IJL),Vandœuvre-lès-Nancy, France
Ayoub.bourjilat@univ-lorraine.fr, djilali.kourtiche@univ-lorraine.fr, frederic.sarry@univ-lorraine.fr,
and mustapha.nadi@univ-lorraine.fr

Abstract— In this work, we present the design of a biosensor for specific DNA sequences detection. Bioimpedance spectroscopy (BIS) is used to characterize DNA hybridization. This theoretical study allows to optimize the geometric parameters of the biosensors by reducing the cell factor (Kcell), where the CoventorWare software was used to modelize the biosensors for design. In addition, the acquisition by a Red Pitaya fast acquisition board for measuring the bioimpedance parameters was studied before the design of the biosensor

Keywords—Biosensor; DNA hybridization; interdigitated electrodes; RedPitaya; bioimpedance spectroscopy

I. INTRODUCTION

In the field of the design of the biosensor, much effort has involved the development of systems for the detection of DNA such as biosensors which incorporate reaction technique polymerase chain reaction (PCR) [1], these developments allow to measure amplification and detection of DNA. The DNA is often characterized by optical detection systems [2], micro gravimetric methods [3], surface plasma resonance (SPR) [4] and electrochemical system detection [5]. However the use of optical detection may prove costly, complex and the electrochemical techniques can provide a low cost, small size and convenience of integration and microminiaturization. Instead a single molecule, the DNA is represented as a pair of molecules closely associated, these two long strands forms a double helix. Generally a base linked to a sugar is called a nucleotide. The four DNA bases are Adenine (A), Cytosine (C), Guanine (G) and Thymine (T), the four bases are attached to the sugar /phosphate to form a complete nucleotide. The double helix of DNA is stabilized by hydrogen bonds. Denaturation and hybridization respectively define the separation and matching of the two complementary DNA strands. Since the strands are interconnected by non-covalent hydrogen bond, they can be separated by increasing the temperature of the DNA. Conversely, a recombination of complementary DNA strands may be obtained by slowly cooling a solution containing the DNA strands [6]. This process is called hybridization. The binding energy between the two strands depends on the percentage CG bonds because they offer a triple hydrogen bond, while AT bonds offer only one double bond hydrogen. During hybridization, the base pairing between the two strands of DNA depends on the complementary base sequences AT and GC. This phenomenon is used as the basis of the DNA probe as the hydrogen bonds between the two complementary DNA strands which are more stable than non-specific adsorption of DNA to the surface of the biosensor. In our study we are going to use the spectroscopy of bioimpedance for the characterization of DNA (hybridization, quantification) using an equivalent circuit model.

II. MODELIZATION AND CONCEPTION

A. Basic concept of dielectric material

The dielectric theory is based on the concept of the electric capacitor. In a capacitor, the dielectric is maintained between two parallel electrode plates, which is characterized by an equivalent circuit with two parameters, the capacitance (C) and conductance (G) (Fig.1).

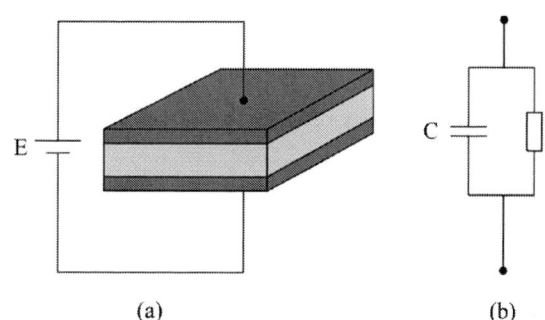

(a)   (b)

Fig. 1.   (a) Dielectric between two metallic plates. (b) Equivalent circuit.

The C and G can be determined by the following equations:

$$C = \frac{A\varepsilon_0\varepsilon_r}{d} \qquad (1)$$

$$G = \frac{A\sigma}{d} \qquad (2)$$

Where; d is the distance between the two metallic planes (μm); A is the surface of the metallic armatures (μm²); ε_0 is the permittivity of vacuum, ε_r the relative permittivity of the dielectric material and σ is the electrical conductivity of the dielectric material.

978-1-4673-8760-6/15 $31.00 © 2015 IEEE

B. Equivalent circuit model

The biosensor to be produced, comprises two interdigitated metal electrodes with : length L, width W and the distance between two adjacent electrodes S (Fig .2.).

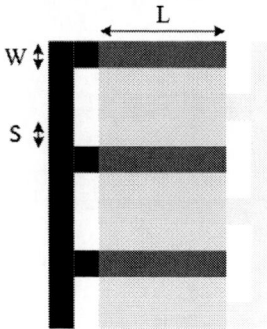

Fig. 2.   Structure of interdigital biosensor.

Figure 3a shows the equivalent circuit of the sensor with DNA strands. Figure 3b corresponds to the equivalent circuit model. R_{SOL} represents the liquid solution in parallel with the DNA resistance (R_{ADN}) [9]. This model does not take into account the diffusion impedance (Warburg impedance). In practice, this phenomenon influences the low frequency impedance. The resistance R_{SOL} shown in the model represents the conductive properties of the medium under the effect of electric field, which is related to the conductivity of the electrolyte medium and the cell factor K_{cell}.

$$R_{Sol} = \frac{K_{Cell}}{\sigma_{Sol}} \tag{3}$$

$$K_{Cell} = \frac{2.K(k)}{L.(N-1).K(\sqrt{1-k^2})} \tag{4}$$

$$K(k) = \int_0^1 \frac{1}{\sqrt{(1-t^2).(1-k^2t^2)}} dt \tag{5}$$

$$k = \cos(\frac{\pi}{2}.\frac{W}{W+S}) \tag{6}$$

With N, the number of electrodes, the function K(k) (5) is the incomplete elliptic integral of the first degree, the constant Kcell depends only on the geometric parameters of the sensor (N,S,W,L).

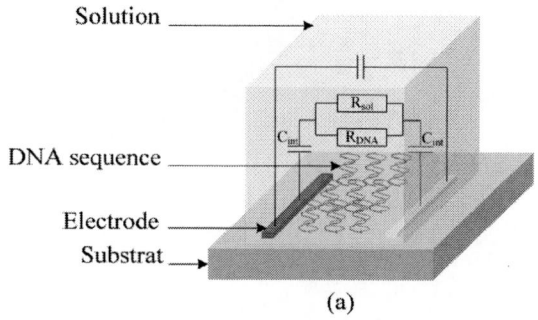

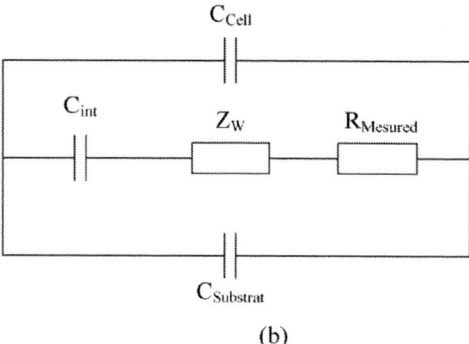

Fig. 3.   Equivalent model including immobilization of DNA strands.

The direct electrode capacitive coupling is represented by the capacity C_{cell}.

$$C_{Cell} = \frac{\varepsilon_0 \varepsilon_{r,Solution}}{K_{Cell}} \tag{7}$$

The impedances representing the double layer phenomena occurring at the electrode electrolyte interfaces Cint, which depend on the electrode material and the electrolyte.

$$C_{int} = C_0 LW \frac{N}{4} \tag{8}$$

The impedance of polarization is determined by the following expression:

$$Z_{int} = \frac{1}{j\omega C_{int}} \tag{9}$$

C. Sensor optimization

The impedance depends on the electrical properties (conductivity and relative permittivity) of the medium and also on the cell factor Kcell [10].

The impedance and the admittance of the biosensor are described by the following expressions:

$$Z = \frac{K_{cell}}{\sigma_{sol} + j\omega\varepsilon_0\varepsilon_r} \quad (10)$$

$$Y = G + j\omega C \quad (11)$$

$$G = \frac{\sigma}{K_{cell}} \quad (12)$$

$$C = \frac{\varepsilon_0\varepsilon_r}{K_{cell}} \quad (13)$$

The impedance and the total admittance in the equivalent electric circuit (see Fig. 3.) are calculated by:

$$Z = \frac{1}{G_{solution} + j\omega C_{solution}} + \frac{1}{j\omega C_{int}} \quad (14)$$

$$Y = G + j\omega C \quad (15)$$

From equations (14) and (15), the total capacity C is:

$$C = \frac{C_{int}G^2_{solution} + \omega^2 C_{solution}C_{int}(C_{int} + C_{solution})}{G^2_{solution} + \omega(C_{int} + C_{solution})^2} \quad (16)$$

When ω reach approximately zero, the value of the total capacitance (C) is equal to C_{int}.

By replacing C in the equation (16) and C_{int} in the formula (8) the capacitance of the double layer becomes:

$$C_0 = \frac{4\varepsilon_0\varepsilon_{r,LF}}{NLWK_{cell}} \quad (18)$$

To reduce the effect of the double layer capacity, decreased Kcell is required. For convenience of design, W and S are set at 5 microns. Figure 4 shows the effects of the variation of L and N on the Kcell cell factor; we used the equations (4), (5) and (6).

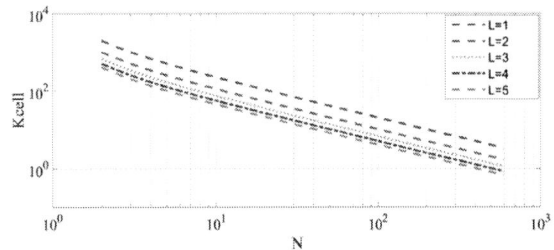

Fig. 4. The cell factor as a function of the number and the width of electrodes

Figure 4 shows the results obtained with the following geometrical parameters: L = 4 mm and a number of electrodes between 80 and 300. To study the influence of geometric

variation of the sensor, we decided to model and design six sensors with optical lithography (table 1)

TABLE I. GEOMETRIC PARAMETERS OF BIOSENSORS

Biosensor	N	L(μm)	W(μm)	S(μm)	Kcell(1/m)
1	500	2.9	5	5	1.389
2	20	1	30	20	90
3	310	3.1	5	5	2.0879
4	80	0.48	3	3	52.7426
5	80	4	3	3	6.32
6	310	4	5	5	1.62

D. Conception of the biosensors

Achieving the sensors was done in Minalor Centre of the University of Lorraine. Glass has been chosen as a substrate for its biocompatibility with the biological media. Its transparency will simplify the visualization of samples deposited on the surface of the sensor. The electrodes are made of gold for stability reasons. Gold has a conductivity of $\sigma_{Au} = 45.2.10^6$ S/m. It is often used in biosensors for DNA hybridization detection by surface functionalization.

III. EXPERIMENTAL RESULTS

Firstly, the characterized sensor from our previous work will be used [10]. This will allow us to test the Red Pitaya board for our future bioimpedance measurement. The Red Pitaya bord is a platform for instrumentation, testing and open source Linux 14.04 LTS measures. It is based on a System On Chip (Soc) Xilinx Zynq 7010 on a Dual ARM cortex A9 MPCore and an FPGA Artix 7. The implementation of a Red Pitaya acquisition board for bio-impedance spectroscopy is performed in a LabVIEW development environment. The main objective of the RedPitaya is to replace expensive and cumbersome measuring. As shown in the figure 5 the program starts with an initialization of the Red Pitaya card, open and short correction, generation of the sweep data in a frequency range from 100 Hz to 1MHz, acquisition of the data after excitation of the DUT (Device Under Test) and finally the Data processing calculates the tested component of the module. For this measurement we need the voltage across the component and the current, the Red Pitaya card allows us to measure only the voltage, a known resistance value is used to deduce the current.

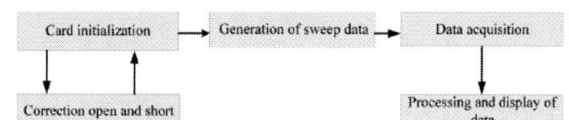

Fig. 5. Red Pitaya Labview diagram flow

A. Measurement with the interdigitated electrodes biosensor

We make measures on standard solutions using an interdigitated electrodes biosensor. The figure 6 confirms the theoretical principle, when the conductivity is low the value of the impedance module is high and when the conductivity is high the value impedance module is low.

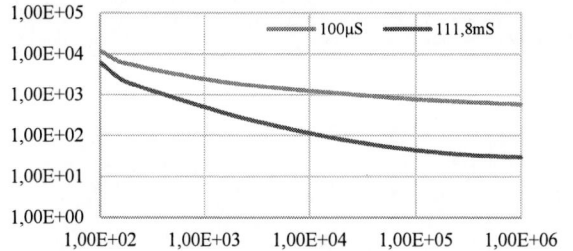

Fig. 6.   The module impedance as a function of frequency for different standard solution conductivity

IV. CONCLUSION

In this work, biosensor's models were realized to reduce the effect of the interface on electrode electrolyte at low frequency. The preliminary results allow us to perform an implementation of bioimpedance architecture measurement in a frequency band (100 Hz to 10 MHz). Our perspective in this work is the design of the biosensors, immobilize DNA for hybridization using the self-assembled monolayer (SAM) [11] and finally proceed the measurement and detection of hybridization using the Red Pitaya fast acquisition board .

REFERENCES

[1]   Z. Gao and Y. Peng, "A highly sensitive and specific biosensor for ligation- and PCR-free detection of MicroRNAs," *Biosens. Bioelectron.*, vol. 26, no. 9, pp. 3768–3773, 2011.

[2]   J. J. Storhoff, S. S. Marla, P. Bao, S. Hagenow, H. Mehta, A. Lucas, V. Garimella, T. Patno, W. Buckingham, W. Cork, and U. R. Müller, "Gold nanoparticle-based detection of genomic DNA targets on microarrays using a novel optical detection system," *Biosens. Bioelectron.*, vol. 19, no. 8, pp. 875–883, 2004.

[3]   R. Gabl, H. D. Feucht, H. Zeininger, G. Eckstein, M. Schreiter, R. Primig, D. Pitzer, and W. Wersing, "First results on label-free detection of DNA and protein molecules using a novel integrated sensor technology based on gravimetric detection principles," *Biosens. Bioelectron.*, vol. 19, pp. 615–620, 2004.

[4]   K. Tamada, F. Nakamura, M. Ito, X. Li, and A. Baba, "SPR-based DNA detection with metal nanoparticles," *Plasmonics*, vol. 2, pp. 185–191, 2007.

[5]   G. Liu, Y. Wan, Z.-Y. Zou, S.-Z. Ren, and C.-H. Fan, "Deoxyribonucleic Acid Molecular Design for Electrochemical Biosensors," *Chinese J. Anal. Chem.*, vol. 39, no. 7, pp. 953–962, 2011.

[6]   W. Cai, J. R. Peck, D. W. van der Weide, and R. J. Hamers, "Direct electrical detection of hybridization at DNA-modified silicon surfaces.," *Biosens. Bioelectron.*, vol. 19, no. 9, pp. 1013–1019, 2004.

[7]   A. Lasia, "Electrochemical Impedance Spectroscopy and its Applications," vol. 32, pp. 143–248, 1999.

[8]   O. A. Sadik, H. Xu, E. Gheorghiu, D. Andreescu, C. Balut, M. Gheorghiu, and D. Bratu, "Differential impedance spectroscopy for monitoring protein immobilization and antibody-antigen reactions," *Anal. Chem.*, vol. 74, pp. 3142–3150, 2002.

[9]   D. Berdat, A. C. Martin Rodríguez, F. Herrera, and M. A. M. Gijs, "Label-free detection of DNA with interdigitated micro-electrodes in a fluidic cell.," *Lab Chip*, vol. 8, pp. 302–308, 2008.

[10]   T. T. Ngo, H. Shirzadfar, A. Bourjilat, D. Kourtiche, and M. Nadi, "A method to determine the parameters of the double layer of a planar interdigital sensor," in *www-ist.massey.ac.nz*, 2014, pp. 2–4.

[11]   K. Kerman, Y. Morita, Y. Takamura, and E. Tamiya, "Label-free electrochemical detection of DNA hybridization on gold electrode," *Electrochem. commun.*, vol. 5, pp. 887–891, 2003.

Self-Charging of Medical Instruments Based on Bioelectric Potentials

Khalifa Elmansouri[*], Rachid latif
Signals System and Computer Sciences Laboratory
ESSI, Morocco
khalifa.elmansouri@outlook.com

Abstract— Today, the designers of medical instruments seek to find more compact standalone recharging systems for the low power medical devices in order that it can be used in isolated and remote communities where electricity is unavailable. In the present study, we will try to show that the electrical activity of the human body can be considered as a source of electrical energy and it can recharge chemical storage batteries. Thus, we have used a recharging a discharged nickel metal hydride battery to accumulate energy based on bioelectric potentials. Promising results have been obtained with a configuration of the offset potentials greater than 10mV.

Keywords—harvesting; portable systems; electrophysiological signal; electrical generators

I. INTRODUCTION

Human power is an attractive energy source and it was studied in many works, such as in [1, 2] where the works have focused on mechanical power and on body's own inertia for generate electricity , whilst the use of the electrical activity of the human body as electrical source remains a challenging issue.

The introduction of bioelectricity dates back to the works of Luigi Galvani in 1787, and his observations that a frog nerve-muscle could be contracted by delivering electrical impulses, this allowed to explain the action of living tissues in terms of bioelectric potentials. Now it is well established that the human body, can be considered as a power station generating multiple electrical signals with two internal sources, namely muscles and nerves. Knowing so, the electric energy which the body has got is approximately 100 W. About 25% of this energy is used by the skeleton and heart, 19% by the brain, 10% by the kidneys and 27% by the liver and spleen [3]. The source of the bioelectricity may be the electrochemical activity of human body or a vibrating structure [2, 4]. So harvesting a part of the body energy is interesting because such power supplies are free, non-polluting and long-lasting [5].

During subsequent years, in spite of the outstanding evolution which paved the way for better understanding of bioelectrical phenomena within the human organs, including the cardiovascular, ophthalmologic and neural systems, the use of these bioelectric potentials as a source of energy was restricted because of its low amplitude.

In this paper, we will show that energy storage based on the electrical activity of the human body is a promising issue to recharge chemical batteries used in the embedded medical systems like pacemaker or pulse generator, which senses the electrical signals activity of the heart and delivers pacing therapy to the patient.

Thus, we divide this manuscript in the following form: we present the electrical activity of the human body; we show experimental setups and battery charging results. Finally, we carry out the conclusion.

II. ELECTRICAL ACTIVITY OF THE HUMAN BODY

Based on the work of Peter Strong in1973, human body is composed of cells of different types; these cells may vary from 1 micron to 100 microns in diameter, from 1 mm to 1 m in length, and have a typical membrane thickness of 0.01 micron [6]. The Cells functioning can be considered similar to a tiny biological battery based on the following principal ions: sodium (Na^+), potassium (K^+) and chloride (Cl^-).

In normal state, the cell's membrane permits the entry of potassium and chloride but impedes the flow of sodium. Consequently, a difference of potential is resulting from the distribution of positive charged ions on the outer surface and negative charged ions inside the cell membrane (Fig. 1).

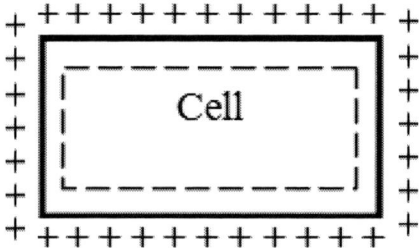

Fig. 1. Cell in normal state

In general, there are three human body's organs characterized by typical bioelectric potentials and they are the heart, brain and skeletal muscle.

The heart has got a group of cells known as the sino-atrial node (SA node is 25 to 30 mm in length and 2 to 5 mm thick) which initiates the heart activity. This system generates impulses at the normal rate of the heart by conducting action potentials through a complex change of ionic concentration across the cell membrane. The potential field generated extends to the other parts of the heart and create a wave called Electrocardiogram (ECG) shown in Fig. 2.

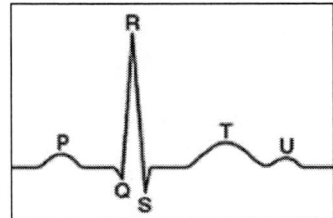

Fig. 2. Electrocardiogram signal

The brain composed by neurons generates rhythmical potentials which appear as a surface waveform known as the Electroencephalogram shown in Fig. 3.

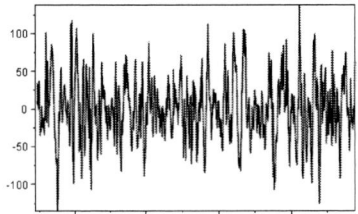

Fig. 3. Electroencephalogram signal

The skeletal muscle generates action potentials such as their summation is known as Electromyogram shown in Fig. 4.

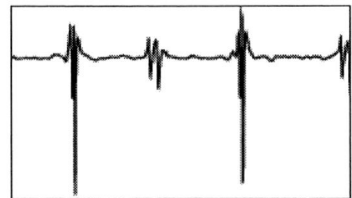

Fig. 4. Electromyogram signal

The bioelectric potentials produced by the coordinated activity of large groups of cells of different body organs tend to migrate through the body fluids and sets up potential differences between various portions of the body. These potential differences can be conveniently picked up by placing two conducting plates at any two points on the surface of the body. This conducting plate is called electrode and it is attached to the surface of the body through an electrolyte (as shown in Fig. 5).

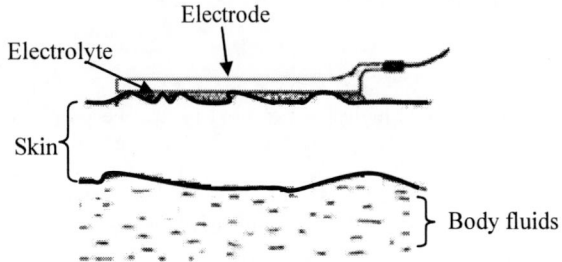

Fig. 5. the electrode and electrolyte

In Fig. 5 we can distinguish two types of contact: Electrode-Electrolyte and Electrolyte-skin contact.

At the first contact, the electrode tends to discharge ions into the electrolyte, and on the other hand, the ions in the electrolyte tend to combine with each electrode. The result is the creation of a difference of potential called electrode potential (or offset potential). The most known electrode potentials are shown on the following table [7]:

TABLE I.   POTENTIAL BETWEEN ELECTROLYTES AND ELECTRODES

Electrode metal	Electrolyte	offset potential
Stainless steel	Saline	10 mV
Silver	Saline	94 mV
Silver-silver chloride	Saline	2.5 mV
Silver-silver chloride	ECG paste	0.47 mV

At the Second contact, the skin acts as a diaphragm arranged between two solutions: electrolyte and body fluids with different concentrations, so that it gives a potential difference composed of biological potential and contact potential.

In practice, the potentials generated by the body's organs are commonly measured by an array of electrodes placed on the body surface. In general, between every two electrodes, we can find the electrical equivalent circuit shown in figure 6 [6]; this circuit presents two electrodes (Electrode 1 and Electrode2) and the voltages presented between them: the contact potential, offset potential and the biological signal.

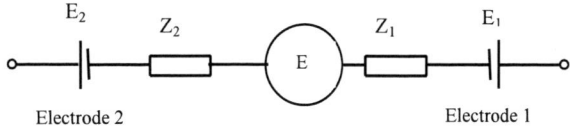

Fig. 6. The electrical equivalent circuit of pair of electrode

E_1 and E_2: Electrode potentials

Z_1: Skin contact Impedance

Z_2: Skin contact impedance plus tissue impedance

E: Bioelectric event plus contact potential

III. EXPEREMENTAL SETUP

A. Experimental signal

The electrophysiological signal used to test the proposed storage device is the abdominal electrocardiogram signal (AECG) presented in Fig. 7; this signal is from physionet database (this database offers a free access to large collections of physiologic signals) with the following characteristics:

- Signals recorded in labor, between 38 and 41 weeks of gestation

- Positioning of electrodes was constant during all recordings

- Ag-AgCl electrodes (3M Red Dot 2271) and abrasive material to improve skin conductance (3M Red Dot Trace Prep 2236)

- Bandwidth: 1Hz - 150Hz (synchronous sampling of all signals)

- Additional digital filtering for removal of power-line interference (50Hz) and baseline drift

- Sampling rate: 1 kHz

- Resolution: 16 bits

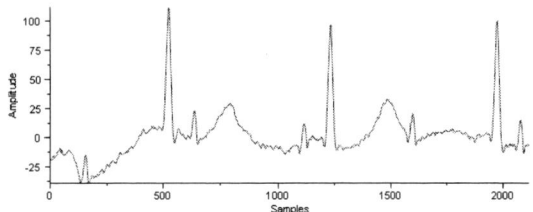

Fig. 7. Real abdominal ECG signal

The AECG signal is continuously generated in a waveform graph using simulator (shown in Fig. 8). The use of a simulator has two advantages, the first one is saving of time, the second one it is removing the difficulties of taking real ECG signals with invasive and noninvasive methods using medical machines.

Fig. 8. The VI which generates AECG continuously

B. Experimental hardware

The experiment is performed with the hardware shown in Fig. 9 and it is explained as follow:

- Computer equipped with Labview software.

- Electronic circuit composed by:
  - The microcontroller 16F628A,
  - The integrated circuit MAX232A (interface RS232C TTL/CMOS, it connects signals TXD and RXD from a

microcontroller to signals RXD and TXD on a PC ; also it convert the voltage levels and invert signals),

- Digital-To-Analog converter.

The computer sends data to the microcontroller (programmed using assembly language with the environment of free development MPLAB IDE of Microchip), once data received; it transmits an instruction towards the digital-to-analog converter.
- 1 quartz 20 MHz (it regulates the rate of transmission of the connection RS232C)

- Loader: it is the circuit constructed to charge the battery; it consists of a full wave rectifier and a capacitor.

- Nickel metal hydride battery (NiMH) is a type of rechargeable battery similar to the nickel-cadmium. The NiMH battery uses a hydrogen-absorbing alloy for the negative electrode instead of cadmium. it can have two to three times the capacity of an equivalent size nickel-cadmium battery; they have a high charge density and they do not require any type of charge controller or voltage regulator to be incorporated into the circuitry. The unit 'mAh' stands for milli-amp-hour and is a measure of the battery's capacity, a 280 mAh capacity means that the batteries will last for 1 h if subjected to a 280 mA discharge current.

- Connectors, resistors, capacitors and wires to connect all the elements.

IV. BATTERY CHARGING RESULTS

The goal of this part is to show that electrophysiological potentials plus contact potential and offset potential can charge batteries used by low power medical devices. As explained above, the maternal composite abdominal electrocardiogram signal is continuously generated by the computer to the electronic circuit for charging the batteries; the electronic circuit is configured to present an electrode potential equal to 10mV and a contact potential equal to 0,6mV. The most important electrical factor of the power supply is that it should be able to provide a fairly significant amount of current. The charge time of a rechargeable battery is directly dependent on the amount of current supplied to it; during the experiments, each battery was charged while the voltage on the battery was measured.

The rechargeable cells used for the test are:

1.2V 280mAh NiMH

1.2V 650mAh NiMH

The time to charge each battery is shown in Table 2, and the plots of the typical batteries charging cycles are shown in Fig. 10.

TABLE II.   TIME REQUIRED TO CHARGING THE TESTED BATTERIES

Battery size (mAh)	charge time (h)
280	11,5
650	28

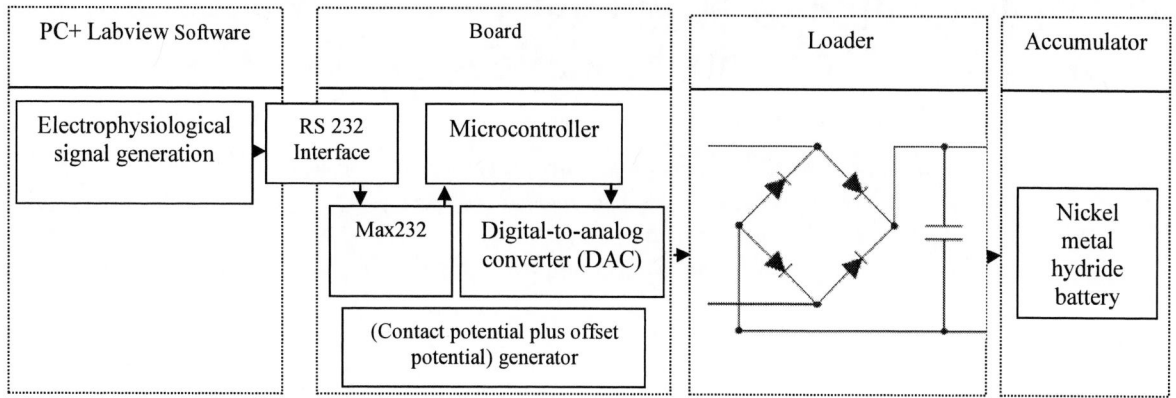

Fig. 9. Block diagram for designed circuit

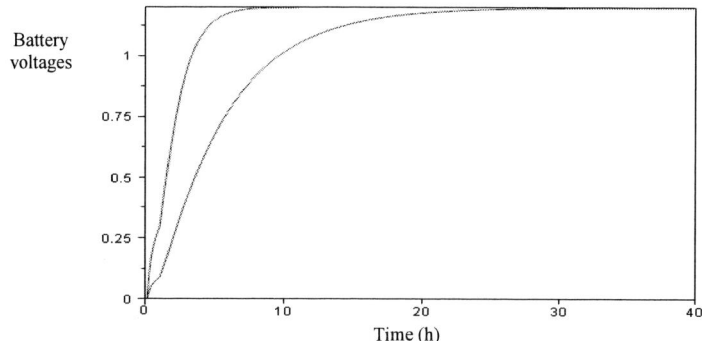

Time (h)

Fig. 10. Charge history of a 280 and 650 mAh batteries

V. CONCLUSION

The need to extend the availability of various energy devices is vital. In this study we have investigated the ability of electrophysiological potentials to recharge various capacities of nickel metal hydride batteries which are generally used by low power medical devices. It is very important to mention that the usual applications of diagnostic and treatment eliminates:

- The offset potential (which depends on species of metal, the type of electrolyte, its concentration and temperature)

- The contact potentials (which depends on the skin and skin preparation)

However in our work these potentials was not eliminated and it was integrated on the designed electronic circuit.

In conclusion, our main contribution it is that we showed that the bio-potentials can reload batteries of low power medical devices. To give an idea of the amount of energy that have been stored in the batteries being tested, a 650 mAh battery contains enough energy to power a rechargeable laptop mouse.

ACKNOWLEDGMENT

The authors wish to acknowledge the anonymous reviewers for their careful readings.

REFERENCES

[1] J. M. Donelan, Q. Li, V. Naing, J. A. Hoffer, D. J. Weber, and A. D. Kuo, "Biomechanical energy harvesting: generating electricity during walking with minimal user effort," Science, vol. 319, pp. 807-10, 2008.

[2] N.S. Shenck, J.A. Paradiso, "Energy scavenging with shoe-mounted piezoelectrics," IEEE Micro., vol. 21, pp. 30–42, 2001.

[3] E. Okuno, I. Caldas, and C.Chow , "Conservação de Energia". In Física para ciências biológicas e biomédicas, vol II, Lisboa: HARBRA, pp. 103-105, 1986.

[4] C.B. Williams, R.B. Yates, "Analysis of a micro-electric generator for microsystems," Sens. Actuat., vol. A 52, pp. 8–11, 1996.

[5] G. Poulin, E. Sarraute and F. COSTA, "Generation of electrical energy for portable devices: Comparative study of an electromagnetic and a piezoelectric system," Sensors and Actuators A: physical, vol. 116, no. 3, pp. 461-471, 2004.

[6] R.S. Khandpur, Handbook of biomedical instrumentation. Tata McGraw-Hill Education, 1992.

[7] G.L. Alexander, B.L. Edward. Principles of applied biomedical instrumentation. John Wiley & Sons, 1975.

Neuro-Space Mapping Technique for microwave nonlinear circuits Modeling

Taj-eddin Elhamadi[#1], Mohamed Boussouis[#2],

[#]*Information system and Telecommunication Laboratory, Faculty of Science,*

Abd Elmalek Essaadi University Tetouan - Morocco

[1]`tajeddinelhamadi@gmail.com`

[2]`m.boussouis@yahoo.fr`

Abstract— **Neural networks are been used for the last years for modeling the nonlinear microwave devices as the Pseudomorphic High Electron Mobility Transistor (pHEMT). In this paper the neuro-space mapping (SMNN)technique is used for modeling the DC characteristic of pHEMT, in this approach the Angelov model is used as the coarse model ,which can be adjusted using the fine model based on measurement. Good results are obtained for the drain-current, the trans-conductance and for the output conductance.**

Keywords— **Neural Networks-Space Mapping-GaAs HEMT-nonlinear modeling.**

I. Introduction

Microwave transistors (MESFET, HEMT, MOSFET and HBT) are used in majority of devices in modern microwave communication systems (satellite systems, mobile systems, etc.). Development of accurate and reliable models of microwave transistors is one of the basic aspects in microwave circuits design.

Different equation-based empirical large-signal models have been proposed for the simulation of field effect transistors (FETs) in nonlinear microwave circuits like amplifiers, filters, mixers and multipliers. For GaAs MESFET, Curtice [1] and Statz [2] models are the most used in simulators; many good models have been developed for HEMT (or pHEMT) devices[3-4]. However, the existing models with closed-form equations, while good for many existing devices, may not fit well with new devices.

Recently, artificial neural networks have been in use to replace empirical equations or compact models for a variety of devices such as the HEMT [5-6] and MOSFET [7]. Artificial neural networks (ANNs) are information processing systems with their design inspired by the studies of the ability of the human brain to learn from observations and to generalize by abstraction [8]. Neural networks are very sophisticated modeling techniques capable of modeling extremely complex functions. Neural networks learn by example. The neural network user gathers representative data, and then invokes training algorithms to automatically learn the structure of the data. The traditional structures of neural networks as the MLP (multi-layer perceptron) and RBF (radial basis function) needs only input-output information without any knowledge about the physical model. But required a high number of examples to be trained, and the model obtained is difficult to use into a standard microwave simulator. While with knowledge based modelling techniques, the neural networks are used to improve and optimized the accuracy of models already exist in microwaves simulators. The most popular used in microwave modelling are: Knowledge based neural network method (KBNN), Prior knowledge input method (PKI) and Space mapping neural network method (SMNN) [9].

In this paper the Space mapping neural network method (SMNN) is used for modeling the DC characteristic of pHEMT. The Angelov model is used as the coarse model, this model is adjusted using the fine model based on measurement. Good results are obtained for the drain-current, the trans-conductance and for the output conductance.

II. Neural Networks

Artificial neural network (ANN) consist of many simple elements called neurons. The neurons interact with each other using weighted connection similar to biological neurons. The network consists of one or more layers between input and output layers [10].

A. Multi-layer Perceptron Structure (MLP)

A standard multilayer perceptron (MLP) as shown in fig.1 consists of an input layer (layer 1), an output layer (layer N_L) and as well as several of hidden layers. Input vectors are presented to the input layer and fed through the network that then yields the output vector. The l_{th} layer output can be written as:

$$z_i^l = \varphi \left(\sum_{j=0}^{N_{l-1}} w_{ij}^l z_j^{l-1} \right) \qquad (1)$$

Where: $l = 2,3,...,L$ and $i = 1,2,3,...,N_l$

z_i^l and z_i^{l-1} are outputs of l_{th} and $(l-1)_{th}$ layer, w_{ij}^l are the weights between the i_{th} neuron in the l_{th} layer and the j_{th} neuron in the $(l-1)_{th}$ layer.

The function φ is an activation function of each neuron, usually the linear function is used for output layer, and the sigmoid function for hidden layers, the sigmoid function is defined as:

$$\varphi(x) = \frac{1}{1 + e^{-x}} \qquad (2)$$

The neural network learns relationship among sets of input output data (training sets).First, input vectors are presented to the input neurons and output vectors are computed. These

978-1-4673-8760-6/15 $31.00 © 2015 IEEE

output vectors are then compared with desired values and errors are computed. The training process proceeds until errors are lower than the prescribed values or until the maximum number of epochs (epoch is the whole training set processing) is reached. Once trained, the network provides fast response for various input vectors (even for those not included in the training set) without additional optimizations.

Many of algorithms can be used for the learning of neural networks [10], the most popular is the back propagation (BP). The Gradient-based optimizations techniques as the Quasi-Newton and Levenberg-Marquardt and the Global optimizations methods as the Genetic Algorithm (GA) can be also adapted for this type of problem because supervised learning of neural network can be viewed as a function optimization problem.

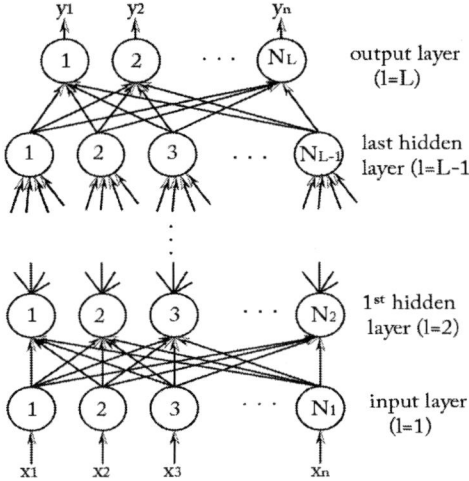

Fig. 1 Multi-layer perceptron structure (MLP)

B. Space Mapping Neural Networks Structure (SMNN)

Space mapping neural network method as shown in Fig.2 establishes a mathematical link between input space of fine and coarse models. Inputs of the coarse model x_c are changed to satisfy more accurate results than coarse model and less computational burden than fine model via input mapping. This mapping is constructed using ANN structure through its nonlinear behavior [9]. The fine model generally consists of EM simulator or direct measurements while the coarse model typically consists of empirical functions or equivalent circuit models. The mapping relationship can be written as:

$$x_c = f_{ANN}(x_f) \qquad (3-a)$$
$$Y_f \cong Y_{SMNN} = f_c(x_c) \qquad (3-b)$$

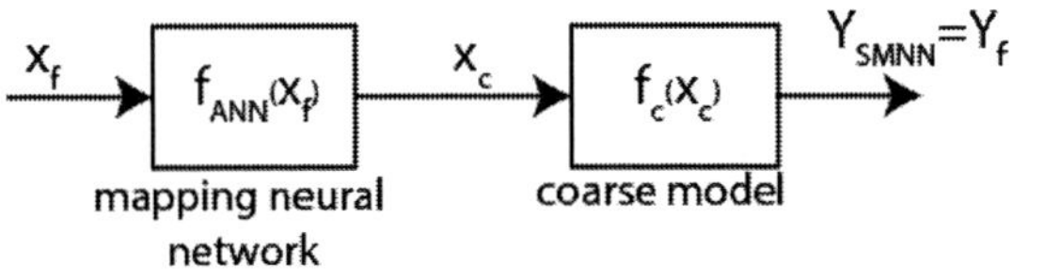

Fig. 2 Space mapping neural network model structure

III. MODELING TECHNIQUE

We propose here the DC modelling technique for a pHEMT. The proposed modelling technique is based on the space mapping neural networks. In this approach, DC-measurement is chosen as fine model. Angelov model of the drain-current $I_{ds} = f_c(V_{gs}, V_{ds})$ is used as coarse model.

A. The Coarse Model (Angelov model)

According to Angelov, drain-source current can be formulated empirically as:

$$I_{ds} = I_{pk}(1 + \tanh(\psi))(1 + \lambda V_{ds})\tanh(\alpha V_{ds}) \quad (4)$$
$$Where:$$

$$\psi = \sum_{i=1}^{n} P_i(V_{gs} - V_{pk})^i$$

I_{pk} and V_{pk} are respectively the drain-current and the gate voltage at the maximum trans-conductance.
The trans-conductance g_m and the output conductance g_d can be computed by:

$$g_m = \frac{\partial I_{ds}}{\partial V_{gs}} \qquad (5-a)$$

$$g_d = \frac{\partial I_{ds}}{\partial V_{ds}} \qquad (5-b)$$

B. SMNN Training Algorithm

Let's denote the input of fine model $V_f = [V_{gs}, V_{ds}]_f$ and the input of coarse model $V_c = [V_{gs}, V_{ds}]_c$. The output can be also denoted respectively I_{ds}^f and I_{ds}^c for the fine and the coarse model. The training algorithm partitioned for two phases: Initialization and Formal Training [11].

1) *Initialization:* The initialization phase making the SMNN model to be just equal to the coarse model. In this phase we initialize the neural networks by a preliminary training to learn unit mapping, where the weights ω are adjusted in order to minimize the mean square error (MSE).

$$\min_{\omega} \frac{1}{2} \sum_{p \in P} \left\| V_f^p - V_c^p \right\|^2 = \min_{\omega} \frac{1}{2} \sum_{p \in P} \left\| V_f^p - f_{ANN}(V_f^p, \omega) \right\|^2 \quad (6)$$

Usually an MLP3 structure is used for resolving this minimization problem, the input is $V_f = [V_{gs}, V_{ds}]_f$ and the output is $V_c = [V_{gs}, V_{ds}]_c$. The training of neural network in this phase is the same as that of the classical neural network. In general, Levenberg-Marquardt training algorithm is used.

2) *Formal Training:* In this phase the mapping neural network will be trained by actual device data in order to exceed the performance of the given coarse model. The coarse model output signal I_{ds}^c should become after receiving the mapped input signal an approximation of the fine output signal.

$$I_{ds}^f \cong I_{ds}^c \big|_{V_c = f_{ANN}(V_f, \omega)} \quad (7)$$

The minimization function can be written for this phase:

978-1-4673-8760-6/15 $31.00 © 2015 IEEE   203

$$\min_{\omega} \frac{1}{2} \sum_{p \in P} \left\| I_{ds,p}^f - I_{ds}^c|_{V_c = f_{ANN}(V_f,\omega)} \right\|^2 \quad (8)$$

The derivative of error function with respect to weight ω_i can be formulated as:

$$\frac{\partial E}{\partial \omega_i} = \frac{\partial E}{\partial I_{ds}^f} \cdot \frac{\partial I_{ds}^f}{\partial \omega_i} = \frac{\partial E}{\partial I_{ds}^f} \cdot \left(\frac{\partial I_{ds}^f}{\partial V_{gs}^c} \cdot \frac{\partial V_{gs}^c}{\partial \omega_i} + \frac{\partial I_{ds}^f}{\partial V_{ds}^c} \cdot \frac{\partial V_{ds}^c}{\partial \omega_i} \right)$$

$$= \frac{\partial E}{\partial I_{ds}^f} \cdot \left(G_m^c \cdot \frac{\partial V_{gs}^c}{\partial \omega_i} + G_d^c \cdot \frac{\partial V_{ds}^c}{\partial \omega_i} \right)$$

$$= \frac{\partial E}{\partial I_{ds}^f} \cdot \boldsymbol{G_c} \cdot \frac{\partial \boldsymbol{f_{ANN}^T}(\boldsymbol{V_f,\omega})}{\partial \omega_i} \quad (9)$$

Where: $\boldsymbol{G_c} = [G_m^c, G_d^c]$ is the conductance matrix of the coarse model, and $\frac{\partial \boldsymbol{f_{ANN}^T}(\boldsymbol{V_f,\omega})}{\partial \omega_i}$ is the first order derivative of ANN output with respect to weight.

IV. RESULTS AND DISCUSSION

A. DC-drain current modeling

The proposed modeling technique is used to modeling a $6 \times 30\mu m$ gate-width pHEMT, the drain-current I_{ds} is measured at 361 bias points (V_{ds} (0V to 5V) and V_{gs} (-2V to 0.6V)). This data measurement (361 sets) is used and partitioned into training (171 sets) and testing (190 sets) data.

Parameters of coarse model are obtained using the curve fitting technique. The optimized Angelov model parameters are given in table 1. For the ψ function we chose $n = 2$.

TABLE I

ANGELOV COARSE MODEL PARAMETERS VALUES

Angelov model parameter	value
V_{pk}	0.1 V
I_{pk}	35.6 mA
P_1	2.204 V^{-1}
P_2	-0.4425 V^{-2}
λ	0.084 V^{-1}
α	2.297 V^{-1}

For the mapping ANN we use an MLP3 structure with 10 hidden layer $[2 - 10 - 2]$. In the initialisation phase, Levenberg-Marquardt training algorithm is used, and for the formal training Back-propagation algorithm with adaptive rate and momentum can be a good chose for this phase. The fig.3 show the results obtained for the drain current before and after the Neuro-space mapping correction.

B. Trans-conductance and output conductance

Using the $(5 - a)$ the trans-conductance can be formulated as:

$$g_m = \frac{\partial I_{ds}^c}{\partial V_{gs}^f} = \frac{\partial I_{ds}^c}{\partial V_{gs}^c} \cdot \frac{\partial V_{gs}^c}{\partial V_{gs}^f} = G_m^c \cdot \frac{\partial V_{gs}^c}{\partial V_{gs}^f} \quad (10 - a)$$

The same for output conductance:

$$g_d = \frac{\partial I_{ds}^c}{\partial V_{ds}^f} = \frac{\partial I_{ds}^c}{\partial V_{ds}^c} \cdot \frac{\partial V_{ds}^c}{\partial V_{ds}^f} = G_d^c \cdot \frac{\partial V_{ds}^c}{\partial V_{ds}^f} \quad (10 - b)$$

The fig.4 shows a comparison betwenn SMNN model and Angelov coarse model obtained before and after the Neuro-space mapping correction.

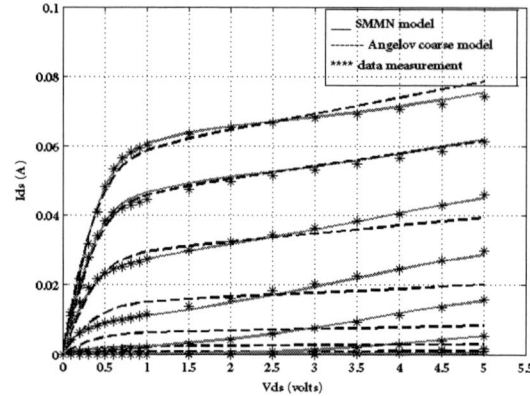

Fig. 3 Drain-current comparison between the SMNN model, Angelov coarse model and the data measurement at various $V_{gs}(-1V \ to \ 0.4V)$

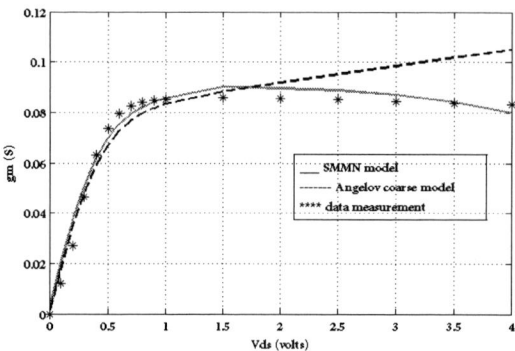

Fig. 4 Trans-conductance comparison between the SMNN model, Angelov coarse model and the data measurement at $V_{gs} = 0.1V$

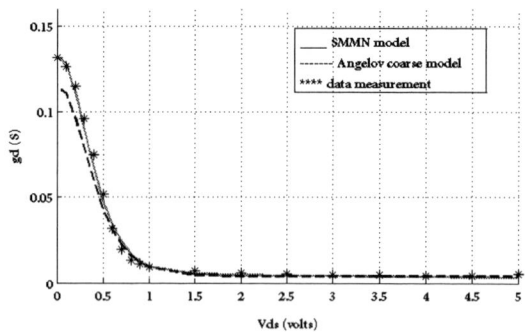

Fig. 5 Output conductance comparison between the SMNN model, Angelov coarse model and the data measurement at $V_{gs} = 0.4V$

V. CONCLUSIONS

The In this paper we propose to use the Space mapping neural network method (SMNN) for modeling the DC characteristic of a pHEMT. To improve the validate of proposed technique the Angelov model of the DC- drain current is modified using a mapping function for the voltage V_{gs} and V_{ds}. The final model present good approximation to the fine model even for the first order derivative g_m and g_d. The presented technique can also be in use with other empirical models as the Curtice and the Statz models. We can also improve the validate o this technique for the small and large signal regime.

ACKNOWLEDGMENT

The authors would like to acknowledge the Department of Communications Engineering at the University of Cantabria, Santander, Spain, for performing the measurements.

REFERENCES

[1] W. R. Curtice, "GaAs MESFET modeling and nonlinear CAD," IEEE Trans. Microwave theory Tech, vol. 36, pp. 220-230, 1988.

[2] H. Statz, P. Newman, I. W. Smith, R. A. Pucel et H. A. Haus, "GaAs FET device and circuit simulation in SPICE," IEEE Trans.Electron devices, vol. 34, pp. 160-169, Feb 1987.

[3] I. Angelov, H. Zirath et N. Rorsman, "A new empirical nonlinear model for HEMT and MESFET devices," IEEE Trans.Microwave theory and techniques, vol. 40, n° :12, pp. 2258-2266, 1992.

[4] T. Tanimoto, "Analytical nonlinear HEMT model for large-signal circuit simulation," IEEE Trans.Microwave Theory Tech, vol. 44, pp. 1584-1586, Sept 1996.

[5] T.Elhamadi, M.Boussouis, N.A.Touhami, "Modeling the drain current of a pHEMT using the artificial neural networks and a taylor series Expansion,"International journal of innovation and applied studies, vol.10, pp.132-137, jan 2015.

[6] M. Hayati and B. Akhlaghi, "An extraction technique for small signal intrinsic parameters of HEMTs based on artificial neural networks," International Journal of Electronics and Communications (AEÜ), vol. 67, pp. 123-129, 2013.

[7] N. Li, X. Li et S. Quan, "An ann-based small-signal equivalent circuit model for MOSFET device," Progress In Electromagnetics Research, vol. 122, pp. 47–60, 2012.

[8] S. Samarasinghe, Neural Network for Applied Sciences and Engineering, Auerbach Publications, 2007.

[9] M. Simsek, Q. J. Zhang, H. Kabir, Y. Cao et N. S. Sengor, "The recent developments in knowledge based neural modeling," Procedia Computer, vol. 1, pp. 1321-1330, 2010.

[10] Q. J. Zhang et K. C. Gupta, Neural network for RF and microwave design, Boston: ARTECH HOUSE, 2000.

[11] L. Zhang, J. Xu, M. Yagoub, R. Ding et Q. J. Zhang, "Efficient Analytical Formulation and Sensitivity Analysis of Neuro-Space Mapping for Nonlinear Microwave Device Modeling," IEEE transactions on microwave theory and techniques, vol. 53, pp. 2752-2767, 2005.

Implementation of a 17 bits Pulse Width Modulation Circuit using FPGA

Lucas Salomon, Robson Moreno and Tales Pimenta
Universidade Federal de Itajuba
Itajuba, Brazil
tales@unifei.edu.br

Abstract — **This paper presents the implementation of a 17 bits digital pulse width modulator (DPWM). Smaller pulses were obtained using internal carry chains that are available in many FPGAs. Our proposed architecture minimizes the critical paths that influence the linearity and induces the creation of carry chains without the use of adders. The frequency of the DPWM is 90.5kHz with a resolution of approximately 80ps, and the clock frequency is 46.34MHz.**

Keywords— Digital pulse width modulator (DPWM); carry chain; field-programmable gate arrays (FPGA); high resolution.

I. INTRODUCTION

Digital Pulse Width Modulation – DPWM is widely used in many applications, such as robotics, movement control, motors and most frequently in control systems and switching converters [1]. The implementation of control circuits using DPWM became popular due to many advantages, such as low sensitivity to parameter variations, programmability, implementation of advanced control approaches, reduction of external passive components, protection of the algorithm and calibration [2]. Some applications demand high precision and high resolution DPWM in which small voltage or current variations are required [3].

DPWM can be implemented using counters and comparators, delay lines, hybrid DPWM, segmented delay lines and dithering [4] [5] [6] [7] [8] [9]. In this work we have used the hybrid architecture, comprised of counters and delay lines in order to achieve higher resolution.

The resolution of a DPWM is given by $(N + n)$ bits, where N is the number of most significant bits (coarse adjustment) and n is the number of least significant bits (fine adjustment) in a duty cycle. The modulator precision is divided into two parts:

- The coarse adjustment, formed by the N bits, comes from a counter. The increments are given by [10]

$$\frac{f_{CLK}}{2^N} \qquad (1)$$

where f_{clk} is the clock frequency.

- The fine adjustment, formed by the n bits, is obtained from a circuit based on special delay lines called "carry chain" [11] [12] [13] [14] that are used for arithmetic operations in most FPGAs. Those lines allow higher resolution due to their short routing, and they are used in some DPWM modulator architectures [15] [16].

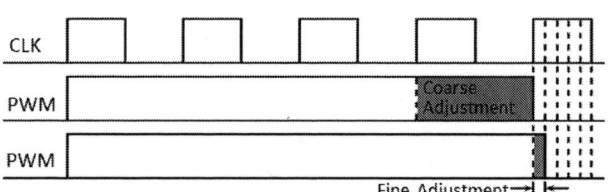

Fig. 1. Coarse and fine adjustments.

The coarse and fine adjustments are shown in the shaded areas in Fig.1. This work presents a 17 bits DPWM implemented in FPGA by using VHDL.

II. DPWM ARCHITECTURE

The DPWM circuit is shown in Fig. 2. The main differences of our work as compared to Liu [15] and Maksimovic [16] are the placement of registers ($F1,F2,...,F_2{}^n$) next to the Chain of Interconnected Logic Gates (CILG) structure in order to minimize the delays, the creation of logic functions that use the carry chain [17] connection, and the use of FFOut flip-flop instead of a SR latch.

Comparators CMP1 compares the counter bits (CNT) to zero, and CMP2 compares CNT to the N bits of the duty cycle. When CNT reaches 0, CMP1 changes its state to high and activates the FFOut register at the rising edge, thus starting the DPWM signal.

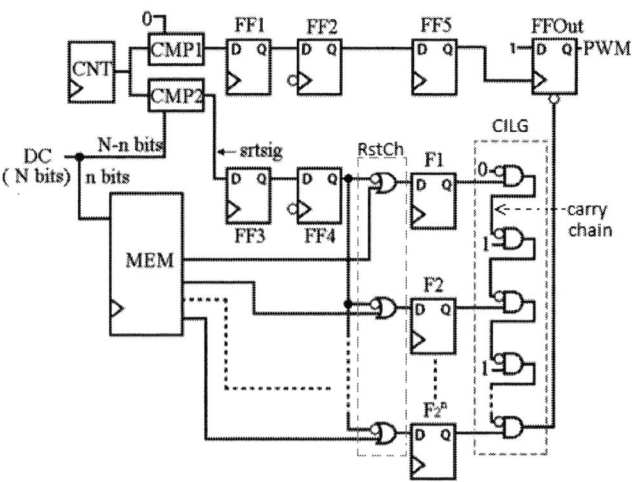

Fig. 2. DPWM circuit.

978-1-4673-8760-6/15 $31.00 © 2015 IEEE   206

When CNT becomes equal to the N bits of the duty cycle, the *srtsig* signal becomes high. The *srtsig* signal is applied to flip-flops FF3 and FF4, and after few clock cycles the CILG is activated, which in turn restarts the register, as can be observed in Fig 2. The structures MEM and *RstCh* are responsible to configure and turn off CILG, respectively.

III. FPGA IMPLEMENTATION

The DPWM circuit was developed for FPGA Altera® Cyclone IV. The duty cycle is comprised of 17 bits – 9 bits of coarse and 8 bits of fine adjustment. It has been proposed a chain of interconnected logic gates with inverters in between them, as shown in Fig. 3, in order to implement the carry chain structure. It was used two delays in the carry chain lines for each increment in the duty cycle (2CCh in Fig. 3), and thus the number of gate is $2 \times 2^n = 512$. The Altera® Cyclone IV series offers D type registers in its internal structure, and consequently, the D flip-flops was used in the FFout structure instead of the SR flip-flops presented in the literature [10] [15] [16]. The FFOut clock signal is responsible to start the DPWM. The FFOut reset signal is responsible to cancel the DPWM.

The CILG structure of Fig. 2 is activated and shut down by a sequence of logic levels applied to the input of the logic gates. As an example, for a chain of 8 gates, the structure is shut down by the sequence 1111. The first increment in the fine adjustment receives 0111. For the second increment, the sequence is 1011, for the third increment it is 1101 and finally 1110 for the fourth increment, as shown in Fig. 3.

Fig. 3.a shows the case where G7 is activated without the carry chain delay. At the restart of DPWM pulse, the CILG structure is shut off, as shown in Fig 3.b. As the sequence is changed to 1011, G5 receives a low level signal and the structure is activated with two carry chain delays, as shown in Fig. 3.c.

As sequence 1101, gate G3 allows *rstffout* signal do be activated after four delays, as shown in Fig. 3.d. The sequence would be 1110 for six carry chain delays. For each PWM period, the chain is turned on and off.

Memory MEM (Fig. 2) is responsible to send the sequence of logic levels to the CILG structure. The determination of the chosen address is performed by the least significant bits of the DC signal, as shown in Fig. 2. Table I presents few samples of the 256 addresses of the memory.

TABLE I. SAMPLES OF MEM.

DC[7:0]	Content
00000000	01111 . . . 1111
00000001	10111 . . . 1111
00000010	11011 . . . 1111
00000011	11101 . . . 1111
00000100	11110 . . . 1111
11111100	11111 . . . 0111
11111101	11111 . . . 1011
11111110	11111 . . . 1101
11111111	11111 . . . 1110

The *RstCh* structure is responsible to turn the CPLI on and off. It is comprised of OR gates with an inverted input. When the *srtsig* is low (as shown in Fig. 2), the OR gates receive the inverted signal and send a high signal to the *CILG*, thus turning the structure off. When the *srtsig* is high, the *RstCg* logic gates allow the transmission of the memory signals, thus activating *CILG*.

Registers *FF1*, *FF2*, *FF3* and *FF4* of Fig. 2 were used to avoid spikes at the comparators output. Flip-flops F_1- F_2^n were implemented in the same *CILG* structure in order to minimize the delay between the *F* flip-flops and the *G* gates. The *What You See Is What You Get* - WYSIWYG library was used for the description of those flip-flops. The gates of the CILG structure were also described using the same library.

The physical proximity between the flip-flops and the gates was achieved by using of the Quartus II Subscription Edition® partition tools, as indicated in Fig. 4.

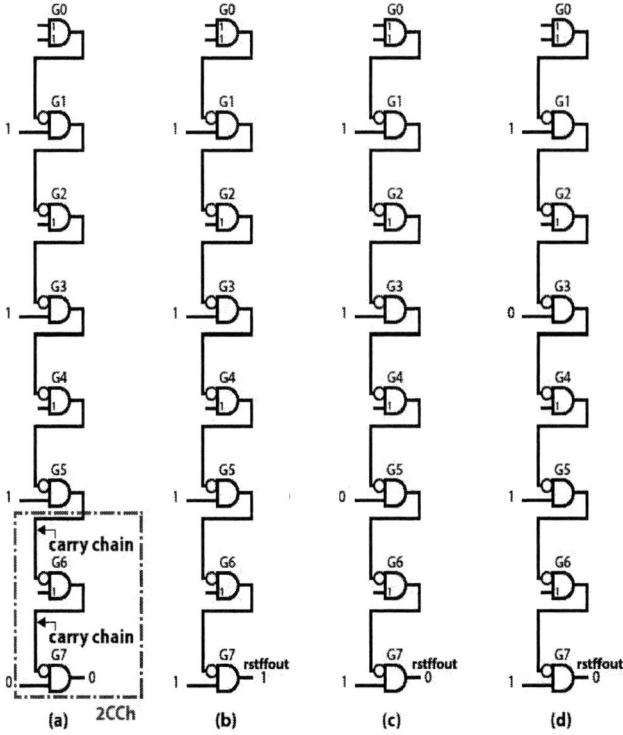

Fig. 3. CPLI structure.

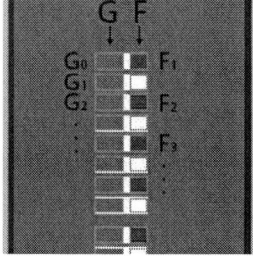

Fig. 4. Physical proximity of flip-flops and CILG structure.

IV. EXPERIMENTAL RESULTS

Experimental tests were conducted on the Cyclone IV EP4CE115F29C7 FPGA available in the DE2-115® board. The measurements were conducted on a 1 GHz MSO4104B® Tektronix oscilloscope. It was connected a 46.34 MHz signal from a generator Tektronix AFG 3252 to the SMA input of the DE2 board, to act as clock signal. Fig 5 shows the DPWM waveforms for the duty cycle range of Table II. The DPWM adjustment signal $-\Delta t_{on}$ is presented in Fig. 6.

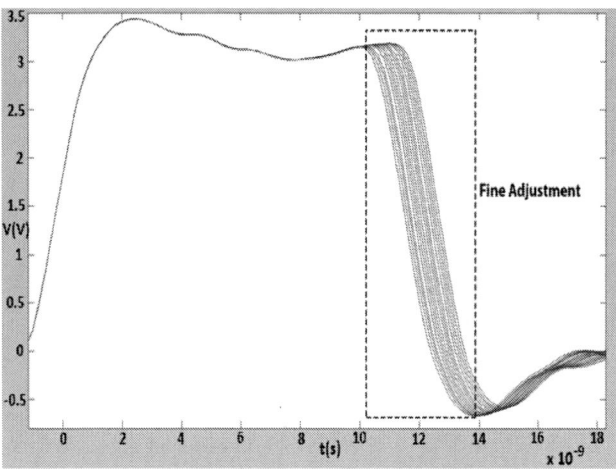

Fig. 5.   DPWM signal with the DC variation form Table II.

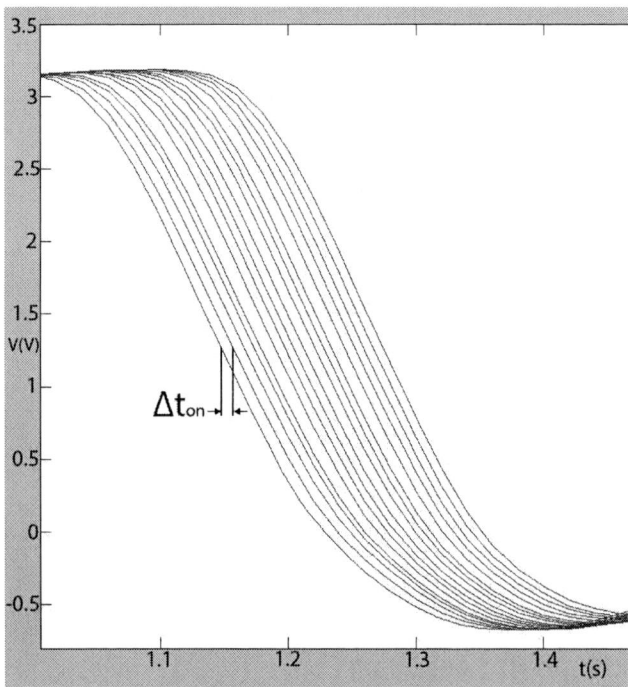

Fig. 6.   DPWM Sinais for 16 duty cycles.

Table II presents few measurements from the DPWM output, and the interval between the positive cycle of a given duty cycle and the negative cycle of the previous duty cycle - Δt_{on}. The larger sample variation was of 112ps while the smallest one was 52ps. The use to carry chain delays allows to obtain a signal monotonicity.

TABLE II.   SELECTED DPWM MEASUREMENTS.

DC	$t_{on}(ns)$	$\Delta t_{on}(ns)$
00000000001111111	11.356	-
00000000010000000	11.445	0.089
00000000010000001	11.531	0.086
00000000010000010	11.610	0.079
00000000010000011	11.662	0.052
00000000010000100	11.774	0.112
00000000010000101	11.853	0.079
00000000010000110	11.921	0.068
00000000010000111	11.995	0.074
00000000010001000	11.088	0.093
00000000010001001	12.154	0.066
00000000010001010	12.214	0.060
00000000010001011	12.319	0.105
00000000010001100	12.399	0.080
00000000010001101	12.470	0.071
00000000010001110	12.544	0.074
00000000010001111	12.621	0.077

In Table III we present a comparison of our work against others presented in the literature [10, 15-19]. As can be observed from the table, our work presented a time resolution of 80ps, comparable to other work. Nevertheless, we obtained a resolution of 17 bits, better than the other works.

TABLE III.   COMPARISON OF OUR WORK AND OTHERS.

	Time resolution	Resolution - bits
[17]	1022.6 ps	10
	1256.4	10
[18]	754 ps	-
[19]	1460 ps	9
[10]	625 ps	11
[15]	70-80 ps	15
[16]	Cyclone® II – 60 ps	8
	Xilinx® Virtex4 – 90 ps	14
This work	80 ps	17

V. CONCLUSIONS

This work described the architecture of a 17 bits DPWM circuit implemented in an Altera® Cyclone IV FPGA. The tests conducted verified the monotonicity of the DPWM signal for many different duty cycles. The carry chain structures were utilized for the fine adjustments, with a Δt_{on} average of 80 ps. The implementation required 1058 logic elements, 535 registers, 20 pins and 65536 bits of memory.

ACKNOWLEDGMENTS

The authors acknowledge CAPES, CNPq and FAPEMIG for their financial support.

REFERENCES

[1] Sabarinath, V.; Sivanandam, K., "Design and implementation of FPGA based high resolution digital pulse width modulator," Communications and Signal Processing (ICCSP), 2013 International Conference on, pp. 410-414, 3-5 April 2013.

[2] Hao Peng; Prodic, A.; Alarcon, E.; Maksimovic, D., "Modeling of Quantization Effects in Digitally Controlled DC–DC Converters," Power Electronics, IEEE Transactions on, vol. 22, n. 1, pp. 208-215, Jan. 2007.

[3] Tanner, L.; Jenni, F., "Digital control for highest precision accelerator power supplies," Particle Accelerator Conference, 2001. PAC 2001. Proceedings of the 2001, vol. 5, pp. 3681-3683, 2001.

[4] Jian Li, Yang Qiu, Yi Sun, Huang, Bin, Ming Xu, Ha, D.S. and Lee, F.C., "High Resolution Digital Duty Cycle Modulation Schemes for Voltage Regulators," Applied Power Electronics Conference, APEC 2007 - Twenty Second Annual IEEE, pp. 871-876, Feb. 25 2007- March 1 2007.

[5] Yan-Fei Liu; Sen, P.C., "Digital control of switching power converters," Control Applications, 2005. CCA 2005. Proceedings of 2005 IEEE Conference on, pp. 635-640, Aug. 2005.

[6] Martin, T.W.; Ang, S.S., "Digital control for switching converters," Industrial Electronics, 1995. ISIE '95., Proceedings of the IEEE International Symposium on, vol. 2, pp. 480-484, Jul1995.

[7] Patella, B.J.;Prodic, A.;Zirger, A.;Maksimovic, D., "High-frequency digital PWM controller IC for DC-DC converters," Power Electronics, IEEE Transactions on, vol. 18, n. 1, pp. 438-446, Jan 2003.

[8] Trescases, O.; Guowen Wei; Wai Tung Ng, "A Segmented Digital Pulse Width Modulator with Self-Calibration for Low-Power SMPS," Electron Devices and Solid-State Circuits, 2005 IEEE Conference on, pp. 367-370, Dec. 2005.

[9] Xiao, J.; Peterchev, A.; Zhang, J.; Sanders, S., "An ultra-low-power digitally-controlled buck converter IC for cellular phone applications," Applied Power Electronics Conference and Exposition, 2004. APEC '04. Nineteenth Annual IEEE, vol. 1, pp. 383-391, 2004.

[10] Navarro, D., Barragán, L.A., Artigas, J.I., Urriza, I., Lucia, O. and Jiménez, O., "FPGA-based high resolution synchronous digital pulse width modulator," Industrial Electronics (ISIE), 2010 IEEE International Symposium on, pp. 2771-2776, July 2010.

[11] Altera®, "Logic Elements and Logic Array Blocks in Cyclone IV Devices," Cyclone IV Device Handbook, vol. 1, November 2009.

[12] Actel Corporation®, "Axcelerator Carry-Connect Macros," Application Note AC163, 2002.

[13] Lattice Semiconductor®, "Design Floorplanning," September 2012.

[14] Xilinx®, "Configurable Logic Block User Guide," November 2014.

[15] L. S. Ge, Z. X. Chen, Z. J. Chen and F. Y. Liu, "Design and Implementation of A High Resolution DPWM Based on A Low-Cost FPGA," PRoc. IEEE Energ. Conv. Cong. Exp., pp. 2306-2311, 2010.

[16] D. Costinett, M. Rodriguez and D. Maksimovic, "Simple Digital Pulse Width Modulator Under 100 ps Resolution Using General-Purpose FPGAs," IEEE Transactions on Power Electronics, vol. 28, n. 10, pp. 4466-4472, 2013.

[17] I. de León, G. Sotta, G. Eirea and J. P. Acle, "Analysis and Implementation of Low-cost FPGA-Based Digital Pulse-width Modulators," Instrumentation and Measurement Technology Conference (I2MTC) Proceedings, pp. 1523-1528, May 2014 IEEE International.

[18] S. C. Huerta, A. de Castro, 0. Garcia and J. A. Cobos, "FPGA based Digital Pulse Width Modulator with Time Resolution under 2 ns," Applied Power Electronics Conference, APEC 2007 - Twenty Second Annual IEEE, pp. 877-881, 1 March 2007.

[19] M. Scharrer, M. Halton and T. Scanlan, "FPGA-Based Digital Pulse Width Modulator With Optimized Linearity," Applied Power Electronics Conference and Exposition, 2009. APEC 2009. Twenty-Fourth Annual IEEE, pp. 1220-1225, 15-19 Feb. 2009.

Low-Power CMOS Variable Gain Amplifier Design in 0.18µm Process

Sawssen Lahiani, Houda Daoud, Samir Ben Salem, Mourad Loulou

Electronic and Communications Group, LETI-laboratory,
National School of Engineers of Sfax, Sfax University, Tunisia
E-mail: sawssenlahiani@yahoo.fr

Abstract— This paper presents a low-power Variable Gain Amplifier (VGA) design in TSMC 0.18µm process. The proposed circuit is composed of tow transimpedance amplifiers and a transconductance amplifier. The VGA control is ensured by using the source degeneration R_S with the feedback resistor R_f. The proposed circuit is designed for low power, low noise and high bandwidth. The studied circuit provides a minimum and a maximum gain of respectively -33 dB and 25.6 dB over more than 133MHz bandwidth. The simulation structure provides less than 20 dB of noise figure (NF). The VGA consumes approximately 40µW under 1V power supply.

Keywords— VGA, CMOS, low power, design.

I. INTRODUCTION

Variable Gain Amplifier (VGA) is an important components in the receiver chain since it ensures the receiver gain regulation [1]. Nowadays, a low power consumption VGA has been designed to achieve high linearity and wide bandwidth with acceptable power dissipation [2, 3]. The VGA circuit is used to maintain the signal power falling to the analog-to-digital converter (ADC) constant while the input one varies randomly [4].

In wireless communication systems, there is a significant amount of the propagation effects and change of the position or direction of the antenna, due to channel fading. These variations are even more numerous when the receiver moves. Therefore, the collaboration with an automatic gain control (AGC), a VGA with fine gain tuning provides relatively constant signal amplitude to maximize the dynamic range of the ADC. The operation principle of the VGA is to compensate these variations by adjusting the input gain. This allows improving the receiver sensitivity. The gain provided by the VGA should be large enough to bring back the level of the weakest signal to a level which enables it to be treated by the ADC [5, 6]. Thus, designing such a block becomes a delicate task.

This paper presents a low power CMOS VGA design using TSMC 0.18µm process. In the first section, differents VGA topologies will be studied and compared. Then, the chosen topology will be studied in the second section. During the third section, simulations measurements will be reported and analysed. Section4 presents some concluding notes.

II. VGA TOPOLOGIES

VGA designers have to think about many considerations: gain, linearity, noise and power consumption. The choice of the VGA topology is strictly related to these parameters as well as specifications required by the standard. There are three commonly used VGA topologies. In this work, these topologies were studied and compared.

The first topology is based on differential pair with diode connected loads. The major drawback of this structure is the trade-off between gain and bandwidth. If the gain increases, the bandwidth decreases considerably. A similar trade-off exists also between linearity and gain [7]. The second structure used in the literature is based on a multiplier pair. It presents a good candidate in term of linearity. However, it is not recommended for low power application. The third VGA topology is based on differential pair with source degeneration. It presents an important gain variation and a good linearity as well as the multiplier structure [7]. The Comparison between these three topologies shows that the two first structures present more limitations in terms of gain, power and linearity. For our application, we deduced that the third VGA topology is the more suitable choice for this design.

A. VGA Block Diagram

The adopted architecture is detailed in [3], [8] and [9]. It is based on a differential pair with source degeneration as presented in figure 1. This structure is obtained by cascading a transconductance amplifier and a transimpedance amplifier. The transimpedance amplifier is performed using a current amplifier with feedback resistors (R_f).

The voltage gain of the VGA is given by formula (1), where G_m and R_m are respectively the transconductance gain and the transresistance gain. The transresistance gain is described by equation (2), where R_{in} and A_i are respectively the input resistance and the current gain of the current amplifier.

$$A_V = G_m R_m \qquad (1)$$

$$R_m = -\frac{(R_f A_i - R_{in})}{(1 + A_i)} \qquad (2)$$

978-1-4673-8760-6/15 $31.00 © 2015 IEEE

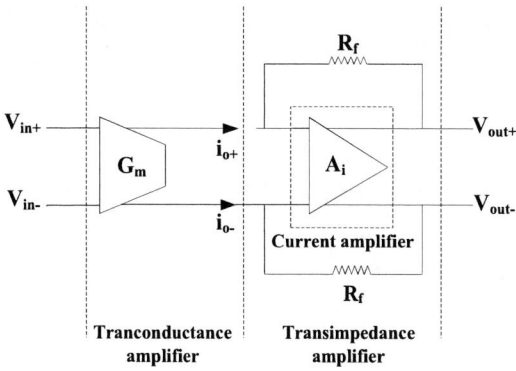

Figure 1. VGA block architecture

B. VGA Electrical Design

The VGA cell is composed by three blocks: two transimpedance amplifiers and the transconductance amplifier as shown in figure 2. The transconductance amplifier is based on the chosen VGA topology described above: differential pair with source degeneration.

The VGA operation can be explained briefly as follows: a differential input voltage is applied at $M_{1a,1b}$ PMOS transistor. It is reproduced across the source degeneration resistor R_S, causing the reproduction of current signal into $M_{2a,2b}$. Current amplification is then ensured by mean of the two transimpedance amplifiers. The amplified current is converted to an output voltage signal via the feedback resistor R_F. The Gain range is ensured via R_S and R_F resistors variations.

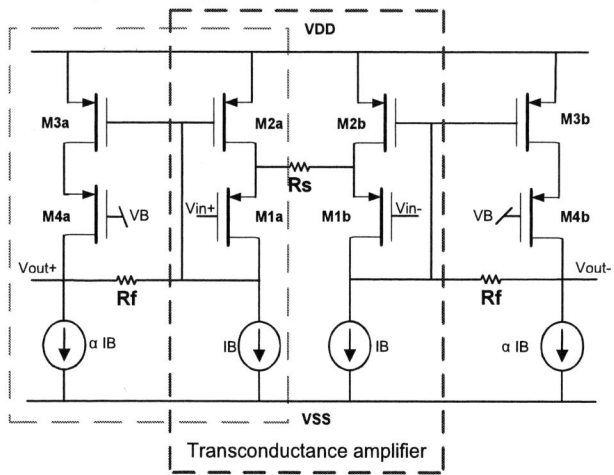

Figure 2. VGA electrical design

Formula (3) describes the DC differential transconductance gain. It can be approximated to $1/R_S$ if $R_S \gg g_{01}/g_{m1}g_{m2}$.

$$G_{md} \approx \frac{1}{(\frac{g_{01}}{g_{m1}g_{m2}} + R_S)} \tag{3}$$

The AC transconductance gain is given by expression (4):

$$G_{mAC} \approx \frac{g_{m1}}{1 + g_{m1}R_S + sR_S C_{gs1}} \tag{4}$$

Where g_{m1} and C_{g1} are respectively the transconductance and the gate-source capacitance of M_1 and M_2 devices. Expression (4) gives the DC transimpedance gain. If we consider that $R_f \gg 1/g_{m3}$, formula (4) can be then approximated to expression (5).

$$R_m = \frac{V_{out}}{I_{in}} = \frac{\frac{1}{g_{m3}} - R_f}{1 + \frac{1}{\alpha}} \approx -\frac{R_f}{1 + \frac{1}{\alpha}} \tag{5}$$

The AC transimpedance gain is approximately given by formula (6), where α is the current gain. g_{m2} and C_{g2} are respectively the transconductance and the gate-source capacitance of M_1 and M_2 transistors.

$$R_{mAC} = \frac{\frac{1}{R_f} - \alpha g_{m2}}{\frac{1}{R_f} g_{m2}(1+\alpha)} * [\frac{1}{1 + s[(\frac{\alpha}{\alpha+1})(\frac{C_{gs1}}{g_{m1}}) + (\frac{C_{gs2}}{g_{m2}})]} + s^2[(\frac{C_{gs1}}{g_{m1}}) + (\frac{C_{gs2}}{g_{m2}})]] \tag{6}$$

III. VGA CIRCUIT SIMULATIONS

The VGA cell is simulated using the Advanced Design System (ADS) tool with TSMC 0.18μm CMOS process parameters under 1V power supply. The NMOS and the PMOS transistor threshold voltages are respectively 0.436 and -0.438 V. The transistors sizing of the VGA circuit are planned in table II. The Current sources are implemented using cascode current mirrors.

TABLE I. TRANSISTORS SIZING

Transistors	W/L (μm/μm)
M_{1a}-M_{1b}	16/0.18
M_{2a}-M_{2b}	3/0.18
M_{3a}-M_{3b}	8/0.18
M_{4a}-M_{4b}	50/0.18

A. Choice of the Bias Current Value

The main objective of this step is to optimize the current (I_B) used for biasing the VGA circuit. Gain and noise figure are the two performances which are closely dependent on the bias current I_B. Accordingly, simulations are accomplished in order to follow up gain and noise figure fluctuation when I_B varies. Figures 3 and 4 present, respectively, the gain and the NF variations versus the bias current I_B. Maximum gain is obtained for $5\mu A < I_B < 15\mu A$ while minimum NF is obtained for $20\mu A < I_B < 40\mu A$. These simulations indicate a trade-off between gain and noise figure. Since noise variation is slightly considerable (between 22dB and 20dB for $5dB < I_B < 20dB$: the gap where the gain is maximum), I_B can be set to 10μA. This value allows a maximum of gain while maintaining an acceptable value for the noise figure.

978-1-4673-8760-6/15 $31.00 © 2015 IEEE

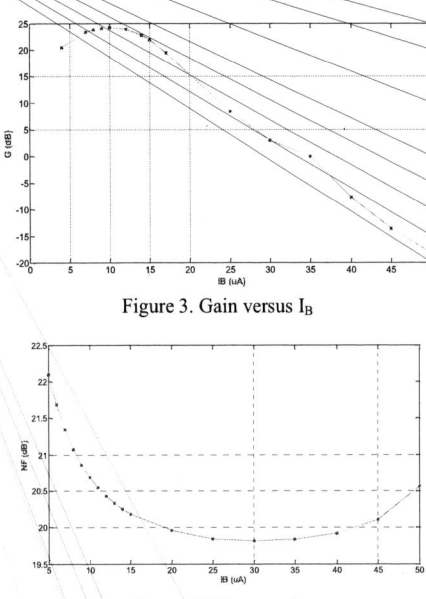

Figure 3. Gain versus I_B

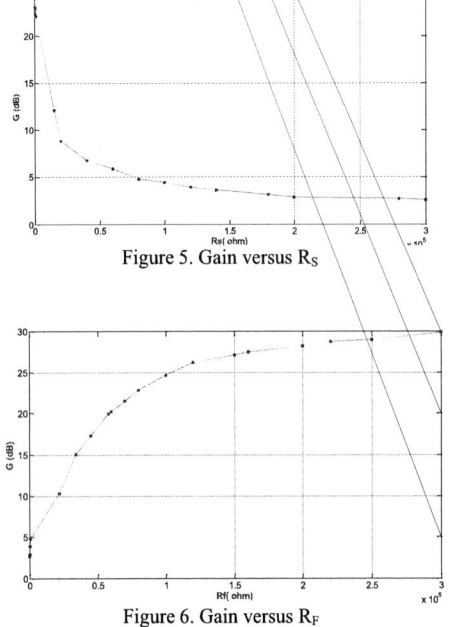

Figure 4. NF versus I_B

B. VGA Cell Simulations

The gain variation of this VGA topology is allowed by varying the R_S and R_F resistors. In the following part, the impact of R_F and R_S variation on gain and noise figure will be presented. As can be seen from figures 5 and 6, the maximum gain is obtained for low values of R_S and high values of R_F. Figures 5, 6, 7 and 8 present the gain and the NF variations versus respectively R_S and R_F. From figures 7 and 8, it is shown that R_F has practically no effect on noise while low noise is obtained for low value of R_S.

The VGA circuit is simulated with a bias current of $10\mu A$ and a voltage supply set at 1V.

Figure 5. Gain versus R_S

Figure 6. Gain versus R_F

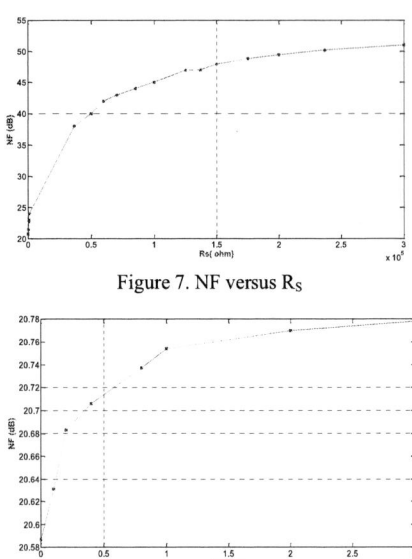

Figure 7. NF versus R_S

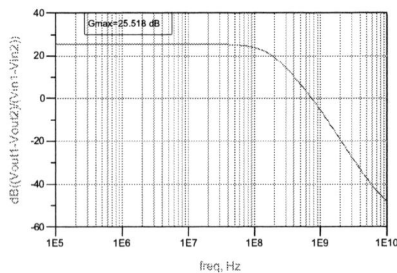

Figure 8. NF versus R_F

The gain range can be adjusted by tuning R_S and R_F resistors. The variations ranges of R_F and R_S are [15KΩ, 100KΩ] and [1Ω, 10kΩ] respectively. Figure 9 and figure 10 present the minimum and the maximum simulated gain obtained for the pair [R_S=10KΩ, R_F=15KΩ] and [R_S=100Ω, R_F=100KΩ] respectively. The frequency bandwidth can be extracted from these two graphs. It is around 133MHz, 1250MHz for the maximum and the minimum simulated gain obtained for the pair [R_S=10KΩ, R_F=15KΩ] and [R_S=100Ω, R_F=100KΩ] respectively. Thus, the VGA circuit can maintain a frequency bandwidth above 133MHz independently of the gain. The corresponding noise figure simulations curves are depicted in figures 11 and 12. Simulations results show that noise figure is around 18dB. Simulations results are summarized in table II.

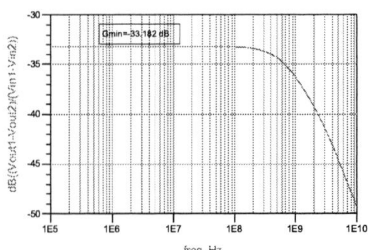

Figure 9. Maximum voltage gain response

Figure 10. Minimum voltage gain response

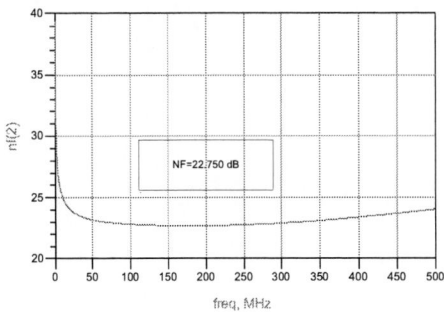

Figure 11. Noise figure response for maximum gain

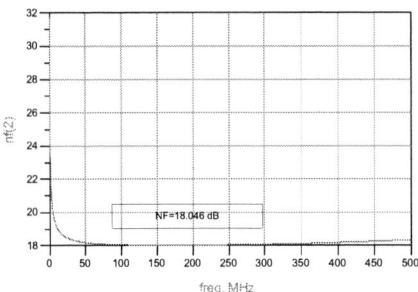

Figure 12. Noise figure response for minimum gain

Table II presents a comparison between the proposed VGA performance and one presented in recent work. One of the main conclusions we can draw is that, the proposed VGA is one of the best in term of dynamic range and power consumption.

TABLE II. COMPARISON WITH RECENT WORK

Parameters	[10]	This work
Gain max (dB)	26	25.6
Gain min (dB)	-7	-33
Bandwith (MHz) (Gain min)	--	1250
Bandwith (MHz) (Gain max)	100	133
NF max (dB)	--	22
NF min (dB)	--	18
Supply Voltage (V)	1	1
Power consumption (μW)	125	40
CMOS Process (μm)	CEDEC 0.18	TSMC 0.18

IV. CONCLUSION

A proposed low power VGA circuit has been designed. The VGA topology combines a source degeneration transconductance amplifier with a transimpedance amplifier.

Simulations results illustrate the very good gain range as well as bandwidth performance. By tuning two resistors R_S and R_F, the gain variation is satisfied. The simulation results using the 0.18μm TSMC process parameters showed a

maximum gain of 25.6dB, a minimum gain of -33 dB and a bandwidth maintained over 133MHz for maximum gain. The circuit consumes only 40μW under 1V power supply.

REFERENCES

[1] Elwan, H., Tekin, A., Pedrotti, K., A Differential-Ramp Based 65 dB-Linear VGA Technique in 65 nm CMOS, IEEE J. Solid-State Circuits, 44(2009), No. 9, 2503-2514.

[2] Apinunt Thanachayanont, "Low-voltage compactCMOS variable gain amplifier," AEU - International Journal of Electronics and Communications, Volume 62, Issue 6, 2 June 2008, Pages 413–420.

[3] Phanumas Khumsat, Piamsuk Anantaseth, Pasin Isarasena, "A Low-Voltage Class-AB CMOS Variable Gain Amplifier", Circuit and systems, 8 Aug 2007, Pages253-256.

[4] Hui Dong Lee, IEEE, Kyung Ai Lee and Songcheol Hong, "A Wideband CMOS Variable Gain Amplifier With an Exponential Gain Control," IEEE Transactions on Microwave Theory and Techniques, vol. 55, no. 6, June 2007.

[5] Hui Dong Lee, IEEE, Kyung Ai Lee and Songcheol Hong, "A Wideband CMOS Variable Gain Amplifier With an Exponential Gain Control," IEEE Transactions on Microwave Theory and Techniques, vol. 55, no. 6, June 2007.

[6] Apinunt Thanachayanont, "Low-voltage compactCMOS variable gain amplifier," AEU - International Journal of Electronics and Communications, Volume 62, Issue 6, 2 June 2008, Pages 413–420.

[7] Lin Chen, "A low power, high dynamic range, broadband variable gain amplifier for an ultra wideband receiver," thesis, Texas A&M University, May 2006.

[8] D. Ayadi, S.Lahiani, S Ben Selem and Mourad Loulou," Variable Gain Amplifier for mobile WiMAX receiver," Journal of Microelectronics, Electronic Components and Materials,Vol. 45, No. 1 (2015), 22 – 28.

[9] S.Lahiani, D.Ayadi, S Ben Selem and Mourad Loulou,"Variable Gain Amplifier in CMOS 0.18μm for WiMAX applicationr," International Conference on Design & Technology of Integrated Systems in Nanoscale Era, April 2015.

[10] M.Idzdihar Idris, M.N.Shah Zainudin, M.Muzafar Ismail, R. Abd. Rahim, Design of a Low Voltage Class AB Variable Gain Amplifier (VGA)", Journal of Telecommunication, Electronic and Computer Engineering, Vol. 4 No. 2 July - December 2012.

A SWOT Analysis of TSV: Strengths, Weaknesses, Opportunities, and Threats

Khaled Salah
Mentor Graphics
Cairo, Egypt
Khaled_mohamed@mentor.com

Abstract—SWOT analysis is one of the most powerful analysis techniques. It is useful in exploring possibilities for new efforts or solutions to problems and making decisions about the best path. In this paper, SWOT analysis for TSV assessment is presented. The aim of this work is to analyze TSV-Based 3D integration from various perspectives and to determine its strengths, weaknesses, opportunities and threats.

Index Terms — Three-Dimensional ICs, Through Silicon Via, TSV, SWOT.

I. INTRODUCTION

Interconnect dimensions and CMOS transistor feature sizes approach their physical limits. Therefore, scaling will no longer be the sole contributor to performance improvement. In addition to trying to improve the performance of traditional CMOS circuits, integration of multiple technologies and different components in a high performance heterogeneous system is a major trend. 3D integration is a key technology level trend to overcome these limitations which summarized in TABLE I [1].

A qualitative comparison regarding the gains introduced by the 3D integration process, compared to existing system design approaches is summarized in TABLE II.

Today, many 3D stacking technologies exist, but TSV is one of the key points and is considered an excellent candidate for 3D integration for different device layers connection [2]-[8].

TSV-based 3D integration is a promising solution to improve the performance, reduce the electromagnetic interference (EMI) effect, where it reduces parasitics and losses and occupies less die area by making use of the z-direction as shown in Fig. 1.

A SWOT (Strengths, Weaknesses, Opportunities, and Threats) analysis is an assessment technique used to guide you to identify the positives and negatives of the studied

Subject [9]-[10]. The SWOT method was originally developed for business and industry, but it is equally useful in other areas and outside the business domain.

SWOT analysis also offers a simple way of communicating and is an excellent way to organize information you have gathered from studies or surveys.

A SWOT analysis focuses on the four elements of the acronym (strengths, weakness, opportunities, threats) as depicted in Fig. 2.

In this paper, SWOT analysis for TSV assessment is presented.

This paper is organized as follows: In Section II, the SWOT analysis of TSV is presented. In Section III, the discussions are introduced. In Section IV, Conclusions are given.

TABLE I
PHYSICAL LIMITATIONS OF SILICON-BASED ICS

Manufacturing Limitations	
Lithography	Lithography technology seems unfit for precise manufacturing beyond 7nm.
Transistor Dimensions	Transistor dimensions are approaching a hard limit that cannot be overcome. That limit is the size of the atom and molecule. Clearly, devices cannot be fabricated smaller than the dimension of a single molecule.
Material Limitations	
SiO_2	If silicon dioxide insulators are reduced to just a few atomic layers, electrons can tunnel directly through the gate.
Design Limitations	
Interconnects Bottleneck	Interconnect delay dominates over gate delay.

TABLE II
COMPARISON BETWEEN ALTERNATIVE DESIGN
IMPLEMENTATIONS

Property	Single chips	SoC	3D
Modular flexibility	High	Low	Medium
System performance	Low	Medium	High
Physical dimension	Large	Medium	Small
Complexity of fabrication	Low	Medium	Medium
Cost	Low	Medium	High
CAD tools	Available	Available	Not deployed yet

978-1-4673-8760-6/15 $31.00 © 2015 IEEE

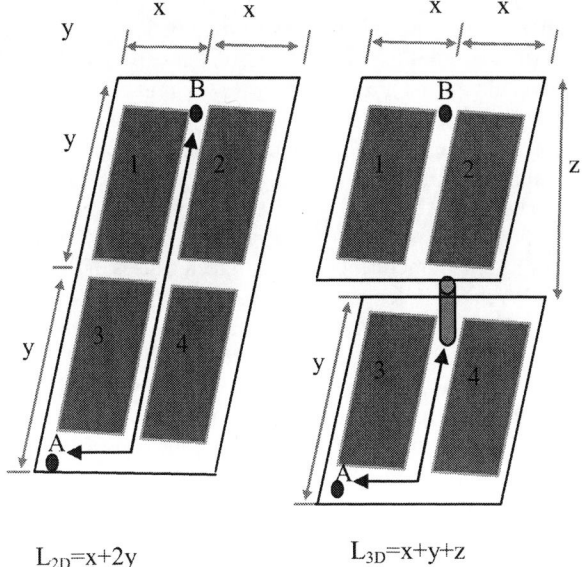

$L_{2D}=x+2y$　　　　$L_{3D}=x+y+z$

Fig. 1 Reduction in wire length, where the original 2-D circuit is implemented in two planes (z < y).

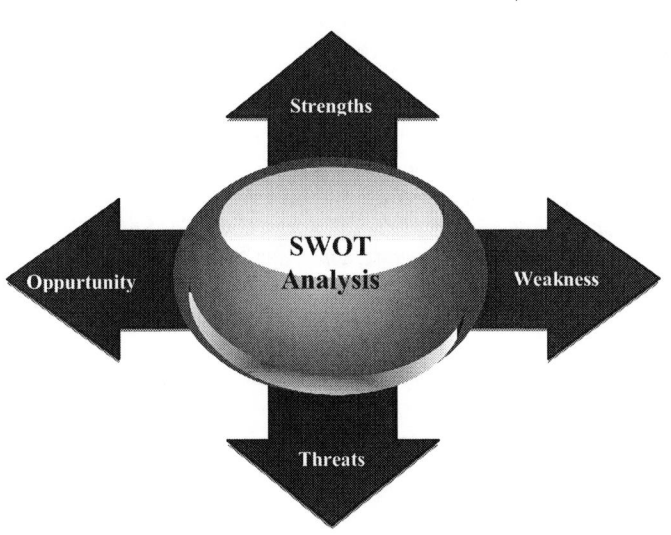

Fig. 2 SWOT Analysis.

II. SWOT ANALYIS OF TSV

In this section, TSV SWOT analysis is discussed. strengths of TSV are discussed in Section II-A, weaknesses of TSV are discussed in Section II-B, opportunities of TSV are discussed in Section II-C, threats of TSV are discussed in Section II-D.

A. Strengths of TSV

Strengths mean positive tangible and intangible attributes. TSV provides short length, high density, less parasitics, and small footprint. Typical values of different interconnection parasitic parameters as compared to TSV parasitics are summarized in TABLE III.

The TSV-based 3D integration also provides opportunities for new circuit architectures for FPGAs, memories (Fig. 3), processes (Fig. 4), NoC and others.

An architecture based on TSV technology for a spiral inductor is possible which provides better quality factor than planar spiral inductor for identical inductance. These good characteristics of the inductor indicate that TSV/3D integration is a promising solution for RF-SoC. Another inductive coupling interface that uses the magnetic near field induced by TSV-based spiral inductor is also available [13].

Applications for TSV-based 3D include image sensors, flash memories, DRAM, processors, FPGAs [14], and power amplifiers. Table VI provides examples of company announcement. The timing for mass production depends on how the TSV compares in terms of cost with existing technologies.

TABLE III
TYPICAL VALUES OF PARASITIC COMPONENTS FOR DIFFERENT CHIP INTERCONNECTION TECHNOLOGIES

Technology	TSV	Wire-bond	Solder bump	Adhesive bumps
Resistance (m Ω)	10-60	30-100	1-3	15-30
Inductance (pH)	5-10	1000-3000	50-100	50-100
Capacitance (fF)	3-10	10-50	4-10	5-10

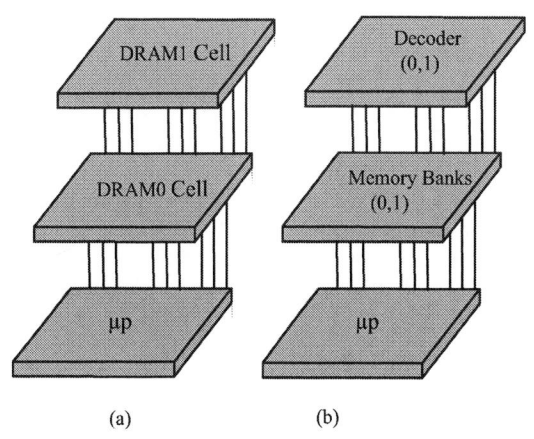

Fig. 3 3D-Memory: (a) complete memory in one layer, (b) split of memory elements in multiple layers.

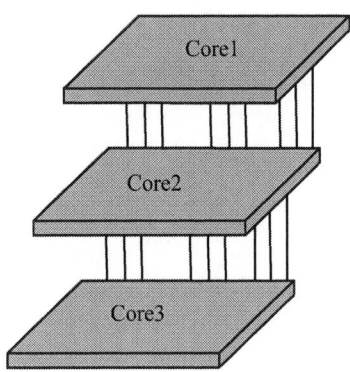

Fig. 4 Multi-core processors stacking in the 3D design, where a 2D chip is divided into a number of different blocks and each one is placed on a separate layer of silicon where each layer is stacked on top of each other

TABLE IV
3D TSV PRODUCTS IN THE NEAR FUTURE

Company	Memory and logic	RAM	Flash	Image sensor	Power Amplifier	FPGA
IBM	✓				✓	
Intel	✓					
Xilinx						✓
Toshiba				✓		
Samsung		✓	✓			
Tezzaron		✓				
Sony	✓					
Sanyo					✓	
Fujikura					✓	
Micron		✓	✓	✓		
ZyCube					✓	
Tessera					✓	

B. *Weaknesses of TSV*

Weakness means in which areas might the organization improve?.

A major difficulty in modeling 3D interconnections results from the need to obtain the entire coupling model of a large number of 3D interconnections. TSV suffers from complex fabrication. Thermal management is one of the important issues of 3D IC integration. The thermal issue of a 3D IC is much severer than that of a 2D IC. The cause is that the ambient environment of the die of a 2D IC is the cooling material, but the ambient environment of a die within a 3D IC may be another die that also generates heat. Therefore, thermal management in 3D ICs is critical for maintaining the required reliability, performance, and power dissipation target.

The addition of a third dimension would require support from more advanced CAD tools. There is a significant amount of work on the floorplanning, placement, and routing for 3D ICs. However, all these tools developed by different groups, using different formats to represent the design data, create barriers for researchers who need to make use of the existing design automation tools to conduct further studies on 3D IC. Also, there is a lack of CAD algorithms and there is only few commercially available EDA tools for 3D integrated circuits. Therefore, current 2D physical design tools (partitioning, placement, routing, timing, extraction, LVS, DRC) must enable 3D designs[11]-[12].

The TSV-based 3D-IC fabrication process can result in several manufacturing defects, which causes 3D IC failure. TSV related defects might occur either in the fabrication of the TSVs themselves such as shorts between TSVs, conductor open defect, and in the bonding of the TSVs to the next tier, or during the life time of the 3D such as mechanical load, and thermal expansion. Solutions are needed to address these challenges before 3D integration technology can be widely used.

3D design methodologies are still needed to reduce coupling between TSVs. Also, more investigations are required for analyzing the effect of TSVs on signal integrity, power integrity and delay.

On-chip power supply noise has worsened in modern systems because scaling of the power supply network (PSN) impedance has not kept up with the increase in device density and operating current due to the limited wire resources and constant RC per wire length, and this situation is worsened in 3D ICs.

The design of a clock distribution network in a digital integrated circuit is challenging in terms of obtaining low power consumption, low waveform degradation, low clock skew and low simultaneous switching noise. In 3D-ICs, the effect of process variations on clock skew and jitter differs from 2D circuits. New schemes for a three dimensional clock distribution network with low skew, low jitter, low power consumption and small area consumption are needed.

The effects of substrate doping density on the electrical performance of TSV should be considered. Moreover, testing multi-layer TSV-based IC is a challenge.

C. *Opportunities of TSV*

An opportunity is a chance for improvement. The weakness points of TSV are opportunities for it. We can overcome electrical, thermal, CAD tools, technological, yield, and test challenges. If this is achieved, so the weaknesses are converted to opportunities.

D. *Threats of TSV*

Threats are barrier or constraint that limits achievement of goals. Threats are external factor that make our choice at risk. There is a competition for TSV-based solution to overcome Moore's law and interconnect limitation by other solutions such as new architectures (SOI, FinFET, Twin-Well), new materials (High-K, metal gate, strained-Si), new interconnect schemes (NoC, optical, wireless), and new devices (molecular computer, biological computer, quantum computer). Fabrication limitation is another threat.

III. RESULTS AND DISCUSSIONS

The SWOT analysis results for TSV are summarized in TABLE V.

TABLE V
TSV SWOT ANALYSIS SUMMARY

Strengths	Weaknesses
• Short length. • High density. • Small footprint. • Less parasitics. • New circuit architectures such as: FPGAs, µps, NoCs. • New applications: spiral inductor, inductive coupling interface. • Many products in the markets.	• Coupling model of a large number of 3D interconnections. • Thermal management. • Advanced CAD tools. • Manufacturing defects (Yield). • Clock and power distribution network. • Substrate doping density effect. • Testing multi-layer TSV-based IC is a challenge. • Manufacturing cost is high.
Opportunities	Threats
• ! Weaknesses.	• New architectures (SOI, FinFET, Twin-Well). • New materials (High-K, metal gate, strained-Si). • New interconnect schemes (NoC, optical, wireless). • New devices such as (molecular-based computer biological-based computer, quantum-based computer). • Fabrication limitation.

IV. CONCLUSIONS

A SWOT analysis identifies your strength, weakness, opportunities and threats to assist you in making decisions. In this paper, SWOT analysis for TSV assessment is presented. The weaknesses of TSV are opportunities for it.

The analysis shows that TSV-based 3D integration is still not mature enough and there are still many opportunities for improvement.

REFERENCES

[1] R. Weerasekera, M. Grange, D. Pamunuwa, H. Tenhunen, and L-R. Zheng, "Compact Modeling of Through-Silicon Vias (TSVs) in Three Dimensional (3-D) Integrated Circuit", in Proc. IEEE International Conference on 3D System Integration (3D IC), 2009, San Francisco, USA, 2009.

[2] J. Burns, L. Mcilrath, C. Keast, C. Lewis, A. Loomis, K. Warner, and P. Wyatt, "Three-Dimensional Integrated Circuits for Low-Power High-Bandwidth Systems on a Chip," In Proceedings of the International Solid-State Circuits Conference (ISSCC'01), pages 268–269, Feb. 2001.

[3] T. Bandyopadhyay, K. Han, D. Chung, R. Chatterjee, M. Swaminathan, R. Tummala, "Rigorous Electrical Modeling of Through Silicon Vias (TSVs) With MOS Capacitance Effects," IEEE Transactions on Components, Packaging and Manufacturing Technology, June 2011.

[4] K. Kanda, D. D. Antono, K. Ishida, H. Kawaguchi, T. Kuroda, and T. Sakurai "1.27-Gbps/pin, 3mW/pin Wireless Superconnect (WSC) Interface Scheme", In Proceedings of the International Solid-State Circuits, 2011.

[5] W. R. Davis, J. Wilson, S. Mick, J. Xu, H. Hua, C. Mineo, A. M. Sule, M. Steer, and P. D. Franzon, "Demystifying 3D ICs: The Pros and Cons of Going Vertical", IEEE Design and Test of Computers, 22(6):498–510, Nov. 2005.

[6] N. Miura, H. Ishikuro, T. Sakurai, and T. Kuroda, "A 0.14pJ/b Inductive-Coupling Inter-Chip Data Transceiver with Digitally-Controlled Precise Pulse Shaping", In Proceedings of the International Solid-State Circuits Conference (ISSCC'07), pages 358–359, Feb. 2007.

[7] N. Miura, D. Mizoguchi, M. Inoue, K. Niitsu, Y. Nakagawa, M. Tago, M. Fukaishi, T. Sakurai, and T. Kuroda, "A 1Tb/s 3W Inductive-Coupling Transceiver for Inter-Chip Clock and Data Link", In Proceedings of the International Solid-State Circuits Conference (ISSCC'06), pages 424–425, Feb. 2006.

[8] S. Saito, Y. Kohama, Y. Sugimori, Y. Hasegawa, H. Matsutani, T. Sano, K. Kasuga, Y. Yoshida, K. Niitsu, N. Miura, T. Kuroda, and H. Amano. "MuCCRA-Cube: a 3D Dynamically Reconfigurable Processor with Inductive-Coupling Link", In Proceedings of the Field-Programmable Logic and Applications (FPL'09), pages 6–11, Sept. 2009.

[9] Department for international development (2002). *Tools for development: A handbook for those engaged in development activity.* Downloaded 1st March from: http://www.unssc.org/web1/ls/downloads/toolsfordevelopment%20dfid.pdf.

[10] European Commission (2004). *Project Cycle Management Guidelines.* Downloaded 1st March from: http://ec.europa.eu/europeaid/qsm/documents/pcm_manual_2004_en.pdf.

[11] C. Chiang, and S. Sinha, "The Road to 3D EDA Tool Readiness, Proc" 2009 Conference on Asia and South Pacific Design Automation, pp.429–436 (2009).

[12] A. Rahman and R. Reif. "System-level performance evaluation of three-dimensional integrated circuits." IEEE Trans. on VLSI Systems, Special Issue on System-Level Interconnect Prediction, 8(6):671–678, 2000.

[13] K. Salah, A. El-Rouby, H. Ragai, Y. Ismail "TSV-Based On-Chip Inductive Coupling Communications" ISCAS, 2013.

[14] K. Salah, M. AbdelSalam "Emerging Reconfigurable Systems: Exploring 3D FPGA Architectures" ICM, 2013.

Shifting the half wave dipole antenna resonance using EBG structure

El Ghabzouri Mohammad, Abdenacer Es-Salhi
LETAS Dept. of physics, Faculty of Sciences, Mohamed I
University
Oujda, Morocco
elghabzouri.mad@gmail.com

Paulo M. Mendes
Dept. of Industrial Electronics, University of Minho
Guimarães, Portugal

Abstract— In this paper, the main motivation is to introduce a new behavior of the electromagnetic band gap (EBG) structures, it is a significant shifting the resonance frequency down of the dipole antennas, this very interesting and useful technique led to low profile, in addition to performance enhancement of dipole antennas, either on return loss or radiation pattern. Also among this EBG structure an investigation on specific absorption rate (SAR) is shown. The used dipole antenna is resonating around 3.5GHz (part of 4G bands), then by using this technique we could shift the working frequency of the same dipole antenna to 2.8GHz worldwide interoperability for microwave access (WiMAX), this new resonance is 80% lower compared with the normal resonance without any structure. The principle of this new technique still valid with other frequency, depending on the frequencies that we would like to shift the working frequency between.

Keywords—Half wave dipole antenna; radiation pattern; electromagnetic band gap (EBG); specific absorption rate (SAR).

I. INTRODUCTION

Many significant works showing the importance of electromagnetic band gap (EBG) structures, firstly in optic area and then in wireless domain. The purpose of the present work is to show how it is efficient to have low profile dipole antenna using those kinds of structures. The Sievenpiper structure which is mushroom like was the beginning of this revolution in optics [1], thereafter one of the famous Sami and Yang works describes how successful it is to reduce a coupling between two patch antennas above this EBG mushroom like structure [2], and some results, recently published present the same behavior of these structures, which manifest in mutual coupling reduction of waveguide slot array antennas (WSAA) [3]. Many of succeeded works try to bring some modifications to this basic Sievenpiper structure, to achieve an enhancement to different antenna parameters [4-5]. Others show specific absorption rate (SAR) investigation by using square and circular patch cells [6-7]. In fact, some recent works have shown that the choice of EBG structures integrated with antennas could be a good alternative to enhance the directivity of monopole, dipole and patch antennas [8].

The present paper illustrate a particular and efficient technique to shift down the half wave dipole antenna resonance frequency, This shifting to low frequency is made by using a standard mushroom like EBG, and the shifted resonance frequency was driven by using our fabricated EBG-CS structure presented in [9]. In fact, the EBG structures can be also used to drive the central frequency of the antenna, by using some specific equations [10]. Also the radiation pattern improvement due to surface wave suppression is showed among this work.

II. CHARACTERISATION BAND GAP METHODS

There are many methods to study this kind of electromagnetic structure, and identifying their band gaps.

We can mention here a dispersion diagram method, which analyze the propagation of electromagnetic waves above the structure by calculating the Eigen values inside the Brillouin zone, the reflection phase method and direct transmission method (DTM). In the following paragraphs, some details describing these methods are detailed. Then we apply the suspended method line to our proposed configuration; cause its practical realization to be compared with feature measurements.

A. Reflection phase methode

In this paragraph, we start with one of the most useable methods. With the reflection phase method we can easily characterize our unit cell and then identify the band gap around the resonance frequency. To extract this reflection phase curve, the unit cell was modeled and excited by a wave guide port at Z_{max} position which is ten time great than the substrate thickness h. We make sure that the boundary conditions X_{min} and X_{max} defined as electric walls, Y_{min} and Y_{max} defined as magnetic walls, and Z_{min} defined as electric walls.

B. Direct transmission line method

This method based on entire modeling of one dimension of the unit cell. The proposed EBG structure is constituted by 6*4 cells, which mean four cells by line. Hence we transmit a planar wave through the line of cells by one side, and we calculate the transmission parameter S_{21} in the other side of the line, in order to determine the prohibited band above our structure. In this method, we have to follow some boundary conditions, because this method considers a line of cells as a TEM wave guide, which mean that the right and the left side are chosen as magnetic walls. In the other hand the bottom and top side are chosen as electric walls. Finally the two rest sides are kept as open boundaries.

C. Dispersion diagram method

This identification technique is more used by programming FDTD (Finite Deference Time Domain) method. But also we can calculate the dispersion diagram by a special solver. Eigen mode solver is another way to determine a band diagram, and then we identify our band gap, after that we optimizing our cell to operate at the work frequency of our antenna. The figure below illustrates the design of the unit cell.

The method consist to define a periodic boundary conditions along U axis (X_{min} and X_{max}) , and same thing to V axis (Y_{min} and Y_{max}), and finally we define W axis (Z_{min} ,Z_{max}) as an $E_t=0$ (tangential electric field) for TE mode calculation or $H_t=0$ (tangential magnetic field) for TM mode calculation.

The main difficulty of the previous three methods is not practical to be realized for measurement and comparison, in opposite to the following method, which is simple and efficient to compute the right S21 of the structure.

D. Suspended line method

The concept of this method is based on designing the whole EBG structure, and then we insert under the structure a transmission line between the two extreme parts of the studied structure, finally we feed the transmission line by two wave guide port to transmit and receive the waves [11].

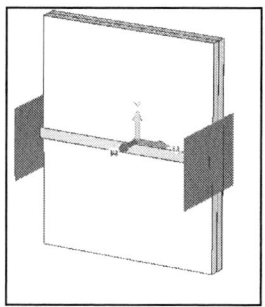

Fig. 1. Design of the suspended line method.

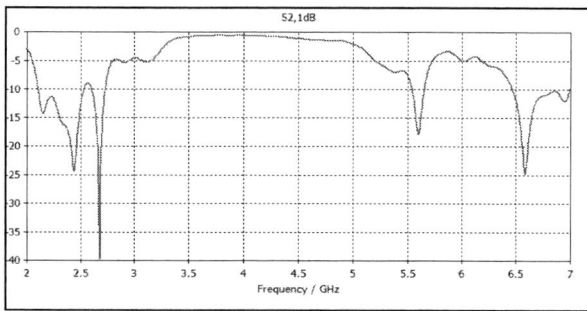

Fig. 2. The transmission coefficient S21 showing the band gap of the used structure

Beside some narrow band gaps, at least less than 7 GHz. the computed transmission coefficient S21 illustrated in Fig.2 shows almost no stop band at -20dB between 2.7 to 3.7 GHz, the dimensions of this EBG structure were chosen to have a transparent structure around these frequencies, which is our zone of interest in this work. Therefore the present structure is modeled to shift the resonance frequency from 3.5 to 2.8 GHz, in favor of some particular application, like a balance between fourth generation band (4G) around 3.5 GHz and WiMAX application at 2.8 GHz for hand cell phones.

III. DIPOLE ANTENNA WITH AND WITHOUT EBG STRUCTURE

A. Return losses results and discussion

In the following section we present some details concerning the half wave dipole antenna design, this dipole antenna modeled to be radiating around fourth generation 3.5 GHz band. The length "l" of dipole antenna is approximately a half wave length l=λ/2, λ present the wave length on free space at the center frequency, the input impedance of the half wave dipole antenna was set at 73ohm. The antenna radiate at 3.5 GHz in free space. Then we will use the antenna in front of the EBG structure background Fig.3, the results in Fig 5 show a very important shifting on the resonance frequency, this new configuration (antenna in front of the EBG mushroom like) shift the resonance frequency down to 2.8 GHz, which the IEEE standard for WiMAX application, the band size of shifting down is 0.7 GHz which is 80% lower from the central frequency of the dipole antenna without EBG structure.

Both the substrate and superstrat material was chosen as Rogers RO4350 with dielectric constant 10.2 with the standard fabricated thickness 1.27 mm. The figures 4 and 5 shows the return losses of both the dipole antenna in free space and mounted with EBG structure.

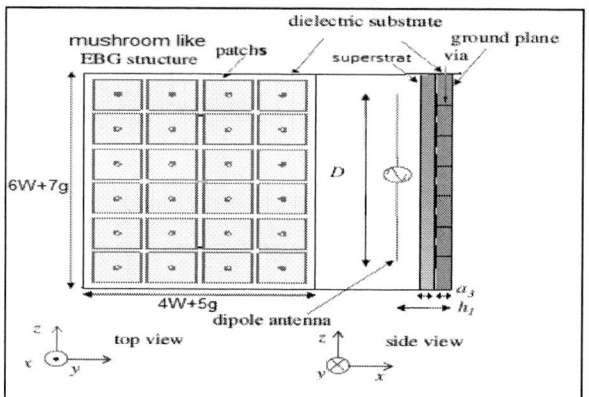

Fig. 3. The design of dipole antenna and EBG structure mounted with superstrat.

W=6.2 mm; g=0.5 mm; a3=1.27 mm; h=2*a3+0.5 mm; D=40 mm

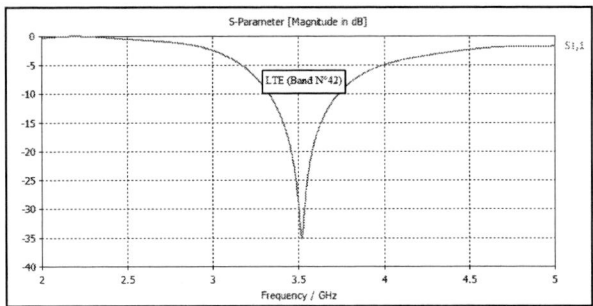

Fig. 4. return loss of the dipole antenna without EBG structure.

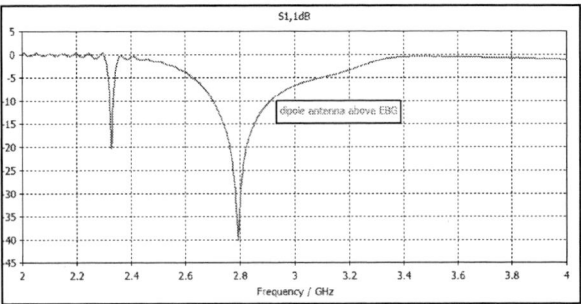

Fig. 5. Return loss of the same dipole antenna with the new working frequency 2.8 GHz when used with the proposed EBG.

B. Rediation pattern results and discussion

The fowling section illustrates the comparison results of the 3D radiation pattern, together with the horizontal and vertical cut. After shifted the antenna working frequency down from 3.5 GHz to 2.8 GHz, when using the EBG structure. Now the results of this new configuration should be compared with a half wave dipole antenna designed to be radiating around 2.8 GHz. The half wave dipole antenna has an Omni-directional diagram with 1.96 dBi of its directivity Fig.2. Therefore the 3D radiation pattern of the proposed configuration of the half wave dipole antenna with EBG structure is showed in Fig.3.

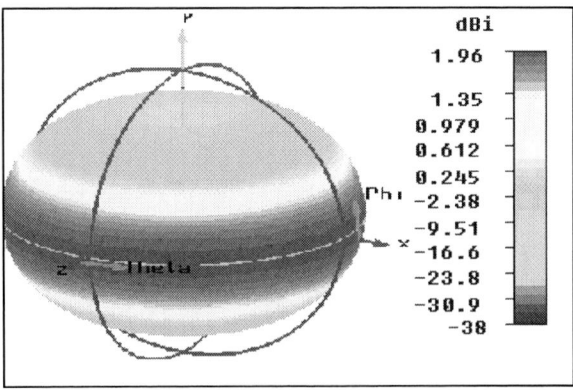

Fig. 6. The far field radiation pattern of the half wave dipole antenna at 2.8GHz

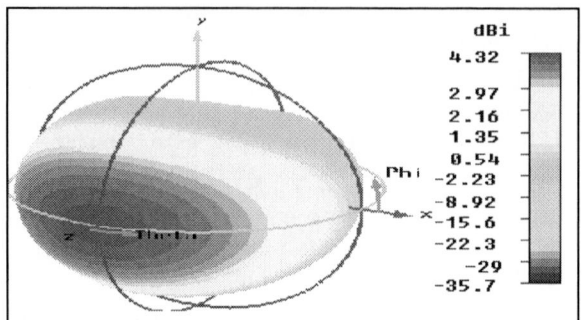

Fig. 7. The far field radiation pattern of the half wave dipole antenna mounted with mushroom like EBG shifted from 3.5 to 2.8 GHz.

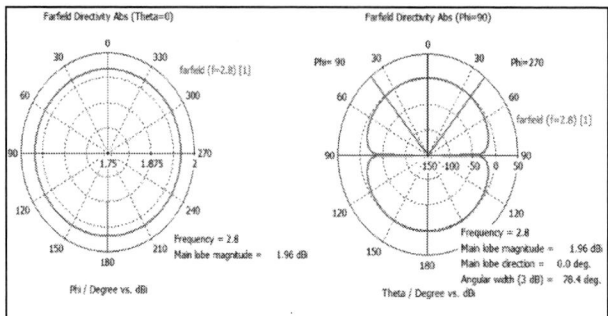

Fig. 8. Vertical and horizontal cut of radiation pattern of dipole antenna.

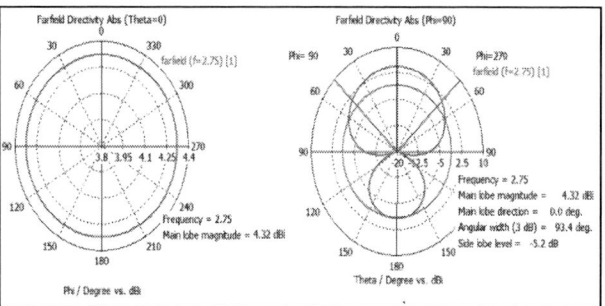

Fig. 9. Vertical and horizontal cut of radiation pattern of dipole antenna with EBG structure.

This proposed configuration, in addition, to low profile achievement, which manifest in shifting the resonance frequency from 3.5 GHz (4G band) to 2.8 GHz (WiMax), led to a very good radiation pattern compared with the reference dipole antenna, as the figures show. the maximum directivity was improved by more than 100% (from 1.96 dBi in Fig.6 to 4.32 dBi in Fig.7), but the improvement was only on the front half space, the directivity on back half space was a bit poor also compared with the reference dipole antenna. Generally the results agreeable than what we were expecting.

IV. SAR INVESTIGATION

The specific absorption rate is a very important parameter to study for in any antenna before its realization, because exposing our bodies in antennas, means that the antennas should satisfy international requirements of SAR. Basically

SAR is presenting the quantity of absorbed power in our biological tissues. The Fig.8, show the far field radiation pattern of the half wave dipole antenna, mounted with the proposed mushroom like EBG, shifted from 3.5 to 2.8GHz, when used with the human head phantom. So we can see that the human head increased the maximum directivity from 4.32 dBi to 6.91 dBi, which is understandable if we take in account the reflected power when used the antenna nearby the human head.

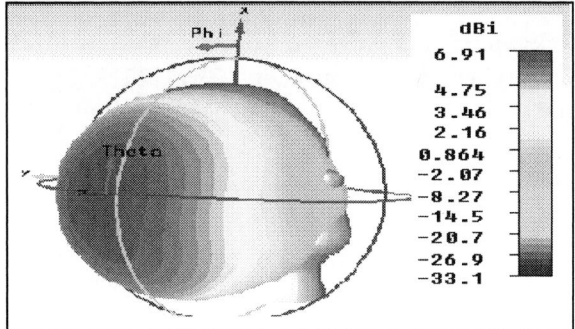

Fig. 10. The far field radiation pattern at 2.8GHz.

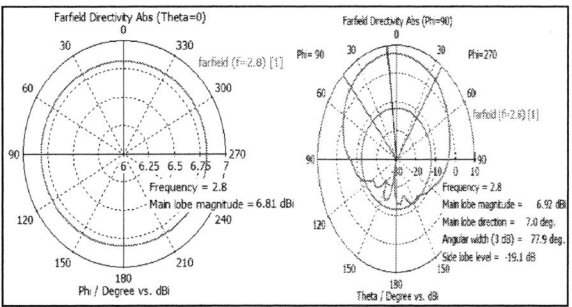

Fig. 11. Vertical and horizontal cut of the influenced radiation pattern of dipole antenna with EBG structure close to the human head.

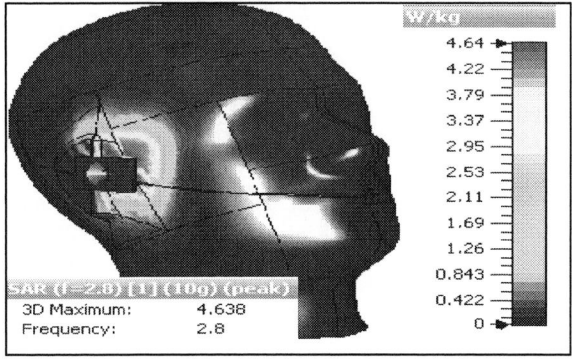

Fig. 12. SAR cartography at 2.8 GHz averaged on 10g.

Finally we have computed the absorbed power (SAR) inside the tissues, the liquid characteristics was chosen to simulate the human tissues at 2.8 GHz, the power was normalized to 1 watt, and the distance between the phantom and the structure was set at 10mm.

V. CONCLUSION

This work resumes another efficient and important use of EBG structure in antenna community, this new technique consist to shift to low frequency the resonating frequency of antennas, this shifting was achieved by using a special design of EBG structure, the half wave dipole antenna was shifted from 3.5 GHz to 2.8 GHz, which present 80% of miniaturization of antenna, which led to the contribution to low profile antenna for cell phones, at the same time we could balance between 4G resonance and WiMAX resonance with only on antenna. Also we have been presented that this technique useable to improve a high efficiency antenna, using the EBG structure. Which manifest in very low return loss of dipole antenna, its 220 MHz bandwidth, in addition to the improvement of the effectiveness of the radiation pattern, by the proposed configuration of EBG structure, finally we have computed the absorbed power (SAR) inside the human head phantom.

References

[1] D. Sievenpiper, L. Zhang, R. F. J. Broas, N. G. Alexópolous, and E. Yablanovitch, "High-Impedance Electromagnetic Surfaces with a Forbidden Frequency Band," IEEE Transactions on Microwave Theory and Techniques, Vol.47, No.11, November 1999, pp. 2059-2074.

[2] Fan Yang, Member, IEEE, and Yahya Rahmat-Samii, Fellow, IEEE "Microstrip Antennas Integrated With Electromagnetic Band-Gap (EBG) Structures: A Low Mutual Coupling Design for Array Applications," IEEE Transactions On Antennas and Propagation, vol. 51,no.10, October 2003.

[3] Siamak Ebadi, and Abbas Semnani, "Mutual coupling reduction in waveguide slot array antennas using electromagnetic band gap (EBG) structures." IEEE Antennas and Propagation Magazine, vol. 56,no.3, June 2014.

[4] N. Elsheakh, H. A. Elsadek, and E. A. Abdallah, "investigated new embedded shapes of electromagnetic bandgap structures and via effect for improved microstrip patch antenna performanced," In Progress In Electromagnetics Research B, Vol. 20, 91_107, 2010.

[5] Seungbae Park, Cheolbok Kim, Youngho Jung, Hosang Lee, Dongki Cho,Munsoo Lee, "Gain enhancement of amicrostrip patch antenna using a circularly periodic EBG structure and air layer," In Int.J.Electron.Commun.(AEÜ) 64(2010)607–613

[6] Kwok-Hung Chan, Member, IEEE, Ryo Ikeuchi, and Akimasa Hirata, Senior Member, IEEE "Effects of phase difference in Dipole phased-array antenna above EBG substrates on SAR," IEEE Antennas Wireless Propag. Lett., vol. 12, 2013.

[7] Ryo Ikeuchi and Akimasa Hirata, Senior Member, "Dipole antenna above EBG substrate for local SAR reduction," IEEE Antennas Wireless Propag. Lett., vol. 10, 2011.

[8] Silvio Ceccuzzi, Lara Pajewski, Cristina Ponti, and Giuseppe Schettini, "Directive EBG Antennas: A Comparison Between Two Different Radiating Mechanisms," IEEE Transactions On Antennas And Propagaton, Vol. 62, NO. 10, october 2014.

[9] Mohammad El Ghabzouri, A. Es Salhi, P. M. Mendes, "Dual band antenna size reduction using EBG structure," Mediterranean Conference on Information & Communication Technologies'2015 May 7-9, 2015 Saïdia, Morocco, unpublished.

[10] Mohammad El Ghabzouri, A. Es Salhi, P. M. Mendes, "Dual band antenna size reduction using EBG structure,"IEEE Mediterranean microwave symposuim 2015 (MMS2015) 'leece, italy, unpublished.

[11] M. S. Alam, M. T. Islam, N. Misran, "A novel compact split ring slotted electromagnetic band gap structure for microstrip patch antenna performance enhancement," Progress In Electromagnetics Research, Vol. 130, 389-409, 2012.

Direct-Elevator: A Modified Routing Algorithm for 3D-NoCs

Maha Beheiry[1], Ahmad Aly[1], Hassan Mostafa[2], and Ahmed M. Soliman[3]

[1]Mentor Graphics, Egypt,

[2,3]Electronics and Electrical Communications Engineering Department, Cairo University, Giza 12613, Egypt,

[2]Center for Nanoelectronics and Devices, AUC and Zewail City of Science and Technology, New Cairo 11835, Egypt.

Emails: maha_beheiry@mentor.com, ahmad_aly@mentor.com, hmostafa@uwaterloo.ca, asoliman@ieee.org

Abstract—In this paper, a Three-Dimensional (3D) flexible routing algorithm is proposed for generic 3D Network-on-Chip (NoC). The proposed approach, Direct-Elevator, is based on the Elevator-First Algorithm which is independent of the network topology, number of interconnects and placement of interconnects. Direct-Elevator is tailored for 3D-NoC structures, offering a lower communication latency than its predecessor.

The proposed approach paves the route for finding the optimal configuration of 3D network topology and different hard and soft router implementations for 3D-NoC based Field Programmable Gate Arrays (FPGA).

Index Terms—3D technology, NoC, 3D Routing Algorithm, 3D NoC.

I. INTRODUCTION

Three-Dimensional technology which groups multiple chips vertically into a single chip with a die-to-die interconnection, provides wide capabilities for system integration [1]. The 3D-integration offers a promising solution with a significant reduction in the power consumption and delay. It also allows the designer to combine different technologies in one fabrication process [2]. Meanwhile, the new 3D trend introduces new challenges such as different interconnect modeling, thermal management, power delivery, cooling techniques, noise anaylsis and fabrication process comp-atibility [3], [4]. On the other hand, Network-on-Chips (NoCs) offer flexibility, re-usability, and reliability in designing networks which can communicate with many intellectual properties (IPs) through efficient communication protocols [5].

The 3D-NoCs combine the benefits provided by the 3D integration technology and NoCs which results in a significant enhancement in the network performance [6]. As the 2D routing algorithms are not applicable in the third dimension, new 3D routing mechanisms are provided to communicate between the different tiers in the 3D-NoCs. The Through-Silicon Via (TSV) interconnection model which is one of the 3D interconnect approaches, offers the highest density of interconnect of all the other interconnecting approaches, but with a high fabrication cost in return [2]. The performance of 3D-NoCs depends on the number and location of the TSVs. Thus, a routing algorithm which guarantees the flexibility in choosing the number and placement of TSVs is a powerful algorithm [7]. The Elevator-First routing algorithm proposed in [7], is also independent of the dimensions of each

tier and the 2D planar topologies. This independence of the shape and size of each tier builds up hierarchical structures of 3D-NoCs. Such flexibility helps to perform variety of future experimental evaluations for the hierarchical structures of the 3D-NoCs. Through these evaluations, a deep analysis can be performed to define the optimal solution for the combination of the network topologies [8], switching techniques and also the count of the TSVs provided in the 3D-NoC.

The paper is organized as follows. In section II, various routing mechanisms for 3D-NoCs are presented. In section III, the Elevator-First algorithm and the Direct-Elevator algorithm are compared to illustrate the differences and the analogy behind each of them. In section IV, the performance analysis between the two algorithms is demonstrated to show the results under various test cases. Section V provides the conclusion of this work.

II. RELATED WORK

Providing new routing mechanisms that can maintain reliability and flexibility offered by 3D-NoCs, is a very challenging research field. Variety of studies adapt new routing schemes for 3D-NoCs. Authors in [9] proposed two routing algorithms for irregular topologies which can be implemented with two different approaches. These algorithms use look up tables for routing the messages over the NoC. In [10], another routing algorithm was proposed, in which virtual channels are used to provide a deadlock free mechanism for irregular networks. This mechanism utilizes compact routing tables to minimize the overhead of the look up tables. The authors in [11] proposed a deadlock and livelock free algorithm. The routing mechanism depends on splitting the network into layers. In [12], a routing algorithm was proposed with lower power consumption and system latency. Other routing schemes are provided for mesh topology only since it is the most popular topology used in the 3D-NoCs [13], [14].

Some studies focus on how to reduce the number of the TSVs to improve the performance. Authors in [8] proposed different 3D topologies in which the number of TSVs is less than other adopted routing algorithms. Mainly, the important aspects that the 3D routing algorithms investigate in are: the number and placement of TSVs in the NoC, the usability of the algorithm for irregular networks, and deadlock and livelock freedom. The authors in [7] provided a routing scheme

978-1-4673-8760-6/15 $31.00 © 2015 IEEE

called Elevator-First having no constraints on the number or placement of TSVs. The Elevator-First scheme is efficient with irregular networks and it is deadlock and livelock free by using two virtual channels. The study is extended in [15] to provide the implementation of the 3D-router that supports the Elevator-First algorithm. The area overhead of the 3D-router implemented is 8% more than the standard 3D-NoC router with same buffer capacity.

III. THE DIRECT-ELEVATOR ROUTING ALGORITHM

The Direct-Elevator algorithm benefits from the advantages and flexibility of the Elevator-First algorithm while optimizing its communication latency. Both distributed algorithms operate on vertically partially connected stacks of 2D topologies.

In the Elevator-First algorithm, adjacent stages are connected through elevator nodes. In order to move a packet from one stage to another, the packet has to go through the elevator node first; hence the algorithm's name. The routing algorithm's logic is distributed over multiple nodes in any tier. To get a better understanding of the algorithm, let's follow a packet path from the source node S1 to the target node D3 shown in Fig.1. Any source issues a packet to destination, it checks whether the destination exists on its tier, accordingly it routes the packet to the destination or the elevator. In case of S1, the destination is on a different tier, hence the packet is forwarded to the elevator E1. Additional information is added to the packet in a temporary header: the elevator's address, a T flag indicating whether the elevator is the final destination and another U flag indicating if the packet will go up or down to the target tier. When the packet reaches the intermediate tier, M acts as the new source and repeats the steps followed by S1. In this scheme, a packet cannot go directly from stage 1 to stage 3; it has to go through multiple nodes in stage 2. The Direct-Elevator algorithm is based on the hypothesis that having smarter elevator nodes connecting the stages will have a positive impact on the communication latency. Those elevator nodes would be able to decide whether the target node exists in their tier or not, then route the packet accordingly.

An example for the Direct-Elevator algorithm is shown in Fig.2. When the source node S1 issues a packet to the target node D3. First S1 adds the elevator address to the temporary header then sets the two flags T and U. The 3D elevator E1 receives the packet and checks the flags then sends the packet to E2. Therefore, E2 resets T flag and checks if the destination node in the same tier or not. If not, it forwards the packet to E3. Otherwise, it routes the packet to D3.

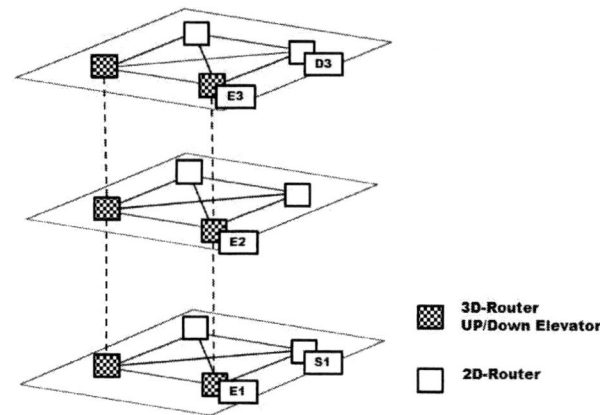

Fig. 2: Direct-Elevator Algorithm

The Direct-Elevator algorithm saves the processing time of adding and removing the 3D elevator address into the temporary header in every tier. So, regardless to where the destination tier is, the adding and removing process of the temporary header is done once.

IV. PERFORMANCE ANALYSIS AND RESULTS

A number of experiments were conducted in order to compare the performance of the Elevator-First and the Direct-Elevator algorithms. Both algorithms are implemented in the same environment to guarantee an efficient and reliable performance evaluation and fair comparison. The two algorithms have been implemented and compiled using C programming language and GNU Compiler Collection (GCC) under Cygwin.

Each experiment measures the throughput of the network in the following cases:

1) Random packet transmission.
2) Transmission over the network's worst path.

The throughput is defined as the number of successful received packets per second, while the worst path is defined as the longest path between two nodes in the NoC.

For each experiment setup the effect of changing the network load, the vertical complexity and the tier complexity on the throughput is measured by varying the number of transmitted packets, the number of tiers and the number of routers per tier, respectively.

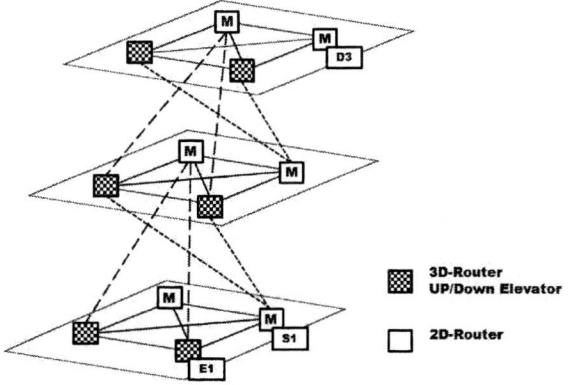

Fig. 1: Elevator-First Algorithm

978-1-4673-8760-6/15 $31.00 © 2015 IEEE   223

A. Network Load

The network load is measured by evaluating the throughput across a range of transmitted packets randomly and over the network's worst path. In these simulations, the number of tiers and the number of routers per each tier is equal to four.

Fig.3 shows the throughput of the two algorithms over a transmitted random packets range which varies from 50,000 to 200,000.

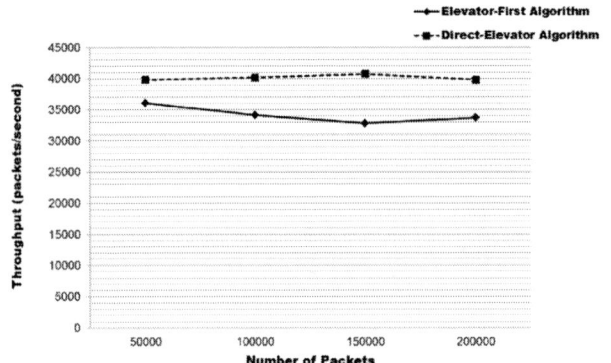

Fig. 3: Throughput Vs. Packets (random test case)

Fig.4 shows the two throughputs over a transmitted packets range over the worst network path which varies from 50,000 to 200,000.

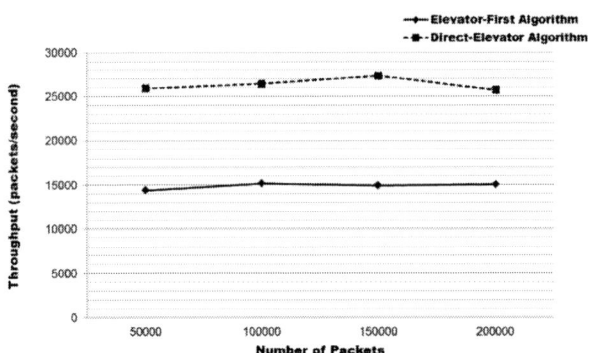

Fig. 4: Throughput Vs. Packets (worst path test case)

In Fig.3 and Fig.4, the throughput of the two algorithms changes slightly over the range of transmitted random or worst path packets. This behavior indicates that tuning the number of packets has no significant effect on the throughput and reflects the performance stability of the two algorithms.

B. Vertical Complexity

The vertical complexity measures the effect of changing the number of tiers in the NoC on the throughput.

Fig.5 shows the throughput of the two algorithms over a range of tiers in the NoC which varies from 3 to 6. In these simulations, the random transmitted packets and routers per tier are 100,000 and 4, respectively.

The throughput of the two algorithms decreases over the range

of tiers as increasing the number of tiers results in new longer vertical paths in the 3D-NoC.

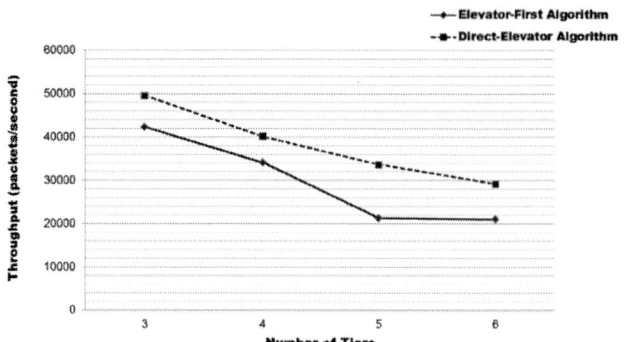

Fig. 5: Throughput Vs. Tiers (random test case)

Fig.6 shows the two throughputs over a range of tiers in the NoC which varies from 3 to 6. Here, the worst path transmitted packets and routers per tier are 100,000 and 4, respectively. The throughput of the two algorithms decreases more than the throughput shown in Fig.5 because in this case the packets are forced to follow the worst path in the 3D-NoC.

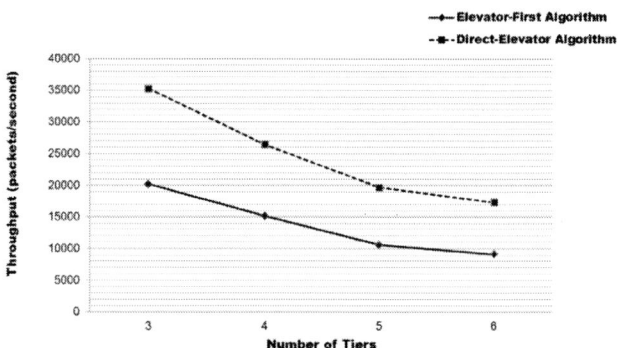

Fig. 6: Throughput Vs. Tiers (worst path test case)

C. Tier Complexity

The tier complexity measures the effect of changing the number of routers per tier in the NoC on the throughput. That can be studied on regular distributed or hierarchical distributed NoCs.

1) Regular Distributed NoC: A regular distributed NoC has the same number of routers in each tier.

Fig.7 shows the two throughputs over a range of routers in the tiers of NoC which varies from 4 to 32. In these simulations, the random transmitted packets and the number of tiers are 100,000 and 4, respectively.

Fig.8 shows the two throughputs over a range of routers per all the NoC tiers which varies from 4 to 32. Here, the worst path transmitted packets and the number of tiers are 100,000 and 4, respectively.

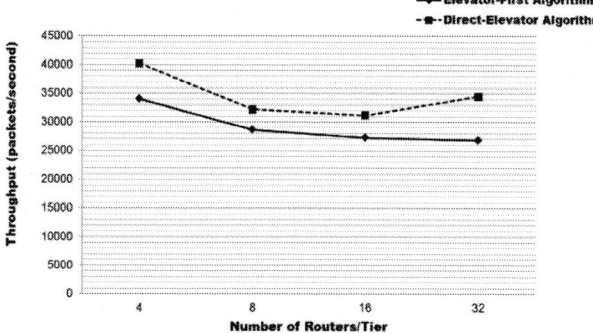

Fig. 7: Throughput Vs. Routers/Tier (random test case)

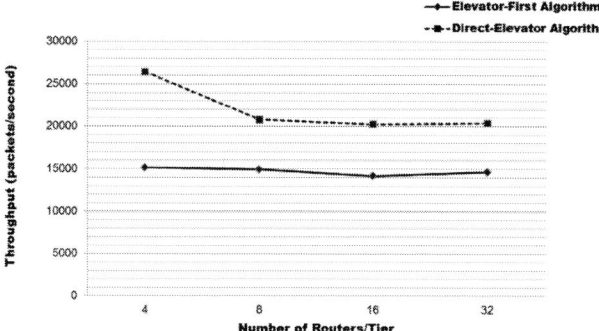

Fig. 8: Throughput Vs. Routers/Tier (worst path test case)

2) Hierarchical Distributed NoC: In the hierarchical structure of a four layered NoC, The routers are distributed as 4 routers in the first layer, 8 routers in the second layer, 16 routers in the third layer and 32 routers in the fourth layer.
In the case of transmitting a random sequence of 100,000 packets, the Direct-Elevator algorithm's throughput is higher than the Elevator-First algorithm's throughput by 5.3%. while in the case of transmitting the packets over the network's worst path,the throughput increase is greater than the random packets test case throughput increase by 40%.

V. CONCLUSION

The experimental evaluations of the Direct-Elevator algorithm show that the throughput of the algorithm is better than the throughput of the Elevator-First algorithm in different 3D-NoCs structures. The Direct-Elevator algorithm saves the processing time taken by the Elevator-First algorithm in the successive operation of adding and removing the elevator address while transmitting the message over the 3D-NoC tiers. The time utilization is better in the case of transmitting packets over the worst path in different combination of tiers per 3D-NoC, transmitted packets and routers per tier. Also, The number and placement of up and down elevators can be adjusted according to the traffic of each application. The Direct-Elevator combines the powerful features of the Elevator-First algorithm, but with better time utilization.

VI. FUTURE WORK

The performance evaluation between the Elevator-First and Direct-Elevator algorithms has been performed from the algorithmic throughput prospective. Future evaluations need to be performed to study the differences between the two algorithms for the area and power aspects.

VII. ACKNOWLEDGEMENT

This research was partially funded by Cairo University, ITIDA, NTRA, NSERC, Zewail City of Science and Technology, AUC, the STDF, Intel, Mentor Graphics, SRC, ASRT and MCIT.

REFERENCES

[1] B. Black, D. Nelson, C. Webb, and N. Samra, "3D Processing Technology and its Impact on iA32 Microprocessors," in VLSI in Computers and Processors, 2004. ICCD 2004. Proceedings. IEEE International Conference on Computer Design, Oct 2004, pp. 316–318.

[2] W. Davis, J. Wilson, S. Mick, J. Xu, H. Hua, C. Mineo, A. Sule, M. Steer, and P. Franzon, "Demystifying 3D ICs: the Pros and Cons of Going Vertical," Design Test of Computers, IEEE, vol. 22, no. 6, pp. 498–510, Nov 2005.

[3] P. Leduca, F. de Crecy, M. Fayolle, B. Charlet, T. Enot, M. Zussy, B. Jones, J.-C. Barbe, N. Kernevez, N. Sillon, S. Maitrejean, and D. Louisa, "Challenges for 3D IC integration: Bonding Quality and Thermal Management," in International Interconnect Technology Conference, IEEE 2007, June 2007, pp. 210–212.

[4] J. Knickerbocker, P. Andry, E. Colgan, B. Dang, T. Dickson, X. Gu, C. Haymes, C. Jahnes, Y. Liu, J. Maria, R. Polastre, C. Tsang, L. Turlapati, B. Webb, L. Wiggins, and S. Wright, "2.5D and 3D Technology Challenges and Test Vehicle Demonstrations," pp. 1068–1076, May 2012.

[5] L. Benini and G. De Micheli, "Networks on Chips: A New SoC Paradigm," Computer, vol. 35, no. 1, pp. 70–78, Jan 2002.

[6] B. Feero and P. Pande, "Networks-on-Chip in a Three-Dimensional Environment: A Performance Evaluation," IEEE Transactions on Computers, vol. 58, no. 1, pp. 32–45, Jan 2009.

[7] F. Dubois, A. Sheibanyrad, F. Pétrot, and M. Bahmani, "Elevator-First: A Deadlock-Free Distributed Routing Algorithm for Vertically Partially Connected 3D-NoCs," IEEE Transactions on Computers, vol. 62, no. 3, pp. 609–615, March 2013.

[8] V. Pavlidis and E. Friedman, "3-D Topologies for Networks-on-Chip," IEEE Transactions on Very Large Scale Integration (VLSI) Systems, vol. 15, no. 10, pp. 1081–1090, Oct 2007.

[9] F. Silla and J. Duato, "High-Performance Routing in Networks of Workstations with Irregular Topology," IEEE Transactions on Parallel and Distributed Systems, vol. 11, no. 7, pp. 699–719, Jul 2000.

[10] W. Jie and S. Li, "Deadlock-free Routing in Irregular Networks Using Prefix Routing Algorithm," Tech. Rep., 1999.

[11] O. Lysne, T. Skeie, S.-A. Reinemo, and I. Theiss, "Layered Routing in Irregular Networks," IEEE Transactions on Parallel and Distributed Systems, vol. 17, no. 1, pp. 51–65, Jan 2006.

[12] A. Ahmed and A. Abdallah, "LA-XYZ: Low Latency, High Throughput Look-Ahead Routing Algorithm for 3D Network-on-Chip (3D-NoC) architecture," in 2012 IEEE 6th International Symposium on Embedded Multicore Socs (MCSoC), Sept 2012, pp. 167–174.

[13] K.-C. Chen, S.-Y. Lin, H.-S. Hung, and A. Wu, "Topology-Aware Adaptive Routing for Nonstationary Irregular Mesh in Throttled 3D NoC Systems," IEEE Transactions on Parallel and Distributed Systems, vol. 24, no. 10, pp. 2109–2120, Oct 2013.

[14] R. Sunkam Ramanujam and B. Lin, "Randomized Partially-Minimal Routing on Three-Dimensional Mesh Networks," IEEE Comput. Archit. Lett., vol. 7, no. 2, pp. 37–40, Jul. 2008.

[15] M. Bahmani, A. Sheibanyrad, F. Petrot, F. Dubois, and P. Durante, "A 3D-NoC Router Implementation Exploiting Vertically-Partially-Connected Topologies," in 2012 IEEE Computer Society Annual Symposium on VLSI (ISVLSI), Aug 2012, pp. 9–14.

978-1-4673-8760-6/15 $31.00 © 2015 IEEE

On Getting Energy for Medical Equipment from Human Body

Shimaa Mahdy Abd Elhalim Marouf [1], Mohamed A.A Eldosoky[1], Yehya H. Ghallab [1,2]

[1]Department of Biomedical Engineering,
Helwan University, Cairo, Egypt.
[2] Center of Nano electronics and Devices (CND), Zewail City of Science and Technology,
Giza, Egypt
Email:shimaa.mahdy@yahoo.com, hm1_eldosoky@hotmail.com;
yghallab@zewailcity.edu.eg

Abstract **– This paper presents a new design of Micro Thermoelectric Generator (µTEG) for human energy conversion. The proposed µTEG uses the body heat for energy generation. Thus, it can be used to convert the temperature difference between the human body and the surrounding environment into an electrical energy to power a rechargeable battery of any medical equipment with lower power. The proposed µTEG uses a bismuth telluride (Bi_2Te_3) thermoelectric material which has an optimum figure of merit at 300°k. The dimensions of the proposed µTEG are 36µm x 20µm x 41µm. It generates an electrical voltage of 9.689µV. Also, it provides 0.01346µv/µm² (1.3VCm⁻²) and 28.5 μWcm^{-2} electrical voltage/area and power/area, respectively.**

I. INTRODUCTION

Human activities can be used to generate energy; this energy may be used as power for an electronic device. Human activities can be categorized into two main types; they are active and passive energy harvesting. The passive energy harvesting means that the energy can be generated as a result of the normal human activities, i.e., someone doesn't have to do special task different than the regular tasks. In other word, the energy is harvested from the human's everyday actions (walking, breathing, body heat, blood pressure, motion, etc.). Once the power is harvested it must be stored by rechargeable batteries.

Most of the current portable products are powered by rechargeable (also called secondary) batteries and they will remain as the main source for this kind of consumer products. However, the disadvantage of batteries is that the batteries need to be either replaced or recharged periodically.

The life time of the portable device is restricted by the duration of the battery. Thus, the principle to use the harvested energy from the human activities is very attractive and useful. One of these harvesting techniques is the usage of the temperature difference between the human body and its surrounding temperature to generate energy. This technique is known as Thermoelectric Generators (TEG), where the human body is used to generate electricity to recharge a secondary battery of some medical devices with low power [1-3].

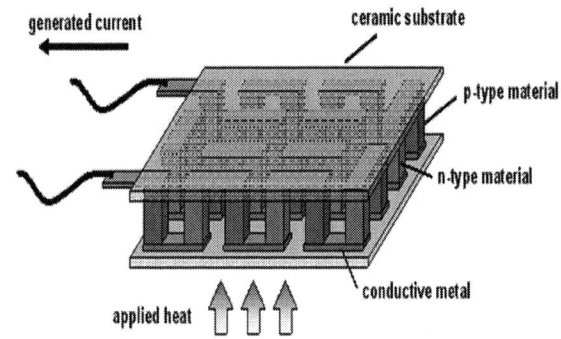

Figure 1: Thermoelectric Generation Principle

TEG involves placing one side of the thermoelectric module into contact with the skin of the human body and the other side in direct contact with ambient air (see Figure 1[4]). The amount of the generated power is associated with the temperature difference between the two sides.

Some medical devices with low power, like a pulse oximeter [5] and an electrocardiography system (ECG) [6], have already been powered by a thermoelectric generator using heat from the skin. This paper is organized as follows: Part II is reviewing the existing used µTEG. The proposed µTEG is presented and discussed in Part III. Simulation results are provided and discussed in Part IV. Part V shows a discussion and comparison between the proposed µTEG and the currently used µTEGs. Part VI application, part VII concludes this paper.

II. EXISTING DEVICES

Various miniaturized micro thermoelectric generators (µ TEG) are developed [7-11]. In [7], NiCu based µTEG that can generate a power up to 2.6 x 10³ µWcm⁻²K⁻². Also, the authors presented a Bi_2Te_3 based µTEG that can generate a power of 0.29µWcm⁻²K⁻². In [8], a µ TEG that has an output voltage of 670µV for the temperature difference of 10°k is developed. Various µTEGs for solar energy conversion are also presented [9, 10]. In [9], a transparent bridge type µTEG is developed which can be integrated with the windows of the building, so that the temperature difference between inside and outside are utilized for the generation of electrical power. This design produces a power of 1.0W for the temperature difference of 10°K. In [10], a µTEG which

978-1-4673-8760-6/15 $31.00 © 2015 IEEE

can generate a power of 8W from 75cm * 75cm area for lighting LED is developed. In [11], a thermo couple is developed to generate output voltage of 9.6868mV and output power of 1.6065µW, the area required to glow 10W, 12V LED has been reduced from 75cm × 75cm to 15cm x 30cm.

III. PROPOSED µTEG

A. Material:

The material that has high thermoelectric energy transformation capacity can be determined by the figure of merit Z, as given in (1)

$$z = \alpha^2 / \rho\lambda \qquad (1)$$

where α is the seebeck coefficient, ρ the resistivity, λ the thermal conductivity of the material. Equation (1) shows that the materials having high seebeck coefficient, low resistivity and low thermal conductivity are considered as thermoelectric materials. But while choosing the apt material, technological aspects also have to be considered. The bismuth telluride (Bi_2Te_3) is used because of its well established technology and high figure of merit (Z) comparing with the other materials at 300°k (see figure 2[4]).

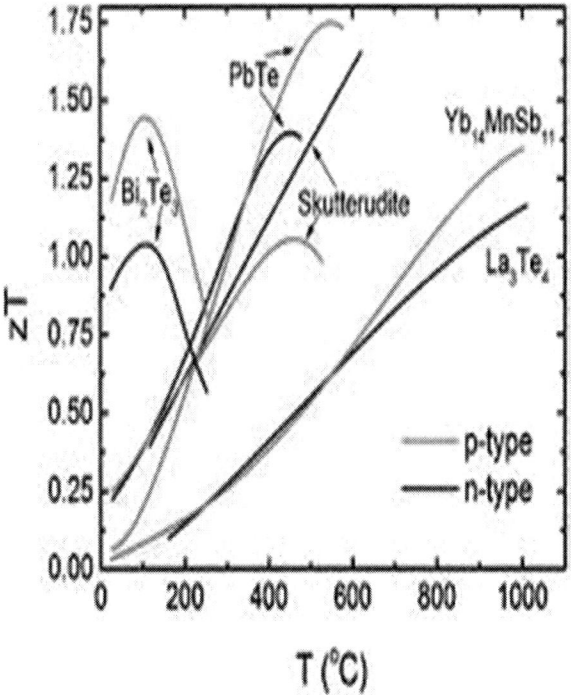

Figure2 Figure of Merit vs. Temperature

B. Material parameters:

The properties of the materials used in the simulation are summarized in Table 1 at 300°k [12].

		TEG material	Electrode (copper)
Seebeck Coefficient	α, V/K	P:200e-6 N:200e-6	6.5e-6
Electric conductivity	σ, S/M	1.1e5	5.9e8
Thermal conductivity	λ, W/mK	1.6	350
Density	ρ, Kg/m	7740	8920
Heat capacity	C, J/Kg/K	154.4	385

Table 1 the properties of the used materials

C. Design:

Figure 3 shows the proposed µTEG, which consists of N and P type (Bi_2Te_3) that are connected by copper for conductivity. The N and P layers are placed between two plates of aluminum (see Fig. 3).

Table (2) shows the dimensions of the used N and P type (Bi_2Te_3), copper and two plates of aluminum (upper and lower plates) (see Table 2).

The dimensions of the proposed µTEG are 36µm x 20µm x 41µm.

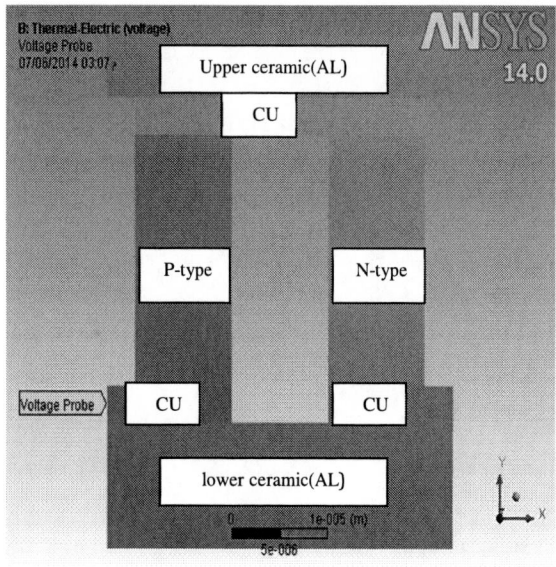

Figure3 the proposed µTEG in ANSYS software

	Height (μm)	Width (μm)	Length (μm)
Upper ceramic(AL)	5	20	36
Upper inter-Connect (CU)	3	20	30
N&P material (Bi$_2$Te$_3$)	20	20	10
Lower ceramic (AL)	10	20	36
Distance between the μTEG legs (μm)	10		

Table 2 Design parameters of the proposed μTEG

IV. Simulation Results

The proposed μTEG is simulated using ANSYS model software. The procedure is as follows:

1-Selection of general governing physics (thermal-electric system).

2-Creation of appropriate sub domains (geometries).

3- Application of appropriate boundary and sub domain conditions.

4- Meshing of the sub domains into an appropriate number of elements.

5- Selecting and utilizing a solver algorithm.

6- Post processing to acquire pertinent information (voltage, current).

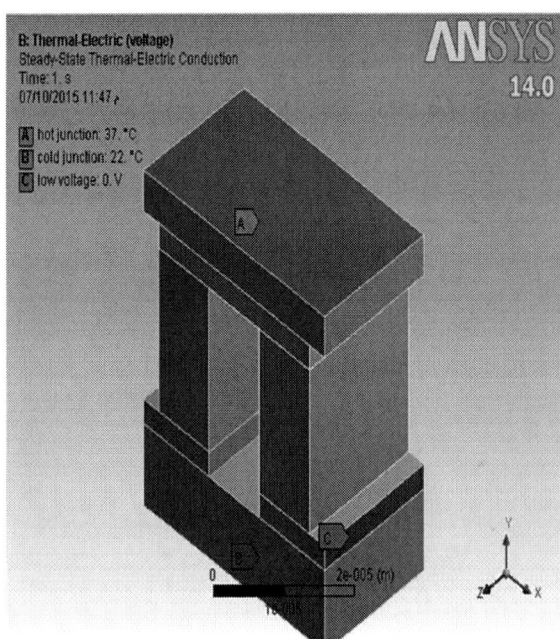

Figure4 steady state thermal-electric conduction of a single TEG

The dimension of the proposed μTEG is 36μm x 20μm x 41μm. It generates an electrical voltage of 9.689μV, and it generates 0.01346 μv/μm^2 (1.34VCm^{-2}) electrical voltage/area and power 2.0587e-10W. The generated power/area is 28.5μWcm^{-2} (where hot junction (Upper ceramic (AL)) is equal to 37°c and cold junction (lower ceramic (AL) is equal to 22°c) see fig.4).

V. Discussion

The proposed μTEG provides better performance compared with other used μTEGs. Table 4 shows a comparison between the proposed μTEG and the currently published μTEGs. The dimension of the proposed μTEG is the smallest compared to all other μTEGs. It also provides better voltage/area and power/area compared to the other μTEGs.

M. Lossec et al. [14] proposed a system that consists of 3300 TEG in 220cm^2 area and in three thermoelectric modules stacked with a heat sink. The number of TEGs is connected in series and material of TEG is bismuth telluride (Bi$_2$te$_3$). The output voltage of TEG is 1.8V and the output voltage/area is 0.0082Vcm^{-2}.

Elena Eet al., [15] proposed a system that consists of a single TEG in 3.4cm^2 and the used material is Bi$_2$Te$_3$.the output voltage of TEG is 5.9929mv and the output voltage/area is 0.001763Vcm^{-2}.

K. Ranjitha et al. [10] proposed a system that consists of a single TEG (15 x 30cm) and the material of TEG is silicon. The output voltage of TEG is 12V and the output voltage/area is 0.02667Vcm^{-2}.

VI. APPLICATION

We can recharge a rechargeable battery of wearable electro cardio graph (ECG) by using number of the proposed μTEG. The rechargeable battery need to 3V to recharge and 40mAh [16] .we will connect the number of proposed μTEG series to reach to a suitable voltage and current to recharge battery where the number of TEG that will used is 10^5 (N=10^5) with area 0.36 cm x 2 cm x 41 μm then this number of TEG connect with DC-DC converter and put number of TEG and dc-dc converter in the cuff around the arm.

VII. CONCLUSION

This paper presents a system that consists of a single μTEG. The proposed μTEG can generate an electrical voltage of 9.689μV from dimension 36 μm x 20 μm x 41 μm, and it can generate an electrical voltage/area of 1.34Vcm^{-2} and it can generate power 2.0587e-10 W and power/area of 28.5μW cm^{-2}.

By comparing with the previous work, the proposed single μTEG generates the highest electrical voltage per area and power per area. The proposed TEG can be used as an electric power source for many applications (medical equipment such as a pacemaker, an artificial heart, wearable ECG) where the number of proposed μTEG can connect series and parallel then this number of TEG connect with DC-DC power converter to reach to a suitable voltage and current to recharge battery of any medical equipment with lower and put number of TEG and dc-dc converter in the cuff around the arm.

References	Number of TEG	Surface area	dimension of single TEG	Type of material	Connection	Number of thermoelectric modules stacked	Power	With heat sink
[14]	3300	220 cm^2	e=3.4mm l=1.5mm L=54mm	Bi$_2$te$_3$	series	3	1.8V (0.0082 Vcm^{-2}) P=7μW cm^{-2}	√
[15]	1	3.4cm^2	e$_p$=1.4cm e$_n$=1cm l=1cm L=3.4cm	Bi$_2$te$_3$	-------	-----	5.9929 mV (0.001763 Vcm^{-2})	----
[10]		15*30 cm^2	e=50 μm l=59μm L=110μm	Si	Series& parallel	-----	10w,12v (0.02667V cm^{-2})	----
Proposed	1	36*20 μm^2	e=20 μm l=26μm L=36μm	Bi$_2$Te$_3$	--------	-----	9.689 μv (1.34 V cm^{-2}) (28.5 μW cm^{-2})	-----

Table 4 The comparison between the proposed μTEG and the currently published μTEGs.
Where: L=side length, l=leg length, e=thickness, e$_n$=thickness of n-type, e$_p$=thickness of p-type

References

[1] T. Starner, IBM Syst. J. 35, p.618, 1996.

[2] A. J. Jansen, A. L. Stevels, "Human Power, a sustainable option for Electronics," in Proc . IEEE Int. Symp. Electron. Environ p. 215, 1999.

[3] D. M. Rowe, CRC Handbook of Thermo electrics, CRC Press, London 1995.

[4] Baranowski, Lauryn L. and Snyder, G. Jeffrey and Toberer, Eric S. and published at journal of Energy & Environmental Science in 2012. [On line]. Available at: http://pubs.rsc.org/en/content/articlelanding/2012/ee/c2ee22248e#!d iv Abstract).

[5] T. Torfs, V. Leonov, R. J. M. Vullers, Sens. Transducers, J. 80, p. 1230, 2007.

[6] V. Leonov, T. Torfs, C. Van Hoof, R.J. M. Vullers, Sens. Transducers, J. 107, p. 165, 2009.

[7] Wulf Glatz, Etienne Schwyter, Lukas Durrer and Christofer Hierold, "Bi$_2$Te$_3$-Based Flexible Micro Thermoelectric Generator With Optimized Design," Journal of Micro electromechanical Systems, vol. 18, no.3, pp. 763-772, June 2009.

[8] Pin-Hsu Kao. Po-Jen Shih, Ching Liang Dai, "Fabrication and Characterization of CMOS-MEMS Thermoelectric Micro Generators," Journal of sensors, vol. 10, no.2, pp. 1315-1325, 2010.

[9] Guan-Ming Chen, I-Yu Huang, et al., "Development of a Novel Transparent Micro-Thermoelectric Generator for Solar Energy Conversion," in Proc. 6th IEEE International Conference on Nano/Micro Engineered and Molecular Systems, pp. 976-979, Feb. 2011.

[10] Y. Jeyashree, Pheba Cherian, A. Vimala Juliet, "Micro Thermoelectric Generator-A source of Clean Energy, " International Conference on Microelectronics, Communication and Renewable Energy (ICMiCR-2013), pp. 1-5, June 2013.

[11] K. Ranjitha et al., "Mono Crystalline Silicon-Based Micro Thermoelectric Generator for Solar Energy Conversion," pp. 1-7, March 2014.

[12] Martin Jaegle, Fraunhofer, "Multi physics Simulation of Thermoelectric Systems-Modeling of Peltier Cooling and Thermoelectric Generation," Excerpt from the Proceedings of the COMSOL Conference 2008 Hannover, 2008.

[13] Diana Davila, "Mono Integration of VLS silicon Nano wires into planer Thermoelectric Micro generators," December 2011.

[14] M. Lossec, et al. "Thermoelectric generator placed on the human body: system modeling and energy conversion improvements", 28 June 2010. Published online: 17 September 2010 – c_ EDP Sciences.

[15] Elena E. Antonova and David C. Looman "Finite Elements for Thermoelectric Device Analysis in ANSYS Inc.,"

[16] Chulsung Park, Pai H. Ying Bai, Robert Matthews, and Andrew Hibbs Chou, "An Ultra-Wearable, Wireless, Low Power ECG Monitoring System," USA.

Enhancing Power Delay Product in DRAMs Using Resonant Tunneling Diode Buffer

Ahmed Lutfi Elgreatly[1], Ahmed Ahmed Shaaban[2]
Dep. of Electrical Engineering
Faculty of Engineering, Port-Said University
Port-Said, Egypt
[1]eng.ahmed_lutfi@hotmail.com, [2]dessouki2000@yahoo.com

El-Sayed M. El-Rabaie
Dep. of Electronics and Communications Engineering
Faculty of Electronic Engineering, Minoufiya University
Minoufiya, Egypt
elsayedelrabaie@gmail.com

Abstract—DRAM chip industry became one of the most researchers' interests nowadays for its simple structure and low power consumption. As the density of DRAM chips increased, many problems occurred that affected the DRAM performance. One of these problems is the increase in the bit-line parasitic capacitance values. These large values slow down the reading operation of the cell and increase the consumed power. This problem gave a great attention to improve the performance of the sense amplifier circuit that is used in the reading operation in the DRAM cell for its great effect on both DRAM access times and overall power consumption.

In this paper, we introduce an alternative circuit architecture for the CMOS sense amplifier. This proposed circuit architecture is a specially designed logic buffer using a Resonant Tunneling Diode (RTD) that can be fabricated in silicon nano-electronics. The proposed design exhibits higher read operation speed, lower power consumption, full noise margin and higher chip density. The Power Delay Product (PDP) is improved by about 40% compared with that in the conventional CMOS sense amplifier and by about 55% compared with that in the conventional RTD-CMOS sense amplifier. The CMOS technology used in this paper is 45nm technology.

Keywords—Circuits and Architecture; CMOS; DRAM; Access time; Resonant Tunneling Diode (RTD); Power Delay Product (PDP).

I. INTRODUCTION

Due to the simple structure of the DRAM cell as shown in Fig.1, it can be easily implemented in arrays which make it widely used in most of recent applications. It consists of a cell capacitance (C_s) and an access transistor (M_{access}) between the bit-line (BL) and the cell capacitance which is enabled by the word-line (WL)[1,2].

A differential sense amplifier is connected to the bit-line to sense the voltage stored in it that results from the charge sharing between the cell capacitance and the bit-line capacitance. According to this value it pulls the bit-line up to the supply voltage (V_{DD}) or down to ground. The voltage shared (sense margin) must be sufficient enough to be sensed correctly by the differential amplifier; it can be expressed as follows [2]:

$$\Delta V = \frac{C_S}{C_S + C_{BL}}\left(V_{CS} - \frac{V_{DD}}{2}\right) \qquad (1)$$

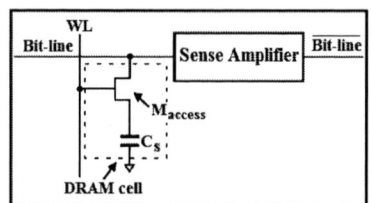

Fig. 1.DRAM cell [1].

where C_S and C_{BL} are the cell and bit-line capacitances respectively, V_{CS} is the voltage stored in the cell and V_{DD} is the supply voltage. Usually C_{BL} is much greater than C_S and (1) can be reduced to [2, 3]:

$$\Delta V \cong \frac{C_S}{C_{BL}}\left(V_{CS} - \frac{V_{DD}}{2}\right) \qquad (2)$$

In this paper, the value of ΔV is assumed to be equal to 5mV (as a worst case) for a very small charge sharing produced. This can be obtained when the cell capacitance (C_s) equals to 10fF and the bit-line equivalent capacitance (C_{BL}) equals to 500fF. This sense margin value (5mV) seems to be confusing if the thermal noise, the bit-line coupling capacitance noise and the process variations are considered. However, this minimum value can be estimated by considering the previously mentioned factors as in [4].

The performance of the memory chip is facing a big problem with the continuing CMOS technology scaling. It results in increasing the parasitic capacitances and increasing the leakage current of the CMOS devices. This leads to a significant increase the power consumed and the read/write access times[3,5].

In this paper, we introduce an effective solution by replacing the sense amplifier with a specially designed RTD logic buffer circuit that is capable of sensing small shared voltages (ΔV) and improves the read access time by a great amount without increasing the consumed power.

The rest of the paper is organized as follows: In section II, we discuss the conventional sense amplifiers including the CMOS sense amplifier and the RTD-CMOS sense amplifier. In section III, we explain the theory of operation for the proposed RTD logic buffer circuit. In Section IV, the simulation results are discussed showing the improvement in the power delay product (PDP). Section V discusses the effect

978-1-4673-8760-6/15 $31.00 © 2015 IEEE

of some important circuit parameters on the proposed design performance. Finally, the conclusion is included in section VI followed by the list of references.

II. CONVENTIONAL SENSE AMPLIFIERS

The CMOS differential sense amplifier [3,5] and the RTD-CMOS differential sense amplifier [6,7] are the most popular circuits used to improve the performance of DRAM memory chip.

A. CMOS Sense Amplifier

The CMOS sense amplifier consists of two identical cross coupled CMOS inverters as shown in Fig.2 [1].Fig. 3, shows the timing diagram of the control signals. The pre-charge signal (PRE) is activated to pre-charge the two bit-lines to half the supply voltage ($V_{DD}/2$). Then by setting the word-line signal (WL) to ($V_{DD}+V_{th}$), charge sharing occurs between the bit-line capacitance (C_{BL}) and the cell capacitance (C_S) resulting in a change in the voltage of the bit-line by ΔV.

In case of the cell capacitance (C_S) holds a logic '1', the bit-line voltage will be slightly greater than the reference voltage ($V_{DD}/2 + \Delta V = 0.505V$). When the sense amplifier is activated by the (SE) signal, the transistor M1 will be activated before M2 pulling the reference bit-line to ground. Similarly, the transistor M3 will be activated at the same time before M4 pulling the bit-line towards V_{DD} and vice versa if the cell capacitance holds a logic '0' (M2 and M4 will be ON).

The characteristics of the control signals are not considered in the scope of this work focusing only on the sense amplifier operation. However, the signal (SE) time period should be taken as small as possible without affecting the reading operation to reduce the consumed static power, especially in the case of RTD-CMOS amplifier circuit since the static power is dominated.

B. RTD-CMOS Sense Amplifier

The RTD-CMOS sense amplifier that introduced by [7] was one of the most promising solutions for enhancing the DRAM cell performance where the PMOS load transistor is replaced with a Resonant Tunneling Diode (RTD) [7] as shown in Fig. 4.This RTD can be fabricated in Nano-scale and exhibit higher operation speed and higher chip density.

The RTD (I-V) characteristic is shown in Fig. 5. It has very small area compared with that for the PMOS transistor and drive much higher current at small voltages (based on the quantum tunneling theory), this makes the RTD the best alternative as a load element to improve the reading access time in the DRAM cell [6, 7].

In this paper, the RTD (I-V) characteristics exhibits a peak-to-valley current ratio (PVCR) of 7.2 with a peak current density of J_P =157.14 A/m^2, peak voltage of V_P = 0.327Vand valley voltage of V_V = 0.55V. These parameters must be considered to insure proper operation of RTD in logic circuits [8].Also, the RTD DC analytical model introduced by [9] is used in our simulations.

Since the RTD as a pull up load can drive more current than an equivalent PMOS transistor, the sensing time during reading logic '1' decreases and consequently the sensing time during

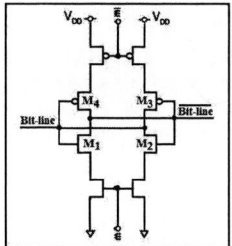

Fig. 2. CMOS sense amplifier [1].

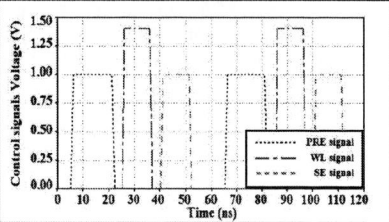

Fig. 3.Timing diagram of control signals

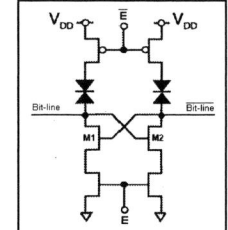

Fig. 4.RTD-CMOS sense amplifier [7].

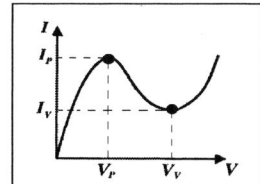

Fig. 5.(I-V) Characteristics of RTD [7].

reading logic '0' from the other bit-line decreases due to the cross coupling connection between the two inverters [7].

An improvement in the reading access time of the conventional RTD-CMOS sense amplifier is observed [7] compared with that in typical CMOS sense amplifier [1, 2]. On the other hand, the conventional RTD-CMOS sense amplifier has an obvious disadvantage of reduced noise margin. When the bit-line holds a logic '0', it holds a small voltage (about 70mV) and doesn't reach 0V as in the conventional CMOS sense amplifier. This results in increasing the static power consumed as shown in Table 1.

The aspect ratios of all NMOS transistors in CMOS sense amplifier circuit are equal to (180/45) each and are equal to (667/45) for each PMOS transistor. The area of each RTD equals to 0.57μm^2, and the aspect ratios of the transistors used in conventional RTD-CMOS circuit are identical to those in the conventional CMOS sense amplifier circuit.

III. THE PROPOSED SCHEME

The proposed RTD-CMOS buffer circuit consists of two cascaded inverter stages as shown in Fig.6. Each stage is specially designed to meet a certain purpose to raise the voltage swing of the logic levels and to improve the power delay product.

The first stage is an RTD-CMOS inverter circuit, it has a sharper transition region than that in the CMOS inverter circuit because it is Bi-stable. This results in a very small sense margin as shown in Fig.7 and Fig.8. This advantage makes the RTD-CMOS very accurate in sensing the small changes in the bit-line voltage produced from charge sharing ($\Delta V = 5mV$ as a worst case). These small changes are difficult to be sensed by the CMOS inverter.

The area of RTD1 and the aspect ratio of the NMOS transistor (M1) are chosen to satisfy the following condition:

$$I_{D_{M1}}(V_{BL} = V_L) < I_{P_{RTD1}} < I_{D_{M1}}(V_{BL} = V_H) \quad (3)$$

where

$$V_{DS_{M1}} = V_{DD} - V_{P_{RTD1}} \quad (4)$$

This condition guarantees that when the bit-line voltage equals to V_H ($V_{DD}/2 + \Delta V = 0.505V$), the output of the inverter is logic '0', and when the bit-line voltage equals to V_L ($V_{DD}/2 - \Delta V = 0.495V$), the output of the inverter islogic'1'.Fig. 9,shows the I-V-characteristics of the first inverter identifying the two logic outputs.

Fig. 10 shows the response of the first inverter stage at different gate widths (W) of the driving transistor M1 at two different RTD areas (A_{RTD1}). The shaded areas include the gate widths values that give the desired response correctly and satisfy (3).So, the condition is necessary and sufficient for the first stage to operate in the way it designed for.

The second stage (output stage) is chosen to be a CMOS inverter circuit over the RTD-CMOS inverter circuit for its full noise margin as shown in Fig. 9. The full output swing helps in decreasing the delay time consumed by the bit-line capacitance (C_{BL}) during charging and discharging which results in reducing both the consumed power and the reading access time. The aspect ratio of the output circuit transistors can be chosen to give the required PDP.

The main advantage of this scheme is that it determines the bit-line logic state voltage directly without differential sensing which decreases the overall parasitic capacitance of the circuit resulting in improving the reading access time of the circuit without increasing the power consumed. Also, no reference voltage is used which improves power consumption compared with the conventional RTD-CMOS sense amplifier discussed above. The feedback connection in the scheme writes back the output voltage of the buffer on the bit-line directly without the need of an access transistor for this write-back operation.

IV. THE PROPOSED SCHEME SIMULATION RESULTS

For the first inverter stage, the area of RTD1and the aspect ratio of transistor M1 should be taken as small as possible and satisfying the condition (3) as they affect the consumed power without a significant change in the delay time.

In this proposed scheme, the area of RTD1and the aspect ratio of transistor M1 are given by the values $0.23\mu m^2$ and (133/45) respectively. The dimensions of the other devices in the circuit are identical to those used in the two previous conventional circuits discussed.

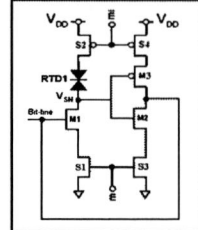

Fig. 6.TheProposed scheme.

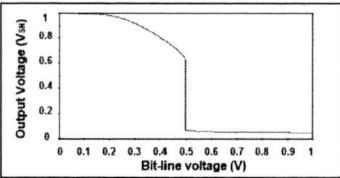

Fig. 7.Voltage Characteristics of the first inverter.

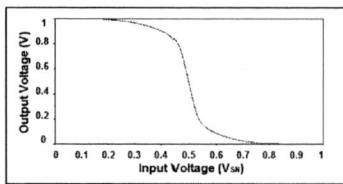

Fig. 8.Voltage Characteristics of the second inverter.

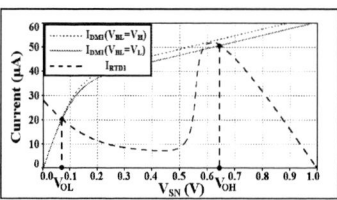

Fig. 9. IV Characteristics of the first inverter.

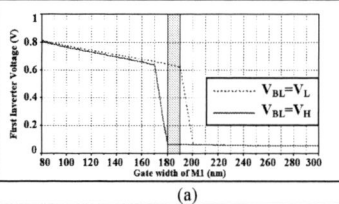

(a)

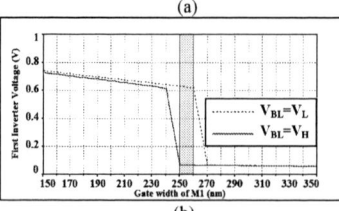

(b)

Fig. 10.Response of the first inverter stage.
(a) $A_{RTD1} = 0.33\mu m^2$(b) $A_{RTD1} = 0.4\ \mu m^2$

Fig.11 shows the reading access time of this scheme compared with the two previous schemes. In this work, The reading access time is defined as the time consumed by the sense amplifier to set the bit-line voltage to logic '1' (0.9 V_{DD}) or to reset it to logic '0' (0.1 V_{DD}). The proposed design exhibits about 40% higher speed than the conventional RTD-CMOS sense amplifier and about 50% higher speed than the conventional CMOS sense amplifier (see Table 1).

The Power Delay Product (PDP) is the most important parameter that measures the performance of the circuit. Table.1 shows the PDP for the proposed scheme (when the bit-line holds logic '1' as the worst case) compared with the two previous schemes. The proposed scheme has the minimum PDP which is about 40% lower than that in the CMOS sense amplifier and about 55% lower than that in the conventional RTD-CMOS sense amplifier.

V. Performance Sensitivity for the Design Parameters

There are two important parameters that significantly affect the performance of the cell operation for the proposed scheme and must be taken in consideration during the design of the sense amplifier. These parameters are: The bit-line capacitance (C_{BL}) and the dimensions of the switching transistors.

A. The bit-line capacitance (C_{BL})

By increasing the value of C_{BL}, it will take longer time while charging (reading logic '1') and discharging (reading logic '0') which increases the reading access time for both cases.

Fig. 12 shows the effect of changing the capacitance (C_{BL}) on the reading access time for the four schemes. The rate of increase of the access time for the proposed scheme is less than that in the conventional schemes; this gives a great advantage for the proposed scheme to be used in large memory arrays.

B. The Dimensions of the switching transistors

The switching transistors used for enabling the RTD-CMOS logic buffer must be taken in consideration during the design. The two switching transistors S1 and S2 are driven to the linear mode to activate the first inverter.

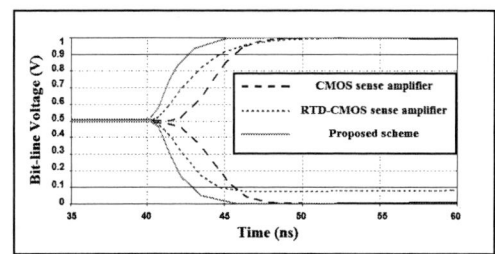

Fig.11. Reading access time of the proposed scheme compared with the previous schemes.

Table.1 Reading access time for the three schemes.

Elements of comparison	Pure CMOS	RTD-CMOS	Proposed
Reading Access time(τ_d)	5.7ns	4.7ns	2.8ns
Power consumed (P_{avg})	9 μW	14.5 μW	10.7 μW
PDP	51.3 fJ	68.2 fJ	30fJ

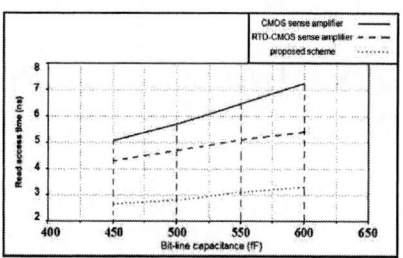

Fig. .12 Reading access time for all schemes at different bit-line capacitances

These switching transistors reduce the voltage supplied to the inverter by twice the drain-source voltage ($2V_{DS_S}$) (assuming all the switching transistors are identical). This voltage drop shifts the peak voltage of the RTD (V_P) by V_{DS} and the value of the drain-to-source voltage (V_{DS}) of the transistor M1 will be equal to:

$$V_{DS_{M1}} = V_{DD} - V_{P_{RTD1}} - 2V_{DS_S} \qquad (5)$$

This equation must be taken in consideration instead of (4).

VI. Conclusion

This paper presents a new design of RTD logic buffer circuit that reads the stored value in the cell and writes it back to the bit-line directly using resonant tunneling diodes. This design gives a great improvement in terms of the power consumed and the reading operation speed compared with the conventional cross coupled sense amplifiers. In this design, the PDP is improved by about 55% compared with that in the conventional RTD-CMOS sense amplifier and by about 40% compared with that in the conventional CMOS sense amplifier.

References

[1] Brent Keeth, R. Jacob Baker, Brian Johnson and Feng Lin, "DRAM Circuit Design: Fundamental and High-Speed Topics," Second edition, Wiley-IEEE Press, December 2007.

[2] Narendra Bahadur Singh, Dharmendra Kumar Rai, Prashant Singh, "Design of Low Power CMOS PSRAM," International Conference on Advanced Electronic Systems (ICAES), Sept. 2013

[3] Karishma Bajaj, ManjitKaur and Gurmohan Singh, "Design and Analysis of Hybrid CMOS SRAM Sense Amplifier," International Journal of Electronics and Computer Science Engineering (IJECSE) Vol.1, Issue 2, April, 2012.

[4] Heribert Geib, Werner Weber, Erdi Wohlrab, and Lothar Risch," Experimental Investigation of the Minimum Signal for Reliable Operation of DRAM Sense Amplifiers," IELE Journal of Solid-State Circuits. Vol. 27. No. 7. July 1992

[5] Abinayadevan. J and Vanitha. N, "Design and Analysis of Hybrid Mode Sense Amplifier for Various SRAM Cells," International Journal of Innovative Research and Studies (IJIRS), Vol. 3, Issue 5, May, 2014.

[6] P. Mazumder, S. Kulkarni, M. Bhattacharya, J.P. Sun and G.I. Haddad, "Digital Circuit Applications of Resonant Tunneling Devices", Proceedings of the IEEE, Vol. 86. No. 4,pp.664-686, April 1998.

[7] Tetsuya UEMURA and Pinaki MAZUMDER, "Design and Analysis of Resonant-Tunneling-Diode (RTD)Based High Performance Memory System," IEICE Trans. Electron.Vol.E82-C, No.9 September 1999.

[8] Christian Pacha, Peter Glosekotter, Karl Goser, Werner Prost, Uwe Auer and Franz-J. Tegude, "Resonant Tunneling Device Logic Circuits," LOCOM technical report, July 1998-1999.

[9] Sherif. F. Nafea, Ahmed. A. S.Dessouki, "An Accurate Large-signal SPICE Model for Resonant Tunneling Diode," International Conference on Microelectronics(ICM), Cairo, 19-22 Dec., pp.507-510, 2010.

Low Voltage Low Power Highly Linear OTA using Bulk Driven Technique

Karima GARRADHI[1], Néjib HASSEN[1], Thouraya ETTAGHZOUTI[1], Kamel BESBES[1,2]

[1]Micro-electronics and instrumentation laboratory University of Monastir, Tunisia
[1,2]Centre for Research on Microelectronics and Nanotechnology of Sousse, Technopole of Sousse, Tunisia
karimagarradhi@gmail.com, nejib.hassen@fsm.rnu.tn, thourayataghzouti@yahoo.fr, Kamel.besbes@fsm.rnu.tn

Abstract— **A new high performance OTA circuit using a bulk-driven differential input stage and a flipped-voltage follower current mirror is presented. The proposed OTA is operated at low supply voltage of ± 0.4V with a reduced power consumption of 0.44mW. All simulations are performed by ELDO technology CMOS TSMC 90nm which is provided a good linearity over the dynamic range, wide bandwidth and an excellent accuracy.**

Keywords— *Classic operational transconductance amplifier (OTA), Bulk driven, Flipped voltage follower, Dynamic range, Low-voltage Low-power.*

I. INTRODUCTION

Low power is one of the key research areas in analog integrated circuits design. Need of low power has created a major pattern shift in the field of electronics where power dissipation is equally important as area, performance etc. In the literature, the Operational Transconductance Amplifier (OTA) is an essential blocks in many mixed-analog systems and sensor interface. It is widely used in both analog and digital applications such as converters (A/D, D/A), Gm-C filters, continuous time oscillators and gyrators [1, 2].

The most significant problem to design the operational amplifier is the power dissipation, the supply voltage, and the limitation of the linearity and of the gain-bandwidth product. To solve this problem, several circuit techniques, crossing-coupling of multiple differential pair [3], adaptive biasing [5], source degeneration using resistors or MOS transistors [7] constant drain-source voltages [10] and super class-AB linear operation [12] can be used. Among the above mentioned techniques, bulk-driven is probably the simplest and most effective solution for realisation of low-voltage and low-power OTA [14]. Furthermore, the bulk-driven differential pair has successfully been used as the input stage of transconductance amplifier to allow a wide input linear range under low power supply voltage operation.

In this context, the implementation of OTA using bulk driven and gate driven structure are proposed. These two architectures have been simulated by ELDO technology CMOS TSMC 90nm which are operated at low voltage with a reduced power consumption .

This paper is organized as follows. In section II the circuit implementation of the proposed OTA using bulk driven and gate driven structure are presented. Simulation results and conclusion are given in sections III and IV, respectively.

II. PROPOSED OTA

A. Description of circuit operation

1) Gate Driven OTA

The conventional structure of proposed OTA using gate driven is shown in Fig. 1, which is achieved with a coupled differential pair (M1, M2) and three simple current mirrors with low operating voltage. The latter are one of the most common building blocks both in analog and mixed-signal VLSI circuits which are very useful elements in the operational amplifier transconductance OTA.

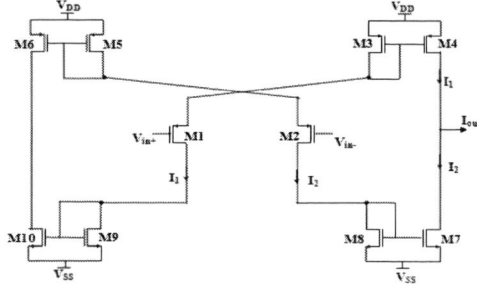

Fig. 1. Gate driven OTA

2) Bulk Driven OTA

In order to improve the previous circuit performance, we replaced the simple current mirrors by current mirrors FVF and we used a bulk driven technique at the input differential stage which is illustrated in Fig. 2.

This topology used bulk-driven technique to achieve highly linear voltage-to-current conversion under low power supply voltage and high bandwidth product and acceptable power supply rejection ratio (PSRR).

The flipped voltage follower (FVF) mirror reaches a low input and a high output impedance that are respectively shown in the following two equations (1, 2) [15].

$$R_{in} = \frac{1}{g_{m15}g_{m16}r_{o15}} \qquad (1)$$

$$R_{out} = r_{o17}r_{o12}g_{m17} \qquad (2)$$

When the input devices of the bulk-driven differential pair in Fig 2 operate in the strong inversion saturated region, their drain current is given by:

978-1-4673-8760-6/15 $31.00 © 2015 IEEE

$$I_d = \frac{1}{2}K(\frac{W}{L})(V_{GS} - |V_{TH}|)^2 \qquad (3)$$

Where, the symbols have their usual meaning and the channel length modulation effect has been neglected. The behavior of the threshold voltage as a function of the bulk-to source voltage V_{BS} of the input devices can be expressed as :

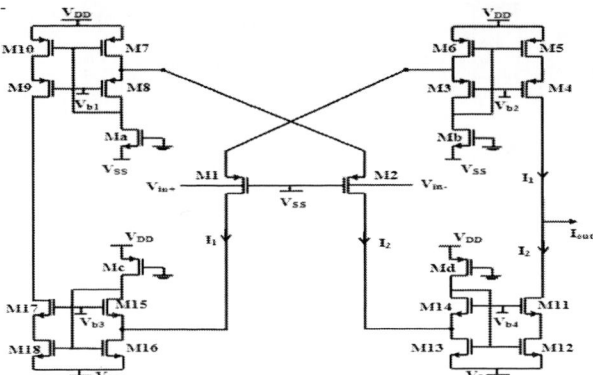

Fig. 2. Bulk Driven OTA

$$V_{TH} = |V_{THO}| + |\gamma|(\sqrt{2|\varphi| + V_{BS}} - \sqrt{2|\varphi|}) \qquad (4)$$

Where, V_{TH0} is the value of the threshold voltage V_{TH} when V_{BS} is zero, δ is the body effect parameter, and φ is Fermi potential. The operation of bulk-driven devices is based on the body effect, that is, on the dependence of V_{TH} on V_{BS} Indeed, from (3) and (4) Id changes when changing V_{BS}, Hence, a transconductance function between the bulk voltage and the drain current is achieved. The transconductance of a bulk-driven MOS transistor may be obtained from (3) and (4) as:

$$g_{mb} = \frac{dI_d}{dV_{BS}} = \frac{|\gamma|g_m}{2\sqrt{2|\varphi| + V_{BS}}} = \frac{|\gamma|}{2\sqrt{2|\varphi| + V_{BS}}}\sqrt{2\beta I_d} \quad (5)$$

Where, g_m is the gate transconductance of the device. If typical values for δ and φ are considered, the value of g_{mb} ranges from 20% to 50% the value of g_m.

To enable body driving, one must first bias the gate to develop a channel inversion layer. Once the inversion layer has been formed there will be depletion region associated with junction of the inversion layer and transistor body. The current is then modulated by varying the body voltage, which in turn modulates the inversion layer width via the depletion region. The drain current versus body voltage for this condition is obtained from (3) and (4):

$$I_D = \frac{1}{2}K(\frac{W}{L})(V_{GS} - |V_{THO}| - \gamma(\sqrt{2|\varphi| + V_{BS}} - \sqrt{2|\varphi|}))^2 \quad (6)$$

Consequently, the principle of the Bulk-driven technique is that; the gate-source voltage is set to a value sufficient to create an inversion layer, and an input signal is applied to the bulk terminal. In this way, the threshold voltage can be either

reduced or removed from the signal path. Therefore, the bulk driven technique devotes many advantages to OTA when compared to traditional OTAs. Firstly, it is used to increase respectively the input common mode voltage range [16], because the Bulk Driven transistor is a depletion type device, it can work under negative, zero, or even slightly positive biasing condition. Since, it has fair low voltage and low power. Secondly, the bandwidth is one of the most important performance parameters. Finally, the Bulk Driven technique is used for the designer of OTA to improve the PSSR characteristic (Power supply rejection ratio); because it is compensated by its load capacitance also it is used to improve the CMRR (common-mode rejection ratio). However, there is one disadvantage of bulk-driven technique, that the transconductance of the bulk-driven MOS transistor g_{mb} is less than the transconductance of the conventional gate-driven MOS transistor g_m. (g_m and g_{mb} are the slopes of its drain current versus its input voltage at the bias point). Hence, when the input differential pair of an amplifier is composed of bulk-driven transistors, the resulting DC gain is relatively low.

B. Dynamic Study of Circuit

The small signal equivalent circuit of proposed OTA is shown in Fig. 3. The differential input voltages V_{in+} and V_{in-} are shown in the following equation:

$$V_{in+} = \frac{1}{r_1 g_{mb1}}[(r_1 I_1 + V_{gs16} - V_{gs6}(1 + g_{m1}r_1 - r_1 g_{mb1})] \qquad (7)$$

$$V_{in-} = \frac{1}{r_2 g_{mb2}}[(r_2 I_2 + V_{gs13} - V_{gs7}(1 + r_2 g_{m2} - r_2 g_{mb2})] \quad (8)$$

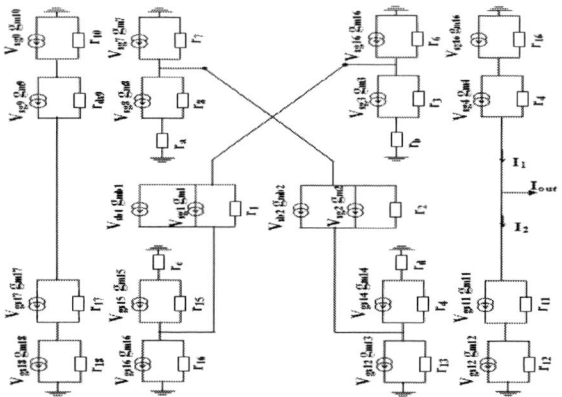

Fig. 3. The small Signal equivalent circuit for the proposed OTA

Where, the gate-source voltage of transistors M16, M13, M6 and M7 are shown respectively in the following equation:

$$V_{gs16} = \frac{r_{16}I_1 - g_{m15}r_{15}V_{gs15}}{1 + \dfrac{r_{15}}{r_c} + \dfrac{r_{16}}{r_c} + r_{16}g_{m16}}, \quad V_{gs13} = \frac{r_{13}I_2 - g_{m14}r_{14}V_{gs14}}{1 + \dfrac{r_{14}}{r_c} + \dfrac{r_{13}}{r_c} + r_{13}g_{m13}}$$

978-1-4673-8760-6/15 $31.00 © 2015 IEEE   235

$$V_{gs6} = \frac{-r_6 I_1 + r_3 g_{m3} V_{gs3}}{1 + \dfrac{r_3}{r_a} + \dfrac{r_6}{r_c} - r_6 g_{m6}} \qquad , \quad V_{gs7} = \frac{-r_7 I_2 + r_8 g_{m8} V_{gs8}}{1 + \dfrac{r_8}{r_a} + \dfrac{r_7}{r_c} - r_7 g_{m67}}$$

Considering the products $g_{mi} r_i$ much greater than 1, the characteristics of the transistors M1 and M2 are identical. Therefore, the output current can be expressed as:

$$I_{out} = \frac{g_{mb1} r_1}{\dfrac{r_1(r_{15} + r_{16} r_c g_{m16}) + r_{16} rc}{r_c r_{16} g_{m16} + r_{15}} + \dfrac{r_6 r_a (1 + r_1 g_{m1} - r_1 g_{mb1})}{r_3 - r_6 g_{m6} r_a}} V_{id}$$

(10)

III. SIMULATION RESULTS

The performance of the proposed CMOS OTA and Conventional OTA are verified by by ELDO technology CMOS TSMC 90nm. This circuit is operated at ± 0.4V supply voltage . Table I gives the dimension; width and lengths, of different transistors, resistor and load capacitor of the presented circuits.

TABLE I. TRANSISTOR SIZES AND BIAS CONDITION FOR ALL CIRCUITS OF OTA

Transistors	Conventional OTA W(μm)/ L (μm)	Proposed OTA W (μm)/ L (μm)
M1 ,M2	100/0.18	100/0.18
M3,M4,M5,M6	80/0.13	50/0.13
M7,M8,M9,M10	50/0.13	50/0.13
Ma,Mb	-	0.7/0.1
Mb,Mc	-	3/0.1
M11,M12,M13,M14	-	50/0.13
M15,M16,M17,M18	-	50/0.13
R(kΩ)	10	10
C(pF)	1	1

The DC transfer characteristic of the proposed OTA is shown in Fig. 4, which is achieved by varying the input voltage from -0.4V to +0.4V. It proves the effectiveness of the bulk driven in improving the dynamic range of the transconductor with nearly the same slope. Differential output current shows a symmetrical and linear behavior over a differential input voltage range of ±0.4V for OTA2 while it is about of ±0.07V for OTA1.

Fig. 5 shows the transconductance (Gm) for all circuits of OTA according to the differential input voltage. A more stable transconductance (G_m) in the interval [-0.3V, 0.3V] is reached by OTA2 and the maximum value of G_m is seen to be equal to 75μS. For OTA1, we noted a faible tansconductance in the interval [-0.07V, 0.07V] which is 65 dB.

Fig.6 shows the open loop frequency response for all circuits of OTA. A open loop gain of 26.53dB with a large GBW of 126.5MHz is reached by OTA2. For OTA1, we noted a low open loop gain closed to 11.85dB and a low GBW of 26.35MHz .

Table II shows a comparison with other low-voltage and low-frequency operational amplifiers. To evaluate this work a figure of merit (FOM) can be defined as [17]:

$$FOM = \frac{Gain(Unit\ gain\ freq)}{(Power\ sup\ ply)(Power\ consumption)} \quad (11)$$

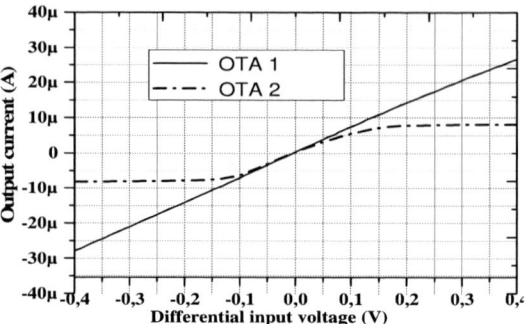

Fig. 4. I-V characteristics of the OTA1, 2

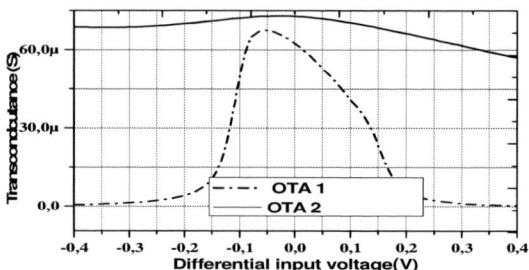

Fig. 5. The simulated Transconductance

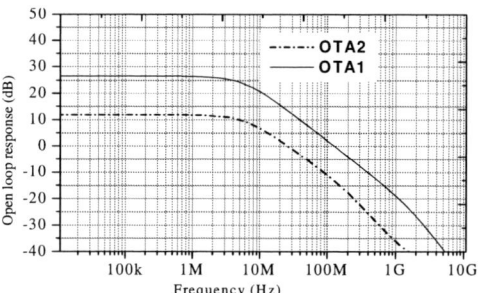

Fig. 6. Open loop frequency response of the OTA1, 2

IV. CONCLUSION

A new ultra-low voltage ultra-low power operational transconductance amplifier design with and without bulk driven technique is described in this paper. Simulation results have been presented to confirm the considerable improvement in unity-gain bandwidth and also the linearity, when compared to other ultra-low power low frequency OTAs conventional. The simulation results show that the open loop gain of the presented amplifier is equal to 26.53dB while achieving unity gain bandwidth of 126.5MHz. The total power consumption of the OTA proposed with bulk driven is as low as 0.44mW and the supply voltage is as low as ±0.4V which are much lower compared to the other presented works.

978-1-4673-8760-6/15 $31.00 © 2015 IEEE

TABLE II.   PERFORMANCE COMPARISON WITH PREVIOUSLY REPORTED WORK

Performance design	Our Circuit		OTA[18]	OTA [19]	OTA [20]	OTA[21]
	Gate Driven OTA	Bulk Driven OTA				
Technology CMOS (m)	90nm	90nm	130nm	0.35μm	50nm	0.35μm
Supply voltage (V)	±0.4	±0.4	0.9	2	0.5	0.9
Power consumption(mW)	0.046	0.44	3.9	0.024	0.0069	0.099
Transconductance (μS)	65	75	-	-	8.24	-
Capacitive load (pF)	1	1	1	1	-	2.5
DC gain (dB)	11.85	26.53	12	76	-	62
Phase margin (degrees)	50°	91 °	107°	-	-	52°
GBW (MHz)	26.53	126.5	1.31	0.736	-	0.54
Linear range (V)	[-0.07;0.07]	±0.4	-	[0.42,1.5]	-	-
PSRR+ (dc) (dB)	37.36	47.68	-	70.12	-	-
PSRR- (dc) (dB)	36.94	50.24	-	70.12	-	-
CMRR (dc) (dB)	33.3	66.44	-	142.24	30.77	129
Positive slew rate (V/μs)	19.84	6.23	0.54	1.23	-	-
Negative slew rate (V/μs)	-11.61	-7.22	-0.226	-1.05	-	-
Input Noise Density (uV/√Hz)	0.5	5	0.8	-	-	-
Output Noise Density (uV/√Hz)	0.001	5	0.4	-	1.12	-
Figure of Merit (dB. MHz/V. mW)	8542.94	9534.21	4.478	1165.3	-	375.75

References

[1] Y. Venkateswarlu, "Design of the 40MHz Double Differential-Pair CMOS OTA with -60dB IM3," International Journal of Engineering Research & Technology, vol. 2, September 2013.

[2] Y.Zheng, "Operational Transconductance Amplifiers For Gigahertz Applications," Department of Electrical and Computer Engineering University in Canada, These, 2008.

[3] Chen J., Sanchez-Sinencio E., Fellow IEEE, and Silva-Martinez J., "Frequency-Dependent Harmonic-Distortion Analysis of a Linearized Cross-Coupled CMOS OTA and its Application to OTA-C Filters," IEEE Transaction on Circuit and Systems, vol. 53, no. 3, pp.499-510, 2006.

[4] Manuel Carrasco-Robles and Luis Serrano, "Obtaining maximally flat transconductances with any number of multiple coupled differential pairs," International Journal of Circuit Theory and Application, Int. J. Circ. Theor. Appl. 2009.

[5] M. G. Degrauwe, J.Rijmenants and E. A. Vittoz, "Adaptive biasing CMOS amplifiers," IEEE Journal of Solid State Circuits, vol. 17, no. 3, pp. 522-528, 1982.

[6] S. Sengupta, "Adaptively biased linear transconductor," IEEE Transactions on Circuits and Systems, vol. 52, no. 11, pp. 2369-2375, 2005.

[7] Ko chi kuo, "A linear mos transconductor using source degeneration and adaptive biasing," IEEE Transactions on Circuits and Systems-ii: analog and digital signal processing, vol. 48, pp. 1057-7130, october 2001

[8] F. A. P. Barúqui ,and A. Petraglia, "Linearly Tunable CMOS OTA with Constant Dynamic Range Using Source Degenerated Current Mirrors," Mirrors' IEEE Transactions on Circuits and Systems II, vol. 53, pp. 791-801, 2006.

[9] Monsuro P, Pennisi S, Scotti G, Trifeletti A, "Linearization technique for source-degenerated CMOS differential transconductors," IEEE Transactions on Circuits and Systems II, vol. 54, no. 10, pp.848–852, 2007.

[10] Ayman A. Fayed and Mohammed Ismail , "A Low-Voltage, Highly Linear Voltage-Controlled Transconductor," IEEE Transactions on Circuits and Systems-II: Express briefs, vol. 52, no. 12, pp. 831-835, 2005.

[11] Ko-Chi Kuo and Hsing-Hui Wu "A Low-Voltage, Highly Linear, and Tunable Triode Transconductor," Electron Devices and Solid-State Circuits EDSSC, pp. 365-368; 2007.

[12] Houda Bdiri Gabboui, Néjib Hassen, and Kamel Besbes, "Low Voltage High Gain Linear Class AB CMOS OTA with DC Level Input Stage," World Academy of Science, Engineering & Technology , vol 5, pp 735-741 , Aug 2011.

[13] M. Laguna, C. De la Cruz-Blas, A. Torralba, R.G. Carvajal, A. Lopez-Martin and A. Carlosena , " A novel low voltage low power class AB linear transconductor," IEEE International Symposium on Circuits and Systems ISCAS, vol.1, pp. 725-728, 2004.

[14] Yasutaka Haga, Hashem Zare-Hoseini, Laurence Berkovi, and Izzet Kale "Design of a 0.8 Volt Fully Differential CMOS OTA Using the Bulk-Driven Technique" Department of Electronic Systems, IEEE,2005

[15] Néjib Hassen, Houda Bdiri Gabbouj and Kamel Besbes, "Low-voltage high-performance current mirrors: Application to linear voltage-to-current converte," International journal of circuit theory and applications, Int. J. Circ. Theor Appl, vol. 39, January 2011

[16] B. J. Blalock, P.E. Allen and G. A. Rincon-Mora, "Designing 1-V Op Amps Using Standard Digital CMOS technology," IEEE Trans. Circuits and Systems – II: Analog and Digital Signal Processing, pp. 769-780, Vol. 45, No. 7, July 1998

[17] E.Kargaran, M.Sawan, Kh.Mafinezhad, H.Nabovati," Design of 0.4V,386nW OTA Using DTMOS Technique for Biomedical Applications," in the proceeding of IEEE 55th International Midwest Symposium on Circuits and Systems (MWSCAS),USA, 2012.

[18] Neha Gupta, Sapna Singh, Meenakshi Sutharand Priyanka Soni, "LOW POWER LOW VOLTAGE BULK DRIVENBALANCED OTA," International Journal of VLSI design & Communication Systems (VLSICS vol 1,2, No.4, December 2011

[19] I. Toihria et T. Tixier "Improved PSRR and Output Voltage Swing Characteristics of Folded Cascode OTA" International Journal of Electronics and Electrical Engineering Vol. 3, No. 4, aout2015

[20] Prabdjot singh et Er.Abdijeet Kumar " Low Voltage and Low Power Bulk Driven O.T.A with Improved Transconductance "IJRIT International Journal of Research in Information Technology, Volume 2, Issue 6, Pg: 22-27 June 2014

[21] Sheng-Wen Pan1, Chiung-Cheng Chuang2, Chung-Huang Yang3, Yu-Sheng Lai "A Novel OTA with Dual Bulk-Driven Input Stage" "Department of Electronic Engineering, Vanung University of Science & Technology, Chung-Li, Taiwan, 2009

Evaluating the Feasibility of Centralized Router for Network on Chip

Mostafa Khamis, Amir Zaytoun
EE Department, Alexandria University
Alexandria, Egypt
amir.h.zaytoun@ieee.org

Ahmed Shalaby
ECE Department, Egypt-Japan University of Science and
Technology (E-JUST), Alexandria, Egypt
ahmed.shalaby@ejust.edu.eg

Abstract— Network-on-chip has been proposed to address the global interconnection problems of System-on-Chip. One of the key factors that affects performance of the network-on-chip is its topology. Mesh is the most popular topology because of its scalability and simplicity in physical implementation. But, their performance degrades with size scaling up due to extra hops. Centralized router is proposed as a candidate to overcome the regular topologies drawbacks. This paper discusses the feasibility of a centralized router to replace distributed network-on-chip. A complete evaluation for centralized router and mesh topology is performed over various network-on-chip sizes and different evaluation parameters. Results show that centralized router outperforms in large scale size at the expense of area and power, while in small sizes it outperforms overall evaluation parameters.

Keywords— System-on-chip (SoC); Network-on-chip (NoC); Crossbar; Batcher-Banyan, router architecture.

I. INTRODUCTION

Evolution in semiconductor technology has enabled integration of hundreds of IP blocks (processor, DSP, memory, etc.) on a single chip Systems-on-Chip (SoC). Network-on-Chip (NoC) is proposed as a promising solution to address the communication challenges of SoC. NoC connects IP blocks through interconnection routers, resembling the computer network, where data is routed through network infrastructure using packet switching rather than global buses [1]. NoC has many advantages compared to bus based architecture like scalability, modularity, reliability and performance. Over the last decade, many techniques have been proposed in order to improve NoC performance in terms of latency and throughput trading-off its area utilization and power consumption [2]. However, many research challenges remain to be solved for NoC to become widely applied and to make very large SoC a reality.

NoC architecture is based on topology, routing algorithm and switching techniques. Topology is one of key elements in NoC architecture and performance. Many interconnecting topologies have been suggested to be suitable for specific hardware and software configurations. Mesh is the most popular NoC topology because of its scalability and compatibility with simple routing algorithms. Moreover, it is more suitable for homogeneous on-chip IP blocks. However, increasing network size increases the number of hops which packet has to pass from source to destination. Consequently, it

leads to performance degradation in terms of latency and power consumption. Therefore, we have to reconsider using other connection schemes like centralized networks and hybrid topologies that combines the advantage of both centralized and distributed networks. In contrast to mesh, centralized networks implement one central router usually to connect heterogeneous on-chip IP blocks. Unfortunately, centralized networks are used to be hungry for power and area.

In this work, we study centralized router compared to mesh topology as the most regular topology for NoC. We provide a precision performance analysis in terms of latency, throughput, area and power consumption. This paper is organized as follows: the next section sheds the light on background and related work. Section III describes the implementation of both centralized and mesh networks. Section IV presents the simulation and synthesis results then a discussion. Finally, section V concludes the paper.

II. BACKGROUND AND RELATED WORK

NoC topology specifies the physical organization of on-chip IP blocks. It defines the connection approach for nodes, switches and links. Various topologies have been proposed for NOC architecture, e.g. mesh, torus, star, ring, butterfly, octagon and irregular interconnections. 2-D mesh is the most popular topology in which all links have the same length. It eases physical design and scalability. In addition, area grows linearly with the number of nodes. However, performance deteriorates as network size increases due to long network diameter and extra hops. Since performance is influenced by the buffers capacity, larger buffer size is proposed to improve mesh performance but this comes at the expense of area and power. As buffers consume 64% of the router leakage power for all process technologies [3], efficient buffer utilization techniques are proposed as another solution for performance degradation. Flexible router is an example for utilizing buffers efficiently [4], whereas the buffer is not confined to its input port but can handle packets from other ports using FIFO Flexibility Controller (FFC).

Centralized router has been proposed as an alternative to the regular distributed networks, as mesh, especially to connect heterogeneous IP blocks. In [5], authors propose two architectures for centralized routers: the first one is FIFO based input buffer while the other is based on Virtual Output Queuing (VOQ) input buffering strategy. However the

978-1-4673-8760-6/15 $31.00 © 2015 IEEE

evaluation lacks the precision and coverage to perform fair analysis to measure the real benefit of proposed architectures. In [6], Passas et al. studied the implementation of a crossbar switching fabric connecting 128 tiles. A detailed study of area and power of the crossbar switching fabric is performed till place and route phase on 90nm CMOS standard-cells. The crossbar is 32 bit width designed to deliver a 32 bit packet in one system clock (no serialization). The results show the area cost of the 128×128 switching fabric as 6% of the total tile area.

III. CENTRALIZED VS. DISTRIBUTED NETWORKS

In order to study the feasibility of centralized router for NoC, centralized and distributed mesh networks are implemented. The following subsections discusses the detailed building blocks of each network.

A. Centralized Router

Our implementation for centralized router is based on [5]. The router main components are: input unit, arbitration unit and switching core as shown in Fig.1 (a). Packet is received and buffered at the input unit. Then the input unit sends a request to the arbitration unit and waits for a grant. Next packet is arbitrated before forwarding it to the switching core. Two different switching cores, Crossbar and Banyan-Batcher (BB), are employed to drive packets to their destination ports from the input units. Both architectures utilize virtual queuing input buffer with Diagonal Propagation Arbitration (DPA) unit, shown in Fig.1 (b).

For input unit, VOQ is implemented whereas the buffer is divided virtually at the run time to queues according to the destination addresses of the ingress packets. For example, if all the input packets are directed to only one output, all the buffer is assigned to it. Whereas, if the input packets are directed to multiple ports, the buffer is divided among them. Consequently, better buffering utilization is achieved. VOQ is performed by dynamic buffer allocation algorithm which reduces packet loss and hence, increase efficiency [8]. A fast and efficient arbitration algorithm is required to handle requests from each virtual queue. and the DPA [9] was the candidate to perform this task. In DPA, a separate arbitration cell is dedicated for each request from every virtual queue to handle its request and grant signals. The Arbitration cells are connected together as shown in Fig. 1(b). Along with a moving activation window used to guarantee fair arbitration between inputs. Packets given grants are directed to the switching core.

Crossbar and Batcher-Banyan switches are implemented. The crossbar is described as an array of cross points to connect a specific input to a specific output. The number of cross points increases exponentially with the number of input/output. On the other hand, Batcher-Banyan switch core consists of two distinct networks: 1) Banyan network, which consists of multi-stage network of switching nodes. Each node is 2×2 self-routing switch. The node tests the destination address of the ingress packet and drives it to one of its outputs. 2) Batcher network, which is multi-stage sorting network used with the Banyan switching core; since it suffers from interconnection contention and needs input packets arrangement. The Batcher

sorting network guarantees that packets are directed to the Banyan network in the order that relates to the destination address.

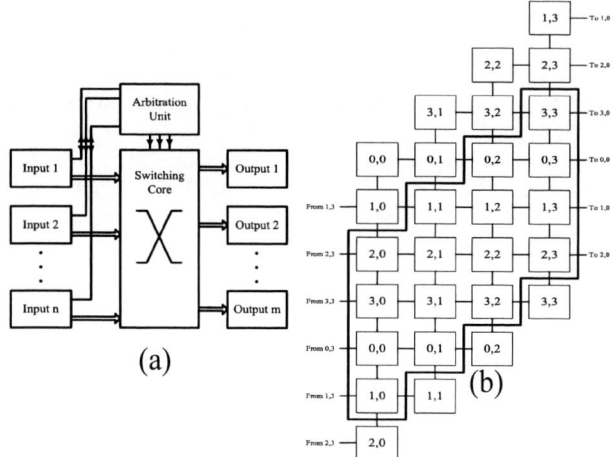

Fig. 1. (a) Centralized router components. (b) DPA for 4×4 VOQ router.

B. Distributed mesh network

Distributed network is based on topology, routing and switching schemes are defined by router which are the core elements in NoC. Dimension-Order routing algorithm is implemented. For fair evaluation, two routers were implemented: base and flexible routers [4]. Flexible router proposes an efficient method for buffering utilization, which can be compared to VOQ employed in the centralized router. The next subsection sheds the lights on basic components of both routers.

1) Base Router

Base router consists of five input/output ports, connected together by intermediate crossbar. The input port has three main components: 1) FIFO buffer: it stores the incoming packets from the upstream router. 2) Input controller: it handles requests and grants from/to upstream router. In addition, it communicates with output ports to transfer the received packets. 3) Routing logic: it applies the routing algorithm and determines the packet destination. Similarly, the output port has three main components: 1) Arbiter: it handles all received requests by the output port, then grants one of them based on round robin algorithm. 2) Output controller: it handles the communication with downstream router and input ports. 3) MUX: it selects which packet passes to the downstream router based on arbiter decision.

Base router operates as follows: It waits a request arising by one of the neighboring upstream routers, and then the input controller checks the FIFO buffer. If the FIFO is full, the input controller waits till at least one slot becomes free. Then, grant signal is sent to the upstream router to state that packet is stored. Next, the destination address field of the packet is checked by the routing function to determine the appropriate output port. After determining the output port, the input controller communicates to that specific output port to transfer the packet to the downstream router. Once the packet is

978-1-4673-8760-6/15 $31.00 © 2015 IEEE   239

transferred, the input port can check another packet in the FIFO and repeat the previous steps.

2) Flexible Router

Flexible router has the same architecture like base router except it has an additional unit, FIFO Flexibility Controller (FFC) at the input port. However, Flexible router operates similarly to the base router till congestion occurs. At that time, flexible router does not wait FIFO to have one or more free slots for incoming packets as the base router does. But FFC unit searches for a free slot at any other FIFO in the router. And once it finds a free slot it grants back the request to the upstream router. Then the packet is transferred to the selected FIFO. After that, the flexible router operates normally like the base router.

IV. SIMULATION, SYNTHESIS AND DISCUSSION

Centralized router is implemented with two different switching cores (crossbar and BB). As well, we implement mesh and torus networks with the two routers (base and flexible), in different sizes (4, 16 and 64) and various buffer size per port (2, 4 and 8). For fair comparison, three different sizes of centralized router are implemented to be compared with various mesh size. Moreover, the total number of buffer slots in the distributed network for all routers is considered to be equivalent to the amount of buffers in the centralized one.

A test environment was developed in Verilog to evaluate the performance of all the implementations (centralized crossbar, centralized BB, mesh base router, mesh flexible router). Each input port has packet injector which injects packets into the network. The injected packets have random destination addresses with uniform distribution traffic pattern. Packets travel through the network and terminate at the packet collector. Packet Injector/Collector logs are saved and analyzed to compute NOC performance.

All designs are written in Verilog/VHDL, tested by ModelSim and synthesized on TSMC 65 nm standard library on Synopsys DC. To calculate average latency and throughput, the simulator is run under different injection rates. Figures 2, 3 and 4 show comparisons between the different routers with different sizes 4, 16 and 64 and various buffer sizes 2, 4 and 8. The power and area are calculated without Injector/Collector. Power is given in mW and area in um².

Figures 2, 3 and 4 present the performance comparison for the four implementations in terms of average latency and throughput. Due to space, figures for network sizes 16 and 64 with buffer size 8 are demonstrated. In general, scaling up network size, degrade performance due to network capacity limitation [10]. On the other hand, increasing the buffer size improves NoC performance and slightly increase injection rate. For the results of small network size 4-ary, the throughputs are the same. Whereas, congestion barely happens in small distributed mesh. The average latency in mesh network is higher than torus network and all are higher than the centralized router.

As shown in Fig. 2, increasing the network size to 16-ary gives an advantage to the centralized networks. Additionally, crossbar centralized router gives lower latency than the

Batcher-Banyan. But both are better than the mesh network. Also we can notice that the flexible router gives lower latency than the base router for higher injection rates.

A great variation in the performance of the centralized networks, which can be noticed for the large network size 64-ary, is shown in Fig. 3. Mesh networks cannot support more than 0.1 packet per cycle injection rate per input/output, whereas the centralized router can achieve more than 0.2 injection rate.

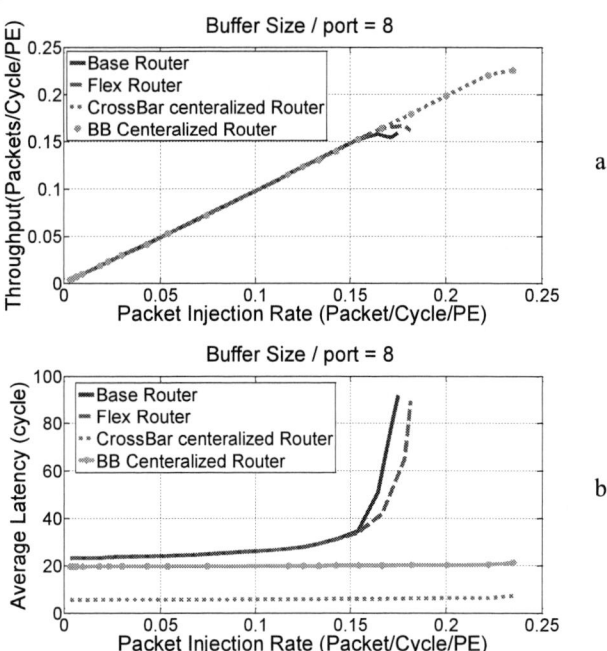

Fig. 2. (a)Throughput and (b) Average latency versus injection rate for 16-ary 2-D mesh with buffer size 8 per port.

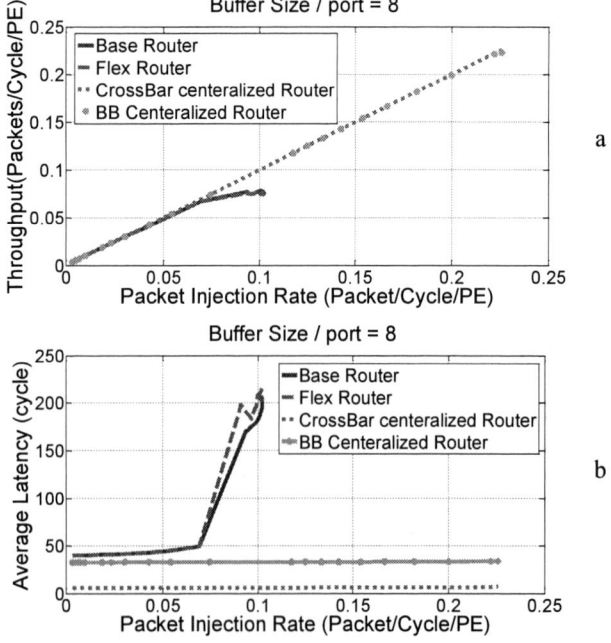

Fig. 3. (a)Throughput and (b) Average latency versus injection rate for 64-ary 2-D mesh with buffer size 8 per port.

Results coincide with our expectation that as network size increases, packets need to move through large number of routers. Thus, it leads to performance degradation at mesh topology. In order to verify our conclusion, we compare with torus topology, which reduces path hop counts due to wraparound links. Fig.3 shows throughput and average latency for centralized network in compare to torus network for 64-ary. It can be noticed that similar results can be achieved, Hence, the centralized network is a way faster than the distributed networks.

Synthesis results are shown in figures 5 and 6. It can be observed that area and power consumption for the small networks below 16-ary is similar. But it varies a lot for larger network sizes. For instance in Fig. 5, for 64-ary, power consumption of base router mesh network is the lowest and slightly increases for the flexible router. However, power consumption is twice over for the Batcher-Banyan centralized router and nearly 120% for the crossbar one. Fig. 6 shows the results for area. We can notice that for network sizes lower than 16-ary, centralized routers have the same area like mesh networks. Even the flexible router occupies larger area than the crossbar one. For the large network size for 64-ary, the areas of the centralized routers is nearly twice the mesh networks and it is expected to increase exponentially for larger networks.

Through all results, it was possible to achieve performance improvement by centralized routers for all evaluation parameters at small size networks, while for large network size, performance improves at the expense of area and power. Overall, results indicate that centralized router has a great potential to improve NoC performance. However, we think that a hybrid network topology with small centralized routers and mesh connecting them can achieve the best performance for large network sizes. This will be studied in our future work.

V. CONCLUSION

In this work, we introduce a feasibility study for centralized routers for NoC. A complete evaluation is performed and a detailed comparison in terms of throughput, average latency, area and power consumption is introduced. Results show that centralized network average latency is a way better than the mesh network. Alike throughput is better for large network sizes. This improvement comes at the expense of area and power consumption which increases exponentially in large network sizes. But on the other hand, it is less than the mesh for small size networks. Overall, our results indicate that centralized router is feasible and applicable for NoC. Hybrid architecture is to be studied in future work.

REFERENCES

[1] L. Benini and G. deMicheli, "Networks on chips: a new SoCparadigm," IEEE Computer, vol. 35, no. I, pp. 70–78, Jan. 2002.

[2] A. Agarwal, C. Iskander and R. Shankar, "Survey of Network on Chip (NoC) Architectures and Contributions," Journal of Engineering, Computing and Architecture Volume 3, Issue1, 2009.

[3] X. Chen and L.-S. Peh, "Leakage power modeling and optimization in interconnection networks," in Proceedings of the international symposium on Low power electronics and design, 2003, pp. 90–95.

[4] M. S. Sayed, A. Shalaby, M. El-Sayed, and V. Goulart, "Flexible router architecture for network-on-chip," Computers & Mathematics with Applications, vol. 64, no. 5, pp. 1301–1310, Sep. 2012.

[5] Amir H. M. Zaytoun, Hossam A. H. Fahmy, Khaled M. F. Elsayed, "Implementation and Evaluation of Large Interconnection Routers for Future Many-core Networks on Chip,"HPCC-ICESS 2012, pp. 524-531.

[6] G. Passas, M. Katevenis, and D. Pnevmatikatos, "A 128 × 128 × 24Gb/s crossbar interconnecting 128 tiles in a single hop and occupying 6% of their area", 4th ACM/IEEE International Symposium on Networks on Chip, NOCS 2010, pp. 87-95.

[7] Goke, L. Rodney, and G. Jack Lipovski. "Banyan networks for partitioning multiprocessor systems.," ACM SIGARCH Computer Architecture News, Vol. 2, No. 4, pp. 21-28, ACM, 1973.

[8] M. Fayyazi, D.R. Kaeli, and Z. Navabi, "Dynamic Input Buffer Allocation (DIBA) for Fault Tolerant Ethernet Packet Switching", in Proc. PDPTA, 2003, pp.819-823.

[9] J. Hurt, A. May, X. Zhu, and B. Lin, "Design and implementation of high-speed symmetric crossbar schedulers", Proc. IEEE International Conference on Communications (ICC'99), Vancouver, Canada, June 1999, pp. 253-258.

[10] Zhou, Xinan. "Performance evaluation of network-on-chip interconnect architectures.," 2009.

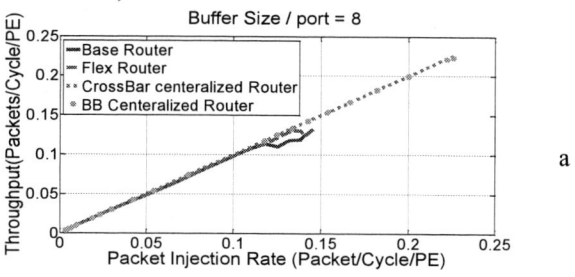

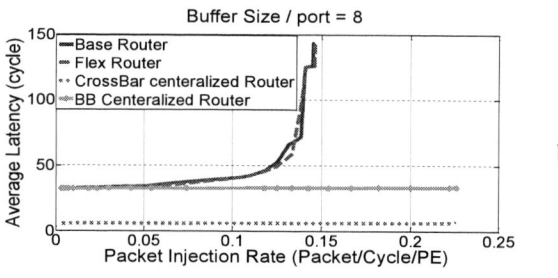

Fig. 4. (a)Throughput and (b) Average latency versus injection rate for 64-ary 2-D Torus with buffer size 8 per port.

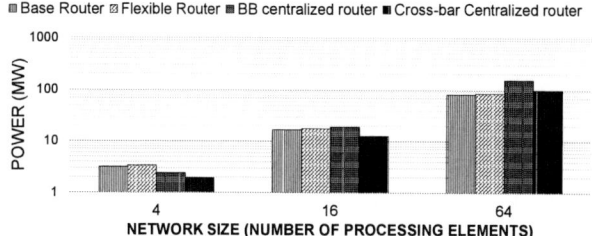

Fig. 5. Power consumption versus number of processor elements.

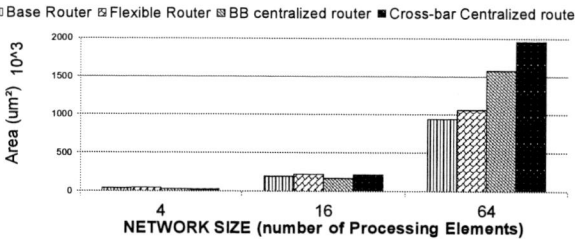

Fig. 6. Area versus number of processor elements.

A Reconfigurable 2-D IDCT Architecture for HEVC Encoder/Decoder

Ahmed Kilany[1], Maher Abdelrasoul[2], Ahmed Shalaby[2], Mohammed S. Sayed[2,3]

[1]Information Technology Institute, Cairo, Egypt, email: eng.ahmed.kilany@gmail.com
[2]ECE Department, Egypt-Japan University of Science and Technology (E-JUST), Alexandria, Egypt
{maher.salem, ahmed.shalaby, mohammed.sayed}@ejust.edu.eg
[3]ECE Department, Zagazig University, Zagazig, Egypt, msayed@zu.edu.eg

Abstract—Recently, HEVC standard have been proposed as a solution for transmitting high quality videos with half bit rate compared to the previous H.264 standard. One of the main properties of the new standard is the variety of the transform unit sizes. In this paper, we propose a new reconfigurable pipelined architecture for Inverse Discrete Cosine transform, which is used in both the HEVC encoder and decoder. Our circuit supports all the transform block sizes with reusability and reconfigurability of the different circuit parts. Our proposed architecture implemented on TSMC 65nm, runs at clock frequency of 500 MHz, and achieves throughput of 1990 Mpixel/sec that is more than the best architecture in the literature, to the best of our knowledge, by about 42%. The proposed architecture can process UHD video resolutions up to 8K with 60 fps.

Keywords—HEVC; DCT; IDCT; reconfigurable architecture.

I. INTRODUCTION

Nowadays, digital video industry advances in a fast manner since the number of electronic devices (smartphones, set-top-box for digital tele- vision, Blu-ray players, etc.) that support digital video increases. Hence video encoders/decoders process high-resolution videos in real time become a demand. This demand drives companies, researchers and standards bodies to provide solutions for the storage and transmission challenges facing digital video industry. Video compression standards were proposed to solve these challenges.

High Efficiency Video Coding (HEVC) standard is proposed as a video coding standard by the MPEG and the ITU bodies [1]. HEVC, when compared to H.264/AVC, provides 50% less bit rate and higher degree of parallelism by adopting a variety of efficient coding tools. The HEVC project was formally launched in January 2010. HEVC is designed along with the successful principle of block-based video coding. A picture is first partitioned into blocks and then each block is predicted by using either intra-picture or inter-picture prediction. In either case, the resulting prediction error, is transmitted using transform coding, scalar quantization of the transform coefficients and entropy coding of the resulting transform coefficient levels. In addition, the HEVC specifies the in-loop filters to be applied in the encoding and decoding processes.

In the transformation module, Discrete Cosine Transform (DCT) is often used to compact the signal energy in a limited number of coefficients. On the other hand, Inverse DCT (IDCT) restores the signal from the coefficients. The transformation is a multiplication between the transformation matrix and the input array. In image/video processing, the transformation is performed for a portion of the image/frame called transform block. This block is a (2D) matrix of pixel values. Therefore, the transformation is performed in two dimensions using two separable 1D transformations. First, the rows are transformed. Next, the result matrix is transposed. Finally, the new rows are transformed resulting in the 2D transformed block.

In HEVC, the transformation matrix is completely integer values. Hence, the DCT/IDCT architectures have to exploit this property efficiently. Many researchers focus on the implementation of the DCT transformation, while only few architectures were proposed for IDCT. In this paper we propose a new reconfigurable architecture for Integer IDCT. The proposed architecture efficiently exploits the reusability of its building blocks to process the different transformation block sizes required by the HEVC standard while the minimum possible area. Moreover, the similarities in the IDCT transformation matrix is exploited and optimized add-shift circuits are implemented instead of multipliers. Therefore, the proposed architecture has a high throughput that it can process UHD video resolutions up to 8K with 60 fps.

The rest of this paper is organized as follows: The related work is reviewed in Section II. Our proposed reconfigurable architecture is presented in Section III. In Section IV, the prototyping results are demonstrated. Finally, in Section V, we conclude the paper.

II. RELATED WORKS

Many architectures were proposed for the computation of DCT/IDCT for H.265/HEVC. All the proposed designs embed (or assume [6]) the full-size transposition buffer (32 × 32 samples) implemented either as a register matrix [3] or memory modules [2 and 7]. As these designs are dedicated for specific throughput, they cannot trade resources for the throughput. Implementation of the full-size transposition buffer involves a large amount of hardware resources, In addition to regular multipliers. To save them, [4] replace constant-multipliers by full-multipliers and constant lookup tables (LUTs) and compute a 1-D IDCT over multiple cycles. This approach is area-efficient when the output contains the number of samples/coefficients equal to the 1D transform size (i.e. 32) [2]. However, this output format is inconvenient and requires additional registers for the adaptation to smaller sizes.

978-1-4673-8760-6/15 $31.00 © 2015 IEEE

III. PROPOSED ARCHITECTURE

Our proposed architecture composed of two main components: 1D-IDCT and transpose memory. Transpose memory acts as intermediate stage between two identical 1D-IDCT units. It is used as a buffer unit to store the output of the first stage 1D-IDCT to be delivered to the next 1D-IDCT unit. Reusability and configurability are main features in the proposed architecture. Where the architecture changes dynamically with the transformation block size. In addition, the proposed architecture is fully pipelined to achieve high throughput. The following subsections discuss the proposed architecture in details:

A. 4-pt IDCT

The 1D 4-point inverse transform is given by the following equation:

$$Y = D_4 X \tag{1}$$

where X= $[X_0 ,X_1 ,X_2 ,X_3]^T$ is the input vector and Y = $[Y_0 ,Y_1 ,Y_2 ,Y_3]^T$ is the output vector and D_4 is the inverse transform matrix of 4x4 block size

$$D_4 = \begin{bmatrix} 64 & 83 & 64 & 36 \\ 64 & 36 & -64 & -83 \\ 64 & -36 & -64 & 83 \\ 64 & -83 & 64 & -36 \end{bmatrix}$$

Our proposed architecture depends on reducing the matrix multiplication into two half-size matrix multiplication followed by one addition or subtraction as follow:
let

$$[Z_0 \quad Z_1] = D_E X_E \tag{2}$$

$$[Z_2 \quad Z_3] = D_O X_O \tag{3}$$

where $D_E = \begin{bmatrix} 64 & 64 \\ 64 & -64 \end{bmatrix}$, $D_O = \begin{bmatrix} -36 & 83 \\ -83 & -36 \end{bmatrix}$,

$X_E = [X_0 \ X_2]$ and $X_O = [X_1 \ X_3]$.

Then, by addition and subtractions as in Eq.4, we can reach the required transformation output

$$Y_0 = Z_0 + Z_3, Y_1 = Z_1 - Z_2, Y_2 = Z_1 + Z_2, Y_3 = Z_0 - Z_3 \tag{4}$$

Fig. 1 shows the proposed architecture for 4-point IDCT. It consists of 4 components: Even Part circuit, Shift Add Unit (SAU), Add Sub Unit (ASU) and Output Add Unit (OAU). 1) The even part circuit implements Eq. 2 by taking (64) as common factor. It ends up with addition and subtraction before the multiplication. In addition, the multiplication by 64 is simply performed by shifting left the input data six times. 2) Shift Add Unit (SAU), this module is a multiple constant multiplication (MCM) circuit, it implements multiplication of the input by D_O matrix using shifters and adders instead of regular multipliers. Simplifying multipliers to SAUs leads to a reduction in the circuit critical path, as a result the throughput increases. There are many algorithms used to generate the shift/add circuit for a multipliers [8]. In this paper, Hcub

algorithm is used to implement SAU. The Hcub is an optimized algorithm in term of delay, which assist in achieving our target to decrease the critical path of transformation circuits, thus reduce its processing time hence increase the circuit throughput. Fig. 2 shows Hcub algorithm implementation for Eq.3 multiplications. 3) Add Sub Unit (ASU) is implemented due to different signs of the multipliers in D_O matrix of Eq. 3. 4) Output Add Unit (OAU) to implement addition and subtraction of Eq.4.

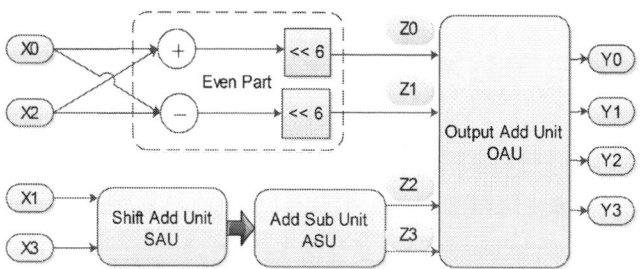

Fig. 1. 4-pt IDCT Architecture

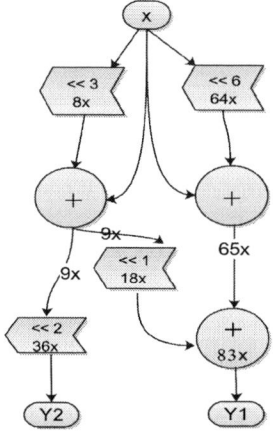

Fig. 2. 4-pt Shift Add Unit (SAU)

B. 8-pt IDCT

Similar to 1D 4-point, the 1D 8-point IDCT is given by the following equation:

$$Y = D_8 X \tag{5}$$

where the input vector X= $[X_0, X_1, X_2, X_3, X_4, X_5, X_6, X_7]^T$, output vector Y= $[Y_0, Y_1, Y_2, Y_3, Y_4, Y_5, Y_6, Y_7]^T$ and the inverse transform matrix of 8x8 block is

$$D_8 = \begin{bmatrix} 64 & 89 & 83 & 75 & 64 & 50 & 36 & 18 \\ 64 & 75 & 36 & -18 & -64 & -89 & -83 & -50 \\ 64 & 50 & -36 & -89 & -64 & 18 & 83 & 75 \\ 64 & 18 & -83 & -50 & 64 & 75 & -36 & -89 \\ 64 & -18 & -83 & 50 & 64 & -75 & -36 & 89 \\ 64 & -50 & -36 & 89 & -64 & -18 & 83 & -75 \\ 64 & -75 & 36 & 18 & -64 & 89 & -83 & 50 \\ 64 & -89 & 83 & -75 & 64 & -50 & 36 & -18 \end{bmatrix}$$

978-1-4673-8760-6/15 $31.00 © 2015 IEEE   243

If we divide D_8 into two matrices D_E, D_O like in the 4-point matrix and the input vector into X_E and X_O where

$$D_E = \begin{bmatrix} 64 & 83 & 64 & 36 \\ 64 & 36 & -64 & -83 \\ 64 & -36 & -64 & 83 \\ 64 & -83 & 64 & -36 \end{bmatrix}$$

$$D_O = \begin{bmatrix} -18 & 50 & -75 & 89 \\ -50 & 89 & -18 & -75 \\ -75 & 18 & 89 & 50 \\ -89 & -75 & -50 & -18 \end{bmatrix}$$

$$X_E = \begin{bmatrix} X_0 & X_2 & X_4 & X_6 \end{bmatrix}$$

$$X_O = \begin{bmatrix} X_1 & X_3 & X_5 & X_7 \end{bmatrix}$$

We can notice that D_E matrix is the same as D_4. Therefore, the 4-point IDCT block is reused in the even part. For the odd part, it can be implemented in a similar way like the odd part in the 4-point IDCT. Correspondingly, 16-point and 32-point IDCT are implemented in a similar manner. In general, N-point IDCT block can be implemented by reusing the even part of N/2-point IDCT.

C. Transpose Circuit

In the proposed architecture, the 2D-IDCT transformation is performed by two identical 1D-IDCT and transpose circuit in between as shown in Fig. 3. The transpose circuit is required to store and reorder the output of first stage 1D-IDCT. In Fig. 4, the transpose element is shown, It consists of a register and a multiplexer. Multiplexer's select signal controls the written data on the register to be transferred to a row or a column. Fig. 5 shows an example for the 4x4 transpose circuit. It works as follows: At the start, the output of first 1D-IDCT unit is written row by row till the end. Then the second 1D-IDCT unit reads column by column. Simultaneously while reading, the first 1D-IDCT writes its new output column wise. After, the second 1D-IDCT reads row wise, and so on. In brief, write and read processes of 1D-IDCT units are alternatively swapped.

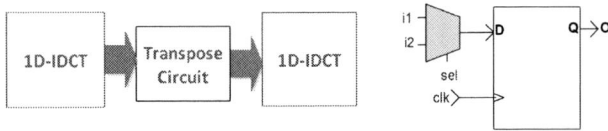

Fig. 3. 2D-IDCT   Fig. 4. Transpose Element

D. Reconfigurable Architecture

The proposed architecture processes different block sizes dynamically in pipeline manner. As shown in Fig.6, the architecture adapts itself according to the input/output block sizes and it is fully pipelinned through six stages pipelinned registers R1 to R6. For instance, if the block size is 4x4 then the outputs O4_0 to O4_3 will be selected by the Output Selection block to be the output after 3 clock cycles.8x8,16x16 an 32x32 will

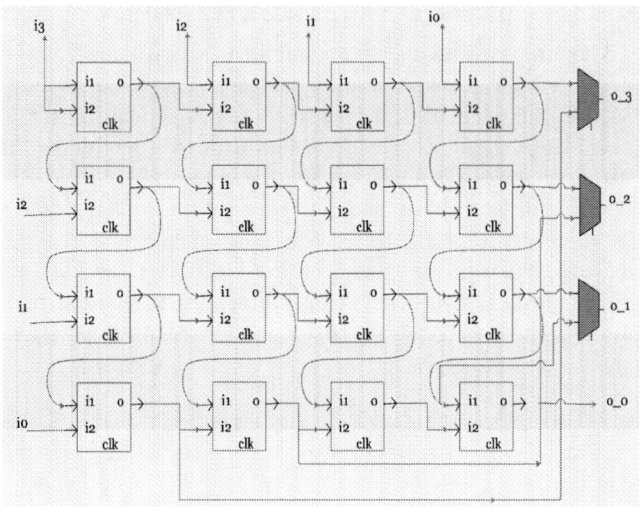

Fig. 5. Transpose Circuit

be calculated in a similar manner but with 4,5 and 6 clock cycles latency. Input/output delay blocks are implemented to adjust the timing between the different modules for different block sizes. For instance, in the case of 32-point IDCT we can notice that the output of the even part from the 16-point IDCT arrives one clock cycle late. So delay blocks are used to guarantee that outputs arrive precisely on time to the OAU.

IV. RESULTS AND COMPARISON WITH RELATED WORK

Our proposed architecture is prototyped using Verilog HDL and mapped to 65nm TSMC standard cell library using Synopsys Design Compiler.The proposed 2D-IDCT reconfigurable architecture supports real-time videos with 8K UHD resolution at 60 fps rate. The maximum achieved frequency for the whole architecture at 32 x32 is 500 MHz and its gate count is 197(2x98.5) K gate.

The comparisons in terms of technology, frequency, area, latency and throughputs of the proposed architecture and several recently published architectures are summarized in Table I. To the best of our knowledge, our proposed architecture achieves the highest throughput with 42% increase over the best implementation in [2]. In addition, significant latency improvement result is achieved, 6 clock cycles for 32x32 block size, compared to 256, 481, and 32 in the architectures presented in [4], [5], and [6] respectively. While, the gate count is comparable to the other designs, where the proposed architecture has less gate count than [2], [3] and [6] and comparable to [5], [7]. Overall, our proposed architecture achieves the highest throughput and the lowest latency at the expense of minor increase in the architecture gate count due to reconfigurability and reusability features applied.

V. CONCLUSION

In this work, we proposed a new reconfigurable architecture of the Inverse Discrete Cosine transform block in the HEVC

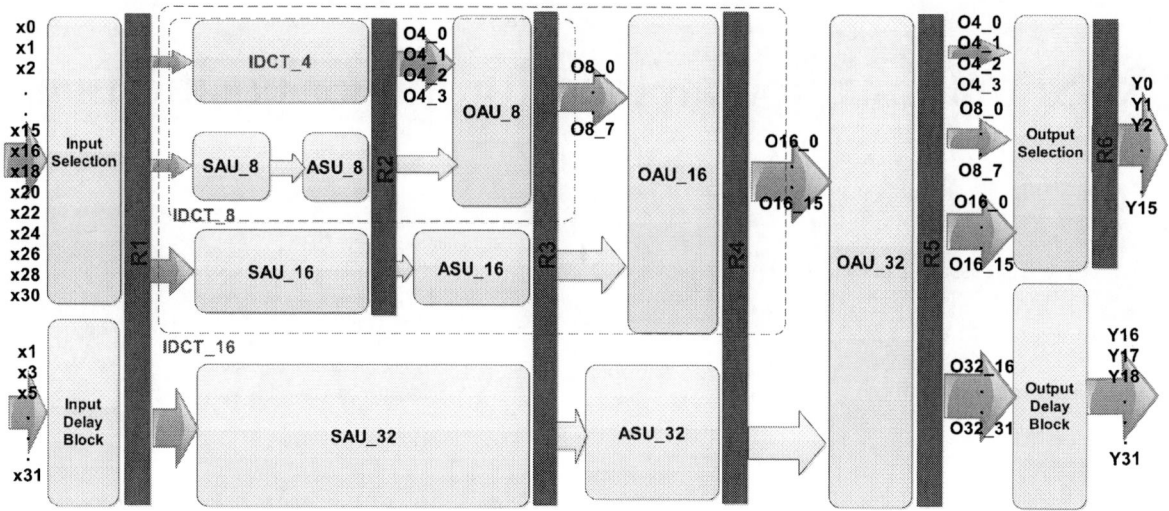

Fig. 6. Reconfigurable Architecture of IDCT_32

TABLE I
COMPARISON WITH RELATED WORK

Design	[2]	[3]	[4]	[5]	[6]	[7]	Proposed
Technology (nm)	130	90	40	180	45	90	65
Max clock (MHz)	350	311	N/A	300	250	312	500
Gate count (K gate)	109.2x2	117.7x2	71	2x52.3	130x2	2x63.8	2x98.5
Latency (clock cycle)	19/68/261	38	256	481	> 32	N/A	3,4,5,6
Max Throughput (Mpixels/s)	1400	1244	249	636	N/A	1248	1990
Transpose Circuit	Memory	Registers/Memory	Memory+36 registers	32x32 registers	not included	4x256x16 bits	32x32x16 bits

encoder/decoder. The proposed architecture is fully pipelined. Moreover, it is totally reconfigurable so that it supports all the necessary transform block sizes. We implemented the proposed architecture using TSMC 65nm process technology. The maximum clock frequency of the proposed architecture is 500 MHz. Furthermore, its throughput is 1990 Mpixel/sec that is higher than the best architecture in the literature, to the best of our knowledge, by about 42%. The area of our architecture is comparable to that of the recently published architectures. Finally, our circuit can process UHD video resolutions up to 8K with 60 fps.

REFERENCES

[1] R.Conceio, J. Cludio Souza Jr., R.Jeske, M.Porto, J.Mattos, L.Agostini."Hardware Design for the 32x32 IDCT of the HEVC Video Coding Standard," In Integrated Circuits and Systems Design (SBCCI), 2013 26th Symposium on, Sept. 2013, pp. 1 - 6.

[2] Shen, S., Shen, W., Fan, Y., and Zeng, X. "A unified 4/8/16/32-point inte- ger IDCT architecture for multiple video coding standards," In Multimedia and Expo (ICME), 2012 IEEE International Conference on, Jul. 2012, pp. 788-793.

[3] Zhu, J., Liu, Z., and Wang, D. "Fully pipelined DCT/IDCT/ Hadamard unified transform architecture for HEVC Codec," In Circuits and Systems (ISCAS), 2013 IEEE International Symposium on, May 2013, pp. 677-680.

[4] Tikekar, M., Huang, C. T., Juvekar, C., Sze, V., and Chandrakasan, A. P, "A 249-Mpixel/s HEVC Video-Decoder Chip for 4K Ultra-HD Applica- tions", IEEE Journal of Solid-State Circuits, vol. 49, no. 1, pp. 61-72, Jan. 2014.

[5] Park, J. S., Nam, W. J., Han, S. M., and Lee, S. "2-D large inverse trans- form (16x16, 32x32) for HEVC (high efficiency video coding)," Journal of Semiconductor Technology and Science, vol. 12, no. 2, June 2012, pp. 203- 211.

[6] M. Budagavi, V. Sze, Unified forward+inverse transform architecture for HEVC. in Proceedings IEEE International Conference on Image Processing (ICIP), (2012), pp. 209–212

[7] Heming Sun, Dajiang Zhou, Jiayi Zhu, Shinji Kimura, and Satoshi Goto,"An area-efficient 4/8/16/32-point inverse DCT architecture for UHDTV HEVC decoder", In Visual Communications and Image Processing Conference, 2014 IEEE on, Dec. 2014, pp. 197 - 200.

[8] http://spiral.net/hardware/multless.html

RSSI optimization method for indoor positioning systems

Youssef AIBOUD[1], Issam ELHASSANI[1][2], Hafid
GRIGUER[1]
[1] SMARTi-Lab
EMSI-Rabat
Rabat, Morocco
Youssef.aiboud@gmail.com

M'hamed DRISSI[2]
[2] IETR
UEB, INSA of Rennes
Rennes, France
Mhamed.drissi@insa-rennes.fr

Abstract— **The aim of this study is to develop a new method for optimizing Received Signal Strength Indicator (RSSI) vaguely used in Indoor Positioning Systems and to compare it to other existing methods. Adaptive filtering, such as Kalman filter, are one of the most used methods for RSSI filtering alongside the median filter. Our approach is to use a new algorithm that restricts the RSSI fluctuation in short periods of time to a changing median range that is calculated based on pervious states. The implementation of the new algorithm is done on Matlab IDE.**

Keywords— *Received signal strength indication, RSSI, adaptive filtering, Median filter, Kalman filter, Indoor positioning system, Trilateration.*

I. INTRODUCTION

The fast development of portable technology affects our daily life, we are becoming more and more dependent on smartphones and wearable devices. Now a GPS chip can guide you through the entire terrestrial globe. But it's still inefficient indoors, that's why a new technology emerged that can guide you through indoor places, such as big malls or football stadiums. There are different methods to elaborate Indoor positioning systems (abbreviated to IPS), from fingerprinting, to triangulation and trilateration.

Our approach is to use the trilateration method, where at least three beacons are installed in a closed room, and using scanned RSSI from each bacon, three distances can be calculated, a position is then deducted [4].

The value of the RSSI, depends strongly on the radio Channel problems where different parameters can be identified, like as: multiple reflections due to metal objects in the closed space, the presence of human bodies in the room, the existence of obstacles between emitter and receiver and reception/emission antenna's polarization.

The use of a brut RSSI is not recommended, as it varies vaguely in short periods of time. Thus, we need to filter the RSSI.

Two of the most used filtering methods are median and adaptive filters such as The Kalman filter [3].

Our paper will be divided into two major sections, the first one describes the method that we developed side by side with

the Kalman filter. In the second section we will compare results of the two methods cited above.

II. METHOD

Received Signal Strength Indicator, is very dependent on, the radio Channel fading and shadowing phenomena, the metallic objects and human bodies existence in the closed space [5], disturb strongly the in time measured RSSI value.

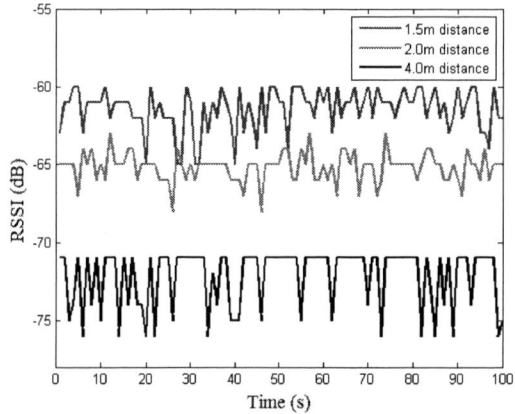

Figure 1. RSSI variance for 3 different distances

In the figure 1, we can observe the variance of the RSSI, it is up to 5dB between two successive values, knowing that the equation for calculating the distance from an RSSI is:

$$d = 10^{\frac{RSSI-A}{-10*n}} \qquad (1)$$

Where:

n: signal propagation constant which is 2 in free space but varies with local settings, for our example and after calcul us we set the n to 2.57

A: is the RSSI value of a *1m* distant device which is around -58dB

For the black signal above the calculated distance using equation (1) varies between *3.21m* and *5.01m*. Which translates to more than *2m* error. There for the use of a filtering method is recommended.

a) *Kalman filter :*

The Kalman filter is an adaptive recursive estimator filter that is capable of estimating data through out a series of incomplete or noisy data [2].

It uses two separate phases, successively, the prediction phase and the update phase. The first one is the estimation of the n-1 state and covariance. The update phase is more of calculating the new state and covariance [6].

The equations for the Kalman filter are [1]:

Prediction phase:

$$\hat{x}_k = A \times \hat{x}_k \qquad (2)$$

$$P_k = A \times P_{k-1} \times A_k^T + Q_k \qquad (3)$$

Update phase:

$$K_k = P_k \times H_k^T (H \times P_k \times H_k^T + R)^{-1} \qquad (4)$$

$$x_k = x_{k-1} + K_k \times (Z_k - H \times x_{k-1}) \qquad (5)$$

$$P_k = P_{k-1} - (K_k \times H \times P_k) \qquad (6)$$

Where:

$A = \begin{bmatrix} 1 & 0.1 \\ 0 & 1 \end{bmatrix}$

Z = Input Signal

K: The Kalman gain

x̂: The estimated state

P: The estimated error covariance

Q and R: noise covariance

b) *The RMM algorithm :*

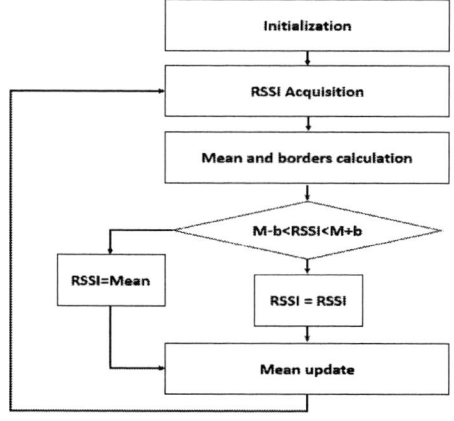

Figure 2. The RMM algorithm

Our approach is based on a basic median filter, at first we set a mean value altogether with its borders. Then we compare every (t) RSSI value to its (t-1) value, if it's over the specified range, the (t) value is then assigned the mean value. Thus, we are preventing peaks over short periods of time. We called our approach the RSSI Moving Median algorithm (RMM).

We predict that the main advantage of the RMM algorithm its simplicity and it's easier implementation in real-time.

III. RESULTS

For our experimentations, we managed to get RSSI from a Bluetooth embedded unit coupled with an Arduino controller using as transmitter, for RSSI's receiver, we have choose smartphone. The RSSI value is then transferred to Matlab using serial communication.

The test that we performed to check the capabilities of the two filters, included two distinct phases, the first one is putting the receiver still at different places and the second one is a moving phase throughout a closed room and with the existence of obstacles and human bodies to emulate a real room activity.

The figure 3 shows the unfiltered RSSI signal compared to the results from the two filtering methods described above.

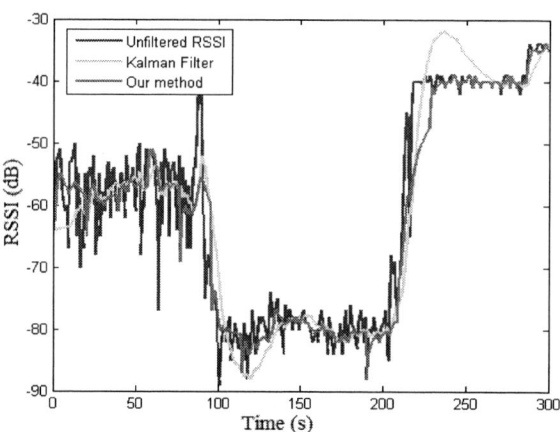

Figure 3. Comparison between unfiltered RSSI and the two filtering methods

After a first examination, we can clearly see that the RSSI signal is very noisy, there for a filtering method is mandatory. In a first hand, we observe that for the stable parts of the signal, the Kalman filter has fewer peaks and is a little more stable than the RMM algorithm. In the other hand, for the movement periods our approach follows in a perfect shape matches to the original signal contrary to the Kalman filter.

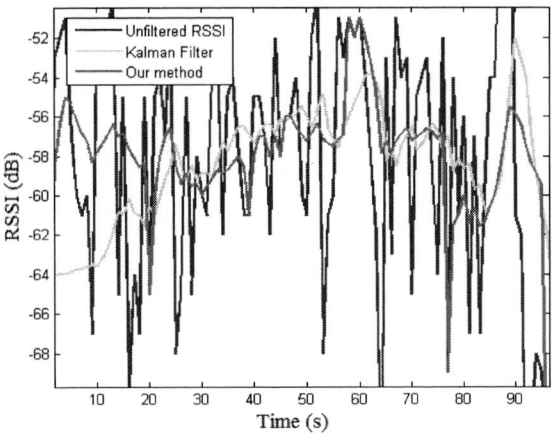

Figure 4. Zoom on the signal *(first stable part)*

The figure 4 shows zooming on the first stable part of the RSSI signal, it is visible that the Kalman filter results are somehow more stable than the RMM algorithm.

The last stable part of the signal indicates that the RMM algorithm has more stability and follows exactly the shape of the unfiltered RSSI signal with elimination of the peak signals as seen in the figure 5.

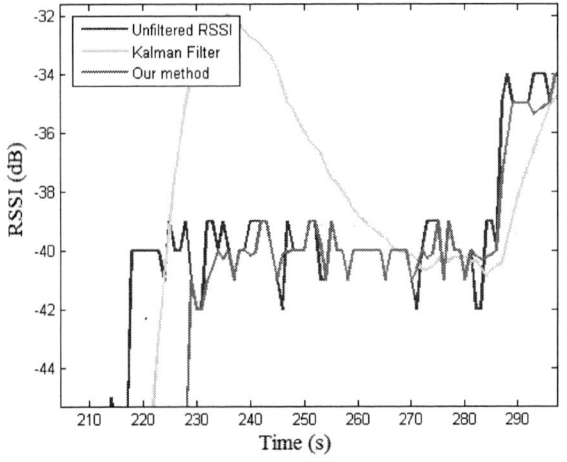

Figure 5. Zoom on the signal *(last stable part)*

For a clearer view on the results we conducted an error comparison, we subtracted the two resulting signals from the original one. As it is shown in the figure 6. This last figure shows that our new optimization gives better results compared to the Kalman filter in a crowded room. The minimal and maximal values of the two signals successively are: *-19.14 dB* and *19 dB* for our method and *-23.59 dB* and *29.01 dB* for the Kalman filter.

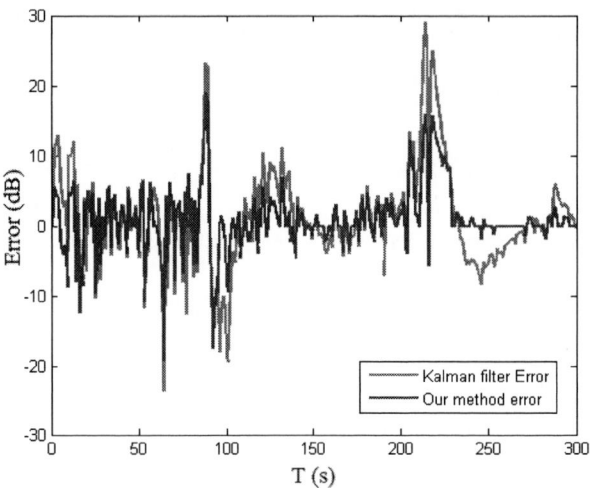

Figure 6. Error comparison

IV. CONCLUSION

RSSI is a very noisy source of data, thus to get the best use of it, a filtering method is mandatory. Our paper describes a newly developed filtering method that uses median filter as a base and then updates the mean value for each sample. We then proceed to compare the results of our method to those of the Kalman filter. The variance of the error signal for the two cited methods are *22.62 dB* for the RMM algorithm and *50.93 dB* for the Kalman filter. The Kalman filter has improved stability in the case of a stable mobile. But unfortunately it's not the case in real world, this is where we observe that the advantage of RMM algorithm is better results in movement. The RMM algorithm is more suitable for a crowded room that includes human presence, and existence of highly moving obstacles and reflectors. The next step is to implement this method into real-time systems.

REFERENCES

[1] Greg Welch and Gary Bishop "An Introduction to the Kalman Filter" ACM technical report, 1995.

[2] Jouni Hartikainen, Arno Solin, and Simo Särkkä "Optimal Filtering with Kalman Filters and Smoothers" Manual for the Matlab toolbox EKF/UKF Version 1.3, August 2011

[3] Sharly Joana Halder, Paritosh Giri, and Wooju Kim "Advanced Smoothing Approach of RSSI and LQI for Indoor Localization System" International Journal of Distributed Sensor Networks, November 2014

[4] B Cook, G Buckberry, I Scowcroft, J Mitchell, T Allen "Indoor Location Using Trilateration Characteristics", LCS 2005

[5] Agustí Corbacho Salas "Indoor Positioning System based on Bluetooth Low Energy" PhD thesis, Universitat Politècnica de Catalunya, June 2014

[6] Y.Aiboud, J.Mhamdi. A.Jilbab. H.Sbaa, "Review of ECG Signal de-noising techniques", WCCS 2015

First specifications of Urban Traffic-Congestion Forecasting Models

Abdellah DAISSAOUI
Moroccan School of Engineering Sciences, Dpt Research
and innovation, Casablanca, Morocco
Daissaoui.abdellah@gmail.com

Azedine BOULMAKOUL
Department of Computer Sciences, LIM/IDS Lab.
Faculty of Sciences and Technologies of Mohammedia
Mohammedia, Morocco

Zineb HABBAS
University of Lorraine
Metz, France

Abstract— **urban traffic congestion and bottleneck has been a global issue for many years due to rapid urbanization. The use of intelligent Transportation System (ITS) in metropolitan cities has become a necessity. In this paper we present traffic congestion prediction models with a novel specifications based on Ant fuzzy model.**

Keywords—Pheromone; ACO; CTM; Traffic modeling; Congestion prediction models

I. INTRODUCTION

Urban traffic congestion and bottleneck has been a global issue for many years due to rapid urbanization. Residents in cities are suffering from the most annoying side-product of urbanization every day. There are more than one billion cars running on all the roads around the world, and the number will double by 2030. Traffic congestion not only causes mental stress in drivers, but also leads to more severe pollution, higher gasoline consumption and huge economic loss [1].

The use of Intelligent Transportation System (ITS) in metropolitan cities has become a necessity. The goal is to manage the evolution of the growing demand for access to urban traffic. One of the technological keys to making these systems more powerful and reactive is to incorporate analysis and prediction models of congestion traffic, which is now regarded as the most important functional component in the ITS, by studying data collected through various sensors (GPS, camera, radar, ...)

For a common sense, traffic congestion is the condition when there is too much traffic in the road. This phenomenon rarely provides enough insight into the overall system, while many approaches of the phenomenon are intuitive and consider the physical flow (capacity, speed, density ...). Also, this phenomenon is considered as a process that evolves in space and time.

An important and abundant number of models that enable congestion traffic analysis and prediction are developed and tested. These models encompass diverse disciplines techniques hence quantitative statistical models, time series, causal methods, methods based on artificial intelligence algorithms, and probabilistic methods.

Various criteria are used to classify developed models, such as both application and independent variables scales, operationalization, processes representation and especially level of detail, which describes the traffic flow. Scale of the independent variables distinguishes between two time scales to describe traffic system's variables which can be either stochastic or determinist.

This paper is organized as follows: In section 2, we define congestion, then the characteristics and measurement are given, state of the art on the traffic modeling and traffic congestion prediction are presented in section 3. Our vision about traffic congestion modeling and prediction is described in section 4, and conclusions are drawn is section 5.

II. CONGESTION DEFINITION, CHARACTERISTICS AND MEASUREMENT

A. Congestion definition

Defining congestion presents a lack of consensus because it is considered both as a physical phenomenon related to the manner in which vehicles impede each other's progression as demand for limited road space approaches full capacity.

There is not any universal definition of "congestion" as the situations where it is experienced vary from person to person and from a place to another. When congestion is defined in a vernacular sense: it is the inability to reach a destination at a satisfactory time due to unpredictable travel speed.

It represents a situation where demand for road space surpasses supply. This definition provides little insight into the multiple, complex and interconnected factors that leads to this mismatch of supply vs demand.

The above definition has led many transportation engineers to solve the problem of congestion by increasing supply which means increasing road space, indeed there is solid evidence that increasing roadway capacity may in many (but not all) circumstances lead to less congestion.

978-1-4673-8760-6/15 $31.00 © 2015 IEEE

B. Congestion characteristics

Traffic congestion can be understood as a factor of the level of traffic, itself a function of how routes are selected by specific roadway users on a road network at a particular time. Therefore, three factors are used to characterize congestion, which are congestion perception by roadway user, streams of road networks and time because of temporal nature of congestion phenomena.

Predictability of travel time is very important when seeking to prioritize congestion policies. An unexpected congestion is perceived much more negatively and is experienced strongly by users than normal congestion situations.

Time reliability is an even more important factor in the user's experience. Travelers undertake most trips in order to arrive at one or several destinations, when the activities at these destinations follow fixed schedules, individuals seek to arrive on time. This has important implications for congestion policy since transport authorities should also demonstrate that the reliability of the roadway network is being addressed through their actions.

Finally, another important factor of congestion is the time because congestion is a temporal phenomenon affecting some periods in time more than others. The affected period is connected to the temporal scale (daily, weekly, monthly or yearly) and to timing of urban activities that are linked to the decisions made by individuals and are firmly related to the purpose of their trips.

We can identify three categories of factors that cause congestion: micro-level factors for example those related to traffic on the roadway ,macro-level factors that are linked to the demand for road usage and a set of other factors linked to patterns and volumes of trip-making and there are also random variables such as weather and visibility (see figure 1).

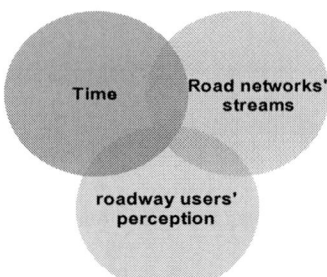

Fig. 1. Congestion characteristics.

C. Congestion measurement

In order to deliver better congestion outcomes, a necessary step is measuring congestion. At micro-level, roadway managers need a congestion measurement that allows them to address operational concerns on specific roadway links. For this purpose, road managers and engineers rely on collected indicators from roadway sensors. These sensors are used to collect both the extent and relative scale and congestion

evolution. Some indicators are strongly relevant for road users such as predictability of travel time and system reliability, while others are relevant to road systems operators namely speed and flow on the network links. However, these metrics are difficult to aggregate and do not directly, address the concerns of roadway system managers or users [2]. System managers need to understand how good the entire network works (see figure 2).

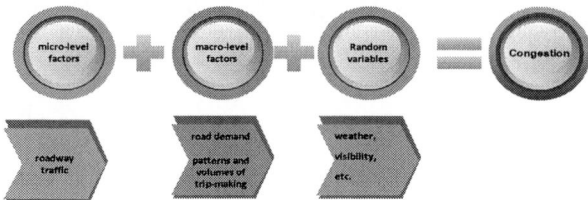

Fig. 2. Congestion causes' factors.

III. RELATED WORKS IN TRAFFIC MODELING AND CONGESTION PREDICTION

A. Traffic Modeling

In planning and operational transportation domain, traffic simulation models are one of the most important and helpful tools. According to the generated level of detail, they are classified as microscopic, mesoscopic and macroscopic. Typically, microscopic simulation models generate further and deeper measures to the macroscopic simulations models, but they require more detailed and refined data in input; however, the mesoscopic simulation models lie between the two.

Several research works have shown the capacity of these models to represent both usual and unusual events such as the situations of evacuations after hurricanes and wildfires [3].

The work leaded to seeks to define an optimal strategy for better mobility in the city, and demonstrated the existence of Macroscopic Fundamental diagram (MFD) in wide urban area.

Cell transmission model (CTM) is a discrete approximation to the LWR model proposed by Daganzo [4-5].

CTM (Fig. 4) is based on the assumption that the road is divided into similar cells whose lengths are equal to the distance traveled by free-flowing traffic in given interval (see figure 3).

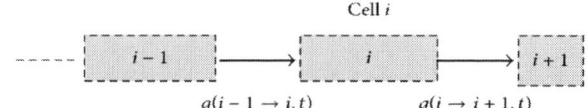

Fig. 3. Link homogeneous discretization.

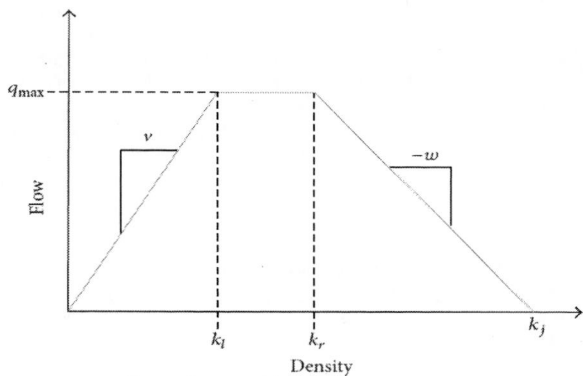

Fig. 4. The trapezoidal fundamental diagram.

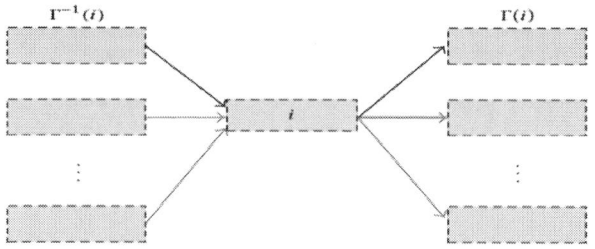

Fig. 5. Generalized extended CTM on network.

CTM has been used in travel time estimate, traffic control, among other. Stochastic cell transmission model has been proposed to model stochastic traffic problems. A novel method that combines CTM and MFD has been proposed to fill the gap due to the modeling of destination networks, able to simulate the traffic in space and time and quantify the effects of spillback in destination networks. The basic idea to associate CTM and MFD is that the accumulation is calculated at each step to determine the rate of inflows and outflows in order to understand the dynamics of congestion in destination networks [3] [6].

An extended CTM model (see Figure 5) allowing constructing an arbitrary network has been proposed in [5].

B. Traffic congestion Prediction models

There are many techniques allowing traffic congestion prediction from network traffic data such as classic statistical model, Bayesian networks, decision trees, linear regression, kernel and support vector machine, cluster and association rules, times series, machine learning, among other [14].

The most frequently used model in the literature for solving this phenomenon is ARMA. Another study has decomposed the network flow as scaling coefficient and wavelet coefficient, and then employed ARIMA model for prediction. An extension ARIMA model was proposed as a prediction method based on alpha smooth information to deal with the long-range burst prediction [7-8].

Other developed model remove the long-term dependence based on experiences using ARMA model for better prediction

accuracy [9]. Furthermore, Gray and Markov models are used for traffic congestion prediction by combining network flow and error predictions [10].

Bilinear regression neural network has been developed to improve the accuracy of prediction. Genetic algorithm was also used for network traffic prediction [11-12].

A comparative study aims to choose the most optimal method in short-term traffic congestion prediction between triple moving average, single exponential smoothing, and double exponential smoothing is given in [13].

In addition to all the above, a traffic bottlenecks analysis model, based on attribute reduction algorithm which is an inductive learning tool was developed. This model allows finding the suitable user parameters to combine with the traffic data [2].

However, it is difficult to say that a model is better than another within a specific context. This finding is related mainly to the use of a minimum data gathered from sensors or simulation. In addition, the accuracy given by a prediction model depends on data quality. In general, neural networks and fuzzy logic have shown promising results [14].

Based on the fact that traffic congestion prediction phenomenon is more complex to be handled by a single model. Recent studies proved the possibility to use multiple models [15].

IV. FUZZY ANT CONGESTION PREDICTION

Swarm societies such as ants perform complex tasks with pheromone communication despite lack of top-down style control. The projected model uses the paradigm of fuzzy theory and ACO meta-heuristic. The bio-inspired character of intelligence in swarms, treat virtual vehicle like a swarm that can act and interact with the components of their environment, while having a collective artificial distributed intelligence. The pheromone deposited by cars represents density of traffic. We propose a method to predict future traffic through a pheromone mechanism using fuzzy potential functions. The collective intelligence enables and predicts the spread and extinction of congestion in the urban transport network.

The pheromone mechanism starts when ants find food and transport it to their nest, deposing attractive sign of pheromone on their return path. The pheromone diffuse into the surrounding environment and its intensity disappear over time. However, pheromone on the shortest path to the food became more intense. Here we talk about the two main characteristics of pheromone: Evaporation and Propagation.

Evaporation: Aggregated pheromone decreases over time due to evaporation.

Propagation: Another fundamental equation describes the propagation received from neighboring places. The transitions function of the propagated amount.

In traffic modeling or congestion prediction, cars are used as ants or bees. Where cars are equipped with some types of sensor allows deposit the pheromone. Therefore, other cars that follow their route would avoid traffic congestion by checking the intensity of pheromone. A car would be able to predict traffic on the road ahead from the information provided by preceding cars. Too many kinds of pheromone models are used in different research fields [18]. A multiple pheromones combination model is proposed to handle traffic congestion prediction. The idea is to predict short-term congestion one minute ahead, without using system traffic control or central server. This distributed system is called center-less probe car system. The three models combined using linear combination and the ratio was turned by uncertainty analysis. The model was evaluated by simulation and the accuracy is calculated by correlation coefficient as error rate indices [16]. In the same topic, a multi-agent coordination mechanism method is presented as congestion forecasting system. It can react to dynamically changing traffic conditions using pheromone communication model. Unlike the model presented above, this congestion forecasting model uses installed sensors in intersection of road to count the number of cars moving through intersections [17].

The solution advocated in this paper in order to ensure urban traffic congestion predictions, refers to ACO paradigm and fuzzy modeling. Thus we propose a fuzzy model using the ant colony works [16-17] where we increase the model by calculating the fuzzy desirability using fuzzy potential functions.

We also propose a distributed architecture for deployment of our proposal (see Figure 6). The context of this work is part of subject concerning urban traffic management for smart cities

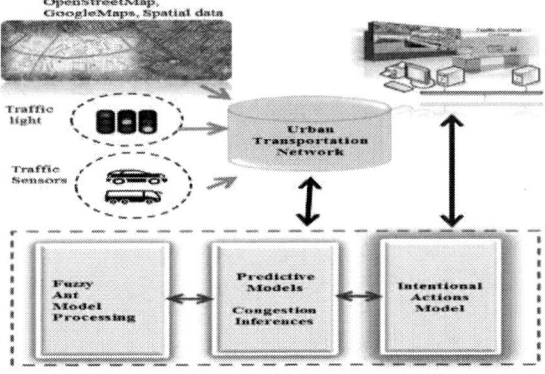

Fig. 6. distributed architecture

CONCLUSION

In this paper we described traffic congestion prediction models used to generate measurement. We also discussed the contribution of ACO models and their highlight relevance to apprehend congestion. In the prospects of this research, we propose to extend the ACO model using fuzzy potential and implement a distributed architecture to deploy this solution. The interest of this research is highly solicited in smart cities project.

REFERENCES

[1] "An assessment of the direct and indirect economic and environmental costs of idling in road traffic congestion to households in the UK, France, Germany and the USA," Report for INRIX, July 2014.

[2] X. Yue, L. Cao, D. Miao, Y Chen, B. Xud, "Multi-view attribute reduction model for traffic bottleneck analysis," Knowledge-Based Systems 86 (2015) 1–10

[3] Z. Zhang, B. Wolshon, V. Dixit, "Integration of a cell transmission model and macroscopic fundamental diagram: Network aggregation for dynamic traffic models," Transportation Research Part C 55 (2015) 298–309, Engineering and Applied Sciences Optimization (OPT-i)

[4] C. F. Daganzo and J. A. Laval, "Moving bottlenecks: a numerical method that converges in flows," Transportation Research Part B: Methodological, vol. 39, no. 9, pp. 855–863, 2005.

[5] A. Boulmakoul, L. Karim, M. Mandar, A. Idri,A. Daissaoui, "Towards Scalable Distributed Framework for Urban Congestion Traffic Patterns Warehousing," Applied Computational Intelligence and Soft Computing Volume 2015, Article ID 578601, 12 pages

[6] A. Sumalee, T. Zhong, N. Indra-Payoong, "Dynamic stochastic journey time estimation and reliability analysis using stochastic cell transmission model: algorithm and case studies," Transportation Research Part C: Emerging Technology, vol 35, 263–285, 2013.

[7] Y. Qiao J. Skicewicz, and P. Dinda, "An empirical study of the multiscale predictability of network traffic", High performance Distributed Computing, Proceedings, 13th IEEE International Symposium, (2004).

[8] W. Yong and G. Zhu, "Prediction of self-similar network traffic with heavy tailness", JournalofHuazhong University of Science and technology (Nature Science), vol. 34, no. 9, (2006).

[9] B. Gao, Q.-Y. ZHANG, Y.-S. Liang, N.-N.. Liu, C.-B. and N. –T. Zhang, "Predicting self-similar networking traffic based on EMD and ARMA", Journal on Communications, vol. 32, no. 4, (2011).

[10] Q.-F. Yao, C.-F. Li, H.-L. Ma and S. Zhang, "Novel network traffic forecasting algorithm based on grey model and Markov chain", Journa;-Zheajiang University –Science Edition, vol. 34, no. 4, (2007).

[11] J.-G Huang, L. Hang, H.-J. Wang and B. Long, "Prediction of Time Sequence Based on GA-BP Neural Net", Journal of University of Electronic Science and Technology of China, vol. 38, no. 5, (2009).

[12] D. C. Park, "Structure optimization of BiLinear Recurrent Neural Networks and its application to Ethernet network traffic prediction", Information Sciences, vol. 237, (2013), pp. 18-28.

[13] Y. Bie, M. Yang, Y. Pei, "Development of Short-term Traffic Volume Prediction Models for Adaptive Traffic Control," International Conference on Advances in Mechanical Engineering and Industrial Informatics (AMEII 2015).

[14] Y. Lv, Y. Duan, W. Kang, Z. Li, F. Wang, "Traffic Flow Prediction With Big Data: A Deep Learning Approach," IEEE TRANSACTIONS ON INTELLIGENT TRANSPORTATION SYSTEMS, VOL. 16, NO. 2, APRIL 2015

[15] X. Ma, H. Yu, Y. Wang, "Large-Scale Transportation Network Congestion Evolution Prediction Using Deep Learning Theory," PLoS ONE 10(3): e0119044. doi:10.1371/journal. pone.0119044, 2015

[16] O. Masutani, H. Sasaki, H. Iwasaki, Y. Ando, Y. Fukazawa, "Pheromone Model: Application to Traffic Congestion Prediction," AAMAS '05 Proceedings of the fourth international joint conference on Autonomous agents and multiagent systems, Pages 1171-1172, 2005

[17] S. Kurihara, "Traffic-Congestion Forecasting Algorithm Based on Pheromone Communication Model," licencee intech, 2013.

[18] D. Teodorovic, p. Lucic, G. Markovic, M. Dell'Orco, "Bee Colony Optimization: Principles and Applications, Neural Network Applications in Electrical Engineering, 2006. NEUREL 2006. 8th Seminar on

Comparison between Active AC-DC Converters For Low Power Energy Harvesting Systems

Ehab Belal[1], Hassan Mostafa[2] and M. Sameh Said[3]

[1, 2,3]Electronics and Communications Engineering Department, Cairo University, Giza 12613, Egypt.
[2]Center for Nanoelectronics and Devices, AUC and Zewail City of Science and Technology, New Cairo 11835, Egypt.
{[1]ehab.belal@gmail.com, [2](hmoustafa@aucegypt.edu, hmostafa@uwaterloo.ca), [3]msmsab@hotmail.com}

Abstract— **Low power energy harvesting systems research is increasing in recent years due to many applications such as wireless sensor node, biomedical implants, highway traffic sensor and demand for increasing the battery life. One of the major challenges of the energy harvesting systems is to convert and store usable power efficiently. This paper presented a survey about several commonly used active AC-DC converter circuits that is based on active diodes such as negative voltage converter with active diode, cross-coupled active full bridge and active bridge voltage doubler. Our study shows that the most efficient rectifying circuit among the surveyed AC-DC converter circuits is the active bridge voltage doubler. It provides a minimum voltage drop compared to other two AC-DC converter circuits. In additions, it provides 2X the amount of electrical power.**

Index Terms—**Energy Harvesting, Active AC-DC, Comparator**

I. INTRODUCTION

AC-DC Converter converts AC signal produced by different energy harvester sensors [1] such as piezoelectric, electromagnetic and electrostatic to DC signal suitable for powering wireless sensors and other systems without the need to external DC power supply. Conventional rectifiers [2] suffer from high diode voltage drop that leads to power loss during the rectification stage. Recently, active diodes are used at low power applications to overcome the voltage drop of the conventional diodes. In this paper, the comparison between active AC-DC converters for low power energy harvesting systems is presented.

The main rectification challenges of junction diodes are the forward voltage drop and the reverse leakage current of junction diodes. On the other hand, active diodes have shown great promise in replacing conventional diodes in low-voltage and low-current applications. As shown in Figure 1, active AC-DC converter is a comparator-controlled switch, whose ON/OFF state is determined by the polarity of the voltage across it, generates low or high voltage at the output terminal.

The behavior of an active diode is similar to an ideal diode, but the comparator requires a power supply. Fortunately, the power consumption of the active diode is quite low. Several commonly used active AC-DC converter circuits are discussed according to efficiency and area. Our study shows that the most efficient rectifying circuit among the surveyed circuits is the active bridge voltage doubler.

The rest of the paper is organized as follows. Section II introduces the surveyed circuits. Section III introduces the simulation results and discussion. Finally, some conclusions are drawn in Section IV.

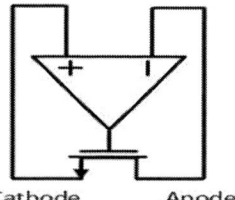

Fig. 1: Schematic of an active diode using PMOS switch [3].

II. SURVEYED CIRCUITS

A. Negative voltage converter with active diode

The proposed active rectifier [3]-[5] is realized using two stages as shown in Figure 2 .The first stage, Negative Voltage Converter "NVC", is used to convert the negative wave of the input AC signal to positive by using two PMOS and two NMOS connected as shown in Figure 3. During the positive wave of the input AC signal ($V_{in1}>V_{in2}$), MP1 and MN1 will be active when the input voltage becomes greater than V_{thn} and V_{thp}. Then, terminal (1) becomes high "V_{in1}" and terminal (2) becomes low "V_{in2}". At the negative wave of the input ($V_{in2}>V_{in1}$), MP2 and MN2 will be active then terminal (1) becomes high "V_{in2}" and terminal (2) becomes low "V_{in1}".

Therefore, terminal (1) is always high "positive" during the positive and the negative cycle of the AC input signal while terminal (2) is always low, the voltage drops of the first stage reach to ($V_{thn}+V_{thp}$) across NMOS and PMOS transistors. Further increasing of the transistor width decreases the resistance to get a smaller voltage drop at the expense of higher power consumption.

The first stage itself cannot control the direction of the current. Therefore, the active diode second stage will be used. This active diode consists of PMOS switch, which is driven by the comparator. The width of the PMOS switch has a great effect on the performance of the active AC-DC converter. Increasing the transistor width decreases the resistance to obtain a smaller voltage drop. On the other hand, this results in larger area and power consumption [6].

978-1-4673-8760-6/15 $31.00 © 2015 IEEE

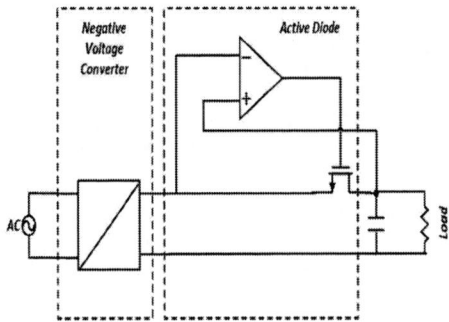

Fig. 2: Negative voltage converter with active diode [5].

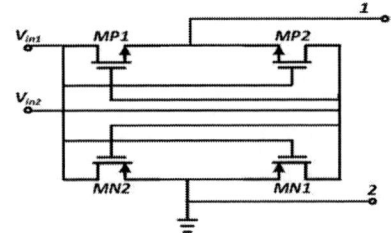

Fig. 3: Negative voltage converter schematic [5].

The comparator should be designed with low power consumption and low input voltage. The circuit implementation of low input voltage comparator , shown in Figure 4, consists of six transistors (M4-M9) work as a comparator to control the gate voltage of the PMOS switch M1. The main drawback of the comparator is the need for an external power supply that can be achieved easily by using current mirroring circuits. Transistors (M10-M12) are used to design two current mirrors for powering the comparator.

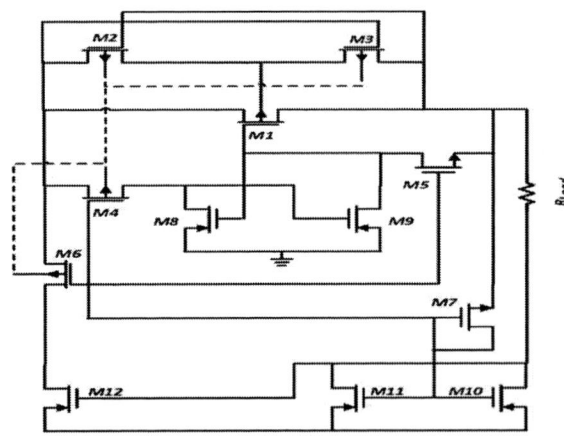

Fig. 4: Schematic of the comparator design [5].

Transistors (M2-M3) work as an active body biasing block to prevent the PMOS switch M1 body-junction from turning ON during start-up for better design reliability. The sizes of

these transistors should be designed as small as possible to provide a low current flow during start up.

The total voltage drop of the negative voltage converter with active diodes is ($V_{thn}+V_{thp}+V_{thp (switch)}$).

B. Cross-coupled active full bridge

The proposed cross-coupled active full bridge [7-8], displayed in Figure 5, consists of two cross-coupled inverters and two active diodes. During the positive wave of the input AC signal ($V_{in1}>V_{in2}$), MP1 will be active when the input voltage becomes greater than V_{thp}. The voltage at the negative input port of the left comparator becomes low "V_{in2}". Correspondingly, the output of the left comparator becomes high which turns the transistor MN2 ON. Thus, the current flowing though (MP1-MN2) charges the loading capacitor CL.

During the negative wave of the input AC signal ($V_{in2}> V_{in1}$) MP2 will be active when the input voltage becomes greater than V_{thp}. While the voltage at the negative input port of the right comparator becomes low "V_{in1}". Correspondingly, the output of the comparator is high which turns the transistor MN1 ON. Thus, the current flowing though (MP2-MN1) charges the loading capacitor CL.

The comparator implementation, shown in Figure 6, consists of three parts (common gate amplifier-current mirror-inverter). Transistors (M8, M10) and transistors (M7, M12) are common gate amplifier while transistors (M9, M13) are used to design current mirror. Transistors (M14-M17) are two inverters driving large gate capacitance for the NMOS switch.

The total voltage drop of the cross coupled active full bridge is ($V_{thn}+V_{thp}$).

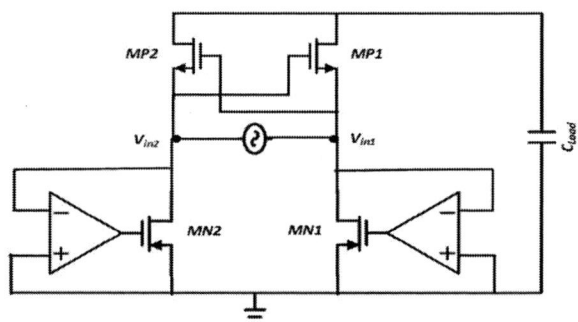

Fig. 5: Cross-coupled with active diode [7].

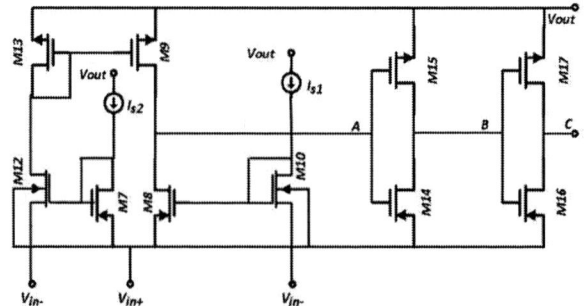

Fig. 6: Schematic of the comparator design [7].

978-1-4673-8760-6/15 $31.00 © 2015 IEEE

C. *Active bridge voltage doubler*

The active bridge voltage doubler [9-10], shown in Figure 7, consists of positive and negative peak detectors that generate DC voltages that track the positive and the negative voltage of the input AC signal.

During the positive wave of the input AC signal the voltage at the negative input port of the comparators is greater than 0V. Correspondingly, the comparator output becomes low which turns the PMOS switch ON and turns the NMOS switch OFF. Thus, the current flowing through the PMOS switch charges the loading capacitor C_{load}. During the negative wave of the input AC signal, the voltage at the negative input port of the comparator is lower than 0V. Correspondingly, the comparator output becomes high which turns the NMOS switch ON and turns the PMOS switch OFF. Therefore, the current flowing through the NMOS switch charges the loading capacitor C_{load}. The total voltage drop of the active bridge doubler is $V_{th(switch)}$. The main advantage of this circuit is providing two positive and negative DC output voltages.

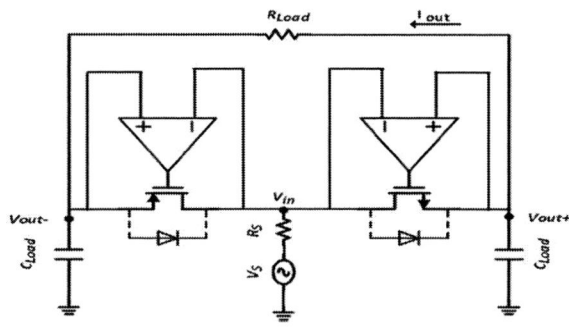

Fig. 7: Active bridge voltage doubler [9].

III. SIMULATION RESULTS

Negative voltage converter with active diode, cross-coupled active full bridge and active bridge voltage doubler are simulated by using Cadence Spectre and hardware -calibrated 130nm CMOS technology provided by UMC. The transient behaviors of the discussed AC-DC converters are shown in Figure 8.

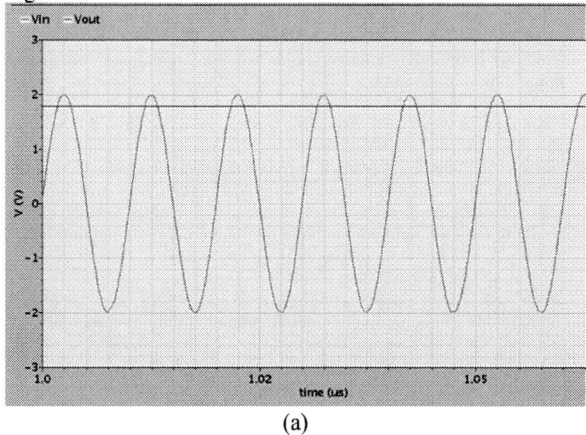

(a)

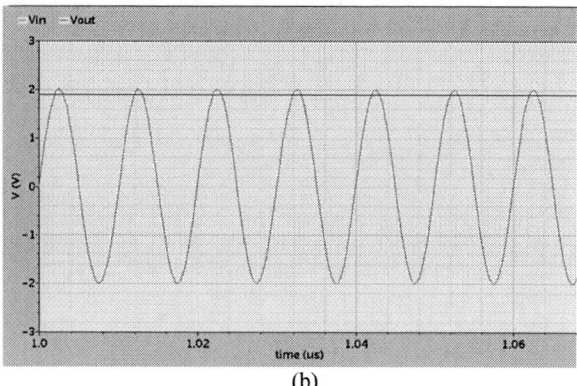

(b)

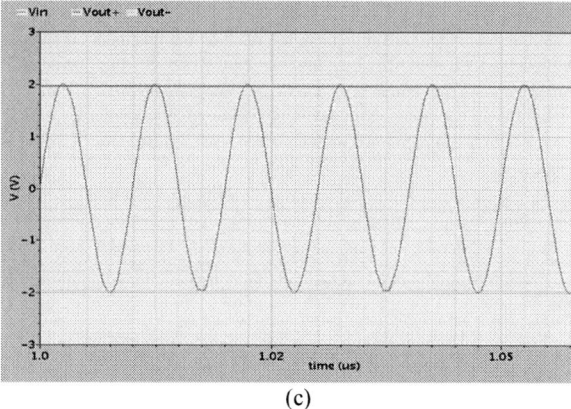

(c)

Fig. 8: Transient behavior of discussed rectifiers (a) Negative voltage converter with active diode (b) Cross-coupled active full bridge (c) Active bridge voltage doubler

It is obvious that the active bridge voltage doubler circuit provides two output voltages (V_{out+}, V_{out-}) with a smaller voltage drop compared with other two AC-DC converters.

The power obtained at the output of the rectifier for the three AC-DC converters portrayed in Figure 9. The active bridge voltage doubler is able to provide 2X the amount of electrical power that was provided by the cross-coupled active full bridge and the negative voltage converter with active diode when a load resistance of 80kΩ load is used.

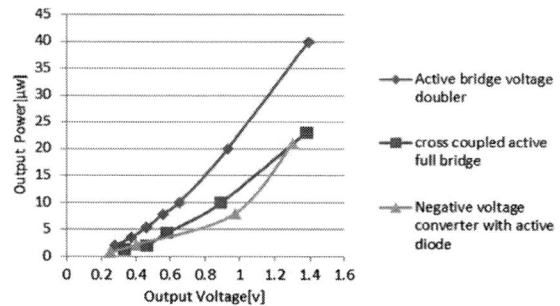

Fig. 9: Simulated power obtained at the output of the Negative voltage converter with active diode, Cross-coupled active full bridge and active bridge voltage doubler.

978-1-4673-8760-6/15 $31.00 © 2015 IEEE

The voltage efficiency versus different input voltage amplitudes "unloaded" for the three AC-DC converters is shown in Figure 10. At an input voltage of 2.5V, the efficiencies of the active bridge voltage doubler, the cross-coupled active full bridge and the active bridge voltage doubler are 98.8%, 98.2%, and 85% respectively.

Figure 11 displays the voltage efficiency versus different input voltage amplitude with ohmic load 80kΩ for the three AC-DC converters. At an input voltage of 2.5V, the efficiencies of the active bridge voltage doubler, the cross-coupled active full bridge and the active bridge voltage doubler are 96%, 94%, and 84% respectively

Table I presents a comparison between the discussed AC-DC converters according to efficiency and area. Where (a): Negative voltage converter with active diode, (b): Cross-coupled active full bridge and (c): Active bridge voltage doubler.

TABLE I
COMPARISON BETWEEN THE DISCUSSED
AC-DC CONVERTERS

	η	Area
(a)	Low	1 comparator+1 switch+4 transistor (Small)
(b)	Moderate	2 comparator+2 switch+2 transistor (Large)
(c)	High	2 comparator+2 switch (Moderate)

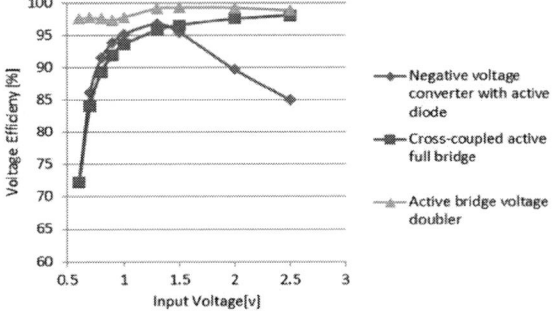

Fig. 10: Simulated voltage efficiency versus input voltage amplitude unloaded

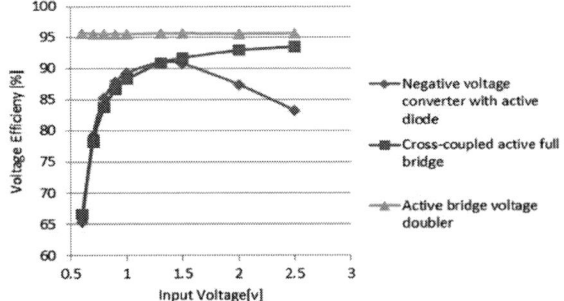

Fig. 11: Simulated voltage efficiency versus input voltage amplitude with 80KΩ ohmic load

It is obvious that the active bridge voltage doubler is the highest efficiency compared with other two AC-DC converters because the total voltage drop of the active bridge voltage

doubler across only the PMOS switch, which is the smallest voltage drop compared with other two AC-DC converters. In additions, it provides two output voltages (V_out+, V_out-). On the other hand, the area of active bridge voltage doubler is a moderate area compared with other two AC-DC converters.

IV. CONCLUSION

This paper presented a survey about several commonly used active AC-DC converter circuits that is based on active diodes such as negative voltage converter with active diode, cross-coupled active full bridge and active bridge voltage doubler to perform signal rectification. Our study shows that the most efficient rectifying circuit among the surveyed AC-DC converter circuits is the active bridge voltage doubler. This is because it provides minimum voltage drop compared to other two AC-DC converter circuits. In additions, it provides 2X the amount of electrical power. Its voltage efficiency reaches 98.8% "unloaded" and reaches 96% with a load of 80KΩ.

V. ACKNOWLEDGEMENT

This research was partially funded by Cairo University, ITIDA, NTRA, NSERC, Zewail City of Science and Technology, AUC, the STDF, Intel, Mentor Graphics, SRC, ASRT and MCIT.

References

[1] Mahammad A Hannan, Saad Mutashar, Salina A Samad, and Aini Hussain, "Energy harvesting for the implantable biomedical devices: issues and challenges," BioMedical Engineering OnLine, vol. 13, no. 79, Jun.2014.

[2] Y.K. Ramadass, and A.P. Chandrakasan, "An Efficient Piezoelectric Energy Harvesting Interface Circuit Using a Bias-Flip Rectifier and Shared Inductor, " IEEE Journal of Solid-State Circuits, vol.45, no.1, pp. 189- 204, Jan. 2010.

[3] D. Niu, Zhangcai Huang, Minglu Jiang, and Y. Inoue, "A sub-0.3V CMOS rectifier for energy harvesting application," Circuits and systems (MWSCAS), 2011 IEEE 54th International Midwest Symposium on, pp. 1- 4, Aug. 2011.

[4] Shuo Cheng, R. Sathe, R. Natarajan, and D.P. Arnold, "A Voltage – Multiplying Self-Powered AC/DC Converter with 0.35-V Minimum input Voltage for Energy Harvesting Applications," Applied Power Electronics Conference and Exposition (APEC) , 2011 Twenty-Sixth Annual IEEE , pp. 1311 -1318, March 2011.

[5] R. Radzuan, M.A.A. Raop, M.K.M. Salleh, M.K. Hamzah, and R.A. Zawawi, "The designs of low power AC-DC converter for power electronics system applications," Computer Applications and Industrial Electronics (ISCAIE), 2012 IEEE Symposium on, pp. 113-117, Dec. 2012.

[6] Nebi Caka, Milaim Zabeli, Myzafere Limani, and Qamil Kabashi, "Influence of MOFSET parameters in its parasitic capacitance and their impact in digital circuits," WSEAS Transactions on Circuits and Systems, vol.6, no.3, March 2007.

[7] Yang Sun, In-young Lee, Chang-Jin Jeong, Seok-kyun Han, and Sang-Gug Lee, "An Comparator Based Active Rectifier for Vibration Energy Harvesting Systems," Advanced Communication Technology (ICACT), 2011 13th International Conference on, pp. 1404-1408, Feb. 2011.

[8] S.S. Hashemi, M. Sawan, and Y. Savaria, "A High-Efficiency Low Voltage CMOS Rectifier for Harvesting Energy in Implantable Devices," Biomedical Circuits and Systems, IEEE Transactions on, vol.6, no. 4, pp. 326-335, Aug. 2012.

[9] Yuan Rao, Arnold, and P. David, "An AC/DC voltage doubler with configurable power supply schemes for vibrational energy harvesting," Applied Power Electronics Conference and Exposition (APEC), 2013 Twenty-Eighth Annual IEEE, pp. 2844-2851, March 2013.

[10] A. Rahimi, O. Zorlu, A. Muhtaroglu, and H. Kulah, "Fully Self-Powered Electromagnetic Energy Harvesting System with Highly Efficient Dual Rail Output, " IEEE Sensors Journal ,vol. 12, no. 6, pp. 2287-2298, Jun. 2012.

A COMPARISON OF MULTI-RESOLUTION AND MULTI-ORIENTATION FOR BREAST CANCER DIAGNOSIS IN THE FULL-FIELD DIGITAL MAMMOGRAM

Abdelaziz Addioui, Faouzia Benabbou and Sanaa El Filali

MITI laboratory
Faculty of sciences Ben Msik
Casablanca, Morocco
addiouiabdelaziz@gmail.com, faouziabenabbou@yahoo.fr
and Elfilalis@gmail.com

Mohamed El Aroussi

LETI
Ecole Hassania des Travaux Publics
Casablanca, Morocco
mohamed.elaroussi@ieee.org

Abstract—**This paper presents a comparative study between Steerable Pyramid, Curvelet, Contourlet, and Gabor for breast cancer diagnosis in the full-field digital mammogram. Using multi-resolution and multi-orientation analysis, mammogram analysis images are decomposed into different resolution levels and orientations. A set of the biggest coefficients from each scale and orientation is extracted. Then a supervised classifier system based on Support Vector Machine is constructed. The experimental results gave a 98.33% classification accuracy rate, which indicate that Steerable Pyramid transformation is a promising tool for analysis and classification of full-field digital mammograms.**

Keywords—**Full-field digital mammogram, Breast cancer diagnosis, Computer-aided diagnostic, Steerable Pyramid, Curvelet, Contourlet, Gabor, SVM.**

I. INTRODUCTION

Breast cancer is the fifth cause of death from cancer. International Agency for Research on cancer (IARC) estimates from 1.67 million new cancer cases diagnosed in 2012 ,522. 000 deaths were women [1].

Digital mammography is a technique for study, detection, diagnostic and control of breast cancer. Reading images is a tiresome task for the radiologists and can fail to detect from 10% to 30% of cancers cases [2].

Computer-aided diagnostic (CADx), helps radiologists to determine a class of abnormal cases detected by radiologist or CADe, which reduce unnecessary biopsies [3].

Eltoukhy et al. presented a comparative study between wavelet and Curvelet transform. Firstly, each mammogram image is decomposed using wavelet and Curvelet transforms separately. The 100 biggest coefficients are extracted from each decomposition level. Then, Euclidian distance is used to construct a nearest neighbor. The classifier achieves 94.07% for Curvelet [4].

Eltoukhy et al. proposed a method using a Curvelet transform at different scales as a per-process for feature extraction and classification of mammogram images, then extracting different ratios of the biggest Curvelet coefficients from each level as a feature vector to be used for classification. Euclidean distance based classifier used for classification process [5].

In addition, A. B. HUDDIN et al (2011) presented Investigation of multiorientation and multiresolution features for microcalcifications classification in mammograms, using steerable pyramid (multiorientation and multiresolution representations). Statistical measures (energy and entropy) were extracted and calculated from each sub bands (wavelet and steerable pyramid decompositions) to form a feature vector that represent each case of image, and selecting best features to represent an image at lower dimension, classifications is made using SVM with radial basis function (RBF) [6].

In [7], explore the steerable pyramid as feature extraction for breast cancer diagnosis in digital mammogram. The proposed approach is compared to some related feature extraction methods as Discrete Wavelet Transform (DWT), linear discriminant analysis (LDA) and principal component analysis (PCA). K-nearest neighbor, Support Vector Machine and Nave Bayesian is used separately to construct a supervised classifier. Experimental results tested on DDSM database. The maximum successful classification with 92%.

This paper introduces a comparative study between Steerable pyramid (SP), Curvelet, Contourlet and Gabor transforms. The motivation is to determine through experimental work, which method is more efficient for diagnosis of breast cancer in full-field digital mammograms. The system aims to differentiate between normal tissue and malignant tumors.

978-1-4673-8760-6/15 $31.00 © 2015 IEEE

The rest of this paper is organized as follows. In Section II, a theoretical background is introduced. The experimental work is treated in Section III. In Section IV, experimental results and the comparative are illustrated. Section V as formulated as a conclusion.

II. THEORETICAL BACKGROUNDS

A. Steerable Pyramid Transform

Steerable pyramid is a multi-resolution transforms similar to the two-dimensional DWT, but with interesting translation and rotation-invariance properties [8].

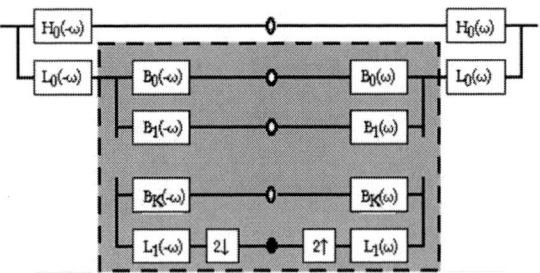

Fig. 1. Tree representation of one-level 2D steerable pyramid transform.

As shown in Fig.1. the input image is firstly decomposed into a high pass subband using a non oriented high-pass filter (ω) and then into a low-pass sub band using a narrow band low-pass filter $L_0(\omega)$. Afterwards, this low pass sub band is decomposed into K-oriented portions using the band pass filters B_k(k = 0,1,..., K-1) and into a low-pass subband.

The decomposition is done recursively by subsampling the lower low-pass subband. The small black boxes represent the decomposed subband images. $2\downarrow$ and $2\uparrow$ indicate down sampling and up sampling by a multiplier of 2 along the rows and columns. Recursive steps extract different directional information at a given scale J [9].

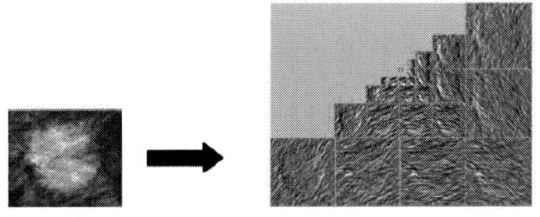

Fig. 2. Detail sub images (image at left) using steerable pyramid filters for image decomposition at 6 orientations and 4 levels of resolutions.

Castleman presented a simple method for designing the finite impulse response filter kernels required to implement the transform [10].

B. Curvelet Transform

Candes and Donoho introduced a new system of multi-resolution analysis called the Curvelet transform in 1999. This system differs from wavelet and related systems. Curvelets takes the form of basic elements, which exhibit a very high directional sensitivity and are highly anisotropic [11]. The Curvelet transform captures curves instead of points as in S-P transform in the continuous domain.

C. Contourlet Transform

Contourlets are an extension of Curvelets, which can be approximated in the discrete domain. Contourlets, however, are defined and derived in the discrete domain from the beginning. They both allow for directionality and anisotropy. At the first stage, we used Laplacian pyramid (LP), and in the second one we used directional filter banks (DFB) [12].

D. Gabor Transform

Gabor filters represent a powerful tool both in image processing and image coding, with the capability to capture important visual features, such as spatial localization, spatial frequency and orientation selectivity [13].

E. Classifying by SVM

The support vector machine (SVM), based on statistical learning theory, is introduced by Vapnik. SVM-based techniques have proven to be powerful in classification They provide a higher performance than that of traditional learning machines. Several kernel functions can be used in SVM, including the linear, polynomial, quadratic.

The aim of an SVM is to find a hyperplane to separate the training data with a maximal margin. This hyperplane, called Optimal Separating Hyperplane (OSH), can minimize the risk of misclassifying examples from the test set [14] [15].

III. EXPERIMENTAL WORK

The database was acquired at the Breast Centre in CHSJ, Porto, under permission of both the Hospital's Ethics Committee and the National Committee of Data Protection. The images were acquired between April 2008 and July 2010 [16].

Original mammograms are almost 50% of the whole image comprised of the background with a lot of noise. Therefore, a cropping operation is applied to the images to cut off the unwanted portions of the images. Regions of Interest (ROI's) 128*128 are cropped [4]. The cropping process was performed manually, where the given center of the abnormality area is selected to be the center of ROI. Thus, almost all the background information and most of the noise are eliminated. 120 regions were extracted from 120 images, including 60 normals and 60 cancers, an example of cropping that eliminates the label on the image and the black background is given in Fig. 3. In Fig .4, some examples ROI's cropped from other mammogram images are represented.

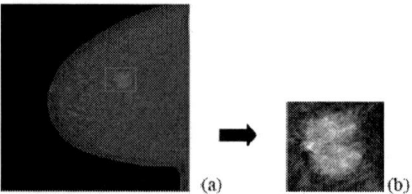

Fig. 3. (a) Original image (22614074). (b) Cropped image (128×128).

978-1-4673-8760-6/15 $31.00 © 2015 IEEE

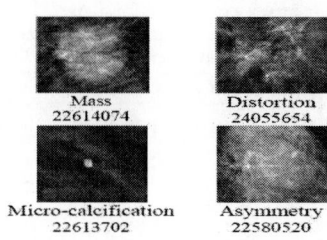

Mass
22614074

Distortion
24055654

Micro-calcification
22613702

Asymmetry
22580520

Fig. 4. ROI's cropped from mammogram images.

Once the images are cropped as described. The SP, Contourlet, Gabor transform is used to represent the ROI's in multi-resolution decomposition levels or scales and multi-orientation. While the available scales are (1, 2, 3, 4) and multi-orientations (1,2,3,4,5,6), we extract the 100 largest coefficients, and then we represent each ROI's by a vector of 600 coefficients. Only in the experimental work the Curvelet is

limited at 4 multi-orientations, then we represent each ROI's by a vector of 400 coefficients. The method of the feature extraction and classification is recapitulated in Fig. 5.

A. Steerable Pyramid results

Steerable Pyramid transform is used as a multi-scale and multi-orientation decomposition to represent mammogram images as a pre-process for mammogram classification. In each orientation, a ratio of the biggest coefficients is used to be the feature vector of the corresponding mammogram.

Table I show, the successful classification rates at scale 2 using SVM (Linear) classificatory. The maximum successful classification average for the 2 classes is 98.33% (Normal: 96.66%, Malign: 100%).

The Steerable Pyramid was also compared to three other multi-resolution and multi-orientation algorithms, Curvelet, Contourlet and Gabor on the same set of mammogram images as discussed in the next section.

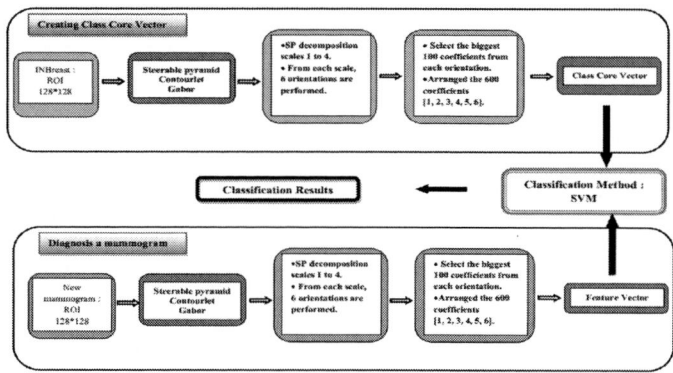

Fig. 5. The proposed system for mammogram diagnosis.

B. Comparative study

Firstly, Curvelet transform is applied to the dataset, by the proposed method for feature extraction. Table II illustrate the performance of the classifier in case of using features provided by Curvelet transform. It shows that the classification accuracy rate reached 95% at scale 4.

Secondly, Table III shows that the average successful classification is 91,66 % by using Contourlet transform and SVM (Linear, Polynomial Order 1) at scale 2.

Finally, Table IV shows the maximum successful classification rate for Gabor is 95 % achieved at scale 4.

TABLE I. THE RECOGNITION ACCURACY OF SP & SVM

	Linear (%)			Quadratic (%)			Polynomial order 1 (%)			Polynomial order 2 (%)			Polynomial order 3 (%)		
	Normal	Malign	Average	Normal	Malign	Average	Normal	Malign	Average	Normal	Malign	Average	Normal	Malign	Average
Scale 1	100	86,66	93,33	93,33	66,66	80	100	86,66	93,33	93,33	66,66	80	0	6	10
Scale 2	96,66	100	98,33	93,33	70	81,66	93,33	70	81,66	93,33	70	81,66	10	23,33	16,66
Scale 3	86,66	90	88,33	90	66,66	78,33	90	66,66	78,33	90	66,66	78,33	3,33	23,33	13,33
Scale 4	86,66	70	78,33	9,33	46,66	70	86,66	70	78,33	93,33	46,66	70	86,66	80	83,33

TABLE II.   THE RECOGNITION ACCURACY OF CURVELET & SVM

	Linear (%)			Quadratic (%)			Polynomial order 1 (%)			Polynomial order 2 (%)			Polynomial order 3 (%)		
	Normal	*Malign*	*Average*	*Normal*	*Malign*	*Average*	*Normal*	*Malign*	*Average*	*Normal*	*Malign*	*Average*	*Normal*	*Malign*	*Average*
Scale 1	100	83.33	91.66	No convergence			100	83.33	91.66	No convergence			No convergence		
Scale 2	100	83.33	91.66	No convergence			100	83.33	91.66	No convergence			No convergence		
Scale 3	96.66	53.33	75	No convergence			96.66	53.33	75	No convergence			No convergence		
Scale 4	96.66	93.33	**95**	No convergence			96.66	93.33	**95**	No convergence			No convergence		

TABLE III.   THE RECOGNITION ACCURACY OF CONTOURLET & SVM

	Linear (%)			Quadratic (%)			Polynomial order 1 (%)			Polynomial order 2 (%)			Polynomial order 3 (%)		
	Normal	*Malign*	*Average*	*Normal*	*Malign*	*Average*	*Normal*	*Malign*	*Average*	*Normal*	*Malign*	*Average*	*Normal*	*Malign*	*Average*
Scale 1	40	76.66	58.33	30	70	50	40	76.66	58.33	30	70	50	26.66	70	48.33
Scale 2	83.33	100	**91.66**	No convergence			83.33	100	**91.66**	No convergence			No convergence		

TABLE IV.   THE RECOGNITION ACCURACY OF GABOR & SVM

	Linear (%)			Quadratic (%)			Polynomial order 1 (%)			Polynomial order 2 (%)			Polynomial order 3 (%)		
	Normal	*Malign*	*Average*	*Normal*	*Malign*	*Average*	*Normal*	*Malign*	*Average*	*Normal*	*Malign*	*Average*	*Normal*	*Malign*	*Average*
Scale 1	66.66	60	63.33	80	56.66	68.33	66.66	60	63.33	80	56.66	68.33	86.66	10	84.33
Scale 2	86.66	73.33	80	63.33	73.33	68.33	86.66	73.33	80	63.33	73.33	68.33	83.33	56.66	70
Scale 3	86.66	86.66	86.66	73.33	76.66	75	86.66	86.66	86.66	73.33	76.66	75	43.44	0	21.66
Scale 4	96.66	9333	**95**	73.33	76.66	75	96.66	93.33	95	76.66	73.33	75	43.33	3.33	23.33

IV. CONCLUSION

This work has explored a case of feature extraction method implemented to diagnostic breast cancer in full field mammogram images. This work focuses on using multi-resolution and multi-orientation representation advantages. The method is based on ranking the feature according to its capability to distinguish between two different classes (Normal or malign). The classification according to SP and SVM has given good results: SP (Scale 2) and SVM has given: **98.33%**.

Acknowledgements

Courtesy of the Breast Research Group, INESC Porto, Portugal.

REFERENCES

[1] GLOBOCAN 2012. Brest cancer Estimate Incidence, Mortality and Prevalence Worldwide in 2012. International Agency for Research on cancer.

[2] Mehul P. Sampat, Mia K. Markey, Alan C. Bovik. "Computer-Aided Detection and Diagnosis in Mammography". In: Bovic AC Handbook of Image and Video Processing Elsevier Academic Press, Amsterdam pp. 1195-1217, 2005.

[3] J. Bozek, M. Mustra, K. Delac and M. Grgic, "A Survey of Image Processing Algorithms in Digital Mammography" in Recent Advances in Multimedia Signal Processing and Communications, Vol. 231, pp. 631-657 Springer, Berlin, 2009.

[4] M.Meselhy Eltoukhy, I.Faye, B. Belhaouari Samir, "A comparison of wavelet and curvelet for breast cancer diagnosis in digital mammogram". Computers in biology and medicine, Computers in Biology and Medicine ,Elsevier vol. 40 (2010)pp. 384–391, 2010.

[5] Mohamed Meselhy Eltoukhy, Ibrahima Faye, Brahim Belhaouari Samir," Breast Cancer Diagnosis in Digital Mammogram Using Multiscale Curvelet Transform", Computerized Medical Imaging and Graphics", Elsevier, 34 (4), pp. 269-276, 2010.

[6] Buciu, I., Gacsadi, A.: Directional features for automatic tumor classification of mammogram images. Biomedical Signal Processing and Control 6(4), 370–378 (2011).

[7] A. Addioui, F. Benabbou, S. El Filali and M. El Aroussi, Breast Cancer Diagnosis in Digital Mammogram Using Steerable Pyramid,(2015) International Review on Computers and Software (IRECOS),10(7),pp.702-709.

[8] E.P.Simoncelli, A rotation-invariant pattern signature, in:Third IEEE International Conference on Image Processing, Laussanne Switzer-land,vol.3, pp.185-188, September 1996.

[9] AyGegl Uar. Color Face Recognition Based on Steerable Pyramid Transform and Extreme Learning Machines. The Scientific World Journal Volume 2014, Article ID 628494, 1-15 pages. Hindawi Publishing Corporation.

[10] K. R. Castleman, M. Schulze, and Q. Wu. "Simplified design of steerable pyramid filters." Presented at the 1998 IEEE International Symposium on Circuits and Systems, Monterey, vol.5, pp329 - 332 California, May 31-June 3, 1998.

[11] E.J.Candes, D.L.Donoho, Curvelets: a surprisingly effective nonadaptive representation for objects with edges,(Online). Available /http://www.Curvelet.org/papers/Curve99.pdfS.

[12] Do and Vetterli, "The Contourlet Transform: An efficient Directional Multi Resolution Image Representation", IEEE Transactions on Image Processing, Vol. 14, pp: 2091-2106, 2005.

[13] M. Elaroussi, "Information Fusion towards a Robust Face Recognition System", Ph.D. Thesis, Lab. Computer Science Research and Telecommunications, Rabat, Morocco, P 38, 2009..

[14] Vapnik, V. "Statistical Learning Theory", WileyInterscience, New York, 1998.

[15] Wen-Jie Wua, Shih-Wei Lina, Woo Kyung Moonb. "Combining support vector machine with genetic algorithm to classify ultrasound breast tumor images". Computerized Medical Imaging and Graphics Vol. 36(8) ,pp. 627-633, 2012.

[16] "INbreast: Toward a Full-field Digital Mammographic Database.Academic Radiology", Vol 19, No 2 February 2012.

Comparison Of Control Structures for Variable Speed Wind Turbine

Salma El Aimani

Ibnou Zohr University - Polydisciplinary of Ouarzazate
P.B. 638 45000

Abstract— this paper presents investigations in the design of wind turbine control systems to maximize the energy capture and their impact onto the grid. The modeling of the wind turbine is presented. Then three different speed control systems are detailed and evaluated through simulations carried out with Matlab Simulink TM. Power fluctuations on the distribution grid are then compared by considering the same wind speed fluctuations.

Keywords—Wind Energy Conversion System, grid, control, Doubly Fed Induction Generator

I. INTRODUCTION

The connection of wind turbines to the electrical distribution grid can induce some power quality problems that may affect users connected onto this grid. The stochastic nature of the wind speed variation leads to voltage and power fluctuations produced by wind turbines, which may cause flicker. The influence of wind power generation on grid power quality will depend on the type of wind turbines installed, the generation level and the characteristics of the grid at the point of common coupling. Flicker produced by power fluctuations and influence of wind turbines on the power quality remains the strongest barrier to the development of wind power generators in the distribution electrical grid (see as example the certification by the International Electrotechnical Commission: 61400-21 standard "Measurement and assessment of power quality of grid connected wind turbines"). Since the last decade new technologies have been introduced in wind power generation systems in order to regulate the power transfers. The main idea is to make the turbine speed variable (and then the wind/electricity conversion) by using power electronic converters and an intermediary dc bus [BAU 02]. However constraints remain and the transferred power is still limited by the capacitor value and the rated voltage of the dc bus [ELA 02].

The purpose of this paper is to present various wind turbine control systems for different objectives as maximum wind power tracking and constant electrical power generation. The first part of this paper presents the wind turbine model, which has been designed to simulate the dynamics of the system from the turbine rotor where the kinetic wind energy is converted to the mechanical energy. Then the principle of the variable speed control is recalled, different control systems are described and compared with the help of simulations.

II. WIND TURBINE MODELING

A. Turbine modeling

The studied horizontal wind power generation system is shown on fig. 1. The power of wind is proportional with the cube of the wind speed:

$$P_{wind} = \frac{\rho}{2}.A.v^3 \qquad (1)$$

where ρ is the air density (approx. 1.22 kg/m^3 in standard atmosphere at 15°C), A is the swept area of the turbine and v is the wind speed. The power coefficient (Cp) represents the aerodynamic efficiency of the wind turbine. It depends on the tip speed ratio (λ) and the blade pitch angle (β).

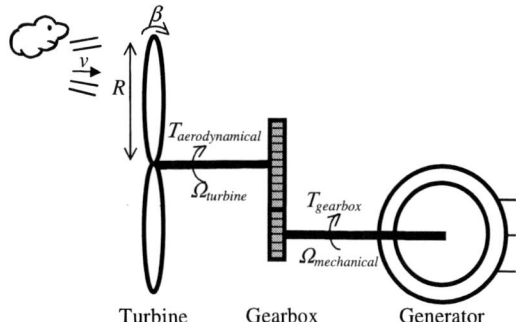

Fig. 1. The horizontal wind power generation system

For the wind turbine used in the study, the following form has been derived from [EZZ 00]:

$$C_p(\lambda,\beta) = (0.3 - 0.00167.\beta)\sin\left[\frac{\pi.(\lambda+0.1)}{(10-0.3.\beta)}\right] - 0.00184.(\lambda-3).\beta \qquad (2)$$

The tip speed ratio is defined as the ratio between the blade tip speed and the wind speed:

$$\lambda = \frac{R.\Omega_{turbine}}{v} \qquad (3)$$

R is the blade radius and $\Omega_{turbine}$ is the angular speed of the turbine.

The characteristic of the power aerodynamic efficiency is shown on Fig. 2.

978-1-4673-8760-6/15 $31.00 © 2015 IEEE

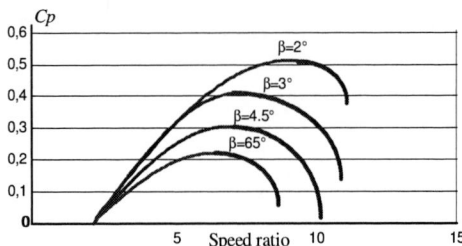

Fig. 2. Power aerodynamic efficiency versus tip speed ratio
For a three bladed turbine

Using Cp, the aerodynamic power $P_{aerodynamcal}$ is determined by:

$$P_{aerodynamical} = C_p.P_{wind} = C_p.\frac{\rho}{2}.A.v^3 \qquad (4)$$

The aerodynamic torque $T_{aerodynamic}$ is determined directly according to

$$T_{aerodynamical} = C_p.\frac{\rho}{2}.A.v^3.\frac{1}{\Omega_{turbine}} \qquad (5)$$

B. Gearbox modelling

The gearbox has the task to transfer the aerodynamical power from the slow rotating rotor shaft to the fast rotating shaft, which drives the generator at the mechanical speed $\Omega_{mechanical}$ (fig.3). It is mathematically described by the following equations:

$$T_{gearbox} = \frac{T_{aerodynamical}}{G} \qquad (6)$$

$$\Omega_{turbine} = \frac{\Omega_{mechanical}}{G} \qquad (7)$$

G is the gear ratio.

C. . Drive train modelling

The drive train is composed of the mass corresponding to the large turbine rotor inertia representing the blades and the hub, and a small inertia representing the rotor mass of the generator. The proposed model only includes the turbine rotor because this part of the wind turbine has the most significant influence on the power fluctuations. The acceleration is governed by the following equation:

$$J.\frac{d\Omega_{mechanical}}{dt} = T_{mechanical} \qquad (8)$$

The total inertia (J), which appears onto the generator rotor, represents the acceleration time constant. Taking into account the electromagnetic torque (T_{em}) and the viscous torque ($T_{viscous}$), the torque balance yields:

$$T_{mechanical} = T_{gearbox} - T_{em} - T_{viscous} \qquad (9)$$

The flexibility of the shaft is modeled as a stiffness (f):

$$T_{viscous} = f.\Omega_{mechanical} \qquad (10)$$

D. Overall system modelling

The wind turbine model output is the aerodynamic torque applied on the gearbox. The inputs of the wind turbine model are the wind speed, the pitch angle and the turbine rotor speed.

The gearbox model transforms the mechanical speed and the aerodynamical torque respectively into the turbine rotor speed and the gearbox torque. The drive train model output is the mechanical speed and it has two inputs: the gearbox torque and the electromechanical torque provided by the generator. Fig.6 shows the corresponding block diagram, which has been simulated under the Matlab Simulink [TM] software.

III. WHY VARIABLE SPEED TURBINE CONTROL?

The stored power (in the inertia) is expressed as the product: $P_{kinetic} = T_{mechanical}.\Omega_{mechanical}$ **(11)**

The turbine speed can be controlled by acting on two inputs: the blade pitch angle and the electromagnetic torque of the generator. The wind speed must be considered as a perturbation input for the system.

In next sections we present various control strategies in order to maximise the electrical power known as Maximum Power Point Tracking (M.P.P.T.). These techniques consist in adjusting the electromagnetic torque of the generator in order to control the speed to the reference value (Ω_{ref}). Other torques, which are applied to the shaft, are then considered as perturbation inputs. As the power must not exceed the rated generator power a pitch actuator is used to shed the aerodynamic power. For this operation, the angle is therefore increased. electromechanical torque is assumed to be equal to the reference value:

$$T_{em} = T_{em\_ref}$$

At constant speed operating, the mechanical torque is null and then the aero dynamical power minus the loss power is converted into electrical power (Fig. 3):

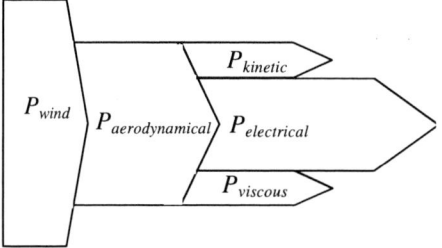

Fig. 3. Conversion transfer of the power

$$P_{electrical} = P_{aerodynamical} - P_{loss} \qquad (12)$$

$$P_{loss} = T_{viscous}.\Omega_{mechanical} \qquad (13)$$

IV. DIRECT CONTROL OF THE WIND TURBINE SPEED

A. Principle

Most of variable speed turbines are controlled by a direct control system, which is now detailed [MUL 01]. In a first step one can neglect the effect of the viscous torque. Moreover, around an operating point, the generator rotor speed varies slowly. Then the dynamic equations (4) and (5) yield:

$$T_{gearbox} = T_{em} \qquad (14)$$

From this equation we can deduce a first control system function to set the reference value of the electromagnetic torque:

$$T_{em\_reg} = T_{gearbox\_estimated} = \frac{T_{aerodynamical\_estimated}}{G} \quad (C1)$$

The aerodynamical torque can be estimated with a wind speed estimation and the measurement of the generator rotor speed:

$$T_{aerodynamical\_estimated} = C_p.\frac{\rho}{2}.A.v^3_{estimated}.\frac{G}{\Omega_{mec}} \quad (C2)$$

To get the maximum of the extracted power from the wind we must set the particular tip speed ratio (λ_{Cp_max}), which corresponds, to $Cp = C_{p\_max}$

$$T_{aerodynamical\_estimated} = C_{p\_max}.\frac{\rho}{2}.A.v^3_{estimated}.\frac{G}{\Omega_{mec}}$$
$$(15)$$

Thus we can deduce an estimation of the wind speed by measuring the mechanical speed and using the expression of the tip speed ratio [THI 93]:

$$v_{estimated} = \frac{R.\Omega_{turbine}}{\lambda} = \frac{R}{\lambda_{C_{p\_max}}}.\frac{\Omega_{mec}}{G} \qquad (C3)$$

For the practical implementation, the reference of the electromagnetic torque is:

$$T_{em\_reg} = k.\Omega_{mec}{}^2 \qquad (16)$$
$$\text{with}$$
$$k = \frac{C_{p\_max}}{\lambda^3_{C_{p\_max}}}.\frac{\rho}{2}.\pi.\frac{R^5}{G^3}$$

V. CLOSED LOOP CONTROL OF THE WIND TURBINE SPEED

A. Principle

The speed is measured and compared with a reference value, and then a corrector adjusts the reference torque in order to reduce this error

$$T_{em\_ref} = C_{ass}\left(\Omega_{mechanical\_ref} - \Omega_{mechanical}\right) \quad (Cass)$$

The speed controller has to achieve two tasks:

- It must set the mechanical speed to its reference value.

- It must damp the action of the wind torque, which represents a perturbation.

To obtain an optimal tip speed ratio, the speed reference value is calculated with the knowledge of the wind speed:

$$\Omega_{mec\_ref} = \lambda_{C_{p\_max}}.\frac{G}{R}.v \qquad (C4)$$

B. Phase delay controller

The first considered speed controller is a phase delay controller (fig. 4):

$$T_{em-ref} = \frac{a_1.s+a_0}{\tau.s+1}.(\Omega_{mechanical-ref} - \Omega_{mechanical}) \qquad (17)$$

a_0, a_1 and τ are parameters, which have to be determined, and s is the Laplace quantity.

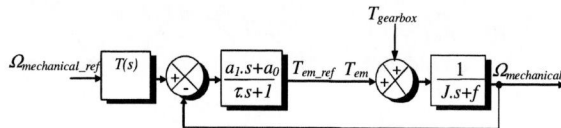

Fig.4. Closed loop control with a phase delay controller

If the electrical machine is suitable controlled then the rotor speed is dependent on the speed reference through the transfer function ($F(s)$) and the gear box torque via a perturbation transfer function ($P(s)$):

$$\Omega_{mechanical} = F(s).\Omega_{mechanical-ref} + P(s).T_{gearbox} \qquad (18)$$

$$\text{with } P(s) = \frac{\tau.s+1}{J\tau.s^2+(f\tau+J+a_1)s+a_0+f}$$

$$\text{and } F(s) = \frac{a_1.s+a_0}{J\tau.s^2+(f\tau+J+a_1)s+a_0+f} .$$

In order to damp the perturbation input ($T_{gearbox}$), the parameter a_0 must be large ($a_0 = 1000$). Then the other parameters (a_1 and τ) are chosen to set a classical second order function with a natural pulsation and a damping ratio given by:

$$\omega_n = \sqrt{\frac{a_0+f}{J\tau}} \quad \text{and} \quad \xi = \frac{\tau f + J + a_1}{a_0+f}.\frac{\omega_n}{2} = 1 .$$

A *50ms* response time of the closed loop control is obtained with an anticipating action, which is given with the following tracking transfer function:

$$T(s) = \frac{J.\tau.s^2 + (f.\tau+J+a_1).s + f + a_0}{(a_1.s+a_0).(\frac{0.05}{3}.s+1)} \qquad (19)$$

Results with the same wind profile are shown in figures 5. A little more power is converted when transient variations of the wind occur, but it is not significant. In steady state an error between the speed and the reference value appears.

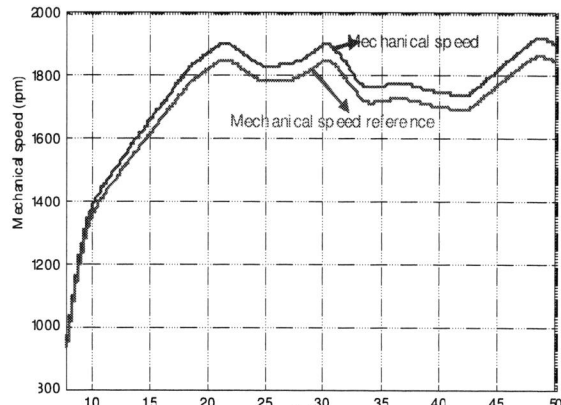

Fig. 5. Obtained results with a phase delay controller

To improve the speed control other controller must be evaluated.

C. Proportional Integrator controller

The second considered speed controller is a proportional integrator controller (fig. 6):

$$T_{em-ref} = b_1 + \frac{b_0}{s}.(\Omega_{mechanical-ref} - \Omega_{mechanical}) \quad (20)$$

where b_1 is the proportional gain, b_0 is the integral gain.

The same mathematical expression of the closed loop is obtained:

$$\text{with } F(s) = \frac{b_1.s + b_0}{Js^2 + (f + b_1).s + b_0}, \quad P(s) = \frac{s}{Js^2 + (f + b_1).s + b_0}$$

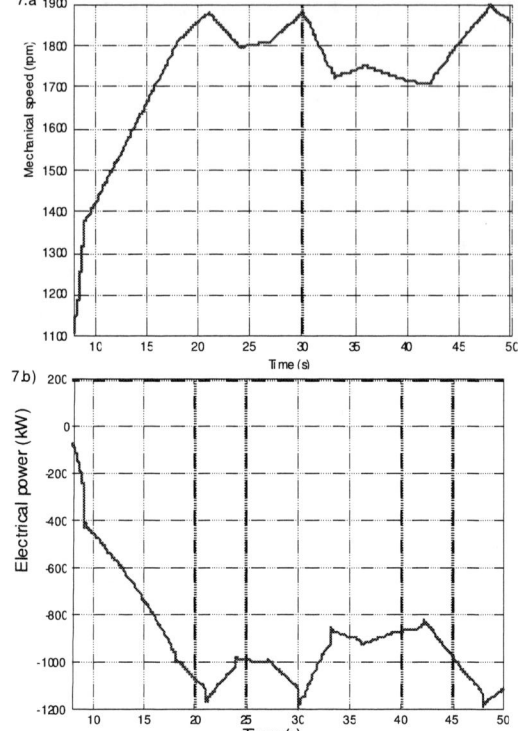

Fig.6. Closed loop control with a PI controller

It is necessary to increase the parameter b_0 to attenuate the action of $T_{gearbox}$. As previously, the natural pulsation and the damping ratio are set by the controller parameters:

$$\omega_n = \sqrt{\frac{b_0}{J}} \quad \text{and} \quad \xi = \frac{f + b_1}{b_0}.\frac{\omega_n}{2}$$

$$T(s) = \frac{Js^2 + (f + b_1).s + b_0}{(b_1.s + b_0).(\frac{0.05}{3}.s + 1)} \quad (21)$$

Simulation results are shown in figure 7. A better control, both in transient and in steady state of the speed is obtained (fig 7.a). This control system is highly dynamic and again more power is converted in transient operation (fig 7.b).

This control system requires the wind speed measurement. In practice it is very difficult to measure the wind speed and optimize the turbine speed accordingly. First the anemometer is located behind the rotor on the nacelle, which distorts the wind speed reading. Secondly, a single anemometer cannot measure the entire wind speed acting since the turbine's rotor sweeps a large area.

Complex filters may be used to smooth these variations [VAN 01].

PARAMETERS

Diameter: 70 m	Number of blades: 3
Hub height: 85 m	Cut-in speed: $\Omega_i = 4$ m/s
Stop wind speed: $\Omega_0 = 25$ m/s	
Viscous coefficient: $f = 7.1e-3$	Inertia: $J = 50$ kg.m^2.

Fig. 7: Obtained results with a proportional integer controller

V. Conclusion

In this paper a simplified modeling of a wind turbine generator has been presented. This simplified model is suitable for modeling large scale wind parks. The direct control system of wind turbines is recalled. However because the system is highly dynamical we show that a control loop of the mechanical speed is required for an optimal response change as the set-point changes. Obtained fluctuations of the electrical generated power with these control systems are then compared. It appears that closed loop control systems of the turbine speed improve the efficiency of the wind/electrical conversion. This requires an accurate wind speed measurement.

REFERENCES

[AME 02] J. L. Rodriguez-Amenedo, S. Arnalte, J. C. Burgos, "Automatic generation control of a wind farm with variable speed wind turbines", IEEE Transactions on energy conversion, vol.17, No.2, June 2002

[BAU 02] P. Bauer, S. W. H.de Haan, M. R. Dubois , "Windenergy and Offshore Windparks:State of the Art and Trends", 10th International Power Electronics and Motion Control Conference: EPE-PEMC 2002, CD, 9-11 September 2002, Cavtat, Croatia

[ELA 2004] S. El Aimani, Modélisation de différentes Technologies d'éoliennes intégrées dans un réseau moyenne tension. Presses Académiques Francophones, ISBN3838170539, 9783838170534, 2012.

[EZZ 00] E. S. Abdin, W. Xu, "Control design and Dynamic Performance Analysis of a Wind Turbine-Induction Generator Unit", IEEE Trans. on Energy conversion, vol.15, No1, March 2000

[MUL 01] E. Muljadi, "Pitch-Controlled Variable-Speed Wind Turbine Generation", IEEE Trans. on Industry Applications, vol.37, No1, Jan./Feb. 2001

[NWCC 01] National Wind Coordinating Committee, "Distributed Wind Power Assesment", Fevbruary 2001, www.nationalwind.org.

Precise electric measurements with temperature using 10-bit embedded system: Application on photovoltaic junctions.

Abdessamad MALAOUI

Laboratoire Interdisciplinaire de Recherche en Sciences et Techniques (LIRST),
Faculté Polydisciplinaire, Université Sultan Moulay Slimane
B.P 592, Béni Mellal, Morocco.
a.malaoui@usms.ma

Abstract— **In this paper, an electronic temperature control system is presented. This contribution aims to provide a not expensive prototype, fast and accurate, which can be used by research laboratories to control the temperature for electrical measurements. This device is based on the implementation of the PID algorithm in an embedded system at 10-bit Analog/Digital Converter (ADC). The electronic schematics and results of this controller are presented and compared with those with an old 8-bits ADC controller and with a standard laboratory regulator. Initial tests are performed on the temperature control of a photovoltaic junction to measure its intrinsic electrical parameters. The obtained results show the range and levels of the advantages of this device.**

Keywords-Embedded system; Arduino; Peltier; Temperature control; PID command.

I. INTRODUCTION

The determination of the intrinsic electrical parameters of photovoltaic junctions, can provide valuable information on the overall behavior of these cells and on the functioning of the solar panels. Knowing precisely these electrical parameters, permit a right choice of electronic components that connect with these cells for optimal operation during energy production. Mathematical models of these junctions show that these parameters (Rs, Rsh, Is and η) clearly depend on the temperature as well as other electrical parameters [1]. In fact, it's necessary to precisely control the junction temperature to have accurate measurements of those parameters.

In this paper, we proposed to use embedded systems, to have a better temperature controller, however inexpensive, accurate and fast. For that, we proposed to study and test an electronic device using Arduino microcontroller and others electronic components to have a desirable objective. This device is improved with on another device which is already realized using an 8-bits ADC microcontroller [2]. This system blends two electronic stages; the first is the power electronic relative to the cooling and heating processes working with thermoelectric (TEC) effect or Pelletier module [3]. The second electronic stage is the programmable Arduino ATmega controller, loaded with PID algorithm. The time constants of two cooling and heating processes are calibrated to eliminate the non-symmetry of this system.

A new architecture of PID controller with Arduino is proposed and compared with others systems to temperature control. The temperature controller is calibrated with an accurate thermometer, and the first tests were carried out on a photovoltaic cell. The obtained results are discussed and analyzed to appraise the proposed controller.

II. DESCRIPTION OF OLD TEMPERATURE REGULATORS

A. Classical device of temperature control.

Several laboratories have conventional devices for the temperature control. In our case we have a thermostat type (Ministat Huber, DIN 12879-KI), and it has a pump of heating and cooling (Fig.1). The thermostat controls the temperature in -20 ° C to +100 ° C. This is a conventional temperature control system, which operates by circulating a thermal liquid around two metal cylinders, through insulated pipes in silicone (2 meters long and 8mm internal diameter). This thermostat is of type RCS6 - LAUDA, it contains a resistance thermometer (PT 500/PID) and a closed-circuit of 10/13 liters per minute, and the suction of 8/9 liters per minute. The system is controlled by computer in the range of temperature [-30 ° C and 150 ° C].

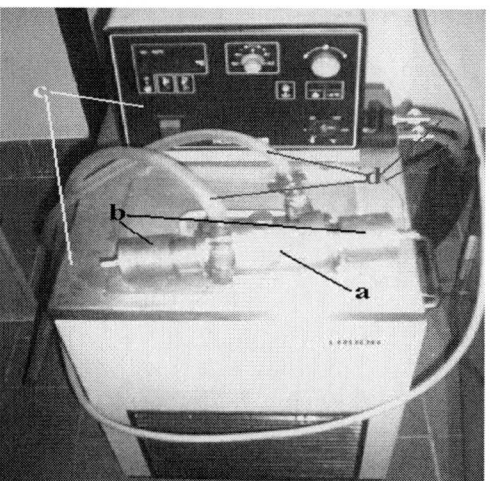

Figure 1. Old laboratory temperature regulator

a- two cylindrical cells, b- connector / support of the test junction, c- temperature servo Ministat, d- liquid circulation pipes.

B. Eelectronic regulator with 8-bits ADC microcontrolor

The first work of our research team in this area, to improve temperature control, was made in 2004. This Work was performed using 8-bit ADC programmed chip. The following diagram shows architecture based on the microcontroller of the ST family, two different processes working with the Joule effect for heating, and with the Pelletier effect for cooling. The performances limit of the programmable component forced us to divide the PID controller into two parts; the first in external hardware of the µC and the second in the program part. Improved previous system was the subject of several works [4]. Indeed, several architectures have been tested and established to improve the performance of the temperature control.

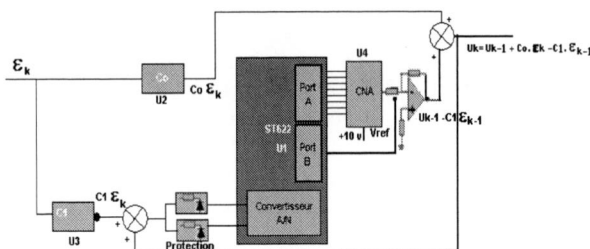

Figure 2.   8-bits ADC temperature regulator

The programmable hardware controller is the ST62xx microcontroller of SGS-Thomson Company. It has a RAM of 64 bytes of data memory, an EPROM of 3872 bytes and 20 pins of A/N converters.

III.   PROPOSED ELECTRONIC REGULATOR WITH ARDUINO

In the proposed electronic system, a 10-bit ADC microcontroller is used to eliminate the shortcomings of the 8-bit ADC µC. The embedded system used in this work is an Arduino Atmega of Atmel Company. It has several advantages in terms of memory size, the processed words type, speed, ports number, addressing modes, and other specifications. These benefits allow changing the regulator architecture imposed by former µC. Indeed, every step of the PID algorithm will be implemented as a program, as shown in the block diagram of Figure 3. Several tests were carried out on the parameters of this regulator. The actions of Proportional (P) and integral (I) are sufficient to give satisfactory results in our case. The transfer function C(s) of a PI analogical controller is given by:

$$C(s) = K_p + \frac{K_i}{s} \qquad (1)$$

Where s is the Laplace variable and K_p and K_i are the proportional and integral gains, they are given by Ziegler-Nichols method to ensure the stability of the system [7]. The discrete form for PI controller is:

$$C(z) = K_p \quad \frac{K_i.T_e}{2} + \frac{K_i.T_e}{1 - Z^{-1}} \qquad (2)$$

Z is the variable of Z-transformation and T_e is the sampling period, used in the following transformation:

$$s = \frac{2}{T_e} \cdot \frac{z-1}{z+1} \qquad (3)$$

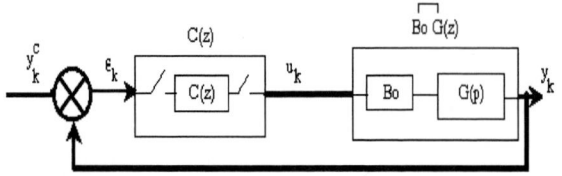

Figure 3.   Block diagram of the controller used by Arduino

The electronic diagram of the controller is given in Figure 4. The embedded system is programmed to control both the Pelletier plates for heating and cooling by the same mechanism. This approach is different than that used by the ST62 µC, cited in Section II. The control of both thermal processes is realized through the set of electronic components (op-amp, TEC transistors, D/A and A/D converters ...). The used sensor of temperature (AD590) has a sensitivity of 1 µA /°K between -55°C and 150°C.

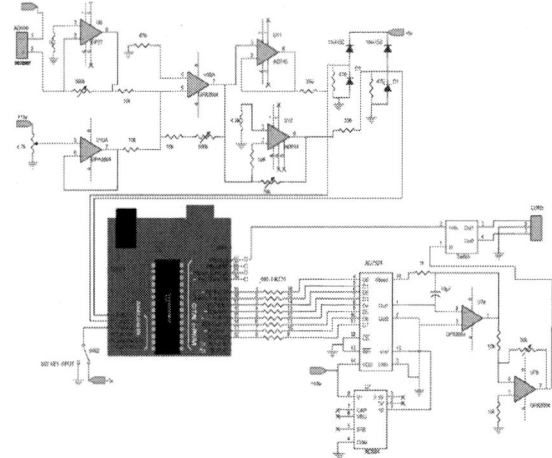

Figure 4.   Electronic schematic diagram of the control assembly

The Power electronic of cooling and heating use the Thermo Electric Cooler or "TEC" component "Pelletier effect – 1834", with reverse the direction of the electrical current. This effect uses a thermoelectric material which is a semiconductor junction "tellurium of bismuth". Pelletier plates require a large input power of the order of 80 Watts is for example 6 amps at 16 VDC. We can use an adjustable switching power supply for this purpose. We use the CP0.8-31-06 model for our system including the following features:

TABLE I.   CARACTERISTIQUES ELECTRIQUE DES PLAQUES DE PELTIER

Max Current (A)	Power (W)	Max Voltage (V)
2.1	4.4	3.75

Each face of the DUT carrier is bonded with four Pelletier plates for heating and 6 Pelletier for cooling plugged in serial. The power of 4 plates is estimated at 16 Watt, the total power is in the order of 100 Watts.

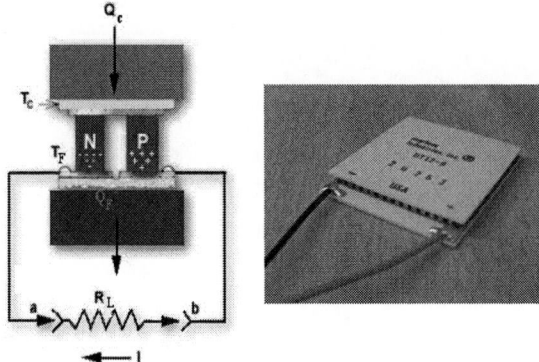

Figure 5. Electronic diagram and image of the used Peltier plate

The voltage in its pins is:

$$U = n.[\alpha.(T_C - T\ ) + R_p.I_p] \qquad (4)$$

$$R_p = \rho.\frac{e_s}{S_e} \qquad (5)$$

Where n, α, ρ, e_s and S_e are constants of the Pelletier plates, T_c it's a hot face temperature. The Thermoelectric junctions of Pelletier are managed by the following laws:

- The cooling capacity (PF) absorbed by the cold face of the Pelletier is expressed as:

$$P_F = n.[\alpha.I.T_F - \frac{1}{2}.R.I^2 - K.(T_C - T_F)] \qquad (6)$$

- The heat capacity (PC) generated by the hot face is:

$$P_C = n.[\alpha.I.T_C + \frac{1}{2}.R.I^2 - K.(T_C - T_F)] \qquad (7)$$

- The electric power (PE) provided to the cell is:

$$P_E = P_C - P_F = n.[\alpha.I.(T_C - T_F) + R.I^2] \qquad (8)$$

- The voltage (U) to the cell terminals:

$$U = n.[\alpha.(T_C - T_F) + R.I] \qquad (9)$$

n : Number of the cell blocks.

α : Seebeck coefficient by paved in volts /°K.

I : Current through the cell.

k : Thermal conductivity in watts/°K.

T_F : Temperature of the cold face in °K.

T_C : Temperature of the hot face in °K.

The output voltage of the generators without load is:

$$V = S.\Delta T \qquad (10)$$

(S =n.α) is the average Seebeck coefficient in volts/°K. If a load is connected to the thermoelectric, the output voltage decreases. The output current is:

$$I = \frac{S.\Delta T}{R_c + R_L} \qquad (11)$$

R_c : Mean internal resistance of thermocouples (Ohms).

R_L : Load resistance (Ohms).

The regulator of thermoelectric current is commanded by voltage to the gate of transistor FET through a closed loop formed by Op Amp. The reversal of the electric polarization of the Peltier modules is ensured by four FETs in H mode, controlled by pins P1, P2, P3 and P4. This polarization reversal, toggles to the cooling or heating by the Peltier plates with the same control system.

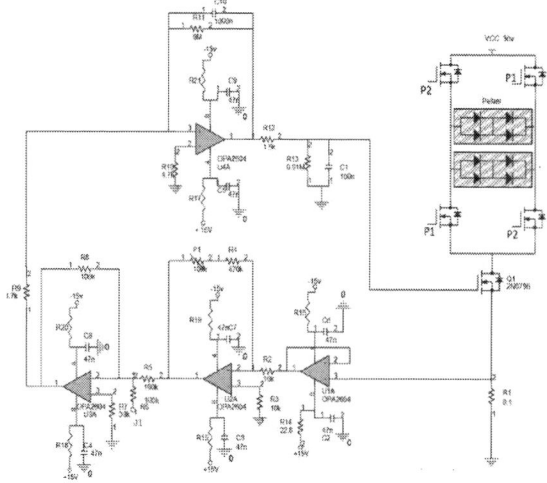

Figure 6. Schematic of heating and cooling power electronics.

IV. RESULT AND DISCUSSIONS

Processes modelling are made by "Strejc" and "Broïda" methods. These are used to find the transfer functions of two processes. The input of thermal system is a step signal with ±10%. The results are the functions of first order:

$$H(s) = K_o.\frac{e^{-T_r.s}}{1 + \tau.s} \qquad (12)$$

Where K_o, T_r and τ are static gain, delay and time-constant respectively.

The parameters of the transfer function of the tree regulators are presented on table II. The comparison shows the improvements made by the electronic system developed with Arduino on several levels. Indeed, an improvement of 10 times is observed on the static error and about 20 times faster compared to the "Ministat".

TABLE II.   PARAMETERS OF THE AIR TRANSFER FUNCTIONS.

Type of regulator	Process	Time constant (seconds)	Static error
Laboratory controller	Heating	2081	10%
	Cooling	2509	
8-bit ADC controller	Heating	161.2	5%
	Cooling	210.8	
Arduino controller	Heating	102	1%
	Cooling	115.4	

Also, it is about 1.5 times faster compared to the 8-ADC controller. The new measuring system brings other advantages over the old system (Ministat). The following table lists the major improvements.

TABLE III.   ADVANTAGE OF PROPOSED CONTROL / MINISTAT

Specification	Improvement (times)
Cost	5
Speed	10
Noise	300
Volume	50
Consumption	8
Weight	34
Stabilty	10

The developed device is tested on the temperature regulation of a photovoltaic cell of the type "H750". Thereafter a series of electrical measurement is performed on the characteristic of that cell, to extract the intrinsic electrical parameters, the series resistance (Rs), the shunt resistance (Rsh), the saturation current (Is) and the factor ideality (η) [5].

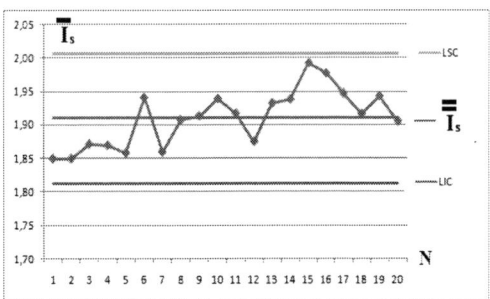

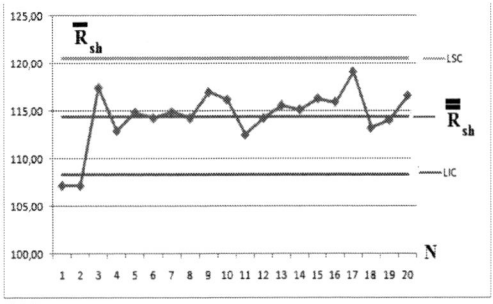

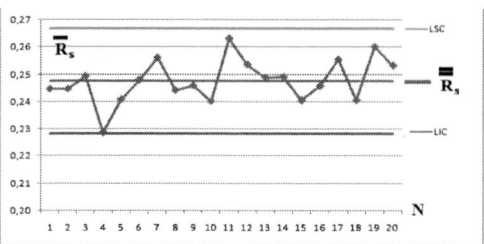

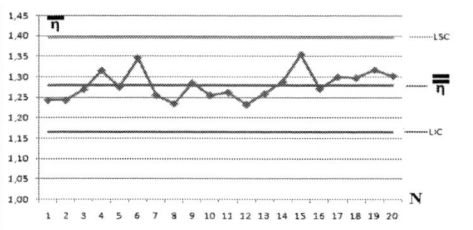

Figure 7.   Electrical measurements of the PV junction by (X,R) chart

We applied the method of control charts (X, R) on these measures to detect likely errors caussées by unstable temperatures [6]. The results of these cards are shown in the following figures, they show that the measurements of 4 intrinsic parameters without temperature control, is considerable fluctuations origin. These fluctuations cause errors on the values of these parameters. However, with thermal regulation, stability is observed on the values of these electrical parameters.

V.   CONCLUSION

The work presented in this paper concerns the description of a programmable electronic tool and low cost, able to control the temperature with an accuracy and fast speed compared to other devices. This system is developed using an 10-bit ADC embedded system to facilitate the integration of the PID algorithm with great flexibility. This system was tested on a photovoltaic cell for extract its intrinsic electrical parameters more accurately.

REFERENCES

[1] Abdessamad Malaoui, Abdelmajid Elmansouri. « Deux nouvelles méthodes complémentaires pour l'extraction optimale des paramètres électriques des jonctions ». Revue CDER, 2010, Vol. 13, N°2.

[2] Malaoui Abdessamad, Quotb Kamal, Ankrim Mohamed, Benhayoun Mohamed. " Accurate temperature control of a ultrasonic cell using a microcontrolor". IEEE International Symposium on Industrial Electronics, ISIE04, 2004, pp. 231-235.

[3] A. Malaoui, "Automatisation en température par microcontrôleur d'un banc de mesure ultrasonore: Applications au contrôle qualité en agroalimentaire.", thèse de doctorat de l'Université de Provence, 2005.

[4] VODA A. A.. Landu I.D. A. "Method of the Auto-calibration of PID Controllers". Automatica. 1995, Vol.31. N 1.pp. 41-53.

[5] Abdessamad Malaoui, "Implementation and tests of an automatic system to improve electrical energy in photovoltaic installations," International Journal of Innovation and Applied Studies, 2014, vol. 8, pp. 328–340.

[6] N. Hadik, A. Malaoui, et al. « A new technique of improvement a dielectric measurements use the statistical method of control: Application on Ba1-x Srx TiO3 ». Physical and Chemical News journal, 2010, Vol 56, pp 26-31.

[7] J.G. Ziegler, N.B. Nichols, "Optimum settings for automatic controllers". Trans. ASME, (1942), 64, pp. 759-768.

Simulation and Control of Takagi-Sugeno Uncertain Model of Buck Converter by Linear Programming

Rkia Oubah, Abdellah Benzaouia and Ahmed El Hajjaji

Abstract— **This paper deals with the problem of stabilization and reference input tracking by fuzzy PDC controller of discrete Takagi-Sugeno (T-S) uncertain model of Buck converter by linear programming (LP) while imposing positivity in closed-loop. The stabilization conditions are derived using the single Lyapunov-Krasovskii Functional (LKF). Simulation results by linear programming methods are given to illustrate the validity of the proposed method.**

Key-Words: T-S fuzzy discrete-time systems, positive systems, Buck converter, Uncertainty Lyapunov-Krasovskii functional, stabilization, Linear programming, PDC controller, reference tracking.

I. INTRODUCTION

The problem concerns a special class of nonlinear systems called Takagi-Sugeno models (T-S) [1], [21], [18]. From the history of this approach, this class can be interpreted as a collection of linear models interconnected by nonlinear functions, called membership functions [3], which are dependent variables. A frequent and inherent constraint in dynamical systems is the nonnegativity of the controls and/or the states. Systems with nonnegative states are important in practice because many physical and chemical processes involve quantities that have intrinsically constant and nonnegative sign: temperatures, level of liquids, concentration of substances, etc, are of course positive or nonnegative. In the literature, systems whose states are nonnegative whenever the initial conditions are nonnegative are referred to be positive [16], [15]. On the other hand, positive systems have been of great interest to researchers in recent years [7], [4], [10], [14], [18]. The class of positive T-S fuzzy systems was considered for the first time in [8]. The obtained results were presented using LMIs.

the chosen model included these characteristics (nonlinearity and positivity) to study is a part of the photovoltaic system is the Buck-converter. A Photovoltaic system is an arrangement of components designed to supply usable electric power for a variety of purposes, using the Sun as the power source, one of importants these components is photovoltaic array, this latter consists of multiple photovoltaic modules, casually referred to as solar panels, to convert solar radiation into usable direct current (DC) electricity. Thus, it uses a DC/DC converter which takes a higher input voltage (Boost converter) or converts it to a lower output voltage (Buck converter).

In this paper, we study the conditions of stabilization by PDC controller of Takagi-Sugeno (T-S) uncertain model of Buck converter, while imposing positivity in closed-loop, by using a different tool called linear programming (LP). Also, we control buck converter system for reach tracking objectif.

The rest of this paper is organized as follows: In Section 2, we give the description of T-S uncertain model of Buck converter, the fuzzy control law based on the PDC structure and conditions of stabilization and positivity in closed-loop applied to T-S Buck converter uncertain model. In Section 3, The results of simulation are presented. Some conclusions are given in Section 4.

II. T-S MODEL OF BUCK CONVERTER

A buck converter is a step-down DC to DC converter. The basic operation of buck converter has the current in an inductor controlled by two switches (usually a transistor and a diode), The basic circuit of the Buck converter is shown in Figure 1.

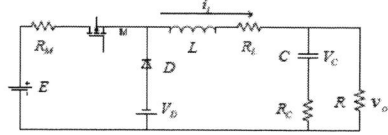

Fig. 1. Circuit of the buck converter

The global state equation of the T-S fuzzy model for the Buck converter is shown as follows:

$$\dot{x}(t) = \sum_{i=1}^{2} h_i(x)(Ac\, x(t) + Bc_i\, u(t)) \qquad (1)$$

where $x(t) = [i_L(t) \quad v_c(t)]^T$,
$h_1 = \frac{-i_L + M_2}{M_2 - M_1}$, $h_2 = \frac{i_L - M_1}{M_2 - M_1}$ is the membership function.

$$A = \begin{pmatrix} -(R_L + RR_c/(R+R_c))/L & -R/L(R+R_c) \\ R/(CR + CR_c) & -1/(CR+CR_c) \end{pmatrix},$$
$Bc_i = [-(R_M M_i - E - V_D)/L \quad 0]^T$

where the duty ratio of the switching tube M is obtained as follows:

$d(t) = u(t) - d_D$ where $d_D = \frac{V_D}{(R_M i_L - E - V_D)}$

Proof: see[5].

We apply the Euler discretization leading to: $A = I_2 + TAc$, $B_i = TBc_i$ where T is the sampling time.

consequently, the discrete-time system is presented as:

Rkia Oubah is with EMSI, Marrakesh, Morocco. rkia.oubah@gmail.com

Abdellah Benzaouia is with LAEPT-EACPI URAC 28, University Cadi Ayyad, Faculty of Science Semlalia, BP 2390, Marrakesh, Morocco. benzaouia@ucam.ac.ma

El Hajjaji is with University of Picardie Jules Vernes (UPJV), 7, Rue de Moulin Neuf 8000 Amiens, France. ahmed.hajjaji@u-picardie.fr

$$x(k+1) = \sum_{i=1}^{2} h_i(x)(A\ x(k) + B_i\ u(k)) \quad (2)$$

Consider the following uncertain discrete-time system:

$$x(k+1) = \sum_{i=1}^{2} h_i(x)(A\ x(k) + B_i\ u(k)) \quad (3)$$

where the matrices A, B_1, B_2 are uncertain with known bounds $\underline{A}, \overline{A}, \underline{B}_1, \overline{B}_1, \underline{B}_2, \overline{B}_2$, such that $\underline{A} \le A \le \overline{A}, \underline{B}_i \le B_i \le \overline{B}_i, i = 1, 2$.

Now we design the fuzzy PDC controller as follows:

$$u(k) = \sum_{j=1}^{2} h_j(x) K_j\ x(k)) \quad (4)$$

where K_j is the feedback gain matrix of the i th subsystem. Substitute (4) into (3), then we can obtain the state function of the entire uncertain system as follows:

$$x(k+1) = \sum_{i=1}^{2}\sum_{j=1}^{2} h_i h_j (A + B_i K_j)\ x(k) \quad (5)$$

where $\underline{A} \le A \le \overline{A}, \quad \underline{B}_i \le B_i \le \overline{B}_i$

Theorem 1: System (5) is asymptotically stable and controlled positive if there exist a vector $\lambda = [\lambda_1 \ldots \lambda_n]^T \in R^n$, and vectors $y_1^j, \ldots, y_n^j, z_1^j, \ldots, z_n^j \in R^m$ for $i, j \in \{1, 2, \ldots, r\}$; $l, s \in \{1, 2, \ldots, n\}$, satisfying the following LPs:

$$\begin{cases} \left[\sum_{i=1}^{r}\sum_{j=1}^{r} (\overline{A}_i \lambda + \overline{B}_i \sum_{s=1}^{n} (y_s^j - z_s^j)) \right] - \lambda \prec 0, \\ \underline{a}_{ls}^i \lambda_s + \underline{b}_l^i (y_s^j - z_s^j) \succeq 0, \\ \lambda \succ 0, \\ y_s^j \succ 0, \\ z_s^j \succ 0 \end{cases} \quad (6)$$

with

$$K_j = \left[\frac{(y_1^j - z_1^j)}{\lambda_1}, \frac{(y_2^j - z_2^j)}{\lambda_2} \ldots \frac{(y_n^j - z_n^j)}{\lambda_n} \right] ; j = 1, \ldots, r.$$

and

$$\underline{A}_i = (\underline{a}^i)_{ls}, l, s = 1, \ldots, n; \underline{B}_i = \begin{bmatrix} \underline{b}_1^i \\ \underline{b}_2^i \\ \ldots \\ \underline{b}_n^i \end{bmatrix}. \quad (7)$$

Proof 1: Let $\lambda \succ 0$, $\underline{A} \le A \le \overline{A}$, $\underline{B}_i \le B_i \le \overline{B}_i$, matrix K can be expressed as the difference of two positive matrices $K_j = K_j^+ - K_j^-$, where

$$K_j^+ = \left[\frac{y_1^j}{\lambda_1}, \frac{y_2^j}{\lambda_2} \ldots \frac{y_n^j}{\lambda_n} \right] ; j = 1, \ldots, r.$$

$$K_j^- = \left[\frac{z_1^j}{\lambda_1}, \frac{z_2^j}{\lambda_2} \ldots \frac{z_n^j}{\lambda_n} \right] ; j = 1, \ldots, r.$$

$$K_j = \left[\frac{(y_1^j - z_1^j)}{\lambda_1}, \frac{(y_2^j - z_2^j)}{\lambda_2} \ldots \frac{(y_n^j - z_n^j)}{\lambda_n} \right] ; j = 1, \ldots, r.$$

So $\underline{B}_i K_j^+ \le B_i K_j^+ \le \overline{B}_i K_j^+$

$$\underline{B}_i K_j^- \le B_i K_j^- \le \overline{B}_i K_j^-$$

$$\underline{B}_i (K_j^+ - K_j^-) \le B_i K_j \le \overline{B}_i (K_j^+ - K_j^-)$$

$$\begin{array}{ccc} (\underline{A} + \underline{B}_i (K_j^+ - K_j^-)) & \le & (A + B_i K_j) & \le \\ (\overline{A} + \overline{B}_i (K_j^+ - K_j^-)) \end{array}$$

we know that, $0 \prec h_i \prec 1 \quad \underline{x} \prec x(k) \prec \overline{x}$ consequently,

$$\sum_{i=1}^{2}\sum_{j=1}^{2} h_i h_j (\underline{A} + \underline{B}_i (K_j^+ - K_j^-)) \underline{x}(k) \le$$

$$\sum_{i=1}^{2}\sum_{j=1}^{2} h_i h_j (A + B_i K_j) x(k) \le$$

$$\sum_{i=1}^{2}\sum_{j=1}^{2} h_i h_j (\overline{A} + \overline{B}_i (K_j^+ - K_j^-)) \overline{x}(k)$$

whether,

$$\overline{x}(k+1) = \sum_{i=1}^{2}\sum_{j=1}^{2} h_i h_j (\overline{A} + \overline{B}_i (K_j^+ - K_j^-)) \overline{x}(k) \quad (8)$$

$$\underline{x}(k+1) = \sum_{i=1}^{2}\sum_{j=1}^{2} h_i h_j (\underline{A} + \underline{B}_i (K_j^+ - K_j^-)) \underline{x}(k) \quad (9)$$

To ensure the stability of system (5), the system (8) must be asymptotically stable.

Now, establish the conditions of stability of system (8) by linear programming approach.

The choice of the Lyapunov-Krasovskii functional is:

$$V(x(k)) = x^T(k)\lambda ; \ \lambda \succ 0 \quad (10)$$

One can deal with the stability of the dual system of (8) given by:

$$\overline{x}(k+1) = \sum_{i=1}^{2}\sum_{j=1}^{2} h_i(x) h_j(x) (\overline{A} + \overline{B}_i (K_j^+ - K_j^-))^T \overline{x}(k) \quad (11)$$

The expression of the rate of increase of the functional (10) becomes:

$$\Delta V(x(k)) = \sum_{i=1}^{2}\sum_{j=1}^{2} h_i(x) h_j(x) \left[x^T(k) (\overline{A} + \overline{B}_i (K_j^+ - K_j^-)) \right]$$

$$-x^T(k)\lambda$$

That is

$$\Delta V(x(k)) \preceq \sum_{i=1}^{2}\sum_{j=1}^{2} x^T(k)\left(\overline{A}+\overline{B}_i\left(K_j^+ - K_j^-\right)\right)\lambda$$
$$-x^T(k)\lambda$$

$$\preceq x^T(k)\left[\left[\sum_{i=1}^{2}\sum_{j=1}^{2}\left(\overline{A}+\overline{B}_i\left(K_j^+ - K_j^-\right)\right)\lambda\right]-\lambda\right]$$

Finally,

$$\left[\sum_{i=1}^{2}\sum_{j=1}^{2}\left(\overline{A}+\overline{B}_i\left(K_j^+ - K_j^-\right)\right)\lambda\right]-\lambda \prec 0 \qquad (12)$$

implies $\Delta V(x(k)) \prec 0$. Now, by letting $K_j^+ = [K_1^{j+}\ K_2^{j+}\ \dots\ K_n^{j+}]$ where K_s^{j+} are vectors in R^m, one has $K_j^+\lambda = \sum_{s=1}^{n}K_s^{j+}\lambda_s = \sum_{s=1}^{n}y_s^j$, with $K_s^{j+}\lambda_s = y_s^j$. The same applies to K_j^-, with $K_s^{j-}\lambda_s = z_s^j$. Consequently, inequality (12) can be written as

$$\left[\sum_{i=1}^{2}\sum_{j=1}^{2}\left[\overline{A}_i\lambda + \overline{B}_i\sum_{s=1}^{n}\left(y_s^j - z_s^j\right)\right]\right]-\lambda \prec 0,$$

To ensure that the system (5) is positive in closed-loop, system (9) must be positive, so matrices $\left(\underline{A}+\underline{B}_i\left(K_j^+ - K_j^-\right)\right)$ must be positive.

Thus

$$\left[\underline{A}_i + \underline{B}_i\left(K_j^+ - K_j^-\right)\right]_{ls} = \underline{a}_{ls}^i + \underline{b}_l^i\left(K_s^{j+} - K_s^{j-}\right),$$

where $\underline{a}_{ls}^i$ are the components of matrix $\underline{A}^i$; $l,\ s\ \in \{1,2,...,n\}$. Therfore, $\underline{a}_{ls}^i\lambda_s + \underline{b}_l^i\left(y_s^j - z_s^j\right) \succeq 0$ implies $a_{ls}^i + b_l^i\frac{\left(y_s^j - z_s^j\right)}{\lambda_s} \succeq 0$, λ being positive. Consequently $a_{ls}^i\lambda_s + b_l^i\left(y_s^j - z_s^j\right) \succeq 0$ implies $\underline{A}_i + \underline{B}_i\left(K_j^+ - K_j^-\right) \succ 0$.

$\square$

III. Simulation of Buck converter T-S uncertain model

In this section, the proposed T-S fuzzy control is applied to the basic buck converter. We noted that the output equation for buck converter is $y(k)=v_o(k)$, then :

$$y(k) = \left[\frac{RR_c}{(R+R_c)}\quad \frac{R}{(R+R_c)}\right]x(k) = G\ x(k)$$

where $G_1 = G_2 = \left[\frac{RR_c}{(R+R_c)}\quad \frac{R}{(R+R_c)}\right]$

One can notice that matrices A and G in this converter are common. The global T-S fuzzy uncertain model of Buck converter is given by:

$$\begin{cases} x(k+1) = \sum_{i=1}^{2} h_i(A_ix(k) + B_iu(k)) \\ y(k) = \sum_{i=1}^{2} h_i G_i x(k) \end{cases} \qquad (13)$$

where $\underline{A} \leq A \leq \overline{A}, \qquad \underline{B}_i \leq B_i \leq \overline{B}_i$.

The objective is to design controllers ensuring stabilization and positivity of systems (5) by using a linear programming technique, using the conditions of Theorem 1.

To achieve our objective of the tracking of a given reference of the voltage capacity, we introduce the following idea. Let us consider the new control given by using the following PDC control

$$u(k) = \sum_{i=1}^{r} h_i(x(k))(K_i\ x(k) + L_i\ x_{ref}) \qquad (14)$$

where x_{ref} is the state to be followed by the closedloop system and the gains L_j to be designed to achieve the tracking problem. Hence, the closed-loop system becomes:

$$\begin{aligned} x(k+1) &= \sum_{i=1}^{r}\sum_{j=1}^{r} h_i(x(k))h_j(x(k))((A+B_iK_j)x(t) \\ &+ B_iL_jx_{ref}) \\ &= G(h)x(t) + B(h)L(h)x_{ref} \end{aligned} \qquad (15)$$

where $G(h) = A + B(h)K(h)$, $B(h) = \sum_{j=1}^{r} h_i(x(k))B_i$. Similar notation is used for $K(h)$ and $L(h)$.

Since the objective is that $x(\infty) = x_{ref}$, one can choose $L(h) = B(h)^+(I - G(h))$, with $B(h)^+$ representing the pseudo inverse of matrix $B(h)$.

The simulation parameters are shown in Table 1.

TABLE I

DATA OF THE BUCK CONVERTER

parameter	value	parameter	value
R	6Ω	R_c	$0.162\ \Omega$
E	30V	L	98.58mH
R_L	$48.5m\Omega$	R_M	0.27Ω
V_D	0.82Ω	C	$202.5\mu F$
M_1	$0A$	M_2	$3A$

and

$$\overline{A}_1 = \overline{A}_2 = \begin{pmatrix} a_{11}+0.01 & a_{12} \\ a_{21}+0.01 & a_{22}+0.015 \end{pmatrix},$$

$$\underline{A}_1 = \underline{A}_2 = \begin{pmatrix} a_{11}-0.01 & a_{12} \\ a_{21}-0.01 & a_{22}-0.015 \end{pmatrix}$$

$$\overline{B}_1 = \left(b_1^1+0.01\quad b_2^1+0.02\right)^T, \quad \overline{B}_2 = \left(b_1^2+0.01\quad b_2^2+0.03\right)^T$$

$$\underline{B}_1 = \left(b_1^1-0.01\quad b_2^1-0.02\right)^T, \quad \underline{B}_2 = \left(b_1^2-0.01\quad b_2^2-0.03\right)^T$$

The obtained solutions of the LP method are as follows:

$$\lambda = \begin{pmatrix} 1.9099 & 293.2065 \end{pmatrix}^T, \quad K_1 = K_2 = \begin{pmatrix} -3.0408 & 0.0353 \end{pmatrix}$$

Matrices in closed-loop are obtained as:

$$\hat{A}_{11} = \begin{pmatrix} 0.0472 & 0.0012 \\ 4.8084 & 0.1986 \end{pmatrix}, \quad \hat{A}_{12} = \begin{pmatrix} 0.0472 & 0.0012 \\ 4.8084 & 0.1986 \end{pmatrix},$$

$$\hat{A}_{21} = \begin{pmatrix} 0.0722 & 0.0009 \\ 4.8084 & 0.1986 \end{pmatrix}, \quad \hat{A}_{22} = \begin{pmatrix} 0.0722 & 0.0009 \\ 4.8084 & 0.1986 \end{pmatrix}.$$

The results of the simulation, with the following data: initial point $x_0 = [0,0]^T$, the trajectory reference

$$\begin{cases} y_r = 5 & 0 \leq k \leq 0.05 \\ y_r = 10 & 0.05 < k \leq 0.1 \\ y_r = 3 & 0.1 \leq k \leq 0.2 \end{cases},$$

and $T=1.e-3$, are obtained as in Figs. 4-5.

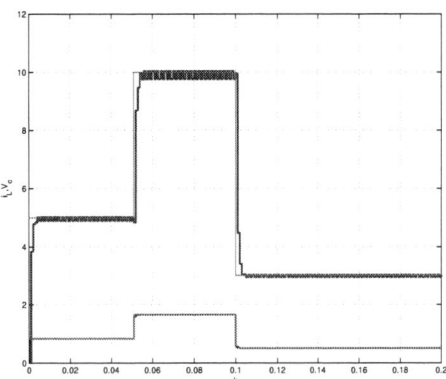

Fig. 2. capacity voltage and inductance current of the DC-DC Buck Converter

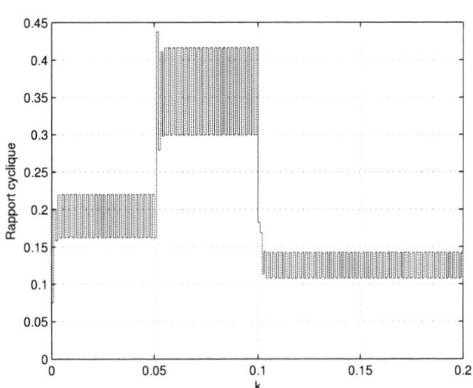

Fig. 3. Duty ratio of the Buck DC-DC Converter.

IV. CONCLUSION

The conditions for stabilization by state feedback of T-S fuzzy systems with constrained state while imposing positivity are presented. The constraints are taken into account during the design phase using an linear programming approach. The membership functions are used to design the global state that respects the state constraints and to achieve the tracking of the reference signal without violating the positivity of the system. The obtained results are applied to the buck DC-DC converter. Simulation results show the effectiveness of the approach. Particularly, the duty ratio respects the physical limitation between 0 and 1 which was shown by Figure 5.

REFERENCES

[1] T. Takagi and M. Sugeno, Fuzzy identification of systems and its application to modeling and control, IEEE Trans on System Man and Cybernetics, Vol.15(1), pp116-132, 1985.

[2] K. Gasso, Identification des systèmes dynamiques non linéaires: approche multimodèles, Thèse de l'Institut National Polytechnique de Lorraine, 2000.

[3] L. X. Wang and J.M. Mendel, Fuzzy basis functions, universal approximation and orthogonal least-squares, IEEE Trans. on Neural Networks, Vol.3(5), pp807-814, 1992.

[4] L. Farina and S. Rinaldi, Positive Linear Systems: Theory and Application, Wiley, New York, 2000.

[5] Yuanlong Li and Zhicheng Ji, T-S Modeling, Simulation and Control of the Buck Converter, Fifth International Conference on Fuzzy Systems and Knowledge Discovery.

[6] A. Benzaouia, R. Oubah, A. El Hajjaji, Stabilization of positive Takagi-Sugeno fuzzy discrete-time systems with multiple delays and bounded controls, Journal of the Franklin Institute, 2013.

[7] M. Ait Rami, F. Tadeo. Lineair Programming approach to impose positiveness in closed-loop and estimated states, In Proc. of the 17th Intern. Symp. on Mathematical Theory of Networks and Systems, Kyoto, Japan, 2006.

[8] A. Benzaouia , A. Hmamed and A. EL Hajjaji, Stabilization of controlled positive discrete-time T-S fuzzy systems by state feedback control. Int. J. Adaptive Control and Signal Processing. Vol. 24, Issue 12, pp.1091-1106, 2010.

[9] A. Benzaouia and A. EL Hajjaji, Delay-dependent stabilization conditions of controlled positive T-S fuzzy systems with time varying delay. IJCIC. Vol. 7, No. 4, 2011.

[10] A. Benzaouia, Saturated Switching Systems, Springer, 2012.

[11] A. Benzaouia, R. Oubah, A. El Hajjaji and F. Tadeo, Stability and stabilization of positive Takagi-Sugeno fuzzy continuous systems with delay, 50th IEEE Conference on Decision and Control and European Control Conference (CDC-ECC) Orlando, FL, USA, December 12-15, 2011.

[12] A. Benzaouia and R. Oubah, Stability and Stabilization by output feedback control of positive Takagi-Sugeno fuzzy discrete-time systems with multiple delays. Nonlinear Analysis: Hybrid Systems, 2013.

[13] L. El Ghaoui, F. Oustry and M. A. Rami, A cone complementarity linearisation algorithm for static output-feedback and related problems, IEEE Transactions on Automatic Control, 1997.

[14] A. Hmamed, A. Benzaouia, M. Ait Rami and F. Tadeo, Memoryless control to Drive States of Delayed Continuous-time Systems within the Nonnegative Orthant, Proceedings of the 18th World Congress, IFAC, Seoul, Korea, July 6-11, 2008.

[15] T. Kaczorek, Positive 1D and 2D Systems, Springer-Verlag, 2002.

[16] T. Kaczorek, Stability of positive continuous-time linear systems with delays. Bulletin of the polish academy of sciences, Technical sciences, Vol. 57, No. 4, 2009.

[17] R. Oubah and A. Benzaouia. Stability and Stabilization by output feedback control of positive Takagi-Sugeno fuzzy discrete-time systems with delay. International Conference on Systems and Control, Marrakech, Morocco, June 20-22, 2012.

[18] M. Sugeno and G. Kang. Structure identification of fuzzy model, Fuzzy Sets and Systems, vol. 28, no. 1, pp. 15-33, 1988.

[19] M. Bolajraf. Robust control and estimation for positive systems. Thesis.

[20] M. Sugeno. Fuzzy Control. North-Holland, 1988.

[21] A. Benzaouia and A. El Hajjaji. Advanced Takagi-Sugeno Fuzzy Systems. Studies in Systems, Decision and Control, Vol. 8, Springer, 2014.

Power optimization of Decode-and-Forward assisted ARQ relaying

Ali Kamouch*, Abdelaali Chaoub[†], and Zouhair Guennoun*
*Laboratory of Electronic and Communication
Mohammadia School of Engineers, Mohammed V-Agdal University, Rabat, Morocco
Email: kamouch.ali@gmail.com, zouhair@emi.ac.ma
[†]Department of Telecommunication, National Institute of Posts and Telecommunications, Rabat, Morocco
Email: chaoub.abdelaali@gmail.com

Abstract—**In this paper we investigate the performances of Decode-and-forward (DF) assisted by automatic repeat request (ARQ) relaying over Rayleigh fast fading channel. We derive a tight asymptotic approximation of the outage probability at high signal-to-noise ratio(SNR) regime. Furthermore we formulate an optimization problem aiming at achieving a target outage probability with an optimum allocation of the power budget depending on the relative position of the relay with regard to source and destination.**

Index Terms—**Decode and forward, Automatic retransmission request, maximum ratio combining, Outage probability, Power allocation, Space-Time Block coding.**

I. INTRODUCTION

Nowadays, relaying concept has ignited an intensive research as an innovative technique to achieve reliable connectivity and guarantee favorable conditions for Qos-sensitive secondary communications [1]. In particular, the source and destination nodes may not always be able to proceed with a direct transmission especially in harsh environments. Thus, the link between source and destination can be bridged by a third party, called relay, with rich spectrum opportunities when there is no available spectrum gaps to establish the direct source-destination link. During the past few years, because of their ability to deliver streams over longer distances with extended network range and faster speeds, the applications of relaying have become quite popular, such as Wireless Sensor Networks (WSN) [2], Vehicular Ad Hoc Networks (VANET) [3] and Wireless Mesh Networks (WMN) [4].

Two axes are considered when designing a relaying network, relaying protocol and resource allocation. In [5] authors analysed the performance of a relaying system consisting of one source, one relay and one destination. the authors assumes the availability of the channel state information CSI at the transmitter and proved that a non-zero delay-limited capacity can be achieved under any given average power constraint. In [6], an opportunistic relaying strategy is proposed where the relay is used only in case of a deep fading in the direct link. The authors proved that with partial CSI available at the transmitter and perfect CSI at the receiver a non-zero delay-limited capacity is achievable given any average power constraint. the authors also showed that opportunistic relaying strategy outperforms the classical multi-hopping and decode

and forward protocols in terms of both achievable delay-limited capacity and maximum outage probability. In [7] The authors derived the closed form expressions for the average power needed for multi-hopping and opportunistic decode and forward strategies to realize a given rate with one-bit feedback CSI at the transmitter. The authors showed that the unavailability of perfect CSI at the transmitter ends in a small degradation in terms of minimum required average power. A variation scheme on the opportunistic cooperation was proposed and analysed in terms of bit error rate in [8]. in [9] authors investigated the improvement provided by opportunistic cooperation in the context of a cognitive network. A secondary user can cooperate with the primary user to increase the latter's transmission success probability. In return, the secondary user can get more opportunities for transmitting its own data when the primary user is idle. In [10], a cooperative system is presented where a threshold-Based criterion is used to switch between the cooperation and non-cooperation modes. The authors showed that with an optimal threshold such a system can perform as better as a system with full channel state information.

In this paper we consider a delay-limited three nodes system with no CSI at transmitter and perfect channel knowledge at the receivers. Actually channel estimation and prediction at receiver is not a demanding task. we provide a semi-analytical method to manage power resources at source and relay depending on the relay position and target outage probability.

The outline of the paper is as follows: in section II system model and relaying protocol are depicted. In section III, we derive a tight approximation of the outage probability and formulate the optimization problem. Numerical results are

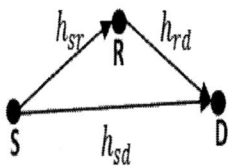

Fig. 1. System model

978-1-4673-8760-6/15 $31.00 © 2015 IEEE

presented in section IV. Conclusions are drawn in section V.

II. SYSTEM MODEL

We consider a three nodes system as depicted in Fig. 1, the system consists of one source, one relay and one destination. We assume a block Rayleigh fading channels constant during one ARQ round but can change independently for the other ARQ rounds. Let $h_{sd}(j)$, $h_{sr}(j)$ and $h_{rd}(j)$ be the channel gains of the source-destination, source-relay and relay-destination links during the jth ARQ round. We assume a distance-dependent path-loss model with power exponent $\nu = 4$, as in a urban environment.The distance source-destination is normalized to 1. The relay is located in the line connecting source to destination in a distance d from the source. The real and imaginary parts of $h_{sd}(j)$, $h_{sr}(j)$ and $h_{rd}(j)$ are assumed to be zero mean Gaussian random variables with variances 1, $d^{-\nu}$ and $(1-d)^{-\nu}$ respectively. In addition transmissions are subject to an AWGN noise with variance N_0 normalized to $1w/Hz$. we consider a delay-limited application with maximum number of retransmission $L-1$. The source first start to transmit the packet in a broadcast manner. The source transmission of the current packet is interrupted at the reception of a positive acknowledgement from either relay or destination or the maximum number of retransmissions is reached. If the relay has correctly decoded the packet before reaching the maximum number of retransmission, it will forward it to the destination until reception of a positive acknowledgement from destination or maximum number of retransmissions is reached.

III. PERFORMANCE ANALYSIS AND OPTIMIZATION PROBLEM FORMULATION

A. Asymptotic outage probability

In this section we derive the outage probability of DF assisted ARQ relaying. We denote by $\{T = k\}$ the event of successful decoding by the relay in exactly k ARQ rounds. The set $S = \{\{T = 1\}, \{T = 2\}, ..., \{T = +infinity\}\}$ forms a partition of a probability space. Hence the outage probability can be expressed as:

$$P_{out} = \sum_{k=1}^{L-1} Pr(T = k)Pr(out|T = k)$$
$$+ Pr(T \geq L)Pr(out|T \geq L)$$
$$= \sum_{k=1}^{L-1} Pr(T = k)Pr(out|T = k)$$
$$+ \left(1 - \sum_{k=1}^{L-1} Pr(T = k)\right)Pr(out|T \geq L) \quad (1)$$

$$Pr(T = k) = Pr\left\{log_2\left(1 + P_s \sum_{j=1}^{k-1} \|h_{sr}(j)\|^2\right) < R,\right.$$
$$\left. log_2\left(1 + P_s \sum_{j=1}^{k} \|h_{sr}(j)\|^2\right) \geq R\right\}$$
$$= Pr\left(\sum_{j=1}^{k-1} \|h_{sr}(j)\|^2 < \frac{2^R - 1}{P_s}\right)$$
$$- Pr\left(\sum_{j=1}^{k} \|h_{sr}(j)\|^2 \geq \frac{2^R - 1}{P_s}\right) \quad (2)$$

$\sum_{j=1}^{k} \|h_{sr}(j)\|^2$ is a non-central chi-square distributed random variable. Hence:

$$Pr(T = k) = P\left(k - 1, \frac{2^R - 1}{P_s d^{-\nu}}\right) - P\left(k, \frac{2^R - 1}{P_s d^{-\nu}}\right) \quad (3)$$

Where $P(k, x)$ is the regularized incomplete gamma function. From (3), we can compute $Pr(T \geq L)$ as follows:

$$Pr(T \geq L) = 1 - \sum_{k=1}^{L-1} Pr(T = k)$$
$$= P\left(L - 1, \frac{2^R - 1}{P_s d^{-\nu}}\right) \quad (4)$$

$$Pr(out|T = k) = Pr\left\{log_2\left(1 + P_s \sum_{j=1}^{k} \|h_{sd}(j)\|^2 +\right.\right.$$
$$\left.\left. P_r \sum_{j=k+1}^{L} \|h_{rd}(j)\|^2\right) < R\right\}$$
$$= Pr\left\{P_s \sum_{j=1}^{k} \|h_{sd}(j)\|^2 +\right.$$
$$\left. P_r \sum_{j=k+1}^{L} \|h_{rd}(j)\|^2 < 2^R - 1\right\} \quad (5)$$

Let $X = P_s \sum_{j=1}^{k} \|h_{sd}(j)\|^2$ and $Y = P_r \sum_{j=k+1}^{L+1} \|h_{rd}(j)\|^2$. As stated before X and Y are non-central chi-squared distributed random variables which probability density functions (pdf) are respectively :

$$P_X(x) = \frac{x^{k-1} \exp\left(\frac{-x}{P_s}\right)}{(k-1)! P_s^k} \quad (6)$$

$$P_Y(y) = \frac{y^{L-k-1} \exp\left(\frac{-y}{P_r(1-d)^{-\nu}}\right)}{(L-k-1)!(P_r(1-d)^{-\nu})^{L-k}} \quad (7)$$

At high SNR regime the pdf of X and Y can be simplified to:

$$P_X(x) = \frac{x^{k-1}}{(k-1)!P_s^k} \tag{8}$$

$$P_Y(y) = \frac{y^{L-k-1}}{(L-k-1)!(P_r(1-d)^{-\nu})^{L-k}} \tag{9}$$

their respective Laplace transforms can be expressed then as :

$$\mathcal{L}_X(s) = \frac{1}{P_s^k} \times \frac{1}{s^k} \tag{10}$$

$$\mathcal{L}_Y(s) = \frac{1}{(P_r(1-d)^{-\nu})^{L-k}} \times \frac{1}{s^{L-k}} \tag{11}$$

Using the fact that Laplace transform of the sum of two random variables is the product of their individual Laplace transforms, we can write Laplace transform of $Z = X + Y$ as: :

$$\mathcal{L}_Z(s) = \frac{1}{P_s^k(P_r(1-d)^{-\nu})^{L-k}} \times \frac{1}{s^L} \tag{12}$$

Inverting the Laplace transform we obtain the pdf of rv Z as:

$$P_Z(z) = \frac{1}{P_s^k(P_r(1-d)^{-\nu})^{L-k}} \times \frac{z^{L-1}}{s^{(L-1)!}} \tag{13}$$

Integrating the pdf of Z we obtain the conditional outage probability $Pr(out|T = k)$ as:

$$Pr(out|T = k) = \frac{(2^R - 1)^L}{P_s^k(P_r(1-d)^{-\nu})^{L-k}L!} \tag{14}$$

For $T \geq L$, the relay will not help in forwarding message toward the destination. The outage probability in this case is :

$$Pr(out|T \geq L) = P\left(L, \frac{2^R - 1}{P_s}\right) \tag{15}$$

From (3), (4), (14) and (15) the asymptotic outage probability is obtained.

B. power allocation problem formulation

In this part we formulate the optimum power allocation problem. The optimization aims at reducing the total average power to the minimum needed to achieve a certain target outage probability P_{out}^{Target} and allowing an efficient management of the power resource between source and relay. Using the same probability space defined in section III-A, we can express the average total power needed to transmit one packet as follows :

$$P_{av} = \sum_{k=1}^{L-1} (kP_s + (L-k)P_r) Pr(T = k) + LP_s Pr(T \geq L) \tag{16}$$

The optimization problem can be stated as :

$$\begin{array}{cc} \underset{P_s, P_r \geq 0}{\text{minimize}} & P_{av} \\ \text{subject to} & P_{out} \leq P_{out}^{Target} \end{array} \tag{17}$$

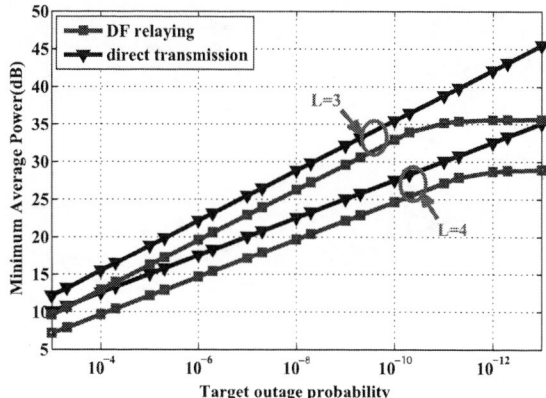

Fig. 2. Optimal total average power vs target outage probability , for $R = 1bit/channel\ use$ and $d = 0.2$

Using a numerical optimization toolbox and expressions derived in section III-A, we solve the problem (17) depending on the position of the relay and targeted outage probability.

IV. NUMERICAL RESULTS ANALYSIS

In Fig. 1, The optimal average power to achieve a target outage probability is plotted for decode-and-forward and direct transmission. From the figure we can notice the energy save provided by relaying especially for applications requiring very small error rate. In Fig. 2, the optimal source power allocation is computed depending on the relay position and target outage probability. It can be seen that equal power allocation is not efficient especially when the relay is near to the destination. This figure also shows that relay need to be allocated more power in other to enhance the system reliability. In Fig. 3, the optimal average power is computed for different target outage probability. the results shows that the optimal relay position to achieve the target outage probability with minimum total power is obtained when $0.6 \leq d \leq 0.8$.

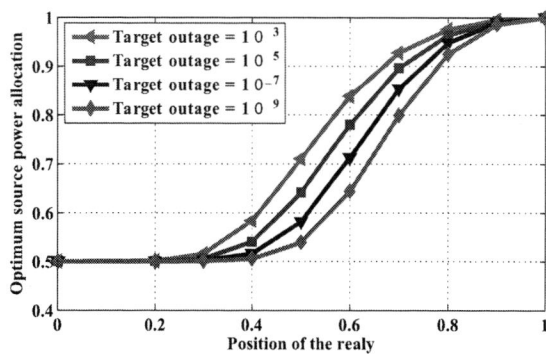

Fig. 3. Optimal power allocation vs relay position, for different P_{out}^{target} and $L = 4$

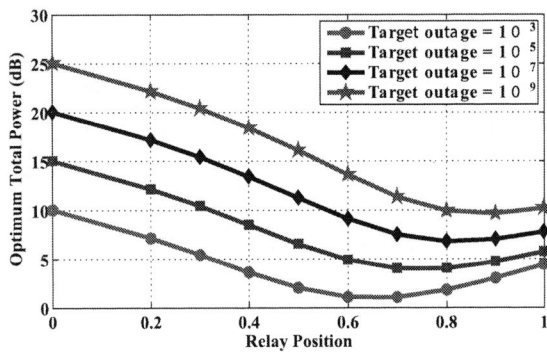

Fig. 4. Optimal total average power vs relay position, for different P_{out}^{target} and $L = 4$

V. CONCLUSION

In this paper we have derived the outage probability of DF relaying assisted ARQ which is a key performance to characterize a communication system. The derived analytic expression provide an effective tool to optimize the network resources and to improve the relaying protocol design. An open issue is to check the impact of the presence of multiple relays in power and bandwidth allocation.

REFERENCES

[1] Sendonaris, A., Erkip, E., Aazhang, B., User cooperation diversity. Part I. System description, *IEEE Transactions on Communications*, 51(11), 1927-1938, 2003.

[2] I.F. Akyildiz, W. Su, Y. Sankarasubramaniam, and E. Cayirci, A Survey on Sensor Networks, IEEE Communications Magazine, vol. 40, no. 8, pp. 102114, Aug. 2002.

[3] H. Hartenstein and K.P. Laberteaux, A tutorial survey on vehicular ad hoc networks, IEEE Communications Magazine, vol. 46, no.6, pp. 164 171, Jun. 2008.

[4] I.F. Akyildiz and X. Wang, A Survey on Wireless Mesh Networks , IEEE Communications Magazine, vol. 43, no.9, pp. S23S30, Sep. 2005.

[5] D. Gunduz, and E. Erkip, "*Opportunistic Cooperation and Power Control Strategies for Delay-Limited Capacity*", Conference on Information Sciences and Systems, The Johns Hopkins University, March, 2005.

[6] D. Gunduz, and E. Erkip, "*Opportunistic Cooperation by Dynamic Resource Allocation*", IEEE transactions on wireless communications, Vol. 6, No. 4, April 2007.

[7] I. A. El-Bakoury, K. G. Seddik and A. Elezabi, "*Opportunistic Relaying with Partial CSI and Dynamic Resource Allocation*", Global Conference on Signal and Information Processing (GlobalSIP), page 899 - 902, December 2013.

[8] Y. Zou, B. Zheng and W. Ping, "*An Opportunistic Cooperation Scheme and its BER Analysis*", IEEE transactions on wireless communications, Pages 4492-4497, Vol. 8, No. 9, September 2009

[9] R. Urgaonkar and M.J. Neely, "*Opportunistic Cooperation in Cognitive Femtocell Networks*", IEEE Journal On Selected Areas In Communications, Pages 607-616, Vol. 30, No. 3, April 2012.

[10] D.K. Jeong and D. Kim, "*Outage-Optimal Threshold-Based Opportunistic Cooperation in AF Relaying Systems with Staying Link Information*", IEEE Asia Pacific Conference on Wireless and Mobile, Page 264-268 August 2014.

Bearing Fault Diagnosis Based on Alpha-Stable distribution feature extraction and wSVM Classifier

B. Chouri[1], M. EL Aroussi[2], M. Tabaa[1], A.Jarrou[1], M. Fabrice[3], A. Dandache[3]

[1]Moroccan School of Engineering Sciences (EMSI), Research and Innovation department, Casablanca, Morocco
[2]GE EHTP, Km 7, Casablanca, Morocco
[3]LGIPM, University of Metz, France

brahimchouri@gmail.com

Abstract

Bearing fault diagnosis has attracted significant attention over the past few decades. In this paper, Alpha-stable distribution was introduced for feature extraction from faulty bearing vibration signals.Such a non-Gaussian model can accurately describe statistical characteristic of bearing fault signals with impulsive behavior. After extracting feature vectors by Alpha-stable distribution parameters, the weighted support vector machine (wSVM) was applied to automate the fault diagnosis procedure. Simulation results demonstrated that the proposed method is a very powerful algorithm for bearing fault diagnosis and has much better performance.

Index Terms–fault diagnosis; machine vibration, weighted support vector machine, alpha Alpha-stable, Bearing Prognostic.

1 Introduction

Bearings are the most commonly used part in a rotary machine [?]. Bearing failures could guide to random productivity losses for manufacture amenities. Consequently, bearing fault diagnosis has attracted important attention from the research and engineering communities over the past decades.

The choice of bearings can be explained by the fact that these components are considered as the most common mechanical elements in industry and are present in almost all industrial processes, especially in those using rotating elements and machines. Moreover bearing failure is one of the foremost causes of breakdowns in rotating machinery and such failure can be catastrophic [?], resulting in costly downtime. Many previously studies[?] have developed theoretical foundation and tools to describe bearing failure modes.

Time-frequency analysis methods, such as the short-time Fourier transform [?], the Wigner Ville distribution [?], and the wavelet transform [?], have been extensively employed to sense bearing faults since they can give plentiful information concerning machine faults. Though, these time-frequency based methods often require many computation time, as they engage a lot of Fourier transforms or convolution operations. Furthermore, due to the factors of clearance and nonlinear inflexibility of bearings, the vibration signals are frequently characterized by nonlinearity. Therefore, these normally used time-frequency analysis techniques may reveal limitations for the reason that their linearity assumption.

In order to defeat this difficulty, numerous nonlinear parameter estimation techniques were applied to take out defect-related features concealed in the measured signals. In [?], Hong and Liang jointed the Lempel-Ziv complexity with the continuous wavelet transform and established that the novel method was more efficient in bearing fault diagnosis.

When a rolling element bearing runs in fault condition, the observed vibration signal from the bearing is a non- Gaussian signal [?]. To the non-Gaussian signal, some statistical analysis approaches were proposed to detect bearing fault, such as rms, crest factor, skew and kurtosis [?]. But these statistical parameters describe only some local statistical characteristic of non-Gaussian and extract the part information about bearing fault. This is incomplete to research of the bearing fault signals. In statistics, the probability density function can completely describe statistical characteristic for statistic data. So the non- Gaussian signal need an accurate description based on probability density function. Considering impulse-like nature of the bearing fault signals, an advanced statistic model based on alpha-stable distribution [?] is a good choice.

In this paper, we introduce stable distribution based wSVM for the bearing fault diagnosis system. After extracting feature vectors by stable distribution, the weighted support vector machine (wSVM) [?] is used as a classifier. The use of this tools is motivated in the one hand by the fact that the $\alpha-$ stable described effectively the vibration data, and because is used for feature extraction in the literature [?]. In the second hand, wSVM is chosen as classifier due to its promising empirical performance, moderate computation complexity and its strong mathematical foundation. Experimental results demonstrate that the proposed stable distribution and wSVM scheme provides a significantly higher accuracy of prediction than the traditional feature extraction methods.the use of wSVM can be justified by the facts that SVM considers all training data points equally in order to establish the decision surface. But wSVM, known as fuzzy SVM (FSVM) or weighted SVM (wSVM) weight all training data points in order to allow different input points to contribute differently to the learning decision surface. The objective of present study is to examine the potential of weighted support vector machines (wSVMs) a classifier and alpha-stable as feature extraction

The remainder of the paper is organized as follows. In Section **??**, the proposed approach is presented, are explained. In Section **??** we give more information about the used data in this work. Also in this section an experimental results are presented and discussed. Finally, in Section **??**, conclusions and future

978-1-4673-8760-6/15 $31.00 © 2015 IEEE

recommendations are given.

2 Approch for Bearing fault detection based on stable distribution and SVM

The method is based on a nondestructive control, and uses the data provided by the sensors installed to monitor the components condition. The acquired signals are first processed to extract features in the form of stable distribution coefficients (μ, c, α and β), which are then used to learn the SVM classifier) for fault identification. In addition, multiple observations, instead of the traditional mono observation approach, are considered for both learning and exploitation phases. The principle of the proposed method relies on two main phases, as shown in Figure ?? : a learning phase, and test phase. In the first phase, conducted offline, the raws data recorded by the sensors are processed to extract the alpha stable distribution. These features are then used to learn several wSVM classifier (one versus all) corresponding to bearing normal state and different bearing fault states.

Indeed, each raw data history corresponding to a given machine states (normal or (inner race, outer race or ball) fault), is transformed to a feature matrix F, by using stable distribution coefficients. In the matrix F, each line vector (of c features at time t) corresponds to a snapshot on the raw signal. The advantage of using several features instead of only one is that a single feature may not capture all the information related to the behavior of the component. The parameters μ, c, α and β of each state are learned by using wSVM classifier. The estimation of machine state is done according to the following steps.

- The first step consists in detecting the appropriate alpha stable coefficient that best fits and represents the on-line observed sequence of vibration data.

- The second step of this procedure concerns the identification of the current machine state's.

2.1 Features Calculation

2.2 Classification by wSVM

The weighted support vector machine (WSVM) is a SVM adaptation to the cost sensitive learning framework. This supervised learning technique has been successfully applicable for imbalanced classification. The term weighted support vector machines (wSVMs) was proposed by Fan and Ramamo hanarao [?] as a synonym for Fuzzy Support Vector Machines (FSVMs) to draw attention to the effective weighting of fuzzy memberships at each FSVM training point. Fan and Ramamo hanarao [?] stated that different input vectors make different contributions to the learning of decision surface. Thus, the important issue in training weighted SVMs is how to develop a reliable weighting model to reflect the true noise distribution in the training data. Fan and Ramamo hanarao [?] developed emerging patterns (EPs) to weight the training data. Lin and Wang [?] developed FSVMs to enhance support vector machine (SVM) abilities to reduce the effects of outliers and noise in data points. While SVMs a recent AI paradigm developed by Vapnik [?] that has been used in a wide range of applications, treat all training points of a given class uniformly, training points in many real world applications bear different importance weightings for classification on purposes. To solve this problem, Lin and Wang [?] applied a fuzzy member to each SVM input point, thus allowing different input points to contribute differently to the learning decision surface. In such time series prediction problems, older training points are associated with lower weights, so that the effect of older training points is reduced when the regression function is optimized.

3 Experiment fault detection Database and results

3.1 Database presentation

In order to validate the capability of the proposed approach, experimental analyses on bearing faults were conducted. All the bearing data we used were obtained from the CWRU Bearing Data Center [?]. The time-domain vibration signals of bearing were collected from the normal case, the ball fault case, the inner race fault case, and the case of the outer race fault at the 6 o'clock position. The sampling frequency is 48 kHz [?]. For all fault conditions, the defect size of point fault is 14 mil in diameter. As shown in Figure ?? , the test stand consists of a 2 hp motor (left), a torque transducer/encoder (center), a dynamometer (right), and control electronics (not shown). In these experiments, the vibration signals collected from different fault conditions are divided into several non-overlapping 2048-point width windows.

Figure 2: Experimental platform : global overview.

To demonstrate the effect of the number of training samples, the experiments were designed by different training set sizes (10%, 20%, 30%, 40%, and 50% of total samples), and the remaining samples are used for prediction. The average accuracy of prediction for each experiment was quantified over 200 tests.

3.2 Results

The average ac-curacies of prediction for different feature extraction methods are presented in Tables 2 to 4. In the experiments, we compare PCA, LDA, α stable feature extraction with SVM classification α-S-SVM and α stable feature extraction with weighted SVM classification $\alpha-$S-wSVM. The parameters C and γ, of the wSVM were 43 and the reciprocal of the feature number,respectively. As presented in Tables 24, in the cases where

978-1-4673-8760-6/15 $31.00 © 2015 IEEE

Figure 1: The principle of the proposed method: a learning phase, and test phase.

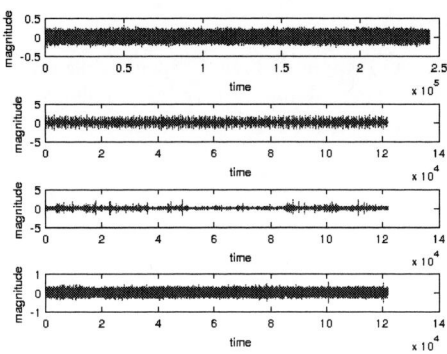

Figure 3: Samples of the used vibration raw signals from an experiment, Normal, Inner Race, Ball Fault and Outer Race with the position relative to load zone (Load Zone Centered at 6 : 00) Centered @6 : 00.

the percentages of training samples are 10%, 20%, 30%, 40%, and 50%, the ac-curacies of the stable distribution feature extraction based fault diagnosis system are all superior to those of the LDA, the PCA and the $\alpha-$S-SVM. As shown by the experimental results, the single scale $\alpha-$S-SVM is not good enough to classify different bearing faults. However, a fault diagnosis system with the accuracy of prediction up to 94% will be obtained if stable distribution features with wSVM. Another advantage of $\alpha-$S-wSVM is that it is more robust on the variation of the training size. The computational cost for the wSVM training procedure can be greatly reduced since a large number of training samples are unnecessary.

Table 1: recognition accuracy of proposed approach compared with other algorithms at 1730 rpm

training %	PCA	LDA	$\alpha-$S-SVM	$\alpha-$S-wSVM
10%	75.23%	80.41%	85.97%	**87.01%**
20%	76.25%	81.25%	86.65%	**88.12%**
30%	77.50%	85.36%	88.04%	**90.00%**
40%	76.25%	84.58%	89.87%	**92.08%**
50%	76.00%	88.00%	92.07%	**94.00%**

4   Conclusions

Stable distribution is an effective way to measure the complexity of chaotic time series, such as the vibration signal of bearings in our experiments. Compared with well-known complexity measures, stable distribution can extract the features with high discriminate capability. Jointed with the wSVM, the simulation results of bearing fault diagnosis show that the proposed framework achieves much accuracies than other methods. Due to the

fact that stable distribution is robust to the training set data size, a large amount of computational cost could be saved in the training process.

References

[1] A. K., Jardine, D. Lin, D. Banjevic, . "A review on machinery diagnostics and prognostics implementing condition-based maintenance" Mechanical systems and signal processing, 20(7), 1483-1510, 2006.

[2] Mehala, N.; Dahiya, R. A comparative study of FFT, STFT and wavelet techniques for induction machine fault diagnostic analysis. In Proceedings of the 7th WSEAS International Conference on Computational Intelligence, Man-Machine Systems and Cybernetics, Cairo, Egypt, 2931 December 2008.

[3] Staszewski, W.J.; Worden, K.; Tomlinson, G.R. Time-frequency analysis in gear box fault detection using the wigner-ville distribution. Mech. Syst. Signal Process. 1997, 11, 673692.

[4] Peng, Z.K.; Chu, F.L. Application of the wavelet transform in machine condition monitoring and fault diagnostics: A review with bibliography. Mech. Syst. Signal Process. 2004, 18, 199221.

[5] Hong, H.; Liang, M. Fault severity assessment for rolling element bearings using the Lempel-Ziv complexity and continuous wavelet transform. J. Sound Vib. 2009, 320, 452468.

[6] UCKUN, Serdar, GOEBEL, Kai, et LUCAS, Peter JF. Standardizing research methods for prognostics. In : Prognostics and Health Management, 2008. PHM 2008. International Conference on. IEEE, 2008. p. 1-10.

[7] D. Vaughan, "The Challenger launch decision", Risky technology, culture, and deviance at NASA. University of Chicago Press, 2009.

[8] Q. Hai, L. Jay, L. Jing, "Robust performance degradation assessment methods for enhanced rolling element bearing prognostics", Advanced Engineering Informatics, 2003, vol. 17, no 3, p. 127-140.

[9] H. Runqing, X.Lifeng, L. Xinglin, "Residual life predictions for ball bearings based on self-organizing map and back propagation neural network methods", Mechanical Systems and Signal Processing, 2007, vol. 21, no 1, p. 193-207.

[10] Y. T. Irem, B. S. Robert, "Mapping function to failure mode during component development", Research in Engineering Design, 2003, vol. 14, no 1, p. 25-33.

[11] D. A. TOBON-MEJIA, K, MEDJAHER, N. ZERHOUNI, "CNC machine tool's wear diagnostic and prognostic by using

dynamic Bayesian networks", Mechanical Systems and Signal Processing, 2012, vol. 28, p. 167-182.

[12] Li, Changning, and Gang Yu. "A new statistical model for rolling element bearing fault signals based on Alpha-stable distribution." Computer Modeling and Simulation, 2010. ICCMS'10. Second International Conference on. Vol. 4. IEEE, 2010.

[13] M. -C., Lee; C. To, Comparison of Support Vector Machine and Back Propagation Neural Network in Evaluating the Enterprise Financial Distress, International Journal of Artificial Intelligence & Applications (IJAIA), Vol.1, No.3, July 2010.

[14] A. Karatzoglou; D.Meyer and K.Hornik support vector machine in R , journal of statistical software April 2006, Volume 15, Issue 9.

[15] J. H., McCulloch, "Simple consistent estimators of stable distribution parameters", Communications in Statistics-Simulations.15(4), 11091136 (1986).

[16] C. Zhenyu; L., Jianping; W., Liwei, "A multiple kernel support vector machine scheme for feature selection and rule extraction from gene expression data of cancer tissue", Artificial Intelligence in Medicine, 2007, vol. 41, no 2, p. 161.

[17] C., Argon; S. G., WU, "Real-time health prognosis and dynamic preventive maintenance policy for equipment under aging Markovian deterioration", International Journal of Production Research, 2007, vol. 45, no 15, p. 3351-3379.

[18] L., Zhijuan; L., Qing, MU, Chundi. "A Hybrid LSSVR-HMM Based Prognostics Approach" In : Intelligent Human-Machine Systems and Cybernetics (IHMSC), 2012 4th International Conference on. IEEE, 2012. p. 275-278.

[19] N. S., JAMMU; K, P, KANKAR, "A Review on Prognosis of Rolling Element Bearings.", International Journal of Engineering Science, 2011, vol. 3.

[20] V., Vapnik, "Statistical Learning Theory", Wiley-Interscience, New York, 1998.

[21] Case Western Reserve University Bearing Data Center Website. Available online:http//csegroups.case.edu/bearingdatacenter/pages/downloaddatafile (accessed on 29 June 2013).

[22] C.C., Chang; C.J., Lin, "LIBSVM: A library for support vector machines" ACM Transactions on Intelligent Systems and Technology 2011, 2, 27:127:27. Software available online: http://www.csie.ntu.edu.tw/ cjlin/libsvm (accessed on 30 July 2011).

[23] C. Li, G. Yu, "A New Statistical Model For Rolling Element Bearing Fault Signals Based On Alpha-Stable Distribution", 2010 Second International Conference on Computer Modeling and Simulation, 22-24 Jan. 2010.

[24] White, M. F. "Simulation and analysis of machinery fault signals." Journal of sound and vibration 93(1): 95-116, 1984.

[25] McFadden, P. D. and J. D. Smith. "Model for the vibration produced by a single point defect in a rolling element bearing." Journal of sound and vibration 96(1): 69-82, 1984.

[26] Nikias. C, M. Shao: Signal processing with alpha-stable distributions and applications. New York: John wiley & Sons Inc. 1995.

[27] Shao, M. and C. L. Nikias. "Signal processing with fractional lower order moments: stable processes and their applications." Proceedings of the IEEE 1983, 81(7): 986-1010, 1993.

[28] Fan, H., & Ramamohanarao, K. (2005, July). "A weighting scheme based on emerging patterns for weighted support vector machines". In Granular Computing, 2005 IEEE International Conference on (Vol. 2, pp. 435-440). IEEE.

[29] Vapnik, V. (2013). "The nature of statistical learning theory". Springer Science & Business Media.

[30] C.F. Lin, S.D. Wang, "Fuzzy support vector machines", IEEE Transactions on Neural Networks 13 (2) (2002) 464471.

[31] Khemchandani, R., & Chandra, S. (2009). "Regularized least squares fuzzy support vector regression for financial time series forecasting". Expert Systems with Applications, 36(1), 132-138.

[32] Brahim chouri,fabrice monteiro,mohamed tabaa, abbas dandache "residual useful life estimation based on stable Distribution feature extraction and svm classifier" Journal of Theoretical and Applied Information Technology 30th September 2013. Vol. 55 No.3

Design rules for RF Micro Energy Harvesting under near Field probing considerations

Hafid GRIGUER[1], Hicham LALJ[1], Mohammed
Amine Benfetah[1] [2]

[1] Smarti-Lab
EMSI-Rabat
Rabat, Morocco
hafid.griguer@gmail.com

M'hamed DRISSI [2]

[2] IETR
UEB, INSA of Rennes
Rennes, France
Mhamed.drissi@insa-rennes.fr

Abstract— **In this paper new method of determining the design rules in a near Field RF energy harvesting system is presented. A novel model of integrated RF near Field energy recycling from smartphone is investigated. The electromagnetic near Field scanning is deployed, the measurement results confirm a high efficiency RF harvesting in near Field zoning within minimum coupling between system antennas.**

Keywords— Micro energy recycling; RF energy harvesting; MPT; RF near Field energy; Near Field Scanning

I. INTRODUCTION

Microwave Power Transmission (MPT) technology is proposed to transmit Electric power under a microwave radiation [1]. Several applications were subject to proposition in literature like a microwave powered satellite. The MPT is actually supposed to be the key technology to provide energy for the Wireless communication components and for the internet of things [2]. The highest challenge of the MPT is to be used as Wireless battery to supply power to the Wireless components like smartphones or Wireless sensors [3].

The MPT is based on the technology of the RF Energy Harvesting; the two major components of the RF energy harvesting schemes are the receiving antenna and the rectifying circuit. In the last few years, researchers have focused on designing new schemes that are specially used at the harvesting of the ambient energy provided by dedicated energy sources [4]. The RF energy in this case, is supported by a plane wave in the far Field context.

Recently, our research group has been interested into the use of the RF energy harvesting to recycle an ambient Wireless energy provided by not-dedicated RF energy sources, like a Smartphone or Wireless sensors antennas [5]. These antennas are supposed to be used in Wireless communication applications, we so have made possible using this antennas as local RF micro energy sources. In This case, the receiving antenna is placed in the near Field environment of the sources Antennas, looking so to maximize the receiving energy.

Unfortunately, in the near Field environment, the radiated wave is not plane, in fact, both of electric and magnetic fields follow an arbitrary propagation law depending on the type of source antenna and its ambient environment. The main challenge of our micro energy recycling solution is providing an efficient RF energy harvesting system without disturbing the native function of the transmission antenna.

In this paper we propose a new rules' design method for positioning the RF harvesting receiving antenna into the near field environment of the Smartphone antenna using as local RF source. This method must verify the geometrical positioning conditions, respecting simultaneously, a maximal receiving signal and a minimal near field coupling between the two antennas.

First, the smartphone antenna will be analyzed under an EM near Field scanner, in this section both of the near Field distribution and the geometrical center of smartphone antenna equivalent surface will be identified at the desired frequency. The conditions of near Field decoupling and maximum signal receiving will be verified. The workflow and the measurement setup of the S matrix under VNA will be presented. In the last section, the measurements results and the positioning conditions will be discussed.

II. ELECTROMAGNETIC NEAR FIELD PROBING

The *Samsung Note 3*©smartphone is used as local RF energy source. To avoid the RF signal intermittent problems that can occur during a real radio communication, the main Smartphone antenna has been disconnected to its inner RF transmitter and has been connected to an external RF synthesizer trough specified 50Ω U-fl connection. GSM center band frequency is chosen to investigate the Smartphone antenna near field behavior. An electromagnetic Near Field scanner (NFS) [6] is used to investigate both of electric and magnetic near Fields around the antenna under test (AUT). As shown in Fig.1 the NFS is composed of 3 axes displacements machine controlled by a PC, a near Field probe and a spectrum analyzer. The total surface of the Smartphone is about 15cm x 7,5cm. Only the near

978-1-4673-8760-6/15 $31.00 © 2015 IEEE

surface around the antenna will be investigated, it's about 4,5cm x 7,5cm. In Z axis the near Field probe is maintained at $rz+$ =1cm from the antenna surface located at the Smartphone bottom surface. Antenna Input power is Pe = 30dBm. The measurement results at 900MHz show that the maximum near field is concentrated nearby the center of the geometrical antenna surface; this point is admitted as the equivalent antenna surface center and chosen as the geometrical reference point for our investigation process.

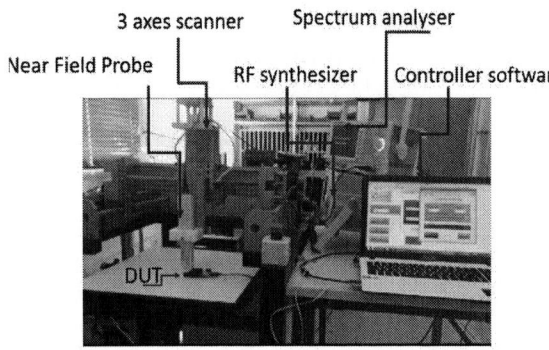

Figure 1. Electromagnetic near Field scanner setup

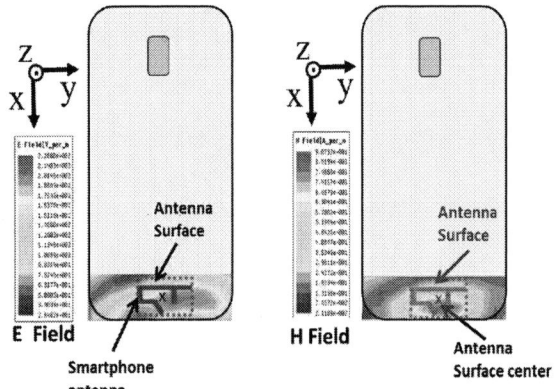

Figure 2. Smartphone antenna Electric and Magnetic Fields measurement results@ **900MHz** & $rz+$ **=1cm**

III.   NEAR FIELD RF HARVESTING MEASURMENT CONDITIONS AND SETUP

In this section, we are interested in studying the electromagnetic coupling phenomena between the smartphone's antenna and our harvesting system antenna. To ensure our investigation, we propose the experimental configuration in Figure 3. The smartphone antenna (SA) is connected to port 1 of the *N522xA Keysight©* Vector Network Analyzer (VNA), via an UF-l / SMA 50Ω transition, a 900 MHz monopole antenna is selected to perform the role of receiving antenna of harvesting system (HRA). The HRA is connected to the port 2 of the same VNA with a similar connection of the SA. The HRA is placed in a 3-

axis machine to ensure its displacement around of the SA. The displacement of the HRA is relative to the reference point which is the SA surface geometric center. (rx, ry, rz) are the displacements' distances from the SA reference point. The displacements are made in positive and negative directions of the x and y axis. The Only positive displacement is provided on the z axis. Δx and Δy are the limitations displacements in x and y axis from the Smartphone perimeter, in z axis $\Delta z.$ =1cm. The $S11$, $S22$ and $S21$ parameters must be measured in a sufficient frequency range covering the GSM 900MHz band.

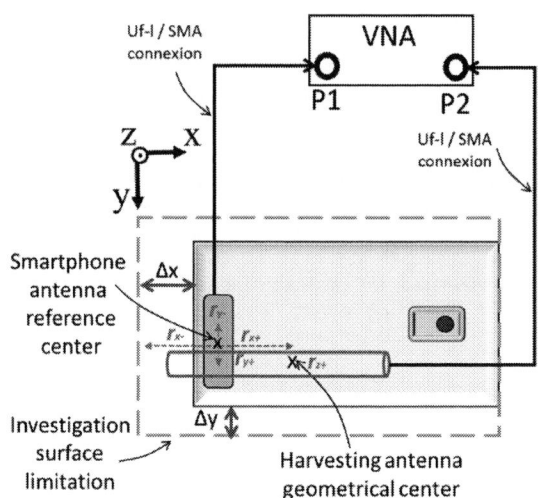

Figure 3. Smartphone & harvesting antennas S matrix measurement setup

As suggested before, our objective is to define the HRA geometrical positioning limits in order to maximize first the received RF power which is then converted into DC power. Since positioning is performed in a near-field area of the SA, this may adversely affect its native operation mode. To characterizing the antennas native operation mode stability as well as maximizing the harvesting RF signal, a set of criteria and its tolerance margins are proposed in Table.1, using: Δfr: Frequency peak dynamic, ΔA: Power peak dynamic and $\Delta S21$: $S21$ Power peak dynamic.

Parameters	Criteria tolerance margin	Limitations	Frequency band
$S11$(dB)	Δfr = +-10 MHz ΔA= +-1dB	-7dB	GSM band
$S22$(dB)	ΔA = +-2dB	-7dB	Up-link GSM band
$S21$(dB)	$\Delta S21$=[-5dB, -10dB]	*******	Up-link GSM band

Table 1. Criteria and process margins

IV. DESIGN RULES RESULTS

In figure 4, we show the S parameters measurement results for the case of ($r_{x+} = 5mm, r_{y+} = 15mm, r_{z+} = 10mm$). The $S11$ peak level of the SA is about -18dB, the peak frequency is near to 925 MHz. With HRA in proximity we notice a slight $S11$ variation within $\Delta fr < +$-10 MHz and input impedance matching improvement. In figure 4(b), the $S22$ of the HRA show a good impedance matching within the GSM band < -7dB. The maximum mutual coupling between antennas is about -9dB, which is suitable since within criteria tolerance margin.

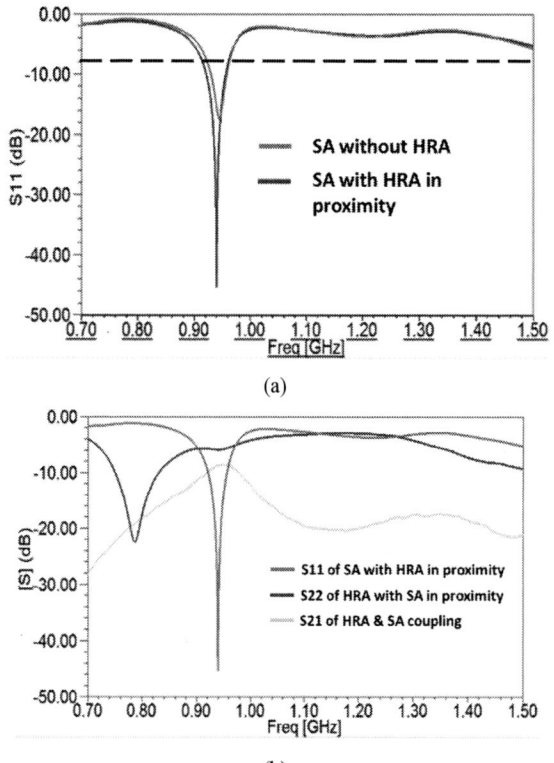

(a)

(b)

Figure 4. (a) $S11$ results comparison for SA with and without HRA in proximity. (b) $S11$, $S22$ and $S21$ of the SA with HRA system.

The HRA displacement on the x and y axis is performed by a step of 5 mm, the x- and y+ displacements were operated within limitations $\Delta x = 10$ mm and $\Delta y = 0$ mm. For each operated position, a measurement of the different criteria is performed. For all cases, moving distance ratios (kn) depending on the distance limit of the near field radiation zone ($r0$) are given in table.2. With, $r0 = (2.D^2) / \lambda$ (D is the biggest dimension of the SA and λ is the wavelength) and $kn = rn/r0$; ($n=x, y$ and z). From the results, we find that the three criteria are met simultaneously when kn varies between 0.5 and 1 in x and y axis. This explains why we can satisfy the criteria of decoupling and maximizing the RF power, while remaining within the near field zone in proximity to the SA. This interesting result will help us later to provide near field integrated association models of RF harvesting circuit near the wireless component radiation systems, where the spatial integration is among the most highlighted objectives.

Parameters	Ratio
kx	[0.5 ; 1]
ky	[0.5 ; 1]
kz	[0.25 ; 0.5]

Table 2. kn ratios depending on the distance limit of the near field radiation zone $r0$

V. CONCLUSION

In this paper we have introduced a new method of determining the design rules in a near Field RF energy harvesting system. Electromagnetic near field scan was performed to determine the effective radiation area of the Smartphone antenna (SA), the results helped us to assume that the geometric center of the SA surface is the reference point of our investigations. The SA was assumed as an ambient RF energy source and a GSM monopole antenna was used as RF energy harvesting antenna (HRA) in proximity of the SA. An experimental configuration has been proposed to measure and verify the operating criteria and tolerance margins for a near field RF harvesting system. The results of the positioning ratio Kn show a freedom of positioning margin within 0.5 and 1. These results gave us the opportunity to offer in the future, new efficient and integrated RF energy harvesting systems associated with small wireless communication systems in there near field area.

REFERENCES

[1] A.P.Sample,D.J.Yeager,P.S.Powledge,A.V.Mamishev,andJ.R. Smith, "Design of an RFID-based battery-free programmable sensing platform,"IEEETrans.Instrum.Meas.,vol.57,no.11,pp.2608–2615, Nov. 2008.

[2] S.Ladan, N.Ghassemi, A.Ghiotto,and K.Wu,"Highly efficient compact rectenna for wireless energy harvesting application," IEEE Microw. Mag.,vol.14,no.1,pp.117–122,Feb.2013.

[3] Z.Popovic,E.Falkenstein,andR.Zane,"Low-power density wireless poweringforbattery-lesssensors,"inProc.IEEERWS,Jan.2013,pp. 31–33.

[4] A.Massa, G.Oliveri, F.Viani, and P.Rocca,"Array designs for long distance wireless power transmission: state-of-the-art and innovative solutions,"Proc.IEEE,vol.101,no.6,pp.1464–1481,Jun.2013

[5] H. Griguer, H. Lalj, and M.Drissi, "Multi-Band and isotropic superstrat for electromagnetic wave absorption and high efficiency energy harvesting," OMPIC Marocain patent, N° MA 37209.

[6] H. Griguer, H. Lalj, " Electromagnetic NearFieldscanningsystem" OMPIC Marocain patent , N° MA30550.

978-1-4673-8760-6/15 $31.00 © 2015 IEEE

Polarization Insensitive Metamaterial Absorber For Energy Harvesting

Hicham LALJ [1], Hafid GRIGUER[1], Mohammed Amine Benfetah[1][2]

[1] Smarti-Lab
EMSI-Rabat
Rabat, Morocco
Lalj.hicham@gmail.com

M'hamed DRISSI [2]

[2] IETR
UEB, INSA of Rennes
Rennes, France
M'hamed.drissi@insa-rennes.fr

Abstract— this paper presents the design of a polarization insensitive microwave absorber based on metamaterial for electromagnetic energy harvesting. The unit cell of the metamaterial consists of eight-fold rotational symmetric electric resonator printed on FR4 substrate with a ground plane to realize both electric and magnetic resonances to achieve efficient absorption of the incident microwave energy.
The electromagnetic simulation results of our absorber demonstrates high microwave absorption up to 92% for different polarized incident electromagnetic waves, from 0 to 90 for both transverse electric wave and transverse magnetic wave.

Keywords—absorber, metamaterial, energy harvesting

I. Introduction

In recent years, left-handed Metamaterials have attracted considerable interest of scientists and engineers working in the field of microwave technology. These Metamaterials exhibit both a negative permittivity and permeability which result in a negative index of refraction, a property not available within any natural material [1-3]. Metamaterials with these unique properties have been applied to create novel microwave devices [4] with better performances and will find more EM applications such as invisibility cloaks, sub-wavelength imaging [5-8].
This property has been successfully utilized to realize EM absorbers with nearly perfect absorption in microwave and terahertz region [9-11].
Theoretically, full absorption can be achieved by tuning μ and ε such that the impedance of the metamaterial is matched to the impedance of free-space.
However, the metamaterial absorbers recently proposed in [12-14] are polarization sensitive, which only works for EM waves with one particular polarization.
In this paper, a polarization insensitive metamaterial absorber is presented based on an array of ELC (Electric-Inductive-Capacitive resonators.
A metamaterial absorber array design is proposed, and finally the absorbing characteristics of metamaterial absorber array for different polarizations of normal incident EM for both transverse electric (TE) and transverse magnetic (TM) waves are investigated.

II. MetaMaterial Absorber Unit Cell

The unit cell geometry of the proposed absorber is illustrated in Figure 1 (a). The unit cell was designed using the microstrip lines to resonate at 10 GHz. The substrate used is a FR4 having the following characteristics (relative permittivity εr = 4.4, loss tangent tg(φ) = 0.001 and thickness h = 0.8 mm).
The size of the unit cell is about 10mm*10mm. It consists of a metallic electric-LC (ELC) resonator in the front side of the dielectric substrate and metallic ground plane in the back side of the substrate. The inner and outer radius of the ELC are 2.1mm and 2.3 mm, respectively, the gaps in the ELC are 0.2mm and the width of the central cross in ELC is 0.3 mm, the size of the back ground plane is 10mm * 10 mm.
The unit cell was designed using the full-wave three-dimensional electromagnetic fields simulator HFSS. The unit cell was placed in a waveguide with Perfect Magnetic and Perfect Electric walls to realize TEM mode excitation with polarization as shown in Figure 1 (b). The incident magnetic field waves will penetrate the space between the ELC and back cross.

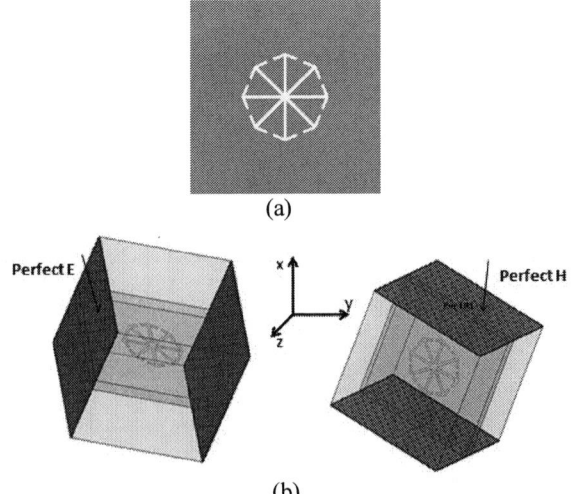

(a)

(b)

Figure 1. (a) Unit cell geometry of the metamaterial absorber
(b): Excitation condition of the metamaterial absorber

978-1-4673-8760-6/15 $31.00 © 2015 IEEE

Using scattering parameters, the absorption of the unit cell is expressed by:

$$A(\omega) = 1 - |S11|^2 - |S21|^2 \qquad (1)$$

Figure 2 shows the simulated transmission, reflection and the absorption at the designed frequency of the unit cell at 10GHz. It can be observed that the transmission and reflection is simultaneously minimized at 10 GHz so that a very good absorption about 92% is achieved at this frequency.

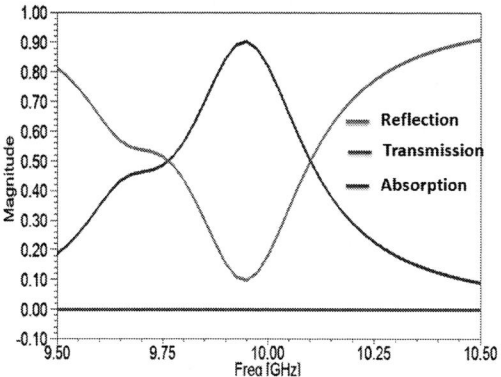

Figure 2. Simulation results of the absorption, reflection, and transmission for the metamaterial absorber

The electric field at the surface of the ELC and the current distribution on the metals for the incident wave at 10 GHz shown in Figure 3 confirm the peak absorption obtained.

The front ELC is excited by the incident electric field so that opposite charges are accumulated at the upper and lower arches of the ELC, which can be determined by opposite directions of electric field at the edge of each arches.

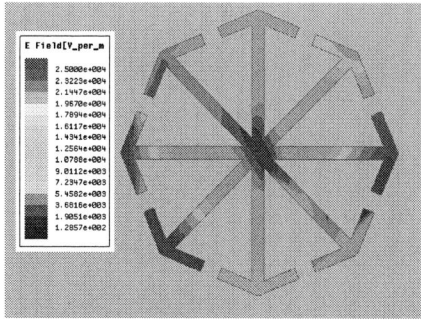

Figure 3. Simulated Electric field surface of ELC at 10 GHz.

III. POLARIZATION INSENSITIVE METAMATERIAL ABSORBER

An array based on the optimized metamaterial absorber shown in Figure 4 is designed used the same FR4 dielectric substrate. The size of the array is 400* 400mm.

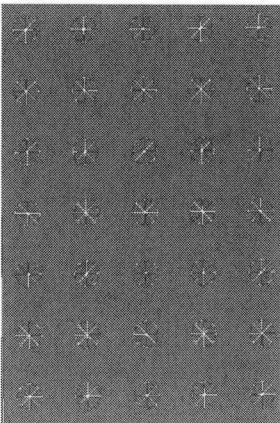

Figure 4. A 40* 40 array of metamaterial absorber

The simulation results of the metamaterial absorber array for both TE and TM waves with different polarizations of normal incidence are illustrated in Figure 5 and Figure 6, respectively. The results show a peak absorption of 92% for different polarizations, which verifies the very good absorption and polarization insensitivity of the proposed design in the case of TE and TM waves.

The transverse field of normally incident EM waves with any polarization can be decomposed into two orthogonally polarized components. Owing to the orthogonal symmetry of the proposed absorber in x-y plane, it couples and responds to each component equally which ensures that this absorber can work for incident EM waves with all polarizations, including circularly polarized waves.

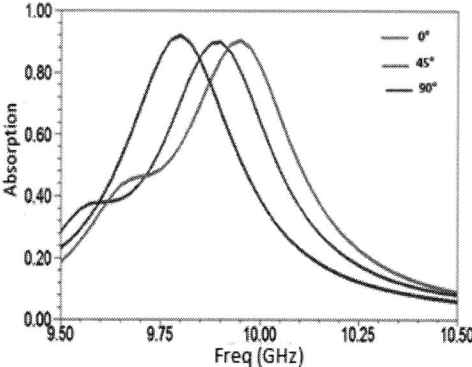

Figure 5. Simulated absorptions at different incident angles for TE modes. The Labels for different curves indicate the incident angles.

978-1-4673-8760-6/15 $31.00 © 2015 IEEE

The absorbing capability of the metamaterial absorber is directly resulted from the resonant structures of the unit cell, so it is possible to tune the absorbing frequency in function of metamaterial absorber dimensions over a significant EM frequency spectrum.

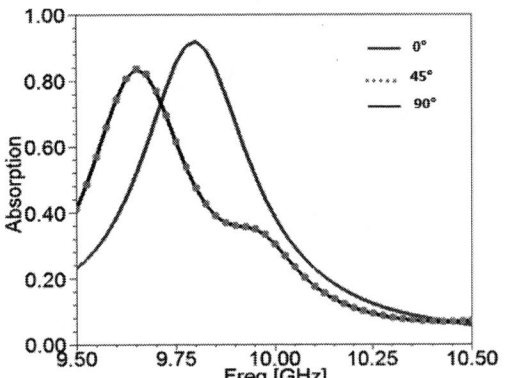

Figure 6. Simulated absorptions at different incident angles for TM modes. The Labels for different curves indicate the incident angles.

IV. CONCLUSIONS

In this paper, a design of a polarization insensitive microwave absorber based on metamaterial for electromagnetic is proposed. Polarization insensitivity has been verified with approximate responses for different polarized EM waves in both TE and TM mode. Peak absorption of 92% has been obtained for normal incident waves. Investigations into the EM field distribution in the unit cell have revealed the working mode of the absorber.

With geometrical scalability, this absorber could be designed to work at other EM frequency range with very good absorption. These remarkable features suggest many potential applications, energy harvesting and stealth technology.

REFERENCES

[1] J. B. Pendry, A. J. Holden, D. J. Robbins and W. J. Stewart, "Magnetism from Conductors and Enhanced Nonlinear Phenomena," IEEE Transactions on Microwave Theory and Techniques, Vol. 47, No. 11, 1999, pp. 2075-2084. doi:10.1109/22.798002

[2] R. Marqués, F. Martin and M. Sorolla, "Metamaterials with Negative Parameters: Theory, Design and Microwave Applications," John Wiley & Sons, Inc., Upper Saddle River, 2007.

[3] R. W. Ziolkowski and N. Engheta, "Metamaterial Special Issue Introduction," IEEE Transaction on Antennas and Propagation, Vol. 51, No. 10, 2003, pp. 2546-2549.

[4] Caloz, C. and T. Itoh, Electromagnetic Metamaterials: Transmission Line Theory and Microwave Applications, Wiley, New York, 2006.

[5] . Schurig, D., J. J. Mock, B. J. Justice, S. A. Cummer, J. B. Pendry, A. F. Starr, and D. R. Smith, \Metamaterial electromagnetic cloak at microwave frequencies," Science, Vol. 314, 977{980, 2006.

[6] Liu, R., C. Ji, J. J. Mock, J. Y. Chin, T. J. Cui, and D. R. Smith, \Broadband ground-plane cloak," Science, Vol. 323, 366{369, 2009.

[7] Pendry, J. B., \Negative refraction makes a perfect lens," Phys. Rev. Lett., Vol. 85, 3966{3969, 2000.

[8] Zhao, J., Y. Feng, B. Zhu, and T. Jiang, \Sub-wavelength image manipulating through compensated anisotropic metamaterial prisms," Opt. Express, Vol. 16, 18057{18066, 2008

[9] Lagarkov, A. N., V. N. Kisel, and V. N. Semenenko, \Wide-angle absorption by the use of a metamaterial plate," Progress In Electromagnetics Research Letters, Vol. 1, 35{44, 2008.

[10] Landy, N. I., S. Sajuyigbe, J. J. Mock, D. R. Smith, and W. J. Padilla, \Perfect metamaterial absorber," Phys. Rev. Lett., Vol. 100, 207402-1{207402-4, 2008.

[11] Tao, H., N. I. Landy, C. M. Bingham, X. Zhan, R. D. Averitt, and W. J. Padilla, \A metamaterial absorber for the terahertz regime: Design, fabrication and characterization," Opt. Express, Vol. 16, 7181{7188, 2008.

[12] Landy, N. I., S. Sajuyigbe, J. J. Mock, D. R. Smith, and W. J. Padilla, \Perfect metamaterial absorber," Phys. Rev. Lett., Vol. 100, 207402-1{207402-4, 2008.

[13] Tao, H., N. I. Landy, C. M. Bingham, X. Zhan, R. D. Averitt, and W. J. Padilla, \A metamaterial absorber for the terahertz regime: Design, fabrication and characterization," Opt. Express, Vol. 16, 7181{7188, 2008.

[14] Wang, J. F., S. B. Qu, Z. T. Fu, H. Ma, Y. M. Yang, X.Wu, Z. Xu, and M. J. Hao, \Three-dimensional metamaterial microwave absorbers composed of coplanar magnetic and electric resonators," Progress In Electromagnetics Research Letters, Vol. 7, 15{24, 2009.

A new FPGA-based DPLL algorithm to improve SAT solvers

Khadija Bousmar[1,2], Fabrice Monteiro[1], Zineb Habbas[3], Sofiene Dellagi[1], Abbas Dandache[1]

[1] Laboratoire de Génie Industriel et de Production de Metz (LGIPM), Université de Lorraine, France
[2] Ecole Marocaine des Sciences de l'ingénieur (EMSI), Département de Recherche et Innovation (DRI), Casablanca, Maroc
[3] Université de Lorraine, France
bousmar.khadija@gmail.com, {fabrice.monteiro, zineb.habbas, sofiene.dellagi, abbas.dandache}@univ-lorraine.fr

Abstract: SAT (SATisfiability of Propositional Formula) is a well-known NP-Complete problem [1][2]. Conventional solvers for SAT based on traditional DPLL algorithm presents serious CPU-Times limitations, especially when addressing large size instances. These last decades, a promising approach has emerged for solving efficiently large size instances by using FPGA architectures. This paper follows this last direction and proposes a new and original DPLL solving algorithm based on FPGA. This new FPGA-based DPLL algorithm will use a new backtrack method to reduce the time of problems resolution by using registers which help to save data from the RAM.

Keywords—SAT, SATisfiability Boolean, NP Complete, DPLL, FPGA, RTL, FPGA, backtrack.

I. INTRODUCTION

SAT problems [3] are a well-studied example of the NP-complete problem family. SAT is used to express and to solve a great class of problems in Artificial Intelligence. Some examples are scheduling, temporal reasoning, graph problems, circuit verification ...etc.

Exact solvers [4] for SAT are based on enumerative search algorithm, such that DPLL (Davis-Logemann-Loveland [5][6], the most used one. Several works and improvements of DPLL have been proposed. However, the facts remain that face to real world problems or large size instances of SAT, the computational cost continues to be prohibitive. In order to improve the processing time of SAT solvers, a promising approach for solving SAT on FPGA [7] has emerged these last decades. Some works present interesting results related to this approach by creating a logic circuit dedicated to solve one problem instance on Field Programmable Gate Arrays (FPGA). This approach becomes feasible due to the advances in FPGA and High Logic Synthesis. In this approach, each SAT instance is automatically analyzed, compiled with the search algorithm and implemented on FPGAs.

This paper follows this new emerging idea. Mainly, it proposes a new DPLL algorithm based on FPGA for improving SAT solvers. The main goal of this first contribution is to propose a general methodology for oriented FPGA solver. DPLL is considered here as an example of a target algorithm.

The idea Behind this approach is using a new backtrack based on logic circuit synthesis, leading to a reduction of the global search space.

To sum up, the important characteristics of this approach is using the RTL programming on the FPGA and the test of this new backtrack method.

The paper is organized as follows: Section II presents formally the SAT problem and DPLL algorithm. Section III concerns the presentation of our main contribution, namely the FPGA-based DPLL algorithm, where we will also develop and explain the diagram block of this new method. Section IV presents the implementation of our approach by using the RTL programming and some first experimental results.

II. PRELIMINARIES

This section presents the SAT problem by defining the type of the used formula, then we will detail the famous algorithm DPLL and we will finish this section by an example which illustrates this algorithm.

A. SAT problem

The SATisfiability problem (SAT) [3][8] is of central importance in computation theory. SAT formalism is used to modelise many academic or real problems like coloration problem, decision support and automated reasoning. Formally, SAT is defined as follows:
Let $V = \{v1, v2 ..., vn\}$ be a set of n boolean variables.
A partial truth assignment for V is a partial function:
$V \rightarrow \{True, False\}$. A clause C is a disjunction of literals, while a literal is a variable vi from V or its negation, noted ¬cvi. A Conjunctive Normal Form (CNF) [9] is a conjunction of clauses. A clause is satisfied when at least one of its literals is set to true. A CNF [10] is satisfied if an assignment of some variables in V satisfies all the clauses. The SAT problem asks for an assignment of some variables in V that satisfies a CNF formula. The problem is said SAT if such an assignment exists and UNSAT otherwise [11].

B. DPLL algorithm

Most of the more successful complete SAT solvers are instantiations of the DPLL algorithm (Davis, Putnam, Logemann, Loveland [4][8][10].

978-1-4673-8760-6/15 $31.00 © 2015 IEEE

This algorithm is an exact enumerative search algorithm [12] for deciding the satisfiability of propositional logic formulae. The principle of DPLL can be summarized by the procedure given by Algorithm 1. SAT enumerates explicitly all possible interpretations of n variables (where n the number of Boolean variables) and assigns a truth value to the variable $v \in V$. In the algorithm 1, all that is written in bold is defined as follows:

- **Empty clause**: An clause is empty if all of its variables have been assigned in a way that makes all the literals false.

-**Unit clause**: A unit clause is a clause including only a single unassigned literal.

-**Unit-progation**: For each unit clause, no choice is necessary. Unit propagation leads to a deterministic cascade avoiding a large part of the search space.

-**Pure-literal-assign**: A literal is pure if it appears with only one polarity (positive or negative). Pure-literal-assign assigns to pure literals boolean values making true all clauses containing them. Such clauses can be deleted.

-**Chosse-literal**: determines, according to a given heuristic, the next literal to be instantiate.

Algorithm 1: DPLL
Input: A set of clauses Φ.
Output: A Truth Value.
Function DPLL (Φ)
If Φ is a consistent set of literals
    Then return true;
If Φ contains an **empty clause**
    Then return false;
For every **unit clause** l in Φ
    Φ = **unit-propagation** (l, Φ);
For every literal l that occurs pure in Φ
    Φ = **pure-literal-assign** (l, Φ);
    l= **choose-literal** (Φ);
return DPLL(Φ ∧ l) or DPLL (Φ ∧ ¬l);

Algo.1.   The basic DPLL algorithm

Example:
Solving following formula:
(a v b) ^ (a v ¬c) ^ (¬a v b) ^ (a v c)   (1)
by using DPLL algorithm gives rise to the following tree search.

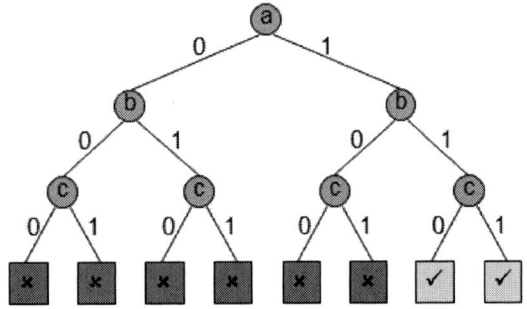

Fig.1.   Example of solving a formula

III.   OUR PROPOSED

This section gives a detailed idea about our new approach of the DPLL algorithm; in the first subsection we will present the principle of the new idea of the algorithm, then we will explain its architecture by presenting the states machines and the block diagram.

A. The FPGA-based DPLL algorithm: principle

This new approach consists to develop the old DPLL algorithm which used to take a lot of time to find solution of a typical problem due to browsing all the branches of the tree search, to a new method that will use a new backtrack but keeping the same steps for the resolution, with some little modification, to be more clear, at each steps the algorithm DPLL picks a variable v and assigns a value to v. The formula is simplified by removing the satisfied clauses and eliminating the false literals. An inference rule may then be applied to assign values to some more variables which are implied by the current assignments. If an empty clause results after simplification, the procedure backtracks and tries the other value for v therefore yields a smaller search space, this details will be discussed in the following subsections.

B. State machine

State machine diagrams specify state machines. This clause outlines the graphic elements that may be shown in state machine diagrams, and provides cross references where detailed information about the semantics and concrete notation for each element can be found.

Every state machine has an arc from "reset". This indicates what state the state machine goes to when a reset is applied. The diagram is worthless without knowing what the initial state is.

Each state is given a name. In this case we are using a type for the states that is an enumerated state type, it provides an easy way to understand and to talk about what and how the state machine works.

Each possible transition between states is shown via an arc with the condition for the transition to occur shown. The condition need not be in VHDL syntax but should be understandable to the reader. Typically logic expressions are given with active high assertion assumed. It should be understood that all transitions occur on the clock edge.

Outputs from the state machine should be listed. The only outputs from this state machine are its present state. Most likely, some other state machine is watching this one's state to determine its next state.

In our case the state machine had a lot of state starting by the first state machine where we begin by the initialization.
The following figure presents the first state machine where you can notice the conversion of any problem to the SAT formalism (conjunction of disjunction), then to solve it we have to save it in the RAM, and after we try to extract a defined number of clause that meets the size of the first register and finally we copied this number of clauses in the

second register to start the evaluation step which is detailed in the block diagram.

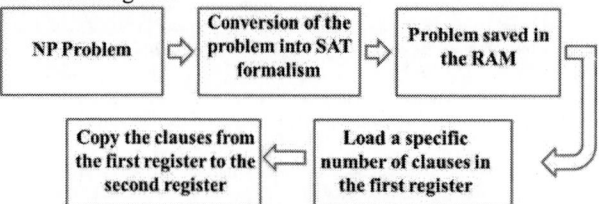

Fig. 2.   The first step of solving a NP problem, for the new DPLL method

C. Block diagram for FPGA-based DPLL algorithm

A block diagram is a specialized, high-level and it is used to design new systems to describe them and improve existing ones. Its structure provides a high-level overview of major system components, key process participants, and important working relationships. The block diagram provides a quick high-level view of the system to rapidly identify points of interest. Because of its high-level perspective, it may not offer the level of details required for more comprehensive planning or implementation.

The block diagram for the new FPGA-based DPLL algorithm will be presented in three blocks as follows:

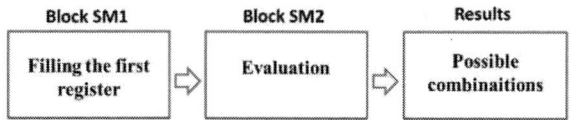

Fig. 3.   The block diagram of the new DPLL method

Block SM1: This block consists to fill the register of the first register with the clauses of the converted problem which was saved from the RAM. The second phase is to save this data in a register to make the evaluation easy.

Block SM2: All the clauses in this block are evaluated by considering all the possibilities for each variable. This step can lead two three possible results:
- The returned result is 1: All ll the clauses of the first register are satisfied. This positive result is saved in the last block.
- The returned result is 0: A backtracks has to be done. This consists of flipping last bit and a restarting of the evaluation process.
- The return result is U: This means Undefined status.
The important point in this new method is, if we finished the test of all the clauses already presented in the register of the first register we do not need to wait for the loading data from the RAM of the following clauses. However to reduce the solution search time we have to continue checking other possibilities with variables and clauses that we already had, by the time the first register is filled to transfer the data to the register of the memory cache that only take a single clock cycles and thus ultimately, the search time solutions will be reduced since we have had no results with which we can continue our research in the following clauses

Results   : This block records all possible combinations that make a problem correct, in two lists:

-The first list stores the first evaluation results that will be used thereafter as and when new terms or new variables are added.
-The second list stores the final results of the problem.

IV.   STRATEGY

This section presents the strategy that we will use to implement this new DPLL algorithm; in the first subsection we will give a short presentation of the programming language adapted to the hardware called the Register-Transfer-Level (RTL), then in the second subsection, we will discuss about the gain theoretical complexity of this new approach.

A.   RTL programming

Register-Transfer-Level (RTL) [13] abstraction is used in hardware description languages (HDLs) like Verilog and VHDL to create high-level representations of a circuit, from which lower-level representations and ultimately actual wiring can be derived. The design at the RTL level is typical practice in modern digital design. This programming language is used in the logic design phase of the integrated circuit design cycle. An RTL description is usually converted to a gate-level description of the circuit by a logic synthesis tool. The synthesis results are then used by placement and routing tools to create a physical layout. And the logic simulation tools may use a design's RTL description to verify its correctness.

B.   Experiments

The results obtained with the old DPLL and the old backtrack are serious CPU-Times limitations, especially when addressing large size instances, because the time of the evaluation of clauses may test a lot of combinations so take many backtrack and then many decision which maximize the resolution time, the figure below give an idea about the steps of this method.

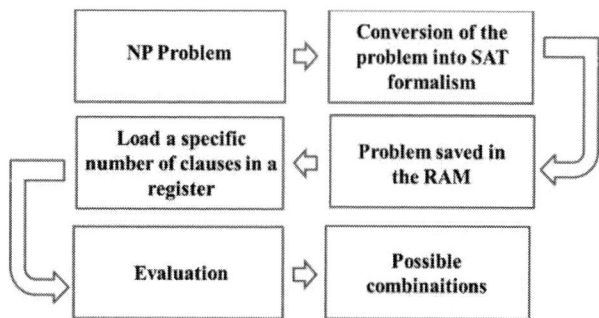

Fig. 4.   The steps of solving a NP problem, for the old DPLL method

Each time that the evaluation of a clauses loaded in the register are ended, we have to wait a long time to load again the same number of clauses from the RAM to the register, then we can start the evaluation again.

But for our proposed, the new DPLL algorithm, as it is was presented in the Fig 2 and the Fig 3. The theoretical results for this approach are positive because of the smart backtrack, and the two registers added after the RAM, this was obtained by removing the waiting time thanks to the second register inserted between the first register which will extract some clauses of RAM and the evaluation step. Then the results obtained with the first tests will be saved on the active list until the first register depletes downloaded clauses, and finally we save the last results in the result list, then if the first register is filled by the clauses from the RAM it will be copied in the second register then we do the same think for the evaluation, if else, we keep the clauses of the second register and do a backtrack on the results saved in the active list, and test them, until the first register receives the clauses from the RAM.

All this is programmed by the RTL language, and this is done first of all by the creation of new types that we use in the DPLL algorithm for the evaluation step.

V. CONCLUSION AND PERSPECTIVES

This paper presents the old DPLL algorithm and gives a detailed description of our new approach methods for solving NP-complete problems converted into the formalism SAT, it present also the different states machine which encompasses the steps of the resolution problems by explaining the most important parts which are the backtrack and the evaluation steps.

The future works will give more details about this idea in addition to this new FPGA-based DPLL algorithm we can also explain the possibility of using new methods of research called heuristics on programmable logic circuits with a suitable logic circuits to optimize more these complex problems and to reduce the resolution time. Indeed, there are still a lot of studies to make improvements to our work as developing the backtrack and make a comparison between the new DPLL and the existing heuristics. Another improvement is to identify and exploit the programmable logic FPGA to achieve the great perspective that is the parallelization of DPLL algorithm and these heuristics we have proposed.

REFERENCES

[1] The P versus NP–complete dichotomy of some challenging problems in graph theory, Celina M.H. de Figueiredo, 2012, Journal Discrete archive, Volume 160 Issue 18, December, 2012, Pages 2681-2693.

[2] Stephen cook, "The P versus NP complet", The P versus NP Problem, Stephen Cook, Manuscript prepared for the Clay Mathematics Institute for the Millennium Prize Problems, April, 2000.

[3] Xili Wang "A novel approach of solving the CNF-SAT problem", Journal CoRR Computer Research Repository Vol. abs/1307.6291, 2013.

[4] Broes De Cat, Bart Bogaerts, Marc Denecker." MiniSAT(ID) for satisfiability checking and constraint solving", September 25, 2014.

[5] João F. Lima, "Boolean Satisfiability Solvers: Techniques, Implementations and Analysis", Electronica e Telecominicaçoes,Vol 5,N°2, JUNHO,2010.

[6] Robert Nieuwenhuis,Albert Oliveras,Cesare Tinelli "Solving SAT and SAT Modulo Theories: From an abstract Davis--Putnam--Logemann--Loveland procedure to DPLL", Journal of the ACM (JACM), November, 2006, New York, NY, USA.

[7] Ling, A.C., Singh, D.P. ; Brown, S.D. "FPGA PLB evaluation using quantified Boolean SATisfiability", Field Programmable Logic and Applications, 2005. International Conference on, Page 19 – 24, ISBN 0-7803-9362-7;

[8] Lintao Zhang, Sharad Malik, "The Quest for Efficient Boolean Satisfiability Solvers", CAV'02 of the 14th International Conference on Computer Aided Verification, Page 17-36, ISBN 3-540-43997-8, Spring-Verlag London, UK 2002.

[9] Calabro C, Impagliazzo R, Kabanets V, Paturi R, The complexity of unique K-SAT an isolation lemma for k-CNF, Journal of Computer and System Sciences, Volume 74, Issue 3, May 2008, Pages 386-393, Computational Complexity 2003.

[10] Schoning, T. "A probabilistic algorithm for k-SAT and constraint satisfaction problems", Foundations of Computer Science, 1999. 40th Annual Symposium on, ISBN 0-7695-0409-4, New York City, NY 17-19 Oct 1999.

[11] Pascal Vander-Swalmen, « Aspects parallèles des problèmes de satisfaisabilité », Thesis of University of Reims Champagne-Ardenne, 2009.

[12] Olivier Fourdrinoy, « Hybridation des méthodes de résolution pour SAT », 7e Rencontres Jeunes Chercheurs en Intelligence Artificielle, RJCIA'2005 à NICE du 31 mai au 3 juin 2005.

[13] Reddy, B.Naresh Kumar, Suresh, N. ; Ramesh, J.V.N. ; Pavithra, T. ; Bahulya, Y.Krupa ; Edavoor, Pranose J ; Ram, S.Janaki ,"An efficient approach for design and testing of FPGA programming using LabView", Advances in Computing, Communications and Informatics (ICACCI),2015. International Conference on 10-13 Aug. 2015, ISBN 978-1-4799-8790-0,Kochi,India.

Real time EEG compression for energy-aware continous mobile monitoring

*Mohamed Adel Serhani, **Mohamed El Menshawy, ***Abdelghani Benharref, *Alramzana Nujum Navaz

*College of Information Technology, UAE University, Al Ain, UAE
**Faculty of Engineering and Computer Science, Concordia University, Canada
***University of Wollongong in Dubai, UAE

Abstract—EEG-based mobile monitoring is recognized to be a resource-constrained activity because of the limited mobile battery, the intermittent network, and the size of EEG signal generated from continuous monitoring. In this paper, we propose a novel approach combining both EEG compression and mobile resource availability evaluation to boost and save energy for longer monitoring episode. The main core of our approach lies in developing and implementing an algorithm, which evaluates on the fly the compression cost and available resources on the mobile device to decide whether to fully/partially compress the input EEG data or not. We experimentally evaluated and tested the effectiveness of our approach using both offline and online data recorded by the Emotiv EEG device. The obtained results show that our approach significantly saves mobile battery and processing power to cope with critical health situations.

Keywords—Mobile monitoring, EEG, compression.

I. INTRODUCTION

The tremendous development in medicine has converted many previously fatal diseases into curable ones or into chronic diseases. While not totally cured, chronic patients can enjoy a relatively natural life when they follow a prescribed life style and have access to efficient medical control. Unfortunately, chronic patients cannot be kept in hospital because of their long-standing diseases and, at the same time, cannot be sent home without proper control and monitoring. In this perspective, mobile and wireless monitoring represents an interesting candidate to balance the dilemma of keeping patients out of hospital premises while undergoing thorough monitoring.

Mobile continuous monitoring is based on the use of sophisticated biosensors and wireless communication technologies. A seamless health monitoring setup generally includes a multidisciplinary sensor that senses vital signs, a visualization screen to inform the patients, the ability to send information to Hospital Information System (HIS), and receives information from the HIS specialists. Unfortunately, the state of the art of sensors can only perform the first task (i.e. vital signs sensing) and have very limited communication capabilities within a very limited range.

The emergence of smartphones seems perfect fit for health monitoring situation. In fact, having a smartphone close to the sensor solves the communication issue. The new configuration will have a sensor getting readings from patient's body, and communicating the data to the smartphone. The smartphone will then take care of the communication with HIS and health professionals back and forth. This communication can happen over Wi-Fi when there is a network; otherwise, the smartphones switch to 3G/4G network if available.

This new configuration is suitable when the size of sensed data over a large period of time (a day for example) is small. For example, monitoring blood glucose requires a few readings per day. Even when each reading consists of many numbers, the overall data exchanged over 3G/4G during that time period is insignificant. This is not the case when the vital signs being sensed generated a tremendous amount of data even during smaller period of times (hours for example).

When monitoring ElectroEncephaloGram (EEG) for example, there should be many sensors used at the same time on different parts of the scalp. Also, to have a representative EEG signal, there is a minimum frequency of readings, which is usually in terms of milliseconds. The combination of all these factors leads to a situation where there will be a huge amount of data that, when sent over 3G/4G, will generate a considerable communication cost and drain extensively the smartphone's battery.

The aim of this paper is to reduce the size of data being exchanged during continuous monitoring of EEG while providing an accurate view of the patient's brain data. We particularly look at the benefits and costs of using compression during the communication between the smartphone and HIS. As compression itself uses additional resources (e.g., processor, memory, battery, etc.), there is a need to evaluate the cost-effectiveness of compression. This decision is quite subjective and depends on the context (e.g., patient critical situation, battery level, and network capacity).

In this paper, we explore various options in using compression during EEG monitoring. In particular, we design, implement, and test an on the fly algorithm to decide if the compression activity is worthily needed or not. This algorithm takes EEG data as input, gets most of its decision factors from the smartphone including battery level, size of data, and urgency of data, and finally engenders an output decision, which is to compress, or not the EEG data. To have a better control over the testing environment, we use text files as input rather than readings coming straight from the sensor. This setup allows instrumentation of the number of sensors, frequency of synchronization, urgency of data, different levels

978-1-4673-8760-6/15 $31.00 © 2015 IEEE

of battery and other mobile device resources (e.g., Memory, CPU).

II. RELATED WORK

Reducing the size of EEG data using compression algorithms has two main benefits: 1) reduce the size of locally stored data and 2) efficiently transmit over a wireless communication network. Thus, the overall cost and power consumption of the monitoring system will be minimized. When EEG applications entail the demand for data transmission and storage algorithms, which have low latency and power along with high fidelity, lossy or near-lossless data compression algorithms can be then used. These algorithms should combine with modern information theories and signal processing tools. These theories and tools include universal coding, universal predication, fast online implementation of multivariate recursive least squares [1], fast discrete cosine transform algorithms [2], and discrete Wavelet transform [3].

Since EEG is widely recognized as an important medical diagnosis and monitoring technique, information, which will affect the examination results must not be lost. In other terms, EEG lossless compression is critical to guarantee exact recovery of EEG data needed for diagnostic purposes. A comprehensive quality review of the lossless compression techniques applied to EEG signals has been reported in [3]. Two-stage lossless compression algorithms incorporating predictors and entropy encoders are introduced in [4, 5]. Specifically, context-based offset bias cancellation is applied to the predictive error in order to enhance the distribution of residues suitable for encoding [4, 5]. In [6], the authors presented lossless EEG data compression using integer wavelet transform to generate coefficients, which in turn use to train the neural network predictors. A compression ratio of 2.99 along with compression efficiency 67% is achieved. The authors in [7] presented hardware and software requirements to implement arithmetic coding, a form of variable length entropy encoding, for lossless compression of EEG data. This technique indeed depends on a probability model to satisfy compression. Xu et al. [3] proposed an algorithm to compress EEG multi-channel signals. The algorithm performs 1-D discrete Wavelet transform to preprocess raw EEG data and then rearranges the generated Wavelet coefficients into 2-D matrix form, and then it applies the No List Set Partitioning in Hierarchical Trees (NLSPIHT) algorithm to compress these coefficients. According to a significant promise of the compressive sensing paradigm in domains such as MRI, the authors in [8] investigated the performance of 18 various implementations of this paradigm when applied to EEG signals. The idea is to embed low power consumption lossy compression algorithm of the raw EEG data into portable and wearable device itself. While the compressive sensing (CS) paradigm contributes in achieving energy consumption, data compression, and device cost, it is "difficult for current CS algorithms to recover EEG with the quality that satisfies the requirements of clinical diagnosis" [9]. This is because EEG is non-sparse enough in terms of both time domain and transformed domains including the Wavelet domain. The

authors in [10] therefore proposed Block Sparse Bayesian Learning to address the CS problem, compress and recover EEG to design EEG tele-monitoring system (e.g., home-based e-Health) via wireless body-area networks. In this system, the compressed EEG data is sent to the nearest smartphone, which transmits it to a remote terminal (e.g., hospital) via the Internet.

The aforementioned techniques in principle suffer from a technical problem stating that while the developed algorithms achieve high performance, they consume high memory and result in high frequency memory access and computation complexity, especially those compression algorithms that use discrete Wavelet transform and neural networks. This problem makes those algorithms unsuitable to employ in mobile EEG-based monitoring systems wherein real-time and low power and memory present essential constraints.

III. MOBILE SYSTEM DEVELOPMENT

A. Mobile EEG monitoring requirements

EEG monitoring is characterized as a heavy process since it handles multi-channel continuous and voluminous data acquisition. It requires sufficient available resources in order to be conducted in an efficient manner. These resources are subject of scarcity, especially once the monitoring is handled on mobile devices wherein resources are constrained by set of parameters. The following requirements then need to be met:

- Energy consumptions/battery drainage: the battery drainage affects considerably the monitoring process and may cause disruption of data collection and transfer. Metering the battery level will allow optimizing the battery drainage for more sustainable monitoring.

- Available CPU and memory: EEG is a heavy signal, requiring CPU capacity and memory space to process continuous EEG signal. Evaluating CPU and memory capacity will help deciding on the fly whether to process it on the mobile device or delegate to a powerful back-end server.

- Network connection and bandwidth: it is a very important requirement to ensure an appropriate connection that supports an acceptable transfer time of voluminous and continuous EEG data.

- Lightweight communication protocol: it will assure reducing the communication overhead and data transfer.

B. Monitoring model

Figure 1 describes the EEG monitoring model we have used to conduct a mobile monitoring service. The model involves three main components: data acquisition module, intelligent module, and back-end server. EEG signals collected from sensors are relayed, processed (e.g., compressed) on a mobile device, and then sent to a back-end server for further processing if required. In the following, we describe the role of each of components involved in the monitoring process.

Data acquisition: EEG data is acquired from wearable sensors via mobile device. Sensors are attached to the head/scalp of the subject under monitoring. Data streams are transmitted using Bluetooth to a mobile device.

Intelligent module: it is a mobile application that implements two main components: resource monitor and the Compression Viability Algorithm (CVA). The resource monitor inspects and measures in a real time the available resources the mobile device. These resources include battery usage, network bandwidth, CPU and memory. The CVA determines whether it is viable to compress EEG data on the mobile device or send raw data to a back-end server. Raw EEG data is transferred to the server using RESTful Web services and JSON. If the CVA algorithm decision is to compress EEG data, then it will be converted into a compressed and sent to a back-end server using WebSockets.

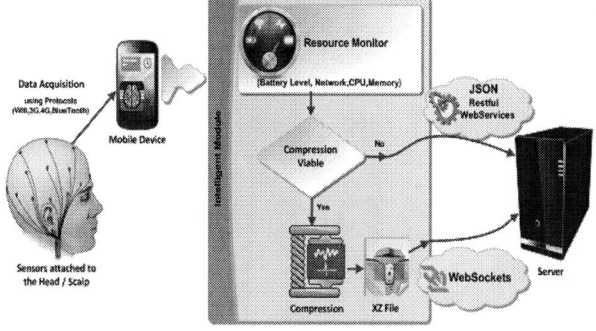

Fig. 1.   The main components of the EEG monitoring model

IV. COMPRESSION OF EEG DATA

In this section, we describe the key requirements of EEG data compression, and the CVA algorithm and its implementation on the mobile application.

A. Our compression algorithm

A considerable number of compression algorithms have been proposed in the literature [1-10]. The purpose of this study is not to develop a compression algorithm. Instead, we used an existing algorithm that is tested to be efficient, lossless, and lightweight to be executed on mobile devices. We particularly used the XZ data compression algorithm [11], which is a lossless data compression program, and file format. The main advantages of XZ are its high compression ratio (around 30%) and low required processing power, which make it very appropriate for our mobile EEG monitoring scheme, especially when devices are resource-constrained.

B. Compression viability algorithm(CVA)

The CVA algorithm gets the EEG data size as input, then meters and evaluates on the fly available resources on the mobile device as well as the network bandwidth in order to evaluate the compression cost. As depicted in Figure 2, it is worth compressing the EEG data when the data size is big enough. The CVA algorithm measures the compression gain in terms of the transfer time and current available resources.

```
Input: EEGData: Collection of EEG readings;   // coming from the EEG sensor
/* The below five inputs are obtained experimentally, beyond the scope
   of this algorithm.                                                  */
Input: N_min: Float;                  // Network bandwidth minimum threshold
Input: M_min: Float;                       // Memory minimum threshold
Input: B_min: Float;                      // Battery minimum threshold
Input: C_max: Float;                         // CPU maximum threshold
Input: S_min: Float;          // minimum size of EEG data to compress
Data: N : Float;                            // Network bandwidth
Data: M : Float;                            // Available memory
Data: B : Float;                        // Available battery level
Data: C : Float;                          // Available CPU level
Data: CompressedEEGData: Collection of bytes ;    // Compressed EEGData

function assessCompressionCost()
begin
    N ← getNetworkBandwidthFromOS();
    M ← getAvailableRAMFromOS();
    B ← getBatteryLevelFromOS();
    C ← getUnusedCPULevelFromOS();
    if (sizeof(EEGData) ≥ S_min) // if EEGData is big enough
    then
        if ((B ≥ B_min) AND (C ≤ C_max) AND (M ≥ M_min) AND (N ≥ N_min))
        then
            /* Compression is viable                                  */
            CompressedEEGData ← compress(EEGData);
            send(CompressedEEGData);
        else
            /* Compression is not viable                              */
            send(EEGData);
    else
        /* EEGData is too small, no need for compression              */
        send(EEGData);
```

/* The routine send() above sends a stream of bytes over the available network. */

Fig. 2.   Our compression viability algorithm

V. EXPERIMENTATION

A. Setup

We used the CHB-MIT scalp EEG dataset, which is publicly available on Physionet [12]. We also used real life data generated from a wearable Emotiv EEG headset [13]. Moreover, our experiments were performed on Galaxy Note SM-N900 running Android KitKat 4.4.2 as front end and an Intel(R) Core(TM) i5-2430M CPU @ 2.40GHz with 6 GB memory running 64-bit Windows 7 as a back-end server. We implemented a set of RESTful Web services to send EEG readings to the server and we used JSON, which is a lightweight communication protocol to communicate with the back-end server.

B. Scenarios

1) In the first scenario, we focused on evalauting the compression ratio of EEG data. We also evaluated the total compression time of scaling EEG data size; this includes the compression time, the transfer time, and the decompression time.

2) In the second scenario, we evaluated the cost of EEG compression for an increasing size of EEG data. This measured based on the mobile battery drainage.

C. Results and discussion

Figure 3 illustrates a steadily high EEG compression ratio, which remains stable in around 50% ratio, then it increases linearly to reach 85% compression ratio at a size of 10 MB EEG data. However, Figure 4 shows that EEG compression reduces significantly the file transfer time (FTT) compared to uncompressed file transfer time. The compression and decompression times (or CT+DT) are much smaller than the file transfer time as shown in Figure 5, thus can be neglected. Figure 6 shows that the mobile battery drains significantly when the size of EEG data increases gradually.

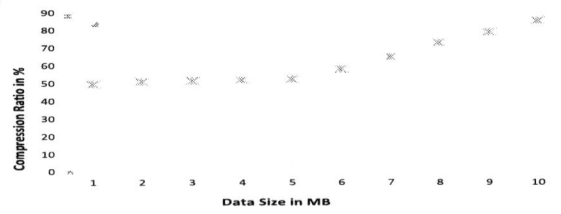

Fig. 3.   EEG compression ratio

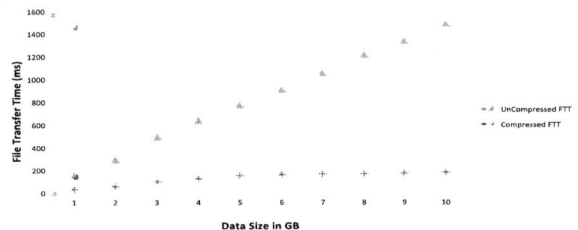

Fig. 4.   Compressed/uncompressed EEG file transfer time

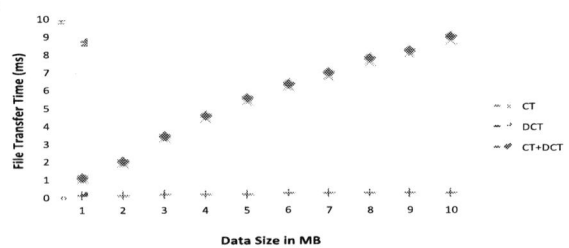

Fig. 5.   Compression/decompression time

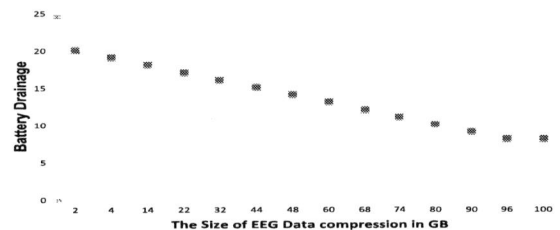

Fig. 6.   Battery drainage versus the size of EEG data compresison

VI. Conclusion

Continuous mobile monitoring is recognized to be a very challenging activity because of the constrained resources of the mobile devices, required to handle and process a continuous data collection. In terms of data size, a continuous EEG monitoring over a relatively large period of time generates a huge amount of data that is challenging to store locally on the device and to transfer to HIS. While compression might provide an interesting compression ratio, current compression algorithms are themselves energy and resources-hungry. The main contribution of this paper lies in developing, implementing and testing an algorithm that decides on the fly whether the compression activity is cost effective or not. Based on the decision of the algorithm, compression might or might not be performed. For the future work, we are planning to extend our CVA algorithm to formally calculate a cost compression function. We are eventually planning to conduct extensive evaluation of our CVA algorithm to cope with different monitoring situations and to consider other QoS rather than only compression cost.

References

[1] D. Birvinskas, V. Jusas, I. Martisius and R. Damasevicius, Fast DCT algorithms for EEG data compression in embedded systems. Computer Science and Information Systems, 2015, 12(1): 49-62.

[2] I. Capurro, F. Lecumberry, A. Martín, I. Ramírez, E. Rovira and G. Seroussi, Low-complexity, multi-channel, lossless and near-lossless EEG compression, proceedings of the 22nd International Conference on European Signal Processing, 2014, pp.2040-2044.

[3] G. Xu, J. Han, Y. Zou and X. Zeng, A 1.5-D Multi-Channel EEG Compression Algorithm Based on NLSPIHT, IEEE Signal Processing Letters, 2015, 22(8): 1118-1122.

[4] G. Antoniol and P. Tonella, EEG data compression techniques, IEEE Transactions on Biomedical Engineering, 1997, 44(2):105-114.

[5] N. Memon, X. Kong and J. Cinkler, Context-based lossless and near-lossless compression of EEG signals, IEEE Transaction on Information Technology in Biomedicine, 1999,3(3):231-8.

[6] N. Sriraam and C. Eswaran, Context based error modeling for lossless compression of EEG signals using neural networks, Journal of Medical Systems, 2006, 30(6):439-48.

[7] N. Sriraam, A High-Performance Lossless Compression Scheme for EEG Signals Using Wavelet Transform and Neural Network Predictors, International Journal of Telemedicine and Applications, vol. 2012, Article ID 302581, 8 pages, 2012. doi:10.1155/2012/302581

[8] D. O'Shea, R. McSweeney, C. Spagnol and E. Popovici, Efficient Implementation of Arithmetic Compression for EEG, Journal of Dostupn`yz WWW:<usb.issc.ie/download/file/288.pdf, 2011.

[9] A. M. Abdulghani, A. J. Casson and E. Rodríguez-Villegas, Compressive sensing scalp EEG signals: implementations and practical performance. Medical and Bioligical Engineering and Computing, 2012, 50(11): 1137-1145.

[10] Z. Zhang, T. Jung, S. Makeig and B. D. Rao, Compressed Sensing of EEG for Wireless Telemonitoring With Low Energy Consumption and Inexpensive Hardware, IEEE Transactions on Biomedical Engineering, 2013, 60(1): 221-224.

[11] XZ data compresison software: http://tukaani.org/xz/

[12] Physionet Database: http://www.physionet.org/pn6/chbmit/

[13] Emotiv Headset: https://emotiv.com

Highly Linear Low Voltage Low Power OTA using source-degeneration Technique and Universal Filter Application

Karima GARRADHI[1], Néjib HASSEN[1], Thouraya ETTAGHZOUTI[1], Kamel BESBES[1,2]

[1]Micro-electronics and instrumentation laboratory University of Monastir, Tunisia
[1,2]Centre for Research on Microelectronics and Nanotechnology of Sousse, Technopole of Sousse, Tunisia
karimagarradhi@gmail.com, nejib.hassen@fsm.rnu.tn, thourayataghzouti@yahoo.fr, Kamel.besbes@fsm.rnu.tn

Abstract— **In this paper, low voltage low power highly linear operational transconductance amplifier (OTA) using source-degeneration technique is presented. Source-degeneration techniques improve the bias current of the input differential pair when large signals are applied, thus, increasing circuit dynamic characteristics without affecting stand-by dissipation. The OTA is designed to operate with a ±0.55V supply voltage and consumes 0.5mW power. All simulations are performed by ELDO technology CMOS TSMC 90nm. The simulation results of this circuit showed a high DC gain of 42.58dB with a large bandwidth of 244MHz. Based on this circuit, a voltage mode universal filter has been implemented. The simulation results are in a good agreement with the theoretical calculations.**

Keywords—Operational transconductance amplifier (OTA), Current mirror, Dynamic range, Low-voltage Low-power, Filter.

I. INTRODUCTION

The Operational Transconductance Amplifier (OTA) is a basic building block in many analog circuit applications, such as including multipliers [1,2], continuous-time - filters [3, 4], voltage-controlled oscillators (VCOs) [5] and continuous-time sigma-delta modulators [6] which is converts an input voltage to an output current by means of transconductance. The most critical design requirement is a highly linearity. It means that voltages transform to currents without any distortion. particularly, in the case of delta-sigma modulators for high resolution Analog/Digital converters, it needs highly linearity transconductor to accomplish the required signal-to (noise +distortions) ratio.

In this paper, the implementation of a highly linear transconductor using a source-degeneration technique which operates at a low supply voltage ±0.55V with a reduced power consumption of 0.56mW is described. Based on this circuit, a voltage mode universal filter with a max central frequency equal to 20.7MHz has been implemented.

II. LOW VOLATGE LOW POWER OTA

A. Description of circuit operation

1) Conventional OTA

Without considering the resistance R, the conventional OTA circuit is presented in Fig .1. It is composed with a differential pair (M1, M2) and three high performance current mirrors with low operating voltage.

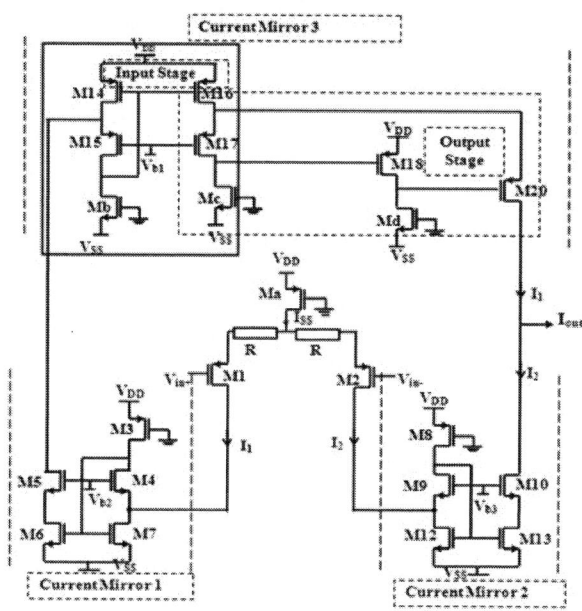

Fig. 1. Operational transconductance amplifier (OTA)

The current mirrors 1, 2 which represent the flipped voltage follower (FVF), reach a low input and a high output impedance that are respectively shown in the following two equations (1, 2) [7].

$$R_{in} = \frac{1}{g_{m14} g_{m17} r_{o14}} \qquad (1)$$

$$R_{out} = r_{o5} r_{o6} g_{m5} \qquad (2)$$

The current mirror 3 is proposed in [8]. Where the input stage has implemented by flipped voltage follower (FVF) and the output stage has implemented by a "super cascode transistor". The OTA uses a super cascode transistor to achieve a high slew rate, a large bandwidth and a high open loop gain [9].

This mirror has a very low input and a high output impedance. Their expressions are shown in Eq. 3 and Eq. 4:

978-1-4673-8760-6/15 $31.00 © 2015 IEEE   295

$$R_{in} = \frac{1}{g_{m17}g_{m16}r_{o17}} \quad (3)$$

$$R_{out} = g_{m18}r_{o18}g_{m19}g_{m15}r_{o15}r_{o14} \quad (4)$$

In this configuration, the differential input voltage between V_{in+} and V_{in-} is converted into differential current between I_1 and I_2. The pair transistors M1 and M2 are functional in the saturated region. Hence, the differential input voltage is shown in the following equation:

$$V_{id} = V_{in+} - V_{in-} = V_{GS1} - V_{GS2} = \sqrt{\frac{I_1}{\beta_p}} - \sqrt{\frac{I_2}{\beta_p}} \quad (5)$$

The differential output current can be expressed as:

$$I_{out} = I_1 - I_2 = \sqrt{2\beta_p I_{SS}} \times V_{id}\sqrt{1 - \frac{\beta_p V_{id}^2}{2I_{SS}}}$$

$$I_{out} \cong \sqrt{2\beta_p I_{SS}} \times V_{id} - \frac{1}{4}\sqrt{2\beta_p I_{SS}}\left(\frac{\beta_p}{I_{SS}}\right) \times V_{id}^3 \quad (6)$$

In equation (6), the higher order terms have been neglected. Therefore, the transconductance can be written as:

$$G_m = \sqrt{2\beta_p I_{SS}} \quad (7)$$

2) Proposed OTA

In order to improve the conventional circuit performance, we used source-degeneration technique by adding resistance (R) at the input differential stage which is illustrated in Fig.1. The differential input voltage is shown in the following equation:

$$V_{id} - RI_{out} = V_{GS1} - V_{GS2} = \sqrt{\frac{I_1}{\beta_p}} - \sqrt{\frac{I_2}{\beta_p}} \quad (8)$$

The differential output current can be expressed as:

$$I_{out} = \sqrt{2I_{SS}\beta_p}\,(V_{id} - RI_{out})\sqrt{1 - \frac{\beta_p(V_{id} - RI_{out})^2}{2I_{SS}}} \quad (9)$$

$$I_{out} \cong \sqrt{2\beta_p I_{SS}} \times (V_{id} - RI_{out}) \quad (10)$$

Therefore, the transconductance G_m can be written as:

$$G_m = \frac{dI_{out}}{dV_{id}} \cong \frac{g_m}{1 + g_m R} \quad (11)$$

When: $g_m = \sqrt{2\beta_p I_{SS}}$ is the transconductance of transistors (M1 and M2).

Consequently, source degeneration using resistors is the simplest topologies to linearize the transfer characteristic of the MOS transconductor. Although, the disadvantage of this configuration is the large resistor value needed to achieve a

wide linear input range. Since in this case, gm ≈ 1/R the obtained transconductance is restricted to small values. Moreover, this technique eliminates the electronic tuning capability of the transconductance because its value is set by the degeneration resistor.

B. Dynamic Study of Circuit

The output current can be expressed as:

$$I_{out} = \frac{r_1 g_{m1} V_{id}}{r_a + r_1 r_a g_{m1} - \dfrac{r_7}{\left(r_7 g_{m7} + \dfrac{r_7}{r_3} + \dfrac{r_4}{r_3} + 1\right)} - (R + r_1 + r_a + r_1 r_a g_{m1})}$$

$$I_{out} \approx -\frac{r_1 g_{m1}(r_7 g_{m7} r_3 + r_4)}{r_7 r_3 + R(r_7 g_{m7} r_3 + r_4)} V_{id} \quad (12)$$

III. SIMULATION RESULTS

The performance of the proposed CMOS OTA and Conventional OTA are verified by by ELDO technology CMOS TSMC 90nm. This circuit is operated at ± 0.55V supply voltage with a capacitive load of 1pF. The transistors aspect ratios of the two OTA are given respectively in Table.I.

TABLE I. ASPECT RATIOS OF THE TRANSISTORS

Transistors	Conventional OTA W(μm)/ L (μm)	Proposed OTA W (μm)/ L (μm)
M1 ,M2	0.5/0.1	100/0.18
Ma	1/0.1	50/0.13
M3,M8	-	0.5/0.1
M8,M3	1/0.1	0.5/0.1
M9, M10, M12, M13 M4,M5,M6,M7	50/0.13	25/0.13
M14, M15, M16, M17	20/0.1	20/0.1
Mb, Mc	0.2/0.1	0.2/0.1
M18	10/0.1	50/0.13
M20	100/0.1	10/0.1
Md	1/0.1	1/0.1

The DC transfer characteristic of the proposed OTA is shown in Fig. 2, which is achieved by varying the input voltage from -0.55V to +0.55V. It is understood that the proposed OTA has a good linearity in the interval [-0.55V, 0.4V].

From this characteristic, the trasconductance G_m of the OTA proposed is plotted and found to be more stable in the interval [-0.45V, 0.45V] and the maximum value of G_m is seen to be equal to 130μS (Fig. 3).

The frequency response of the two OTA is shown in Fig. 4. The proposed OTA reaches an open loop gain of 42.58 dB with transition frequency of 247.3MHz. For OTA conventional, a relatively small gain of 20.54 dB with a transition frequency of 13.65 MHz is noted.

978-1-4673-8760-6/15 $31.00 © 2015 IEEE

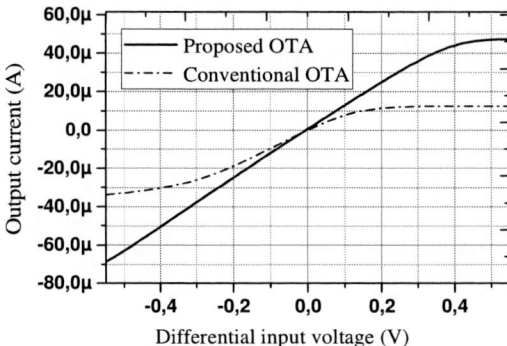

Fig. 2. DC Transfer characteristic

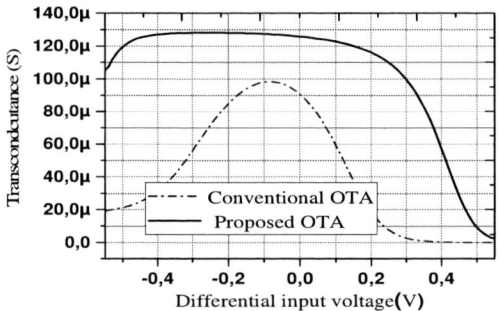

Fig. 3. The simulated Transconductance

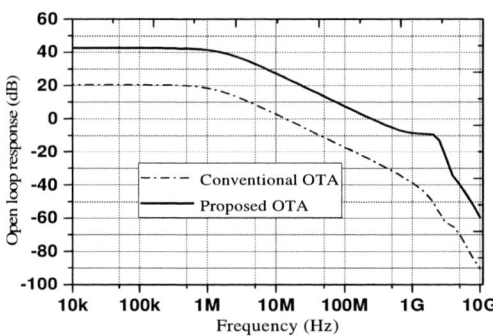

Fig. 4. Open loop frequency response of the OTA

In Table III, the simulated performances of the proposed OTA and the conventional OTA along with some of the recent works are summarized. To evaluate this work a figure of merit (FOM) can be defined as [10]:

$$FOM = \frac{Gain \ (GBW)}{(Power \ \text{sup} \ ply) \ (Power \ consumption)} \quad (13)$$

IV. APPLICATION

A. Theoretical Result

The range of the transconductance should be increased to achieve a good tuning range of the filter, therfore the filter performance is largely determined by the G_m value of the OTA. The topology adopted by the voltage mode universal filter with multi inputs and a single output is shown in Fig.5.

This designed voltage mode universal filter uses two proposed OTA which are controllable by three voltages. By controlling the inputs to the three input voltages terminal, low pass, high pass and band pass filters can be implemented. The differential output voltage can be expressed as:

$$V_0 = \frac{s^2 C_1 C_2 V_C + s \ g_{m2} C_1 V_B + g_{m1} g_{m2} V_A}{s^2 C_1 C_2 + s g_{m2} C_1 + g_{m1} g_{m2}} \quad (14)$$

Thus, the circuit can be used for different filters by setting the values of V_A, V_B and V_C as describe below.

If, $V_A = V_{in}$ and $V_B = V_C = 0$, a low pass filter (LP) is obtained.
If $V_C = V_{in}$ and $V_A = V_B = 0$, a high pass filter (HP) is obtained.
If $V_B = V_{in}$ and $V_A = V_C = 0$, a band pass filter (BP) is obtained. If $V_A = V_C = V_{in}$ and $V_B = 0$, a reject pass filter (BP) is obtained. and if $V_A = V_C = V_B = V_{in}$, a all pass filter (BP) is obtained.
Where, the pole frequency (ω_0) and the pole quality factor (Q_0) of these transfer functions can be given as follows:

$$\omega_0 = \sqrt{\frac{g_{m1} g_{m2}}{C_1 C_2}} \qquad Q_0 = \sqrt{\frac{C_2}{C_1}}$$

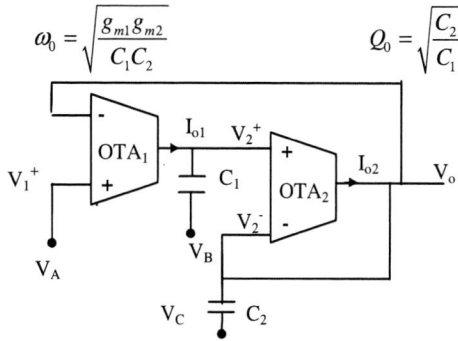

Fig. 5. Basic Circuit of a Universal Filter

B. Simulation results

The proposed filter operates at a low supply voltage of ±0.55V with a reduced power consumption of 1mW. After having used $C_1 = C_2 = 1pF$, a center frequency of to 25.2 MHz (Fig.6) has been obtained. The theoretical calculation allows us to obtain a centre frequency of 20.7 MHz, wherever $C_1 = C_2 = 1$ pF and $g_{m1} = g_{m2} = 130\mu S$ are used. The simulation results are in good agreement with the theoretical calculations.

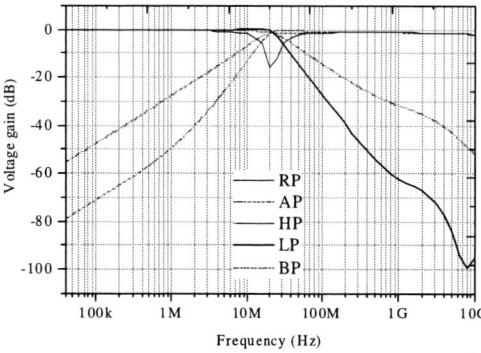

Fig. 6. Fequency responses of universal filter with varied controlled inputs

978-1-4673-8760-6/15 $31.00 © 2015 IEEE

TABLE II. PERFORMANCE COMPARISON WITH PREVIOUSLY REPORTED WORK

Performance design	Our circuit		OTA[11]	OTA [12]	OTA [13]
	Conventional OTA	Proposed OTA			
Technology CMOS	90nm	90nm	0.35um	0.5 um	0.18 um
Supply voltage (V)	±0.55	±0.55	3.3	2.5	±0.8
Power consumption (mW)	0.55	0.5	-	24	0.45
DC gain (dB)	20.54	42.58	-	78,2	73.6
GBW (MHz)	13.65	247.3	5	-	50.19
Phase margin (degrees)	76	92			84.5
Transconductance (µS)	98	130	40	-	350
Linear range (V)	[-0.1; 0.05]	[-0.55; 0.4]	-		±0.55
PSRR+ (dB)	47.34	79.68	-		65.88
PSRR- (dB)	49.35	98.43	-		63.33
CMRR (dB)	39.13	57.48	-	-	86.53
Input Noise Density (uV/√Hz)	10	5u	-		28.4
Output Noise Density (mV/√Hz)	0.5	0.1			-60.81
Positive slew rate (V/µs)	13.9	18.84	-		695.48
Negative slew rate (V/µs)	-10.31	-26	-		-438.90
Charge capacitive CL (pF)	1	1	-	-	10
Figure of Merit (dB.MHz/V.mW)	926.84	3829	-	-	-

V. CONCLUSION

This paper discusses the approach to low voltage and low power OTA design using a source degeneration technique to improve the linearity of the conventional differential pair balanced transconductance amplifier. The latter offers a large differential input voltage capability and a wide gm adjustment range. The resulting topology achieves a good input range of 1Vpp with a high GBW of 247,3MHz in ±0.55V supply voltage. Based on this circuit, a voltage mode universal filter has been proposed and a center frequency of 20.7 MHz has been obtained.

References

[1] M. Ismail and T. Fiez, Analog VLSI Signal and Information Processing. New York: McGraw-Hill, 1994.

[2] S. R. Zarabadi, M. Ismail, and C. C. Hung, "High performance analog VLSI computational circuits," IEEE J. Solid-State Circuits, vol. 33, no. 4, pp. 644–649, Apr. 1998.

[3] C. C. Hung, K. A. Halonen, M. Ismail, V. Porra, and A. Hyogo, "A low-voltage, low-power CMOS fifth-order elliptic GM-C filter for baseband mobile, wireless communication," IEEE Trans. Circuits Syst. Video Technol., vol. 7, no. 4, pp. 584–593, Aug. 1997.

[4] T. Y. Lo and C. C. Hung, "A wide tuning range G -C continuous time analog filter," IEEE Trans. Circuits Syst. I, Reg. Papers, vol. 54, no. 4, pp. 713–722, Apr. 2007.

[5] J. Galan, R. G. Carvajal, A. Torralba, F. Munoz, and J. Ramirez-Angulo, "A low-power low-voltage OTA-C sinusoidal oscillator with a large tuning range," IEEE Trans. Circuits Syst. I, Reg. Papers, vol. 52, no. 2, pp. 283–291, Feb. 2005.

[6] J. van Engelen, R. van de Plassche, E. Stikvoort, and A. Venes, "A sixth-order continuous-time bandpass sigma–delta modulator for digital radio IF," IEEE Solid-State Circuits, vol. 34, no. 12, pp. 1753–1764, Dec. 1999.

[7] Néjib Hassen, Houda Bdiri Gabbouj and Kamel Besbes, "Low-voltage high-performance current mirrors: Application to linear voltage-to-current converte," International journal of circuit theory and applications, Int. J. Circ. Theor Appl, vol. 39, January 2011.

[8] A.Torralba, R.G. Carvajal, J. Ramirez-Angulo and F. Munoz, "Output stage for low supply voltage high performance CMOS current mirrors" Electronics letters, vol. 38, pp.1528-1529, 2002.

[9] Manoj K. Taleja, and Manoj Kumar, "Bias Current Effect on Gain of a CMOS," IEEE International Conference on Advanced Computing & Communication Technologies, pp. 396-397, 2011.

[10] E.Kargaran, M.Sawan, Kh.Mafinezhad, H.Nabovati," Design of 0.4V,386nW OTA Using DTMOS Technique for Biomedical Applications," in the proceeding of IEEE 55th International Midwest Symposium on Circuits and Systems (MWSCAS),USA, 2012.

[11] Ko-Chi Kuo "A Linear MOS Transconductor Using Source Degeneration and Adaptive Biasing" IEEE TRANSACTIONS ON CIRCUITS AND SYSTEMS—II: ANALOG AND DIGITAL SIGNAL PROCESSING, VOL. 48, NO. 10, OCTOBER 2001

[12] Fernando A. P. Barúqui and Antonio Petraglia,Linearly Tunable CMOS OTA With Constant Dynamic Range Using Source-Degenerated Current Mirrors IEEE TRANSACTIONS ON CIRCUITS AND SYSTEMS—II: EXPRESS BRIEFS, VOL. 53, NO. 9, SEPTEMBER2006

[13] Karima GARRADHI, Néjib HASSEN,Kamel BESBES " Low Voltage Low Power Analog Circuit Design OTA Using signal attenuation Technique in Universal Filter Application " «12th international Multi-conference on Systems, Signals & Devices» 2015.

A Novel High Bandwidth Current Mode Instrumentation Amplifier

Zineb M'HARZI, Mustapha ALAMI
CSE Laboratory
INPT, 2 Avenue Allal El Fassi, Madinat Al Irfane
Rabat, MOROCCO
{Mharzi, Alami}@inpt.ac.ma

Farid TEMCAMANI
QUARTZ Laboratory
ENSEA, 6 Avenue Ponceau, 95014
Cergy-Pontoise Cedex, FRANCE
Farid.temcamani@ensea.fr

Abstract— **A novel structure of current mode instrumentation amplifier based on the second generation current conveyor is introduced. This circuit offers several advantages over the architectures of instrumentation amplifiers described in the literature. It presents a wide bandwidth with high current controlled gain and a high common mode rejection ratio. The second generation controlled current conveyor used in this topology has a very simple structure and operates at high frequency with a low supply voltage. PSPICE simulations results are included to support the proposed theory. These results are compared to those of instrumentation amplifier topologies proposed in the literature to prove the benefits of the new circuit.**

Keywords— *Second Generation Controlled Current Conveyor (CCCII); Instrumentation Amplifier (IA); Current Mode Instrumentation Amplifier (CMIA); Voltage Mode Instrumentation Amplifier (VMIA); Common Mode Rejection Ratio (CMRR); Integrated Circuit (IC).*

I. INTRODUCTION

Instrumentation amplifier (IA) is a very important analog block used in a wide range of application such as: medical instrumentation, read-out circuit of biosensors, data acquisition, and signal processing [1], where it is necessary to amplify the small differential signals in the presence of large common mode interference [2].

In voltage mode, the conventional instrumentation amplifier is realized with three operational amplifiers and seven resistors as shown in Fig. 1 [3]. The common mode rejection ratio (CMRR), which is one of the most important parameters of IA, is limited by the problem of resistor mismatching. In addition, the IA cannot operate at high frequencies due to the constant gain-bandwidth product of the classical operational amplifiers [1], [3]-[7]. Therefore, high common mode rejection ratio with high frequency is difficult to conceive in VMIA [3].

To overcome these drawbacks, much attention has been paid to the design of instrumentation amplifiers in current mode [1], [3]-[15].

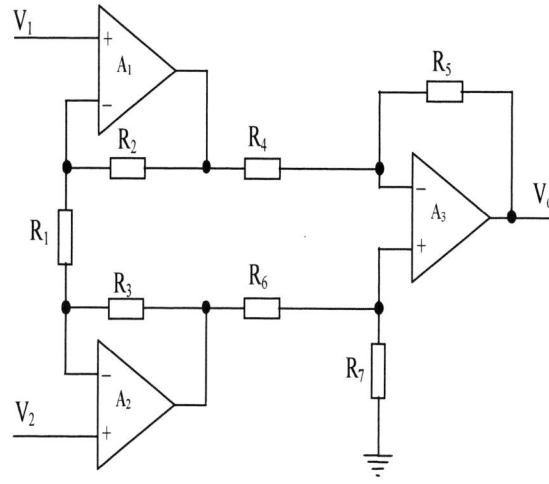

Fig. 1: Conventional voltage mode instrumentation amplifier using three operational amplifiers

These circuits have wide bandwidth independent of gain, great linearity, and large dynamic range [13]-[14]. The CMRR of these structures is high and independent of any resistor matching condition [3]-[7].

In the literature, certain CMIA structures using current conveyors as basic element have been reported [3]-[11]. This element has been chosen due to these advantages features of simplicity, wide bandwidth, wide dynamic range, and low power consumption, compared to other circuits used in current mode [4]-[5].

The second generation controlled current conveyor (CCCII) is a universal element used in a great number of analog functions in voltage mode and current mode. It comprises principally three ports, designated Y, X, and Z. A fourth terminal can be added to represent the bias current (Fig. 2) [16]-[20].

978-1-4673-8760-6/15 $31.00 © 2015 IEEE

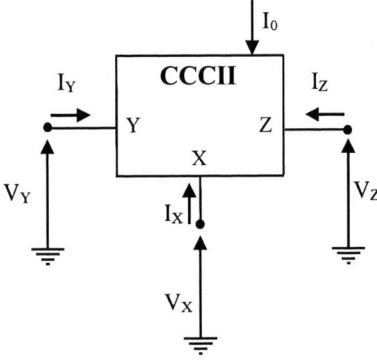

Fig. 2: Symbol of CCCII conveyor

In addition, it is characterized by parasitic impedances Z_X, Z_Y, and Z_Z at terminals X, Y, and Z respectively, as well as the voltage and current transfer gains $\beta(s)$ and $\alpha(s)$ respectively, as described in the matrix follows [19]:

$$
\begin{pmatrix} i_Y \\ v_X \\ i_Z \end{pmatrix} = \begin{pmatrix} \frac{1}{Z_Y} & 0 & 0 \\ \beta(s) & Z_X & 0 \\ 0 & \alpha(s) & \frac{1}{Z_Z} \end{pmatrix} \begin{pmatrix} v_Y \\ i_X \\ v_Z \end{pmatrix} \tag{1}
$$

The current conveyor provides a wide bandwidth independent of gain. It has simultaneously the properties of voltage and current followers. And, it present an intrinsic resistance R_X, in X terminal, controlled by the bias current [16]-[20].

This paper describes a novel approach of a current mode instrumentation amplifier which consists of three CCCIIs. This topology has particular advantages over other instrumentation amplifiers presented in the literature, namely: high CMRR, wide bandwidth, current controlled gain, high input impedance, simple circuit with no passive resistor, and low supply voltage, which make it attractive for integrated circuit implementations. The proposed circuit has been designed in BiCMOS 0.35 μm technology of ST [21] and has been compared to the previous proposed solutions.

II. INSTRUMENTATION AMPLIFIER

The proposed current mode instrumentation amplifier is shown in Fig. 3. This amplifier requires no external resistor to function. It consists of three second generation controlled current conveyors. The first one is a conveyor with positive current transfer (CCCII+), while the two others are current conveyors with negative current transfer (CCCII-).

The CCCII- used in this structure is simple. It composes of two NPN transistors for transferring different signals, and four current sources, of the same value, for the bias circuit (Fig. 4) [20]. In the real case, these sources will be replaced by CMOS current mirrors.

This conveyor is characterized by the resistance R_X related to the bias current by the following equation [20]:

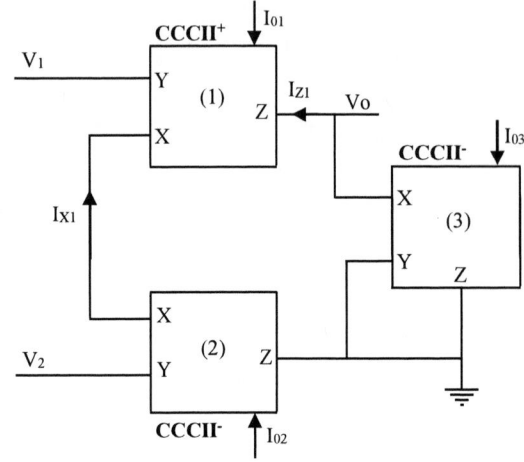

Fig. 3: Proposed current mode instrumentation amplifier

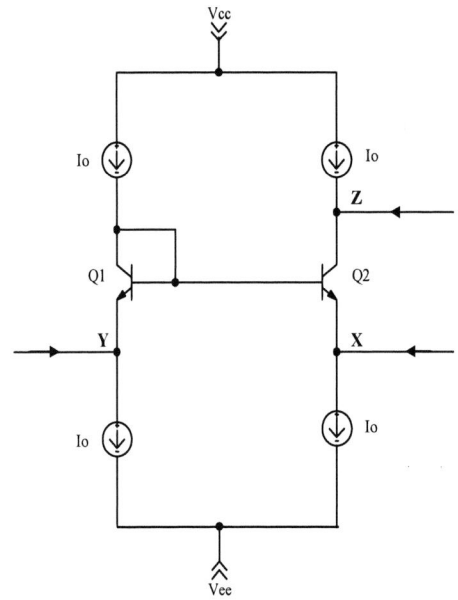

Fig. 4: Schematic implementation of CCCII-

$$
R_X = \frac{V_T}{I_0} \tag{2}
$$

With V_T the thermal voltage ($\approx$26 mV at 27 °C).

The CCCII+ is directly deducted from CCCII- by adding two buffer cells to reverse the current in the Z terminal and to keep the same electrical characteristics of the CCCII-.

In figure 3, the voltages V_1 and V_2, applied at high impedance terminals Y_1 and Y_2 respectively, are transferred to the low impedance terminals X_1 and X_2 by producing a current across the resistors R_{X1} and R_{X2}. This current is conveyed at the output terminal Z_1 and it is injected at the terminal X_3 of the third conveyor.

978-1-4673-8760-6/15 $31.00 © 2015 IEEE   300

The voltage output of the instrumentation amplifier can be written as:

$$V_0 = \frac{R_{X3}}{R_{X1} + R_{X2}}(V_1 - V_2) \tag{3}$$

Equation (4) represents the differential gain of the instrumentation amplifier:

$$A_d = \frac{V_o}{V_1 - V_2} = \frac{R_{X3}}{R_{X1} + R_{X2}} \tag{4}$$

From equations (2) and (4), we can see that the gain is controlled by the bias current of the conveyors.

Taking into account the voltage and current transfer gains $\beta(s)$ and $\alpha(s)$ of the conveyors and the parasitic capacitance at terminal Z_1, the output voltage becomes:

$$V_0 = \frac{\alpha_1 R_{X3}}{R_{X1} + R_{X2}} \frac{\beta_1 V_1 - \beta_2 V_2}{1 + sR_{X3}C_{Z1}} \tag{5}$$

The differential gain is given by:

$$A_{dm} = \frac{V_0}{V_{dm}} = \frac{1}{2} \frac{\alpha_1 R_{X3}}{R_{X1} + R_{X2}} \frac{\beta_1 + \beta_2}{1 + sR_{X3}C_{Z1}} \tag{6}$$

From this equation, we can see that the bandwidth depends on $R_{X3}C_{Z1}$.

The common mode gain is obtained with $V_1 = V_2 = V_{cm}$, and expressed as:

$$A_{cm} = \frac{V_0}{V_{cm}} = \frac{\alpha_1 R_{X3}}{R_{X1} + R_{X2}} \frac{\beta_1 - \beta_2}{1 + sR_{X3}C_{Z1}} \tag{7}$$

The CMRR is obtained from equations (6) and (7), and written as:

$$CMRR = \frac{A_{dm}}{A_{cm}} = \frac{1}{2} \frac{\beta_1 + \beta_2}{\beta_1 - \beta_2} \tag{8}$$

It is clear that an infinite value of CMRR can be easily obtained without any condition of matching resistances, and it is independent of bias current.

III. SIMULATION RESULTS

The proposed instrumentation amplifier is simulated for differential voltage gain of 10, 15 and 20 with the bias currents as $I_{03} = 10$ μA and $I_{01} = I_{02} = 200$ μA, 300 μA and 400 μA respectively.

The variation of the differential gains as a function of frequency is shown in Fig. 5. It is clear that the simulated gains 19.6 dB, 23 dB and 25.6 dB, for the above bias currents, are in good agreement with the theoretical values and they have a wide bandwidth equal to 90 MHz and independent to gain variations.

The CMRR frequency response of the current mode instrumentation amplifier for different values of gain is shown

in Fig. 6. In this figure, we see that the CMRR has a high value equal to 114.6 dB independent of bias current.

A comparison between the proposed circuit and the structures of instrumentation amplifiers presented in the literature is given in Table 1.

Our amplifier has remarkable advantages over other circuits at the wide bandwidth, and the low supply voltage while maintaining a simple structure and a current controlled gain.

Table 1 shows that the CMRR of the proposed circuit is higher than those of [1], [6], [9] and [13] but it is smaller than those of [10] and [11].

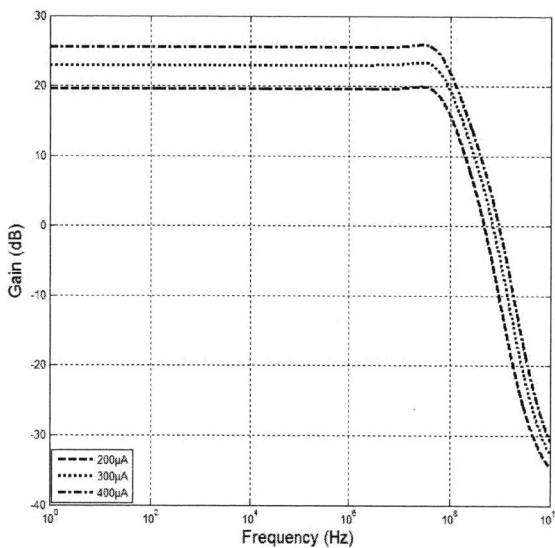

Fig. 5: Variation of gains in function of the frequency for three values of bias currents

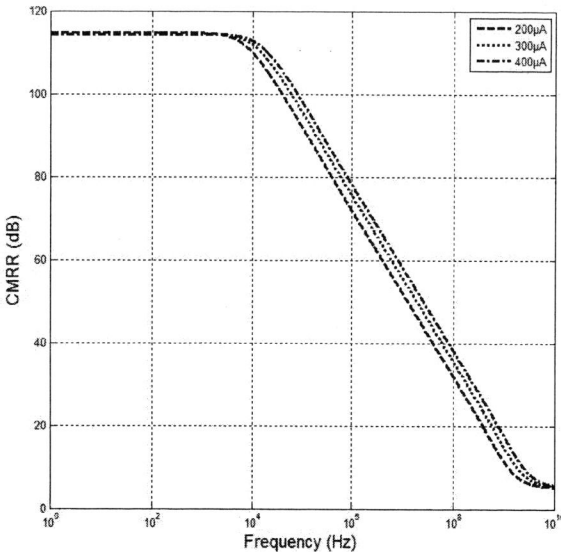

Fig. 6: CMRR frequency response of the circuit for different gains

TABLE I.   COMPARISON BETWEEN SOME INSTRUMENTATION AMPLIFIERS

Study	Vs (V)	Gain	Bandwidth (MHz)	CMRR (dB)	Control function	Technology	Number of building blocks used
[1]	-	10	1.2	76	Passive resistor	BJT	2 OFCCs and 2 passive resistors
[6]	-	29	0.592	95	Passive resistor	BJT	2 CCCIIs and 2 passive resistors
[9]	± 2.5	-	70	102	Current	BJT	2 CCCIIs and an active resistor
[10]	± 2.5	-	11	147	Current	BJT	3 CCCIIs
[11]	± 3.3	45	70	142	Voltage or current	CMOS	2 CCCIIs and an active resistor
[13]	± 1.5	-	10	64.5	Passive resistor	CMOS	3 OTRAs and 5 passive resistors
Proposed circuit	± 0.75	15	90	114.6	Current	BiCMOS	3 CCCIIs

IV. CONCLUSION

In this paper, we have presented a novel structure of current mode instrumentation amplifier. This CMIA is constituted of three simplified current conveyors and required no external resistors which make it an ideal choice for IC implementation.

The simplified topology and the low supply voltage of the circuit have not limited their performances. It has a large bandwidth equal to 90 MHz, a current controlled gain and a high CMRR equal to 114.6 dB.

The comparison between our instrumentation amplifier and the previous proposed solutions shows the superiority of our circuit in terms of high CMRR and wide bandwidth with low supply voltage. This circuit can be used in several applications in telecommunication and medical instrumentation.

REFERENCES

[1] Y. H. Ghallab, W. Badawy, K. V. I. S. Kaler, and B. J. Maundy, "A Novel Current-Mode Instrumentation Amplifier Based on Operational Floating Current Conveyor", IEEE Transactions on Instrumentation and Measurement, Vol. 54, No. 5, pp. 1941–1949, 2005.

[2] A. Worapishet, A. Demosthenous, and X. Liu, "A CMOS Instrumentation Amplifier with 90-dB CMRR at 2-MHz using Capacitive Neutralization: Analysis, Design Considerations, and Implementation", IEEE Transactions on Circuits and Systems I: Regular Papers, Vol. 58, No. 4, pp. 699-710, 2011.

[3] K. Koli and K. A. I. Halonen, "CMRR Enhancement Techniques for Current-Mode Instrumentation Amplifiers", IEEE Transactions on Circuits and Systems I: Fundamental Theory and Applications, Vol. 47, No. 5, pp. 622–632, 2000.

[4] B. Babaei and S. Mirzakuchaki, "High CMRR, Low Power and Wideband Current-Mode Instrumentation Amplifier", 24th IEEE Norchip Conference, pp. 121-124, 2006.

[5] B. Wilson, "Universal Conveyor Instrumentation Amplifier", Electronics Letters, Vol. 25, No. 7, pp. 470-471, 1989.

[6] S. J. Azhari and H. Fazalipoor, "A Novel Current Mode Instrumentation Amplifier (CMIA) Topology", IEEE Transactions on Instrumentation and Measurement, Vol. 49, No. 6, pp. 1272–1277, 2000.

[7] C. Galanis, and I. Haritantis, "An Improved Current Mode Instrumentation Amplifier", in Proceedings of ICECS, pp. 65–68, 1996.

[8] S. J. G. Gift, "An Enhanced Current-Mode Instrumentation Amplifier", IEEE Transactions on Instrumentation and Measurement, Vol. 50, No. 1, pp. 85–88, 2001.

[9] S. A. Tekin, H. Ercan, and M. Alçi, "Electronically Adjustable Wide Bandwith Instrumentation Amplifier", 14th National Biomedical Engineering Meeting, pp. 1-4, 2009.

[10] S. Maheshwari, "High CMRR Wide Bandwidth Instrumentation Amplifier Using Current Controlled Conveyors", International Journal of Electronics, Vol. 89, No. 12, pp. 889–896, 2002.

[11] H. Ercan, S. A. Tekin, and M. Alçi, "Voltage And Current-Controlled High CMRR Instrumentation Amplifier Using CMOS Current Conveyors", Turk Journal Electrical Engineering and Computer Science, Vol. 20, No. 4, pp. 547–556, 2012.

[12] T. M. Hassan and S. A. Mahmoud, "New CMOS DVCC Realization and Applications to Instrumentation Amplifier and Active-RC Filters", International Journal of Electronics and Communications, Vol. 64, No. 1, pp. 47–55, 2010.

[13] R. Pandey, N. Pandey, and S. K. Paul. "Electronically Tunable Transimpedance Instrumentation Amplifier Based on OTRA", Journal of Engineering 2013.

[14] C. Chanapromma, C. Tanaphatsiri, and M. Siripruchyanun, "An Electronically Controllable Instrumentation Amplifier Based on CCCCTAs", International Symposium on Intelligent Signal Processing and Communication Systems (ISPACS), 2008.

[15] W. Surakampontorn, V. Riewruja, C. Surawatpunya, and S. Yodladd, "Instrumentation Amplifiers Using Operational Transconductance Amplifiers", International journal of electronics, Vol. 71, No. 3, pp. 511-515, 1991.

[16] A. Fabre, O. Saaid, F. Wiest, and C. Boucheron, "High Frequency Applications Based on a New Current Controlled Conveyor", IEEE Transactions on Circuits and Systems I: Fundamental Theory and Applications, Vol. 43, No. 2, pp. 82–91, 1996.

[17] A. Fabre, O. Saaid, F. Wiest, and C. Boucheron, "Low Power Current-Mode Second-Order Bandpass IF Filter", IEEE Transactions on Circuits and Systems II: Analog and Digital Signal Processing, Vol. 44, No. 6, pp. 436–446, 1997.

[18] F. Seguin and A. Fabre, "2 GHz Controlled Current Conveyor In Standard 0.8 µm BiCMOS Technology", Electronics Letters, Vol. 37, No. 6, pp.329 -330, 2001.

[19] F. Seguin, B. Godara, F. Alicalapa, and A. Fabre, "2.2 GHz All-n-p-n Second Generation Controlled Conveyor in Pseudoclass AB Using 0.8 µm BiCMOS Technology", IEEE Transactions on Circuits and Systems II: Express Briefs, Vol. 51, No. 7, pp. 369–373, 2004.

[20] Z. M'harzi, M. Alami, and F. Temcamani, "Improvement of Current Mode Controlled Amplifier Using Current Conveyors," in Proceedings of WCSIT, pp. 2-4, 2014.

[21] STMicroelectronics, "0.35 µm SiGe BiCMOS", Grenoble (France), 1994.

Performance Enhancement Of $0.18\mu m$ CMOS On Chip Bandpass Filters Using H-Shaped Parasitic Element

Nessim Mahmoud[1], Anwer S. Abd El-Hameed[1,2], Adel Barakat[2], Adel B. Abdel-Rahman[1], Ahmed Allam[1], and Ramesh K. Pokharel[3]

[1]Egypt-Japan University of Science and Technology, Alexandria, 21934, Egypt, nessim.mahmoud @ejust.edu.eg

[2]Electronics Research Institute, Giza, 12622, Egypt, adel.barakat@eri.sci.eg

[3]Kyushu University, Fukuoka, 819-0395, Japan, pokharel@ed.kyushu-u.ac.jp

Abstract— A design of an improved open loop resonator on-chip bandpass filter for 60 GHz millimeter-wave applications using 0.18 μm CMOS technology is presented. The proposed on-chip BPF employs H-shaped parasitic structure inserted between two open loop coupled resonator. The adoption of a two open loop coupled resonators BPF and the utilization of two transmissions zero located at 48 and 80 GHz permit a compact size and high selectivity of the BPF. In addition, the parasitic H-shaped structure increases the capacitance between the two resonators, which enables a further reduction of the physical length of the filter and enhances the coupling between the resonators which improve the filter insertion loss .The proposed BPF has a center frequency of 60 GHz, an insertion loss of -2 dB, a 3dB band width of 13 GHz, and a core size 160×480 μm² with total chip size 680×280 μm² (including bonding pads).

Keywords—60GHz band; Bandpass; CMOS; Open loop Resonators

I. INTRODUCTION

Filters constructed with CMOS technology in modern RF wireless communication systems have great importance for enabling the integration of the millimeter -wave system on chip (SoC). The widespread use of the passive bandpass filters (BPFs) makes it necessary for enhancing the performance of the front-end circuits in radio communication system. Designing a bandpass filter (BPF) that can be easily integrated and has a high performance, is currently of a great interest. Although several related studies [1-7] have focused on continuous improvements of on-chip passive bandpass filters in the unlicensed 60 GHz frequency band, these filters have various disadvantages, e.g. high insertion loss, poor selectivity, wide band width, or large chip size [2, 8 and, 9] .The lumped component BPF [9] showed excellent selectivity response and small size, however, the 6.9 dB insertion loss is too high. In [8], the dual-mode loop resonator BPF presented good selectivity, nevertheless, the 4.9 dB insertion loss was still high, and moreover, the special input position and the large chip size restricted its application. The folded microstrip line BPF [3] successfully fabricated at 60 GHz with insertion loss of 2.7 dB, however it is difficult to be used in the 60 GHz band because it has more than 30 GHz bandwidth.

Many techniques were proposed to enhance the pass band insertion loss. In [10], the H-shaped defected ground structures (DGS) was used under coupled area of the BPF to increase the slow wave factor which improve the insertion loss. In [6], a folded open loop structure was used on patterned ground shields in first two stage open loop resonator.

In this paper, a compact size, low insertion loss and sharp-rejection on-chip bandpass filter is proposed and simulated in 0.18μm CMOS technology. The proposed BPF employs H-shaped parasitic structure inserted in the middle of two open-loop coupled resonators to reduce the size and to enhance the insertion loss of the proposed filter.

II. FILTER GEOMETRY AND DESIGN

The filter configuration comprised of two square-shaped microstrip open loop resonators; each has a perimeter about half-wavelength of the guided waves at the center frequency of 60GHz [10]. Fig.1 shows the geometry and dimensions of the BPF designed using 0.18μm CMOS technology.

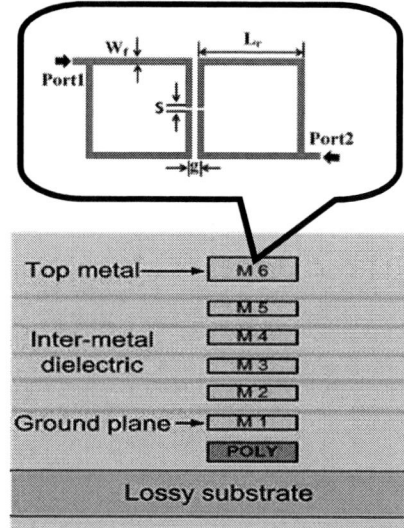

Fig.1. Structure of levels of metal in CMOS process and layout of a 60-GHz CMOS millimeter-wave square-shaped two open loops bandpass filter.

The square coupled resonators have a width W_f of 6 μm for 50 ohms impedance and length L_r of 345μm. The value of the open section S is 1.5 μm while, the coupling gap between resonators g has the value of 6.7 μm. The two open loop resonators are constructed on the top metal layer (M6) while, the ground plane is constructed on the bottom layer (M1). The standard thickness h_d of SiO$_2$ layer; distance between M1 and M6 is 8μm. To investigate the performance of the proposed BPF in terms of achieving good operations, the commercially available simulation software ANSFT HFSS® was used for numerical analysis. Fig. 2 shows the simulated results of S_{11} and S_{21} versus frequency of the BPF which illustrates that the insertion loss of the BPF is more than 4 dB.

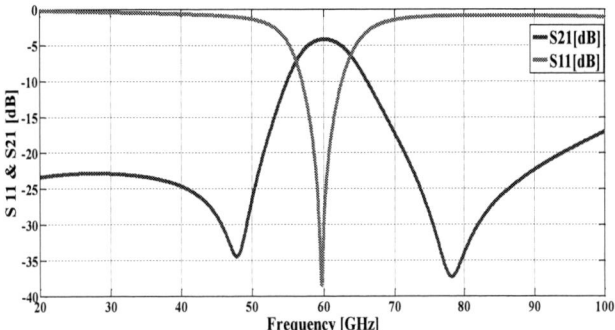

Fig.2. S-Parameters of the square-shaped BPF

Due to the close proximity of the ground plane to the coupled resonators at M6, a strong coupling path exists between M1 and M6. Accordingly, part of the transmitted power is coupled from the coupled resonators to the ground plane, which leads to degradation in the filter insertion loss.

To study the effect of SiO$_2$ thickness h_d on the insertion loss, a parametric study is conducted. Fig.3 presents a comparison between insertion loss of the proposed BPF at three different values of h_d considering the optimization of the filter dimensions to guarantee the same bandwidth and same coupling the. The results show that as h_d increases from 8 μm to 24 μm, the insertion loss is improved from -4.2 dB to -1.49 dB. This can be interpreted due to the increase of SiO$_2$ thickness, the coupling between the resonators and the ground plane decreases and hence less power is coupled to the ground plane.

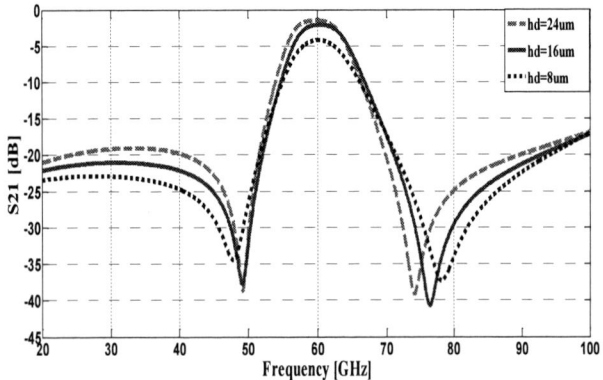

Fig. 3. S_{21} versus frequency at different SiO$_2$ thicknesses

TABLE I.   INSERTION LOSS AND RETURN LOSS AT DIFFERENT SIO2 THICKNESSES

h_d (μm)	g (μm)	L_r (μm)	S_{11} (dB)	S_{21} (dB)
8	6.7	340	-38	-4.216
16	10.2	355	-36	-2.09
24	13.6	363	-36	-1.49

III.   BPF INSERTION LOSS IMPROVEMENT USING RECTANGULAR SHAPE RESONATOR

To reduce the coupling between the coupled resonators and the ground plane and hence improve the insertion loss , the shape of the coupled resonators was modified; where the square-shaped resonator was replaced by a rectangular-shaped resonator with the same half-wavelength perimeter as shown in Fig.(4).

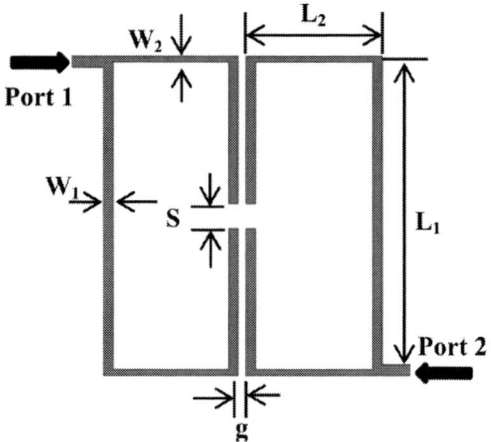

Fig. 4. The layout of the rectangular-shaped BPF

By increasing the arm length L_1, the coupling between the two resonators has been enhanced and by reducing the horizontal arm L_2, the coupled power to the ground plane was diminished which lead to further improvement in the insertion loss. The dimension of the coupling arm L_1 and the horizontal L_2 were optimized to L_1= 500 μm, L_2=200 μm, S=40 μm, w_1= 8 μm, w_2= 5 μm, with a coupling gap g =5 μm. The Simulated S_{11} and S_{21} versus frequency of the rectangular-shaped resonator are shown in Fig.5. The insertion loss rose to -2.5dB, while the upper frequency transmission zero moved to the right with a bandwidth of 14GHz.

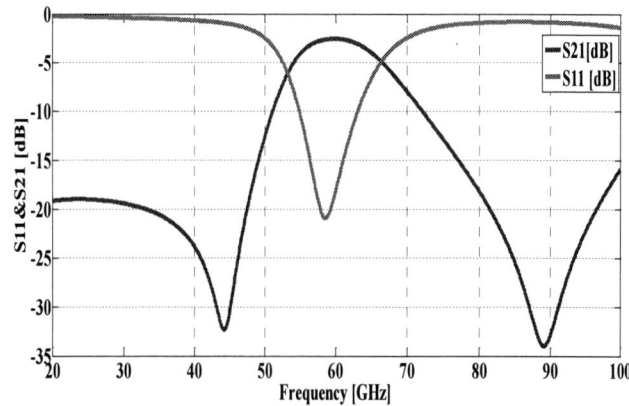

Fig. 5. Simulated S-Parameters of the rectangle-shaped BPF

978-1-4673-8760-6/15 $31.00 © 2015 IEEE   304

III. BPF WITH H-SHAPED ELEMENT

In this section, we discuss the final design of the proposed filter; H-shaped parasitic element was inserted between the coupled resonators in the same top metal layer (M6) as shown in Fig. 7. The existence of the parasitic element increases the capacitance between the two coupled lines which leads to a reduction in the physical dimension of the BPF. The length of the coupling arm L_1 is reduced to 480 μm while the length of the horizontal arm L_2 is reduced to 80 μm which in turn reduce the total filter size by 20%. This parasitic H-shaped consists of two rectangular lines of length 460 μm and width 10 μm is connected by a transverse rectangular line of width 10 μm and length 25.2 μm symmetrically.

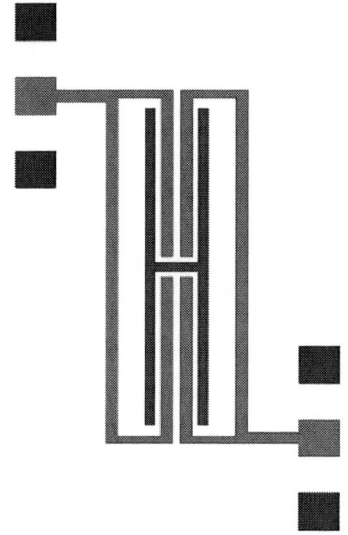

Fig. 7. The proposed BPF including pads

Moreover, by adding the parasitic element between the two resonators; a new coupling path was created through the parasitic element. Therefore, better passband performance and more enhancements in the insertion loss were achieved.

Fig.8 shows the simulated S_{11} and S_{21} versus frequency for the rectangular-shaped resonator including the H-shaped element. The insertion loss rose clearly from -4.2dB to -2dB, which is more than 2 dB enhancement in the BPF's insertion loss. Besides, a 13GHz bandwidth was obtained which imply more selectivity.

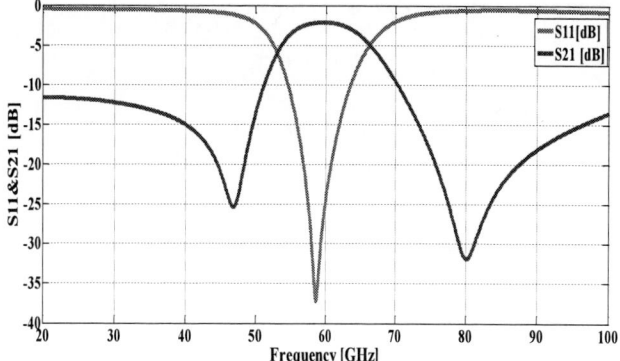

Fig. 8. Simulated S-parameters of the proposed BPF

The overall performance of the proposed BPF was summarized and compared with other recent publications in Table 2 .The simulated insertion loss and return loss of the proposed BPF are 2 dB and greater than 35 dB respectively, which represents a significant improvements over recently published CMOS BPFs.

IV. CONCLUSION

In this paper, an open loop bandpass filter is designed at 60 GHz in 0.18 μm CMOS technology for millimeter-wave applications with insertion loss of -2dB, bandwidth of 13GHz and a chip area of 0.2mm². By using rectangular-shaped resonator less power is coupled to the ground plane which in turn improves the insertion loss of the BPF. The parasitic H-shaped element inserted between the two coupled resonators further improves the insertion loss and reduce the filter size significantly.

TABLE II.   SUMMARY OF RESULTS AND PERFORMANCE

Ref	CMOS	Frequency	Core Size (mm²/ λ_g^2)	Insertion Loss	3-dB FBW (%)
[1] 2008	0.18 μm	70 GHz	0.65 × 0.67 (=0.436 mm²)/ 0.3 × 0.3 (=0.09 λ_g^2)	3.6 dB	25.7
[2] 2010	0.18 μm	65.25 GHz	0.36 mm ×0.21 mm (=0.08 mm²)	3.86 dB	36
[4] 2011	0.18 μm	77 GHz	0.31 × 0.26 (=0.08 mm²)/ --	3.9 dB	22
[5] 2011	0.18 μm	57.5 GHz	with pads: 0.91 ×0.65 (=0.59) mm²/0.34 × 0.24(=0.08) λ_g^2	2.8 dB	24
[6] 2012	0.13 μm	60 GHz	0.39 ×0.74 (=0.29) mm²/ --	4.1 dB	17
[7] 2013	0.18μm	60 GHz	with pads: 0.45 × 0.41 (=0.18) mm²/--	4 dB	38
[11] 2014	0.18 μm	77 GHz	0.056 × 0.353 (=0.02 mm²) / 0.03 × 0.18 (=0.0054 λ_g^2)	2.9 dB	27
[12] 2015	0.18 μm	56.0 GHz	With pads : 0.651 mm²	2.682	19
This work	**0.18 μm**	**60 GHz**	**0.16×0.48(=0.07 mm²)/0.064×0.192(=0.0112× λ_g^2)**	**2**	**21.6**

978-1-4673-8760-6/15 $31.00 © 2015 IEEE

ACKNOWLEDGMENT

The authors would like to thank the ministry of higher Education (MoHE)-mission department, and Egypt-Japan University of Science and Technology (E-JUST) for funding our work, in addition, this work was partly supported by a Grant-in-Aid for Scientific Research (B) from JSPS.KAKENHI (Grant no. 23360159).

REFERENCES

[1] K. Ma, S. Mou, and K. S. Yeo, "Miniaturized 60-GHz on-chip multimode quasi-elliptical bandpass filter," Electron Device Letters, IEEE, vol. 34, pp. 945-947, 2013.

[2] C.-Y. Hsu, C.-Y. Chen, and H.-R. Chuang, "70 GHz Folded Loop Dual-Mode Bandpass Filter Fabricated Using 0.18 m Standard CMOS Technology," Microwave and Wireless Components Letters, IEEE, vol. 18, pp. 587-589, 2008.

[3] Y.-C. Hsiao and C.-H. Tseng, "Design of 60 GHz CMOS bandpass filters using complementary-conducting strip transmission lines," in Microwave Symposium Digest (MTT), 2010 IEEE MTT-S International, 2010, pp. 1712-1715.

[4] S.-C. Chang, Y.-M. Chen, S.-F. Chang, Y.-H. Jeng, C.-L. Wei, C.-H. Huang, and C.-P. Jeng, "Compact millimeter-wave CMOS bandpass filters using grounded pedestal stepped-impedance technique," Microwave Theory and Techniques, IEEE Transactions on, vol. 58, pp. 3850-3858, 2010.

[5] Y.-M. Chen and S.-F. Chang, "A ultra-compact 77-GHz CMOS bandpass filter using grounded pedestal stepped-impedance stubs," in Microwave Conference (EuMC), 2011 41st European, 2011, pp. 194-197.

[6] R. K. Pokharel, X. Liu, R. Dong, A. Dayang, H. Kanaya, and K. Yoshida, "60GHz-band low loss on-chip band pass filter with patterned ground shields for millimeter wave CMOS SoC," in Microwave Symposium Digest (MTT), 2011 IEEE MTT-S International, 2011, pp. 1-4.

[7] A.-L. Franc, E. Pistono, D. Gloria, and P. Ferrari, "High-performance shielded coplanar waveguides for the design of CMOS 60-GHz bandpass filters," Electron Devices, IEEE Transactions on, vol. 59, pp. 1219-1226, 2012.

[8] C.-y. Hsu and H.-r. Chuang, "A 60-GHz Millimeter-Wave Bandpass Filter Using 0.18-µm CMOS," in Technology," IEEE Electron Device Lett, 2008.

[9] B. Dehlink, M. Engl, K. Aufinger, and H. Knapp, "Integrated bandpass filter at 77 GHz in SiGe technology," Microwave and Wireless Components Letters, IEEE, vol. 17, pp. 346-348, 2007.

[10] S. K. Parui and S. Das, "Performance enhancement of microstrip open loop resonator band pass filter by defected ground structures," in Antenna Technology: Small and Smart Antennas Metamaterials and Applications, 2007. IWAT'07. International Workshop on, 2007, pp. 483-486.

[11] L.-K. Yeh, Y.-C. Chen, and H.-R. Chuang, "A novel ultra-compact and low-insertion-loss 77. GHz CMOS on-chip bandpass filter with adjustable transmission zeros," in European Microwave Conference (EuMC), 2014 44th, 2014, pp. 1056-1059.

[12] D. Mat, R. Pokharel, K. Yoshida, and H. Kanaya, "Comparison of 60GHz CSRRs Ground Shield and Patterned Ground Shield On-chip Bandpass Filters Designed for 0.18 µm CMOS Technology."

A60-GHz double-Y balun-fed on-chip Vivaldi antenna with improved gain

Anwer S. Abd El-Hameed[1,2], Adel Barakat[2,3], Adel B. Abdel-Rahman[1], Ahmed Allam[1], and

Ramesh K. Pokharel[3]

[1]Egypt-Japan University of Science and Technology, Alexandria, 21934, Egypt, anwer.sayed@ejust.edu.eg

[2]Electronics Research Institute, Giza, 12622, Egypt, adel.barakat@eri.sci.eg

[3]Kyushu University, Fukuoka, 819-0395, Japan, pokharel@ed.kyushu-u.ac.jp

Abstract—A60-GHz double-Y balun-fed exponential tapered slot Vivaldi antenna-on-chip (AOC) is designed using standard 0.18μm six metal-layer CMOS technology. A double-Y balun feeding structure is used to make transition from coplanar to slot line. Three methods are developed for improving antenna radiation properties. First, an impeding longitudinal rectangular slits on the backed edge of the Vivaldi antenna are used to enhance the gain. Second, loading circular metal-strips are used as additional director into the slot area of Vivaldi antenna on M6. Finally, a planar arc reflector is used to inhibit the back lobe, contributing to the enhancement of gain and efficiency. The overall antenna size is very compact and equal to 700um×940μm. The influence of the antenna position on the radiation properties is also studied. The proposed antenna offers a simulated peak gain and a radiation efficiency of -1.9 dBi and 24%, respectively.

Keywords—60 GHz; antenna-on-chip; slits; balun; reflector.

I. INTRODUCTION

In recent years, an opening of 60GHz spectrum as an unlicensed band has increased the demand of the millimeter wave in applications such as broadband radio communication, radiometry and radio astronomy. This millimeter wave technology has created challenges to the design of the front end systems such as antennas. With the considerations of size miniaturization and RF- system-on-chip (SoC) integration, the commercial silicon CMOS process will be very effective for high-frequency circuit integration. However, it is very challenging to achieve either high gain or high efficiency on-chip antenna using CMOS technology due to the low resistivity and high permittivity of silicon substrate [1]. Artificial Magnetic Conductor (AMC) can be used for improving radiation characteristics in the broadside antennas [2, 3].

End-fire antenna plays an important role in communication and radar [4]. Tapered slot antenna (TSA) and quasi Yagi Uda antennas are two popular end-fire antennas reported in the literature. Vivaldi antennas which are in TSA category have received considerable attention due to their high gain, relatively wide band, simple structure, and wide use in radar applications. Their small lateral dimensions and simple integration make them excellent candidates for array development.

Almost always Vivaldi antennas are excited by a slotline; thus, feeding a TSA requires building a transition between the slot-line and other transmission media. In the case of microstrip or coplanar waveguide, the transition is a balun. Double-Y balun is a six-port transition balun with three balanced port and three unbalanced. It has various realizations with the same equivalent circuit as the transmission lines differ. Previous researches have investigated double-Y balun theoretically [5], [6] and practically [7-11].

To our knowledge, no work was proposed for Vivaldi antenna on 0.18 μm CMOS technology. However, many works have been reported on the Yagi antenna which focuses on the study of geometry, antenna characteristics and some application specific designs. Previous 60-GHz Yagi antenna designs using CMOS technology have been restricted in the possible application because of their very poor radiation efficiency of about 5%-16.5% [1, 12, 13]. Typically the low gain on-chip antennas could be used in chip-to-chip wireless communications to replace the metal interconnects between chips [14].

In this paper, a 60 GHz exponential tapered slot Vivaldi antenna with double Y balun-fed is developed using standard 0.18μm CMOS technology. The radiation characteristics are improved using different techniques. By inserting a longitudinal rectangular slits on the backed edge of the TSA antenna the gain is increased. On the other hand, loading circular metal-structures as an additional director into the slot area of the TSA shows significant improvement in the gain and efficiency. Besides, the planner arc reflector has the potential to inhibit the back lobe, contributing to the enhancement of antenna radiation properties. The effect of the antenna position on its performance is also studied.

II. ANTENNA DESIGN

In this section, the design of the proposed antenna is presented. The proposed design is constructed by widening the slot line in exponential tapered structure to produce Vivaldi shape, Fig.1. The antenna is designed using standard 0.18μm six metal-layer CMOS technology. Ten circular metal strips are placed on the slot area of the Vivaldi antenna. Furthermore, a planner arc reflector is constructed using metal vias between the two layers (M1 and M6) to reflect the waves travel in to the back direction.

TABLE I. GEOMETRICAL PARAMETERS OF EXPONENTIAL SLOT VIVALDI ANTENNA (μm).

W_v	L_v	W_f	R	d_x	d_y	L1	W1	L2	L3	L4
640	350	10	10	50	40	80	60	50	157.5	157.5
L_{slit}	W_{slit}	Wc	R_o	R_i	s	L_y	W2	R1	W_{step}	
50	15	20	350	270	4	30	30	24	520	

TABLE II. TABLE II: COMPARISON WITH PREVIOUSLY REPORTED 60-GHZ CMOS YAGI ANTENNAS

	Process Technology	Antenna size	Frequency	Gain	Efficiency
Quasi-Yagi[1]	Post-back-end-of-line	NA	65 GHz	-12.5 dBi (Measured)	5.6%
Yagi [12]	0.18-μm CMOS	1.1 mm× 0.95mm	60 GHz	-10.6 dBi (Measured)	10%
Yagi[13]	0.18-μm CMOS	2.45mm×1.8mm	60 GHz	-2.64 dBi (Simulated)	16.8%
Our work	0.18-μm CMOS	0.7 mm×0.94mm	60 GHz	-1.9 dBi (Simulated)	24%

The coplanar waveguide feeding technique in millimeter wave circuits is used to fit to the Ground-signal-Ground (GSG) probe which is used in the measurement phase. Consequently, a double-Y balun is used to couple the CPW to the slotline input of the TSA antenna. An expanded view of the double-Y balun junction is illustrated in Fig.1 (b). The pads dimensions are constructed to be 50 Ohms and the input and output impedances of the double-Y balun are also chosen to be 50 Ohms. The proposed antenna is designed on 5×5 mm² substrate at distance d=2mm from the center of substrate as it will be demonstrate in section IV.

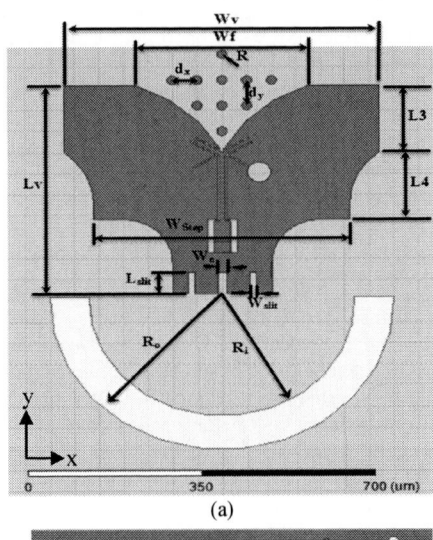

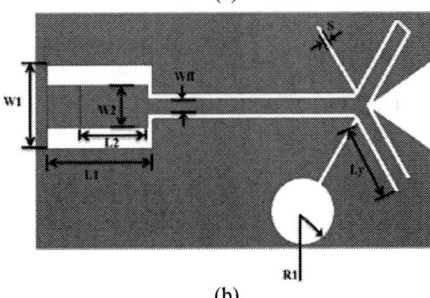

(b)

Fig.1. Antenna geometry: (a) view of the proposed double-Y fed exponential tapered slot Vivaldi on-chip antenna, (b) Expanded view of double Y junction structure.

To investigate the performance of the proposed Vivaldi antenna in terms of achieving good gain and efficiency, the commercially available simulation software ANSFT HFSS® was used for numerical analysis. Fig.2, 3 and 4 show that the proposed antenna operate at 60 GHz and cover all unlicensed band from 55 GHz to more than 65 GHz with a peak gain at the

end fire direction of -1.9 dBi and radiation efficiency of 24%. It also features both physically and electrically small dimensions; 700μm ×940 μm. Optimized dimensions of the proposed Vivaldi antenna are introduced in Table I.

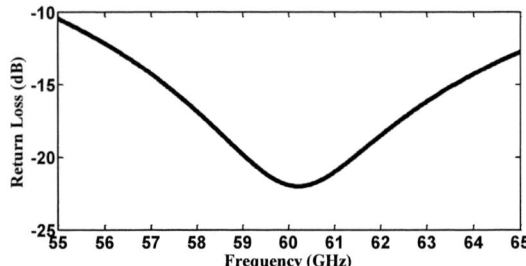

Fig.2. Simulated return loss versus frequency.

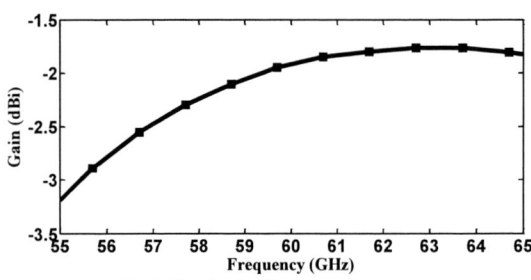

Fig.3. Simulated gain versus frequency.

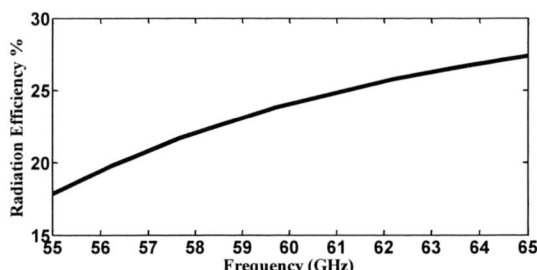

Fig.4. Simulated radiation efficiency versus frequency.

Compared with the widely used Yagi Uda antennas; this antenna design provides a fairly small size, high gain and high radiation efficiency. Table II demonstrate a comparison between the proposed Vivaldi antenna and other Yagi Uda designs.

III. DESIGN PROCEDURES

As a preliminary study and to track the effect of every modification on the antenna performance, a set of simulations were carried out; the results of each are detailed below. For

978-1-4673-8760-6/15 $31.00 © 2015 IEEE

clarifying the modification process, four modification steps are defined as follows: In step.1, the Vivaldi antenna without any addition structures shown in Fig.5 (a) was simulated. In step.2, three rectangular slits were inserted on the backed edge of Vivaldi antenna, Fig.5 (b). In step.3, circular metal-strips were impeded as additional director into the slot area of the antenna on M6, Fig. 5(c). In step4, a planner arc reflector was placed back to the antenna, Fig.5 (d).

The simulated curves of the return loss, gain and radiation efficiency for all steps are plotted in Fig. 6, 7, and 8 respectively. For the four prototypes, it is clear that the return loss is matched at 60GHz with more than 10 GHz bandwidth, as shown in Fig.6. After adding rectangular slits in step.2, the gain is increased from -4.92dBi to -3.69dBi due to the rectangle slit effect on the backed radiation. Moreover, the radiation efficiency is improved slightly from 16.5% to 17.5%. In addition, the effect of the planner arc reflector is also observed; the gain and efficiency are reached to -2.5dBi, 20.5% respectively.

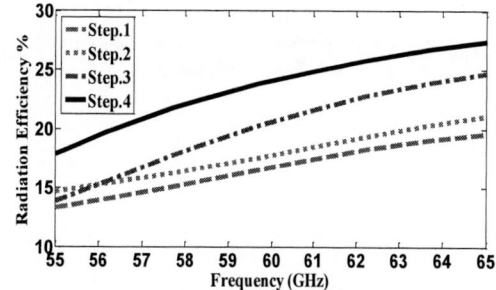

Fig.8: Simulated radiation efficiency versus frequency of four modifications.

Later, by setting the metallic strips as a grating into the slot-area of Vivaldi antenna in step.4, a 0.6dBi increment of gain at 60 GHz is achieved. Besides, the radiation efficiency is increased by 3.5%. The overall effect of these modifications on the gain and the efficiency are 3.02dBi and 7.5% enhancement, respectively.

IV. EFFECT OF ANTENNA POSITION ON ITS PERFORMANCE

For an integrated SoC, the typical substrate size is about 5mm×5mm. Thus, the effect of the antenna position on the antenna performance is required. Extensive parametric analysis has been done to obtain the optimum position of the antenna on the substrate to achieve the best performance. The Si substrate has a permittivity 11.9 and high conductivity of 10 S/m. For clarification, a comparison between the optimum location (position 2) and the center location (position 1) is presented. Both antennas are matched at the bandwidth of interest as shown in Fig. 10.

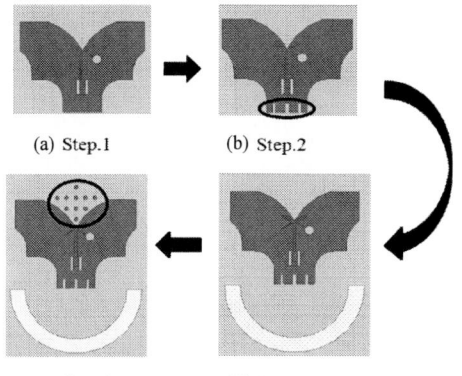

Fig. 5. Modification steps of designed antennas.

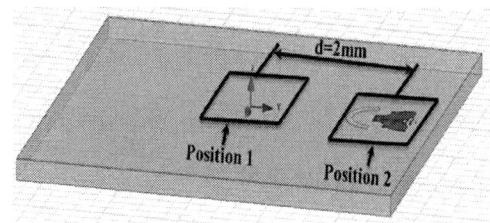

Fig.9. proposed antenna design in different locations.

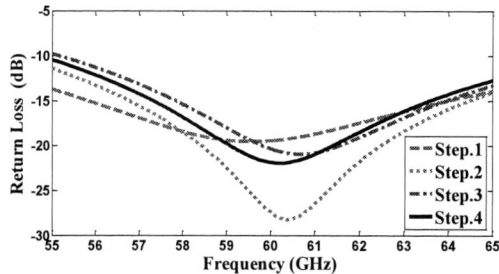

Fig.6. Simulated return loss versus frequency of four modifications.

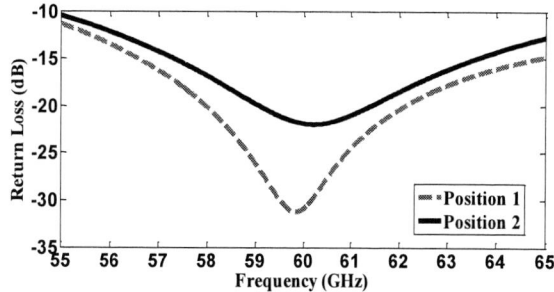

Fig.10. Simulated return loss versus frequency for two locations.

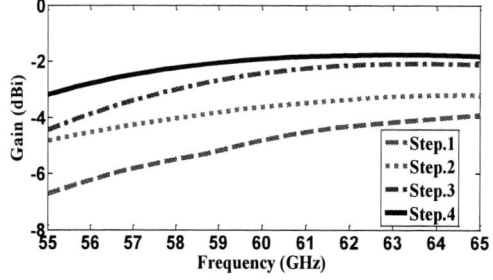

Fig.7. Simulated gain versus frequency of four modifications.

However, as depicted in Fig.11, the gain of the antenna located in position 2 is the higher than the located in position 1 due to better coupling with air. Owing to the small area of lossy substrate front to the antenna in position 2, the radiation

978-1-4673-8760-6/15 $31.00 © 2015 IEEE   309

efficiency is greatly superior to the other, Fig. 12. The antenna has an end-fire radiation pattern, Fig.13. The poor front-to-back ratio is due to the lossy substrate configuration.

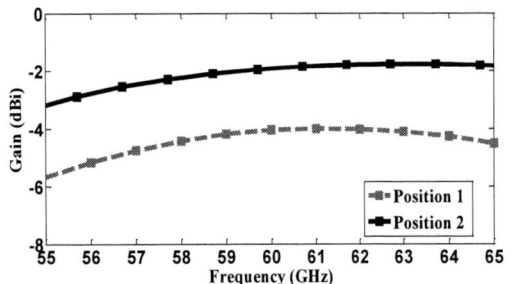

Fig.11. Simulated gain versus frequency for two locations.

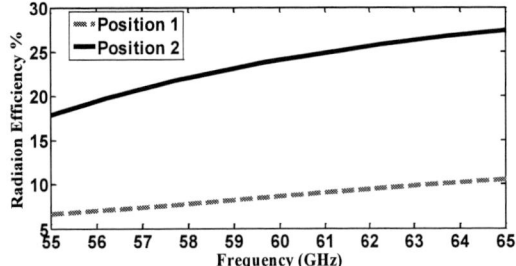

Fig.12. Simulated radiation efficiency versus frequency for two locations.

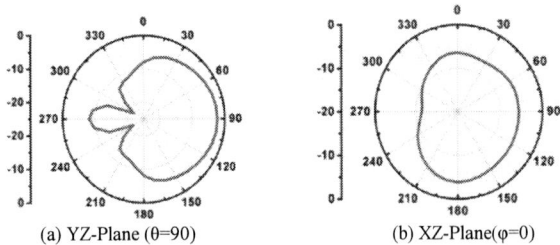

(a) YZ-Plane (θ=90)   (b) XZ-Plane(φ=0)

Fig.13. Simulated radiation pattern at 60GHz.

V. CONCLUSION

A double-Y balun-fed exponential tapered slot Vivaldi antenna on-chip with improved gain has been designed using standard 0.18 μm six metal-layer CMOS technology. Different techniques have been used to improve performance such as, loading of rectangular slits, grating elements and planar reflector. The antenna radiation properties have been presented in step by step manner for better understanding of the antennas operation. The overall size of the antenna is slightly affected by the techniques used, therefore, the overall size of the antenna remain compact of 700um×940um. The proposed antenna covers the FCC defined mm-wave band and has more than 10GHz of bandwidth (from 55GHz to more than 65GHz). The effect of the antenna position on the radiation characteristics is also studied. The proposed antenna has a simulated peak gain and a radiation efficiency of -1.9 dBi and 24% at 60 GHz, respectively. An array of the proposed antenna element is suitable for mm wave radar and microwave imaging applications.

ACKNOWLEDGMENT

The authors would like to thank the ministry of higher Education (MoHE)-mission department, and Egypt-Japan University of Science and Technology (E-JUST) for funding our work, in addition, this work was partly supported by a Grant-in-Aid for Scientific Research (B) from JSPS.KAKENHI (Grant no. 23360159).

REFERENCES

[1] Y. P. Zhang, M. Sun, and L. Guo, "On-chip antennas for 60-GHz radios in silicon technology," Electron Devices, IEEE Transactions on, vol. 52, pp. 1664-1668, 2005.

[2] A. Barakat, A. Allam, H. Elsadek, A. B. Abdel-Rahman, S. M. Hanfi, R. K. Pokharel, "Miniaturized 60 GHz triangular CMOS Antenna-on-Chip using asymmetric artificial magnetic conductor," IEEE, Silicon Monolithic Integrated Circuits in RF Systems (SiRF) meeting, 2015, pp.92-94.

[3] A. Barakat, A. Allam, H. Elsadek, H. Kanaya, R.K.Pokharel, "Small size 60 GHz CMOS Antenna-on-Chip: Gain and efficiency enhancement using asymmetric Artificial Magnetic Conductor," Microwave Conference (EuMC), 2014, pp.104 - 107.

[4] S.-G. Kim and K. Chang, "A low cross-polarized antipodal Vivaldi antenna array for wideband operation," in National radio science meeting, 2004, pp. 2269-2272.

[5] N. Marchand, "Transmission-line conversion transformers," Electron, vol. 17, pp. 142-146, 1944.

[6] B. Jokanović and A. Marinčić, "Microwave circuits based on six-port junction," in Telecommunications in Modern Satellite, Cable and Broadcasting Service, 2003. TELSIKS 2003. 6th International Conference on, 2003, pp. 218-222.

[7] B. Jokanovic and D. Markovic, "Wideband microstrip-to-waveguide transition using double-Y balun," Electronics Letters, vol. 42, pp. 1043-1045, 2006.

[8] J. Venkatesan, "Novel version of the double-Y balun: microstrip to coplanar strip transition," Antennas and Wireless Propagation Letters, IEEE, vol. 5, pp. 172-174, 2006.

[9] J. B. Venkatesan and W. R. Scott Jr, "Design of the Double-Y Balun for use in GPR Applications," in Defense and Security, 2004, pp. 383-398.

[10] B. Schiek and J. Köhler, "An improved microstrip-to-microslot transition (letters)," Microwave Theory and Techniques, IEEE Transactions on, vol. 24, pp. 231-233, 1976.

[11] M. Ruyu and F. Jiahui, "Microstrip to coplanar strip double-Y balun with very high upper frequency limitation," in Antennas and Propagation (APCAP), 2014 3rd Asia-Pacific Conference on, 2014, pp. 1402-1405.

[12] S.-S. Hsu, K.-C. Wei, C.-Y. Hsu, and H. Ru-Chuang, "A 60-GHz millimeter-wave CPW-fed Yagi antenna fabricated by using 0.18-CMOS technology," Electron Device Letters, IEEE, vol. 29, pp. 625-627, 2008.

[13] X. Bao, Y. Guo, and S. Hu, "A 60-GHz Differential on-chip Yagi antenna using 0.18-μm CMOS technology," in Antennas and Propagation (APCAP), 2012 IEEE Asia-Pacific Conference on, 2012, pp. 277-278.

[14] B. T. S. Rappaport, J. N. Murdock, and F. Gutierrez, "State of the art in 60-GHz integrated circuits and systems for wireless communications," Proceedings of the IEEE, vol. 99, pp. 1390-1436, 2011.

Design and Performance Analysis of Energy Conversion Chain, from Multilevel Inverter until the Grid

Chirine Benzazah, Loubna Lazrak, Mustapha Ait lafkih

Laboratory of Automatic, Energy Conversion and Microelectronics (LACEM),
Electrical Engineering Department, University of Sultan Moulay Slimane, Faculty of Sciences and Technology,
B.P: 523 Beni-Mellal 23000, Morocco
ch.benzazah@gmail.com ; lazrakfst@yahoo.fr ; idoimad@yahoo.fr

Abstract—**In this paper, a complete and comprehensive study has been explored on the design and performance analysis of grid side multi-level inverter in a high power renewable energy. The LCL-Filter and dq-PLL structure plus a multi-variable band-pass filter (FMVPB) must be used as the appropriate topologies to comply with the grid interconnection requirements. Details mathematical models and simulation using Matlab-Simulink/SimPower Systems for all these components have been given and evaluated. By providing future users a simulation tool for performance analysis and control design of any type of electrical power units installed with renewable energy production.**

Keywords—*control design; high power; LCL-filter; NPC three-level inverter; phase locked loop; renewable energy*

I. INTRODUCTION

In the last years, demand for high-power has increased in the renewable energy units, which requires new concepts in the design conversion systems to achieve the maximum efficiency and good performances. The appropriate technology and simplest alternative used in high-power applications is the three-level Neutral-Point Clamped - NPC inverter [1, 4].

The power quality injected into the grid must comply with interconnection standards [5]. However, the inverter produced a modulated output voltage that must be filtered to reduce harmonics. Compared with the L- and LC filters, the LCL filter has excellent harmonic suppression capability [6], [7], but the inherent resonant peak of the LCL filter may introduce instability in the whole system. Therefore, a passive damping due to its simple implementation must be added in series with the filtering capacity [8, 12], to improve the control of the system.

Other challenge met in the grid-connection is, to synchronize the converter with the grid to operate properly. Synchronous Reference Frame PLL (dq-PLL) is the most extensively utilized technique for the grid synchronization in three-phase system [13, 15], but its performance is not good when unbalanced faults and voltage distortion are happened. By adding the multi-variable band-pass filter (FMVPB) block to the dq-PLL structure [16, 17], the system will have certainly a good robustness against utility voltage distortions and unbalances. In this paper, a three-level grid side converter is designed for a 7MW renewable energy source. The general control scheme of the three levels grid side converter is given in section I. The comprehensive and detailed design procedure for the LCL filter is discussed in section II. Section III deals with the grid synchronization issue, dq-PLL and FMVPB-PLL are described, their performances are evaluated under general disturbance utilization grid conditions. Before the conclusion, the measurement results are presented and analyzed in Section IV.

II. OVERALL SYSTEM DESCRIPTION

The overall system studied is shown in Fig.1. It consists of an NPC three-level converter connected to the grid via an LCL filter. The control of this system is realised by an internal control loop current and other outer loop voltage DC, which maintains a constant value of the DC link voltage and provides the reference to the loop current. And an additional closed-loop for the control of neutral-point voltage deviation is necessary to avoid the problem of unbalance the neutral-point. A grid synchronization method (PLL) is used to synchronize the control with the grid phase angle. And finally, the modulation SPWM block calculates the proper states of the switches in order to obtain the reference input voltage.

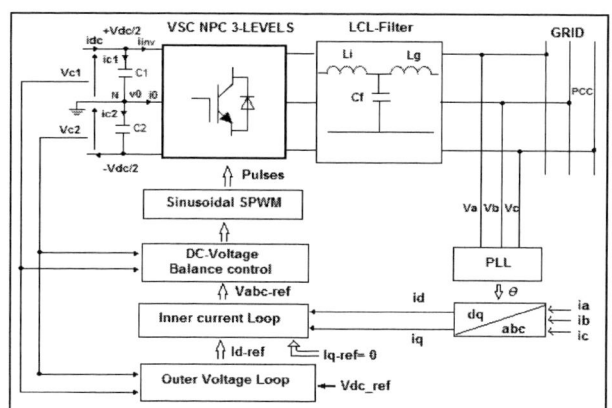

Fig. 1. Block diagram of the overall system

III. DESIGN OF THE FILTER CONNECTED TO THE GRID

The procedure for choosing the LCL-filter parameters requires the power rating of the converter, the grid frequency

978-1-4673-8760-6/15 $31.00 © 2015 IEEE

and the switching frequency as inputs. Detailed algorithm of LCL-filter design is shown in Fig. 2 [18].

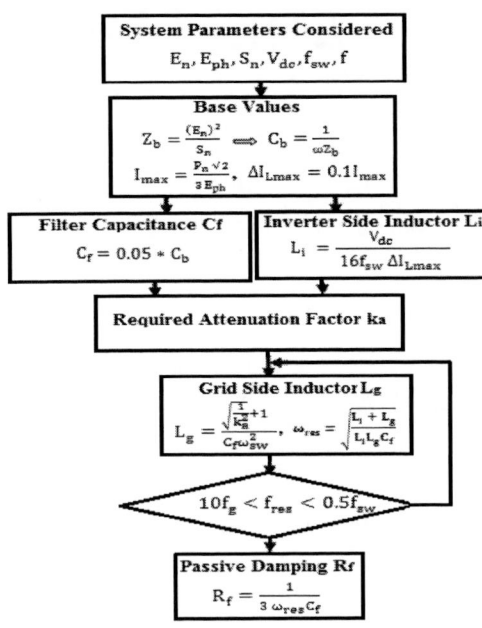

Fig. 2. LCL filter design algorithm

By application of the algorithm above, the different values of the filter components and system parameters considered are presented in the table below.

TABLE I.   PARAMETERS OF THE CONSIDERED SYSTEM AND VALUES OF FILTER COMPONENT

Values of the Considered System Parameters		Values of Filter Component	
Grid Line to line voltage	E_n=1380V	Inverter side inductor	L_i =0.443 mH
NPC DC-Link voltage	V_{dc}=2200V	Grid side inductor	L_g= 0.0645 mH
Output Power of the Inverter	S_n=7MVA	Capacitor filter	C_f =589.45 µF
Grid Frequency	f=50Hz	Damping Resistor	R_f =0.103 Ω
NPC Switching Frequency	f_{sw}=2000Hz	Resonant frequency	f_{res}=873.9 Hz
Required Attenuation Factor	ka= 20%	Resonant pulsating	ω_{res}=139 rad

IV. GRID SYNCHRONIZATION (PHASE LOCKED LOOP, PLL)

The most commonly used method today is the phase locked loop (PLL) used in the rotating dq-reference frame synchronous dq-PLL, shown in Fig. 3.

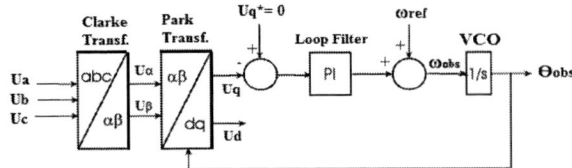

Fig. 3. Block diagram of classical dq-PLL

The dq-PLL structure is made by Clarke and Park transformation, a PI controller as a loop filter [13-15] and an integrator as that the voltage controlled oscillator (VCO). This system offers a good synchronization between the signals varying in time. But it is very sensitive to unbalance and distortion of grid voltage. A solution to these problems is adding a multi-variable filter, which ensures decoupling between sensitivity to disturbance and dynamic performance. So, a reliable detection of the frequency and phase will be achieved when voltage unbalances and distortions occur. The overall structure of the MVBPF+ dqPLL is shown in Fig. 4.

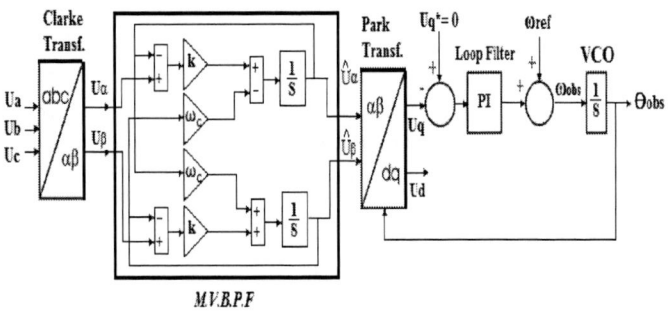

Fig. 4.   Block diagram of the new PLL structure with MVBPF

The design of the MVBPF+dq-PLL gain (k) is a critical point within this process. From the point of view of dynamic systems, high gains will imply higher dynamics [16], but stability may become unacceptable. The closed loop transfer function of the dq-PLL H(s) is similar to the transfer Function of second order G(s) as shown by Eq. 1 [17]:

$$H(S) = \frac{K_P S + K_i}{s^2 + K_P S + K_i}; \quad G(S) = \frac{2\xi\omega_0 S + \omega_0^2}{s^2 + 2\xi\omega_0 S + \omega_0^2} \quad (1)$$

Where (K_P, K_i) are the proportional and integral gains, ω_0 the natural angular frequency and ξ is the damping factor. By the identification [17]:

$$K_i = \omega_0^2 \quad ; \quad K_p = 2\xi\sqrt{K_I} \quad (2)$$

Where: $\xi = \sqrt{\frac{\ln(P/\pi)^2}{1+\ln(P/\pi)^2}}$; $\omega_0 = \frac{4.6}{\xi T_{set}}$

T_{set}: the settling time
P: the overshoot

The transfer function of the multi-variable band-pass filter is given by equation [16]:

$$H(s) = \frac{\hat{V}_{S\alpha\beta}(S)}{V_{S\alpha\beta}(S)} = \frac{\hat{V}_{S\alpha}(S) + j\hat{V}_{S\beta}(S)}{V_{S\alpha}(S) + V_{S\beta}(S)} = K\frac{(S + K) + j\omega_c}{(S + K)^2 + \omega_c^2} \quad (3)$$

Then, the final writing of the structural form of MVBPF which is introduced into the dq-PLL:

$$\begin{cases} \hat{V}_{S\alpha}(s) = \frac{K}{s}[V_{S\alpha}(S) - \hat{V}_{S\alpha}(S)] - \frac{\omega_c}{s}\hat{V}_{S\beta}(S) \\ \hat{V}_{S\beta}(s) = \frac{K}{s}[V_{S\beta}(S) - \hat{V}_{S\beta}(S)] - \frac{\omega_c}{s}\hat{V}_{S\alpha}(S) \end{cases} \quad (4)$$

So, the overall block diagram of MVBPF +dq-PLL is shown in Fig. 4.

This filter allows, at the frequency of 50Hz:

- The preservation of the input and output signals amplitude by the band pass filter.

- The conservation of the input and output signals phase regardless of the value of k.

- The best selectivity of the filter when k decreases.

V. SIMULATION RESULTS

On the basis of the design above, the simulation of the entire system is done in Matlab Simulink/SimPowerSystems to investigate more the robustness of the developed control system for 3-Level NPC inverter connected to grid via a LCL filter. It was simulated for 0.2 seconds with a sampling period of 5μs. Where, the system rated power is 7MW, the line voltage is 1380V, the switching frequency is 2 KHz, the fundamental frequency is 50Hz.

A. Output waveforms for both cases: with and without LCL-Filter (FFT analyses)

The Fig.5 and 6 show the output Phase-to-Phase (Vab) and Output current (Ia) before and after using the LCL-filter circuit.

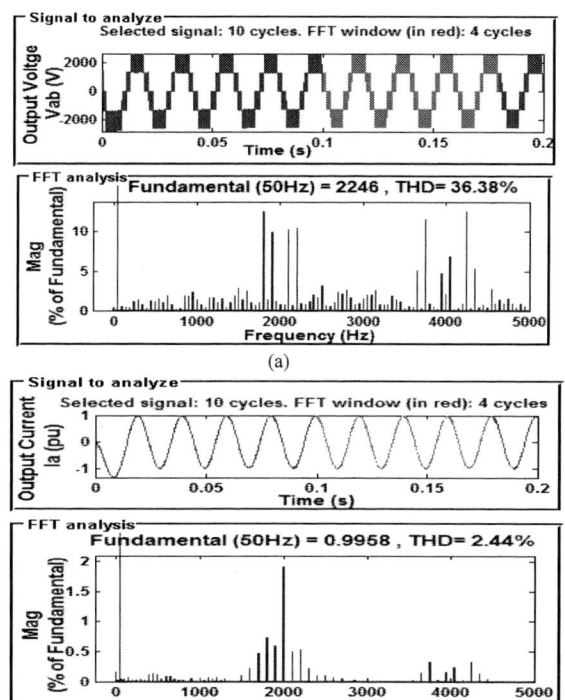

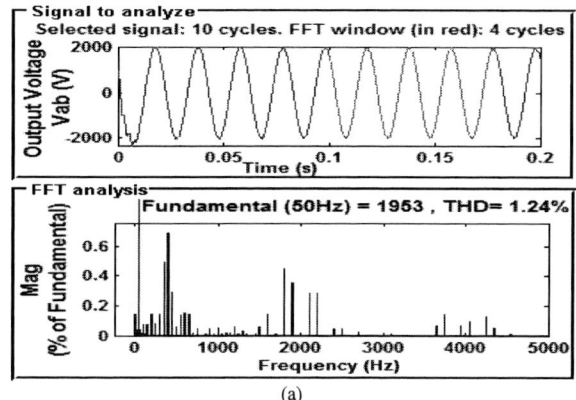

Fig. 5.   (a) FFT analysis of Output Phase-to-Phase voltage (Vab) waveform; (b) FFT analysis of Output current (Ia) waveform (without LCL-filter).

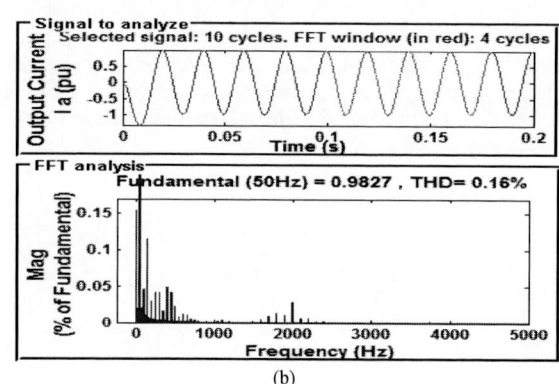

Fig. 6.   (a) FFT analysis of Output Phase-to-Phase voltage (Vab) waveform; (b) FFT analysis of Output current (Ia) waveform (with LCL-filter)

The waveform of Fig.5a shows that the output is square wave which is due to the presence of the harmonics; the THD in this case is equal to 36.38%. When LCL-filter was introduce (Fig.6) the output of the inverter seems to almost sinusoidal, and the THD is less than 2%. All these results affirm that the parameters of LCL-filter are well calculated.

On another side, It is found that the design meets the standards (IEEE Std 519-1992) keeping the THD within the limited range.

B. Outputs active and reactive power injected to the grid, voltage and current waveforms

The Fig. 7 compares the demanded active/reactive powers with the ones measured.

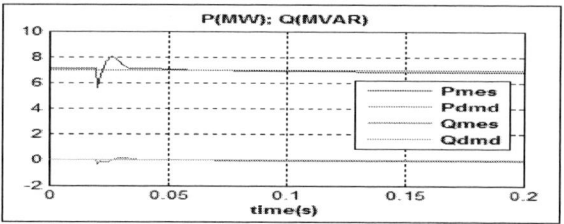

Fig. 7.   Demanded and Measured active/reactive powers.

The Fig. 8 shows the DC-link voltage during power flow.

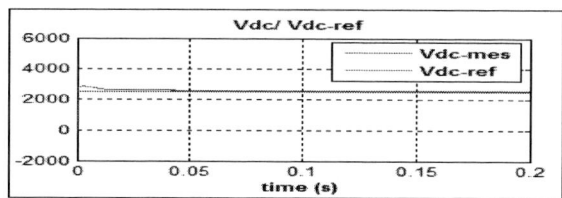

Fig. 8.   DC-link voltage during power flow

At the initial time, there is a low variation of DC-voltage. Therefore, there are small perturbations in Fig. 7, but this transient state has a very limited time and minimal overshoot. However, in steady state, the both of measured active/reactive powers follow closely that asked by the grid.

As clearly shown with these simulation results, the grid side converter is well controlled, fulfilling the necessary requirements, to send only the active electrical energy to the grid. The Fig. 9 represents the line current and tension for the phase (a) of grid side.

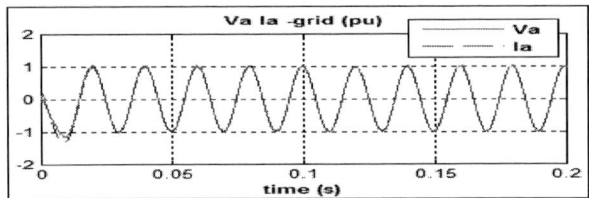

Fig. 9.   Line Current Ia and tension Va of grid side.

As shown in Fig. 9, the line current is in phase with the grid voltage when the reactive power is null.

C. Performance of grid synchronization under harmonic distortions

The Fig. 10 shows the results of application of the new PLL for voltage signals containing in addition to the fundamental term, the harmonics.

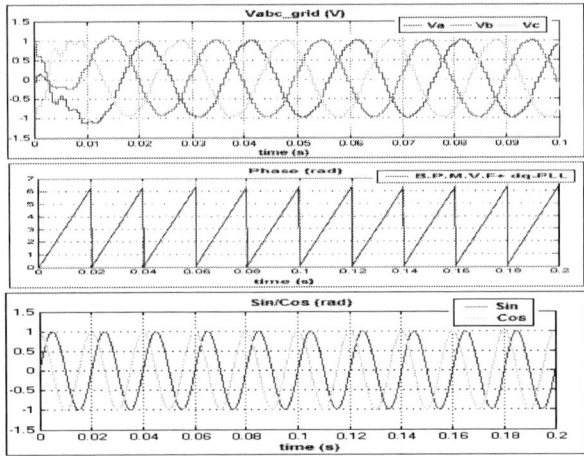

Fig. 10.   Simulation results of the FMVBP-PLL voltage source for balanced three phase containing harmonics

It is noted from the Fig. 10 that the new PLL structure, allows to obtain a phase angle non-oscillatory and linear periodically, and that the presence of multi-variable filter improves the quality of voltage signals.

VI. CONCLUSION

This paper presented the design and performance analysis of different basic components in the grid side 3-levels NPC inverter such as the LCL-Filter and grid synchronization, aiming to enlarge the grid compatibility with a high power renewable energy units. The proposed system acquires a good robustness against utility voltage distortions by adding a multi-variable band-pass filter (FMVPB) block to the classical structure dq-PLL.

In addition, the simulation results showed that the inverter control algorithm is successful in converting renewable energy DC power to AC power with suitable THD level according to IEEE Recommendation for supplying power to grid as well.

REFERENCES

[1] LG. Franquelo, J. Rodriguez, JI. Leon, S. Kouro, RC. Portillo, and MAM. Prats, "The age of multilevel converters arrives," IEEE Ind. Electron. Mag., vol. 2, no. 2, pp. 28–39, 2008.

[2] J. Rodriguez, LG. Franquelo, S. Kouro, JI. Leon, RC. Portillo, MAM. Prats, and MA. Perez, "Multilevel converters: An enabling technology for high-power applications," Proc. IEEE, vol. 97, no. 11, pp. 1786–1817, 2009.

[3] J. Rodriguez, JS. Lai, and FZ. Peng, "Multilevel inverters: A survey of topologies, controls, and applications," IEEE Trans. Ind. Electron., vol. 49, no. 4, pp. 724–738, 2002.

[4] C. Benzazah, M. Ait Lafkih, and L. Lazrak, "Comparative study between two topologies three-phase inverters, conventional 2-level and NPC 3-level with two methods different of control, SPWM and SWM," International Journal of Innovation and Applied Studies, vol. 9, no. 2, pp. 841–852, 2014.

[5] TM. Blooming and DJ. Carnovale, "Application of IEEE Std 519-1992 harmonic limits," In Pulp and Paper Industry Technical Conference, 2006. Conference Record of Annual, pp. 1-9. IEEE, 2006.

[6] H. Cha and Tk. Vu, "Comparative analysis of low-pass output filter for single-phase grid-connected Photovoltaic inverter," In Applied Power Electronics Conference and Exposition (APEC), 2010 Twenty-Fifth Annual IEEE, pp. 1659-1665. IEEE, 2010.

[7] J. Lettl, J. Bauer and L. Linhart, "Comparison of different filter types for grid connected inverter," PIERS Proceedings, Marrakesh, Morocco (2011).

[8] TCY. Wang, Y. Zhihong, S. Gautam, and Y. Xiaoming, "Output filter design for a grid-interconnected three-phase inverter," In Power Electronics Specialist Conference, vol. 2, pp. 779-784, IEEE 34th Annual, 2003.

[9] W. Zhao, and C. Guozhu, "Comparison of active and passive damping methods for application in high power active power filter with LCL-filter," In Sustainable Power Generation and Supply, 2009. SUPERGEN'09. International Conference on, pp. 1-6. IEEE, 2009.

[10] M. Liserre, F. Blaabjerg, and S. Hansen, "Design and control of an LCL filter-based three-phase active rectifier," IEEE Trans. Ind. Appl., vol. 41, no. 5, pp. 1281–1291, 2005.

[11] C. Zhang, D. Tomislav, CV. Juan, and MG. Josep, "Resonance damping techniques for grid-connected voltage source converters with LCL filters—A review," In Energy Conference (ENERGYCON), 2014 IEEE International, pp. 169-176. IEEE, 2014.

[12] DK. Choi and KB. Lee, "Stability improvement of distributed power generation systems with an LCL-filter using gain scheduling based on grid impedance estimations," Journal of Power Electronics, vol. 11, no. 4, pp. 599-605, 2011.

[13] EM. Adžic, MS. Adžic, and VA. Katic, "Improved PLL for Power Generation Systems Operating under Real Grid Conditions," Electronics, vol. 15, no.2, pp. 5-12, 2011.

[14] LGB. Rolim, DR. Costa, M. Aredes, "Analysis and Software Implementation of a Robust Synchronizing PLL Circuit Based on the pq Theory," IEEE Transactions on Industrial Electronics, vol. 53, no. 6, pp. 1919-1926, 2006.

[15] V. Kaura, V. Blasko, "Operation of a phase locked loop system under distorted utility conditions," In Applied Power Electronics Conference and Exposition, 1996. APEC'96. Conference Proceedings 1996., Eleventh Annual, vol. 2, pp. 703-708. IEEE, 1996.

[16] GX. Qiang, WY. Wu, and HR. Gu, "Phase locked loop and synchronization methods for grid-interfaced converters: a review," Przegląd Elektrotechniczny, vol. 87, no. 4, pp. 182-187, 2011.

[17] EM.Adžic, MS.Adžic, and VA.Katic, "Improved PLL for Power Generation Systems Operating under Real Grid Conditions," Electronics,Vol.15, no.2, pp. 5-12, 2011.

[18] T.Ngo and S.Santoso, "Grid-connected photovoltaic converters: Topology and grid interconnection," Journal of Renewable and Sustainable Energy 6, " no. 3 (2014): 032901.

ParametersIdentification of a Thin-FilmPhotovoltaic Panels

Abdellatif OBBADI*, Youssef ERRAMI, AbdelkrimELFAJRI, Mustapha AGUNAOU, Mohammadi BENHMIDA, Smail SAHNOUN

Laboratory: Electronics, Instrumentation andEnergy, Team: ExploitationandProcessingof Renewable Energy
Faculty of ScienceUniversity,ChouaibDoukkali,Department of Physics
RouteBenMaachou, 24000El-Jadida,Morocco
obbadi.a@ucd.ac.ma, errami.y@ucd.ac.ma, elfajri@hotmail.com, mostaf_agn@yahoo.fr, benhmida@gmail.com,
sahnoun.s@ucd.ac.ma

Abstract—In this paper, our work consist to identify the parameters of both complete (5-parameter) and simplified (4-parameter) single-diode PV models by non-iterative and iterative methods.The objective isto predict the behavior of a Thin-Film module under real environmental conditions. A new parameter Series/Parallel Ratio (SPR) ranking photovoltaic modules is defined. According tothe value SPR, we can neglect the series or shunt resistance of single-diode model without affect the accuracy. The results obtained with non-iterative and iterative methods are compared with experimental data. Simulations are performed in the MATLAB/Simulink environment.

Keywords-single-diode; Thin-Film;Photovoltaic (PV)module; parameter identification; iterative method; non-iterative method.

I. INTRODUCTION

The new generation Thin-Film Photovoltaic Module(amorphous silicon, CIS (copper indium diselenide) and other technologies, CdTe(cadmium telluride), etc.), gradually appearing on the market. While these modules have performances (in terms of efficiency) lower than the conventional modules (monocrystalline or polycrystalline) [1]. They therefore require larger capturing area for the same installed power.However, their manufacturing techniques show great potential for a significant drop in PV cost. However, the use of these modules, and therefore their industrial development on a large scale is still relatively limited, partly because of the uncertainty of users for these new technologies. One of the obstacles to this development is the lack of reliable procedure for the simulation module and therefore the difficulty to evaluate the performance as compared to a PV system equipped with monocrystalline or polycrystalline modules [1- 4]. There is currently no consensus in the scientific community of PV on a physical model [1, 4].However, for fast and reliable PV system design, simulator efficient, fast and accurate is essential. This simulation tool can be used to predict the behavior of Maximum Power PointTracker (MPPT), to estimate the efficiency of the PV system and to study the interaction between the power converter and PV generators [1-4]. The iterative methods are often slow and their convergences to an accurate solution are not always assured. Specifically, the convergence to the exact solution depends on the initial value, the accuracy and efficiency of the algorithm used [1]. Therefore, iterative methods cannot be used intracking applications of MPPT for on-line [4]. In this article, a non-iterative method of parameter estimation is used to reduce the

complexity of the PV model. And a new simplified 4-parametersingle-diode PV model is proposed. The model aims to reduce computation time and complexity without losing accuracy. Finally, the objective of this article is to compare the non-iterative and iterative methods are: Newton-Raphson [3], Halley's method [5] to estimate solar modules of Thin-Film parameters. The simulation results are compared with experimental data to validate the different methods used. We then found that the Thin-Film module behaves as a crystalline module and perfectly obeys to the standard model.

This work is organized as follows.Section II discusses in brief the modeling of photovoltaic module based on a single-diode model.Section III describes simulation, validation and comparative analysis.Finally, conclusions are given.

II.MATHEMATICAL MODELING OF PHOTOVOLTAIC MODULEBASED ON A SINGLE-DIODE MODEL

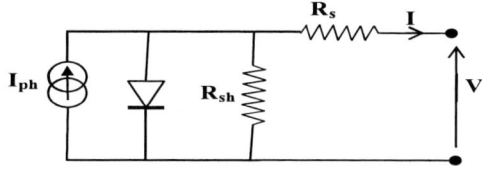

Figure 1.   Model of a PV cell

A single diode model consists of a current source, diode,series, and shunt resistances, as shown in Fig. 1. Thecurrent equation is given as follows:

$$I = I_{ph} - I_0\left[e^{\left(\frac{V+IR_s}{N_sV_t}\right)} - 1\right] - \frac{V+IR_s}{R_{sh}} \qquad (1)$$

With:
$$V_t = \frac{AKT}{q} \qquad (2)$$

This is a nonlinear and implicit function, which will be denoted as: $I = f(I, V)$ (3)

Whereqelectron charge ($1,6 \times 10^{-19}$ C);KBoltzmann constant ($1,38 \times 10^{-23}$ Nm/K);TPV module temperature in Kelvin;I_0reverse saturation current of diode;Adiode ideality constant of diode;I_{ph}light generated current of PV module in amperes;$R_s$series resistance of PV module;R_{sh}shunt resistance

---

978-1-4673-8760-6/15 $31.00 © 2015 IEEE   315

of PV module;$N_s$number of PV cells connected in series;$V_t$the junction thermal voltage;Voutput voltage of PV module;Icurrent of PV module in amperes.PV manufacturers typically provide values of open-circuit voltage (V_{oc}), short-circuit current (I_{sc}), and the maximum power point (V_{mp}, I_{mp}) at Standard Test Conditions (STC) as shown in table I[7,8].The following set of five equations is to be solved in order to estimate all the model's parameters:

$$I_{sc} = f(I_{sc}, 0);\ f(0, V_{oc}) = 0;\ I_{mp} = f(I_{mp}, V_{mp});\ \left.\frac{dp}{dV}\right|_{\substack{I=I_{mp}\\V=V_{mp}}} =$$

$$0;\ \left.\frac{dI}{dV}\right|_{\substack{I=I_{sc}\\V=0}} = -\frac{1}{R_{sh}} \qquad (3a, b,c,\ d,\ e)$$

According to [1,3, 5, 6], we can rewrite (3) to obtain a system of three equations with three variables (R_s, R_{sh}, V_t),equation(4a) isobtained from(3d), (4b)of the inverse of(3e)and(4c)from(3c) as shown below:

$$g(R_s, R_{sh}, V_t) = 0;\ h(R_s, R_{sh}, V_t) = 0;\ V_t = \frac{l(R_s, R_{sh})}{N_s} \quad (4a, b, c)$$

With [5]:

$$g(R_s, R_{sh}, V_t) = I_{mp} -$$

$$V_{mp}\frac{\frac{[I_{sc}(R_s+R_{sh})-V_{oc}]}{N_sV_t}e^{\left(\frac{I_{mp}R_s-(V_{oc}-V_{mp})}{N_sV_t}\right)}+1}{R_s\left(\frac{[I_{sc}(R_s+R_{sh})-V_{oc}]}{N_sV_t}e^{\left(\frac{I_{mp}R_s-(V_{oc}-V_{mp})}{N_sV_t}\right)}+1\right)+R_{sh}}(5)$$

$$h(R_s, R_{sh}, V_t) = \frac{R_s\left(\frac{[I_{sc}(R_s+R_{sh})-V_{oc}]}{N_sV_t}e^{\left(\frac{I_{sc}R_s-V_{oc}}{N_sV_t}\right)}+1\right)+R_{sh}}{\frac{[I_{sc}(R_s+R_{sh})-V_{oc}]}{N_sV_t}e^{\left(\frac{I_{sc}R_s-V_{oc}}{N_sV_t}\right)}+1} - R_{sh}$$

$$(6)$$

$$l(R_s, R_{sh}) = \frac{I_{mp}R_s-(V_{oc}-V_{mp})}{\ln\left(\frac{(I_{sc}-I_{mp})(R_s+R_{sh})-V_{mp}}{I_{sc}(R_s+R_{sh})-V_{oc}}\right)} \qquad (7)$$

Where$g(R_s, R_{sh}, V_t)$derivative of output power respect to voltage;$h(R_s,R_{sh},V_t)$tolerance on the computed value of R_{sh};$l(R_s,R_{sh})$expression of thermal voltage for a single PV cell; And two additional equations (8) and (9),(8) are obtained by combining(3a) and(3b)andassume that [1,3,5]:$e^{\frac{V_{oc}}{N_sV_t}} \gg 1$:

$$I_0 = \left(\frac{I_{sc}(R_s+R_{sh})-V_{oc}}{R_{sh}}\right)e^{\left(-\frac{V_{oc}}{N_sV_t}\right)} \qquad (8)$$

$$I_{ph} = \frac{I_{sc}(R_s+R_{sh})}{R_{sh}} \qquad (9)$$

A. Non-iterative method of parameter identification

The resolution of theequationsystem (4) to determine R_s, R_{Sh}, and V_t that can be substituted in (8) and (9) for determining I_0 and I_{ph} respectively.The systemoffiveequations

(3)can be reduced toa system of twoequations in two variables$R_s$andR_{sh}, as shownbelow [6]:

$$m(R_s, R_{sh}) = 0;\ n(R_s, R_{sh}) = 0 \qquad (10a,b)$$

Where$m (R_s, R_{sh})$derivative of output power respect to voltage,with implicit dependence on V_t;$n (R_s, R_{sh})$tolerance on the computed value of R_{sh}, withimplicit dependence on V_t;$m (R_s, R_{sh})$ and $n(R_s, R_{sh})$ whoserespectiveexpressions(11) and(12)[5] are obtained by replacing(7)in(5) and(6).The resolution of(4)is madeaccording to the followingconditions: a)draw curves$s(R_s, R_{sh})$ and $t(R_s, R_{sh})$,defined by the intersectionbetweenthezeroaxisand$m$ (point P_1) or n(point P) respectively; b)find theonlycommon point P betweencurvess and t.This can be achievedby using the "*contour*" function ofMATLAB, the result isshown in Figs. 2(a) and 2(b)for bothPVmodules, representing twotypical behaviorsthat may be encountered; this has already been observed experimentally with several case studies [5, 6].Theirreference valuesatSTC, taken from the manufacturers'datasheets, are summarized in Table I.s curve always has a horizontal asymptote R_s^*on the R_{sh}-R_s plane, the sign can be positive, as shown in Fig. 2(a), or negative, as shown in Fig. 2(b).In the latter case, the curve's intercepts the horizontal axis in R_{sh}^*. P_1 represents either s asymptote (Fig.2(a))of the s curve or the intersection with the horizontal axis.

$$M(R_s, R_{sh}) = I_{mp} - V_{mp}\frac{\frac{(I_{sc}-I_{mp})(R_s+R_{sh})-V_{mp}}{l(R_s, R_{sh})}+1}{R_s\left(\frac{(I_{sc}-I_{mp})(R_s+R_{sh})-V_{mp}}{l(R_s, R_{sh})}+1\right)+R_{sh}}$$

$$(11)$$

$$n(R_s, R_{sh}) = -R_{sh} +$$

$$\frac{R_s\left(\frac{[I_{sc}(R_s+R_{sh})-V_{oc}]}{l(R_s, R_{sh})}\left(\frac{(I_{sc}-I_{mp})(R_s+R_{sh})-V_{mp}}{I_{sc}(R_s+R_{sh})-V_{oc}}\right)e^{\left(\frac{I_{sc}R_s-V_{oc}}{R_sI_{mp}-(V_{oc}-V_{mp})}\right)}+1\right)+R_{sh}}{\frac{[I_{sc}(R_s+R_{sh})-V_{oc}]}{l(R_s, R_{sh})}e^{\left(\frac{I_{sc}R_s-V_{oc}}{l(R_s, R_{sh})}\right)}+1}$$

$$(12)$$

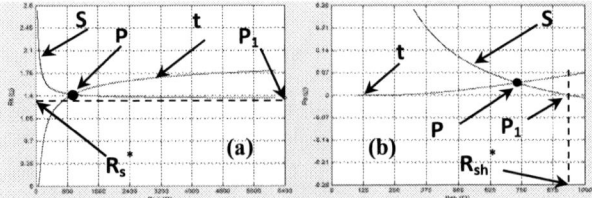

Figure 2. Contour curves on R_{sh}-R_s plane for,(a) SHELL ST40 PV panel, SPR=3,69686, Rs=1,3582 Ω, Rsh = ∞, (b) SPR-200-BLKPV panel, SPR=0,847733, Rs=0, Rsh= 933,876Ω.

Therefore, whenthesolutionisanegativeasymptotethe solution of (4) can be approximatedwith the intersectionbetweenthecurve sandthe horizontal axis,that is,$R_s=0$Ωand$R_{sh}=R_{sh}^*$. On the contrary, when the asymptote is positive solution of (4) can be approximated with the asymptote R_s^*, namely, $R_s = R_s^*$ and $R_{sh}=\infty$. In both cases, the value of V_t can be calculated from (4c) and (7). As usual, the determined values of R_s, R_{sh}, and V_t

may be substituted in (8) to find the value of I_0. Finally, I_{ph} can be calculated from (9). It has been shown in [5] that is no difference between the solutions given by the points P_1 and P.

TABLE I.   REFERENCE VALUES OF SAMPLE PV MODULES AT STC [7,8]

Brand	Model	Reference values at STC
Shell	ST40	V_{mp}=16,6V, I_{mp}=2,41A, V_{oc}=23,3V, I_{sc}=2,68A, N_s=42.
SUNPOWER	SPR-200-BLK PV	V_{mp}=40V, I_{mp}=5A, V_{oc}=47,8V, I_{sc}=5,4A, N_s=72.

B. Expressions allowing non-iterative parameter identification of the single-diode model

Solving the equationsfor identificationthe parameters byanon-iterative methodis to identifythe equation of thecurve sinthe R_{Sh}-$R_s$plane. For this, thezero setting(11)can express(13) [5]:

$$R_s = \frac{1}{I_{mp}}\left[V_{mp} - l(R_s, R_{sh})\frac{I_{mp}(R_s+R_{sh})-V_{mp}}{(I_{sc}-I_{mp})(R_s+R_{sh})-V_{mp}}\right] \quad (13)$$

Given theparametersγ_i=I_{mp}/I_{sc},andγ_v=V_{mp}/V_{oc}, andsubstitute(7) into (13)and using(14),(15)we canobtain(16) and (17) [6]:

$$u = R_s + R_{sh} \quad (14)$$

$$q(u) = \ln\frac{(1-\gamma_i)I_{sc}u-\gamma_v V_{oc}}{I_{sc}u-V_{oc}} \quad (15)$$

$$R_s =$$

$$\frac{V_{oc}}{I_{sc}}\frac{u\left[\frac{\gamma_v}{\gamma_i}(1-\gamma_i)q(u)+(1-\gamma_v)\right]-\frac{\gamma_v V_{oc}}{\gamma_i I_{sc}}[(1-\gamma_v)+q(u)\gamma_v]}{u[(1-\gamma_i)q(u)+\gamma_i]-\frac{\gamma_v V_{oc}}{I_{sc}}[1+q(u)]}$$

$$(16)$$

$$R_{sh} = u - \frac{V_{oc}}{I_{sc}}\frac{u\left[\frac{\gamma_v}{\gamma_i}(1-\gamma_i)q(u)+(1-\gamma_v)\right]-\frac{\gamma_v V_{oc}}{\gamma_i I_{sc}}[(1-\gamma_v)+q(u)\gamma_v]}{u[(1-\gamma_i)q(u)+\gamma_i]-\frac{\gamma_v V_{oc}}{I_{sc}}[1+q(u)]} \quad (17)$$

TheasymptoteR_s^*can be obtained bycalculating thelimit of thescurvewhenR_{sh}orutends to infinity. The resultis [6]:

$$R_s^* = \frac{V_{oc}}{I_{sc}}\frac{\frac{\gamma_v}{\gamma_i}(1-\gamma_i)\ln(1-\gamma_i)+(1-\gamma_v)}{(1-\gamma_i)\ln(1-\gamma_i)+\gamma_i} \quad (18)$$

With R_s^* series resistance computed as the asymptote of curve s (R_s, R_{sh}). To calculateR_{sh}^*, $R_s$is replacedby zero in(13), anduse the equations(7), (14),(15),(19),(20),(21)and (22)obtaining (23)as shown in[6]:

$$r = \frac{\gamma_i(1-\gamma_v)}{\gamma_v(1-\gamma_i)} \quad (19)$$

$$\lambda_1 = \frac{(1-\gamma_v)}{(1-\gamma_i)}\cdot\frac{2\gamma_i-1}{\gamma_v+\gamma_i-1} \quad (20)$$

$$\lambda_2 = \frac{\gamma_v}{1-\gamma_i} \quad (21)$$

AndtheSerial/Parallel Ratio (SPR) can be computed as:

$$SPR = \frac{1-\gamma_i}{e^{-r}} \quad (22)$$

Ifa PV modulehasan SPR>1, thecurvehas apositiveasymptoteas shown inFig. 2(a);thePVmodule can bemodeled using(R_s=R_s^*; R_{sh}=∞). On the contrary,ifa PV moduleSPR<1, the curve hasanegativeasymptotesas shown inFig.2(b);thePVmodule can bemodeled using(R_s= 0;R_{sh}=R_{sh}^*).Then,by a simple calculationit is possibleto deducethe expression of R_{sh}^* [6] following:

$$R_{sh}^* = \frac{V_{oc}}{I_{sc}}\cdot\frac{\lambda_2 W(-SPR.\lambda_1 e^{-\lambda_1})+\lambda_1}{W(-SPR.\lambda_1 e^{-\lambda_1})+\lambda_1} \quad (23)$$

With R_{sh}^*shunt resistance computed as the intersection between curve s (R_s, R_{sh}) and the R_{sh}-axis. It has been shownin[6]thatthesingle branchW(denotedW$_{-1}$) [5,6] to be taken intoconsiderationwhen evaluating(23) isthat which corresponds totherange(exp(-1), 0)andverifying theinequality(24)follows:

$$\lambda_2\frac{V_{oc}}{I_{sc}} \le R_{sh}^* \le +\infty \quad (24)$$

The non-iterative method usedin this workis toapproximateW$_{-1}$(x) by theequation(25)with an errorof no more than0,025%[6].

$$W_{-1} = -1 - \sigma - \frac{2}{M_1}\left(1 - \frac{1}{1+\frac{(M_1\sqrt{\sigma/2})}{\left(1+M_2\sigma e^{(M_3\sqrt{\sigma})}\right)}}\right) \quad (25)$$

With: σ=-1-ln(-x); M_1=0,3361; M_2=-0,0042; M_3=-0,0201

III. SIMULATION VALIDATION AND COMPARATIVE ANALYSIS

The estimated parameters are already shown in Table II. The small difference between the parameters has no influence on the IV characteristics (Figure 3 (a), 4 (a) and 5 (a)). The effectiveness of the method is evaluated by the value of the error as shown in Figures 3 (b), 4(b) and 5 (b). The reference values of the PV module at the irradiance measured are summarized in Table III and those of the measured temperature are summarized in Table IV. We also notice a small difference between the experimental results and the results rebuilt with experimental data as shown in Figures 3, 4 and 5. This discrepancy may be the result of noise in the experimental data and inaccurate determination of the parameters I_m and V_m. By against the values of I_{sc} and V_{oc} are almost insensitive to changes in temperature as shown in Figure 5. The small gap can be attributed either to the operation of the photovoltaic module itself, or the method of calculation (Villalva [3]). Against by monitoring near the maximum power point remains dependent on the precision of

978-1-4673-8760-6/15 $31.00 © 2015 IEEE

experimental data. Therefore, the Fill-Factor FF defined as the ratio of the maximum power to the product of the short circuit current and the open circuit voltage decreases at non STC. Figure 5 compares the different methods for a simple diode module. The results clearly show that the proposed methods provide the most accurate results. Accordingly, the error values of the proposed methods are greatly reduced compared to other models (Villalva [3]). The good agreement with high irradiance 1000W/m² with the new single-diode model can be explained by the fact that, in these conditions the traps may be involved in recombination are saturated and then the diffusion process becomes predominant as in concentration. In the end, users of such methods to be very attentive to experimental conditions and approximations involved. It is also noted that the new model 4-parameter combination between the simplicity, accuracy, and speed of calculation in a real situation and shows the choices we believe the most interesting.

I. CONCLUSION

The purpose of the extraction parameters of thin-film solar panels is obviously describing their behavior under all conditions of use. The four methods we have outlined are used to calculate the parameters of a photovoltaic panel's single-diode Thin-Film from the manufacturer's data. The proposed non-iterative method is most suitable for proper simulation of the operation of photovoltaic thin-film modules in real situations.

TABLE II.   PARAMETERS ESTIMATED BY DIFFERENT METHODS OF MODELING.

Paramet er (SHE LL ST40)	4-parameter (point P_1) SPR=3,69686	5-parameter (point P)			5-parameter
		Halley's method	Non-iterative method	Lambert W function	Villalva [3]
A	1,384108	1,287775	1,295894	1,295350	1,134544
$R_s(\Omega)$	1,358196	1,423932	1,415954	1,415954	1,494
$R_{sh}(\Omega)$	∞	949,6259	965,5860	949,6259	257.165987
$I_0(A)$	$4,473393.10^{-07}$	$1,3813.10^{-07}$	$1,5345.10^{-07}$	$1,5237.10^{-07}$	$1,4137.10^{-08}$
$I_{ph}(A)$	2,680000	2,684018	2,683930	2,683996	2,695670

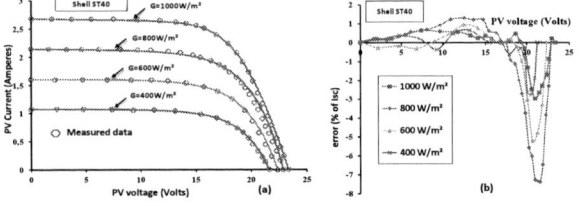

Figure 3.   (a) Reconstructed I–V curves using measured data for different irradiance levels (T = 25 °C). (b) Plot of ([($I_{experimental}$-$I_{theoretical}$)/I_{sc}]*100) for different levels of irradiance (T = 25 °C).

TABLE III.   REFERENCE VALUES FOR DIFFERENT IRRADIANCE LEVELS FOR EXPERIMENTAL VALIDATION

Irradiance (W/m²)	Reference values (N_s=42) (SHELL ST40)			
	$V_{mp}(V)$	$I_{mp}(A)$	$V_{oc}(V)$	$I_{sc}(A)$
1000	17,1235	2,36148	23,3545	2,6788
800	17,4045	1,82989	22,8799	2,1452
600	16,9906	1,38472	22,3734	1,6062
400	16,6068	0,928608	21,614	1,0726

TABLE IV.   REFERENCE VALUES FOR DIFFERENT TEMPERATURES FOR EXPERIMENTAL VALIDATION

Temperature (°C)	Reference values (N_s=42)(SHELL ST40)			
	$V_{mp}(V)$	$I_{mp}(A)$	$V_{oc}(V)$	$I_{sc}(A)$
20	17,7366	2,34631	23,7501	2,6793
40	15,9137	2,30095	21,8439	2,712
60	13,9345	2,28099	19,9064	2,7283

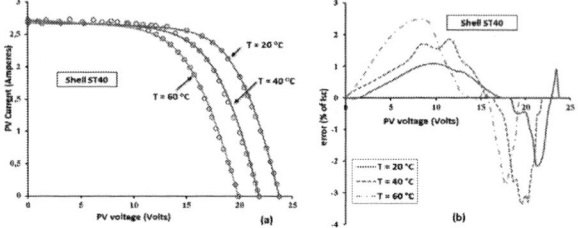

Figure 4.   (a) Reconstructed I–V curves using measured data for different temperatures (G = 1000W/m²). (b) Plot of ([($I_{experimental}$- $I_{theoretical}$)/I_{sc}]*100) for different temperatures (G = 1000W/m²).

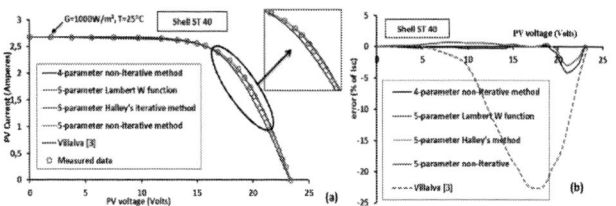

Figure 5.   (a) Comparison of measured data with the I–V characteristics generated from parameters estimated with different method. (b) Plot of ([($I_{experimental}$- $I_{theoretical}$)/I_{sc}]*100).

REFERENCES

[1]   V.J. CHIN, Z. SALAM, K. ISHAQUE, "Cell modelling and model parameters identification techniques for photovoltaic simulator application: A review", Applied Energy, vol. 154, 2015, pp. 500-519.

[2]   M. TALI, A. OBBADI, A. ELFAJRI, Y. ERRAMI, "Passive filter for harmonics mitigation in standalone PV system for non linear load", Renewable and Sustainable Energy Conference (IRSEC), International. IEEE, 17-19 Oct. 2014, pp.499,504.

[3]   M. G. Villalva, J. R. Gazoli, E. R.Filho, "Comprehensive approach to modeling and simulation of photovoltaic array", IEEE Trans. Power Electron., vol. 24, no. 5, May 2009, pp. 1198–1208.

[4]   G. PETRONE, G. SPAGNUOLO, "Parameters identification of the single-diode model for amorphous photovoltaic panels",Proc. IEEE 5th International Conference on Clean Electrical Power (ICCEP), 16-18 June 2015, pp. 105-109.

[5]   S. Cannizzaro, M. C. Di Piazza, M. Luna, G. Vitale, "Generalized classification of PV modules by simplified single-diode models", Proc.IEEE International Symposium on Industrial Electronics (ISIE'14), June1-4, 2014, pp. 2262-2269.

[6]   S. Cannizzaro, M. C. Di Piazza, M. Luna, G. Vitale, "PVID: An interactive Matlab application for parameter identification of complete and simplified single-diode PV models," Control and Modeling for Power Electronics (COMPEL), 2014 IEEE 15th Workshop, 22-25 June 2014,pp.1-7.

[7]   ShellSolarProductInformationSheet[Online].Available at: http://www.solarcellsales.com/techinfo/technical_docs.cfm.

[8]   Twenty 200W SunPower SPR-200-BLK panels, which are a monocrystalline technology. [Online]. available: http://www.rectifier.co.za/Solar/sunpower/pdf/sp_200blk_en_a4_p_ds.pdf.

Validation of a multi-exponential alternative model of solar cell and comparison to conventional double exponential model

R. Bendaoud, S. Yadir, C. Hajjaj, Y.Errami,
S. Sahnoun, M. Benhmida
Laboratory of Electronics, Instrumentation and Energetic
Faculty of Sciences, Chouaïb Doukkali University,
B.P 20, El Jadida, Morocco
benhmida@gmail.com

M. El Aydi
Mathematics Department
Regional Center for Education and Training
El Jadida, Morocco

Abstract—The modeling of a solar cell is useful for any operation of efficiency optimization or of diagnostic of the photovoltaic generator. Usually, a solar cell is represented by an equivalent electrical circuit which the physical parameters can be determined experimentally from the (current-voltage) characteristic. For the classical two exponential model, the equivalent circuit contains two diodes, a photogenerated current, a series resistance and a conductance. In the alternative model proposed in [5-7], this circuit is substituted by a Thevenin generator and two resistances in series with each of the two diodes. In this study, we express analytically the physical parameters of the alternative model depending to those of the classical model, considering the equivalence of the two circuits. It is then possible to reduce the number of equations and extract physical parameters easily. The reliability of both models is evaluated by comparing them to experimental (current-voltage) characteristics of a real solar cell.

Keywords— Solar cell alternative model, solar cell physical parameters extraction, double-exponential solar cell model.

I. INTRODUCTION

A photovoltaic solar cell is a semiconductor device which permits converting solar energy into electricity. Various models have been proposed to describe its electrical behavior [1-7]. Solar cell can be represented by an equivalent electrical circuit with a simple diode [1, 2], or two diodes, that's taking into consideration separately diffusion and generation-recombination mechanisms in the cell [3, 4]. Recently, a model with N diodes *Fig.1*, has been proposed by D.C. Lugo-Muñoz and by A. Ortiz-Conde and al [5-7]. Description of the equivalent electrical circuit was modified by application of Thevenin's theorem *Fig.2*, [6]. The Electrical circuit corresponding to this model so called alternative [5, 6], is represented on *Fig.3*.

Experimental data used in this study were obtained by taking of experimental characteristics (I-V) with a rectangular polycrystalline solar cell (8cm x 5cm).

II. PHYSICALS PARAMETERS EXTRACTION

To simplify the writing of the characteristic equation describing solar cell electrical equivalent circuit with N diodes

Fig.1, A. Ortiz-Conde and al proposed [5-7] an alternative model represented by the electrical circuits of *Fig.2* and *Fig.3*.

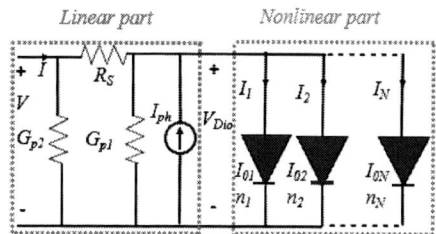

Fig.1. Single Solar cell equivalent circuit corresponding to conventional model with N diodes [5-7]. R_S: Serie resistance, G_{p1}, G_{p2}: Parallel conductances ; I_{ph} : Photogenerated current ; I_{0i} : Reverse saturation current of diode i ; n_i : Ideality factor of the diode i ; V_{Dio} : Terminal voltage of the diodes.

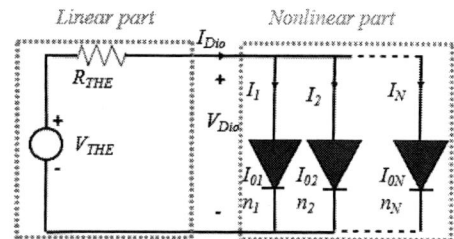

Fig.2. Equivalent circuit corresponding to modified conventional model with N diodes. R_{THE}: Thevenin equivalent resistance; V_{THE}: Thevenin equivalent generator; I_{Dio}: Total current in the N diodes.

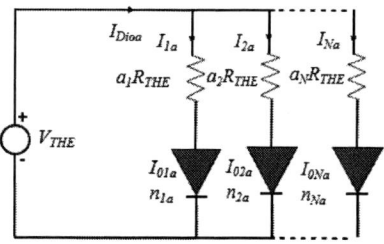

Fig. 3 Equivalent circuit corresponding to simplified alternative model with N diodes. I_{0ia}: Reverse saturation current of diode i; n_{ia}: Ideality factor of the diode i; I_{Dioa}: Total current in the N diodes; a_i: Multiplicative coefficients.

978-1-4673-8760-6/15 $31.00 © 2015 IEEE   319

Current intensity delivered by solar cell represented by the equivalent circuit with two diodes *Fig.1*, is written as [6]:

$$I = \left\{ \sum_{i=1}^{2} I_{0k} \left[\exp\left(\frac{V - R_S \left(I - G_{p2}V\right)}{n_k . v_t} \right) - 1 \right] \right\} - I_{ph}$$
$$+ G_{p2}V\left(1 + G_{p1}.R_S\right) + G_{p1}(V - R_S.I) \tag{1}$$

By using alternative model with two diodes and the LambertW function [6, 8], the current intensity I, when the solar cell is irradiated, can be written as:

$$I = \left\{ \frac{n_{1a}v_{th}}{a_1 R_S} W_0 \left[\frac{a_1 R_S I_{01a}}{n_{1a}v_{th}\left(1 + R_S.G_{p1}\right)} \times \exp\left(\frac{V + I_{ph}R_S + a_1 R_S I_{01a}}{n_{1a}v_{th}\left(1 + R_S.G_{p1}\right)} \right) \right] \right.$$
$$- \frac{I_{01a}}{\left(1 + R_S.G_{p1}\right)} \right\} + \left\{ \frac{n_{2a}v_{th}}{a_2 R_S} W_0 \left[\frac{a_2 R_S I_{02a}}{n_{2a}v_{th}\left(1 + R_S.G_{p1}\right)} \right. \tag{2}$$
$$\times \exp\left(\frac{V + I_{ph}R_S + a_2 R_S I_{02a}}{n_{2a}v_{th}\left(1 + R_S.G_{p1}\right)} \right) - \frac{I_{02a}}{\left(1 + R_S.G_{p1}\right)} \right\} + \frac{VG_{p1} - I_{ph}}{\left(1 + R_S.G_{p1}\right)}$$

The equivalence between conventional and alternative models is discussed by comparing the extracted physical parameters of the solar cell, by applying the two models to its experimental characteristic.

Considering the equivalence of resistive losses in the two models, we can reduce the number of searched physical parameters and simplify their extraction. When expressing the equality of the current intensities through the equivalent circuits with two diodes *Fig.2* and *Fig.3*, we have:

$$\left(I_{1a} + I_{2a}\right)^2 = a_1 I_{1a}^{\;2} + a_2 I_{2a}^{\;2} \tag{3}$$

and

$$I_{Dio} = \left[\frac{a_1 \pm \sqrt{a_1 + a_2 - a_1 a_2}}{a_1 - 1} \right] I_{2a} \tag{4}$$

I_{Dio} expressed in the conventional model is written as:

$$I_{Dio} = I_{1a} + I_{2a}$$
$$= I_{01a}\left(\exp\left(\frac{V_{THE} - a_1 R_{THE}.I_{1a}}{n_{1a}V_T} \right) - 1 \right) \tag{5}$$
$$+ I_{02a}\left(\exp\left(\frac{V_{THE} - a_2 R_{THE}.I_{2a}}{n_{2a}V_T} \right) - 1 \right)$$

with :

$$V_{THE} = \left(\frac{V}{R_S} + I_{ph} \right) R_{THE} \tag{6}$$

and

$$R_{THE} = \frac{R_S}{1 + R_S.G_{p1}} \tag{7}$$

The effect of G_{p2} on the total current is accounted when I_{Dio} is calculated [6].

When considering the equivalent circuits of *Fig.1* and *Fig.2*, we have [6]:

$$I = G_{p1}.V_{Dio} - I_{ph} + I_{Dio} \tag{8}$$

$$V_{Dio} = V_{THE} - I_{Dio}.R_{THE} \tag{9}$$

The combination of (4), (5), (6), (7), (8) and (9) permits rewriting the intensity of total current corresponding to the conventional model with two diodes, according to the parameters introduced in the alternative model:

$$I = I_{02a}(1 - G_{p1}R_{THE})\left(\frac{a_1 \pm \sqrt{a_1 + a_2 - a_1 a_2}}{a_1 - 1} \right) \times$$
$$\left(\exp\left(\frac{V_{THE} - \dfrac{a_2 R_{THE}(I - G_{p1}V_{THE} + I_{ph})(a_1 - 1)}{(1 - G_{p1}R_{THE})\left(a_1 \pm \sqrt{a_1 + a_2 - a_1 a_2}\right)}}{n_{2a}V_T} \right) - 1 \right) \tag{10}$$
$$+ G_{p1}V_{THE} - I_{ph}$$

Using LambertW function, I can be written as:

$$I = \left(\frac{a_1 \pm \sqrt{a_1 + a_2 - a_1 a_2}}{a_1 - 1} \right) \times$$
$$\left(\frac{n_{2a}.V_T}{a_2.R_S}.LambertW\left(\frac{a_2.R_S.I_{02a}}{n_{2a}.V_T.(1 + R_S.G_{p1})} \right. \right.$$
$$\left. \left. \times \exp\left(\frac{V + I_{ph}.R_S + a_2.R_S.I_{02a}}{n_{2a}.V_T.(1 + R_S.G_{p1})} \right) \right) - \frac{I_{02a}}{(1 + R_S.G_{p1})} \right) \tag{11}$$
$$+ \frac{V.G_{p1} - I_{ph}}{(1 + R_S.G_{p1})}$$

III. RESULTS AND DISCUSSION

The extraction of solar cell physical parameters was carried out using iterative injection method [9, 10]. In this case, thereof involves injecting parameters test values of a_1 and a_2 in characteristic equations (10) and (11).

This allows generating a set of probable characteristics. The error rate parameter "%$\Delta Area$" [11, 12], defined below permits to evaluate the relative magnitude difference between the generated characteristic (I-V) and that obtained experimentally:

$$\%\Delta Area = \frac{\text{area difference between the extracted and experimental curves}}{\text{total area of the experimental characteristic}} \times 100$$

The retained values of a_1 and a_2 are those corresponding to the %$\Delta Area$ minimum value.

To retrieve physical parameters values of Table. I, we assigned to R_S, G_{p1} and I_{ph} respectively the values 0.15(Ω), 1/233 (Ω^{-1}) and 0.178(A), found in a previous study [9].

TABLE. I shows that by introducing the equivalence of resistive losses, physical parameters extracted within the framework of conventional and alternative models are quite identical. Moreover, we note that the error rate parameter in this case is twice more important than of starting alternative model (2).

978-1-4673-8760-6/15 $31.00 © 2015 IEEE   320

This result is also illustrated by *Fig. 4*, *Fig.5* and *Fig.6*, and can be explained by the fact that in the adjustment process between the experimental and extracted characteristics, only the most preponderant parameters are taken into account in the expression of resulting characteristics equations.

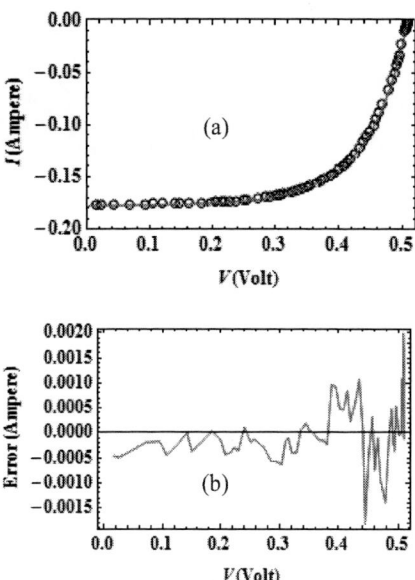

Fig. 4. (a) Characteristics (I-V): (O) Experimental data; (▬) with alternative model using iterative extration method.
(b) Current deviation between experimental and extracted characteristics

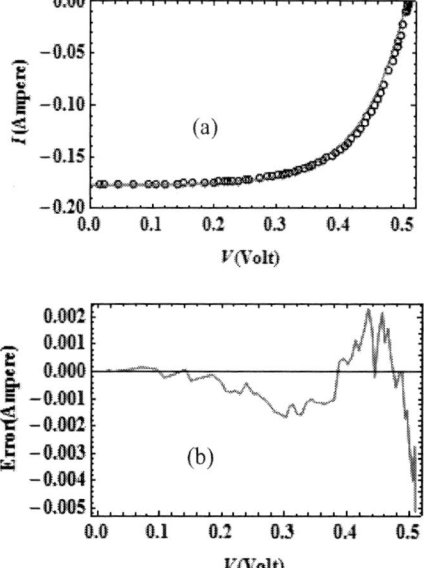

Fig. 5. (a) Characteristics (I-V): (O) Experimental data; (▬) with alternative model considering the equivalence of resistive losses, using iterative extraction method.
(b) Current deviation between experimental and extracted characteristics.

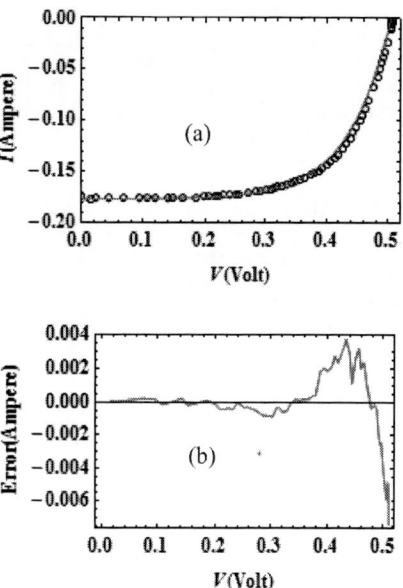

Fig. 6. (a) Characteristics (I-V): (O) Experimental data; (▬) with conventional model considering the equivalence of resistive losses, using iterative extraction method.
(b) Current deviation between experimental and extracted characteristics.

TABLE. I. $(I_{01a}, I_{02a}, n_{1a}, n_{2a}, a_1$ and $a_2)$ EXTRACTED BY ITERATIVE INJECTION METHOD.

Models	Parameters						
	$I_{01a}(A)$	$I_{02a}(A)$	n_{1a}	n_{2a}	a_1	a_2	%ΔArea
Alternative model	$6\ 10^{-11}$	$6\ 10^{-5}$	0.94	2.5	1.7	1.3	0.21
Alternative model *	-	$6\ 10^{-6}$	-	2.4	1.3	1.2	0.50
Conventional model*	-	$6\ 10^{-6}$	-	2.4	1.3	1.2	0.50

* Conventional and alternative models considering the equivalence of resistive losses.

IV. CONCLUSION

The results obtained by considering solar cell conventional and alternative models with two diodes and the assumption of the resistive losses equivalence are comparable to those obtained using the originally alternative model proposed by D. C. Lugo-Muñoz, J. Muci, A. Ortiz-Conde, F. J. García Sánchez, M. de Souza, and M. A. Pavanello [5, 6].

TABLE. I illustrates that in the three models, the compatibility with the experimental results, characterized by the parameter %ΔArea, is quite acceptable.

The reliability of the two alternative models should be tested by comparison of results to those obtained with the conventional model and that for different radiation intensities and for different solar cells temperatures. In addition, it would be interesting to test the reliability of the alternative model with three or even four diodes, in order to obtain a better identification of the physical mechanisms acting within the cell.

REFERENCES

[1] J. P. Charles, M. Abdelkrim, Y. H. Muoy and P. Mialhe, "A practical method of analysis of the current-voltage characteristics of solar cells", Solar Cells 4, 169 (1981).

[2] D. S. H. CHAN, J. R. Phillips and J. C. H. Phang, "A comparative study of extraction methods for solar cell model parameters", Solid-State Electronics Vol. 29, No. 3. pp. 329-337, 1986.

[3] C.T. Sah, R.N. Noyce, W. Shockley, "Carrier generation and recombination in P-N junctions and P-N junction characteristics", Proc IRE; vol. 45, pp. 1228–1243, Sept. 1957.

[4] C. T. Sah, "Effect of surface recombination and channel on p-n junction and transistors Characteristics", IRE Transactions on Electron Devices, ED- 9, pp. 94-108. Jan. (1962).

[5] D. C. Lugo-Muñoz, J. Muci, A. Ortiz-Conde, F. J. García Sánchez, M. de Souza, and M. A. Pavanello, "An explicit multi-exponential model for semiconductor junctions with series and shunt resistances", Microelectron. Reliabil, vol. 51, pp. 2044–2048, 2011.

[6] A. Ortiz-Conde, D. Lugo-Muñoz and F. J. García Sánchez, " An explicit multi-exponential model as an alternative to traditional solar cell models with series and shunt resistances", IEEE Journal of Photovoltaics, vo.2, pp.261-268, July 2012.

[7] A. Ortiz-Conde, F.J. García Sánchez, A. Terán Barrios, J. Muci, M. de Souza, M.A. Pavanello "Approximate analytical expression for the terminal voltage in multi-exponential diode models", Solid-State Electronics, vol. 89, pp. 7-11, Nov. 2013.

[8] R.M. Corless and all, "On the Lambert W function", Advances in Computational Mathematics 5(1996)329-359.

[9] S. Yadir, " Extracting physical parameters of a solar cell. Study of their influence on the efficiency. (Extraction des paramètres physiques d`une cellule solaire. Etude de leur influence sur le rendement)", Thesis , Chouaib Doukkali University, 2014.

[10] S. Assal, " Extracting physical parameters and influence of the technological factors on photovoltaic cells efficiency. (Extraction des Paramètres physiques et influence des Facteurs Technologiques sur la Rendement d'une Cellule Photovoltaïque)", Thesis , Chouaib Doukkali University, 2015.

[11] S.Yadir, S. Benhmida, M. Sidki, M. Assaid, E. Khaidar, M. "New method for extracting the model physical parameters of solar cells using explicit analytic solutions of current-voltage equation", The International IEEE Conference on Microelectronics (ICM), 19-22 Dec, 2009, pp. 390 – 393.

[12] S.Yadir, S. Assal, M. Benhmida, M. Sidki, M. Khaidar, O. Aomari, A. Malaoui, E.Bendada, M. Mabrouki, "Extraction of physical parameters of solar cell model with double exponential from illuminated I-V experimental curve", Global Journal of Physical Chemistry. Volume 2, Issue 2 (2011) pp. 236-240.

Physical parameters extraction by a new method using solar cell models with various ideality factors

S. Yadir, H. Amiry, R. Bendaoud, A. El Hassnaoui,
A. Obbadi, M. Benhmida
Laboratory of Electronics, Instrumentation and Energetic
Faculty of Sciences, Chouaïb Doukkali University,
B.P 20, El Jadida, Morocco
benhmida@gmail.com

M. El Aydi
Mathematics Department
Regional Center for Education and Training
El Jadida, Morocco

Abstract—Increasing solar cells efficiency is considered as a critical factor in reducing the production costs of photovoltaic cell manufacturers. Modeling, simulation and design are key tools achievement of high-efficiency solar photovoltaic cell. The efficiency is closely connected to the physical parameters associated with the solar cell model adopted. They represent an essential analysis tool for understanding cell operating. The model with two diodes permits to better take into account the physical processes involved in the solar cell. Nonetheless, in which case both exponential terms included within the characteristic equation make it difficult to numerically manage.

In this paper, we developed a numerical method for extracting the physical parameters of solar cell models. Results are compared to those obtained by application of two other classical methods in the case of a single diode model and two diodes model with two constant or various ideality factors. The comparison of results highlight that the two diodes models are more appropriates to describe physical phenomena acting within the solar cell. The proposed technique permits to have a better theoretical curve compatibility with the experimental characteristic (I-V). Furthermore, it requires less computational power.

Keywords—Solar cell, Physical parameters extraction, solar cell efficiency.

I. INTRODUCTION

The solar cell physical parameters are closely related to the internal physical mechanisms acting within the solar cell [1-6].There are tools for characterization which permit to understand not only cell operating but above all take controls the parameters limiting the cell's performance [1-5]. Their determination is therefore important. Given the difficulty to find exact analytical solutions of solar cell characteristic equation, it is usual to resort to their numerical resolution [1-5]. Typically, the extraction methods proposed can be classified, according to their approximation, into three main groups [5,7]:

➢ Graphical methods: they represent approximate procedures and product higher errors method than those of measurements.The physical parameters are determined from several experimental characteristics or even by using selected aspects of the characteristics where some parameters may be negligible;

➢ Method using Newton's algorithm [1-3]: they product fewer errors than graphical methods. Writing the characteristic

equations for particular points (I ,V) permits to deduce a system of nonlinear equations whose solutions are carried out using a Newton algorithm;

➢ Numerical methods using the whole experimental characteristic (I-V) with no recourse to specific approximations [2-4]. The accuracy of physical parameters values, in this case, depends on the experimental measurements results.

II. SOLAR CELL EQUIVALENT ELECTRICAL CIRCUIT MODELS

The solar cells are electronic devices that directly convert sunlight into electrical current I_{ph}. Several models are used to express the current-voltage characteristic (I-V) of a solar cell. Two equivalent electrical circuits *Fig.1*, *Fig. 2*, are often used [8-13] to modeling these characteristics:

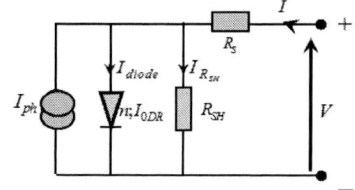

Fig. 1. Single diode equivalent electrical circuit model including a series (R_S) and a shunt (R_{SH}) resistances.

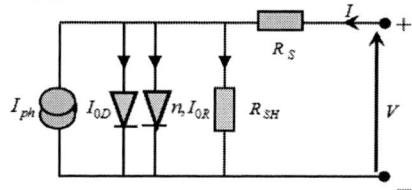

Fig. 2. Two diodes equivalent electrical circuit model including a series (R_S) and a shunt (R_{SH}) resistances

✓ **Single diode equivalent electrical circuit model (SEM) [1,2,12,13] :**

The schematic model is consisting of a current generator associated with single diode. The mathematical description of the circuit *Fig.1*, is given by the following equation:

$$I(V) = I_0 \left(\exp(\frac{V - R_S I}{n V_{th}}) - 1 \right) + \frac{V - R_S I}{R_{SH}} - I_{ph} \qquad (1)$$

Where I_0 , n , R_S and R_{SH}, are respectively, the saturation currents of diode, the ideality factors, the series resistance which takes account of total ohmic losses, and the shunt resistance which takes account of parasitic currents of the solar cell [1,2,12,13].

$V_{th}=k_B T/e$, with e is the electron charge, k_B is the Boltzmann constant and T is the absolute temperature.

The total number of physical parameters related to the model is of five: I_0, I_{ph}, R_S, R_{SH}, and n.

✓ **Double diode equivalent electrical circuit model (DEM) [8-11] :**

The equivalent circuit includes two diodes D_1 and D_2, a current generator I_{ph}, R_S, and R_{SH}. The first diode D_1, with a ideality factors n_1 and a saturation currents I_{O1}, depends mainly on the diffusion current in the quasi-neutral region. The second diode D_2, with a ideality factors n_2 and a saturation currents I_{O2}, is related to the generation-recombination current mainly in the depletion region.

According to the considerations about the ideality factor, the following models can distinguished:

✓ **Double diode model with two variables ideality factors (2VDEM) [8-11] :**

$$I(V) = I_{OD}\left(\exp(\frac{V-R_SI}{n_1 V_{th}})-1\right) + I_{OR}\left(\exp(\frac{V-R_SI}{n_2 V_{th}})-1\right) + \frac{V-R_SI}{R_{SH}} - I_{ph} \quad (2)$$

The total number of physical parameters related to the model is of seven: I_{OD}, I_{OR}, n_2, n_1, R_S, R_{SH} et I_{ph}.

✓ **Double diode model with one variable ideality factor (1VDEM) [8-11] (with n_1=1):**

$$I(V) = I_{OD}\left(\exp(\frac{V-R_SI}{V_{th}})-1\right) + I_{OR}\left(\exp(\frac{V-R_SI}{n_2 V_{th}})-1\right) + \frac{V-R_SI}{R_{SH}} - I_{ph} \quad (3)$$

The total number of physical parameters related to the model is of six: I_{OD}, I_{OR}, n, R_S, R_{SH} et I_{ph}.

✓ **Double diode model with two constants ideality factors n_1=1 and n_2=2 [11,14-17] (CDEM):**

$$I(V) = I_{OD}\left(\exp(\frac{V-R_SI}{V_{th}})-1\right) + I_{OR}\left(\exp(\frac{V-R_SI}{2V_{th}})-1\right) + \frac{V-R_SI}{R_{SH}} - I_{ph} \quad (4)$$

The total number of physical parameters related to the model is of five: I_{OD}, I_{OR}, R_S, R_{SH} et I_{ph}.

III. ITERATIVE INJECTION METHOD

The difficulty of finding analytical solutions to (4) is due to its nonlinearity and the presence of R_S, n_1 and n_2 within the two exponentials. The determination of R_S, n_1 and n_2 independently from the others physical parameters makes easy to solve (4). The injection of test values of R_S, n_1 and n_2 in (4) permits to generate a set of probable numerical characteristics. The assessment of the absolute error rate between generated and experimental characteristics *Fig.3*, is defined by the parameter, denoted "%ΔArea", given by the following equation [2,5,12]:

$$\%\Delta Area = \frac{\text{area difference between theoretical and experimental curve}}{\text{total area of the experimental characteristic}}\%$$

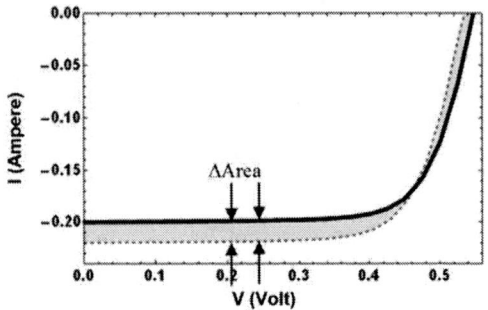

Fig. 3: Area error rate (%Area) between experimental (- -) and generated (–) characteristics curves.

% ΔArea depends on the deviation of R_S, n_1 and n_2 from their optimal value. The theoretical deviation value reaches zero when (I-V) generated characteristics coincide exactly with the experimental characteristic (% ΔArea = 0). Solar cell physical parameters extraction consists of finding R_S, n_1 and n_2 values for which % ΔArea is minimal. R_{SH}, I_{ph}, I_{01} and I_{02} are then easily deduced for all set of R_S, n_1 and n_2 values. The extraction is performed in the following way:

✓ Injection of various n_1, n_2 and R_S test values in the (I-V) characteristic equation;

✓ Determination of R_{SH}, I_{ph}, I_{OR} and I_{OD} for each set of R_S, n_1 and n_2 values, followed by generation of a set of probable numerical characteristics ;

✓ Determination of n_1, n_2 and R_S values corresponding to %Δarea minimum value.

We tested the reliability of this proposed extracting method, in the case of the 1VDEM model, by generating a set of (V,I) points using (2), for: R_S= 0.1 (Ω), R_{SH}= 100(Ω), I_{ph}= 0.2 (A), I_{01}= 10^{-10} (A), I_{02} = 10^{-6} (A) and n_2 = 2.5 ; the parameters are determined in the following steps:

✓ Injection of various R_S test values in (2), for each n_2 value ;

✓ Determination of R_{SH}, I_{ph}, I_{OR} and I_{OD} and %ΔArea for each R_S values. The retained R_S value correspond to the minimum value of %ΔArea (ΔArea$_{min}$) ;

✓ Similarly, the retained n_2 value corresponds to the minimum value of ΔArea$_{min}$.

Fig. 4, shows %ΔArea variations as a function of introduced test values of n_2 and R_S.

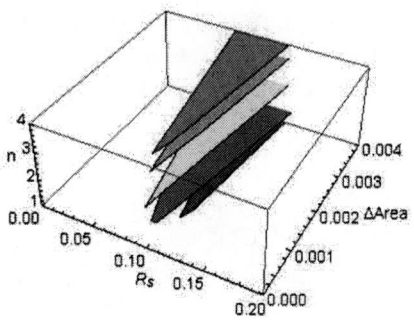

Fig. 4 (a): %ΔArea plot obtained using the proposed extraction method, for various introduced test values of R_S and ideality factor (n).

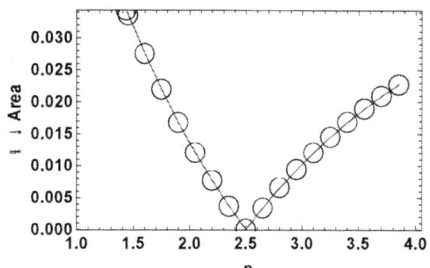

Fig. 4 (b): %ΔArea plot using the proposed extraction method, for various introduced test values of n.

As seen on *Fig. 4* (a, b), %ΔArea minimum value correspond to R_S and n values initially introduced ($R_S = 0.1\Omega$, $n = 2.5$) used to plot the studied (I–V) characteristic.

IV. RESULTS AND DISCUSSION

We use this technique in the case of the MDE2V model to calculate R_S, n_2 and %ΔArea$_{min}$ minimum values for each injected test value of n_1, in the same way as in the case of the 1VDEM model TAB. I.

Similarly, the retained R_S, n_2 and n_1 values are those corresponding to the minimum value of %ΔArea$_{min}$. For (MDEC) model, the injection is limited to the R_S test values, with n_2=2 and n_1= 1 TAB. I. Similar steps are used in the case of SEM and 1VDEM models TAB. I.

Carrying out of this extraction method to the solar cell experimental characteristic, in the case of the MDE2V model, permits to obtain *Fig. 5*, : R_S=0.196 (Ω), n_1=0.84 and n_2=2.75.

Current deviation *Fig.6*, represents the algebraic difference between experimental currents intensities and that generated with the iterative injection method using representative solar cell models. As shown on *Fig.6*, the better agreements between the experimental and the rebuilt characteristics are obtained for 2VDEM and 1VDEM models.

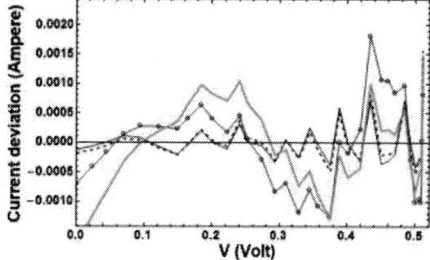

Fig.6: Current deviation between experimental and extracted characteristics using the four models:
SEM (-●-●-●), CDEM (—), 1VDEM (- - - -), 2VDEM (—).

Fig. 7 (a) and (b), present current deviation obtained using SEM and CDEM models by three methods: the graphical method consisting of injecting approximate R_S value [3], five-point method [2,3] and the iterative injection method. These figures show that the better agreement between experimental and extracted (I-V) characteristics is achieved through the iterative method.

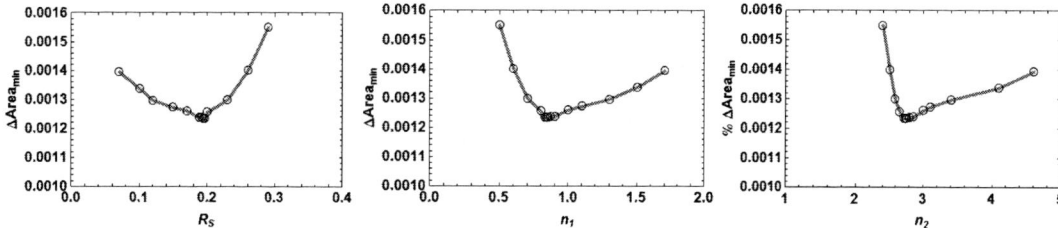

Fig. 5: %ΔArea plot as a function of R_S, n_1 and n_2

TAB. I: EXTRACTED PHYSICAL PARAMETERS VALUES USING THE ITERATIVE INJECTION METHOD FOR REPRESENTATIVE SOLAR CELL MODELS: 2VDEM, 1VDEM, CDEM AND SEM.

	$R_S(\Omega)$	n_1	n_2	$I_{01}(A)$	$I_{02}(A)$	$I_{ph}(A)$	$R_{SH}(\Omega)$	%ΔArea
Double-diode model with two variables ideality factors	0.196	0.84	2,75	$1,34.10^{-11}$	$9.38 10^{-5}$	0,1774	434	0,12 %
Double-diode model with one variable ideality factor	0.162	1.00	2,83	$50,40.10^{-11}$	$11.2 10^{-5}$	0,1776	666	0,13 %
Double-diode model with two constants ideality factors	0.093	1.00	2,00	$0,86.10^{-11}$	$1.18 10^{-5}$	0,1793	50	0,37 %
Single-diode model	0.063	---	2,10	----	$2.06 10^{-5}$	0,1781	84	0,29 %

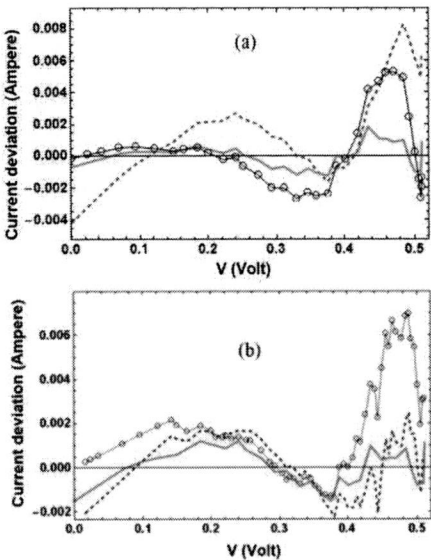

Fig.7: Current deviation between experimental and extracted characteristics obtained using SEM (a) and CDEM (b) models for three methods: five-point method (o), graphical method (- - -) and iterative injection method (—).

TAB. II, shows that the %ΔArea$_{min}$ values are laying between 1.1% to 0.8% by using SEM and CDEM models, for all methods, except the case of the iterative method where this value reaches 0.37%, for MDEC model. It's interesting to note that %ΔArea$_{min}$ is just of about 0.1% for the iterative injection method using 1VDEM and 2VDEM models. Thus, both models (VDEM) appear more appropriate to describe cell electrical characteristic evolution.

TAB. II: %ΔArea$_{min}$ VALUES OBTAINED BY FOUR MODELS USING THREE EXTRACTION METHODS: GRAPHICAL METHOD (SEM, CDEM), FIVE-POINT METHOD (SEM, CDEM) AND ITERATIVE INJECTION METHOD (SEM, CDEM, 1VDEM, 2VDEM).

	%ΔArea$_{min}$			
	Single-diode equivalent circuit model (SEM)	Double-diode equivalent model (CDEM)	Model (1VDEM)	Model (2VDEM)
graphical method	1.06	1.10	- - - - -	- - - - -
five-point method	1.06	0.80	- - - - -	- - - -
Iterative injection method	0.90	0.37	0.13	0.12

Physical parameters' values obtained with MSE model are globally comparable to those related to the generation-recombination current of VDEM models. This finding confirms that, in the case of solar cell, diffusion current is negligible compared to recombination-generation current.

V. CONCLUSION

In this paper, we highlight the reliability of the so called iterative injection extraction method used especially in the frame of representative solar cell models with variables

ideality factors **(1VDEM, 2VDEM)**. It will be interesting to test this method by operating for different intensities of radiations, various temperature conditions and type of solar cells. The evolution of cell physical parameters should permit to deduce their influence on the mechanisms acting inside thereof.

REFERENCES

[1] S. Yadir, S. Aazou, N. Maouhoub, K. Rais, M. Benhmida, and E. Assaid, "Illuminated solar cell physical parameters extraction using mathematica," in IEEE/ACS International Conference on Computer Systems and Applications, 2009. AICCSA 2009, 2009, pp. 63–64.

[2] S. Yadir, M. Benhmida, M. Sidki, E. Assaid, and M. Khaidar, "New method for extracting the model physical parameters of solar cells using explicit analytic solutions of current-voltage equation," in 2009 International Conference on Microelectronics (ICM), 2009, pp. 390–393.

[3] S. Yadir, S. Assal, M. Benhmida, M. Sidki, M. Khaidar and O. Aomari, A. Malaoui, E.Bendada, M. Mabrouki, "Extraction of physical parameters of solar cell model with double exponential from illuminated I-V experimental curve," Global Journal of Physical Chemistry, vol. 2, no. 2, pp. 236–240, 2011.

[4] A. Ortiz-Conde, F. J. García Sánchez, and J. Muci, "New method to extract the model parameters of solar cells from the explicit analytic solutions of their illuminated I–V characteristics," Sol. Energy Mater. Sol. Cells, vol. 90, no. 3, pp. 352–361, Feb. 2006.

[5] S. Yadir, S. Assal, A. El Rhassouli, M. Sidki, and M. Benhmida, "A new technique for extracting physical parameters of a solar cell model from the double exponential model (DECM)," Opt. Mater., vol. 36, no. 1, pp. 18–21, Nov. 2013.

[6] O. Aomari, A. Malaoui, M. Mabrouki, E. Bendada, A. Elmansouri, S. Yadir, S, Assal, M. Benhmida, "Implementation of a new analytical technique to determine the electrical parameters of junction models," Global Journal of Physical Chemistry, vol. 2, no. 2, pp. 68–72.

[7] Jean-Pierre Charles et al., "La Jonction, du Solaire à la Microélectronique," Rev. Energ. Ren., vol. 3, pp. 1–16, 2000.

[8] M. Wolf, G. T. Noel, and R. J. Stirn, "Investigation of the double exponential in the current #8212;Voltage characteristics of silicon solar cells," IEEE Trans. Electron Devices, vol. 24, no. 4, pp. 419–428, Apr. 1977.

[9] D. S. H. Chan and J. C. H. Phang, "Analytical methods for the extraction of solar-cell single- and double-diode model parameters from I-V characteristics," IEEE Trans. Electron Devices, vol. 34, no. 2, pp. 286–293, Feb. 1987.

[10] U. Stutenbaeumer and B. Mesfin, "Equivalent model of monocrystalline, polycrystalline and amorphous silicon solar cells," Renew. Energy, vol. 18, no. 4, pp. 501–512, Dec. 1999.

[11] K. Kurobe and H. Matsunami, "New Two-Diode Model for Detailed Analysis of Multicrystalline Silicon Solar Cells," Jpn. J. Appl. Phys., vol. 44, no. 12R, p. 8314, Dec. 2005.

[12] D. S. H. Chan, J. R. Phillips, and J. C. H. Phang, "A comparative study of extraction methods for solar cell model parameters," Solid-State Electron., vol. 29, no. 3, pp. 329–337, Mar. 1986.

[13] K. Rajkanan and J. Shewchun, "A better approach to the evaluation of the series resistance of solar cells," Solid-State Electron., vol. 22, no. 2, pp. 193–197, Feb. 1979.

[14] M. Haouari-Merbah, M. Belhamel, I. Tobías, and J. M. Ruiz, "Extraction and analysis of solar cell parameters from the illuminated current–voltage curve," Sol. Energy Mater. Sol. Cells, vol. 87, no. 1–4, pp. 225–233, May 2005.

[15] J. M. Zhu, W. Z. Shen, Y. H. Zhang, and H. F. W. Dekkers, "Determination of effective diffusion length and saturation current density in silicon solar cells," Phys. B Condens. Matter, vol. 355, no. 1–4, pp. 408–416, Jan. 2005.

[16] K. Taretto, U. Rau, J. H. Werner, "Method to extract diffusion length from solar cell parameters - application to polycrystalline silicon," Journal of Applied Physics, vol. 90, no. 9, pp. 5447–5455.

[17] . Zerga, F. Benyarou et B. Benyoucef, "Optimisation du Rendement d'une Cellule Solaire NP au Silicium Monocristallin," Rev. Energ. Ren. : Physique Energétique, pp. 95 – 100, 1998.

IEEE
445 Hoes Lane
Piscataway, NJ 08854-4141

ISBN 978-1-4673-8760-6